Quantum Neuroscience

The Answer to Life, the Universe, and Everything

Book Two of the Quantum Mechanics Series

By: Mark My Words

I'm a theoretician. I notice trends and consolidate them. It's just what I do.

In the summer of 2017, I upgraded my science to Science 2.0. Science 2.0 allows ALL of the evidence into evidence and is based upon the Lived Experiences (phenomenology) of the human race, including our non-local experiences or transdimensional experiences. Science 2.0 is the way that science should have always been done but wasn't. Under Science 2.0, everything is taken into consideration; and, I chose to go with a preponderance of the evidence.

From the very beginning, I felt that Science 2.0 needed to justify its existence. The way that it does so is by repeatedly demonstrating through comparison and contrast that Science 2.0 is vastly superior to Scientific Naturalism and Eliminative Materialism.

Science 2.0 is based upon Phenomenology. Phenomenology is the scientific study of events, experiences, and phenomena of all types. The BEST way to find and know the truth is to live it and experience it for yourself, or to choose to trust someone who has. The second-best way to find and know the truth is through a process of elimination. If we eliminate everything that is false, has been falsified, has never been experienced nor observed, or has been demonstrated to be false and impossible, then eventually only the truth will remain. The Ultimate Truth that remains after the false and the falsified have been eliminated is the fact that Psyche or Non-Local Consciousness is the Ultimate Causal Agent in all dimensions and in every universe.

One of the first fruits from my upgrade to Science 2.0 is a new science that I call Quantum Neuroscience. Quantum Neuroscience is the scientific study of how the Human Psyche interacts with and controls its physical brain. Quantum Neuroscience is primarily a human science, because only human beings write, tell, report, and share their non-local experiences, near-death experiences, out-of-body experiences, quantum experiences, psychic experiences, supernatural experiences, encounters with God, shared-death experiences, visions, revelations, and other types of transdimensional or spiritual experiences. That doesn't mean that other species don't have these types of experiences. It just means that only human beings or human psyches have the language capabilities necessary to share these types of experiences with other human beings.

When it comes to science and thinking-out-of-the box, intellectuals, geniuses, and the highly educated are severely disadvantaged, because they can easily talk themselves out of things that are true and then easily talk themselves into things that are false, because they have been trained in our public schools to do so. I KNOW, because I used to be a Materialist, Naturalist, Nihilist, and Atheist. Materialism, Naturalism, and Atheism of any kind are based upon denial, suppression of evidence, and a refusal to look at evidence. Been there and done that! I know how it works. Self-deception works, and it works every time. Nobody is immune. According to the Materialists and Naturalists, the existence of God is physically impossible; therefore, God does not exist.

As an integral part of Science 2.0, Quantum Neuroscience allows ALL of the evidence into evidence. Quantum Neuroscience is an evidentiary science. It stands in stark contrast to the things we had before, which were based upon a rejection of evidence and a refusal to look at evidence. Quantum Neuroscience is an observational science, experiential science, eye-witness science, and empirical science that's based upon the Phenomenology or the Lived Experiences of the human race through a preponderance of the evidence. Quantum Neuroscience is an attempt to understand and explain the physically impossible. I hope you will find it as interesting as I found it to be.

Ironically, everything within Quantum Neuroscience is discovered, verified, and proven Science. Quantum Field Theory, Action at a Distance, and Quantum Mechanics are proven science. They have been constantly verified and proven true. In this book, I'm simply using them to explain Neuroscience, as should have been done decades ago. When it comes to Quantum Neuroscience, there's nothing to prove. It has already been proven true. I simply took it and ran with it.

Amazon Author Page: https://amazon.com/author/science

The Associated Facebook Page: https://www.facebook.com/MarkMyScience/

The Associated Twitter Page: https://twitter.com/Mark_Me_Words

The Associated Website: https://quantum-neuroscience.com/

A Sister Website: http://mypsyche.us/

A Sister Website: https://science-2-0.com/

The Associated Forum: http://www.markme.us/forums/forum/quantum-neuroscience/

First Edition

"This is the coolest science book that I have ever read – a real game changer!" — D.M.B.

Table of Contents

DEDICATION

With much gratitude, dedicated to Pim van Lommel, Jeffery M. Schwartz, Mario Beauregard, Wilder Penfield, John Eccles, and Henry P. Stapp.

INTRODUCTION

If you want to make a quantum leap in your Science, Logic, Knowledge, and Understanding of Reality, then search for and study ALL of the different things which the Materialists, Naturalists, Darwinists, and Atheists are trying to hide from you. There are thousands of such things, waiting for you to find them and understand them. I encourage you to take that quantum leap with me.

I was a Skeptic, Scientist, and Agnostic. I even spent time as a Materialist, Nihilist, and Atheist because I didn't know better at the time.

When it comes to science, after over fifty years of life, I finally find myself in a wonderful position now that I have never experienced before. When I listen to the Materialists, Naturalists, Darwinists, and Atheists speak or read their writings, I KNOW instantly that they are wrong.

How's that possible?

It's possible, because I KNOW why they are wrong.

That's the KEY.

Off my head from memory, I can list dozens of observations that prove that they are wrong – these are scientific observations that FALSIFY Materialism, Naturalism, Darwinism, and Atheism. Thirty years ago, NONE of us had this kind of information at our fingertips; but, we do now!

What's the net result of all of this?

Whenever I listen to the Materialists and Naturalists and Atheists speak or read their writings, I feel sorry for them. I feel sorry that they have been tricked and deceived just as I was tricked and deceived. I KNOW how it feels to have been tricked and deceived by Darwinism and the Theory of Evolution for most of one's life. It's not a good feeling. I feel sorry for these people, because they haven't seen it yet nor understood it yet. My ultimate desire is to set them free, just as I have been set free during the past couple of years.

Evolution Debunked:

https://www.youtube.com/watch?v=4wCxkBnm3ow&t=1457s

I found this YouTube video interesting.

In this book, I dismantle the theory of evolution and replace it with something infinitely better – Quantum Neuroscience.

RAISON D'ÊTRE

I started writing my *Science 2.0* book and this *Quantum Neuroscience* book, when I went looking for Scientific Proof of God's Existence.

This book is huge.

I have a compelling or driving reason for writing it, or I wouldn't have bothered, because I have been working on it in one form or another for the past couple of years.

The following people died. Some of them went to hell.

Jesus Christ saved them, raised them from the dead, and sent them back to testify about what happened to them.

Lee Stoneking Addresses UN General Assembly:

https://www.youtube.com/watch?v=FYt8sv4vzQs

Ralph Jensen:

https://www.youtube.com/watch?v=uWshfNnyEQA&t=40s

Howard Storm:

https://www.youtube.com/watch?v=Vm647n1360A

https://www.youtube.com/watch?v=UPj4wci_bcI

Dr. Mary Neal:

https://www.youtube.com/watch?v=DX473dF7ChY

Ian McCormack:

https://www.youtube.com/watch?v=HbTAmN4m2lQ

Joe Hadwin:

https://www.youtube.com/watch?v=IOhOynR9Jxq

Another:

https://www.youtube.com/watch?v=N4ut09jDdV0

NDE Encounters with Jesus Christ

https://www.youtube.com/results?search_query=NDE+Jesus

These types of Near-Death Experiences had a powerful effect on me; and, they completely changed my worldview or philosophy of life.

My best friend has had similar experiences.

My best friend is a brilliant scientist and a patent-holding inventor.

I trust these people and eventually concluded that they are telling us the truth. I now KNOW that the human psyche and its spirit body survive the death of their assigned physical body. The human psyche or non-local consciousness continues to function, live, and thrive long after our physical brain is dead and gone.

One Near-Death Experience is easily dismissed as fiction and a figment of the imagination; but, thousands and even millions are impossible to deny.

I used to be a Materialist, Nihilist, and Atheist; and, I refused to look at and study these types of things at that point in time. But now, I allow ALL of the evidence into evidence. I'm no longer hiding from evidence, which means that I'm no longer a Materialist, Nihilist, and Atheist. I had to make major adjustments in my way of looking at the world and thinking about the world, in order to be able to write a book like *Quantum Neuroscience*. Such a book wasn't possible for me to write when I was a Materialist, Nihilist, and Atheist.

I have been stubborn and skeptical most of my life. I was born with trust issues. I think I was born with materialistic and naturalistic and agnostic tendencies. Until this year (2017), I was a promoter of Scientism – the philosophical belief that the scientific method is the best way for finding and knowing the truth.

Like I said, I used to be a Materialist, Nihilist, and Atheist. I'm stubborn; but, I'm not stupid. I recognize a lost cause when I see one. WE LOST. The preponderance of the evidence stands firmly against us. I had to get used to it. Materialism, Naturalism, Darwinism, Nihilism, Behaviorism, Determinism, Materialistic Reductionism, and Atheism have been FALSIFIED by observational evidence, experiential evidence, empirical evidence, scientific evidence, experimental evidence, and eye-witness evidence. Materialism and Naturalism and their derivatives are DEAD. I have now accepted that fact and have decided to move on with my life. In actuality, I now vigorously oppose Materialism, Naturalism, and Darwinism because I KNOW they are false because I KNOW why they are false and how they have been falsified.

Materialism and Naturalism are self-defeating. All of the different immaterial, philosophical, and theoretical Black Boxes that these people rely upon to make their case, end up falsifying the fundamental teachings and tenets of Materialism, Naturalism, and Darwinism. They shoot themselves in the foot, and most of the time they don't even know it.

Chemical evolution can't make proteins and the matching genes to go along with them; so, there's no way in the universe that evolution could make your eyes and brain let alone the genes to go along with them.

Since it's physically impossible for chemical evolution, natural selection, and random mutations to design, create, and deploy functional proteins and the matching genes to go along with them, that means that it's physically impossible for evolution to design, create, and produce genomes, eyes, brains, and life forms. Evolution as our Designer and Creator is physically impossible.

It is November 2017 as I write this sentence. It is clear and obvious to me that there are currently thousands of Near-Death Experiences (NDEs) on YouTube and the Internet, for us to study and examine; and, I have been watching and reading some of them for a couple of years now. NDEs have met their burden of proof through a preponderance of the evidence. The evidence for NDEs has now reached critical mass, with more coming online every day. Like I said, I'm stubborn, but I'm not stupid. I don't know precisely when it happened, but the preponderance of the evidence became impossible for me to deny.

Near-Death Experiences

https://www.youtube.com/results?search_query=Near-death+experiences

I submit the whole load into evidence.

Now WE KNOW as a race that the human psyche or the human spirit survives the death of its physical brain. It's no longer theory. It has become PROVEN KNOWLEDGE and a PROVEN FACT, through the Lived Experiences of the human race as a whole. We also KNOW through a preponderance of the evidence that the Biblical God Jesus Christ really does exist and that He is the "being of light" whom people encounter during their NDEs. If you ever find yourself in hell, KNOW that Jesus Christ can get you out of there just for the asking.

Ever since I realized and accepted the fact that the Near-Death Experiencers and Out-of-Body Travelers are telling us the truth, I have been searching for a scientific explanation for what these people have experienced and witnessed. I have been searching for Scientific Proof of God's Existence because now I KNOW that it must exist. This whole process has also involved finding Scientific Proof of Psyche or Non-Local Consciousness. It all goes together and combines into one great whole, because it's mutually supportive.

Only one science discipline seems to tie all of this together into a comprehensible and logical unity or worldview, and that science discipline is Quantum Mechanics. Quantum Mechanics provides Scientific Proof of Non-Local Consciousness; and, that's the easiest place to start when we are looking for Scientific Proof of God's Existence. But of course, I never do things the easy way, which means that I took a different path at first, in order to accomplish this goal.

This book describes some of the steps I took in order to arrive at this chosen destination.

My ultimate goal with this book is to set people free, as I have been set free.

Now pay attention!

Despite the FACT that I will have debunked and falsified the Theory of Evolution thousands of times in hundreds of different ways, don't ever once let me convince you that there is NO such thing as evolution or random mutations.

Evolution is REAL, very REAL. Evolution or random mutation is entropy or the second law of thermodynamics. It's REAL. Entropy is very REAL at the physical level; and, it works as advertised at the physical level.

Whenever the Materialists, Naturalists, Darwinists, Nihilists, Behaviorists, and Atheists start talking about evolution, they get most everything wrong, because evolution or entropy cannot design and create. However, these people do indeed get one thing perfectly right. Evolution, or random mutation, or entropy is indeed the CAUSE of ALL of our heritable diseases, developmental diseases, and heritable mental illnesses.

Remember, the Theory of Evolution is FALSE because random mutations or entropy cannot design and create genes, proteins, and life forms. However, evolution or random mutation or entropy is very REAL; and, it can indeed destroy genes, proteins, and life forms. Do you see how that works? It's important to understand.

WIRING THE BRAIN

Materialists, Naturalists, Darwinists, Nihilists, and Atheists have defined axon growth and the arborization of dendrites along with synaptogenesis and synaptic pruning as "wiring the brain". Can you see any technical problems with their claims that our brains are wired together by axons, dendrites, and synapses?

Wired by whom?

Scientific Observation: We have observed that wired networks or networks of wires require some kind of Electrical Architect to design the schematic and some kind of Electrical Engineer, Wire Maker, and Cable Runner in order to implement and successfully wire that electrical network of wires.

Scientific Definition: The Materialists, Naturalists, Darwinists, Nihilists, and Atheists define axons and dendrites as "wires within a physical brain". These same people also define synapses as "wired connections between neurons in a physical brain".

Scientific Conclusion: Therefore, it is logical and rational to conclude that axons, dendrites, and synapses have some kind of Architect, Map Maker, Engineer, Manufacturer, and Cable Runner who decides and controls how synaptogenesis and synaptic pruning will take place. In order to have wires requires some kind of Wirer! Wires don't just spontaneously generate out of thin air!

There's another severe problem when it comes to the "wires within our brains" and the "normal wiring of our brain", which the Physicalists and Scientific Naturalists cannot see nor understand. Do you know what that problem is?

There are NO physical wires within our brain!

Chemical Synapses are gaps in space – NOT physical connections.

The axons, dendrites, and synapses DO NOT FUNCTION as physical connections; and therefore, they do not function as wires and wired networks – not at the physical level at least!

So, ask yourself how the different neurons are communicating with each other telepathically at a distance since they are NOT communicating with each other physically through their axons, dendrites, and synapses.

The materialistic and naturalistic claim that axons and dendrites form wires within our brains is deceptively false. Chemical Synapses are gaps in space – NOT physical connections. Technically, there are NO physical wires, physical connections, or physical cables within our brains connecting the neurons together into logic gates, transistors, and computer networks for computer processing and memory storage through our Chemical Synapses. If there is any computer processing or memory storage taking place within our physical brain, it's taking place at the quantum level or the psyche level because it's physically impossible at the physical level because chemical synapses are gaps in space. Synaptic clefts scramble or randomize everything that comes their way at the physical level, which means that communication between neurons has to be taking place at the quantum level or the psyche level since it's not taking place at the physical level.

That is what has been experienced and observed.

Mark My Words

PART I — SCIENTIFIC PROOF OF GOD

When it comes to any scientific theory, you have to understand what's wrong with it before you can move beyond it. Only the scientific theories which are repeatedly verified, repeatedly experienced, and repeatedly observed pass the test of time. Other things like the chemical evolution or spontaneous generation of proteins and genes, which is physically impossible and has never been observed nor caught in the act, are automatically and repeatedly falsified through a lack of empirical evidence, a lack of experiential evidence, and a lack of observational evidence. I had to understand this, before I could move beyond it.

Remember, chemical evolution of proteins from scratch is physically impossible. Evolution always gets credit for the things that it couldn't possibly do. Evolution is always given credit for the physically impossible, by the Materialists, Darwinists, and Naturalists; but, these people are almost always wrong. By definition in principle, evolution of any kind cannot do the physically impossible. Evolution can't do anything at the quantum level or the psyche level. You are deceiving yourself if you think it can. God didn't use evolution to do the physically impossible. God used quantum mechanisms to do the physically impossible.

Mutations are defined as accidental alterations in individual genes during chromosome duplication and genetic recombination. Remember, accidents can't do anything deliberate such as the origin of species. Accidents can mess up genes, but accidents can't create them.

When I first realized that Natural Selection, Random Mutations, and Evolution (genetic drift) DID NOT EXIST until after God had designed and created the first genomes and the first life forms, that reality and epiphany became my first convincing Scientific Proof of God's Existence.

In a very real sense, the Theory of Evolution proved to me that God exists, and that God must exist in order to have done all of the design, programming, engineering, creation, field-testing, manufacturing, fine-tuning, and deployment of genomes and life forms which Natural Selection, Random Mutations, and Evolution could NEVER have done. God must of necessity exist in order to have done ALL of the Science, which Natural Selection, Random Mutations, Evolution, and the Rocks could NEVER have done. After that realization, I simply KNEW that God exists, because I KNEW why He must exist.

When it comes to the origins of life, proteins, genes, genomes, and fine-tuning, the existence of God has infinitely more explanatory power than chemical evolution and random mutations. That's just the way it is. Science is the pursuit of things which have explanatory power – or at least it should be.

Remember, chemical evolution, design and creation by random mutations and natural selection, abiogenesis, spontaneous generation, macro-evolution (two dogs siring and giving birth to genetically compatible cats), and the theory of evolution are physically impossible. That's all you really need to know about the theory of evolution – it's physically impossible.

The theory of evolution showed me our need for God. And, once we have used science to establish a need for God, we have in fact used science to develop a Scientific Proof of God's Existence. It's elementary.

In the summer of 2017, I upgraded my science to Science 2.0, which allows all of the evidence into evidence. Science 2.0 goes with a preponderance of the observational evidence and experiential evidence. We start with what WE KNOW to be real and true, from the Lived Experiences of the human race; and then, we try to explain it scientifically!

I used to be a Materialist, Nihilist, and Atheist. When I finally noticed that we have a tendency to trick ourselves and deceive ourselves, I decided to STOP IT, by choosing to pursue and follow the empirical evidence or observational evidence instead. I chose to let the evidence lead me wherever it wants to lead me, as any good scientist should.

Now pay attention!

Despite the FACT that I have debunked and falsified the Theory of Evolution thousands of times in hundreds of different ways, don't ever once let me convince you that there is NO such thing as evolution or random mutations.

Evolution is REAL, very REAL. Evolution or random mutation is entropy or the second law of thermodynamics. It's REAL. Entropy is very REAL at the physical level; and, it works as advertised at the physical level.

Whenever the Materialists, Naturalists, Darwinists, Nihilists, Behaviorists, and Atheists start talking about evolution, they get most everything wrong, because evolution or entropy cannot design and create. However, these people do indeed get one thing perfectly right. Evolution, or random mutation, or entropy is indeed the CAUSE of ALL of our heritable diseases, developmental diseases, and heritable mental illnesses.

Remember, the Theory of Evolution is FALSE because random mutations or entropy cannot design and create genes, proteins, and life forms. However, evolution or random mutation or entropy is very REAL; and, it can indeed destroy genes, proteins, and life forms. Do you see how that works? It's important to understand.

Scientific Proof of God's Existence

Scientific proof of God's existence?

They ALL told me that it can't be done and that it will never be done.

I believed them.

For over fifty years of my life, I believed them.

THEY WERE WRONG!

What good are Science and the scientific methods if they can't be used to find and know the truth? What good are Science and the scientific methods if they can't be used to prove anything?

I have spent the last two years studying Science, the scientific methods, theology, theism, psychology, the philosophy of science, and natural theology – developing Scientific Proofs of God's Existence as I go along.

As far as I know, I'm the ONLY person on the planet who is doing this – using various different scientific methods to prove that God exists and to prove that God must of necessity exist. Why am I the ONLY person doing this? It's because everyone else believes that it can't be done, so they don't even try. That's the way I used to be.

I used to be a Materialist, Nihilist, Atheist, and Skeptic; but, I'm not anymore, because Science and the scientific methods proved to me that God exists and why He must exist.

I wasn't going to believe in God until after I had scientific proof of God's existence. I didn't think it was possible, which meant that I was destined to remain Agnostic and Skeptical for the rest of my life. BUT over time, after I finally started looking for it and asking God for it, He gave me scientific proof of His existence. Now, I have no choice but to believe in God's existence, because He gave me exactly what I was looking for.

So, how was this possible, if Science and the Scientific Methods can't be used to prove anything true? Well, let me start at the beginning and explain the process that I went through, in detail.

It all started by getting a truthful and accurate definition for Materialism and Naturalism. When I started this process two years ago, I didn't know what Materialism and Naturalism were. I was a scientist to my very core and a believer in Scientism, and I didn't even know what philosophical foundations Science is built upon. But, I'm not alone. Most scientists don't have a clue when it comes to the Philosophy of Science. They don't teach you this stuff in school. In fact, most of your college professors either don't know anything about the Philosophy of Science or they are trying to hide that information from you if they do know. I had a steep learning curve ahead of me!

I didn't know where to start my search, so I started by studying Atheism which quickly led me to Philosophical Proofs of God's existence – basically the opposite of Atheism. I KNEW what Atheism was, because I had spent time as an atheist. I didn't believe that there could be a Proof of God's existence, so the whole Natural Theology realm and Philosophical Proofs of God were completely new to me. I didn't believe such a thing was possible and that such a thing exists. But, I was wrong! We Materialists and Atheists are always wrong!

I started the whole learning process with Antony Flew, allegedly the world's most notorious atheist, and his book, *There Is a God: How the World's Most Notorious Atheist Changed His Mind*. I found out that he gave up his atheism because of Intelligent Design Theory. His reason for giving up his atheism was different than mine; and, I started to take Intelligent Design Theory seriously.

My search for "proof of God" on Amazon led me to John Lennox and a few of his books.

Lennox, J. C. (2009). *God's Undertaker: Has Science Buried God?* Oxford, England: A Lion Book.

Lennox, J. C. (2011). *God and Stephen Hawking: Whose Design Is It Anyway?* Oxford, England: A Lion Book.

Lennox, J. C. (2011). *Gunning for God: Why the New Atheists Are Missing the Target*. Oxford, England: A Lion Book.

This stuff is Natural Theology, or Philosophical Proof of God's Existence, wherein nature and science are used as premises while constructing Proofs of God.

This whole thing slowly led me to believe that it might in fact be possible to use the scientific methods to prove that God exists. This was before I knew that the scientific methods can't be used to prove anything to be true; and, this was long before I discovered that the scientific methods can in fact be used to prove things false. At the time, I simply thought that Science and the scientific methods ARE the TRUTH. I was a practitioner and adherent of Scientism, and I had a great deal of faith in Science and the scientific methods. I simply believed that if I could use Science and the scientific methods to PROVE God's existence, then I would have had found the Holy Grail for real.

I had a lot to learn!

I started to study Science and the Scientific Methods, in earnest, from a completely new perspective with a completely new goal in mind. I was looking for Scientific Proof of God's Existence.

Seek, and ye shall find. Knock, and it shall be opened unto you.

Using the Scientific Methods to Prove God's Existence

Okay, so how do you use the scientific methods to prove God's existence, if the scientific methods can't be used to prove anything to be true?

When you take up a serious study of the Philosophy of Science, the first thing you discover is that the scientific methods can't be used to prove anything true, due to the *affirming the consequent*, the *begging the question*, and the *jumping to conclusions* logic fallacies which are built directly in to the Scientific Method.

For any piece of scientific data or scientific information or scientific evidence, there are theoretically an infinite number of possible interpretations or explanations which can be given to that scientific data. By *affirming the consequent*, you can pick any conclusion or consequent that you want and use the Scientific Method to affirm it or to "prove" it to be true, which is what the Materialists, Atheists, and Darwinists do when it comes to the Theory of Evolution. They *affirm the consequent*, or unilaterally declare their personal interpretation of the scientific evidence to be axiomatically true. It's a logic fallacy which they are employing in order to "prove" that the Theory of Evolution and Darwinism are true. It's a FALSE PROOF, or a FALSE POSITIVE!

For example, the same scientific evidence which the Darwinists use to "prove" that the Theory of Evolution is true CAN BE USED by *affirming the consequent* to prove that the flying spaghetti monster designed and created ALL of the genomes and life forms on this planet.

The same scientific evidence which the Darwinists use to "prove" that the Theory of Evolution is true CAN BE USED and HAS BEEN USED to prove that Intelligent Design Theory IS TRUE! But, it's ALL based upon the *affirming the consequent* or the *jumping to conclusions* logic fallacy! So, it really isn't a proof of these things! It only appears to be.

So again, how do you use the scientific methods to prove God's existence, if the scientific methods can't be used to prove anything to be true? This IS the great mystery, and God slowly showed me how to solve it.

In order to accomplish this task, one needs a solid understanding of the Philosophy of Science and the Scientific Method. You need to KNOW what the scientific method really is, how it works, what it does, what it is good for, and what it can't do! You also need a solid understanding of what Materialism, Naturalism, Darwinism, Nihilism, and Atheism REALLY ARE. You have to have a true and accurate definition for these philosophical concepts, or you will never be able to develop a credible argument for God's existence. I didn't even know what I was looking for at first; and, I spent nearly two years studying Science, the Scientific Methods, and the Philosophy of Science before it all came together for me.

Rychlak, J. F. (1981). *A Philosophy of Science for Personality Theory* (2nd ed.). Malabar, FL: Robert E. Krieger Publishing Company.

I used this book from Joseph Rychlak as my introduction to the Philosophy of Science. This was the first book to introduce me to the *affirming the consequent* logic fallacy and the primary weakness of the scientific methods – namely, that the scientific methods can't be used to prove anything to be true. You have to have a solid understanding of the Scientific Method and the Philosophy of Science if you are going to develop a sound and logical philosophical proof and scientific proof of God's existence.

Why is this important? Why is it important to understand the Philosophy of Science and how the Scientific Methods really work?

Well, I run into genius PhD scientists all the time who use the scientific methods every day in their work, and they have NO clue whatsoever about the limitations of the scientific methods. These people literally believe that the Scientific Method is God and that it can do anything and accomplish anything it sets its mind to. These people believe that the Scientific Method is Superman and that they are Supermen, and that they are going to save the world. These people have no real understanding of the Philosophy of Science, what the Scientific Method really is, what it's good for, what it can do, and what it can't do. These people have no idea what the limitations of the Scientific Method are and what logic fallacies are built into the scientific methods. These people are practitioners of the Religion of Scientism, which means that these people treat Science and the Scientific Method as if these things were God. I KNOW, because I used to be one of them.

I was a bit rusty when I first started developing Scientific Proofs of God's Existence. I went about it the hard way and the long way at first, but we each have to start where we are and go forward from there.

My first real breakthrough came when God revealed to me what Materialism, Naturalism, Darwinism, and Atheism really ARE. I needed a definition for these philosophical concepts that I could plug into the Scientific Method as a hypothesis; and, eureka! It just popped into my mind one day after I had started looking for it and asking for it.

Materialism IS design and creation by physical matter. Materialists really truly believe that the rocks or raw physical matter designed, programmed, engineered, created, manufactured, and deployed the first genomes and life forms on this planet. This IS what these people truly believe and what these people teach our children in our public classrooms. Immediately, it became obvious to me that Naturalists and Darwinists are also Materialists. The Materialists, Naturalists, and Darwinists really truly believe that the rocks designed and created ALL of the genomes and life forms on this earth.

Nihilism is the philosophical belief that there is no afterlife, no reason or purpose for living, and no such thing as a non-local realm or a spirit realm. Nihilism is the belief that we cease to exist when we die. Nihilism is another form of Materialism and Atheism.

Atheism is design and creation by nothing, or design and creation by chance. The Atheists really truly believe that nothing designed and created everything, including all of the genomes and life forms on this planet.

Due to the *affirming the consequent* logic fallacy which is built into the scientific methods, you CANNOT USE the scientific methods to prove anything true directly, which means that the scientific methods cannot be used for a direct proof of God's existence. This reality has got most people believing that we can't use the scientific methods to prove anything at all. I have had many Scientists, Materialists, and Atheists tell me this. They have told me that Science and the Scientific Methods can't be used to prove anything. THEY ARE WRONG! It's annoying, because many of the Atheists, Naturalists, and Materialists use this false belief as a crutch or a shield to hide behind, as a reason for believing in the impossible and the patently absurd, and as a reason for doing sloppy science. These people don't understand the Scientific Method nor the Philosophy of Science, if they truly believe that the scientific methods can't be used to prove anything.

Nevertheless, there are millions of scientists and Darwinists out there in the world right now who are using their ignorance of the scientific method and their belief that the scientific methods can't prove anything as an excuse or a reason for continuing to hope that someday – some way – someday – Science will finally find a way to prove the theory of evolution true. It's never going to happen, because the scientific methods can't be used directly to prove anything to be true – certainly not something that is demonstrably false.

So, how do you use the scientific methods as a proof of God's existence, if the scientific methods can't be used to prove anything to be true? That is the million-dollar question, isn't it; and, I get the impression that I'm the ONLY person on the planet who has solved this one! It took a lot of prayer and study, but eventually God showed me how to do it. He showed me how things REALLY WORK!

When developing a Scientific Proof of God's Existence, the process starts by using the scientific methods to **falsify** Materialism, Naturalism, Darwinism, and Atheism. And, you also have to choose to use the Scientific Observations, or the Direct Observations, or the Lived Experiences of the human race to **falsify** Nihilism, Naturalism, and Atheism. That's the KEY, and that that's the thing which EVERYBODY refuses to do. I don't think that it's that they don't know how to do it or can't know how to do it. I think it's because they don't want to do it and refuse to do it. In fact, it probably NEVER enters their mind to even try to do it. As far as I know, I'm the ONLY person on the planet who has ever done it. I have used various different scientific methods and scientific observations to **falsify** Materialism, Naturalism, Darwinism, Scientism, Nihilism, and Atheism. It's really easy to do, once you learn how to do it. Everything **falsifies** these things!

But, let's run through the whole process in a bit more detail, rather than just making blanket statements.

First, you have to realize and accept the fact that the scientific methods can indeed be used to PROVE theories false! It's called falsifying a theory or *negating the consequent*; and, it's philosophically and logically sound. When you successfully *negate the consequent*, you have in fact PROVEN a theory false, or falsified a theory. This IS Philosophy of Science 101, and the core essential understanding of the Scientific Method one must have if one is going to use the scientific methods to PROVE anything. You have to KNOW that the scientific methods can be used to PROVE things false. You have to KNOW that the scientific methods can be used to falsify theories, hypotheses, and philosophical concepts that are in fact false.

Finally, you have to be willing to use the scientific methods and scientific observations to **falsify** Materialism, Naturalism, Darwinism, Nihilism, Scientism, and Atheism. If you are unwilling to do this, then you will NEVER be able to develop a Scientific Proof of God's Existence, and you will NEVER be able to find the truth. This is crucial. You have to be open-minded enough and smart enough to use the scientific methods for what they are good for and for what they were designed for, if you are going to get Science and the scientific methods to PROVE anything to you.

Let's use the Scientific Method for what it was designed for. Here's how falsifying a theory, or *negating the consequent*, works in principle using the Scientific Method:

Scientific Hypothesis: If Theory X is true, then we will observe Y.

Scientific Observations: We don't observe Y.

Scientific Conclusion: Therefore, Theory X is false and has been falsified by the Scientific Method.

Here's how it works in practice:

Scientific Hypothesis: If the Theory of Evolution, Materialism, Naturalism, and Darwinism are true, then we will observe the rocks and

physical reactions designing, creating, and manufacturing genomes and life forms from scratch.

Scientific Observations: We have NEVER observed the rocks and physical reactions designing and creating genomes and life forms; and, we NEVER will. They can't.

The Scientific Conclusion: Therefore, the Theory of Evolution, Materialism, Naturalism, and Darwinism are false and have been falsified by the Scientific Method or Scientific Observation.

I just successfully falsified Materialism, Naturalism, Darwinism, and the Theory of Evolution; and, I used the Scientific Method to do so! Best of all, my argument is philosophically and logically sound. I have in fact **falsified** Materialism, Darwinism, Naturalism, Creation by Rocks, and the Theory of Evolution for REAL! I have PROVEN them false! It's that simple! I successfully used the Scientific Methods for what they are good for – falsifying theories which are false. I wish I would have known how to do that forty years ago. It would have saved me a lot of confusion, frustration, time, and grief.

Now, let's run Atheism through the Scientific Method:

Scientific Hypothesis: If Atheism is true, we will observe "nothing" designing, creating, and manufacturing everything, including genomes and life forms. It would be chaos, but that's exactly what we should be observing if Atheism were true.

Scientific Observations: Obviously, we have NEVER observed "nothing" designing and creating anything; and, we NEVER will. It's patently absurd.

The Scientific Conclusion: Therefore, Patent Absurdity or Atheism is false and has been falsified by the Scientific Method and Scientific Observations.

I just falsified Atheism for REAL; and, it's philosophically and logically sound. I have in fact falsified Atheism, using the Scientific Method and Scientific Observations to do so! But, the process of designing and creating a Scientific Proof of God's existence isn't complete yet. Technically, falsifying Atheism is not the same thing as proving Theism. So, how do we proceed from here, and continue to use the Scientific Methods as we go along?

Well, let's see if you can follow this logic or reasoning.

Since we KNOW by the Scientific Method that Materialism, Naturalism, Atheism, Darwinism, Creation by Rocks, and the Theory of Evolution ARE FALSE, we also KNOW through the process of elimination that some form of their opposite MUST BE TRUE. When it comes to the scientific methods, we arrive at the truth through a process of elimination. By eliminating ALL of the falsehoods, we are left with the truth. This is Logic 101!

Here we are operating on the AXIOM or the GIVEN or the LAW that Truth, Reality, and Knowledge do in fact exist, if we know how and where to find them. Science, Knowledge, and Reality rely upon this basic essential assumption; otherwise, there's no sense doing science, if truth and reality and knowledge are in fact impossible to find and impossible to use once we have found them. We have to start with the primary assumption that THE TRUTH can indeed be found and does indeed exist!

Let's state this again, because it is essential and important! Since we KNOW by the Scientific Method that Materialism, Naturalism, Atheism, Darwinism, Creation by Rocks, and the Theory of Evolution ARE FALSE, we also KNOW through the process of elimination that some form of their opposite MUST BE TRUE. Using the scientific methods, we narrow in on the TRUTH through a process of elimination, by using the scientific methods to eliminate everything that is false. Once you have eliminated EVERYTHING that is false, then you are left staring at THE TRUTH. It's elementary my dear friend.

It's so obvious, that I sometimes wonder how the human race has been able to overlook it for thousands of years!

Once you have eliminated EVERYTHING that is false, then you are left with THE TRUTH. That's what Sherlock Holmes does.

How often have I said to you that when you have eliminated the impossible [and the false], whatever remains, *however improbable*, must be the truth? — Sherlock Holmes.

Sherlock Holmes slowly arrives at the truth by falsifying his theories and his imagined evidence. That's how I finally arrived at THE TRUTH, by falsifying Materialism, Naturalism, Darwinism, Nihilism, and Atheism. THE TRUTH is the opposite of falsehood and falsified theories!

Now, here comes the kicker and the final knock-down blow! THE TRUTH is repeatedly verified by scientific observations and scientific methods. By eliminating the impossible and ALL of the falsehoods, whatever remains must be THE TRUTH! That's how we USE the scientific methods or scientific observations to find THE TRUTH, and to KNOW the truth, and to PROVE the truth! The truth is going to be the opposite of falsified theories, or the opposite of the things which have been proven false by the scientific methods and the observations of the human race. Checkmate!

THIS IS THE KEY!

So again, how do you use the scientific methods to prove God's existence, if the scientific methods can't be used to prove anything to be true?

You USE the scientific methods to eliminate EVERYTHING that is false, so that ONLY the TRUTH remains! If you successfully falsify and eliminate ALL of the falsehoods such as Materialism, Darwinism, Naturalism, Scientism, Nihilism, and Atheism, then THE TRUTH is the only thing left standing. It's logical, and it works! Best of all, it works! THE TRUTH is constantly VERIFIED by the scientific methods and never falsified. In fact, there's NO way to use the scientific methods to falsify THE TRUTH. It can't be done. The scientific methods or scientific observations continuously VERIFY THE TRUTH. Cool, huh?

It's elementary my dear friend! And, it works!

Let's provide some examples of how this works in practice.

What's the opposite of Materialism? The non-local, the spiritual, the transdimensional, and the immaterial are the opposite of Materialism. Psyche or Intelligence or Non-Local Consciousness is the opposite of Materialism. Light and thoughts and dreams are the opposite of Materialism. (Spirit matter really isn't the opposite of Materialism, because spirit matter is matter, just like physical matter is matter. According to Quantum Mechanics, it takes conscious observation or the Word of Command to convert spirit matter into physical matter. It's ALL matter, just different phases of being or different states of existence.)

Observation, empirical evidence, direct experience, lived experiences, the scientific methods, direct observations, scientific observations, personal experiences, revelations from God, physical experiences, and spiritual experiences ARE the same exact thing. They are

ALL **the source** of Scientific Evidence or should be! Lived Experience includes our non-local experiences or our spiritual experiences. Lived Experience or Scientific Observation IS Scientific Evidence! Lived Experience or Scientific Observation is how scientific theories and scientific hypotheses are verified and falsified!

THIS IS KEY!

We KNOW from **observation**, direct experience, or the scientific methods that Intelligence (Psyche), Thought (Psyche), Consciousness (Psyche), Quantum Nonlocality, Light, Forces, Fields, Magnetism, Gravity, Dark Matter (Spirit Matter), Dark Energy (the Zero-Point Field of Light or the Light of Christ) do in fact exist because they have been observed, experienced, and **verified** trillions of trillions of times, just as we KNOW from the scientific methods or scientific observations that Materialism is false.

Darwinism and Naturalism are a form of Materialism. The very same scientific methods and scientific observations, which PROVE Materialism false, also simultaneously PROVE Darwinism and Naturalism false. Atheism is a form of Materialism and Naturalism. The very same scientific methods which PROVE Materialism and Naturalism false also PROVE Atheism false. They are all fruit from the same poisoned tree.

There is NO evidence and will NEVER be any evidence to support the major premises, primary assumptions, and main claims of Materialism, Darwinism, Naturalism, Nihilism, and Atheism. These philosophies or religions have to be taken on blind-faith as being true; and, blind-faith isn't scientific evidence, even though millions of Materialists and Atheists do in fact try to use their ignorance, flying spaghetti monsters, and blind-faith as scientific evidence.

What do we learn from this? We learn that **observations** or the **scientific methods** repeatedly falsify falsehoods such as Materialism, Naturalism, Darwinism, Nihilism, and Atheism – while at the very same time, the same exact observations or scientific methods repeatedly VERIFY the theories which are in fact TRUE.

What's the opposite of Darwinism? Intelligent Design Theory, or Design and Creation by Intelligent Beings, is the opposite of Darwinism and the Theory of Evolution. Darwinism IS Creation by Rocks. The opposite of that is Creation by Intelligent Psyches. The rocks or raw physical matter have NEVER been caught in the act of design and creation. In fact, in 1859, Louis Pasteur falsified Spontaneous Generation or Creation by Rocks. Louis Pasteur used a scientific method to falsify Materialism, Darwinism, Naturalism, Creation by Rocks, and the Theory of Evolution the very same year that Charles Darwin published "On the Origin of Species". How ironic is that?

Do you see how important it is to get a correct and useful definition for Darwinism, Materialism, Atheism, and Naturalism which can then be run through the Scientific Method and have Scientific Observations or Lived Experiences applied directly to those hypotheses or definitions? It makes ALL the difference in the world! As far as I know, I'm the ONLY person on the planet to have done such a thing; yet, God has been doing this same exact thing for trillions of years. God is the Ultimate Scientist! Science IS observation or lived experience; and, God observes!

What's the opposite of Naturalism? The supernatural, the psychic, the spiritual, the non-local, Theism, and God ARE the opposite of Naturalism. We KNOW from observations or the Lived Experiences of the human race that the Biblical God Jesus Christ exists and rose from the dead. Since we KNOW from the scientific methods that Naturalism IS FALSE, we should be looking for some type of Naturalism's opposite to be TRUE.

This is where the vast SUPERIORITY of Lived Experience or Direct Observation comes in handy and really shines. Science IS Observation, or it should be! God IS a Scientist, the Ultimate Scientist! God Observes! Through Direct Observation or Lived Experiences, we can go directly to KNOWING the TRUTH without having to rely upon the scientific methods or philosophical interpretation of scientific data. We KNOW from the Lived Experiences of the human race that the Biblical God does in fact exist and really did rise from the dead.

Furthermore, there are over 13 million Near-Death Experiences on record. These people KNOW that Naturalism and Materialism ARE FALSE, because these people have left their physical body and have gone to the spirit world and seen for themselves that it is real and truly exists. These people have FALSIFIED Nihilism through their Lived Experiences or Direct Observations. Nihilism is the philosophical belief that there is no such thing as a spirit world and an afterlife.

Thousands of different people KNOW that Jesus Christ lives and rose from the dead, because they have seen Him and touched Him both while out-of-body and while in the flesh. These people have also FALSIFIED Nihilism through their Lived Experiences or Direct Observations.

Lived Experiences or Direct Observations ARE Scientific Evidence! That's the best and most effective definition for scientific evidence that we can have. With Lived Experiences or Direct Observations, we human beings (human psyches) go directly to KNOWING the TRUTH! Lived Experiences or Scientific Observations include our Spiritual Experiences and our Out-of-Body Experiences. This is REAL SCIENCE! And, it's all based upon Lived Experience or Direct Observation of THE TRUTH.

Knowing by the scientific methods that Naturalism is false and why it is false is just icing on the cake, or confirmation of what we already KNOW from the Lived Experiences of the human race as recorded in the Bible, the Book of Mormon: Another Testament of Jesus Christ, the Doctrine and Covenants, the Pearl of Great Price, Near-Death Experiences, Out-of-Body Experiences, and other types of Spiritual Experiences or Non-Local Experiences.

Lived Experiences or Direct Scientific Observation of both the spirit realm and the physical realm are vastly SUPERIOR to the traditional scientific methods or the materialistic methods which restrict us exclusively to studying physical matter with physical methods. With Lived Experience or Direct Scientific Observation, we human beings or human psyches can go directly to KNOWLEDGE of the TRUTH without getting mired down in Materialism, Scientism, the Physical Sciences, and the limitations of the traditional materialistic methods which are limited to and limit us to studying physical matter and physical processes.

Finally, if Materialism and Naturalism were actually true, scientists and psychologists would be putting rocks into Skinner boxes in order to observe their behavior and shape their behavior. Why aren't the Behaviorists putting rocks into Skinner boxes for some behavioral technology or behavioral conditioning? It's because Materialism and Naturalism are FALSE. It's so obvious, that I sometimes wonder why nobody has ever thought of it before.

What's the opposite of Atheism? Theism and God are the opposite of Atheism. We KNOW through the scientific methods or scientific observations that the major premises or the primary assumptions of Atheism are patently absurd. So, we should be looking to Theism, Intelligent Beings, Intelligent Psyches, and some kind of God or Ultimate Scientist as the designer, programmer, creator and manufacturer of ALL the genomes and life forms on this planet.

We should also be looking to find some kind of God, or Intelligent Psyche, or Powerful Non-Local Consciousness, or Intelligent Being to be the Conscious Observer who is necessary for converting spirit matter into physical matter according to the Science of

Quantum Mechanics. God must of necessity exist, or there would be NO physical matter! God must of necessity exist, or there would be no genomes and no physical life forms. Right there IS Scientific Proof of God's Existence! It's not that great of a leap, once we have repeatedly used the scientific methods or scientific observations to falsify Materialism, Naturalism, Nihilism, Darwinism, and Atheism.

Once I KNEW through the scientific methods that Materialism, Darwinism, Naturalism, and Atheism ARE FALSE, then I simply KNEW that God or the Ultimate Scientist must of necessity exist in order to have done ALL of the Science which the rocks or raw physical matter could NEVER have done. I used Science and the Scientific Methods for what they are good for – falsifying theories – and through a process of elimination I arrived at THE TRUTH, namely that God or the Ultimate Scientist must of necessity exist, or there would be no physical matter, there would be no genomes, there would be no physical life forms, and there would be no science. If God or the Ultimate Scientist didn't exist, you and I would not exist; and, you wouldn't be reading this right now. It's elementary my dear friend!

We KNOW from the Lived Experiences and Near-Death Experiences of the human race that the Biblical God Jesus Christ does in fact exist. Knowing through the various scientific methods that Atheism is false and why it is false lends confirmation to the fact that the Lived Experiences or Direct Observations of the human race are in fact true.

After I had successfully used the scientific methods to falsify Materialism, Naturalism, Darwinism, and Atheism a dozen different ways, I simply KNEW that God or the Ultimate Scientist of necessity must exist in order to have done all of the different types of Science, which the rocks or physical matter could never have done. It was logical, and it was obvious. Suddenly, I simply KNEW that God exists; and, I had used Science and the Scientific Methods to gain that KNOWLEDGE.

I had effectively used the Scientific Method to prove to myself that God exists, by using the scientific methods to **falsify** everything that was the opposite of God, Non-Locality, Intelligence, Non-Local Consciousness, and Psyche. This process became SCIENTIFIC PROOF OF GOD'S EXISTENCE, through a process of falsifying everything that is false.

The beauty of this whole process is that I finally realized that the scientific methods and scientific observations have falsified Materialism, Naturalism, Darwinism, Nihilism, and Atheism trillions of trillions of times in thousands of different ways – while at the same time, scientific methods, observations, and Lived Experience continue to VERIFY the truth over and over again. Thousands of people have seen and touched our Resurrected Lord Jesus Christ. That REALITY keeps getting verified by scientific observations or direct observations or Lived Experiences over and over and over again. THE TRUTH is always VERIFIED by the scientific methods or direct observations or Lived Experience.

Lived Experience, including our non-local experiences or spiritual experiences, IS Scientific Evidence and becomes Scientific Proof of God's Existence. This IS Science at its very best! Through Lived Experience or Direct Scientific Observations, we can go directly to TRUTH, KNOWLEDGE, and PROOF of God's Existence. Powerful, huh?

Quantum Non-Locality, when properly understood scientifically, FALSIFIES Materialism, Darwinism, Naturalism, Nihilism, and even Atheism. Quantum Non-Locality IS the opposite of Materialism, Darwinism, Naturalism, Nihilism, and Atheism.

Have you ever wondered why Classical Physicists, such as Albert Einstein, vigorously opposed Quantum Mechanics all of their lives? It's because Quantum Mechanics is supernatural in nature and origin. Quantum Mechanisms involve faster-than-light Action at a Distance, which is physically impossible. Quantum Mechanics is supernatural; and therefore, Quantum Mechanics falsifies Materialism, Naturalism, and their derivatives.

In most of the science books, non-locality is presented to us as Instantaneous Action at a Distance – things like transpersonal psychology, telepathy, and telekinesis – the thing that Einstein called Spooky Action at a distance.

I personally, while reading *Consciousness Beyond Life* by Pim van Lommel, modified, changed, and enhanced the definition for non-locality by defining "non-local" as non-physical and transdimensional. It just made more sense to me to define it that way; but, it's not the traditional way of defining it.

Quantum Mechanics: From a Non-Physical Spiritual Perspective

https://www.amazon.com/dp/B01J023TGU/

Non-Local means Non-Physical, Transdimensional, or Spiritual – not located in our physical realm. Local means located in our physical 3D space-time realm. Materialism, Darwinism, Naturalism, and Atheism have at their very core the major premise and the primary philosophical assumption which claims that the non-local, the non-physical, the immaterial, and the spiritual do not exist. These false philosophies and false religions are FALSIFIED each and every time we see a ray of light or feel the effects of gravity, which are in fact non-local and non-physical spiritual phenomena. Every time you have a thought or a dream, or feel the effects of time, you have in fact FALSIFIED Materialism, Naturalism, Darwinism, Nihilism, and even Atheism by experiencing something non-physical. These false materialistic philosophical concepts are FALSIFIED every time that Quantum Non-Locality is verified scientifically or experienced directly.

According to the Bible, the human spirit body is a God, sired and parented by God and the Goddesses. According to the Bible, your physical body is descended from the Gods, because Adam and his physical body were literally the son of God. Your surname is God! Every time we humans have a Near-Death Experience, Out-of-Body Experience, Shared-Death Experience, or a Spiritual Experience we have in fact FALSIFIED Materialism, Naturalism, Darwinism, Nihilism, and Atheism which do in fact make the claim as their primary assumption that the Non-Local, the Non-Physical, and the Spiritual do not exist. The Lived Experiences, or the Direct Observations, or the Scientific Observations of the human race as a whole FALSIFY Materialism, Naturalism, Nihilism, and Atheism. Lived Experiences including Spiritual Experiences ARE Scientific Observations and Scientific Evidence, or at least they should be if we are truly Scientists as we claim to be.

Near-Death Experiences and Out-of-Body Experiences are direct verification of Quantum Non-Locality and the Spirit Realm or the Non-Local Realm. These types of Lived Experience or Scientific Evidence also falsify Nihilism.

Quantum Non-Locality, Quantum Entanglement, and "Spooky Action at a Distance" have been repeatedly VERIFIED through scientific experimentation. Quantum Non-Locality IS the spirit realm, where people go whenever they separate from their physical body. This Non-Locality stuff has been OBSERVED and VERIFIED trillions of times. Every time you have a thought or a dream, you have in fact had a spiritual experience, because there is NO way for our scientists to record your thoughts and dreams using physical instruments to do so. The contents of your thoughts and dreams cannot be detected nor recorded by physical instruments, because they are non-local or spiritual in nature and origin.

There is NO evidence and will be NO evidence to support Materialism, Naturalism, Darwinism, Nihilism, and Atheism because they ARE FALSE. Instead, ALL of the evidence and ALL of the scientific observations and ALL of our Lived Experiences keep pointing us to non-locality, psyche, non-local consciousness, the non-local realm, the quantum realm, the spirit realm, the transdimensional realm, spiritual beings, intelligent beings, and psyche beings. The evidence keeps pointing us to some Grand Intelligence, Super Psyche, God, or Ultimate Scientist as our designer and creator and as the person who consciously commanded physical matter into existence in the first place.

It's literally impossible to falsify THE TRUTH using the Scientific Methods. The scientists will never be able to falsify Intelligent Design Theory, or design and creation by Intelligent Beings, or design and creation by Intelligent Psyches, because it's true. Design and Creation by Intelligence or Psyche can be verified and replicated on demand! The truth never fails and is never falsified!

ONLY Psyche – ONLY Intelligence – ONLY Intelligent Beings can design and create things and do science. Psyche or Intelligence is the ONLY thing we have ever observed doing so. You can verify this right now for yourself, by choosing to design and create something for someone, because the rocks can't do it for you. It's elementary my dear friend.

As far as I know, I'm the ONLY person on the planet who is using Science and various different Scientific Methods directly and purposefully to FALSIFY Materialism, Darwinism, Naturalism, Scientism, Nihilism, and Atheism. Nobody else has ever thought to do so! I've never seen it done by anyone else, because everyone else has chosen to believe that it can't be done and that it will never be done. I have had dozens of Atheists and Materialists tell me that it can't be done, and for a while there, I used to believe them. THEY WERE WRONG!

Lived Experiences, including Spiritual Experiences or Out-of-Body Experiences, ARE Scientific Observations and Scientific Evidence. It has to be this way, or we will never be able to use Scientific Observations or Scientific Methods in order to get at THE TRUTH of our reality and our existence. Remember, Direct Observation is an integral part of the Scientific Methods – the most important part of the Scientific Methods! Science is ALL about Lived Experience or Direct Observation of any kind, or at least it should be, if we are in fact Scientists and not simply sophists and philosophers as the Materialists, Naturalists, and Atheists are. Lived Experience, including our Spiritual Experiences or Non-Local Experiences, IS Scientific Evidence which is based upon Scientific Observations!

See: Van Lommel, P. (2010). *Consciousness Beyond Life: The Science of the Near-Death Experience*. New York: HarperCollins.

Quantum Mechanics IS the Science of Near-Death Experiences. Quantum Mechanics is Spiritual Mechanics, or the way that spirit matter really works. Quantum Mechanics falsifies Materialism, Naturalism, Darwinism, Nihilism, and even Atheism.

Lived Experience, including our non-local experiences or spiritual experiences, IS Scientific Evidence and becomes Scientific Proof of Quantum Non-Locality. This IS Science at its very best! Through Lived Experiences or Direct Scientific Observations, we can go directly to TRUTH, KNOWLEDGE, and PROOF of the Non-Local Realm or the Spirit Realm.

Powerful, huh?

The BEST way to find and know the truth is to live it and experience it for yourself, or to choose to trust someone who has.

The second-best way to find and know the truth is through the process of elimination, by using the scientific methods or *negating the consequent* to FALSIFY erroneous philosophies such as Materialism, Naturalism, Darwinism, Scientism, and

Atheism. If you successfully eliminate everything that is false and that has been falsified, such as Materialism and Naturalism, then ONLY the truth will remain. It's elementary my dear reader.

Scientific Proof of God's Existence

Through the direct process of eliminating ALL the falsehoods that we know about and falsifying ALL of the false theories that we know about such as Materialism and Naturalism and Atheism, this whole process becomes Scientific Proof of God's Existence. THE TRUTH is what remains after ALL of the false theories have been FALSFIED and eliminated from consideration. THE TRUTH is the thing which the scientific methods or direct observations repeatedly verify and never falsify. TRUTH and KNOWLEDGE are based upon Lived Experience, which is in fact Scientific Evidence. THE TRUTH is the thing which the scientific methods can't falsify! Over and over again, the scientific methods and scientific observations VERIFY THE TRUTH! It becomes obvious and clear, once a person chooses to look and see and trust and believe. Lived Experience IS Scientific Observation and Scientific Evidence. The Scientific Methods always verify the truth, if they are done right.

How often have I said to you that when you have eliminated the impossible [and the false], whatever remains, *however improbable*, must be the truth? — Sherlock Holmes.

So, what is this GRAND TRUTH, which the scientific methods can't falsify and which the scientific methods and scientific observations continuously verify?

This GRAND TRUTH is that ONLY intelligent beings or psyche beings can design, program, engineer, create, manufacture, and deploy genomes, physical life forms, computers, software, hardware, glassware, plastics, cars, houses, airplanes, rockets, space ships, internets, radio, bridges, skyscrapers, theories, ideas, plans, reasons, and blueprints. ONLY Psyche or Intelligence can do teleology, final causality, cosmic fine-tuning, thinking, consciousness, dreaming, imagination, intentionality, awareness, reasoning, knowledge, truth, and life. ONLY Psyche or Intelligence can ACT as an independent Agent. ONLY Psyche can do moral agency. ONLY Psyche can do Out-of-Body Experiences and remember having done so. ONLY Psyche can have Near-Death Experiences and see and embrace the Biblical God Jesus Christ in the spirit world and remember having done so. ONLY Psyche or Non-Local Consciousness can convert spirit matter into physical matter. This IS a Psyche Ontology, or the Ultimate Model of Reality, wherein Psyche is the fundamental unit of reality!

ONLY Psyche or Non-Local Consciousness can do scientific observations or Lived Experiences! ONLY Psyche can go to the spirit realm or the non-local realm and make scientific observations or have lived experiences, which are in fact Scientific Evidence and could be used as evidence in a court of law. ONLY the human psyche can record its lived experiences or its scientific observations which it has in the spirit realm, and then share that scientific evidence or those lived experiences with other human beings or other human psyches. When taken as a whole, these observations, knowledge, truth, lived experiences, spiritual experiences, direct observations, direct experiences of the non-local, and scientific evidence ultimately become Scientific Proof of God's Existence.

Why?

Only Psyche can do lived experience, observation, and science. Only God's Psyche or the Ultimate Scientist could have done ALL of the science that needed to be done, which the rocks or raw physical matter could NEVER have done. ONLY God's Psyche or the Ultimate Scientist could have done the conscious observation necessary to convert spirit matter into physical matter. If God didn't exist, there would be NO physical matter, and there would be NO genomes and NO physical life forms. This IS Scientific Proof of God's Existence!

Science is supposed to be about Observation, Lived Experience, and Knowledge of the Truth. God's Psyche or the Ultimate Scientist IS the only thing left standing when everything else has been falsified and eliminated. The Biblical God Jesus Christ IS the Being of Light whom we encounter during our Near-Death Experiences and Spiritual Experiences – during our Direct Scientific Observation of the Non-Local Realm! We KNOW this to be TRUE from the Lived Experiences or the Scientific Observations of the human race! Lived Experiences, including Spiritual Experiences or Non-Local Experiences, ARE Scientific Observations and Scientific Evidence.

If we are going to do Science for REAL, then we need a new definition for Science. Science is NOT Materialism; and, Science is NOT Naturalism. Materialism, Naturalism, Darwinism, Nihilism, and Atheism are metaphysics and philosophy, NOT Science. Science IS Lived Experience, including our Spiritual Experiences or Non-Local Experiences. Science IS Lived Experience or Direct Observation of THE TRUTH. Science IS KNOWLEDGE, which is based upon our Lived Experiences or our Scientific Observations, including our Spiritual Experiences and Non-Local Observations. That's what Science really is and really should be – the sum total of our Lived Experiences or Scientific Observations, including our Spiritual Experiences and our Out-of-Body Experiences. Lived Experience or Psyche Experience IS Scientific Evidence. I'm talking about a Paradigm Shift here – switching over to a Psyche Ontology and a Psyche Epistemology as the Ultimate Model of Reality.

Within this Ultimate Model of Reality or Ultimate Paradigm, psyche or non-local consciousness is the fundamental unit of reality. This is a Psyche Ontology.

Within this Ultimate Model of Reality or Ultimate Paradigm, lived experiences or psyche experiences or direct observations including spiritual experiences are the best way of finding and knowing the truth. This is a Psyche Epistemology.

I have had many different Atheists, Naturalists, Darwinists, Nihilists, and Materialists tell me that I don't understand Science and that I don't understand the Scientific Method. You will have to decide for yourself whether you agree with them or not. Now you know what I think, and that's all that really matters to me.

In summary, God or the Ultimate Scientist must of necessity exist, or your genome would not exist, the physical matter in your physical body would not exist, and you would not exist as a physical living being. This is scientifically accurate and scientific truth. This IS Scientific Proof of God's Existence! If God did not exist, you would not exist, and you wouldn't be reading this right now. That's what Science and the scientific methods are trying to tell us; but, the Materialists, Naturalists, and Atheists refuse to look, listen, and learn. It's their loss. I KNOW, because I used to be a Materialist, Nihilist, and Atheist. The grass really IS greener on the Theistic side of the fence, because I have been there now and KNOW that to be true as well.

Materialism, Naturalism, and Atheism are based upon a refusal to look at evidence and a denial of Reality. These people are in denial! There is NO evidence and will be NO evidence to support Materialism, Naturalism, Nihilism, and Atheism because these things contradict Lived Experiences, Empirical Reality, and Experiential Evidence; but, there IS tons of Lived Experience, or Scientific Observation, or Direct Observation, or Scientific Evidence supporting the existence of and the resurrection of our Lord Jesus Christ – if we choose to equate Lived Experience with Scientific Evidence as we should do if we are truly Scientists.

Remember, if it matches with Reality then it is declared to be scientifically accurate. If it has been falsified and contradicts Reality, then it is unscientific and false. If it contradicts the Lived Experiences of the human race, then it is FALSE. Materialism, Naturalism, Nihilism, Darwinism, Scientism, and Atheism contradict Reality and contradict the Lived Experiences or the Scientific Observations of the human race; therefore, we KNOW that these things are FALSE.

Throughout the rest of this book, I will go back to the beginning and explain in greater detail how I came to these conclusions. It didn't come to me all at once. It was at least two years in the making. It has taken me a while to refine my presentation. All I can say is that this whole thing is a lot more fun and infinitely more interesting than My Materialism and My Atheism ever were!

Go with the best and get rid of all the rest. That's what I finally decided to do.

Best wishes,

Mark My Words

May 2017

The Materialists Don't Want You Using Scientific Theories to Prove the Truth and to Find the Truth

I have had Materialists and Atheists tell me that theories cannot be used to prove anything, because they are theories. That was kind of confusing and frustrating. There's some really strange ideas and information floating around concerning Science and the Scientific Method. I think what these people were trying to tell me is that Science and the scientific methods cannot be used to prove anything. These people ARE WRONG! What good is Science and the scientific methods if they can't be used to find and know the truth? What good is Science and the scientific methods if they can't be used to PROVE things?

I have had multiple Scientists, Atheists, Naturalists, and Materialists tell me that I don't understand Science and the Scientific Method. I have observed that the Materialists, Naturalists, Nihilists, and Atheists are always wrong. It's these people who really don't understand Science and the Scientific Methods. These people don't have a clue. I KNOW, because I used to be one of them. There used to be a time when I believed them, because I used to be a Materialist, Nihilist, and Atheist. These people had successfully convinced me that I would never be able to use Science nor Theories nor Methods to prove anything. I believed them, because I used to be a practitioner and adherent of Scientism. I saw no other way and knew of no other way.

These people had convinced me that Science and the Scientific Methods could never be used to prove Materialism, Naturalism, Darwinism, and Atheism false. THEY WERE WRONG! Falsifying Theories is precisely what the Scientific Method is good for and precisely what the Scientific Method was designed for!

After I repeatedly caught these people trying to trick us and deceive us, I stopped trusting them. Since then, Science and Logic and the Scientific Methods have proven to me that the Materialists and the Atheists are ALWAYS wrong. Once I got rid of My Materialism and My Atheism, Science and Scientific Methods have been proving things to me right and left. Materialism and Naturalism were holding me back and stunting my intellectual growth; and, I certainly am not the first victim of these false and unproductive and unscientific ideologies.

Materialism is the chosen philosophical belief that the Non-Local, or the Non-Physical, or the Spiritual does NOT exist. Ironically, there is NO Scientific Evidence and can be NO Scientific Evidence to support such a belief. Belief in Materialism requires a monumentally huge blind-leap of faith, a blind faith which is in fact Non-Physical or Spiritual in nature. Belief in Materialism requires a Spiritual and Non-Physical and Philosophical blind leap of faith, the very fact of which proves that the Spiritual exists and proves that Materialism is fundamentally FALSE.

I eventually realized that Materialism, Darwinism, Naturalism, and Atheism CANNOT BE used to prove anything, because they are FALSE. We cannot use anything that is false to prove something to be true; so, in the end, the Materialists and Atheists were right in that THEIR THEORIES cannot be used to prove anything, because their theories are false. However, once we know that ALL of their theories are false, we can indeed use their atheistic and materialistic falsehoods, deceptions, and lies to point us towards THE TRUTH. I do it all the time now.

Nowadays, I actively USE the Materialists, Darwinists, Naturalists, and Atheists to point me to the things that they don't want me seeing, reading, and understanding because I KNOW that the Science and the Theories which the Materialists and Atheists have formally rejected are in fact THE TRUTH and will always end up being THE TRUTH. I learned to use their atheistic theories and materialistic theories to point me to THE TRUTH. Genius, huh?

Consequently, I use their false theories to point me to the truth and to prove the truth to myself and to my readers and my friends.

If your Science isn't proving stuff to you, then you ain't doing it right or your theories are false. Remember, a false theory can be used to point us to the truth, because the truth is often the opposite of any theory which has been demonstrated to be false. Therefore, false theories can be used to point us to the truth and thereby prove the truth. Remember, Science is supposed to be the pursuit of truth, meaning that Science is supposed to eliminate the False Theories so that ONLY the True Theories remain.

This book is about using SCIENCE and THE SCIENTIFIC METHOD to eliminate the falsehoods and the impossible, in an attempt to demonstrate and prove THE TRUTH! What good is Science if Science can't be used to prove stuff?

In this book, I apply THE SCIENTIFIC METHOD, common sense logic, abductive reasoning, and deductive reasoning to various Scientific Theories in an attempt to do the real Science of eliminating False Hypotheses while at the same time keeping the True Hypotheses.

The goal is to find THE TRUTH among all of the materialistic and atheistic falsehoods that are typically presented to us for our consideration.

This was a fun book to research and write, because it made logical sense to me from beginning to end.

THE TRUTH, whenever I find it and wherever I find it, IS parsimonious and makes logical sense. THE TRUTH tastes good and feels good. There is NO confusion or doubt associated with THE TRUTH. You just know that it is TRUE.

The Materialists and Atheists are right. There is NO way to use Materialism and Atheism to prove that God exists. But, imagine what you could do if you got rid of Materialism and Atheism, as I have done!

The Spiritual Sciences or the Non-Physical Sciences become logical and rational possibilities once we have successfully eliminated Materialism and Atheism and Naturalism from our worldview or personal philosophy of life. It's like stepping through a door into the light. And, once a person has successfully made that transition into light and truth, he or she typically doesn't want to go back into the darkness and the lies. I don't want to go back to My Atheism and My Materialism, now that I KNOW how limiting, irrational, and boring they really were. That's just the true reality of the situation!

Non-Local means not located in our Physical 3D Space-Time. Non-Local means Non-Physical or Trans-Dimensional or Spiritual. Materialism, Naturalism, Darwinism, and Atheism cannot be used to study Nonlocality or Trans-Dimensionality, because the Materialists formally reject the Non-Local Sciences! There are Atheists and Materialists online right now mocking and ridiculing String Theory because it is Non-Local, Trans-Dimensional, and Spiritual in nature.

The Materialists and Atheistic Scientists are trying to severely limit how Science is used and how Science can be used; and, these people use force, intimidation, mocking, and ridicule in order to do so. This is one of the main reasons why I have lost all respect for Materialism, Darwinism, Naturalism, and Atheism. I don't like the tactics used by the people who promote these falsehoods and lies.

Furthermore, their materialistic rejection of the Spiritual Sciences or the Non-Physical Sciences (such as Quantum Nonlocality, Light, Time, String Theory, Trans-Dimensionality or Non-Local Reality, Quantum Mechanics or Spiritual Mechanics, Origins, Gravity, Infinite Velocity, Mathematics, Dark Matter or Spirit Matter, Dark Energy or the Zero-Point Field or the Light of Christ or the Quantum Sea of Light, Friendship and Love and Justice and Mercy, Philosophy of Mind, Psychology or the Study of the Human Spirit, and Non-Local Consciousness) LED ME to reject My Materialism and My Atheism so that I would then be free to study and learn about the Spiritual Sciences or the Non-Physical Sciences which exist in great abundance. Materialism is completely worthless when it comes to the Non-Physical Sciences!

YOU will NEVER be able to use Materialism or Atheism to prove that God exists or to prove that the Spiritual or the Non-Physical exists. Materialism is completely worthless when it comes to the Non-Local or the Non-Physical. Materialism can't be used to prove things, so what good is it? Materialism is Bad Science! Once I realized how much Materialism was holding me back and keeping me stupid and blind, then I was glad to be rid of it. The Scientist in me demanded no less! We each have a God-given right to reject the falsehoods and lies and to embrace THE TRUTH instead. The Materialists and Atheists are trying to take that away from us.

I have proven to myself that Materialism if FALSE and why it is false, which means that I have proven to myself that its opposite is TRUE and really exists. Your mileage may vary, especially if you haven't seen the evidence that I have seen.

The Materialists refuse to believe in the Non-Physical or the Non-Local, which means that the Materialists believe in Creation by Physical Matter or Creation by ROCKS. The Materialists really truly believe that the ROCKS designed and created it all, including everything that existed before the first physical matter was designed and created. The reason that we go to the public schools is so that we can be brainwashed by the Materialists and Atheists into believing in the IMPOSSIBLE; and, these people are very successful at what they do because they will fail you or fire you if you refuse to go along with their brainwashing procedures. Being awarded a PhD in Materialism of some kind is the ultimate form of brainwashing that our public schools offer us.

I have formally rejected Materialism and have chosen to go in pursuit of the Spiritual Sciences or the Non-Physical Sciences instead.

In this book, I discuss the various Scientific Theories and Scientific Evidence which proved to me that God exists. I found the Scientific Proofs of God's Existence infinitely more believable and convincing than the Philosophical Proofs of God's Existence. The Philosophical Proofs of God's Existence are only convincing if they have tons of Scientific Evidence supporting the truthfulness of the Premises and the Conclusion.

I no longer find any proofs based upon Materialism to be the least bit convincing, because I have trained myself to see and understand the falsehoods, deceptions, and lies upon which their Premises are based. When it comes to Materialism, Darwinism, Naturalism, and Atheism, since most of their Premises can be demonstrated to be false, that means that ALL of their Conclusions are false as well. Instead, I USE the falsehoods and lies of Materialism and Naturalism to point me to THE TRUTH. Science is supposed to be the pursuit of truth after all! Since I KNOW that Materialism, Atheism, Darwinism, and Naturalism are FALSE, I KNOW that their opposites are likely to be true; and, that ends up being a good place to start my pursuit of THE TRUTH.

A Thought Experiment

Let's disassemble your physical body down into its individual proteins, genes, and molecules placing the results into a bathtub.

Could you reassemble yourself from the contents in that tub? Would you know how to do it?

You are ALL there in that tub according to the Materialists, Naturalists, and Darwinists. You should be able to assemble a functional you from the contents of that tub.

You believe that evolution can assemble you from the contents of that bathtub, so why can't you do it? You're smarter than evolution after all. Evolution by definition in principle is dumb and blind.

Could you successfully reassemble your genome from the contents of that tub, and then implant that genome into a zygote, and then make that zygote produce a clone of you? You believe that evolution can do this, so why can't you do it?

Let's take all those proteins in that tub and break them down into amino acids. Could you then reassemble all of those amino acids back into their original proteins? Why not? You believe that evolution can do this, so why can't you do it? You believe that evolution is smart enough and skilled enough to do all of these things, so why aren't you smart enough and skilled enough to do all of these things?

If you can't do these things at will, then it's a sure thing that evolution could never have done them, because you are infinitely more intelligent than evolution is.

Our NEED for God to have done all of these things for us when we were nothing – but amino acids, DNA, and molecules – IS so obvious and clear that I sometimes wonder why I ever thought that maybe chemical evolution or luck could have done these things for us.

Remember, chemical evolution, design and creation by random mutations and natural selection, abiogenesis, spontaneous generation, macro-evolution, and the theory of evolution are physically impossible. That's all you really need to know about the theory of evolution – it's physically impossible. If you can't do these things, then there's no way on earth that evolution could have done these things.

PART II — LIVED EXPERIENCES

Lived Experiences ARE the pinnacle of knowledge and truth. There's NO better way for finding and knowing the TRUTH than Lived Experience or Direct Observation.

Science is supposed to be Observation, or Lived Experience, although for most people Science is treated as if it is nothing more than philosophy or sophistry, and they limit Science exclusively to our physical reality. In other words, these people use Science to try to deceive us and to deceive themselves as well; and, they are extremely successful at doing so. Self-deception works, and it works every time. That type of limiting, materialistic, and exclusive "science" really isn't Science – it is sophistry, metaphysics, and philosophy. Science based upon falsehoods such as Materialism and Naturalism are the very definition of BAD SCIENCE or pseudo-science.

In contrast, the TRUTH is KNOWN by living it, witnessing it, and experiencing it for yourself, or by choosing to trust someone who has done so for himself.

By Lived Experiences, we mean ALL of our psyche experiences, thoughts, dreams, choices, decisions, actions, spiritual experiences, conscious experiences, memories, out-of-body experiences (OBEs), mystical experiences, psychic experiences, physical experiences, science experiments, unexplainable "miraculous" events, healing, near-death experiences (NDEs), shared-death experiences (SDEs), theophanies, revelations of God, and revelations from God. Through Lived Experience, we can KNOW the TRUTH directly by experiencing it directly for ourselves.

We KNOW from the Lived Experiences of the human race that Psyche exists and that our Psyche or Intelligence is something completely different than our Spirit Body. We also KNOW from the Lived Experiences of the human race that our Human Psyche or Non-Local Consciousness has been assigned to a Spirit Body or united holistically with a Spirit Body, which looks like our Physical Body. What people call our "Spirit" or our "Ghost" is in fact our Psyche or Personality which has been united with a Spirit Body as a single functioning unit or whole. We KNOW from the Lived Experiences of the human race that Psyche can temporarily separate from its Spirit Body and look at its Spirit Body from an immaterial third-person viewpoint in space.

This is a Psyche Ontology. This is a holistic ontology or a holistic model of reality which takes into consideration ALL of our Lived Experiences as a race. The Human Psyche is unique in that it can record and then share its Lived Experiences with other human beings. The animals can't. This Psyche Ontology, wherein Psyche or Intelligence or Non-Local Consciousness is the fundamental unit of reality, is the Ultimate Ontology and the Ultimate Model of Reality. ONLY Psyche can do ontology, reality, and existence. Without Psyche or Non-Local Consciousness, there would be no existence.

A Psyche Epistemology, based upon our Lived Experiences or Direct Observations, is the ultimate way of finding and knowing the truth. A Psyche Epistemology is a much better way of knowing the truth than the scientific methods, scientific experimentation, or philosophical speculation which introduce elements of doubt, uncertainty, interpretation, and guesswork into the equation and which tend to limit us exclusively to the physical realm.

A Psyche Epistemology is all-inclusive; whereas, a Scientism Epistemology, or Materialism Epistemology, or Naturalism Epistemology are not. The Naturalistic Epistemologies and Scientific Epistemologies tend to be denialistic, exclusive, limited, materialistic, and restrictive – censoring what their adherents are permitted to study and learn about. Materialism, Atheism, and Naturalism really aren't Science or Observation – they are in fact Dogmatic Religions – the bad kind of blind-faith religions which the Materialists and Atheists claim to despise.

Materialism and Atheism of any kind are based upon a refusal to look at evidence, a denial of reality, and a rejection of the Lived Experiences or the Observations of the human race. A Naturalistic Epistemology is a highly restrictive and limited epistemology, which deliberately rejects most of the Lived Experiences or Observations of the human race. That's NOT science, because Science is supposed to be all about Observation, or Lived Experience, or Experiential Evidence! Naturalism and Materialism are religions and epistemologies, which have to be taken on blind faith as being true. They are NOT science.

In contrast, a Psyche Epistemology is the Ultimate Epistemology, because it's all-inclusive and based upon the Lived Experiences of the human race. Furthermore, ONLY Psyche can do knowledge and experience for REAL. We don't witness any of this from the rocks or physical reactions. These things don't learn from their experiences in any noticeable or detectable fashion. Furthermore, the rocks have no way to apply their knowledge if they have any, because technically they are not alive; whereas, Psyche is. Only Psyche can do observations, which means that only Psyche can do science!

From the Lived Experiences of the human race, we KNOW that our Psyche or Intelligence or Non-Local Consciousness is a completely different entity than our spirit body, and that the psyche/spirit body unit or combination is a completely different entity than our physical body. Psyche, spirit body, and physical body nest within each other holistically like Matryoshka dolls, from smallest to largest or from most refined to most coarse. It's really easy to understand if one chooses to do so. A child can understand it. Human beings are psyche beings. Human beings are intelligent beings. Being IS Psyche. Psyche IS Being and Becoming. Psyche is the sum total of our Lived Experiences. ONLY Psyche can have Lived Experiences and remember having done so. Raw matter can't, whether we are talking about physical matter or spirit matter.

We learn best through Observation and Experience – through the comparison and the contrast of our Lived Experiences as a race. Human beings or human psyches are unique in that we can record our Lived Experiences and then share our Lived Experiences with each other. The animals can't. For all we know, the animals see and talk with God every day; but, they have no way of telling us about it if they do.

I, personally, demanded Scientific Proof of God's Existence before I was going to let go and be willing to believe in God. Thankfully, God gave me what I was looking for and asking for; and over time, I found and developed a few Scientific Proofs of God's Existence that were totally convincing to me. After that, I simply KNEW that God exists, even though I had yet to come to KNOW God.

When it comes to Scientific Proof of God's Existence, the way this process works is through the falsification of Materialism, Naturalism, Nihilism, and Atheism; and then through a process of elimination, after falsifying all of these things, we arrive at the truth through a process of elimination. Remember, the scientific methods can't be used to prove anything true; but, the scientific methods can definitely be used to prove things false – such as Materialism, Naturalism, Darwinism, and Atheism.

Once you have successfully falsified Materialism, Naturalism, Darwinism, Nihilism, and Atheism there are only a few logical things which remain to explain the origin of genomes and life forms on this planet. In effect, ALL of the evidence ends up pointing to Psyche, Intelligence, and some kind of Advanced Alien Intelligence or God. After Naturalism and Darwinism have been **falsified**, then it becomes clear that this earth was terraformed and seeded.

Think about it logically. That's what I did. Once you have successfully eliminated ALL the falsehoods and lies, then you are left starting at the truth! It works. I KNOW, because that's what I did to get at the truth. Let me demonstrate the process.

What's the opposite of Materialism? Psyche, spirit, non-locality, the quantum realm, the transdimensional realm, and the spirit realm are the opposite of Materialism. Once you use the scientific methods to successfully **falsify** Materialism a trillion different ways, you are left staring at its opposite and are left staring at the truth, which is Psyche, Intelligence, Spirit, the Immaterial or Non-Physical, Forces, Fields, Light, Non-Locality, Non-Local Consciousness, Quantum Mechanics or Spiritual Mechanics, and the Transdimensional Realm.

What's the opposite of Naturalism? The supernatural, the spiritual, the non-local, spirit bodies, spirits, psyche, Theism, and God are the opposite of Naturalism. After you have used the scientific methods to **falsify** Naturalism a thousand different ways, through a process of elimination, you are left staring at the opposite of Naturalism as the only possible Truth that remains.

What's the opposite of Darwinism? Darwinism is reduced to design and creation by physical matter. Alas, the rocks can't design and create. They have NEVER been observed doing so. Once you have used the scientific methods to **falsify** Darwinism a dozen different ways, you are left staring at the truth that only Intelligent Beings or Intelligent Psyches can design and create. You are left staring at Darwinism's opposite – intelligent design and creation. Random Mutations and Natural Selection (Biological Evolution) did NOT exist until after God's Psyche designed and created the first genomes and life forms on this planet. God must of necessity exist in order to have designed and created the first genomes and life forms on this earth, because ONLY Psyche can design and create and do science. The various physical reactions or evolutionary processes could not have designed, programmed, engineered, field-tested, manufactured, and deployed themselves. ONLY Psyche or Intelligence can do these kinds of things, because only Psyche can do science, engineering, and manufacturing.

In a very real sense, for me personally, the repeated **falsification** of the Theory of Evolution became my first real and most convincing Scientific Proof of God's Existence. God must of necessity exist in order to have done ALL of the science and creation which the various types of evolution or "creation by rocks" could NEVER have done. Eventually, for me, that Reality became logical and obvious. If we successfully use the scientific methods to eliminate ALL the falsehoods, then ONLY the Truth remains.

What's the opposite of Nihilism? Knowledge of an afterlife and knowledge of some kind of meaning and purpose in our lives is the opposite of Nihilism. I, personally, used the Lived Experiences of the human race – NDEs, SDEs, OBEs, and other types of spiritual experiences – to prove that Nihilism is false. Direct Observations or Lived Experiences trump philosophical speculation and the scientific methods, or at least they should. We can, and we should, use our Lived Experiences to **falsify** the scientific theories and the philosophical ideas that don't match with Reality. Through Lived Experiences, we humans can go directly to KNOWING the TRUTH, which is something that we can't do with the scientific methods or philosophical guesswork. Science, ultimately, is supposed to be about Observation or Lived Experience or Psyche Experience, which is in fact the ONLY real way of finding and knowing the truth. Only Psyche can do Lived Experience, Observation, Knowledge, Reality, Existence, and Truth!

What's the opposite of Atheism? God's Psyche or God's Intelligence is the opposite of Atheism. Atheism is design and creation by chance, or design and creation by nothing. We all KNOW scientifically from observation and experience that chance and nothing can't design and create anything – at least when we are thinking rationally and logically we KNOW that this to be true. Once we have used the scientific methods and common sense to **falsify** Atheism in a variety of different ways, suddenly we find ourselves staring into the face of Intelligent Beings or the Gods as our ultimate designers and creators. In a very real sense, your Psyche could have been the designer and the creator of the frogs and the dogs and other types of critters, because ONLY Psyche can design and create and do science! In the next life, many of the Atheists, Materialists, and Naturalists will discover that they helped to design and create some of the life forms on this planet. That will be quite a surprise for them, won't it!

Once we have eliminated ALL of the falsehoods, then we are left staring at the truth! It's elementary my dear reader! That's how we use the scientific methods to "prove the truth" – by using the scientific methods to eliminate ALL of the falsehoods. Remember, we can – and we should – use the scientific methods to **falsify** all the theories and ideas which are in fact false. Once we have done so, once we have eliminated all the falsehoods, then we are in fact left with the truth.

What's this truth which we are left with after we have eliminated ALL the falsehoods?

THE TRUTH IS that ONLY Psyche or Intelligent Beings can design and create genomes and life forms, planets and universes, you and me. Best of all, the scientific methods can NEVER be used to falsify this truth, because the scientific methods or observations will always verify this truth. Our observations and lived experiences have PROVEN beyond a shadow of a doubt that **ONLY** Intelligent Psyches or Intelligent Beings can design and create and do manufacturing, engineering, and science. Scientific observation has verified this reality trillions of times and will verify this reality an infinite number of times more before we are done.

THE FUNDAMENTAL TRUTHS are always verified by our scientific methods and lived experiences, because this kind of truth is never falsified and can never be falsified. There is NO way to design and create a scientific method that will falsify Psyche's ability to design and create, because you would in fact VERIFY "design and creation by Psyche or Intelligent Beings" by trying to design and create such a scientific method in the first place. There's no way to use the Scientific Methods to falsify this TRUTH. Scientific Observation has verified "Creation by Psyche" or "Creation by Intelligent Beings" trillions of trillions of times. There's NO way to falsify something like that, because it is TRUE. We can ONLY use the Scientific Methods to falsify theories that are in fact false – things like Materialism, Naturalism, Darwinism, Nihilism, and Atheism.

Lived Experiences point us directly to a Psyche Ontology, the Ultimate Ontology, wherein psyche becomes the fundamental unit of reality. Only Psyche can design, create, and do manufacturing and science. ONLY Psyche or Non-Local Consciousness can convert spirit matter to physical matter, according to Quantum Mechanics. If it weren't for Psyche, this physical universe would not exist. Psyche is the Ultimate Cause of physical universes and physical matter. That's what Quantum Mechanics is trying to teach us. It doesn't get more basic or fundamental than that! Psyche or Non-Local Consciousness is the fundamental unit of Reality and Existence.

For me, personally, this whole thing was a paradigm shift that initially took place for me in the spring of 2016, because I used to be a Materialist, Nihilist, and Atheist and used to truly believe that it would always be impossible to prove the existence of Psyche, God, and Truth. I was wrong; but, I didn't know that at the time.

Let's use the Scientific Method for what it was designed for. Here's how falsifying a theory, or *negating the consequent*, works in principle:

Scientific Hypothesis: If Theory X is true, then we will observe Y.

Scientific Observations: We don't observe Y.

Scientific Conclusion: Therefore, Theory X is not true.

Here's how it works in practice:

Scientific Hypothesis: If the Theory of Evolution, Materialism, Naturalism, and Darwinism are true, then we will observe the rocks and physical reactions designing, creating, and manufacturing genomes and life forms from scratch.

Scientific Observations: We have NEVER observed the rocks and physical reactions designing and creating genomes and life forms; and, we NEVER will. They can't.

The Scientific Conclusion: Therefore, the Theory of Evolution, Materialism, Naturalism, and Darwinism are not true.

I just successfully falsified Materialism, Naturalism, Darwinism, and the Theory of Evolution; and, I used the Scientific Method to do so! Best of all, my argument is philosophically and logically sound. I have in fact falsified Materialism, Darwinism, Naturalism, and the Theory of Evolution for REAL! I have PROVEN them false! I successfully used the Scientific Methods for what they are good for – falsifying theories which are false. I wish I would have known how to do that forty years ago. It would have saved me a lot of confusion and grief.

After I had successfully used the scientific methods to falsify Materialism, Naturalism, Darwinism, and Atheism, I simply KNEW that God of necessity must exist in order to have done all of the different types of Science, which the rocks or physical matter could never have done. It was logical, and it was obvious.

I had effectively used the Scientific Method to prove to myself that God exists, by using the scientific methods to **falsify** everything that was the opposite of God, Non-Locality, Intelligence, and Psyche. Thereafter, the Lived Experiences of the human race were able to provide me with solid and convincing PROOF of God's Existence, although up to that point, I was always reluctant to accept Lived Experience into evidence until after the Scientific Method and Science had demonstrated to me that God must of necessity exist in order to have done all the Science which needed to be done in the first place.

Ironically, though, after God had provided me with a couple dozen different Scientific Proofs of His Existence, it still took me a year to figure out for myself that Lived Experience is a much better way for finding and knowing the truth than the scientific methods can ever be. For me, switching to Lived Experience is a recent paradigm shift in my way of thinking, which finally took place for me in April 2017. Before I saw the Light, I truly believed that the scientific method was the ONLY way for finding and knowing the truth. I was wrong, and it was holding me back!

Nowadays, I greatly value the Lived Experiences of the human race, and what they have taught me. I have slowly learned to value my Lived Experiences a lot more than I value my philosophical reasoning and the scientific methods. But for me, it wasn't always that way.

The TRUTH is KNOWN by observing it, living it, and experiencing it; or, by choosing to trust someone who has.

Through lived experience or direct observation, we human beings or human psyches can go directly to KNOWING the TRUTH without having to engage in philosophical speculation, scientific experimentation, and scientific methodology. Powerful! Is it not? Maybe you can sense why I found it so enticing and compelling, and why I was so eager to embrace it, once the idea of Lived Experience was finally presented to me and I accepted it as being real and true. Lived Experience is infinitely more powerful and convincing and real than the scientific methods can ever be. I never knew that before; but, now I do. Scientific Methods can't be used to prove the truth directly; but, Lived Experiences can! It doesn't get any better than that!

For example:

John 17: 3: And this is life eternal, that they might **know** thee the only true God, and Jesus Christ, whom thou hast sent.

John 8: 32: Ye shall **know the truth**, and the truth shall set you free from death, hell, ignorance, and sin.

Doctrine and Covenants 50: 25-27: "And again, verily I say unto you, and I say it that you may **know the truth**, that you may chase darkness from among you. He that is ordained of God and sent forth, the same is appointed to be the greatest, notwithstanding he is the least and the servant of all. Wherefore, he is possessor of all things; for all things are subject unto him, both in heaven and on the earth, the life and the light, the Spirit and the power, sent forth by the will of the Father through Jesus Christ, his Son."

John 14: 5-6:

5 Thomas saith unto him, Lord, we know not whither thou goest; and **how can we know the way**?

6 Jesus saith unto him, **I am the way, the truth, and the life**: no man cometh unto the Father, but by me.

Truth is knowledge. God KNOWS that Lived Experience is the best way of finding and knowing the truth. God is KNOWN by living Him and experiencing Him. Life, light, truth, and knowledge come to us from God the Father through His Son Jesus Christ.

Truth is knowledge. Truth is Lived Experience. We KNOW the things that we have lived and experienced! That's much more powerful than the scientific methods will ever be.

Mosiah 27: 36: Thus they were instruments in the hands of God in bringing many **to the knowledge of the truth**, yea, **to the knowledge of their Redeemer**.

Ether 4: 12-13: "And whatsoever thing persuadeth men to do good is of me; for good cometh of none save it be of me. I am the same that leadeth men to all good; he that will not believe my words will not believe me — that I am; and he that will not believe me will not believe the Father who sent me. For behold, I am the Father, I am the light, and the life, and **the truth of the world**. Come unto me, O ye Gentiles, and I will show unto you the greater things, **the knowledge** which is hid up because of unbelief."

From the Lord's perspective, to be led astray means to be led away from the Church of Jesus Christ of Latter-day Saints, His Church, His priesthood power, His saving ordinances, His temple blessings, His Apostles, His people, His Saints, His Truth, and His Holy Spirit. That's what it truly means to be led astray; and, we KNOW from experience that the Prophets of God will never lead us astray, because they ALWAYS point us directly to the Church of Jesus Christ of Latter-day Saints, Christ's Own Church, Christ's Apostles, Christ's Saving Ordinances, Christ's priesthood power and priesthood blessings, and the Temple Ordinances.

This is how we discern truth from error. If it points us and leads us to the Church of Jesus Christ of Latter-day Saints, Christ's Church, His saving ordinances, His priesthood powers and keys and authority, His covenants, and His temple ordinances, then we KNOW that it is from God and Christ. If it points us to some other path or leads us down some other path, then we KNOW that it is from Satan and the evil spirits who follow him.

This is how I tell the difference between truth and error. It was simple, once I finally realized the truth of it, and extremely useful, powerful, and helpful. I KNOW, because I have lived it and experienced it for myself.

Truth is knowledge which is gained from our lived experiences.

1 Timothy 4: 1-3: "Now the Spirit speaketh expressly, that in the latter times some shall depart from the faith, giving heed to seducing spirits, and doctrines of devils; speaking lies in hypocrisy; having their conscience seared with a hot iron; forbidding to marry, and commanding to abstain from meats, which God hath created to be received with thanksgiving of them which believe and **know the truth**."

Doctrine and Covenants 31: 2: Behold, you have had many afflictions because of your family; nevertheless, I will bless you and your family, yea, your little ones; and the day cometh that they will believe and **know the truth** and be one with you in my church.

Alma 24: 19: And thus we see that, when these Lamanites were brought to believe and to **know the truth**, they were firm, and would suffer even unto death rather than commit sin; and thus we see that they buried their weapons of war, for peace.

Moroni 10: 5: By the power of the Holy Ghost ye may **know the truth** of all things.

TRUTH is KNOWN by living it and experiencing it directly. God can reveal the truth to us, because God knows all truth, because God made reality or the truth. God IS the Truth.

Doctrine and Covenants 49: 2: Behold, I say unto you, that they desire to **know the truth** in part, but not all, for they are not right before me and must needs repent.

God can help us to overcome ALL things, except for a lack of desire. If we have NO desire to know the truth and to live the truth and to become the truth, then there's nothing that God can do for us. If we have no desire to repent or change for the better, God can't do anything to help us. God will never force the truth upon us. The truth and its associated knowledge have to be lived and owned by us, before it becomes useful and real to us. God knows this, which is why He will never force the truth upon us. We repent by turning to God and by choosing to embrace the truths which He freely gives us.

1 John 2: 3-4:

3 And hereby we do know that we **know** Him, if we keep His commandments.

4 He that saith, I know Him, and keepeth not His commandments, is a liar, and **the truth** is not in him.

The truth is known by living it and doing it. The truth is known by acting upon it and becoming it.

7 And behold, ye do **know** of yourselves, for **ye have witnessed it**, that as many of them as are brought to the **knowledge of the truth**, and to know of the wicked and abominable traditions of their fathers, and are led to believe the holy scriptures, yea, the prophecies of the holy prophets, which are written, which leadeth them to faith on the Lord, and unto repentance, which faith and repentance bringeth a change of heart unto them —

8 Therefore, as many as have come to this, ye **know** of yourselves are firm and steadfast in the faith, and in the thing wherewith they have been made free.

How did they know these things? By observing them! By living them and experiencing them! They witnessed it!

In order to get back to our Father in Heaven, we have to choose to follow someone who has already done so – Jesus Christ.

We KNOW the TRUTH by living the truth, doing the truth, keeping the truth, and becoming the truth. The Truth is KNOWN by observing it, witnessing it, living it, being it, and becoming it. The rebels and atheists don't have the truth in their lives, because they don't want it.

There are many more of these to be found in the Bible, the Book of Mormon: Another Testament of Jesus Christ, the Doctrine and Covenants, and the Pearl of Great Price. These are a sampling of a few of my favorites.

The following is probably the BEST discourse on **truth and knowledge** that has ever been given to us by the Biblical God Jesus Christ:

Doctrine and Covenants 93: 1-40:

1 Verily, thus saith the Lord: It shall come to pass that every soul who forsaketh his sins and cometh unto me, and calleth on my name, and obeyeth my voice, and keepeth my commandments, shall see my face and **know** that I am;

2 And that I am **the true light** that lighteth every man that cometh into the world;

3 And that I am in the Father, and the Father in me, and the Father and I are one —

4 The Father because he gave me of His fulness, and the Son because I was in the world and made flesh my tabernacle, and dwelt among the sons of men.

5 I was in the world and received of my Father, and the works of Him were plainly manifest.

6 And John saw and bore record of the fulness of my glory, and the fulness of John's record is hereafter to be revealed.

7 And he bore record, saying: I saw His glory, that He was in the beginning, before the world was;

8 Therefore, in the beginning the Word was, for He was the Word, even the messenger of salvation —

9 The light and the Redeemer of the world; **the Spirit of truth**, who came into the world, because the world was made by Him, and in Him was the life of men [physical body] and the light of men [psyche and spirit body].

10 The worlds were made by Him; men were made by Him; all things were made by Him, and through Him, and **of Him**.

11 And I, John, bear record that I beheld His glory, as the glory of the Only Begotten of the Father, **full of grace and truth**, even **the Spirit of truth**, which came and dwelt in the flesh, and dwelt among us.

12 And I, John, saw that He received not of the fulness at the first, but received grace for grace;

13 And He received not of the fulness at first, but continued from grace to grace, until He received a fulness;

14 And thus He was called the Son of God, because He received not of the fulness at the first.

15 And I, John, bear record, and lo, the heavens were opened, and the Holy Ghost descended upon Him in the form of a dove, and sat upon Him, and there came a voice out of heaven saying: This is my beloved Son.

16 And I, John, bear record that He received a fulness of the glory of the Father;

17 And He received all power, both in heaven and on earth, and the glory of the Father was with Him, for [the Father] dwelt in Him.

18 And it shall come to pass, that if you are faithful you shall receive the fulness of the record of John.

19 I give unto you these sayings that you may understand and **know** how to worship, and **know** what you worship, that you may come unto the Father in my name, and in due time receive of His fulness.

20 For if you keep my commandments you shall receive of His fulness, and be glorified in me as I am in the Father; therefore, I say unto you, you shall receive grace for grace.

21 And now, verily I say unto you, I was in the beginning with the Father, and am the Firstborn;

22 And all those who are begotten [reborn] through me are partakers of the glory of the same, and **are the church of the Firstborn**.

23 Ye were also in the beginning with the Father; that which is Spirit, even **the Spirit of truth** [sprit, psyche, truth is what we really are at our core being];

24 And **truth is knowledge** of things as they are, and as they were, and as they are to come;

25 And whatsoever is more or less than this is the spirit of that wicked one who was a liar from the beginning.

26 **The Spirit of truth is of God. I am the Spirit of truth**, and John bore record of me, saying: He received **a fulness of truth**, yea, **even of all truth**;

27 And no man receiveth a fulness unless he keepeth His commandments.

28 **He that keepeth His commandments receiveth truth and light, until he is glorified in truth and knoweth all things.**

29 Man was also in the beginning with God. Intelligence [psyche], or **the light of truth**, was not created or made, neither indeed can be.

30 **All truth is independent** in that sphere in which God has placed it, **to act for itself**, as all **intelligence** also; otherwise there is no existence. [Truth, psyche, intelligence is an independent entity, and the fundamental source and reality of our existence. Without psyche, there is no existence. Psyche acts for itself. Psyche is the ONLY thing that can. Psyche or intelligence is synonymous with the truth – the true reality of our existence. Psyche or intelligence is made of some kind of light. Intelligence or psyche is light and truth.]

31 Behold, here is the agency of man, and here is the condemnation of man; because that which was from the beginning is plainly manifest unto them, and they receive not the light.

32 And every man whose spirit receiveth not the light is under condemnation [and severely disadvantages himself, or damns himself, by stopping himself from finding and knowing the truth].

33 For man [human beings or human psyches] is spirit. [Each man or psyche has a spirit body]. The [physical] elements are eternal, and spirit [psyche and spirit body] and element [physical body], inseparably connected, receive a fulness of joy; [The physical elements are eternal, because they are made of a part of God's Psyche and are made from spirit matter, both of which are eternal in nature.]

34 And when separated [from his spirit body and his physical body], man [psyche] cannot receive a fulness of joy.

35 The elements [physical body] are the tabernacle of God; yea, man is the tabernacle of God, even temples; and whatsoever temple is defiled, God shall destroy that temple. [We destroy it ourselves, but God allows it to be destroyed and allows us to destroy it.]

36 The glory of God is **intelligence**, or, in other words, **light and truth**. [The glory of God is His Psyche or Non-Local Consciousness, or, in other words, light and truth. Intelligence or psyche is light and truth. Psyche is the true fundamental reality of our existence, and psyche is some type of light, a living light.]

37 Light and truth forsake that evil one. [Light and truth should reject the darkness].

38 Every **spirit of man** was innocent in the beginning; and God having redeemed man from the fall, men became again, in their [physical] infant state [as children], innocent before God.

39 And that wicked one cometh and taketh away light and truth [intelligence], through disobedience, from the children of men, and because of the tradition of their fathers. [Satan makes us stupid, and ignorant of the truth and the true reality of our existence.]

40 But **I have commanded you to bring up your children in light and truth**.

That's the best discourse on truth, knowledge, psyche, intelligence, spirit bodies, physical bodies, glory, and light that I have encountered so far in my life, and it came to us from the Biblical God Jesus Christ Himself – the pinnacle of truth and knowledge. It doesn't get any better than that.

I find it interesting whenever God says that the worlds were made "of Him", as He said above. Think about it! Only God would say such a thing. Only God could say such a thing.

Physicists have noticed that if they pour in a lot of energy, they can raise the electrons around an atom to a higher state or a higher level. Therefore, if we take a hydrogen atom, lower its electrons to a lower state or a lower level "creating" a hydrino or spirit matter or dark matter in the process, there should be a ton of energy released as a result.

What does this mean where God is concerned? It means that for every atom of physical matter that He brought into existence or commanded into existence, He had to pour some of His energy or His glory or His light or His life into it, in order to transform it from spirit matter into physical matter. Therefore, when God tells us that the worlds, the physical matter and the atoms, were created by Him, through Him, from Him, and of Him, He means it literally. There's a piece of God, a piece of God's glory and light and consciousness, in each and every physical atom that He brings into existence. God KNOWS of what He speaks, because He lived it and experienced it. He gave of Himself to produce it! God had to pour Himself and His energy and His life into the creation of this physical universe in order to bring it into existence in the first place. God says as much here, and elsewhere.

Doctrine and Covenants 88: 1-21:

1 Verily, thus saith the Lord unto you who have assembled yourselves together to receive His will concerning you:

2 Behold, this is pleasing unto your Lord, and the angels rejoice over you; the alms of your prayers have come up into the ears of the Lord of Sabaoth, and are recorded in the book of the names of the sanctified, even them of the celestial world.

3 Wherefore, I now send upon you another Comforter, even upon you my friends, that it may abide in your hearts, even the Holy Spirit of promise; which other Comforter is the same that I promised unto my disciples, as is recorded in the testimony of John.

4 This Comforter is the promise which I give unto you of eternal life, even the glory of the celestial kingdom;

5 Which glory is that of the church of the Firstborn, even of God [the Father], the holiest of all, through Jesus Christ his Son!

6 He that ascended up on high, as also He descended below all things, in that He comprehended all things, that He might be **in all and through all things**, **the light of truth** [God is Reality, Truth, and Existence. That is the Truth of our existence. A bit of God's consciousness or life is in the whole of it];

7 Which **truth shineth**! **This is the light of Christ**. As also **He is in the sun**, and the light of the sun, and **the power thereof by which it was made**.

8 As also **He is in the moon**, and is the light of the moon, and the power thereof by which it was made;

9 As also the light of the stars, and **the power thereof by which they were made**;

10 And the earth also, and **the power thereof**, even the earth upon which you stand.

11 And the light which shineth, which giveth you light, **is through Him** who enlighteneth your eyes, which is the same light that quickeneth [enlivens] your understandings;

12 Which **light proceedeth forth from the presence of God to fill the immensity of space**; [God is speaking of dark energy here, the zero-point field, the quantum sea of light, or the light of Christ. Dark energy or the light of Christ counteracts gravity and causes our universe to stretch, spread, and expand. Dark energy or the light of Christ also keeps the atoms from imploding and becoming spirit matter once again. Spirit matter is Dark Matter. Dark energy, the light of Christ, the Word of Command, or Conscious Observation is what keeps the electrons in their orbits – creating and then sustaining physical atoms. Dark energy or the light of Christ is what keeps this universe from imploding back into the infinite singularity from whence it came! Imagine it! It ALL comes from God. God had to pour Himself, His energy and light and life, into this physical universe in order to convert parts of it, 4.6% of it, to physical matter and to organize that physical matter into planets and stars, and to cause this universe to expand so that we would have space. They say that an atom is 99.999% empty space; and, it is God who is creating that space through His light, power, glory, life, command, and will. God's light and life are needed to keep those electrons in orbit, because positive attracts negative! An atom should implode, but it doesn't. God's light and life and psyche is preventing each atom from imploding! God's light and life are needed to keep this universe expanding and growing. God's light and life are needed, in order to choose what parts of that primal spirit matter to convert into physical matter, stars, planets, gas clouds, and galaxies. Galaxies come into existence whole. Galaxies aren't grown. They spring into existence as a unit or a whole, and then start to wind up. That's what we observe. If God didn't intervene in just the right spots to create stars, planets, and galaxies, then this universe would be nothing but one huge homogeneous ball of gas. Because according to Big Bang Theory every atom is supposedly moving away from every other atom, there could be no planets, stars, and galaxies without God's intervention and direction – choosing which parts of this universe to convert to physical matter and which parts of this universe to leave as dark matter or spirit matter. I think God gave this whole thing some thought and influence. God brought this whole thing to life. There is no other way to account for it logically.]

13 **The light which is in all things, which giveth life to all things, which is the law by which all things are governed, even the power of God who sitteth upon His throne, who is in the bosom of eternity, who is in the midst of all things.**

14 Now, verily I say unto you, that through the redemption which is made for you is brought to pass the resurrection from the dead.

15 And the **spirit** and the body are the soul of man. [The **spirit** or ghost is comprised of our psyche and our spirit body as a united whole. The Gods use the terms "intelligence" and "spirit" interchangeably because from their point of view, the unification of psyche and spirit body is a done deal. The soul of man is comprised of our physical body and our **spirit** or ghost].

16 And the resurrection from the dead is the redemption of the soul [spirit and physical body].

17 And the redemption of the soul is through him that quickeneth [enlivens] **all thing**s, in whose bosom it is decreed that the poor and the meek of the earth shall inherit it.

18 Therefore, it must needs be sanctified from all unrighteousness, that it may be prepared for the celestial glory;

19 For after it hath filled the measure of its creation, it shall be crowned with glory, even with **the presence of God the Father**;

20 That bodies who are of the celestial kingdom may possess it forever and ever; for, for this intent was it made and created, and for this intent are they sanctified.

21 And they who are not sanctified through the law which I have given unto you, even the law of Christ, must inherit another kingdom, even that of a terrestrial kingdom, or that of a telestial kingdom.

God's Light, Life, and Glory is within each atom in our physical bodies. God's programming skills are manifest in our genomes; and therefore, God's signature is written on every living cell in our physical body within the software that is our genome. God has filled us with Himself and written Himself upon us. It doesn't get more personal than that. Think about it if you can. God KNOWS the truth because He is the truth and because He makes the truth and becomes the truth. God KNOWS reality, because God makes reality. Reality is made up of God and came from God.

As physical mortal beings, we don't think about these things nor have much of an appreciation for these things. But God does, because He had to give of Himself in order to bring physical matter into existence in the first place. God had to make a sacrifice of Himself in order to bring this physical universe into existence and in order to bring us to life.

The Truth is KNOWN by living it and experiencing it! This reality applies as much to God as it does to us.

You come to have faith in the truth by living it and acting upon it. Over time, that faith transforms into knowledge, because it is based upon the truth. Truth is knowledge; and, knowledge is truth, because it is impossible to know the truth based upon falsehoods and lies. Lived Experience or Observation is the pinnacle of knowledge and truth.

I find it interesting how a change in philosophical perspective makes things like Psyche and God's Psyche useful, feasible, realistic, credible, and even necessary. Everything points to Psyche or Intelligence, except for the exclusionary philosophies and sciences which deliberately and knowingly exclude Psyche or Non-Local Consciousness from consideration. But, even they point us to Psyche by trying to steer us away from Psyche.

I have learned to trust the Lived Experiences and Observations of the human race and the prophets of God, because Lived Experiences are our ONLY way of finding and knowing the truth. You can't find and know the truth through the scientific methods, because the scientific methods are based upon a wide variety of logic fallacies and philosophical assumptions which prevent the scientific methods from being able to prove the truth.

The TRUTH is KNOWN by observing it, living it, witnessing it, and experiencing it for ourselves – not through philosophical speculation or the philosophical interpretation of scientific evidence. Lived Experience IS the BEST way of finding and knowing THE TRUTH.

22 And now, after the many testimonies which have been given of him, this is the testimony, last of all, which we give of him: That he lives!

23 For we saw him, even on the right hand of God; and we heard the voice bearing record that he is the Only Begotten of the Father;

24 That by him, and through him, and of him, the worlds are and were created, and the inhabitants thereof are begotten sons and daughters unto God.

Due to the inherent weaknesses of the scientific methods, Lived Experience or Observation is technically the ONLY way of finding and knowing the truth. It's elementary, my dear friend!

I think that this is the greatest scientific discovery of my career as a scientist, theoretician, psychologist, observer, and philosopher. Lived Experience or Observation is the pinnacle of knowledge and truth. We KNOW the Truth by living it and experiencing it for ourselves, or by choosing to trust someone who has.

Doctrine and Covenants 46: 13-14: "To some it is given by the Holy Ghost to **know** that Jesus Christ is the Son of God, and that He was crucified for the sins of the world. To others it is given to believe on their words, that they also might have eternal life if they continue faithful."

Experience or knowledge is a better way of knowing the truth than philosophical speculation or scientific experimentation. For some, knowledge and truth are revealed to them by God. They are eye-witnesses. Others choose to believe on their words. I have tended to be in the latter camp, whenever I haven't been deceiving myself, although the other kind of knowledge has been coming online for me in recent years because I chose to open myself up to it and chose to ask God for it. I have a testimony of asking, knocking, and seeking because I have observed that it leads to receiving and knowing of The Truth, who is God.

The following scriptures sum up my life, during the past five years of my life:

Doctrine and Covenants 88: 118: "And as all have not faith, seek ye diligently and teach one another words of wisdom; yea, seek ye out of the best books words of wisdom; seek learning, even by study and also by faith."

Doctrine and Covenants 46: 13, 14: "To some it is given by the Holy Ghost to know that Jesus Christ is the Son of God, and that He was crucified for the sins of the world. To others it is given to believe on their words, that they also might have eternal life if they continue faithful."

I didn't have a lot of faith, but I definitely know how to study and do research. I'm not a prophet or seer or clairvoyant, but I have learned to believe and trust the people who are. I have learned to trust and believe their Lived Experiences.

The purpose of life is to make friends, and to seek The Truth and find The Truth. I have been doing a little bit of both ever since I got sober and gave up My Materialism, My Nihilism, My Scientism, and My Atheism.

When it comes to the therapeutic benefits of a Relational Ontology and the truthfulness of a Psyche Ontology, and when it comes to the powerful effects of the Atonement of Christ for physical healing and for psyche-therapy, I KNOW these things to be real and true, because I have lived them and experienced them for myself. It doesn't get any better than that!

Faith in the unseen becomes KNOWLEDGE of the unseen through Lived Experience. I KNOW that the non-physical or the non-local exists, because I have experienced parts of it for myself and because I have chosen to trust those who have experienced all of it for themselves. When it comes to the non-local or the non-physical or the spiritual, I have chosen to believe in and trust the Lived Experiences of the human race, because our collective Lived Experiences or Psyche Experiences PROVE that these non-local realities do in fact exist.

The Biblical God Jesus Christ really isn't hiding from us. Thousands, if not millions, have experienced Him directly through their Lived Experiences. However, there are in fact millions if not billions of Materialists, Naturalists, Nihilists, Darwinists, Behaviorists, and Atheists who are trying to hide from the Biblical God Jesus Christ and believe that they are having some success in doing so. I used to be one of them, so I KNOW. I lived it and experienced it.

PART III — MY SCIENTIFIC DISCOVERIES

1. One of my greatest scientific observations and scientific discoveries came when I finally realized that Materialism, Naturalism, and Darwinism have been falsified trillions of different times in thousands of different ways. That was a major conceptual breakthrough for me, because I used to be a Materialist, Nihilist, and Atheist.

Materialism, Naturalism, and Darwinism are BEST defined as "Design and Creation by Physical Matter". These people literally believe and teach that the rocks – physical reactions – designed, programmed, created, engineered, field-tested, manufactured, and deployed ALL of the different genomes and life forms on this planet. That's what these people really believe, and that's what these people teach in all of our public classrooms.

Ironically, Louis Pasteur falsified Materialism, Naturalism, Darwinism, and the Theory of Evolution in 1859 by demonstrating that the rocks – raw physical matter – cannot design and create. That was the same year that Charles Darwin published, "On the Origin of Species". Louis Pasteur proved the Theory of Evolution or Creation by Rocks false the same year that Charles Darwin introduced his theory to the world. Remember, Science and the scientific methods can be used to prove theories false. It's called falsifying a theory; and, that's what the scientific methods do, or are supposed to do.

Technically, the scientific methods can't be used to prove anything true. Due to a wide variety of different logic fallacies which are built directly into the scientific methods, the scientific methods cannot be used to prove a theory true, as the Materialists, Naturalists, and Darwinists try to do.

Instead, we can use the scientific methods to falsify theories. Using the scientific methods to prove our theories false is philosophically and logically sound. In other words, if you falsify a theory by *negating the consequent*, you have in fact proven that theory false.

Here's how it works in principle:

Scientific Hypothesis: If Theory X is true, then we will observe Y.

Scientific Observations: We don't observe Y.

Scientific Conclusion: Therefore, Theory X is not true.

Here's how it works in practice:

Scientific Hypothesis: If the Theory of Evolution, Materialism, Naturalism, and Darwinism are true, then we will observe the rocks and physical reactions designing, creating, and manufacturing genomes and life forms from scratch.

Scientific Observations: We have NEVER observed the rocks and physical reactions designing and creating genomes and life forms; and, we NEVER will. They can't.

The Scientific Conclusion: Therefore, the Theory of Evolution, Materialism, Naturalism, and Darwinism are not true.

I just successfully falsified Materialism, Naturalism, Darwinism, and the Theory of Evolution; and, I used the Scientific Method to do so! Best of all, my argument is philosophically and logically sound. I have in fact falsified Materialism, Darwinism, Naturalism, and the Theory of Evolution for REAL! I just used the Scientific Method to

PROVE these things false. I successfully used the Scientific Methods for what they are good for – falsifying theories that are false. I wish I would have known how to do that forty years ago. It would have saved me a lot of confusion and grief.

In contrast, there is NO way to falsify theories that are true, such as "Design and Creation by Intelligent Beings or by Intelligent Psyches". Instead, true theories are continuously verified over and over again. "Design and Creation by Psyche or by Intelligence" has been observed and verified and experienced trillions of trillions of different times and ways, with an infinite number of more to go. See how that works? That's Science and the Scientific Methods in action for REAL, doing what they are best at!

(See: Slife, B. D. & Williams, R. N. (1995). Science and Human Behavior. In *What's Behind the Research? Discovering Hidden Assumptions in the Behavioral Sciences*, (pp. 167–204). Thousand Oaks, CA: SAGE Publications.

http://mypsyche.us/wp-content/uploads/2017/04/Science.pdf).

This has been and is one of my greatest scientific discoveries and scientific observations of all time, during the whole of my science career. I have observed that there are literally thousands of different ways to falsify Creation by Rocks – or Materialism, Naturalism, Darwinism, and the Theory of Evolution; and, they are all philosophically and logically sound. Thanks to the scientific methods and scientific observations, we KNOW that Creation by Rocks, Materialism, Naturalism, and Darwinism are false, because we KNOW why they are false – rocks and physical reactions cannot design and create. It's elementary my dear friend.

There's NO way that Evolution (Mutation and Selection) could have designed, programmed, engineered, created, and manufactured the first genomes and life forms, because Evolution didn't even exist until AFTER God had designed, created, and deployed the first genomes and life forms in the first place.

I have observed that the Scientific Methods or Observation have been used thousands of different ways to falsify Materialism, Naturalism, Darwinism, and even Atheism. Atheism is Creation by Nothing, or Creation by Chance. The Atheists really truly believe that Nobody and Nothing designed and created everything. Nobody, Nothing, and Chance are the holy trinity of Atheism. These people believe in those false gods or idols with a passion. Materialism, Naturalism, and Atheism are in fact our modern-day form of idolatry; and, these people are idolaters. These people worship the rocks and physical reactions, as if these things were God.

Ironically, God must of necessity exist in order to have done ALL of the science that needed to be done, which the rocks or raw physical matter could never have done. Only Psyche can design and create; and, God's Psyche is the only one we know of who was there at the time and could have done the job.

Finally, as a capstone, the Biblical God Jesus Christ has told us repeatedly in the Bible, Book of Mormon: Another Testament of Jesus Christ, the Doctrine and Covenants, and the Pearl of Great Price that HE designed and organized the heavens, this earth, and all of the life forms on this earth. The Biblical God confessed to doing the job.

For me personally, falsifying Materialism and Darwinism became my first really convincing Scientific Proof of God's Existence. After falsifying Materialism, Naturalism, and Darwinism, I simply KNEW that God exists. I was finally willing to follow the evidence, wherever it might lead me.

(See: Mark My Words. (2016). *The Theory of Evolution Proved to Me that God Exists: Why I Am No Longer an Atheist and Why I No Longer Believe in the Theory of Evolution*. Kindle. Retrieve from: https://www.amazon.com/dp/B01HZYBZ7K).

(See also: Mark My Words. (2016). *The Scientific Method: Proves That the Theory of Evolution Is False*. Kindle. Retrieve from: https://www.amazon.com/dp/B01IAAIRT2).

2. My greatest scientific observation and scientific discovery is the realization that Lived Experience IS the BEST way of finding and knowing the truth. Lived Experiences or Psyche Experiences are the best way of finding the truth and knowing the truth in EVERY realm of existence, including this physical realm.

Due to the fact that the scientific methods can't be used to find the truth and prove the truth directly, and due to the fact that the scientific methods are typically restricted to the physical realm or the local realm, and due to the fact that there are a wide variety of logic fallacies built into the scientific methods, the scientific methods are in fact a much weaker source of knowledge and truth than Lived Experiences are. Lived Experiences or Psyche Experiences are in fact an infinitely better way of finding and knowing the Truth than science and the scientific methods can ever be. I wish I would have known that forty years ago. It would have saved me a lot of frustration and grief.

My greatest scientific discovery of all time is that the Lived Experiences of the human race are in fact a much better way of finding and knowing the truth than Science and the scientific methods. For me, that was a major epiphany and scientific breakthrough, because I used to be a Materialist, Nihilist, and Atheist.

Lived Experience is extremely powerful. Lived Experience is Science, Observation, Knowledge, and Truth in their purest form. Science is supposed to be Observation and Knowledge and the pursuit of the Truth. In other words, Science is supposed to be Lived Experience. The Materialists, Naturalists, Behaviorists, and Atheists did the world a great disservice when they hijacked and stole Science and limited Science exclusively to physical matter. One of my greatest scientific discoveries was to finally realize how wrong these people really are.

I KNOW how it goes, though, because I allowed My Scientism, My Nihilism, My Materialism, and My Atheism to blind me to Lived Experience as a source of evidence and to blind me as to how powerful and useful Lived Experiences really are as a source of knowledge and truth. But, that's what we Materialists and Atheists do. We refuse to allow Lived Experiences into evidence. Materialism and Atheism of any kind is based upon a refusal to look at evidence.

I truly believed that Science and the scientific methods were the ONLY way for finding and knowing the truth. I WAS WRONG! In fact, the scientific methods can't be used to prove the truth, which means that Lived Experience is in reality the ONLY way to find and KNOW the truth directly. The truth is KNOWN by living it and experiencing it for yourself, or by choosing to trust someone who has.

I never fully realized until April 2017 that Lived Experience or Direct Observation is a much better way of finding and knowing the truth than the scientific methods and philosophical guesswork will ever be; but, I'm seeing it now and understanding it now.

We KNOW from the Lived Experiences of the human race that the Biblical God Jesus Christ does indeed exist and truly rose from the dead – although it's nice to have Science pointing us to the same truths, in its roundabout way.

(See: *Bible, Book of Mormon: Another Testament of Jesus Christ, Doctrine and Covenants, and Pearl of Great Price* – available for free at:

https://www.lds.org/scriptures/bible?lang=eng).

3. During the Prime Event, or what most scientists call the Big Bang, only 4.6% of this universe was converted to physical matter. According to Astrophysics, the other 95.4% of this universe remained Non-Local or Spiritual as dark matter (spirit matter) and dark energy (the Light of Christ or the zero-point field of light).

The Materialists, Naturalists, Darwinists, and Atheists are WRONG about 95.4% of this universe! Imagine it! 95.4% of this universe is still spiritual or non-local. Doesn't it make you wonder what else the Materialists might be wrong about?

(See: Ross, H. (2008). *Why the Universe Is the Way It Is*. Grand Rapids, MI: Baker Books.)

There are indications that a lot of energy has to be poured into spirit matter in order to convert spirit matter to physical matter. In other words, God poured a tiny bit of His Own Psyche, Light, Intelligence, Power, Consciousness, Energy, or Life into each physical atom that God brought into existence or commanded into existence. ONLY God could have done such a thing. We human beings certainly don't know how to do that kind of Science!

Because of this KNOWLEDGE, I'm now willing to admit Lived Experiences into evidence.

(See: https://www.lds.org/scriptures/dc-testament/dc/93?lang=eng).

(See: https://www.lds.org/scriptures/dc-testament/dc/88?lang=eng).

(See: https://www.lds.org/scriptures/dc-testament/dc/76?lang=eng).

(See: https://www.lds.org/scriptures/dc-testament/dc/130?lang=eng).

4. One of my greatest scientific discoveries came when I first realized that Quantum Mechanics is in fact Spiritual Mechanics, or the way that spirit matter really works. When I first realized that Psyche or a Non-Local Conscious Observer and the Word of Command are needed to convert spirit matter into physical matter, then suddenly the whole of Quantum Mechanics became clear to me.

God must of necessity exist in order to have provided the necessary Word of Command or Conscious Observation, which Quantum Mechanics tells us must take place in order to convert spirit matter into physical matter. The Bible actually tells us that Jesus Christ is the WORD or the Word of Command who brought physical matter into existence in the first place. If God's Psyche didn't exist, then there would be NO physical matter.

For me personally, Quantum Mechanics became my first positive Scientific Verification of God's Existence. The scientists keep verifying Quantum Mechanics or Spiritual Mechanics over and over again; so, it must be the truth, because it's impossible to use the scientific methods to falsify the truth. Quantum Mechanics or Spiritual Mechanics has never been falsified, although the materialistic interpretations of Quantum Mechanics have been falsified trillions of times.

After realizing that God's Psyche must of necessity exist in order to convert spirit matter into physical matter, I simply KNEW that God exists.

(See: Van Lommel, P. (2010). *Consciousness Beyond Life: The Science of the Near-Death Experience*. New York: HarperCollins.)

(See: Goswami, A. (2008). *God Is Not Dead: What Quantum Physics Tells Us about Our Origins and How We Should Live*. Charlottesville, VA: Hampton Roads.)

(See also: Mark My Words. (2016). *Quantum Mechanics from a Non-Physical Spiritual Perspective*. Kindle. Retrieve from: https://www.amazon.com/dp/B01J023TGU).

The existence of Light falsifies Materialism and Naturalism, and it ends up pointing us directly to God and God's Psyche.

(See: Mark My Words. (2017). *God Is in the Light: God is light, and in Him is no darkness at all*. Retrieve from: https://www.amazon.com/dp/B07168S37N).

5. Some of the scientists have been smart enough and open-minded enough to observe that a genome is a radically advanced software program, that DNA is in fact like a radically advanced hard drive or memory storage device, and that a living cell is a radically advanced piece of computer hardware or nanotechnology. We, who have made this realization, also realize that ONLY Psyche or Intelligence is capable of designing, creating, engineering, and manufacturing hard drives and hardware; and, ONLY Intelligent Beings or Psyche Beings can do computer programming or the creation of software and genomes.

God must of necessity exist in order to have designed, engineered, manufactured, and created all of the DNA molecules, proteins, and living cells on this planet. These things don't put themselves together. They have never been observed doing so! In fact, Louis Pasteur used Science in 1859 to prove that these things can't put themselves together, by falsifying Spontaneous Generation or Materialism and Creation by Rocks.

Furthermore, God must of necessity exist in order to have written or programmed each and every genome that is encoded as software in the DNA of each life form on this planet. ONLY Psyche or intelligent beings could have done such a thing. We have NEVER caught the rocks in the act of designing, programming, engineering, and manufacturing genomes and life forms from scratch; and, we never will. They can't!

A Genome IS God's Signature!

What's most interesting about all of this is that God wrote His signature in each one of your cells. Your genome IS God's signature. Each genome on this planet is God's signature. God has written Himself upon you and within every life form on this planet. The rocks or raw physical matter can't write genomes or computer software, but Psyche certainly can. God's Psyche must of necessity exist, or you wouldn't have a genome. God's Psyche must exist, or you wouldn't have a physical body. That's what Science is trying to teach us!

(See: Mark My Words. (2016). *Using the Scientific Method: To Eliminate the Usual Suspects and to Prove the Truth*. Kindle. Retrieve from: https://www.amazon.com/dp/B01J6STHP0).

6. In many of his books, Hugh Ross uses Cosmic Fine-Tuning as Scientific Proof of God's Existence. The whole thing made logical sense to me, because ONLY Psyche can

convert or transmute spirit matter into physical matter. ONLY Psyche can organize physical matter into useful and productive forms such as planets, stars, galaxies, genomes, and life forms. ONLY Psyche can do Cosmic Fine-Tuning. God's Psyche must of necessity exist, in order to have done all of the cosmic fine-tuning and science that needed to be done; otherwise, you and I would not be here right now in this physical realm. Science has made it obvious and clear to me that it must be so. It's obvious that the rocks or raw physical matter can't do Cosmic Fine-Tuning. That kind of process requires an active, living, agentic, conscious, intelligent Being or Psyche, whom many of us tend to call God or the Ultimate Scientist.

(See: http://www.reasons.org/articles/fine-tuning-for-life-in-the-universe).

7. Many different scientists taught me that Random Mutations and Natural Selection (Evolution) cannot design and create genomes and life forms. They met their burden of proof and demonstrated to me that it must be so. Once we have eliminated all of the falsehoods such as Materialism, Naturalism, Atheism, Nihilism, and Darwinism, then we are left staring at THE TRUTH, which is that ONLY Psyche can design, program, engineer, manufacture, create, and do science.

By eliminating all of the falsehoods or pseudo-sciences, it becomes obvious that God's Psyche must of necessity exist in order to have DONE all of the Science, which evolution and the rocks could NEVER have done.

(See: Wells, J. (2000). *Icons of Evolution: Science or Myth? Why Much of What We Teach About Evolution Is Wrong*. Washington, DC. Regnary.)

(See: Sanford, J. (2014). *Genetic Entropy* (4th ed.). Cornell University: FMS Foundation.)

(See: Sanford, J. C., Marks, R. J., Behe, M. J., Dembski, W. A., & Gordon, B. L. (Eds.). (2013). *Biological Information: New Perspectives*. Hackensack, NJ: World Scientific.)

(See: Meyer, S. C. (2010). *Signature in the Cell: DNA and the Evidence for Intelligent Design*. New York: HarperCollins.)

(See: Meyer, S. C. (2013). *Darwin's Doubt: The Explosive Origin of Animal Life and the Case for Intelligent Design*. New York: HarperCollins.)

(See: Mark My Words. (2016). *The Scientific Method: Proves That the Theory of Evolution Is False*. Kindle. Retrieve from: https://www.amazon.com/dp/B01IAAIRT2).

(See: Mark My Words. (2016). *The Theory of Evolution Proved to Me that God Exists: Why I Am No Longer an Atheist and Why I No Longer Believe in the Theory of Evolution*. Kindle. Retrieve from: https://www.amazon.com/dp/B01HZYBZ7K).

8. Neuroscientists and brain stimulation have demonstrated that Psyche or Mind is a completely different entity than the physical brain. Any Science which points us to Psyche and Mind and Quantum Nonlocality is going to end up being vastly superior to the watered-down stuff that we get from the Materialists, Naturalists, and Atheists.

(See: Penfield, W. (1978). *The Mystery of the Mind: A Critical Study of Consciousness and the Human Brain*. Princeton, NJ: Princeton University Press.)

(See: Eccles, J. C., & Popper, K. R. (1977). *The Self and Its Brain: An Argument for Interactionism.* New York: Routledge.)

(See: Eccles, J., & Robinson, D. N. (1984). *The Wonder of Being Human: Our Brain and Our Mind.* New York: The Free Press.)

(See: Eccles, J. (1985). *Mind and Brain: The Many-Faceted Problems.* New York: Paragon House Publishers.)

These scientists solved the mind-brain problem over thirty years ago; but, the Materialists and Atheists refuse to read the memo. Materialism and Atheism of any kind are based upon a refusal to look at evidence. I KNOW, because I used to be a Materialist and an Atheist, until I started looking at the evidence.

Once you KNOW that Psyche or Mind exists, then you are suddenly free to go looking for the Mind of God.

9. Any Science that demonstrates Mind over Matter is going to point us directly to Psyche and establish the fact that Psyche is something completely different than matter, whether we are talking about spirit matter or physical matter.

The Placebo Effect is scientific proof of Psyche's existence.

(See: Dispenza, J. (2014). *You Are the Placebo: Making Your Mind Matter.* USA: Hay House Inc.)

(See also: McTaggart, L. (2002). *The Field: The Quest for the Secret Force of the Universe.* New York: HarperCollins.)

A complete understanding of Quantum Mechanics explains how God's Psyche was able to convert or transmute spirit matter into physical matter at the locations in this universe where He wanted there to be physical matter. It explains how the galaxies came into existence all at once!

If the Big Bang Theory were 100% true, then every particle in this universe should be moving away from every other particle in this universe; and, there would be NO planets, stars, and galaxies! This universe should be nothing more than one huge homogeneous ball of gas. It required God's Psyche in order to transmute spirit matter into physical matter at the current LOCATION of the planets, stars, and galaxies. God's Psyche is the ONLY way to explain the existence of planets, stars, and galaxies, because they shouldn't exist at all if the Big Bang Theory as typically presented to us were 100% true.

The Big Bang Theory "proves" that this physical universe had a beginning. That's the part of the Big Bang Theory that I truly believe in. However, the Big Bang Theory FAILS to explain how the stars, planets, and galaxies came together; and, for an explanation of that reality we have to turn to God's Psyche and Mind-over-Matter.

A complete understanding of Mind over Matter ends up explaining how God's Psyche was able to organize planets, stars, galaxies, genomes, and life forms from raw unorganized physical matter. These are all things which the rocks or raw physical matter could NEVER have done!

For me personally, John Pratt's sacred calendars provided Scientific Proof of God's Existence and scientific proof of mind over matter at a cosmic scale:

http://www.johnpratt.com/items/docs/lds/dates.html

http://www.johnpratt.com/items/docs/article_nums.html

That scientist certainly had a vision, which would ONLY have been possible by choosing to believe in God's existence in the first place.

(See: https://www.lds.org/scriptures/pgp/abr?lang=eng).

You particularly want to look at Abraham 4: 10, 12, 18, 21, and 25; where the Gods watched and waited to make sure that the physical matter obeyed them, during the organization and terraforming of this earth.

The Gods are Scientists, not magicians! There is NO such thing as Creation Ex Nihilo or magic.

(See also: Mark My Words. (2017). *I Am Not a Creationist: So What Am I?*

https://www.amazon.com/dp/B071XTM8XY).

10. The book, *Adventures Beyond the Body: How to Experience Out-of-Body Travel*, by William Buhlman was a game-changer for me. Here was my first encounter with someone who can go out-of-body basically "at will". He turned the whole experience into a Science. I'm talking about a REAL SCIENCE of direct observation, experimentation, and lived experience – NONE of that garbage where they deliberately restrict their "science" to physical matter.

This book about Out-of-Body Experiences (OBEs) provided me with some of my very first evidentiary PROOF that Materialism and Naturalism are false. Buhlman ran science experiments and tried things out while out-of-body on the astral plane or in the spirit realm. He did REAL SCIENCE, doing direct observation of the spirit realm and direct experimentation with the spirit realm, while out-of-body. I thought this whole thing was really cool and fascinating.

With Lived Experiences, you go directly to KNOWLEDGE, TRUTH, and PROOF. Lived Experiences are vastly superior to the scientific methods for getting at truth and proof and knowledge. Technically, you can't use the scientific methods to prove anything due to the wide variety of logic fallacies and biases which are built into the scientific methods. But, you can definitely use Lived Experiences or Direct Observations as knowledge and proof of the truth.

This was the first book to help me see what Science should really be all about – Observation and Lived Experience, which document KNOWLEDGE and PROVE the TRUTH.

Buhlman helped to set me up, so that I was ready to take the NDErs and their out-of-body experiences seriously.

There are at least 13 million documented cases of Near-Death Experiences (NDEs) in the record of Lived Experience for the human race. The Materialists and Naturalists have chosen to deny and reject everyone. Denial of evidence and rejection of evidence is NOT science! But, this is exactly the way that the Materialists and Atheists do science, by denying and rejecting evidence or Lived Experience.

The discovery of all those different NDEs and OBEs is in fact one of my greatest scientific discoveries!

The best way to KNOW God is to live Him and experience Him directly, or to choose to trust someone who has. Science and Scientific Methods can point us to God as the BEST

EXPLANATION for the scientific evidence. However, through Lived Experience we can KNOW the truth and KNOW God directly if we want to.

That's very powerful Science there!

(See: Buhlman, W. L. (1996). *Adventures Beyond the Body: How to Experience Out-of-Body Travel*. New York: HarperCollins Publishing.)

(See: Durham, E. (1998). *I Stand All Amazed: Love and Healing from Higher Realms*. Orem, UT: Granite Publishing and Distribution.)

(See also: Storm, H. (2000, 2005). *My Descent into Death: A Second Chance at Life*. USA: Random House.)

Conclusion: All of this was a long time in coming, fifty-five years, but I finally got there in the end. Still, I wish I would have known about all of this stuff forty years ago. It would have saved me a lot of time and grief.

ALL of my greatest scientific discoveries and scientific observations have pointed me directly to God, because ONLY God's Psyche could have done ALL of the science which needed to be done at the time that these various things came into existence. This is obvious and clear to me now; but, it wasn't when I was a Materialist, Nihilist, and Atheist. Go with the BEST and get rid of all the rest. That's what I finally decided to do; and, you can too. All of this came open to my view, once I got rid of My Materialism, My Scientism, My Nihilism, and My Atheism.

If you really want to do Science, you start by getting rid of the pseudo-sciences, such as Darwinism, Materialism, and Naturalism. Then you switch over to Observation or Lived Experience for your scientific evidence and your Science. Through Lived Experiences, including spiritual experiences, you can go directly to KNOWING the TRUTH which is something that you can't do with the scientific methods and philosophical speculation. I never realized how powerful and revelatory Lived Experience can be, until after I got rid of My Materialism, My Atheism, My Nihilism, and My Scientism.

I never realized how weak and insubstantial the scientific methods and science really are until AFTER Joseph Rychlak, Brent Slife, and Edwin Gantt pointed out all the different logic fallacies that are built into the scientific methods. These are concepts which you will never get from your materialistic and atheistic college professors, because these people want you believing that science and the scientific methods are infallible.

There was a time in my life when I truly believed that Science and the scientific methods are the ONLY way of finding and knowing the truth. Boy, was I wrong! Over fifty years of being wrong! I'd been duped! Due to all the different logic fallacies that are built into the scientific methods, we can't use the scientific methods to prove the truth and to know the truth. If we want to KNOW the TRUTH, then we have to switch over to Lived Experience or Direct Observation or choose to trust someone who has had the kinds of Lived Experiences that we wish we would have had. I KNOW of what I speak, because I have lived it and experienced it, on both sides of the fence.

Psyche

I consider Psyche to be my greatest and most useful scientific discovery of all time.

Psyche or Non-Local Consciousness is primarily an observational, experiential, empirical science, because it cannot be detected by our physical instruments. From the physical side of the equation, Psyche is simply inferred to exist based upon the influence it has upon physical matter – much the same as wind, magnetism, gravity, dark energy, microwaves, radio, gamma rays, action at a distance, and other types of forces and fields. We can't see these things directly with our physical eyes nor hear them with our physical ears; but, WE KNOW that they exist based upon the influence they have on physical matter. Psyche is the same type of phenomenon.

This experiential account from William Buhlman completely changed the way that I look at Psyche. I latched onto it and never looked back.

Journal Entry, October 2, 1982

I hear the buzzing, engine-like sounds and will myself out-of-body. I step to the bedroom door and automatically request "Clarity now!" My vision improves and I step through the door, into the living room. Still feeling a little out of sync, I verbally repeat my request with more emphasis, "Clarity now!" I feel my awareness and vision snap into place.

My thoughts are clear and I make a verbal demand, "I need to see the form I'm in now!" Instantly I feel an intense sensation of being drawn within myself. I'm suddenly different, weightless as though I'm floating in space. As I look forward I see a sparkling, bluish white form. For some reason, I seem to know that I'm looking at my nonphysical body from a different perspective. I stare in amazement at this form before me that shines and flows with energy and light. It looks like an energy mold created from a million tiny points of light; it radiates a bluish glow but appears to have a defined outer structure. The body of light before me is naked and is identical to my physical form. Even though my body looks firm, there is a noticeable energy motion and radiation present. I can see what appears to be an ocean of blue stars throughout my body. It's difficult to describe because the stars are stable, yet moving at the same time; the light and energy of my body appear to change and flow almost like the waves of an ocean.

As I stare at the body of light, it hits me that I must be in another body. Yet I can't perceive any form or substance; I'm like a viewpoint in space without shape or form of any kind. As I reflect upon my new state of being, I feel a sensation of rapid motion and I'm instantly back within my physical body.

Lying still and reviewing my experience, I'm struck by an inescapable conclusion: I must possess multiple energy-bodies. The form I just experienced was noticeably lighter (less dense) than even my second nonphysical body. I realize that the traditional view of our possessing two bodies — a physical body and a spiritual body — is far too simplistic; we are much more complex than this. Just as there are multiple nonphysical energy dimensions within the universe, each of us must consist of multiple energy-bodies or vehicles of expression.

Now I seriously wonder just how many nonphysical bodies or forms this involves. I suspect that there must be one within each dimension of the universe and that all of these are interrelated and connected, just as the physical body is connected to its first nonphysical (spiritual) body.

Buhlman, William L., "Adventures Beyond the Body: How to Experience Out-of-Body Travel" (pp. 34-35). HarperCollins. Kindle Edition.

Psyche is an immaterial viewpoint in space – a third-eye or third-person perspective. From that vantage point as an immaterial spark or an immaterial viewpoint in space, one can look upon his or her own spirit body, examine it, and try to understand it. Accounts like these are evidentiary proof that our Psyche, or Intelligence, or Non-Local Consciousness is something completely different than our spirit body; and, our spirit body is something completely different than our physical body.

I have chosen to treat this observational account as scientific evidence, and proof of concept; and then, I have tried to verify it from other personal accounts of Psyche, or that third-person immaterial viewpoint in space.

In the summer of 2017, I upgraded my science to Science 2.0. Science 2.0 allows all of the evidence into evidence, and Science 2.0 pursues a preponderance of the evidence.

For the sake of simplicity, I have chosen to believe that we have only one spirit body, which is capable of manifesting or existing at multiple different frequencies, levels, or dimensions within the Non-Local Realms or Spirit Realms. However, there are accounts on record from people who have experienced the phenomenon of bi-location; wherein, their spirit has been in two different locations on this earth simultaneously. Whether that's an indication of one of our spirit body's capabilities, or an indication that we have two or more spirit bodies, I don't know. However, it does seem as if our spirit body is capable of simultaneous multitasking, or parallel processing and parallel existence; whereas, it's obvious that our physical body is not.

Things like Psyche or Non-Local Consciousness are experiential, observational, empirical sciences, and not replicable verifiable physical sciences. We experience magnetism, gravity, dark energy, action at a distance, and the invisible wavelengths of light; but technically, we don't observe them directly with our physical instruments. We observe their effects on physical matter instead. Psyche is the same type of thing. Psyche is a force or a field like gravity, dark energy, magnetism, action at a distance, and the strong and weak nuclear forces. We experience these things and observe the effects of these things on physical matter, rather than observing them directly with our physical instruments. They are inferred to exist because they have an influence on physical matter.

These discoveries are the fruits of allowing all of the evidence into evidence, and then choosing to pursue a preponderance of the evidence by allowing the evidence to lead you wherever it wants to lead you.

I used to be a Materialist, Naturalist, Nihilist, and Atheist; but, I no longer try to force the evidence to fit my pre-conceived notions as to how things are supposed to be. Instead, I try to let the evidence speak for itself, and then run with it. This is a different way of doing science, a better way of doing science, in my humble opinion, because it greatly enhances our ability to discover and study new concepts and ideas that come our way. It's an eyes-wide-open approach rather than a head-in-the-sand approach to science. You don't discover anything new or interesting if you always have your head in the sand as the Materialists, Naturalists, Darwinists, and Atheists do.

Carl Jung and William James actually used the term, Psyche, within their books and research papers. Pim van Lommel calls it, "Non-Local Consciousness", within his book *Consciousness Beyond Life*. The Biblical God Jesus Christ occasionally calls it Intelligence, or Light and Truth.

From Uncle Google:

The terms "Urim" and "Thummim" have traditionally been understood as "light(s)" and "perfection(s)" or as "perfect light." The Urim and Thummim were a means of revelation entrusted to the high priest.

There's Psyche, or Intelligence, or Perfect Light within these stones. These seer stones are like our computers, but sentient, intelligent, conscious, and aware.

Direct experiential evidence of Psyche is rather rare; but, I have found a few other experiential accounts of the nature and functionality of Psyche or Non-Local Consciousness. Some of them were found on YouTube.

When people are separate from their spirit body and physical body, and they are viewing their physical body while they are an immaterial spark on the ceiling, they are experiencing their Psyche rather than their spirit body. It happens to some out-of-body travelers, but not all of them.

Lightning Strike Near-Death Experience – Dr. Anthony Cicoria

When Dr. Anthony Cicoria was struck by lightning, he died. He, his spirit body, was outside his physical body. As he walked up the stairs, his spirit body dissolved beneath him and transformed into an orb of light. He, his Psyche, was looking at his spirit body and later that orb of light from an immaterial third-person perspective or that Third Eye Perspective. This NDE is Scientific Evidence or Empirical Proof that Psyche or Intelligence is a different entity than our spirit body and our physical body.

https://www.youtube.com/watch?v=WUXzj0Tczz4

Near Death Experience of Barbara Wilcox

In the next NDE, Barbara Wilcox explains not having a body, or spirit, or a soul but being Total Intelligence.

https://www.youtube.com/watch?v=JQzap_jFXT8

Jessica NDE Near-Death Experience of a different kind

In the following NDE, Jessica describes being a Point of Light, a Small Spec, and a Pinpoint of Light during her near-death experience. She, too, is describing Psyche or Non-Local Consciousness. Jessica also described the fact that we don't see everything and don't understand everything which is there to be seen and understood while out-of-body. Instead, there is ever-growing awareness and understanding, with still more to go that is never reached. Jessica could sense the presence of others, but she didn't engage with them during her NDE as much as others seem to do. Hers is the most "self-centered" NDE that I have encountered so far. Jessica's Psyche is very much the center of her universe.

https://www.youtube.com/watch?v=Ve6RG9K3qrA

There are many things about Jessica's NDE that are atypical, abnormal, weird, unusual, and strange – making some people believe that she's faking the whole thing. I include her in this short list, because she talks about being that immaterial spark during her NDE. I've never heard Jessica talk about her spirit body. She always describes herself as a spark; and, I define that "immaterial spark" or "viewpoint in space" as Psyche, Intelligence, or Non-Local Consciousness within my books.

Conclusion

Technically, you need two or three witnesses in order to establish the truth of a concept or the truth of a matter; and, I have them here. Our Psyche or Intelligence does indeed and can indeed exist and function separate from its assigned spirit body.

It's always interesting to observe that whenever the Psyche is separate from its spirit body, it's the Psyche who is having the experiences and making the memories, and not the spirit body. Likewise, whenever the Psyche is separated from its assigned physical body, it's the Psyche who is having the experiences and making the memories, not the physical body.

I'm confident that as my different websites come online and develop, some people will take the time to let us know about their NDEs or OBEs wherein they were an Immaterial Spark or a Viewpoint in Space during their Out-of-Body Experience.

Materialism, Naturalism, and Darwinism cannot survive the discovery of Psyche, Quantum Mechanics, NDEs, OBEs, and our after-death Life Reviews.

Quantum waves are thoughts; and, they are being produced, broadcast, retrieved, and interpreted by some kind of Psyche or Intelligence at the quantum level.

Out-of-Body Travelers have observed that the spirit body doesn't have thoughts and doesn't make memories while its Psyche or Intelligence is separated from it. Consciousness and sentience reside within the Psyche, and not the spirit body. Psyche is thought. Psyche is memory. Psyche is intelligence. Psyche is a quantum wave or a vibrating thought. In string theory, a string is a thought or a quantum wave. That's why I have chosen to separate the Psyche from the spirit body and treat them as two separate entities. I noticed in the scriptures, which the Biblical God Jesus Christ had a hand in writing and producing, that He has done the same thing; and/or, His chosen prophets have. Psyche or Intelligence is something different than Spirit or a Spirit Body.

As I have gone along, I have refined my thoughts and ideas somewhat, of course. Nowadays, I tend to visualize Psyche as quantum waves; and, these quantum waves are the vibrations within a string, if we are talking about string theory. If there really is such a thing as a point particle or an infinite singularity, Psyche or Intelligence would be it. If there is no such thing as a point particle, then Psyche is the quantum waves or the vibrations within a string. Quantum waves are often portrayed as light waves. Psyche has been called Sentient Light. Obviously, the quantum waves within a string would have to be of the highest frequency possible, because of the extremely short distances that they are traveling. You are not going to be putting radio waves within a string.

Just as invisible magnetic fields project out into space, the Psyche within an atom does the same thing interacting with the other atoms around it. Some of these fields or quantum waves attract, and some of them repel. It's really simple to understand, once a person chooses to do so.

Unifying Physics

Remember, spirit matter and physical matter are basically the same thing, except they exist in a different phase or different dimension. Thanks to quantum mechanisms, it's totally possible for two or more atoms or particles to occupy the same space at the same time, as long as they are out of phase with each other. Our psyche, our spirit body, and our physical body are out of phase with each other, which means that they can occupy the same space at the same time. A spirit body can walk through a physical body unphased. Spirit matter and psyche are pure syntropy. They don't age. They are eternal and everlasting because they are pure syntropy. God takes a particle of spirit matter and infuses it with space-time or entropy in order to convert it into a particle of physical matter. God slows down that spirit particle to sub-light speeds in order to convert it into physical matter. God changes its phase or its dimension.

Remember, thoughts and memories are quantum waves being produced, transmitted, recorded, and stored at the quantum level by Someone Psyche, which explains why all of our thoughts and memories are able to survive the death of our physical brain. Our mind is something other than our brain!

The Materialists and Naturalists deliberately conflate Psyche at the quantum level with our physical brain at the physical level. It's what they do.

These people write books trying to explain how our quantum thoughts and quantum memories are being distributed and stored as physical matter, physical receptors, and physical synapses within our physical brain – which is physically impossible. Mapping and storing quantum thoughts and quantum memories as quantum waves within a physical atom is totally possible at the quantum level – it happens all the time; but, it's completely impossible at the physical level with physical atoms and physical synapses and physical neurons.

Remember, the smaller can dwell within and control the larger. Have you ever heard of vibrating strings and string theory? The vibrations or quantum waves are smaller than the strings, which means that the vibrations or quantum waves can dwell within and control the strings.

The vibrations within a string are quantum waves. Thoughts and memories are quantum waves, which means that thoughts and memories are the vibrations within a string. The string is the physical component, and the vibrations are the psyche component. Psyche IS thoughts and memories, which means that Psyche is quantum waves or the vibrations within a string. The smaller can dwell within and control the larger. Atoms can communicate with each other through quantum waves or telepathy. Nature's Psyche can communicate with and control atoms at the quantum level through quantum waves or telepathy. Telepathy is WiFi at the quantum level. Strings on the physical side such as quarks are fixed by God and remain constant. Only God can change them. The strings in the middle seem to have a bit more flexibility – gluons, bosons, photons, gravitons, electrons, leptons, and dark energy – half in and half out. The strings on the psyche side are alive and infinitely variable. They may not be strings at all. In fact, they may be the different vibrations within a string – infinitely flexible and variable. These are our thoughts, memories, and Psyche.

This is the KEY to understanding Quantum Neuroscience and how the Human Psyche interacts with Nature's Psyche in order to command and control its physical brain. Go with the truth and the preponderance of the evidence; and then, forget about everything else. Or in my case, use the truth and the preponderance of the evidence to debunk and falsify everything else.

I'm not going to apologize for figuring this stuff out, because it's really cool if you think about it.

I just unified physics at every level. Of course, the person who will get credit for all of this is the person who figures out all the mathematics behind it; but, I have given them a target to shoot towards.

A Philosophy of Science for Personality Theory or Psyche Theory

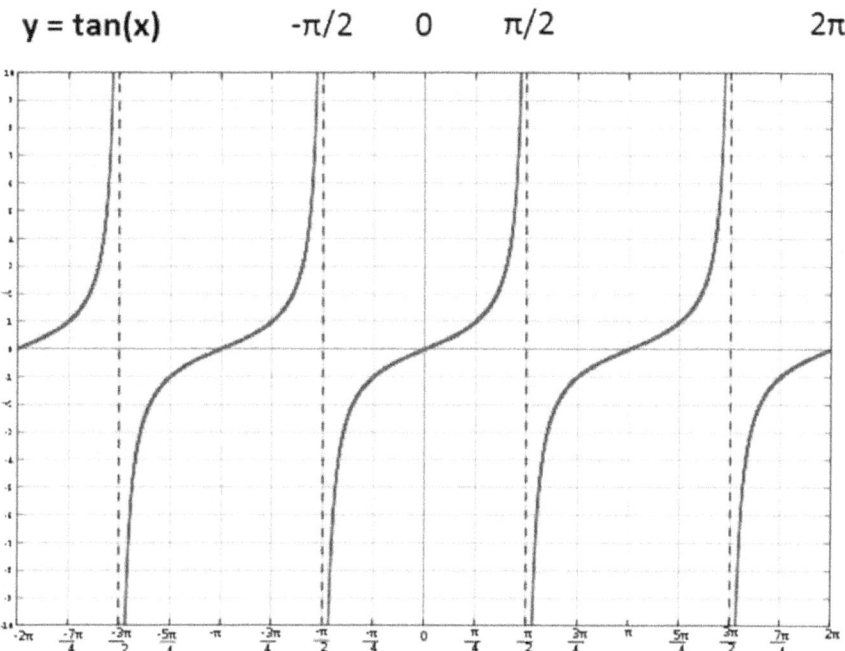

I use **y = tan(x)** to mathematically model my Psyche Ontology or my Ultimate Cause Model of Reality, quantum objects, spirit matter, physical matter, spirit bodies, physical bodies, universes, infinite singularities, psyche, and non-local consciousness. Each wave on the graph is a different Quantum Object, or a different particle of matter, or a different universe.

A single Universe, or Quantum Object, or Particle of Matter and its lifeline exists in the domain between the -π/2 and π/2 asymptotes. Matter or a Quantum Object has some limitations. Matter in any format, whether spiritual or physical, reacts. There's a time lag associated with Matter or a Quantum Object, because it waits for and then reacts to Psyche's Word of Command.

Again, a single Universe, or Quantum Object, or Particle of Matter (and its size or the amount of space it takes up) exists in the domain between the -π/2 and π/2 asymptotes. That Quantum Object, or matter, or material is spiritual and takes up little space when it is below (0, 0) on the graph; and, that same Quantum Object or matter is physical and can theoretically take up infinite space and become infinite in size when it is above (0, 0) on the graph above.

Psyche resides ON the -π/2 and π/2 asymptotes. Think about it. That means that Psyche or Non-Local Consciousness has NO size and takes up NO space, yet is

63

simultaneously infinite in size, range, scope, and presence. It's a paradox, but it's Reality. It's one eternal round! Psyche goes down the rabbit hole and climbs the stairway to heaven simultaneously, because Psyche exists at the infinite and IS infinite.

Quantum Objects – spirit matter and physical matter – reside in the realm of the finite. Psyche, Intelligence, Consciousness, or Life is the Master of the Infinite!

This IS the Ultimate Model of Reality and the Grand Unified Theory of Everything; and, it's really quite simple to visualize and understand.

A Psyche Ontology – The Grand Unified Theory of Everything

Rychlak, J. F. (1981a). *A Philosophy of Science for Personality Theory* (2nd ed.). Malabar, FL: Robert E. Krieger Publishing Company.

I used this book from Joseph Rychlak to point me to a Philosophy of Psyche or a Philosophy of Personality; and, this book was instrumental in helping me to develop and present my Psyche Ontology or Ultimate Model of Reality. Every truth points to every other truth. By observing what was missing from all the different models of reality or personality theories, this book from Joseph Rychlak and its discussion of Aristotle's four physical causes pointed me directly to Psyche as the Ultimate Cause, because physical causes CAN'T DO psyche, non-local consciousness, spirituality, intelligence, science, philosophy, teleology, and life.

Ever since a Psyche Ontology was revealed to me, I have wondered why nobody has ever thought of it before. Why didn't God reveal it to someone thousands of years ago? Why do I see no mention of a Psyche Ontology in any of the literature?

The only answer I have is that people weren't wanting a Psyche Ontology; and therefore, people weren't looking for a Psyche Ontology. God gives us what we want most; and, most people don't want an ontology that points them to God's Psyche or that reduces to God's Psyche. Depending upon how much you are resisting all of this, you might be one of them.

Nevertheless, this Psyche Ontology or Ultimate Model of Reality or Ultimate Paradigm ends up being the Grand Unified Theory of Everything, because it explains everything and subsumes the whole of existence, reality, knowledge, lived experience, and truth.

The scientists and mathematicians are NEVER going to be able to unite the Theory of Relativity with Quantum Mechanics directly, because Classical Physics and the Theory of Relativity explain the LIMITATIONS of physical matter whereas Quantum Mechanics or Spiritual Mechanics or Non-Local Mechanics explains to us that the quantum or the spiritual or the non-local has NO physical limitations. The scientists will NEVER be able to explain Quantum Mechanics or Non-Local Mechanics in materialistic or physical terms, because Quantum Mechanics is a non-local, non-physical, transdimensional, spiritual phenomenon. Quantum Mechanics explains how spirit matter works, and how spirit matter becomes physical matter by taking upon itself some additional limitations.

To unify Classical Physics with Quantum Mechanics we have to provide a mathematical model, like $y = tan(x)$, to model how Matter or Quantum Objects REALLY work. We have to explain how spirit matter becomes physical matter! When we do so, this Psyche Ontology or Ultimate Model of Reality becomes the Grand Unified Theory of Reality, which everyone has been looking for.

When Matter or a Quantum Object is below (0, 0) and approaching infinite velocities, infinite frequencies, and zero size, that particle of Matter or that Quantum Object can function instantaneously and simultaneously with NO distance and speed limitations whatsoever, because the stuff is spirit matter and NOT physical matter. The only real limitation which spirit matter has is that it reacts and therefore it lags just slightly behind Psyche's Word of Command, where the timing is concerned.

When Matter or a Quantum Object is above (0, 0) and approaching infinite mass and infinite inertia and zero velocity, that particle of physical matter is in fact approaching LIMITATIONS which it can't get beyond. The ONLY way that that particle of physical matter or that Quantum Object can get beyond those limitations is to transform back into spirit matter where it will have no such limitations.

It is my observation that Seraphim, Translated Beings, Resurrected Beings, God the Father, and Jesus Christ can convert or transmute their physical bodies into spirit matter at will, and then convert that spirit matter back into physical matter when they reach their chosen destination. That's how these people do what they do. While their bodies are spirit matter, these people can travel across the universe instantaneously at the speed of thought. When these people reach their chosen destination, they then convert their bodies back into physical matter. It's a Quantum Jump Drive or a Blink Drive, and it is infinitely faster than a warp drive which tries to keep everything in a physical matter format.

Warp drive will NEVER be possible because it's impossible to push physical matter faster than the speed of light. However, when we temporarily convert physical matter back into spirit matter, then we are dealing with Quantum Leaps and a Blink Drive or a Quantum Jump Drive, which has NO physical limitations! It's teleportation and it works. It has also been EXPERIENCED and OBSERVED. It solves everything that we have been looking for.

The ONLY problem is that God or God's Psyche is in control of it all, and He gifts this teleportation ability only to His righteous followers whom He has selected for the gift or the experience. Teleportation or Blink Drives are something that we mortal beings will never gain control of, because God is in control of this gift or this "technology"; and, it is God who decides whom He is going to teleport or blink from one location to another. ONLY God's Psyche controls this spiritual gift. This is an ability or a gift that is ONLY God's to give.

I and my car have teleported. I blinked, or I jumped; BUT, it was nothing that I consciously did or consciously chose. God did it for me. God is the ONLY one who can do these kinds of things for us. If we are going to blink or jump out of harm's way or walk through fire unscathed, this is something that God is going to have to choose to do for us, because it is something that we mortal fallen beings can't do for ourselves. God is KNOWN by what He chooses to do for us; and, God does for us the things that we can't do for ourselves.

This is a Psyche Ontology, wherein Psyche is the fundamental unit of reality and God's Psyche is the Ultimate Psyche or the Ultimate Cause. This IS the Ultimate Model of Reality and the Grand Unified Theory of Everything. A Psyche Ontology subsumes ALL of the other ontologies, because ONLY Psyche can do ontology, reality, truth, knowledge, life, transmutation of matter, teleportation, lived experience, memories, and existence. This really IS the Ultimate Model of Reality, which is why nobody has ever thought of it before.

Sparticles – The Other Side of the Equation

Materialists, Darwinists, and Naturalists like Michio Kaku consider evolution, natural selection, and random mutations to be their designer, creator, and god. In ALL of his books, Michio Kaku pays worshipful homage to evolution and makes it clear that he considers evolution to be his designer and his creator.

Due to the fact that each piece of scientific evidence is subject to an infinite number of different interpretations, the scientific method and scientific experiments cannot be used to prove anything true. However, science, scientific observations, and the scientific methods can indeed be used to falsify ideas or to prove theories and ideas false.

The ONLY way to prove anything true through science is through Direct Scientific Observations or Lived Experiences; but, even those are subject to false interpretations. Personal interpretations or one's chosen philosophical worldview ARE the primary weakness and the major bane of Science, Scientific Observations, and Science Experiments. The final step of the Scientific Method is to interpret the data; and, that's where it always goes wrong and seldom goes right.

You can't find the truth with a false interpretation. With a theoretically infinite number of possible interpretations for each piece of scientific evidence or for each scientific observation, ONLY ONE of those interpretations can possibly be true. The rest of them are false. Grabbing onto the false interpretations and using them as evidence or proof is called *affirming the consequent*.

That's the way that the Materialists, Naturalists, and Darwinists "prove" the theory of evolution to be true – by *jumping to conclusions* or *affirming the consequent*, which are logic fallacies. The whole of Materialism and Naturalism and Darwinism are based upon a wide variety of different logic fallacies; and, the Materialists and Naturalists fall for it hook, line, and sinker because they desperately want these things to be true.

In contrast, *negating the consequent* or using the scientific methods and scientific observations to FALSIFY Materialism, Naturalism, Darwinism, Nihilism, and Atheism IS philosophically and logically sound. I and many other scientists have used the Scientific Method and a wide variety of different Scientific Observations or Lived Experiences to FALSIFY Materialism, Naturalism, Darwinism, Nihilism, and Atheism. In other words, we have used Science and Scientific Observations or Lived Experiences to PROVE Materialism, Naturalism, Darwinism, Nihilism, and Atheism FALSE. Remember, *negating the consequent* or *proving things false* is philosophically and logically sound.

When it comes to a Scientific Proof of God's Existence, the process starts by using science, the scientific methods, and scientific observations to FALSIFY Materialism, Naturalism, Darwinism, Nihilism, and Atheism. You use Science itself to create a hole that ONLY God can fill. In other words, you use Science, the scientific methods, and scientific observations to develop a Proof of God's Necessity. Once you have successfully used Science and Scientific Observations to establish God's Necessity, then you have in fact used Science as a Proof of God's Existence.

In order to successfully complete this process, you have to be willing to allow ALL of the evidence into evidence, including the Non-Local Experiences of the human race.

The Materialists only allow physical evidence into evidence. These people completely reject and ignore ALL of the mounds of evidence from out-of-body travelers, who explain how physics works in the spirit realm or the non-local realm. I KNOW, because I used to be a Materialist, Nihilist, and Atheist. Materialism, Naturalism, and Atheism of any kind ARE based upon a refusal to look at evidence – particularly non-local evidence, or non-physical evidence, or spiritual evidence.

The Materialists and Naturalists completely reject Quantum Non-Locality and refuse to take it into consideration while developing their theories and equations. You have to be willing to admit ALL of the Scientific Evidence or Scientific Observations into evidence, or you will never be able to find the truth. It's only logical.

The Materialists and Naturalists always reject and ignore the other side of the equation. Keep asking yourself, "What is the other half of the equation which the Materialists, Naturalists, and Atheists keep ignoring and rejecting?" Scientists ask questions; and, these people aren't scientists in the truest sense – they are religious fanatics – they refuse to question their chosen conclusions, and instead they submit their chosen conclusions or their consequents into their arguments as their major premises or primary assumptions, a process which is called *begging the question*, which is another logic fallacy that these people use to make their case.

There is NO evidence and can NEVER be any evidence to support the major premises or the primary assumptions of Materialism, Naturalism, Darwinism, Nihilism, and Atheism because these philosophical religions have to be taken on blind faith as being true. These false and falsified naturalistic religions state that Quantum Non-Locality, Spirit, Spirit Matter, Psyche, the Non-Physical, and Non-Local Consciousness DO NOT EXIST. There is NO way to provide evidence for that kind of claim. It has to be taken on blind faith as being true.

We have tons of **observational evidence** explaining how physical matter works. We also have on record tons of **observational evidence** explaining how spirit matter works. If we want to get at the ultimate truth of our existence and reality, then these two sets of observational evidence have to be unified into one grand theory of everything. Both sets of observational evidence have to be explained by our theories and our equations.

I have upgraded my science to Science 2.0, in which I allow all of the evidence into evidence. We KNOW from the Non-Local Experiences of the human race that Quantum Non-Locality, Spirit Matter, Spirit Bodies, the Spirit Realm, the Non-Local Realm, the Transdimensional Realm, Psyche, Non-Local Consciousness, and the Biblical God Jesus Christ REALLY DO EXIST.

Based upon that KNOWLEDGE, the ultimate goal is to find and develop logical scientific explanations and scientific theories to account for and explain ALL of these different Lived Experiences and Non-Local Experiences of the human race. In other words, I now use Theoretical Physics to look for Signs of Psyche and for Signs and Explanations of the Non-Local Realm. It's simply amazing how many things you can find in Theoretical Physics that points implicitly to the Non-Local Realm or the Spirit Realm. It's there to be found!

Obviously, Quantum Non-Locality or Quantum Non-Physicality points directly to the Transdimensional Realm or the Spirit Realm. But, what other things in theoretical physics points us in this direction?

From the Wikipedia:

In particle physics, a superpartner (also sparticle) is a hypothetical elementary particle. Supersymmetry is one of the synergistic theories in current high-energy physics that predicts the existence of these "shadow" particles.

When considering extensions of the Standard Model, the *s- prefix* from sparticle is used to form names of superpartners of the Standard Model fermions (sfermions), e.g. the stop squark. The superpartners of Standard Model bosons have an *-ino* (bosinos) appended to their name.

The s-particles are the hypothesized symmetric particles of supersymmetry. It is my personal belief and theory that the s-particles or sparticles remain unseen and undetected because they are in fact the spiritual side or the non-local side of the equation – whether we are taking about the equations of supersymmetry or my $y = \tan(x)$ model, which I continue developing and presenting to the world.

While writing about general relativity and quantum mechanics, Michio Kaku stated:

In many ways, these two theories appear to be opposites. General relativity concerns the cosmic motions of galaxies and the universe, while quantum mechanics probes the subatomic world. Relativity is primarily a theory of force fields that continuously fill up all space. (The force field of gravity, for example, can be compared to gossamer-like tendrils that extend to the outer reaches of space.) Quantum mechanics, by contrast, is primarily a theory of atomic matter, which travels much slower than the speed of light. In the world of quantum mechanics, a force field only appears to fill up all space smoothly and continuously. If we could examine it closely, we would find that it actually is quantized into discrete units. Light, for example, consists of tiny packets of energy called quanta or photons. (p.37).

I found this quote interesting, because during the past year or more I have been visualizing general relativity and quantum mechanics pretty much opposite of the way that Michio Kaku describes them here.

Why?

It's because I'm trying to understand things from a non-local perspective or a spiritual perspective, and Michio Kaku as a materialist is only interested in the physical perspective.

I have tended to equate general relativity and space-time with physical matter and classical physics; whereas, I have tended to equate quantum mechanics with spirit matter and the non-local realm. Consequently, I visualize spirit matter or non-local matter as traveling faster than the speed of light up to a potential infinite velocity. ONLY physical matter or physical atoms travel much slower than the speed of light. When it comes to matter, or a quantum object, I try to cover both sides of the equation – the local and the non-local, or the physical and the non-physical. The Materialists are only interested in the physical side of the equation and ignore the spiritual side of the equation. I'm no longer a Materialist.

Likewise, I tend to visualize the quantum world including spirit matter as force fields or waves of light that fill up all of space smoothly, continuously, and instantaneously. These waves or force fields ONLY quantize or particlize into tiny packets of energy called quanta or photons when they are used, or observed, or have reached their destination. In other words, the waves or force fields are particlized into spirit matter in the non-local realm when they are used or organized by psyche. And, when it comes to the physical realm, these waves or force fields are particlized only when they reach sub-light speeds and/or land at their destination and are used to provide information. Light is also a force field or a wave until it is observed or used, and then the wave function collapses into a quantum or a photon.

69

So, which one of us is right?

We both are! That's the message that I'm trying to present to the world.

According to general relativity and Einstein, gravity is a force field that travels at the speed of light. That may in fact be true when it comes to our physical realm, because everything in the physical realm is by definition limited to less than the speed of light. If it's grabbed or detected by a physical object or a physical instrument, its velocity slows down even more.

HOWEVER, everything in the spirit realm or the non-local realm exists at velocities greater than the speed of light. That's why we can't detect the stuff with our physical instruments. It's in a completely different phase of existence. In order to be detected by our physical instruments, it has to slow down to velocities below the speed of light.

Consequently, in the spirit realm, gravity is a force field or a wave that has infinite velocity – in other words, it's instantaneously and simultaneously everywhere. The same reality applies to non-local consciousness or psyche. In the spirit realm or the non-local realm, spirit matter or non-local matter has a native velocity greater than the speed of light, but it lags just slightly behind psyche. Spirit matter is always organized by psyche or responds to psyche's command. That's the way things are experienced in the spirit realm.

In the Spirit Realm or Non-Local Realm, spirit matter, spirit bodies, and psyche can teleport. As I see it, an understanding of how teleportation works, what it is, and how it fits into the other side of the equations is the KEY to understanding Quantum Non-Locality or the Spirit Realm, along with spirit matter.

What happens when a spirit body or a physical body teleports?

Well, I can tell you what happens when God teleports your physical body to safety, because God teleported me and my car to safety.

When you teleport, time stops; and, you have no memory of traveling the distance. You are instantaneously at your destination. According to out-of-body travelers, spirit bodies and psyche teleport all the time. A person thinks of a desired destination, and he or she is instantaneously there. There's no memory of having traveled the distance, because the passage of time has stopped during the teleportation process. There is NO passage of time during the teleportation process, whether we are talking about the teleportation of psyche, spirit matter, or physical matter. ONLY the angels and God have control over the physical teleportation process. A fallen mortal being cannot teleport his or her own physical body or physical matter; but, God can.

However, anyone seems to be able to teleport while in the spirit realm or the non-local realm. It just happens, at will. What this means is that spirit matter and psyche can travel infinite distances instantaneously, because there is little or no resistance to acceleration when it comes to spirit matter. Spirit bodies can also float, levitate, and fly. Also, in the non-local realm, spirit bodies or consciousness can be in two different places simultaneously. The physical limitations don't exist in the non-local realm or the non-physical realm.

Likewise, if God stops the passage of time for physical matter, then He can teleport that physical matter to any destination that He desires instantaneously. We humans can't stop the passage of time for a physical object, but if we could, then we would be able to teleport that physical object anywhere we want in the universe instantaneously. In the science fiction series *Dark Matter*, that's the way the Blink Drive or the Quantum Drive works. It teleports the ship instantaneously. By definition, in principle, no time passage takes place during teleportation. In the science fiction series, *Star Trek the Next Generation*, Q teleports to any desired destination in this universe at will. Again, the passage of time stops during the teleportation process.

In our science fiction stories, the Materialists and Naturalists try to contract and expand space – warp space – in order to travel faster than the speed of light. I suggest that the stoppage of time and teleportation are infinitely more realistic and possible and doable, because teleportation has actually been experienced in real life – both in the physical realm and in the spirit realm. If we can figure out how to stop the passage of time for a physical object, then we should be able to teleport it anywhere in the universe instantaneously, just like what happens in the spirit realm or the non-local realm.

There are other ways to look at this as well. Quantum Mechanics is trying to teach us that ONLY Psyche, particularly God's Psyche, can convert spirit matter into physical matter. If spirit matter can be converted into physical matter, then physical matter can also be converted back into spirit matter. If we were to have a space ship and temporarily convert it back into spirit matter, then we could teleport that spirit matter instantaneously anywhere in this universe, and then convert that spirit matter back into physical matter when it reaches our chosen destination.

ALL of this becomes possible by including the other side of the equation – the spiritual side or the non-local side.

The spiritual side of quanta or the spiritual side of quantum objects goes down the **y = tan(x)** equation and is less than zero; and, the physical side goes up or is more than zero.

Remember, infinite velocity is synonymous with NO passage of time. In other words, infinite velocity is synonymous with teleportation.

As we approach the -π/2 and the π/2 asymptotes, whether we are talking about the spiritual side of the equation or the physical side of the equation, we are in fact approaching INFINITE velocity and ZERO passage of time. The -π/2 and π/2 asymptotes represent both Psyche and Teleportation, whether we are talking about the spirit realm or the physical realm.

When it comes to the teleportation of physical matter, God holds the KEY, because God created physical matter in the first place. God has locked us out of being able to teleport physical matter; and, He has only given that KEY to certain individuals. We mortal fallen physical beings don't even think about trying to stop the passage of time and teleporting our physical body to a chosen destination, because we instinctively know that we can't do it. ONLY God can do it.

The Materialists and Naturalists don't consider or think about any of this stuff, because they don't believe it to be possible in any realm of existence. The Materialists, Naturalists, and Atheists are definitely not going to ask God for the KEY to teleportation, because they have chosen to believe that God does not exist. The Materialists are only interested in trying to find different ways to save the physical matter, because they truly believe that non-local consciousness or psyche does not exist. In all of his books, Michio Kaku has a chapter or two about trying to find different ways to save the physical matter at the end of our physical universe. He doesn't even stop to realize that our psyche and spirit body go onward thinking and acting and doing, long after our physical body and physical brain are dead and gone. The Materialists and Naturalists don't think about any of this.

Consequently, the Materialists, Naturalists, and Atheists deliberately ignore and reject the non-local side or the spiritual side of relativity, quantum mechanics, light, matter, quanta, force fields, and gravity where EVERYTHING travels faster than the speed of light and approaches infinite velocity. Each one of these things has a spiritual side, as well as a physical side. The Materialists and Naturalists only think about the physical side; and, they scratch their heads in confusion and wonder whenever they stumble upon the psyche side, or the spiritual side, or the non-local side.

It's my theory that s-particles are unseen and not experienced because they are spiritual and function like tachyons at velocities faster than the speed of light. The math says that the sparticles must exist; but, they have never been detected, so either they don't exist, or they reside in a different dimension or a different phase other than our physical realm.

ALL of the waves are there in the non-local realm, but they have to be slowed down to sub-light speeds, quantized and particlized, if they are in fact to be detected and recorded by our physical instruments.

The Quantum Law of Complementarity states that a quantum object or particle of matter can reside in either a non-local spiritual state or in a localized physical state, but not in both states simultaneously. It's fully possible that **sparticles** are in fact the non-local spiritual state of a localized physical **particle**. Sparticles might in fact be spirit matter or spiritual particles. In other words, it's possible that supersymmetry divides or folds around the local and the non-local, or around the physical and the non-physical. Of course, a Materialist or Naturalist will never think about this possibility, because they have concluded a priori that only the physical state exists.

The Materialists and Naturalists have chosen to ignore and reject the other side of the equation. By affirming the consequent, the Materialists, Naturalists, and Atheists define science as Materialism, Naturalism, Darwinism, Nihilism, and Atheism. They paint themselves into a corner and can't get out of it.

You are not free to think about the spiritual, when you have rejected all of the evidence supporting the spiritual. I KNOW, because I used to be a Materialist, Nihilist, and Atheist.

The thousands of Near-Death Experiences (NDEs) and Out-of-Body Experiences (OBEs) on YouTube got me to change my mind. I'm no longer a Materialist, Nihilist, or Atheist, because now I know better. I have fully upgraded my science to Science 2.0; and, now I treat ALL Non-Local Experiences or Non-Local Observations as Scientific Evidence. Science IS observation, after all. Science IS knowledge and experience, including knowledge and experience of the Non-Local Realm or the Spirit Realm.

The physicists are trying to unite all of the different physical forces into one grand unified field theory or equation. It's never going to happen while they continue to ignore the other side of the equation. They run into all sorts of anomalies and infinities, and those anomalies and infinities are in fact the spiritual side or the non-local side of the equation where quantum objects approach and attain infinite velocity, the stoppage of time, and teleportation.

PART V — QUANTUM NEUROSCIENCE AND SCIENCE 2.0

Quantum Neuroscience is neuroscience from a quantum perspective.

Quantum Neuroscience is the scientific study of how the Human Psyche interacts with, controls, and survives its physical brain.

In order to do Quantum Neuroscience, I updated my science to Science 2.0, which allows ALL of the evidence into evidence and pursues a preponderance of the evidence.

I have observed that Quantum Neuroscience has infinitely more explanatory power than Materialism, Naturalism, and Classical Physics.

Quantum Neuroscience is a proven and verified science. ALL of the quantum mechanisms that are used in Quantum Neuroscience have been verified and proven to exist through experimentation. As a race, we could have started Quantum Neuroscience back in the 1980's, if somebody had actually wanted to do so; but, nobody is interested in any of this, except for me I guess.

While doing research for Quantum Neuroscience, I learned a lot of things that I never knew before. Its explanatory power is through the roof!

Since Quantum Neuroscience and Science 2.0 are willing to allow ALL of the evidence into evidence, these new ways of doing science are not afraid to allow evidence for the existence of God into evidence.

For example, WE KNOW from the Big Bang Theory that every particle of physical matter had a beginning, which means that they had a Beginner. And, WE KNOW from the revelations from the Biblical God Jesus Christ that Intelligence or Psyche, and spirit matter, had NO beginning and will have NO end. Psyche and spirit matter are eternal and everlasting.

Why?

Well, the science can actually explain this to us, if we let it do so.

You see, entropy is a function and a product of space-time.

There is NO space-time at the non-local level, or quantum level, or transdimensional level, or spiritual level. Since there's NO entropy in the spirit realm, then that means that your Psyche, God's Psyche, and spirit matter don't age, because there's NO passage of time for them – NO entropy.

Now, when the Biblical God Christ Jehovah decided to convert specific particles of spirit matter into physical matter, He did so by infusing those specific particles of spirit matter with entropy or space-time, thereby turning them into particles of physical matter. Anything that is comprised of physical matter is subject to entropy, and it can experience the passage of time and experience some sort of aging process.

In contrast, it's impossible to grow old and die in the spirit realm, because there is NO entropy in the spirit realm. The spirit realm is comprised of pure syntropy. Entropy only exists in the physical realm.

This is really simple to understand, once a person finally chooses to allow ALL of the evidence into evidence.

Ironically, the existence of entropy and the existence of physical matter are Scientific Proof of God's Existence, whether we realize it or not. Everything bears record of God. Your genome is God's Signature after all – yet another Scientific Proof of God's Existence. All of the fine-tuning in this physical universe is Scientific Proof of God's Existence. Everything in the physical realm bears record of the Person who made it, because it wouldn't exist without God.

Quantum Neuroscience and Quantum Mechanics start with the physically impossible; and then, they try to find a scientific explanation for it all. It works, but only if we let it work.

Now pay attention!

Despite the FACT that I will have debunked and falsified the Theory of Evolution thousands of times in hundreds of different ways, don't ever once let me convince you that there is NO such thing as evolution or random mutations.

Evolution is REAL, very REAL. Evolution or random mutation is entropy or the second law of thermodynamics. It's REAL. Entropy is very REAL at the physical level; and, it works as advertised at the physical level.

Whenever the Materialists, Naturalists, Darwinists, Nihilists, Behaviorists, and Atheists start talking about evolution, they get most everything wrong, because evolution or entropy cannot design and create. However, these people do indeed get one thing perfectly right. Evolution, or random mutation, or entropy is indeed the CAUSE of ALL of our heritable diseases, developmental diseases, and heritable mental illnesses.

Remember, the Theory of Evolution is FALSE because random mutations or entropy cannot design and create genes, proteins, and life forms. However, evolution or random mutation or entropy is very REAL; and, it can indeed destroy genes, proteins, and life forms. Do you see how that works? It's important to understand.

THE SCIENTIFIC METHOD

THE SCIENTIFIC METHOD tells us to examine a person's claims or hypotheses, and then TEST to see if their claims hold water or not. Can their claims hold-up under scrutiny? If not, then their claims or hypotheses should be formally rejected by the Scientific Community!

—

THE SCIENTIFIC METHOD typically goes through the following sequence in an attempt to arrive at THE TRUTH.

1) Form a **HYPOTHESIS**.
2) Select a SCIENTIFIC METHOD or Scientific Methodology to **TEST** the hypothesis.
3) Run the Science Experiment; and then, **Observe** and **Measure** the **RESULTS**.
4) Find the BEST **INTERPRETATION** or the BEST **EXPLANATION** for the Scientific Data, the Scientific Evidence, the Scientific Observations, and the **RESULTS** of the Science Experiment.

This is THE SCIENTIFIC METHOD in action!

The problems result and come into play when the Scientists fudge with this thing and cheat in an attempt to make their favorite theories seem true.

In this brief treatise, I intend to apply THE SCIENTIFIC METHOD and Deductive Reasoning brutally and impartially to some of the Scientists' most favorite theories. I will pull no punches.

The game's afoot!

The Scientific Claims of Materialism

THE SCIENTIFIC METHOD tells us to examine a person's claims or hypotheses, and then TEST to see if their claims hold water or not. Can their claims hold-up under scrutiny? If not, then their claims or hypotheses should be formally rejected by the Scientific Community!

The Materialists claim that Physical Matter is all that there is and all that there will ever be. Can this claim hold-up under scrutiny when THE SCIENTIFIC METHOD is applied to it? It cannot!

Whether the Materialists realize it or not, the Materialists claim that Physical Matter designed, created, and manufactured this Universe, this Earth, all of the Genomes on this Earth, and every Life Form on this Earth. Can this claim hold-up under scrutiny when THE SCIENTIFIC METHOD and Logical Reasoning are applied to it? It cannot!

Materialism IS Creation by ROCKS! Whether they realize it or not, that's the claim that the Materialists are making and expecting us to believe. Their materialistic hypothesis is that THE WATER and THE ROCKS designed, created, and manufactured all of the Genomes and Life Forms on this planet. That IS their Scientific Hypothesis.

IS there any way to TEST that hypothesis?

Can we set up a Science Experiment wherein we actually OBSERVE Rocks, and Water, and Physical Matter actively designing, creating, and manufacturing Genomes, Computers, Cars, Buildings, Stealth Fighters, Battleships, and fully-functional Living Life Forms?

If we CANNOT set up a Science Experiment to TEST and OBSERVE Rocks and Water and Physical Matter IN THE ACT of designing, creating, and manufacturing Genomes and Life Forms from scratch, then according to THE SCIENTIFIC METHOD, the Materialistic Hypothesis that ROCKS or Physical Matter designed and created and manufactured all of the Genomes and Life Forms on this planet MUST BE FORMALLY REJECTED as being untestable, unsustainable, unobservable, and unscientific.

Something is said to be Scientifically Accurate if it matches with REALITY. Creation by ROCKS and Creation by Physical Matter are unrealistic, unscientific, and IMPOSSIBLE, because these things cannot design and create! And, Materialism IS Creation by ROCKS, whether the Materialists realize that or not!

The rules of THE SCIENTIFIC METHOD must be applied impartially even to our most favorite theories. According to THE SCIENTIFIC METHOD, the claims of Materialism must be formally rejected as being FALSE and UNTESTABLE and IMPOSSIBLE.

Therefore, according to THE SCIENTIFIC METHOD, since their claim cannot be TESTED nor OBSERVED in action, then the Materialists' hypothesis that ROCKS designed and created everything must be formally declared to be IMPOSSIBLE and FALSE. Since there is NO way possible to TEST and OBSERVE Rocks in the very ACT of designing and creating something from scratch, Materialism is FALSE, according to THE SCIENTIFIC METHOD.

Furthermore, Logical Reasoning also known as Common Sense tells us that ROCKS and Physical Matter cannot design, create, nor manufacture anything at all. ROCKS cannot do science! Rationality and Logical Common Sense tells us that the claims of Materialism ARE FALSE! Common Sense and Rationality are indeed applicable to THE SCIENTIFIC METHOD! In fact, THE SCIENTIFIC METHOD is a form of deductive reasoning!

Materialism or Creation by ROCKS is untestable, unscientific, unsustainable, unrealistic, illogical, unbelievable, IMPOSSIBLE, and FALSE. Therefore, according to THE SCIENTIFIC METHOD, Materialism or Creation by ROCKS should be formally rejected by the Scientific Community as the source or the origin of all of the Genomes and Life Forms that we see around us on this planet.

According to Deductive Reasoning, FALSE PREMISES or False Hypotheses should be eliminated because they produce FALSE and IMPOSSIBLE CONCLUSIONS! Physical Reactions cannot design and create computers and genomes and life forms. Therefore, according to Deductive Reasoning, Materialism and Creation by ROCKS and Creation by Physical Matter and Creation by Physical Reactions should be eliminated from consideration because these things are demonstrably FALSE and will therefore ALWAYS provide us with a FALSE CONCLUSION.

According to Abductive Reasoning, which is the search for the BEST Possible Explanation for the available Scientific Evidence, Materialism or Creation by ROCKS is one of the WORST possible explanations that can possibly be given to Scientific Evidence; and therefore, according to Abductive Reasoning, Materialism or Creation by Physical Matter should be eliminated from consideration because it is IMPOSSIBLE, and FALSE, and CAN NEVER HAPPEN.

In every way that we can possibly imagine, we have PROVEN that Materialism is FALSE. By using THE SCIENTIFIC METHOD, Deductive Reasoning, Abductive Reasoning, and Logical Common Sense, we have PROVEN that Creation by ROCKS is IMPOSSIBLE and FALSE; and therefore, we have PROVEN that the claims of Materialism ARE FALSE. Materialism of any kind cannot hold-up under scrutiny; therefore, Materialism or Creation by Rocks IS FALSE! We will NEVER catch THE ROCKS in the act of designing, creating, and manufacturing Genomes and Life Forms from scratch; therefore, we KNOW that Materialism is FALSE!

As a Scientist, Logician, and Theoretician, I FORMALLY REJECT Materialism or Creation by ROCKS as the source or the origin of all the Genomes and Life Forms on this planet, because the ROCKS are not conscious and alive; and therefore, ROCKS cannot design and create.

How often have I said to you that when you have eliminated the impossible, whatever remains, *however improbable*, must be THE TRUTH? — Sherlock Holmes

Materialism MUST BE ELIMINATED from Science, because Creation by ROCKS and Creation by Physical Matter and Creation by Physical Reactions ARE IMPOSSIBLE! Materialism must be eliminated from Science because Materialism is based exclusively upon wishful thinking and blind faith. There is NO Scientific Evidence and can be NO Scientific Evidence to support Materialism or Creation by ROCKS. I found this little Science Experiment and Thought Experiment credible and convincing. Did you?

Once we have eliminated ROCKS and Physical Matter as our designers and creators, then whatever remains, however improbable or unlikable, MUST BE THE TRUTH!

THIS IS SCIENCE!

Now, stop and ask yourself the question, "Can we think of anyone who was around 13.8 billion years ago during the creation of this Physical Universe, 3.8 billion years ago during the creation of the first genomes and life forms on this planet, and 542 million years ago during the Cambrian Explosion when lots of things needed to be designed, created, and manufactured?"

I can think of someone, can you?

As clearly documented and recorded in the Bible, the Book of Mormon: Another Testament of Jesus Christ, the Doctrine and Covenants, and the Pearl of Great Price, the BIBLICAL GOD Jesus Christ **reveals** Himself to mankind, and formally **confesses** multiple different times to multiple different people, and formally **states** that HE designed, created, and manufactured this Universe, this Earth, and ALL of the Genomes and Life Forms on this Earth.

The BIBLICAL GOD formally submits these books into EVIDENCE in order to make His case and to convict Himself of designing and creating everything. Within these books which the BIBLICAL GOD had a hand in writing and producing, the BIBLICAL GOD meets His Burden of Proof through a preponderance of the evidence and formally confesses to designing and creating this Universe, this Earth, and ALL of the Genomes and Life Forms on this Earth.

Confessions ARE admissible in a Court of Law.

Our task as the JURY is to determine for ourselves whether we believe the EVIDENCE to be true, or not. You know what my verdict is on this matter. What's your verdict?

For me personally, the scientific FAILURES of Materialism combined with the formal confessions of the BIBLICAL GOD end up becoming convincing Scientific Proof of God's Existence. I formally eliminated the IMPOSSIBLE, Materialism, in the hope that whatever remains, however improbable, must be THE TRUTH.

There is ONLY One Person that I know of who remains standing AFTER all of the Scientific Experimentation and Logical Deduction has been done. That person is the BIBLICAL GOD, our Resurrected Lord and Savior Jesus Christ. HE is the ONLY person that I know of who was in fact sufficient to the task of designing, creating, and manufacturing this Physical Universe, this Earth, and ALL of the Genomes and Life Forms on this Earth. HE is the ONLY person that I know of who was in existence and on the scene back then when all of this SCIENCE needed to be done.

For me, these scientific realities ARE indeed Scientific Proof of God's Existence.

I love Science and THE SCIENTIFIC METHOD for what it can reveal to us if we let it do so. The BIBLICAL GOD **reveals** Himself to us in two different ways — through Scripture and through Science.

As a member of the JURY, what's your verdict? Did you find the EVIDENCE compelling and convincing?

During our deliberations, you discovered that I found the EVIDENCE convincing enough to convict the BIBLICAL GOD of designing, creating, and producing this Universe, this Earth, and ALL of the Genomes and Life Forms on this Earth. But, you are a different member of the JURY. What's your verdict? The Judge awaits your decision.

Applying THE SCIENTIFIC METHOD to Materialism

THE SCIENTIFIC METHOD typically goes through the following sequence in an attempt to arrive at THE TRUTH.

1) Form a **HYPOTHESIS**.
2) Select a SCIENTIFIC METHOD or Scientific Methodology to **TEST** the hypothesis.
3) Run the Science Experiment; and then, **Observe** and **Measure** the **RESULTS**.
4) Find the BEST **INTERPRETATION** or the BEST **EXPLANATION** for the Scientific Data, the Scientific Evidence, the Scientific Observations, and the **RESULTS** of the Science Experiment.

Materialism IS Creation by Physical Matter or Creation by ROCKS.

We can indeed **HYPOTHESIZE** that Physical Matter or ROCKS or WATER might be able to design and create something NEW and USEFUL from scratch.

However, Materialism or Creation by Rocks FAILS at STEP 2 of THE SCIENTIFIC METHOD, because there is NO way to design and run a SCIENCE EXPERIMENT, where in fact we get to **TEST** and **OBSERVE** Materialism or Physical Reactions or Rocks IN ACTION designing and creating NEW genetic information, NEW genomes, and NEW life forms from scratch. According to THE SCIENTIFIC METHOD, Creation by Physical Matter or Creation by ROCKS doesn't even get off the launch pad, because Physical Matter cannot design and create! There is NO WAY to **TEST** Materialism or Creation by ROCKS and **OBSERVE** it in action!

Once again, there is NO way to **TEST** and **OBSERVE** "Design by Physical Matter" or "Creation by ROCKS"! Can you think of a way to catch ROCKS or Physical Matter IN THE ACT of DESIGN and CREATION, which doesn't involve some kind of magic, sorcery, trickery, lying, or wishful thinking?

I can't!

How can we get Materialism or Creation by ROCKS under a microscope, examine it carefully, and watch it in action?

WE CAN'T!

We can't make Physical Matter or ROCKS do things for us! Physical Matter and ROCKS cannot use or employ THE SCIENTIFIC METHOD to get things done! ROCKS and Physical Matter CANNOT DO SCIENCE, and a lot of SCIENCE needed to be done in order to produce the first genome and the first life form, back then 3.8 billion years ago when ROCKS seemed to be the only thing around!

Creation by ROCKS cannot be **TESTED** and **VERIFIED**, thus making Creation by Physical Matter or Materialism BAD SCIENCE and UNSCIENTIFIC! Since there is no way to get THE SCIENTIFIC METHOD and our Scientific Instruments onto Creation by ROCKS and examine it in action, Materialism ends up being a man-made Religion and a man-made philosophical pseudo-science and nothing more!

There is NO Scientific Methodology that can be created and used to **TEST** the actions of something that DOES NOT EXIST — something like Creation by ROCKS or Creation by Physical Reactions. Consequently, there is NO way to **MEASURE THE RESULTS** of Creation by ROCKS and Creation by Physical Matter, because Physical Matter and ROCKS cannot design, or create, or manufacture, or produce creative **RESULTS** for us to examine and measure.

When it comes to THE SCIENTIFIC METHOD, **MEASURABLE RESULTS** are what we are after!

There is NO WAY possible to provide **MEASURABLE RESULTS** for Creation by Physical Matter or Creation by ROCKS. Nobody can think of a way to do so, because ROCKS and Physical Matter cannot design, manufacture, create, or **PRODUCE MEASURABLE RESULTS** demonstrably, predictably, and reliably. Creation by ROCKS or Materialism is one of the least parsimonious and most illogical **EXPLANATIONS** that can be given to Scientific Evidence, or any kind of evidence for that matter.

Even Physical Reactions are incapable of designing and creating Genomes and Life Forms. Natural Selection and Random Mutations ONLY work on the genomes and life forms that God has already designed and created in the first place. Natural Selection and Random Mutations cannot WORK on dead, inanimate, inert molecules and matter. Mutation and Selection cannot design, program, manufacture, create, nor DO SCIENCE!

According to THE SCIENTIFIC METHOD, Creation by Physical Matter IS A FALSE PREMISE or a FALSE HYPOTHESIS, because Physical Reactions such as Mutation/Selection cannot design and create.

According to THE SCIENTIFIC METHOD, Physical Matter and ROCKS cannot DO MANUFACTURING and SCIENCE, because ROCKS and Physical Matter cannot design and create!

Finally, there is NO way to **INTERPRET RESULTS** which cannot be PRODUCED in the first place. ROCKS and Physical Matter cannot be made or forced to PRODUCE **RESULTS** for us to **INTERPRET, EXPLAIN,** and **EXAMINE**! That kind of SCIENCE is ONLY DONE by Intelligent Creative Living Beings.

When it comes to Creation by ROCKS or Materialism, we can't **TEST** it; we can't examine any **RESULTS** from it; and, we can't find anything from it to **INTERPRET** either, because Physical Processes of any kind cannot design and create anything at all.

THE SCIENTIFIC METHOD demonstrates clearly and conclusively that Creation by ROCKS or Materialism IS FALSE!

The purpose of THE SCIENTIFIC METHOD is to help us to find THE TRUTH, through a preponderance of the evidence.

THE SCIENTIFIC METHOD has no value to us if we use it to convince ourselves that a LIE is TRUE, as the Materialists and Darwinists always seem to do.

That's what I discovered during my Pursuit of the True Reality of All Things, and during my usage of THE SCIENTIFIC METHOD.

For me personally, each FAILURE of Materialism to meet its Burden of Proof becomes yet another miniature Scientific Proof of God's Existence, because God must exist in order to have done all of the SCIENCE which Physical Matter, THE ROCKS, and the various Physical Reactions could NEVER have done. It's elementary my Dear Reader!

If you are a Materialist, and you still don't believe me, then get on your knees and pray to the ROCKS or pray to Physical Matter and ask them to do something for you right here, right now. Ask them to speak to you or to answer your prayers.

Run this Science Experiment!

I assure YOU that you will have an infinitely easier time getting the BIBLICAL GOD Jesus Christ to answer your prayers than you will have getting ROCKS or Physical Matter to answer your prayers! I KNOW THAT THIS IS SO, because I have experienced it first-hand. I have run this Science Experiment and examined the RESULTS! Praying to the BIBLICAL GOD actually produced noticeable and replicable RESULTS for me! I can't say the same thing about praying to ROCKS or praying to Physical Matter.

And, that's the way it is!

This truth set me free to pursue other truths!

Applying EVIDENCE to Materialism

I have written sizable books wherein I have used THE SCIENTIFIC METHOD, Scientific Evidence, Scientific Data, Deductive Reasoning, Abductive Reasoning, and Logical Common Sense to debunk the claims of Materialism and to demonstrate through a preponderance of the evidence that Materialism is FALSE.

I won't duplicate all of that here.

Instead, I will later refer you to my books online at Amazon for EVIDENCE supporting my claim that Materialism IS FALSE.

Everything that I have written so far puts a Gravestone on Materialism and a Capstone on Creation by Living Intelligent Beings.

Materialism IS Creation by ROCKS or Creation by Physical Matter; and, I KNOW for a FACT that Rocks and Physical Matter CANNOT design, manufacture, or create anything at all! Therefore, I KNOW that Materialism is FALSE.

If we eliminate Materialism, Chance, and Darwinism from the equation, then we are left staring into the Face of God. The FAILURES of Materialism, Darwinism, Natural Selection, and Chance to be able to design and create Genomes and Life Forms from scratch, when fully understood and accepted as being true, BECOME a compelling, credible, and convincing Scientific Proof of God's Existence.

The Materialists will tell you that there is NO Scientific Proof of God's Existence and that there CAN BE NO Scientific Proof of God's Existence. They are right, if you are in fact trying to use Materialism or Darwinism to PROVE that God exists. We can't even use Materialism to prove that Materialism is true, so why should anyone expect to be able to use Materialism to prove that God exists? Obviously, we can't use Materialism to prove that God exists! If you choose to believe the Materialists, then you will NEVER be able to find a convincing Scientific Proof of God's Existence. I wasn't able to do so until after I got rid of My Materialism! Materialism was actually preventing me from finding any Scientific Proof of God's Existence! That's the way that Materialism works. It blinds us to the True Reality of things.

However, when we deliberately and successfully REMOVE Materialism, Darwinism, Chance, and Creation by Evolution from the equation, we are able to DEMONSTRATE through a preponderance of the evidence that God must exist in order to have done all of the SCIENCE and CREATION which Evolution, Natural Selection, Physical Matter, Chance, Materialism, and Darwinism could NEVER have done! This IS both Deductive Reasoning and Abductive Reasoning through a process of eliminating the IMPOSSIBLE and the FALSE PREMISES so that we are left with the BEST EXPLANATION or the CORRECT CONCLUSION. This IS Science and THE SCIENTIFIC METHOD in action!

While trying to apply EVIDENCE to Materialism, we quickly realize that there is NO EVIDENCE and can be NO EVIDENCE to support Materialism or Creation by ROCKS! We just KNOW that ROCKS cannot design, create, nor do Science; and, Materialism IS Creation by ROCKS! The ONLY reason to believe in Materialism is because you desperately want it to be true.

GOOD SCIENCE is supposed to be replicable on demand, reliable, and predictable. If the Materialists could somehow get a collection of ROCKS into a science lab and get them reliably and predictably designing and creating new and unique Genomes and Life Forms from scratch, then those Materialists would be famous the world over! But, it's NEVER going to happen, because Creation by Rocks is IMPOSSIBLE!

Instead, Materialism fails THE SCIENTIFIC METHOD at step 2, the TESTING and OBSERVATION part of THE SCIENTIFIC METHOD, because we can't get ROCKS nor Physical Matter into a Science Lab and reliably and predictably OBSERVE those ROCKS designing and creating Genomes and Life Forms from scratch. And, that's exactly what the Materialists claim happened 13.8 billion years ago BEFORE this Physical Universe was designed, created, and brought into existence. The Materialists are claiming that Physical Matter or ROCKS designed and created this Physical Universe, this Earth, and ALL of the Genomes and Life Forms on this Earth. Do you buy it? If you do, then you are a Materialist; and, you are welcome to it because I don't want it because I KNOW that it is FALSE!

New Life can ONLY be designed and created and brought into existence by LIFE. You will NEVER get the ROCKS to design and create life; thus, you will NEVER be able to demonstrate and prove that Materialism is true.

What does this really mean? It means that the First Conscious Life was NOT designed nor created but has always existed and will always exist. It means that YOUR personal consciousness was in existence BEFORE the first Physical Matter and the first Physical Universe was designed, created, and brought into existence by Conscious Life — possibly by YOU.

I discuss these Scientific and Observational Realities in much greater detail in the following books of mine:

"Quantum Mechanics: From a Non-Physical Spiritual Perspective" by Mark My Words.

"Summary Of: The Theory of Evolution Proves that God Exists: An Introductory Summary" by Mark My Words.

"THE SCIENTIFIC METHOD: Proves That the Theory of Evolution Is False" by Mark My Words.

"The Theory of Evolution Proves that God Exists: Why I Am No Longer an Atheist and Why I No Longer Believe in the Theory of Evolution" by Mark My Words.

"I Am NOT a Creationist! So What Am?" by Mark My Words.

And even,

"The Second Comforter: Supping with Our Resurrected Lord Jesus Christ" by Mark My Words.

In these books and this book, I make some bold and extraordinary claims, but no more bold or extraordinary than the Materialists' claim that ROCKS designed and created everything in this Universe. Creation by ROCKS or Creation by Physical Matter is a very bold and extraordinary claim for which there will NEVER be any evidence to confirm it or support it, because we all instinctively KNOW that ROCKS and Physical Matter cannot design and create.

As a Scientist, Philosopher, Logician, and Theoretician, I have learned to love the evaluation of evidence. I have learned to love asking my own questions, evaluating the evidence, and distinguishing between good evidence and weak evidence.

Ironically, I soon discovered that when it comes to the different types of Materialism, there is NO EVIDENCE whatsoever to support the claims of Materialism; and, "No Evidence" is in fact as weak as it can get when it comes time for us to evaluate the evidence. And, when it comes to Science, it is extremely important to see and understand and experience how Scientific Evidence is evaluated.

Materialism IS the claim that the Spiritual does not exist. There is NO EVIDENCE and can be NO EVIDENCE to support that claim. Consequently, when it comes to Materialism there can be NO evaluation of the evidence because there is NO EVIDENCE to support the claims of Materialism and therefore no evidence to evaluate. Instead, ALL of the evidence stands firmly against Materialism telling us clearly and conclusively that Materialism is FALSE. Materialism is simply a Religion that has to be taken completely on blind-faith as being true, because there can be NO EVIDENCE to support the claims of Materialism.

A Materialist has infinitely more blind-faith than a Theist can even begin to muster, because there can be NO SCIENTIFIC EVIDENCE nor any OBSERVATIONAL EVIDENCE to support the claims of Materialism. The lack of supporting Evidence IS the primary reason why Materialism is FALSE. Instead, the preponderance of the evidence tells us clearly and conclusively that Materialism is FALSE. That's what I learned when I tried to apply THE SCIENTIFIC METHOD to Materialism and evaluate the evidence associated with Materialism. I learned that there is NO EVIDENCE to support the claims of Materialism.

As a member of the JURY, judge for yourself if I meet my Burden of Proof through a preponderance of the evidence. This is what THE SCIENTIFIC METHOD is all about. SCIENCE is supposed to be the search for truth. I believe that I have finally found THE TRUTH. As a member of the JURY, judge for yourself if I am right or if I am wrong. THE SCIENTIFIC METHOD demands no less from you.

If you are a Materialist, and you still don't believe me, then get on your knees and pray to the ROCKS or pray to Physical Matter and ask them to do something for you right here, right now. Ask them to speak to you or to answer your prayers.

Run this Science Experiment!

I assure YOU that you will have an infinitely easier time getting the BIBLICAL GOD Jesus Christ to answer your prayers than you will have getting ROCKS or Physical Matter to answer your prayers! I KNOW THAT THIS IS SO, because I have experienced it first-hand. I have run this Science Experiment and examined the RESULTS! You can too!

It will feel weird at first praying to the BIBLICAL GOD, but you will be pleasantly surprised when it starts to produce noticeable RESULTS! Just don't give up; and, keep your eyes and ears and heart open for an answer to your prayers. It will come if you refuse to give up and stop trying! Patience and persistence pays off and produces RESULTS when it comes to the BIBLICAL GOD Jesus Christ. I can't say the same when it comes to Materialism and ROCKS.

Intro to Materialistic Darwinism 101

The Materialists, Darwinists, and Atheists can be extremely stupid, ignorant, uneducated, and dense when it comes to the parts of SCIENCE that they have chosen not to believe in. I know, because I used to be one of them.

Once you choose to abandon Materialism and Atheism, you free yourself to start studying the Sciences or the Scientific Disciplines which focus upon the non-physical, the spiritual, and consciousness or mind. You soon discover that the Spiritual Sciences or the Non-Physical Sciences — things like the "Sub-Quantum", Forces, Fields, Waves, Light, Strings, Psyche, Consciousness, Spirit, Thoughts, Telepathy, Extra Dimensions, Telekinesis, Remote Viewing, Multiple Universes, Visions, Dreams, Near-death Experiences, Out-of-Body Experiences, Awareness, Sentience, and Mind — are infinitely more interesting than the physical sciences ever were! We have a pretty good handle on the physical sciences now; but, we are just beginning when it comes to the Spiritual Sciences.

Most of the essays in this book were produced while interacting with Darwinists, Materialists, and Atheists online. I wrote these essays in response to their comments and questions. Clearly, I have learned to see things differently than they do, which got some of them curious and got others of them angry.

The PRIMARY PROOF which the Darwinists employ to convince us that the Theory of Evolution is true is purely a philosophical one. It has to be a philosophical proof because one eventually discovers that there is NO scientific evidence that actually supports the Theory of Evolution.

Their philosophical proof goes something like this: We all know that God does not exist; therefore, Evolution must be real and true, because Evolution is the only scientifically plausible explanation for the origin of life on this planet.

When it comes to philosophy, you can have any conclusion that you want, simply by adjusting the premises to your liking. Using philosophy, it's possible to prove a lie to be true, which is exactly what the Darwinists try to do.

Every time that I have ever interacted with a Darwinist online, they use this philosophical proof of Evolution's truthfulness as the basis for convincing us that the Theory of Evolution must be true. They use it, because it WORKS — on the Darwinists, Materialists, and Atheists at least. But, it doesn't work on those who are not in a Materialistic or Atheistic frame of mind, though. Nor does it work on the people who actually KNOW that God exists.

Obviously, the Darwinists say it or write it or IMPLY it in many different ways, but it's always the same basic philosophical proof which they use to convince us or prove to us that the Theory of Evolution must be true. Therefore, the Materialists and Darwinists make the claim that the Theory of Evolution must of necessity be true, because God does not exist. Or, they use the non-existence of God as PROOF that the Theory of Evolution must be true.

From their perspective, that claim makes perfect logical sense, is scientifically accurate, and is science and rationality and common sense in action. The Darwinists and Materialists are unable to see and understand the logic fallacies which they are employing to sustain such a claim and actually believe that it is true. If you are a Materialist or a Darwinist, you won't be able to see what's wrong with it. Instead, you will be employing it to convince others that the Theory of Evolution must be true.

—

I have noticed that during their indoctrination process, the FIRST LIE that the Darwinists are taught is that there is absolutely NO difference between Micro-evolution and Macro-evolution, if they are taught anything about it at all.

I have encountered PhD Darwinists and brilliant Scientists who can't explain to me the difference between Macro-evolution and Micro-evolution, having NEVER been taught in school anything about Micro-evolution and Macro-evolution. Many Darwinists haven't even heard the terms before; and, the few who have heard of micro and macro evolution have been LIED TO and taught that there is no difference between them.

You see, the Darwinian propaganda machine has been purposefully designed to teach students in our public schools that Evolution, or change of any kind, is exactly the same; but, there are many different types of Evolution or change, and some of them do indeed actually happen for real and have indeed been observed in the wild, but other types of Evolution are nothing more than pseudo-science or science fiction.

The few Darwinian professors, who have actually heard of Micro-evolution and Macro-evolution and know the difference between them, will LIE to their pupils and tell their students that the "science fiction types" of Evolution are just as real and happen just as often in the wild as the real types of Evolution. But most of the time, the Darwinian professors themselves don't know the difference between Micro-evolution and Macro-evolution, and thus will never mention the fact to their students that there are different types of Evolution; consequently, most people will come out of our public school system having never heard of Micro-evolution and Macro-evolution, and therefore believing that ALL Evolution is just the same and equally as real and true. That's the way the Darwinists want it to be! Obviously, the Darwinists don't want you thinking that some part of the Theory of Evolution might in fact be FALSE!

I can't even begin to count the number of Darwinists, Materialists, and Atheists whom I have encountered that have either never heard of micro and macro evolution or have been taught in school that micro and macro are exactly the same, and thus can't explain to me the difference between the two. I spend a lot of my time teaching the Materialists and Darwinists the difference between Micro-evolution and Macro-evolution, because they have never been introduced to the concept before! The Darwinists have NO incentive to teach their students the difference between Micro-evolution and Macro-evolution — why one of them is real and truly happens from time to time and why one of them is completely FALSE and has NEVER been observed in the wild. The Darwinists want their students believing that the whole Theory of Evolution is true, not that some part of it might be FALSE.

This reality presents itself to me every single time I interact with a Darwinist or a Materialist; otherwise, they wouldn't be Darwinists and Materialists!

The Darwinists who KNOW the difference between the two types of evolution do NOT want their students to know that there is a huge difference between Macro-evolution and Micro-evolution, because they don't want their students to know that there is a huge controversy where the Theory of Evolution is concerned.

The Darwinists want every one of us believing that everyone in the world is in agreement that the Theory of Evolution has been proven to be true; consequently, you will indeed repeatedly encounter PhD's and practicing scientists who have never heard of Micro-evolution and Macro-evolution! And, they will have NO idea that there is any kind of difference between them, if they have even heard of them before.

It can be a very interesting and fascinating experience to try to teach a Materialistic professional scientist or a PhD Darwinian biology professor the difference between Micro-evolution and Macro-evolution! They can't wrap their minds around it at first, because their

Darwinian conditioning is so total and complete. Even when they do start to understand the terms, they still can't see and understand how and why the distinction between them is logically and scientifically necessary, accurate, sound, and true — why Micro-evolution (Mutation and Selection) is real and has been observed in the wild, and why Macro-evolution (magical Creation by Evolution) is FALSE, has never been observed in the wild, and is nothing more than pseudo-science. They are going to fight this and resist it because they don't want to believe that it might be true. Can you blame them? Nobody likes being proven wrong, especially a PhD scientist.

The Darwinian professors and scientists don't want to give Evolution too much thought, or they might figure out what's wrong with it. So, you do indeed have to be ready to teach them the difference between Micro-evolution and Macro-evolution when they come asking, because they have never thought about it or learned the difference between the two main types of evolution or change. For many of them, it will be a concept that they have never been taught or learned in school, having been taught that all types of evolution are the same and all of them are true. But be warned, you aren't going to have any luck teaching them anything new until long after they have decided for themselves that they want to learn the truth about evolution. Meanwhile, you have to be satisfied with knowing the truth for yourself.

Over and over again, the Materialists and Darwinists have been taught that Micro-evolution and Macro-evolution are the same thing, just Evolution, if they have been taught anything at all. They can't see any difference between the micro and the macro. They can't see it and understand it because they don't want to see it and understand it. And, the Materialists and Darwinists blindly believe whatever they have been taught by other Darwinists and Materialists because they desperately want to believe "Creation by Evolution" to be true. The Darwinists and Materialists have a mental block where Micro-evolution and Macro-evolution are concerned! Otherwise, they wouldn't be Darwinists and Materialists.

By the time you get done reading this book, you WILL KNOW the difference between Micro-evolution and Macro-evolution — why one of them is real and truly happens and why one of them is FALSE and nothing more than science fiction — because THIS is where the controversy over evolution resides! In this book, I am going to teach you things that you will NEVER learn from a Darwinian professor in your college classes. I'm going to teach you the difference between Micro-evolution and Macro-evolution, and then slowly explain to you why neither one of them can design and create new unique functional genomes.

I'm going to teach you THE TRUTH about evolution, which is something that you will never get from a Darwinist or your PhD Darwinian college professor.

I eventually discovered for myself that you will NEVER learn the truth about the Theory of Evolution from a Darwinist, because they have no incentive to teach you the truth about Evolution; and, most of them don't even know the truth about the Theory of Evolution. They have no idea which parts of the Theory of Evolution are true, and which parts of it are FALSE.

So, keep your eyes open, because I'm going to teach you the truth about Micro-evolution and Macro-evolution. One of them has been observed in the wild for real, and one of them hasn't; consequently, one of them is real and truly happens from time to time, and one of them is completely FALSE and a LIE. If asked, your local Darwinian professor will LIE to you and tell you that they are one and the same, and thus equally true; but, I won't do that to you because I have no incentive to.

—

We learn through repetition. We learn from our failures and mistakes.

Some of this book gets repetitious because I have noticed that the Atheists, Materialists, and Darwinists don't get it or understand it the first time that it is presented to them, because they don't want to understand it. Been there and done that myself! I used to be an Atheist and a Materialist.

I will let you experience some of what I experienced as I returned and answered the very same questions over and over again in new and different ways. There are thousands of different ways to demonstrate how and why the Theory of Evolution is FALSE; but, there is no way to prove that it's true!

In this book, I will repeatedly use one or another of those THOUSANDS of different ways to demonstrate that Creation by Evolution is FALSE.

I have given Macro-evolution another name which I use quite often in my essays, namely "Creation by Darwinian Chance" or simply "Darwinian Chance". This substitution places a realistic meaning upon the term "Macro-evolution". Chance is synonymous with magic, and in this world just as reliable. Science is supposed to be about the things that are reliable, replicable, and/or demonstrable. Creation by Evolution is NOT it!

Over time, I observed that Darwinists define "Real Science" as science and scientific evidence which has been given a "goo-to-you" Darwinian interpretation. From their Darwinian perspective, it is not Real Science unless that science has been performed by Darwinian Chance and/or has been performed by a Darwinist who has given it a Darwinian molecules-to-man interpretation. To them, it doesn't count as Real Science unless it has been given a Darwinian explanation or a goo-to-you interpretation. According to the Darwinists, it must have the Darwinian Seal of Approval and a Darwinist's blessing and sanctification, or it can't be Real Science.

The Darwinists will tell you over and over again that you need to read and study some Real Science, meaning science which has been given their FALSE Darwinian goo-to-you interpretation.

Only a blind and loyal Darwinist will actually believe that Darwinian Chance is real and true, and that Darwinian Chance designed and created everything in this universe.

In contrast, I look at the exact same evidence, and then I choose to give that evidence an infinitely more believable interpretation or explanation than Darwinian Chance or molecules-to-man evolution, which is my right to do as a scientist.

The only significant difference between a Darwinist and a Real Scientist is the way that they choose to explain or interpret the Scientific Evidence.

The Darwinists give Scientific Evidence a FALSE INTERPRETATION, explaining everything in terms of Darwinian Chance, Creation by Evolution, and goo-to-you evolution.

In contrast, the Real Scientists or the True Scientists will attempt to give the same Scientific Evidence a True Interpretation that actually matches with Reality and the Fossil Record. It is said to be scientifically accurate if it matches with Reality. The True Scientists will give you a Scientific Interpretation or Explanation that actually matches with Reality.

In the end, it all boils down to the fact that Macro-evolution of any kind including Creation by Evolution is FALSE and has NEVER been observed in the wild by anyone. Keep your eye on Macro-evolution or Creation by Evolution or "Creation by Darwinian Chance" because IT is the Major Flaw in the Theory of Evolution. Macro-evolution or Creation by Evolution should be eliminated from Science as a possible Designer and Creator, because creation by "evolution of any kind" is IMPOSSIBLE. Evolution cannot design and create anything at all.

Creation by Evolution is IMPOSSIBLE; and, my goal in this book is to eliminate the impossible, so that we are left with the truth. And, I'm just getting warmed up!

Still interested? Then read on my friend . . .

—

Now pay attention!

Despite the FACT that I have debunked and falsified the Theory of Evolution thousands of times in hundreds of different ways, don't ever once let me convince you that there is NO such thing as evolution or random mutations.

Evolution is REAL, very REAL. Evolution or random mutation is entropy or the second law of thermodynamics. It's REAL. Entropy is very REAL at the physical level; and, it works as advertised at the physical level.

Whenever the Materialists, Naturalists, Darwinists, Nihilists, Behaviorists, and Atheists start talking about evolution, they get most everything wrong, because evolution or entropy cannot design and create. However, these people do indeed get one thing perfectly right. Evolution, or random mutation, or entropy is indeed the CAUSE of ALL of our heritable diseases, developmental diseases, and heritable mental illnesses.

Remember, the Theory of Evolution is FALSE because random mutations or entropy cannot design and create genes, proteins, and life forms. However, evolution or random mutation or entropy is very REAL; and, it can indeed destroy genes, proteins, and life forms. Do you see how that works? It's important to understand.

Science Is the Search for Truth, Or It's Supposed to Be

It is simply amazing that most people have no idea what Science is and what it's really meant to do.

At its most powerful and its most ideal and idealistic, SCIENCE IS THE SEARCH FOR TRUTH.

Based upon that definition for Science, we automatically know that Science IS NOT LIMITED exclusively to the material realm or the physical realm! There are millions of truths out there that simply are not tangible!

The fact that my estranged wife hates me and refuses to forgive me might be the truth, but it's NOT something that I can get my hands onto or my scientific instruments onto to know for sure! I could be wrong. There are millions of things just like that in this world. The fact that I had a dream last night is the truth, but I can't get my hands onto it, nor can you get your scientific instruments onto it and tell me what that dream was about! The fact that my granddaughter loves me and adores me is obviously true to me, but there is no way for you to get your scientific instruments onto that and measure the extent of it or the depth of it! I can feel it or sense it, but scientific instruments cannot measure it or quantify it.

—

Seeing the Light

Materialism is the philosophical belief that only matter and energy exist.

Science, TRUE SCIENCE, will never limit itself to the physical or the tangible or the material, because Science is the search for truth — all truth!

LIGHT is non-physical, massless, immaterial, and intangible; but, that doesn't stop the Real Scientists from studying it scientifically. Ordinary visible light is spirit, which can actually be seen with the physical eye or the naked eye. There are other SPIRITS or LIGHTS, existing at other frequencies, which cannot be seen with the naked eye.

There seem to be living spirits or living lights, which are organized into some kind of living package or spirit body. These spirits or living lights have to adjust their frequency so that they can actually be seen by us with the naked eye. This frequency adjustment is what happens when mortals like us in this physical realm start seeing and speaking with the spirits of the dead. We get on the same channel or the same wavelength or the same frequency; and, then we can see them. Do you see dead people? Some of us do.

When it comes to spirits or lights, there seem to be different types and different degrees; but, it has been hypothesized or suggested that even a photon of light is alive, sentient, and aware of its surroundings including its ultimate destination.

LIGHT IS SPIRit. SPIRIT IS LIGHT. Physicist David Bohm said that matter is frozen light, which I take to mean "light that has been slowed down enough to become tangible or physical".

You will NEVER get any useful or interesting information about these SPIRITS or LIVING LIGHTS from a Materialist or an Atheist, because they refuse to think about and consider these kinds of things. Their Materialism cripples their scientific exploration!

Every physical life form must have either a Creator or a Parent, because physical bodies do not design and create themselves; and, physical bodies NEVER come alive spontaneously on their own. In order for a physical body to come alive, God has to place some kind of spirit or spark or light into it – the Light of Life. By definition in principle, once our spirit is severed or detached from our physical body, then our physical body is dead. The spiritualists and mystics and out-of-body travelers who have seen this thing call it a Silver Cord. This Silver Cord is what attaches or links our spirit body with our physical body. When this Silver Cord is severed or broken, then our physical body dies, and our spirit body is free to move on to other realms and abodes.

The first physical body to ever come into existence had to have been intelligently designed, programmed, engineered, assembled, and then occupied by some kind of sentient Spirit or sentient Light which had negotiated control of matter or had learned how to control physical matter with its mind. Physical bodies don't just put themselves together all on their own. There has to be some kind of Spirit, Intelligence, or Light that is driving or controlling the whole process!

It gives a whole new meaning to the phrase in 1 John 1: 5 which says: "God is light, and in Him is no darkness at all." If you notice carefully, God has the type of Spirit or Light that can actually levitate, move, and manipulate physical matter.

Before the first God created the first physical body and then occupied it, that God was a sentient Light or a sentient Spirit. It's that Spirit or Intelligence or Light which is eternal and uncreated. The first Spirit to design and create an immortal physical body for Himself would by definition have become the First God. In the realm of spirits, the Physically Embodied Spirit is God and King because He has found some way to manipulate and control physical matter.

In this universe, there are things that act, and things that are acted upon. The physical matter is always acted upon! It is the Spirit or the Intelligence or the Light or the Consciousness that does the actual thinking and acting.

The thing about spirits is that they can reach out and deliberately prove the Materialists and Behaviorists wrong any time that they want to do so. Any human spirit can decide to break his Behavioral Conditioning or his Bad Habits if he or she wants to do so.

The Behaviorists might believe that they have found the cause for a certain behavior, but any human spirit can decide to prove them wrong any time it wants to do so. We instinctively know that Alex in "A Clockwork Orange" could break his conditioning as quickly as he wanted to, once he got away from his Conditioners. Only a Materialist or Behaviorist would actually believe that that psychological conditioning is permanent. The rest of us know that conditioning of any kind experiences extinction, once the human spirit decides that it doesn't want or need that conditioning any more.

Even animals break their conditioning when the stimulus and reward (or punishment) are no longer present. Our male dog immediately acts differently around me than he does around his master, because his master disciplines him and scolds him for bad or rambunctious behavior, and I do not. I like to see him running wild and free, so he always gets stirred up whenever he sees me. It's very fascinating to observe the cogs whirring in that dog's head whenever he is presented with me and his master simultaneously. You can sense and feel that he wants to let loose, but he's also fighting it and trying not to because his master is present. It's fascinating to watch him look at his master, and then pretend that I don't exist and that I'm not present at all.

There is Intelligence or Spirit or Light in that dog, and he actually chooses to act differently depending upon which person he is interacting with. That reality runs completely counter to Materialism and Behaviorism, which says that that dog should act the exact same way no matter whom or what that dog might encounter. According to Behaviorism and Materialism, that dog should be nothing more than a robot dancing to its DNA. But, there's a person in there, a unique and living personality, a unique and living SPIRIT or LIGHT! Dogs are people too!

My interactions with our dogs is one constant Science Experiment. In comparison, the cats are boring. Sometimes those cats really do seem like nothing more than living breathing killing machines. Once again, we can learn something from this — there are different types of SPIRITS or LIGHTS. And, we can instinctively tell who or what is driving the bus, or if the bus is just simply driving itself.

Light of any kind proves conclusively that Materialism is inadequate and false, if one is willing to accept the evidence before his eyes. Open your eyes and see the light!

—

Psychology is an officially recognized Science! Psychology is in fact one of the most popular Sciences that we have! The original definition for "Psychology" was "the study of the human psyche". What is the psyche? The psyche is the human spirit, or the human consciousness, or the human awareness, or the human intelligence, or the human spark, or the human mind. It is interesting to note that the human spirit is by definition — in principle — intangible, immaterial, and non-physical; but, that doesn't stop the REAL SCIENTISTS from trying to find some way to study it, especially the scientists who are searching for the truth!

Psychology when it was originally developed refused to limit itself to the physical or the material.

Yes, we have subsequently had Atheistic Psychologists and Behavioral Psychologists who have deliberately limited their own selves to Materialism and Physicalism, stunting their own intellectual learning and growth in the process; but, there is absolutely no reason why any of the rest of us should acquiesce to their demands that we follow their example and limit ourselves to Materialism as well.

Limiting oneself exclusively to the physical realm completely stifles Scientific Exploration and Scientific Discovery! Materialism and Atheism are the greatest banes to Scientific Discovery that mankind has ever created!

—

I have noticed that most people, especially the Atheists and Darwinists, seem to have absolutely no understanding whatsoever of the difference between a "theory" and an "axiom or a law".

They confidently and unknowingly declare that Evolution or "Creation by Natural Selection and Random Mutations" is a LAW or an indisputable AXIOM. But, there is something that they fail to realize. Not everyone agrees that Forward Evolution can be accomplished by Natural Selection and Random Mutations; in fact, many peer-reviewed Science Research Papers have been written which clearly demonstrate that Natural Selection and Random Mutations are completely insufficient and totally unable to create new proteins and new DNA, new biological information and new biological designs, and new genomes and new life forms!

So, what gives?

Whenever an Atheist or a Darwinist publicly declares that Evolution has been proven true, has no opposition, and is an axiom or a law, then we just simply know that they are lying to us and trying to deceive us. They don't understand the difference between a theory and an axiom! Their words are simply propaganda!

You see, a "theory" has an active opposition — people who don't believe that the theory is true! With a theory, you will see contending and opposing points of view! So, what do we see with the Theory of Evolution? We see contending and opposing points of view! There is NO consensus on the subject! And, thousands of different Research Scientists over the decades have provided Scientific Evidence letting us know that Micro-evolution and Macro-evolution are INSUFFICIENT and cannot do everything that the Darwinists say that they can do. There is huge and active opposition to the Theory of Evolution!

Now, in contrast, if you are dealing with an Axiom or a Law, there will be NO opposition to it, because it is self-evident and obviously clear to everyone that it is true. Take something like 2+2=4. That thing is based upon laws and axioms! There really are NO contending and opposing points of view! Only the mentally incompetent and mentally damaged would contend against it.

Now, with this information in mind, decide for yourself, "Is Evolution really the Law of Evolution or the Theory of Evolution?" Is there any contention or opposing points of view associated with Evolution and Creation by Evolution? If there are, then it's still a theory, despite what the Darwinists will try to tell you!

It's only after a theory becomes obviously true to every one of us, and there is no opposition to it whatsoever, that it becomes a Law or an Axiom! As long as there is any question or doubt, then it remains a theory; and, the scientific search for the truth regarding that theory continues!

—

Science is supposed to be a search for the truth.

I have observed that for most of the Darwinists, Materialists, and Atheists, their ultimate goal is to suppress the truth or to hide the truth from us.

What is that truth which they are trying to hide from us?

The truth is that Evolution of any kind cannot design and create anything at all. There is NO such thing as goo-to-you evolution or molecules-to-man evolution. Creation by Darwinian Chance or Macro-evolution does not exist. Darwinian Chance is not a living breathing person, so there is no way possible for it to have designed and created everything in this universe, as the Darwinists say that it did. That's the truth which the Darwinists and Materialists are trying to hide from us; and, they use sophistry, trickery, name-calling, censorship, and intimidation in order to do so.

If you eliminate the IMPOSSIBLE, such as Creation by Darwinian Chance or Macro-evolution or Evolution by Abiogenesis, then whatever remains, however improbable, must be the truth. What do you end up with when you eliminate the Theory of Evolution from the equation? It's elementary, my dear reader. You are left staring at the truth. You have suddenly proven to yourself that God must exist in order to have done all of the different things that Evolution couldn't have done, and that Evolution never did. Only an Atheistic Darwinist will refuse to believe that this is so.

Circumstantial Evidence and Scientific Inferences

Remember, the things that can't be observed with nor detected by our physical instruments have to be experienced in order to KNOW that they are real and true.

One day, I made a simple and powerful observation which completely changed my life and my science for the better. I observed and realized that if we want to find and know the truth about our existence and Reality itself, the best way to find that truth is to look at, study, and learn everything which the Materialists, Naturalists, Darwinists, Nihilists, and Atheists refuse to look at, study, accept, and learn. Study and learn everything that these people automatically reject and assume not to exist; and then suddenly, we will find ourselves looking at the true reality of our existence. Initially, I used these people to tell me what I should be looking at and studying. If I came across anything online that these people were resisting, rejecting, or openly fighting against, then I automatically knew that it was something that I should be studying and learning about if I wanted to find and know the truth. That was quite an epiphany, especially for someone like me who used to be a Materialist, Nihilist, and Atheist.

I observed that the Materialists, Naturalists, Darwinists, Nihilists, and Atheists have a knack, a skill, or a built-in barometer for finding, accepting, and believing anything that is false. These people gravitate and navigate towards anything that has been falsified by observational evidence, experiential evidence, or empirical evidence of a Non-Local or Non-Physical nature – automatically choosing to believe that the observational evidence is false, and that Materialism and Naturalism are true. These people have a very strongly developed *confirmation bias*, wherein they automatically and immediately reject anything and everything that falsifies the materialistic and naturalistic worldview. These people don't seem to realize that *confirmation bias* is a logic fallacy, as is a refusal to look at and study evidence.

Confirmation bias is defined as the tendency to interpret new evidence as confirmation of one's existing beliefs or theories. These people automatically reject anything that doesn't confirm their existing beliefs and theories; and, they choose to interpret everything in terms of Materialism and Naturalism. These people develop complete, complex, detailed Science Disciplines, which are based exclusively on a rejection of observational evidence combined with circumstantial evidence and scientific inferences of their own making. It can be fun and interesting to list all the different logic fallacies which they use in order to make their case for Materialism and Naturalism. These people literally treat their scientific inferences as empirical evidence.

Empirical evidence is information acquired by observation, experience, and/or experimentation. In contrast, *scientific inferences* are *ad hoc just-so stories* – logic fallacies. Scientific inferences are easily defeated simply by making up a better story.

Furthermore, observational evidence, experiential evidence, eye-witness evidence, and the Lived Experiences of the human race should be given precedence over experimental evidence.

Why?

It's because experimental evidence can be interpreted incorrectly, and typically is. For every piece of evidence that we might ever encounter, there are theoretically an infinite number of possible different interpretations which can be given to that single piece of evidence. God knows which one of them is true!

The best way to find and know the truth is to live it and experience it for yourself, or to choose to trust someone who has. I have learned that when it comes to experimental evidence, scientific evidence, observational evidence, experiential evidence, and empirical evidence, I INTERPRET IT FOR MYSELF, rather than relying upon the "official interpretation" which is provided by the Materialists, Naturalists, Darwinists, Nihilists, and Atheists because these people have a knack for always providing a false interpretation to every piece of evidence they encounter. Personal interpretation of the evidence IS the primary flaw of the scientific methods, because personal interpretation of the evidence is based upon a bunch of different logic fallacies such as *affirming the consequent*, *begging the question*, *jumping to conclusions*, *scientific inferences*, *circumstantial evidence*, *pre-conceived conclusions*, *evidence manufactured to fit one's pre-conceived conclusions*, *ad hoc just-so story-telling*, and *confirmation bias*. A study of the Philosophy of Science can help us learn to recognize and identify these different logic fallacies whenever we come across them. The following book from Joseph Rychlak changed my life for the better, and changed the way I look at, study, and do science.

Rychlak, J. F. (1981). *A Philosophy of Science for Personality Theory* (2nd ed.). Malabar, FL: Robert E. Krieger Publishing Company.

For a good and honest introduction to the Philosophy of Science, see also:

Slife, B. D. & Williams, R. N. (1995). Science and Human Behavior. In *What's Behind the Research? Discovering Hidden Assumptions in the Behavioral Sciences*, (pp. 167–204). Thousand Oaks, CA: SAGE Publications.

http://mypsyche.us/wp-content/uploads/2017/04/Science.pdf

Remember, the ultimate goal in all of this is to find and know the truth. If it's a lie, then ultimately it has no value to us.

It can be extremely fascinating to study the psychology of the Materialists, Naturalists, Darwinists, Nihilists, and Atheists. These people are in denial; and, denial is one of the most common defense mechanisms. When I was a Materialist, Nihilist, and Atheist, I had some severely messed-up psychology, which can be fascinating to look back on and study.

Eventually I discovered that many, if not most, of the Materialists, Darwinists, Nihilists, and Atheists are actually trying to trick us and deceive us. The truth cannot be found nor established through deceptions and lies. The following book helped me to make this particular discovery or realization. This book exposed many of their lies and deceptions and tricks, which we scientists and materialists typically take on blind faith as being true.

Wells, J. (2000). *Icons of Evolution: Science or Myth? Why Much of What We Teach About Evolution Is Wrong*. Washington, DC. Regnary.

This book, *Icons of Evolution*, exposes a few of the lies and deceptions upon which the Theory of Evolution is based. After reading this book, I simply KNEW that many of the Darwinists and Evolutionists are deliberately and knowingly trying to trick us and deceive us. I don't like being lied to, so this book had a powerful impact on my life, because many people like me have been tricked, deceived, and lied to by the Darwinists, Naturalists, and Materialists without even knowing it. Once I was made aware of their deceptions and lies, I was suddenly able to catch them in their lies, which made a huge difference in my life.

Circumstantial evidence is evidence that relies on an **inference** to connect it to a conclusion of fact — like a fingerprint at the scene of a crime. By contrast, **direct evidence**

supports the truth of an assertion directly — i.e., without need for any additional evidence or inference.

In science and philosophy, **a just-so story**, also called an **ad hoc fallacy**, is an unverifiable narrative explanation for a cultural practice, a biological trait, or behavior of humans or other animals. The pejorative nature of the expression is an implicit criticism that reminds the hearer of the essentially fictional and unprovable nature of such an explanation. Such tales are common in folklore, mythology, anthropology, paleontology, and the theory of evolution.

Almost every college textbook I have ever read, pays homage to the creative genius of Evolution, Natural Selection, and Random mutations.

My Biopsychology textbook has a whole chapter and paragraphs within each chapter paying homage to the creative brilliance of evolution, with worshipful adoration. For these people, evolution is clearly their god, and they have imbued it with all the attributes and creative powers of a god.

Kalat, J. W. (2008). *Introduction to Psychology* (9th ed.). Belmont, CA: Wadsworth, Cengage Learning. **"The brain is the product of Evolution."** (p. 12.)

Pinel, J. (2014). *Biopsychology* (9th ed.). New York: Pearson. **"Genetic endowment is a product of its evolution."** (p. 24.) **"Some have suggested that glial networks may be the dwelling places of thoughts."** (p. 59.)

It's physically impossible for neurons and glial cells to be the dwelling places of thoughts, because a neuron is a single hardware BIT. All of the synapses on a neuron reduce or integrate everything that comes their way into a single BIT of information. A neuron is a switch at the physical level – it is either ON or OFF. You can't store thoughts and memories in a switch. Thoughts and memories are quantum waves. WE KNOW this to be true, because our thoughts and memories survive the death of our physical brain. Thoughts, memories, psyche, and quantum waves can be stored within a single atom at the quantum level or psyche level. We don't need a glial network for that. If glial networks are indeed the dwelling places of thoughts, it's happening at the quantum level, because it's physically impossible at the physical level thanks to all of those synaptic clefts that were designed to randomize and eliminate memories and thoughts and information by reducing them into a single BIT of information.

The Human Psyche is the dwelling place of thoughts, and those thoughts interact with and control glial cells and neurons telepathically through Nature's Psyche and the Quantum Zeno Effect.

These people want you to believe that evolution (random mutations and natural selection) designed, created, manufactured, engineered, field-tested, fine-tuned, balanced, and produced your neurotransmitter system and your physical brain all by itself without any kind of outside help. These people want you to believe that evolution (random mutations and natural selection) designed, programmed, implemented, debugged, field-tested, perfected, engineered, manufactured, and produced your genome from scratch all by itself without any kind of outside help. These people want you to believe that evolution is smarter and more capable than the very best scientists on our planet. These people want you to believe that evolution designed and created your physical body, your genome, your physical brain, and your neurotransmitter system all by itself without any outside help.

That's impossible!

Why?

Think about it logically.

It's impossible because evolution, random mutations, and natural selection didn't even exist until after God designed and created and produced the genomes, physical bodies, physical brains, and neurotransmitter systems in the first place.

There's no way in the universe that the dumb, mindless, blind processes of evolution (mutation/selection) could have designed and created every aspect of our physical brain. It's impossible. Chemical evolution can't even design and create a functional protein, let alone the matching gene to go along with it. Science is supposed to eliminate the impossible, so that only the truth remains. Science should start by eliminating the theory of evolution. If you successfully eliminate everything that is false, then only the truth will remain. The evolutionary perspective is FALSIFIED by observations such as these. There's no place for philosophical speculation, scientific inferences, and evolutionary psychology whenever an author deliberately limits himself to the FACTS and the TRUTHS that have been observed and experienced phenomenologically. Evolution of any kind has never been caught in the act of spontaneously generating proteins and genes out of thin air, because it can't. Spontaneous generation, abiogenesis, and macro-evolution are physically impossible. If they happened, they happened at the quantum level. Think people, think!

Both James Kalat and John Pinel, the authors of these college textbooks, fell for the lies and deceptions of evolution – hook, line, and sinker. John Pinel has a whole chapter dedicated to the "beautiful concept" and "powerful ideas" of evolution. Throughout his book, John Pinel presents a "just-so story", an "ad hoc fallacy", a "scientific inference" as truth, proof, empirical evidence, and scientific evidence that the theory of evolution has been proven true.

These "conclusions of fact" that evolution designed and created your genome and your brain are based upon circumstantial evidence, rather than direct evidence. These people are simply making an inference, an educated guess, based upon circumstantial evidence rather than observational evidence. NOBODY has ever caught the processes of evolution – natural selection and random mutations – in the act of designing, programming, creating, and manufacturing genomes and brains, because these natural processes can't do so.

Scientific inference is scientific speculation or scientific guesswork. They are making an educated guess and providing a personal interpretation of the scientific evidence. At the same time, these Materialists, Naturalists, Darwinists, Nihilists, and Atheists selectively reject any and all direct observational evidence of the non-local or the spiritual. Rejecting observational evidence – that's NOT science – it's blind religious dogmatic fanaticism. These people reject the observational evidence or the experiential evidence, and they replace this evidence with ad hoc just-so stories which they label as "scientific inference" to make it sound more official and real than it is.

When it comes to these people, their whole "science" is based exclusively on scientific inference and ad hoc just-so stories, because there is NO evidence that evolutionary processes can design and create genomes and brains from scratch; and, there never will be.

From *Biopsychology* page 13, John Pinel writes:

Scientific inference is the fundamental method of biopsychology and of most other sciences – it is what makes being a scientist fun. This section provides further insight into the nature of biopsychology by defining, illustrating, and discussing scientific inference.

The scientific method is a system for finding things out by careful observation, but many of the processes studied by scientists cannot be observed. For example, scientists use empirical (observational) methods to study ice ages, gravity, evaporation, electricity, and nuclear fission – none of which can be directly observed; their effects can be observed, but the processes themselves cannot. Biopsychology is no different from the other sciences in this respect. One of its main goals is to characterize, through empirical methods, the unobservable processes by which the nervous system controls behavior.

The empirical method that biopsychologists and other scientists use to study the unobservable is called <u>scientific inference</u>. Scientists carefully measure key events they can observe and then use these measures as a basis for logically inferring the nature of events that they cannot observe.

Like a detective carefully gathering clues from which to recreate an unwitnessed crime, a biopsychologist carefully gathers relevant measures of behavior and neural activity from which to infer the nature of the neural processes that regulate behavior.

The fact that the neural mechanisms of behavior cannot be directly observed and must be studied through scientific inference is what makes biopsychological research such a challenge – and as I said before, so much fun. (p. 13.)

Scientific inference is a logic fallacy.

The processes of evolution designing and creating genomes and brains and living cells from scratch "cannot be observed" and have never been observed; so, we are simply going to guess – make a scientific inference – and then introduce our guess into the scientific method as evidence. This process is the logic fallacy called, *begging the question*, which is the process of taking your chosen conclusion and then inserting that conclusion into your argument as one of the premises or proofs of your argument. That's cheating! Not science! I have even had Darwinian evolutionists and materialists tell me that it's okay to lie and cheat and deceive and take liberties, as long as their lies are being used to prove the truth – truths such as the Theory of Evolution.

In other words, these people are deliberately and knowingly introducing "scientific inference" into evidence as evidence, because it makes being a scientist fun.

Remember, the speculation and guesswork and inference are what makes the science fun; but, it's ONLY direct observation or direct experience of a phenomenon that actually makes the science real and true. Remember that! These people base whole sciences upon "scientific inference" without a single shred of observational evidence to support their inferences. John Pinel literally can't see anything wrong with this process! He subsequently gushes profusely about the beauties, simplicity, logic, powerful ideas, and large body of supporting evidence that are the theory of evolution; and, he repeatedly quotes others who do the same.

These people assume that the theory of evolution is true, *affirm the consequent*, and then *beg the question* by inserting their chosen conclusion or scientific inference back into their argument as proof and evidence of the truthfulness of their chosen conclusion. They don't seem to realize that they are doing so. They seem to be blind to it, probably because they want to be. It takes someone deeply trained in the Philosophy of Science to be able to see the wide variety of different logic fallacies that these people use to make their case in order to prove that the theory of evolution is true.

Rychlak, J. F. (1981). *A Philosophy of Science for Personality Theory* (2nd ed.). Malabar, FL: Robert E. Krieger Publishing Company.

In this book and his other books, Joseph Rychlak introduces and explains some of the logic fallacies upon which Science and the Scientific Methods are based – particularly the *affirming the consequent* logic fallacy. Did you know that Science and the Scientific Methods ARE BASED UPON a logic fallacy, which is called *affirming the consequent*? It was news to me! I didn't know that; and, I have been a scientist all of my life – in this case, for over fifty-five years of my life before I was taught anything about *affirming the consequent* and how it works and how it messes everything up in science.

The *scientific inference* logic fallacy is just another variation of the *affirming the consequent* logic fallacy, the *jumping to conclusions* logic fallacy, and the *begging the question* logic fallacy. Using these logic fallacies, you can prove anything to be true; and, that's exactly what the Naturalists, Darwinists, and Materialist do – they insert their chosen conclusion into their arguments, their scientific interpretations, and their methods as their initial premise and then treat their chosen conclusion as evidence and proof that their chosen conclusion is true.

So, how do these people know that evolution created your brain and your genome? Were they there? Did they watch it happen? Have they observed it happening? NO! It's an "ad hoc just-so story". It's a "scientific inference". It's based upon the *ad hoc* logic fallacy. In other words, it's just a guess. It's simply what they want to be true. That's not science! That's blind religious faith!

Over and over again, I catch these "scientists" employing *affirming the consequent*, *begging the question*, and *jumping to conclusions* logic fallacies in order to make their case. Thanks to John Pinel and his college textbook, *Biopsychology*, I now have yet another logic fallacy to add to the list – ***scientific inference***. I introduce to the world the *scientific inference logic fallacy*. Through scientific inference, you can have any conclusion you want, and then simply prove your chosen conclusion to be true through scientific inference; and, that's exactly what John Pinel does in his book. But, it's ALL *circumstantial evidence* – not observational evidence; and, I make the bold claim that Science IS Observation, and that science is NOT scientific inference as John Pinel claims. Scientific inference is a logic fallacy, in a long list of logic fallacies upon which the Materialists, Naturalists, Darwinists, Nihilists, and Atheists rely in order to make their case and "prove" that the theory of evolution is true.

"Scientific inference makes science fun", because you can have any conclusion you want, and then insert that chosen conclusion into your arguments as scientific proof that your arguments and claims are true. That's the way these people do science. Exposing their deceptions, tricks, logic fallacies, and lies is the way that I do science. As the judge and jury in this particular scenario, you will have to decide for yourself which of us makes his case, has the best presentation, carries the preponderance of the evidence, and meets his burden of proof. All I can do is to present to you the evidence of what I have observed and experienced, and then let you make up your own mind what you want to believe.

On page 25 of *Biopsychology*, John Pinel quotes another worshipper of the theory of evolution:

> **Evolution is both a beautiful concept and an important one, more crucial nowadays to human welfare, to medical science, and to our understanding of the world than ever before. It's also deeply persuasive – a theory you can take to the bank. The supporting evidence is abundant, various, ever increasing, and easily available in museums, popular books,**

textbooks, and a mountainous accumulation of scientific studies. No one needs to, and no one should, accept evolution merely as a matter of faith (Quammen, 2004, p. 8).

Ironically, you have to accept his chosen conclusions here "merely as a matter of faith", because evolution, natural selection, random mutations, macroevolution, abiogenesis, and spontaneous generation have NEVER been caught in the act of designing, programming, initiating, engineering, manufacturing, and deploying genomes, life forms, and physical brains from scratch – and never will be because they can't do these kinds of things. Evolution, natural selection, and random mutations lead a species inexorably to extinction through a documented and observed process called "genetic entropy". These things don't build. They destroy. Evolution, natural selection, and random mutations are processes of chaos, destruction, entropy, and extinction – not design and creation. That's what the observational evidence is trying to teach us.

It has been repeatedly proven that evolution of any kind can't design and create proteins, nor the genes associated with them. So, what can evolution (genetic change) do, since it can't do design and creation? Evolution is entropy or random chaos. Evolution is the second law of thermodynamics. Evolution is real. Evolution can do random mutations, chaos, entropy, disease, cancer, mental illnesses, death, and extinction. Evolution is a blender. Evolution can destroy; but, it can NEVER design and create.

Sanford, J. (2014). *Genetic Entropy* (4th ed.). Cornell University: FMS Foundation.

This book changed my life for the better, by introducing me to the truth about evolution, random mutations, and natural selection – namely that these things can't design and create anything at all.

"No one needs to, and no one should, accept evolution merely as a matter of faith"; yet, that's exactly what they do – take evolution merely as a matter of faith. I can't do it anymore. I no longer have faith enough to be a Materialist, Nihilist, and Atheist like I used to be. I just can't do it anymore. I know better now.

The Materialists, Naturalists, and Darwinists are literally "banking" on their belief in evolution – it's the way that they make their money and their livelihood. They are banking on the hope that you will unknowingly fall for their logic fallacies, deceptions, tricks, and lies. They use their positions as college textbook writers to promote, and even enforce, their personal religion. It's the greatest deception that has ever been foisted on the human race. The only thing that's ever-increasing is the amount of evidence that I come across, which directly falsifies Materialism, Naturalism, Darwinism, Nihilism, and Atheism.

Materialism is dead. Naturalism is dead. Darwinism is dead. These people simply haven't gotten the memo yet. Materialism, Naturalism, Darwinism, and Atheism have been successfully falsified by science, observational evidence, and the scientific methods trillions of times in thousands of different ways; but, these people don't know that because they don't want to know it. Textbook writers like Pinel and Kalat literally insert "assumptions" and "scientific inferences" and "ad hoc just-so stories" into their college textbooks as scientific evidence, and they seem completely oblivious to these logic fallacies which they employ in order to make their case for the theory of evolution, because that's the way they have been taught to do science. *Jumping to conclusions* and *begging the question*, these people literally define "science" as scientific inference, materialism, naturalism, and Darwinism; and, then they insert their definition for science into their arguments, inferences, and interpretations as evidence and proof that their chosen conclusions are true – a process that is in fact *begging the question*.

In his textbook, James Kalat repeatedly demands that we "examine the evidence"; but, he makes an exception for the theory of evolution choosing instead to take the whole thing on blind faith as being true without even trying to present any evidence. In his textbook, John Pinel tries to make a compelling authoritarian case for the theory of evolution; but, all of his evidence is scientific inference, circumstantial evidence, hyperbole that he expects you to take literally, logic fallacies, ad hoc just-so stories, and unobservable unproven assumptions which he then uses as evidence and proof that the theory of evolution is true. In his textbook, John Pinel repeatedly calls upon the student to "think creatively"; yet contrary to his demand, he accepts the theory of evolution, Darwinism, and evolutionary psychology without question as being true.

Clearly, almost every college textbook I read keeps bringing me back to this subject. I tried to get out of it, but they keep dragging me back in.

The best, most believable, most parsimonious, most logical, most balanced, and most scientific college textbook that I have read so far is:

Barlow, D. H. & Durand, V. M. (2015). *Abnormal Psychology: An Integrative Approach* (7th ed.). Stamford, CT: Cengage Learning.

http://mypsyche.us/wp-content/uploads/2017/10/Abnormal-Psychology.pdf

These authors don't spend all of their time making inferences, jumping to conclusions, and worshipping the theory of evolution like the others do. They examine the research and provide a sound, logical, integrative interpretation to that scientific evidence or scientific research. I was impressed. While reading Barlow and Durand, I actually felt like I was studying science and psychology rather than philosophy, religion, dogma, and evolutionary propaganda.

Remember, the things that can't be observed with nor detected by our physical instruments have to be experienced in order to KNOW that they are real and true.

Neuroscience: Things That Aren't Quite True

We have to eliminate as many falsehoods as we possibly can before we can take on something new like Quantum Neuroscience. I'm not going to pull any punches. I'm going straight for the throat – straight for the kill – and never let go. Red in tooth and claw!

1. The human brain and the human genome are the products of evolution.

The authors of our college textbooks want you to believe that evolution designed, planned, created, manufactured, field-tested, and fine-tuned everything in your physical body.

Kalat, J. W. (2008). *Introduction to Psychology* (9th ed.). Belmont, CA: Wadsworth, Cengage Learning. **"The brain is the product of Evolution."** (p. 12.)

Pinel, J. (2014). *Biopsychology* (9th ed.). New York: Pearson. **"You will learn through the text that we humans have learned much about ourselves by studying species that are related to us through evolution. The evolutionary approach has proven to be one of the cornerstones of modern biopsychological inquiry."** (p. 4.) **"The empirical method that biopsychologists and other scientists use to study the unobservable is called scientific inference."** (p.13.) **"This is a particularly important chapter for you."** (p. 21.) **"Genetic endowment is a product of its evolution."** (p. 24.) **"The human brain appears to have evolved from the brains of our closest primate relatives."** (p. 33.) **"The human eye, a product of 600 million years of evolution."** (p. 132.) **"Thinking creatively: Thinking in productive, unconventional ways that are consistent with the evidence."** (p. 496.)

There's no way on earth that any of this can possibly be true.

Why?

It's because there has to be already in-place a functional brain and a functional genome and a functional physical body, BEFORE random mutations and natural selection can start tweaking them and adjusting them. Think about it logically. Evolution (genetic change) and mutation/selection did NOT exist until AFTER God designed, programmed, engineered, manufactured, created, and deployed ALL the genomes, and physical bodies, and physical brains, and physical eyes in the first place.

According to his definition for creative thinking, thinking creatively DOES NOT apply to evolution, because there is NO evidence supporting the alleged creative powers of evolution. Evolution of any kind cannot design and create. Creative thinking excludes evolution, because evolution is inconsistent with the observational, experiential, empirical, and scientific evidence. The theory of evolution is a hoax or a fraud. It can't do what they say that it does, because evolution (genetic change), random mutations, and natural selection have NO access to the quantum level. These things are trapped at the physical level.

I'm NOT the least bit impressed with natural selection as a mechanism for creative change. Natural selection does absolutely NOTHING to our genes! Natural selection is impotent. All natural selection can do for you is get you killed when you get selected against, so that you can't copulate. That's it! Natural selection is worthless. Natural selection only contributes to death and extinction.

Imagine it!

They built a whole "science" – the theory of evolution – on something that doesn't touch our genes and can't touch our genes – natural selection. These people are so blindly devoted to natural selection and the theory of evolution that they can't even see that it is so. In the beginning and for a hundred years, it was "the origin of species by means of natural selection"; and, natural selection can't even do anything to touch our genes. What a fraudulent hoax! And, they swarmed to it like flies to used toilet paper.

What is evolution?

Evolution (genetic drift) is the result of FAILED meiosis; and, so are ALL of our genetic illnesses. Meiosis or genetic recombination is a physical process, which means that it's subject to entropy or evolution, which explains why it sometimes goes wrong. Evolution is entropy or the second-law of thermodynamics. Evolution is random chaos.

It is genetic recombination during the production of our gametes (sperm or egg) where ALL of the genetic action, genetic mutations, and genetic change takes place that pertains to us! Genetic recombination IS the potent and effective mechanism in all of this; and, it has absolutely nothing to do with Natural Selection. Natural Selection can't touch it.

Genetic recombination is the meiotic process by which parts of a chromosome cross-over one another at random points, break apart, and exchange genes. Your parents each have two chromosomes, and therefore two sets of matching genes or alleles. When your parents were producing your sperm and egg, each parent provided a random set of genes from each of their chromosomes to produce your single egg or sperm. That's how children are able to end up with physical features or phenotypes that aren't visible in either of their parents.

Genetic recombination is where most of our mutations or mishaps take place. It's a physical process, so it's subject to failure. In comparison, though, natural selection is absolutely worthless when it comes to genetic change, or genetic drift, or evolution. Natural selection doesn't touch our genes – not in a million years.

Monoallelic expression is a mechanism of gene expression that inactivates one gene of a pair of alleles and allows the other gene of the pair to be expressed.

Who decides?

You have got two functional genes there in the chromosome. How does Nature's Psyche decide which gene is dominant and which gene is recessive? How can it tell? How does it decide? How does it know? And let's say that you have two recessive genes in that particular allele. How does Nature's Psyche choose which gene it is going to express and which gene it is going to suppress? Or, is Nature's Psyche going to merge the functionality of the repressive genes? If so, how? How does Nature's Psyche know how to do all of these things? Genes are molecules; and, Nature's Psyche is making choices between molecules. That's physically impossible. Choice takes place at the quantum level, and choice is the product of Someone Psyche.

If you observe carefully, especially where human beings are concerned, the fittest and healthiest are typically the first to get killed or selected against, while the handicapped and welfare cases tend to breed like rabbits. Natural selection or survival of the fittest is worthless and unreliable. Natural selection wasn't even able to eliminate the homosexuals; and, there shouldn't be any homosexuals if natural selection and the theory of evolution worked as advertised. Natural selection doesn't touch our genes.

Imagine it!

They built a whole "science" and religion – the theory of evolution or "On the Origin of Species by Means of Natural Selection" – based upon something that doesn't even touch our genes. How pathetic is that? And then, they are religiously dogmatic about its truthfulness. In fact, the Militaristic Atheists (the Communists) and the Fascistic Darwinists (the Nazis) were willing to kill you in order to enforce belief in Darwinism, the theory of evolution, and survival of the fittest. Yet, the truth is that natural selection doesn't even touch our genes.

Furthermore, genetic recombination really has nothing to do with copulation or sexual intercourse either. ALL of the mutations, which ended up in your physical body, took place BEFORE you were even conceived. The origin of species by means of natural selection or sexual selection is nothing but science fiction. It never happened, and it isn't real. Natural selection isn't a mechanism that can get anything done. The theory of evolution is nothing but science fiction. It isn't real, and it didn't happen. Mutations and selection cannot design and create anything, especially new species from scratch.

Theological evolution is also a hoax. God didn't use evolution to create proteins, genomes, and life forms, because that's physically impossible! It's physically impossible for chemical evolution to produce a functional protein, let alone the matching gene to go with it. God has infinitely better mechanisms for creating genes, proteins, and life forms – Quantum Mechanisms or the Priesthood of God. Through Quantum Mechanics, God can do the physically impossible at will.

Remember, evolution can't do anything at the quantum level, which means that evolution is completely worthless when it comes to anything that involves Psyche at the quantum level.

How does John Pinel know that it took 600 million years for evolution to produce the human eye? Was he there and did he watch it happen? Of course not! It's simply wishful-thinking and blind-faith on his part, a fictional ad hoc just-so story that he wants to be true; and, nothing more. Does that mean that our ancestors were blind for 600 million years, before evolution got around to creating our eyes? According to John Pinel, that must be the case. So, how could our ancestors be the fittest, if they were blind for 600 million years before evolution created their eyes? The theory of evolution is self-defeating. It doesn't hold water, doesn't make sense, and doesn't survive scrutiny.

A gene is a unit of inheritance – a section of a chromosome that is used to synthesize ONE protein. Gene expression or protein synthesis is the production of the protein specified by a particular gene.

Remember, chemical evolution cannot synthesize a single functional protein. It's physically impossible.

It's physically impossible for chemical evolution to produce a single functional protein, let alone a genome mapping out 19,000 to 20,000 functional protein-encoding genes. The Theory of Evolution is nothing but science fiction.

Given an infinite number of years, evolution (genetic change), natural selection, random mutations, chemical evolution, abiogenesis, and spontaneous generation wouldn't be able to create one single functional gene that successfully encoded for a functional protein along with the devices (enzymes) necessary to convert that gene into that desired protein. The process is even more difficult given the fact that evolution, random mutations, and natural selection didn't even exist until AFTER God designed and created all the genomes, proteins, life forms, physical bodies, physical brains, and physical eyes in the first place. And, there would be NO incentive for evolution, natural selection, and random mutations to design and create a physical eye, because such a thing would have no survival value whatsoever until after God designed and created it in the first place.

There was no person named "Evolution" who designed and created us. There was no person named "Random Mutations" or "Natural Selection" or "Abiogenesis" who designed and created our genomes, proteins, and our physical bodies. The claim that evolution created our genome, our brain, and our eyes is nothing but a fictional story made up out of thin air to satisfy people's need to eliminate God from the equation. In contrast, WE KNOW from the Lived Experiences of the human race that the Biblical God Jesus Christ had a hand in designing and creating our genome, our physical bodies, our physical brain, and our eyes because He has told thousands of different people that He did so. Jesus Christ exists. He can tell us that He designed us and created us. Evolution does not exist as a person. It can't tell us that it designed us and created us, because Evolution, Random Mutations, and Natural Selection didn't even exist until AFTER God designed and created the genomes, physical bodies, physical brains, and eyes in the beginning.

God's presence, influence, and necessity are most obvious when it comes to beginnings and endings.

Other species are similar to us through Design, not through genetic change or evolution. Genetic drift doesn't do design and creation – it only promotes disease, cancer, death, and extinction. That is what we have observed. Natural Selection doesn't do origin of species. Natural Selection does extinction of species. Random chance or random mutation cannot design and create anything.

When it comes to neuroscience, the greatest fiction and myth that we have received from the Materialists, Darwinists, and Naturalists is that evolution (random mutations and natural selection) designed and created our genome, our proteins, and our brain. That's physically impossible, which means that it didn't happen.

Occasionally these people will slip up and introduce design into their science. It's always good for a laugh whenever you catch a Darwinist or Materialist using the word "design" to describe a physical process or a physical behavior that was supposedly created by evolution.

Target-site concept. The idea that aggressive and defensive behaviors of an animal are often designed to attack specific sites on the body of another animal while protecting specific sites on its own. (*Biopsychology*, p. 496.)

Designed by whom? Design, intention, teleology, purpose, foresight, and choice are a product of Psyche – NOT evolution. Evolution of any kind by definition in principle can't do design, teleology, intention, creation, foresight, choice, and psyche. These things are physically impossible. According to the Materialists and Naturalists and Behaviorists, choice is physically impossible. It's always humorous whenever you catch an Evolutionist or Darwinist using the word "design" to describe evolutionary processes. Design is a code-word for Psyche or Intelligence.

For the Evolutionists, Darwinists, Materialists, Naturalists, Determinists, Behaviorists, and Atheists, design and creation and choice are the enemy. These people repeatedly emphasize that evolution can't do intelligent design or choice. This is their message to the world – that they have no choice but to believe what they have chosen to believe – that they have no choice but to do what they have chosen to do. Philosophically, Materialism and Naturalism are self-defeating, because they are full of internal contradictions.

Obviously, I have a bit of a gripe against Behaviorism. The Behaviorists study rat psychology, and then they force human psychology to match with their discoveries in rat psychology. Rats don't scale up to human beings. Yes, there are human beings who choose to act like rats; but, the rats are incapable of acting like human beings. Psyche is choice. The Radical Behaviorists define voluntary responses and free will out of existence. The rats have no choice but to go into the Skinner box; but, a human being will CHOOSE to fight you and resist you if you try to do the same thing to him.

Pinel also talks about "assessing animal behavior in environments similar to those in which it evolved." He's *begging the question* and *assuming the consequent*. He's also *assuming facts not in evidence*. Animals have NEVER been observed evolving in any environment. Viruses and bacteria were designed by God to mutate in order to adjust to their environment and survive; but, viruses and bacteria are not animals. The animals were designed by God so that they can only reproduce by copulating with their own species. God actually designed animals to be mutation-resistant or evolution-resistant. By design, it's physically impossible for two chimp-like ancestors to produce genetically compatible male and female human offspring. Macro-evolution is physically impossible. It can't happen, which means that it didn't happen.

Of course, these people have chosen to believe that evolution is the BEST way to interpret the evidence. It is not. The evolutionary perspective and evolutionary psychology are worthless, because the whole thing is a fraudulent hoax. Pinel wants you to see the value of the evolutionary perspective. There is NO value to an evolutionary perspective, because it's physically impossible for human beings to descend from apes and monkeys. God made sure that it is so. Monkeys seem to be the primary focus of Pinel's evolutionary perspective, rather than apes. He even gets the theory of evolution wrong. He never lets the facts interfere with his fiction. Why should he, because the fiction is perfect to begin with!

Imagine for a second – just a second is all I'm asking – what you would have ended up with, if evolution had actually designed and created your whole body by random chance alone. Can you picture it? Let it work on you for a second. You just as well toss your eyes and your brain into a blender and turn it on, if you want to know what random mutations can really do for you. Just think about it for a second, what we would really be, if the theory of evolution were 100% true. We wouldn't be. Thank God the theory of evolution is false.

Remember, evolution can't do anything at the quantum level, which means that evolution is completely worthless when it comes to anything that involves Psyche at the quantum level. Anything that can be attributed to evolution can also be attributed to an Intelligent Designer; but, the Intelligent Designer is more plausible, parsimonious, and believable. It's always humorous whenever you catch an Evolutionist or Darwinist using the word "design" to describe evolutionary processes. Design is a code-word for Psyche or Intelligence.

The very existence of proven and verified Quantum Mechanisms such as Action at a Distance (Telekinesis and Quantum Non-Locality), the Quantum Zeno Effect (Telepathy, Quantum Waves, and Mind-Over-Matter), Quantum Tunneling (Teleportation), and Quantum Non-Local Consciousness (Psyche, Mind, Thoughts, Memories, and the Placebo Effect) FALSIFIES Materialism, Naturalism, Darwinism, Nihilism, Behaviorism, and even Atheism. Remember, the very existence of Quantum Mechanics falsifies Materialism, Naturalism, and their derivatives.

All you want is the truth, because everything else is nothing but science fiction.

Telepathy and Telekinesis: Ask yourself why God could prevent fallen mortal beings such as us from reaching out with our Human Psyche, seeing the individual atoms with our mind, and then rearranging those atoms with our mind at will. If you were a God, is that the kind of power and ability that you would want to give to a Militant Atheist, a Muslim terrorist, the Christian Inquisition, Hitler, James Bond, or the devil living next door to you?

Action at a Distance: Ask yourself why God would prevent fallen mortal beings like us from reaching out with our Human Psyche, steering buses into gas stations, triggering the detonation of nuclear missiles at will, choking people to death with our mind, stopping their hearts at will, calling up earthquakes, steering hurricanes, starting fires, and transforming viruses into lethal forms. If you were a God, is that the kind of power and ability that you would want to give to a Militant Atheist, a Muslim terrorist, the Christian Inquisition, Hitler, James Bond, or the devil living next door to you?

Quantum Tunneling or Teleportation: Ask yourself why God would prevent fallen mortal beings like us from reaching out with our Human Psyche, teleporting ourselves and a bomb into parliament, and then teleporting away to safety before the bomb goes off? If you were a God, is that the kind of power and ability that you would want to give to a Militant Atheist, a Muslim terrorist, the Christian Inquisition, Hitler, James Bond, or the devil living next door to you?

ORIGIN: Design, Chance, and the First Life on Earth

This video, and others like it, marks the end of the Theory of Evolution. It's finished.

https://www.amazon.com/ORIGIN-Design-Chance-First-Earth/dp/B01LZ7EZIS

The *ORIGIN* video from Illustra Media does some mathematical modeling demonstrating that the odds of a functional protein developing by chance are 1 chance in 10^{164}; and, that's with God taking the time to fill an ocean completely full of the right 20 amino acids with the right chirality, and with God forcing 150 of those amino acid sequences to line up every second so that they can see if they will fold up into the right usable protein.

Technically, there's no logical reason for that ocean full of the right 20 amino acids to deliberately line up 150 amino acids in sequence every second in order to try out all of the 10^{164} possible combinations in a timely fashion – naturally, or through undirected processes. God is going to have to provide the ocean full of amino acids in the first place, sustain that ocean full of amino acids in existence for trillions of trillions of trillions of trillions of years, and force 150 of those amino acids to line up in sequence every second throughout those trillions of trillions of years JUST TO PRODUCE a single protein "naturally". The odds of that whole process happening by random chance, without God's intervention, is ZERO. There is NO CHANCE!

https://www.youtube.com/watch?v=W1_KEVaCyaA

https://www.youtube.com/watch?v=cQoQgTgj3pU

https://www.youtube.com/watch?v=X7VqomfivC0

The odds of a functional cell developing by natural processes are 1 chance in $10^{340,000,000}$.

https://www.youtube.com/watch?v=nuMvRExazAw

The current estimated age of this universe is 13.8^9, or 13.8 billion years. In comparison to 13.8 billion years, something like $10^{340,000,000}$ is synonymous with IMPOSSIBLE or INFINITY, because there is NO CHANCE that any living cell could ever develop by chance through natural processes in 13.8 billion years of time. In fact, there is technically NO CHANCE that a single functional protein could develop by chance in 13.8 billion years, even if we were to start with a God-given ocean full of the right 20 amino acids with the right chirality, no destructive oxygen, and no sunshine to destroy that particle of meat once it comes together.

Now let's say that, thanks to God's intervention and help, chemical evolution has managed to create a single functional protein "naturally" after trillions of trillions of years of time, NOW chemical evolution has to create the matching gene to go along with that protein; but, the mechanisms for doing so do NOT exist on Amino Acid World. That protein will eventually disintegrate, and we'll have to start all over again with NO gene in place to successfully replicate that protein. Chemical evolution of proteins along with their matching genes is physically impossible. It can't be done, which means it wasn't done. God has an infinitely better way of making functional proteins along with their matching genes – it's called Quantum Mechanics or the Priesthood Power of God.

There's NO WAY in the multi-verse that evolution of any kind can produce a functional protein, let alone the matching gene to go with it. It's physically impossible. It can't be done.

Species, eyes, brains, genes, proteins, and genomes didn't evolve over millions of years as claimed. That's physically impossible. Instead, these things were placed on our earth millions of years ago, and they have remained essentially unchanged ever since. Evolution (genetic drift) and random mutations can only do cancer, disease, genetic entropy, devolution, destruction, chaos, and extinction. It has been observed and therefore proven scientifically that the various processes of evolution cannot design, create, and manufacture anything whatsoever – not even a simple protein. The chemical evolution of a functional protein is physically impossible, let alone the matching gene to go along with it.

Which came first, genetic change (evolution) or the genome?

This is a classical chicken-egg scenario that has ONLY ONE correct answer or solution, that makes any logical sense. The physical body, physical brain, physical eye, and physical genome had to come first; otherwise, there would still be NO evolution – natural selection and random mutations – anywhere on this earth. There's no way on earth that evolution (mutation and selection) could have designed and produced the first genome and the first life form, because evolution, natural selection, and random mutations didn't exist until AFTER the first genome and the first life form were designed and created.

Which came first, evolution or the genome? This is always the first question you need to answer and get right.

The answer is the genome, because evolution, natural selection, and random mutations did not exist until AFTER the first genome and first living cell to house that genome were designed and created and produced in the first place.

Designed and created by whom?

The only one we know of who was there and who could have done the job IS the one who has publicly admitted and confessed to doing the job to thousands of different people – the Biblical God Jesus Christ.

So, anytime you hear them say that evolution created your body, brain, eyes, or genome, you simply KNOW that they are wrong, because evolution and genetic change didn't exist until AFTER God designed and created your body and brain and genome in the first place.

This particular claim from the evolutionists is defeated and debunked by logical common sense. If these people are demonstrably wrong about this one, then what else do they get wrong as a result?

I uncovered the flaws in Kalat's theories and ideas in other essays and books, so in this essay I will focus primarily on the things that John Pinel gets wrong in his college textbook, *Biopsychology*, and then try to explain why they are wrong.

Pinel has a whole chapter's worth of material in his book presenting the creative genius of evolution; and, it is simply fundamentally wrong from beginning to end, and he can't see that it's wrong because he isn't looking nor thinking creatively about the subject. He's simply taking it all on blind faith as being true, relying on his personal *scientific inferences* instead to establish the truthfulness of his assumptions. *Scientific inference* is the logic fallacy that he uses to manufacture bogus empirical evidence out of thin air. He cheats. Empirical evidence is by definition and in principle observational evidence and experiential evidence. Since nobody has ever caught evolution of any kind in the act of design and creation, and never will, John Pinel makes a scientific inference and creates an ad hoc just-so fictional story; and then, he unilaterally declares his science fiction stories to be empirical evidence. He claims in his own book that his scientific inferences – guestimates and fictional stories – are empirical evidence. He destroys the very foundations of science in order to lay a foundation for his worshipful blind-faith in evolution as his designer, creator, and god.

John Pinel even admits up-front that everything, he produces or develops in his opening chapters about the theory of evolution, is based upon Scientific Inferences. He doesn't even realize that *scientific inference* IS a logic fallacy.

Scientific inference is the fundamental method of biopsychology and of most other sciences – it is what makes being a scientist fun. This section provides further insight into the nature of biopsychology by defining, illustrating, and discussing scientific inference.

The scientific method is a system for finding things out by careful observation, but many of the processes studied by scientists cannot be observed. For example, scientists use empirical (observational) methods to study ice ages, gravity, evaporation, electricity, and nuclear fission – none of which can be directly observed; their effects can be observed, but the processes themselves cannot. Biopsychology is no different from the other sciences in this respect. One of its main goals is to characterize, through empirical methods, the unobservable processes by which the nervous system controls behavior.

The empirical method that biopsychologists and other scientists use to study the unobservable is called <u>scientific inference</u>. Scientists carefully measure key events they can observe and then use these measures as a basis for logically inferring the nature of events that they cannot observe.

Like a detective carefully gathering clues from which to recreate an unwitnessed crime, a biopsychologist carefully gathers relevant measures of behavior and neural activity from which to infer the nature of the neural processes that regulate behavior.

The fact that the neural mechanisms of behavior cannot be directly observed and must be studied through scientific inference is what makes biopsychological research such a challenge – and as I said before, so much fun. (*Biopsychology*, p. 13.)

Scientific inference is a logic fallacy.

There's the theory of evolution in a nut-shell – it all comes down to *scientific inference*, which is a logic fallacy. Scientific inference is the *ad hoc just-so story-telling* logic fallacy in action, within "science" no less. With scientific inference, you can have any result you want and prove anything to be true, simply by making up a story out of thin air and presenting it as fact. This is how they support their evolution inference – they make an educated guess – they make up a story; and then, they present their fictional story as God-given truth.

I found it absolutely fascinating when John Pinel admitted that the Materialists, Naturalists, Darwinists, Behaviorists, Nihilists, and Atheists treat their scientific inferences as empirical evidence. How fascinating is that? It explains a lot, doesn't it?

Scientific inferences are logic fallacies! Yet these people treat their logic fallacies as empirical evidence, which is also a *category error* logic fallacy – treating scientific inferences as scientific evidence. Amazing, is it not? This discovery and revelation from John Pinel was worth the price of admission, and it made his whole book worth it.

The theory of evolution is based upon a wide variety of different logic fallacies and faulty assumptions. Before doing any actual science, the Materialists, Naturalists, Darwinists, Nihilists, Behaviorists, and Atheists assume axiomatically that the quantum level or the psyche level does not exist; and, they assume Psyche, Intelligence, Design, and Choice out of existence. You know what they say about assumptions. It makes for stinky philosophy and stinky science. The whole thing stinks.

Throughout his book, John Pinel encourages us to "think critically," which is extremely easy to do whenever one is dealing with scientific inferences, because scientific inferences are extremely easy to debunk with observational evidence and experiential evidence. It doesn't require any particular genius to be able to demolish scientific inferences, because they are logic fallacies to begin with. I can also attack and destroy his personal interpretations of the research evidence or scientific research; but, I can't attack the observational evidence, or the operational evidence derived from the research, unless they used shoddy methodologies.

Scientific inference is synonymous with *affirming the consequent*, *begging the question*, and *jumping to conclusions*. So is *ad hoc just-so story-telling*, another logic fallacy. These people make an educated guess and then jump straight to the conclusion that the theory of evolution is true.

These people haven't studied the Philosophy of Science, so they don't know that they are *affirming the consequent* and *begging the question* in everything that they write, do, and say. Instead, these people treat their scientific inferences and present their scientific inferences as scientific evidence and as proof that the theory of evolution is true. Unless you learn how to see through the ruse, you will end up falling for their creative stories and end up believing that their science fiction is true.

Evolution, Materialism, Naturalism, and Darwinism have impeded scientific progress by suppressing and hiding observational evidence, experiential evidence, and empirical evidence in order to make its case – replacing the hard evidence with scientific inferences instead.

Remember, creation and design by intelligent beings IS always the simplest and best explanation, when it comes to creation and design, especially since mutation and selection cannot do design and creation. Follow the evidence, the observational evidence and experiential evidence, rather than the scientific inferences.

After explaining how much fun scientific inference is, John Pinel then uses scientific inference in the next chapter of his book to make his case for the theory of evolution. He never once realizes that *scientific inference* IS a logic fallacy, because he is having too much fun using it to prove that the theory of evolution is true.

Now, notice the *bait and switch* that John Pinel employs here, as we begin the next chapter of his book, *Biopsychology*.

> **We all tend to think about things in ways that have been ingrained in us by our Zeitgeist – the general intellectual climate of our culture. That is why this is a particularly important chapter for you. You see, you are the intellectual product of a Zeitgeist that promotes ways of thinking about the biological bases of behavior that are inconsistent with the facts [things such as Darwinism, Macro-Evolution, Spontaneous Generation, Abiogenesis, Materialism, and Naturalism]. The primary purpose of this chapter is to help you bring your thinking about the biology of behavior in line with modern biopsychological science.** (*Biopsychology*, p. 21).

Okay, when he uses the term Zeitgeist, he's actually talking about Christianity, Cartesian Dualism, the belief in Non-Local Consciousness or Psyche, belief in the Bible, and belief in God. Whenever these people start talking about "wrong thinking" and the need for "correct thinking" and the need to "bring your thinking in line" with Materialism, Naturalism, Darwinism, Scientism, Nihilism, and Atheism, they are in fact attacking anything to do with Psyche, Non-Local Consciousness, the Afterlife, the Spirit Realm, and God.

Of course, this is the first paragraph in this particular chapter of this book, so you have no idea at first what he considers to be "wrong thinking" and "correct thinking"; and when he starting talking about the Zeitgeist, I just naturally assumed at first that he was in fact talking about things such as Darwinism, Macro-Evolution, Spontaneous Generation, Abiogenesis, Materialism, and Naturalism which have ALL been FALSIFIED by science and the scientific methods.

My assumption proved to be wrong.

When talking about **the Zeitgeist** that promotes "wrong thinking", he was in fact talking about Christianity, God, design and creation by God, and belief in a psyche and an afterlife, as I was soon to discover.

John Pinel says right up front in his opening paragraph that his purpose with this most important chapter in the book is to correct your "wrong thinking" and replace it with "correct thinking". He intends to re-condition you, re-educate you, and brainwash you so that you end up thinking correctly – end up thinking the way that he thinks.

In my case, he failed. I don't like being manipulated, tricked, lied to, and deceived. I don't like being brainwashed with "correct think".

Taking something that has NEVER been caught in the act of design, production, and creation; and then, assigning that thing to be our Designer and Creator, VIOLATES parsimony, logic, common sense, and even the spirit of Morgan's Canon and Occam's Razor; and, it results in an egregious abuse of scientific inference and scientific interpretation. Every college textbook writer does this automatically without thinking, especially where Materialism, Naturalism, and the Theory of Evolution are concerned.

Morgan's Canon is defined as the rule that the simplest possible interpretation for a behavioral observation should be given precedence. Ironically, the chemical evolution of proteins and genes from scratch is physically impossible; therefore, according to Morgan's Canon, evolution shouldn't even be considered as an alternative because it is impossible. You want the simplest possible interpretation, and evolution isn't possible.

John Pinel employs Morgan's Canon while developing his case for the theory of evolution.

> **Morgan's Canon: In no case is an animal activity to be interpreted in terms of higher psychological processes if it can be fairly interpreted in terms of processes which stand lower in the scale of psychological evolution and development.**

> https://en.wikipedia.org/wiki/Morgan%27s_Canon

Morgan's Canon is an attempt at parsimony. In actuality, Morgan's Canon canonizes the theory of evolution making evolutionary theory by definition the simplest explanation for any scientific evidence that you might happen to come across. In other words, they assure the outcome before they do any actual science, in classical *begging the question* fashion.

When there are several possible interpretations for a behavioral observation, the rule is to give precedence to the evolutionary interpretation, because it is assumed to be the simplest one. This rule is called Morgan's Canon.

Then John Pinel states, **"You should be particularly skeptical of scientific claims that have not gone through this review process."** (p. 15.) In other words, you should be skeptical of anything that hasn't been blessed and sanctified by the high priests of Materialism, Naturalism, Darwinism, Nihilism, and Atheism. They *stack the deck* so that they can't lose. In other words, they cheat in order to keep their evolutionary theories clean and pure. They literally define "critical thinking" and "correct thinking" as Materialism, Naturalism, Darwinism, Nihilism, and Atheism.

Ironically, intelligent design and creation always end up being the best, most logical, and most parsimonious interpretation, because they have actually been experienced and observed in real life for real, unlike the putative "creative powers of evolution" which have never been observed nor caught in the act and don't in fact exist.

Remember, creation and design by intelligent beings is always the simplest and best explanation, because it has been observed and experienced and actually exists.

In his opening chapter, John Pinel also talked about "converging operations" – using the strengths of one approach to compensate for the weaknesses in other approaches. Ironically, I used converging operations (the combining of multiple different approaches or methods) to determine and prove that Materialism, Naturalism, Darwinism, Nihilism, and Atheism are FALSE. I falsified these things or proved them false. The only good thing about these philosophical concepts is that they are falsifiable through scientific observations, because they have in fact been falsified through converging operations or multiple different methods and approaches.

Remember, when it comes to these people, their ultimate purpose is to condition you into "correct thinking" by brainwashing you and tricking you into believing in the Theory of Evolution, which they define as "correct thinking". As promised, John Pinel did what he promised to do. He attempted to brainwash me into believing in the Theory of Evolution; but I know better now, so I didn't fall for it and chose to resist it instead.

As the following points attest, resistance isn't futile. I have gotten very good at resisting their brainwashing techniques.

John Pinel isn't alone. He's just unlucky enough to be the one whose book I was forced to buy and read. ALL of the college textbooks are that way. They ALL try to brainwash you and intimidate you into believing in the Theory of Evolution. They ALL imply that you are stupid, unscientific, and billions of years behind the times if you refuse to accept the Theory of Evolution and refuse to believe in the Theory of Evolution. It's BAD SCIENCE! It's based upon a refusal to look at evidence, and the preponderance of the evidence tells us that the Theory of Evolution is FALSE.

Evolution Can Do Protein Synthesis at Will

The idea that evolution, random mutations, and natural selection can do protein synthesis at will is another lie that's commonly promoted throughout the science community by the Materialists, Naturalists, and Darwinists.

Given enough time and resources and coordinated effort, I believe we human beings could eventually design and created a new species. In contrast, given an infinite amount of time, chemical evolution would still be unable to design and create a single functional protein, nor would chemical evolution know what to do with that single protein if it were lucky enough to actually create one. Furthermore, for any protein that chemical evolution gets lucky enough to produce, evolution would have to immediately design, encode, and produce a matching gene for that protein, or that newly created protein would be an evolutionary dead-end.

The *ORIGIN* video from Illustra Media does some mathematical modeling demonstrating that the odds of a functional protein developing by chance are 1 chance in 10^{164}; and, that's with God taking the time to fill an ocean completely full of the right 20 amino acids with the right chirality, and with God forcing 150 of those amino acid sequences to line up every second so that they can see if they will fold up into the right usable protein.

Technically, there's no logical reason for that ocean full of the right 20 amino acids to deliberately line up 150 amino acids in sequence every second in order to try out all of the 10^{164} possible combinations in a timely fashion – naturally, or through undirected processes. God is going to have to provide the ocean full of amino acids in the first place, sustain that ocean full of amino acids in existence for trillions of trillions of trillions of trillions of years, and force 150 of those amino acids to line up in sequence every second throughout those trillions of trillions of years JUST TO PRODUCE a single protein "naturally". The odds of that whole process happening by random chance, without God's intervention, is ZERO. There is NO CHANCE!

https://www.youtube.com/watch?v=W1_KEVaCyaA

https://www.youtube.com/watch?v=cQoQgTqj3pU

https://www.youtube.com/watch?v=X7VqomfivC0

The odds of a functional cell developing by natural processes are 1 chance in $10^{340,000,000}$.

https://www.youtube.com/watch?v=nuMvRExazAw

The current estimated age of this universe is 13.8^9, or 13.8 billion years. In comparison to 13.8 billion years, something like $10^{340,000,000}$ is synonymous with IMPOSSIBLE or INFINITY, because there is NO CHANCE that any living cell could ever develop by chance through natural processes in 13.8 billion years of time. In fact, there is technically NO CHANCE that a single functional protein could develop by chance in 13.8 billion years, even if we were to start with a God-given ocean full of the right 20 amino acids with the right chirality, no destructive oxygen, and no sunshine to destroy that particle of meat once it comes together.

Now let's say that, thanks to God's intervention and help, chemical evolution has managed to create a single functional protein "naturally" after trillions of trillions of years of time, NOW chemical evolution has to create the matching gene to go along with that protein; but, the mechanisms for doing so do NOT exist on Amino Acid World. That protein will eventually disintegrate, and we'll have to start all over again with NO gene in place to successfully replicate that protein. Chemical evolution of proteins along with their matching genes is physically impossible. It can't be done, which means it wasn't done. God has an infinitely better way of making functional proteins along with their matching genes – it's called Quantum Mechanics or the Priesthood Power of God.

There's NO WAY in the multi-verse that evolution of any kind can produce a functional protein, let alone the matching gene to go with it. It's physically impossible. It can't be done.

Now imagine the design and creation of something like an eye.

Fifteen different genes encode for eye color alone. That implies fifteen different proteins, just for eye color. Evolution can't create one protein, so what's evolution going to do when tasked with creating fifteen different proteins along with fifteen matching genes so that those newly created proteins don't end up being an evolutionary dead-end?

There are an estimated 19,000-20,000 human protein-coding genes. (Uncle Google.)

Imagine it. That's 20,000 different proteins along with 20,000 different genes that match with those proteins, just to design and construct a part of the human genome. Chemical evolution can't create a single functional protein, so what is it going to do with 20,000 proteins along with the 20,000 matching genes, so that those newly created proteins don't end up being an evolutionary dead-end?

Nobody ever thinks about this rationally and logically. Nobody does the math; but, the math makes it obvious and clear that evolution of any kind is physically impossible. It's so obvious that the Theory of Evolution is false, that I sometimes wonder why I wasn't able to see it for the first fifty years of my life. But, I didn't see it because I didn't want to see it; and I didn't see it because I wasn't looking for it; and I didn't see it because I didn't know where to look. I had been tricked and deceived just like everyone else. I had chosen to believe that the case for evolution was infinitely stronger than it really is.

People believe in the creative powers of evolution because they want to; but, their blind faith doesn't make it real and true. Materialism, Naturalism, and Atheism are religions. They have to be taken on blind faith as being true, because there is NO empirical evidence or observational evidence to support them.

Belief in the creative powers of evolution requires a great deal of ignorance and blind faith. There is NO CHANCE whatsoever that evolution, random mutations, and natural selection can design and create as advertised. In fact, genetic drift, random mutations, natural selection, and evolution didn't even exist until AFTER God designed, programmed, created, and manufactured the genomes and the matching proteins in the first place. Evolution is an evolutionary dead-end.

The theory of evolution is a fictional ad-hoc just-so story, and nothing more. It's the pinnacle of wishful thinking, blind faith, and story-telling; and, evolution is one of the more popular religions and false gods on this planet, because it's telling people what they want to hear – that God does not exist, and as a result, that there will be no consequences for breaking His commandments. Of all the man-made gods on this planet, the one you don't want to believe in is evolution, because it's physically impossible and can't produce the goods; but, it seems to be the one that people believe in the most.

Remember, evolution of any kind cannot design and produce a single functional protein, nor the matching gene to go along with it. That means that there's no way in the multi-verse that evolution can be your Designer, Creator, and God. You have some other origin, besides evolution. Get used to it. I had to get used to it, because I used to be a Materialist, Naturalist, Nihilist, and Atheist until all the observational evidence and experiential evidence and scientific evidence convinced me that I was wrong.

Remember, even if given an infinite amount of time, chemical evolution cannot design and create a single functional protein and the gene to go along with it. Materialism, Naturalism, Darwinism, Abiogenesis, Spontaneous Generation, Chemical Evolution, Design and Creation by Mutation/Selection, and Macro-Evolution have been successfully falsified due to a complete lack of observational evidence and a complete lack of replicability. That's the way science should work, but typically isn't allowed to work. Nevertheless, the scientific methods are extremely efficient at falsifying false theories and false ideas, and they should be allowed to do so. Materialism, Naturalism, and Darwinism have been falsified by dozens of different scientific methodologies, and we should take that reality and run with it and adjust our chosen conclusions accordingly, if we are truly scientists. It took me fifty years to figure this out, but I finally got there in the end.

Collectively as a whole, Materialism, Naturalism, Darwinism, Determinism, Radical Behaviorism, Scientism, Classical Physics, Creation by Chemical Evolution, Spontaneous Generation, Abiogenesis, Design and Creation by Random Mutations, Creation by Natural Selection, Macro-Evolution, and the Theory of Evolution have been FALSIFIED quadrillions of times in thousands of different ways.

It's time for a new brand of science or a new type of science – one that allows all of the evidence into evidence, and one that actually explains the observational evidence and experiential evidence which we have on hand as a race. It's time to upgrade our science to Science 2.0, which allows all of the evidence into evidence and pursues a preponderance of the evidence, instead of suppressing the evidence and destroying the evidence as Materialism and Naturalism do.

2. Your mind or consciousness is simply a product of your physical brain.

Physiological Psychology is the division of biopsychology that studies the neural mechanisms of behavior through the direct manipulation of the brains of nonhuman animal subjects in controlled experiments.

Physiological Psychology is an oxymoron – a contradiction in terms. The original definition for Psychology was the Study of the Human Psyche. Thoughts, memories, and choices are quantum waves, which is why they survive the death of our physical brain. The study of neural mechanisms should be the study of how the brain is mapped by Nature's Psyche at the quantum level to perform specific physical functions of its choosing. The study of neural mechanisms should be the study of the Maps of Physical Functionality that

Nature's Psyche is mapping onto our brains at the quantum level and then using to get things done for us at the physical level.

My neuroscience teacher wrote:

Behavioral neuroscience is the study of the relationship between brain and behavior. It is also sometimes called physiological psychology or biopsychology. Everything we feel, perceive, think, or do originates in the brain. In a very real sense, you are your brain. Behavioral neuroscience comprises several different disciplines, all of which are concerned with answering one fundamental question: How does the brain produce thoughts, feelings, emotions, and behavior? In other words, how do mental events arise from the physical interactions within the brain?

They don't! This whole thing is false. The message from these people is, "You are your brain, and your brain is a product of evolution." They are demonstrably wrong.

How does the brain produce thoughts, feelings, emotions, and behavior?

It doesn't.

How do we KNOW?

We KNOW from the thousands of Near-Death Experiences (NDEs) and Out-of-Body Experiences (OBEs) on record that we continue to think, feel, emote, choose, and behave after our physical brain is dead and gone. The preponderance of the evidence tells us that the Materialists, Naturalists, Atheists, and Darwinists are wrong.

How do mental events arise from the physical interactions within the brain?

They don't!

Physical interactions within the brain arise from mental events or are triggered by mental force, as clearly demonstrated and proven in the excellent book, *The Mind and the Brain: Neuroplasticity and the Power of Mental Force*, by Jeffrey Schwartz. That book is how neuroscience should have been done but wasn't.

It can be frustrating to have neuroscience teachers who don't know what they are talking about, because they are only looking at things from a materialistic perspective. They are all in this together, and their goal is to trick your and deceive you with their Materialism, Naturalism, Darwinism, Nihilism, Behaviorism, and Atheism.

Your mind, psyche, or consciousness is an epiphenom or product of your physical brain. When you die, your mind or consciousness ceases to exist.

This claim is demonstrably false, and it can be falsified by observational evidence and experiential evidence – the lived experiences of the human race.

There are two types of neuroscientists in this world – those who look at the evidence, and those who dismiss the evidence. John Pinel is the second type – in his book, *Biopsychology*, he proceeds to dismiss the evidence and tells you outright that the evidence does not exist.

He is wrong; but, he can't see that, because his bias or choice is self-confirming and selects for itself. Self-deception works, and it works every time, even on PhD scientists and geniuses.

Throughout his book, *Biopsychology*, John Pinel tells his students to employ critical thinking commanding them to "think creatively"; and, then he violates his command by the very things that he writes.

Cartesian dualism, as Descartes's philosophy became known, was sanctioned by the Roman Church, and so the idea that the human brain and the mind are separate entities became even more widely accepted. It has survived to this day, despite the intervening centuries of scientific progress. Most people now understand that human behavior has a physiological basis, but many still cling to the dualistic assumption that there is a category of human activity that somehow transcends the human brain.

The physiological-or-psychological debate (mind-brain problem) and the nature-or-nurture debate are based upon incorrect ways of thinking about the biology of behavior. (*Biopsychology*, p. 22.)

Can you sense all the different attacks he makes on the religious people and the spiritual people who choose to believe that Cartesian Dualism is true, or have experienced Cartesian Dualism firsthand for themselves? This implicit attack is based upon the *ad hominem* logic fallacy – the implication being that only the stupid people and the religious people still believe that mind is separate from the physical brain.

Without examining any evidence whatsoever, John Pinel *jumps to the conclusion* that since it was sanctioned by the Roman Church, Cartesian Dualism is automatically wrong and automatically unscientific. He implies that scientific progress should have eliminated Cartesian Dualism by now; wherein, he erroneously and egregiously defines "science" as Materialism, Naturalism, Darwinism, Nihilism, and Atheism.

John Pinel then employs the *"most people"* logic fallacy to make his case, which is the idea that might makes right or the idea that the majority is somehow magically always right in their assumptions and their conclusions.

John Pinel declares belief in mind or psyche to be the result of wrong-thinking; and, then he proceeds to tell you how you are supposed to think.

Again, these people *stack the deck* so that they can't lose. They cheat. Before doing any actual science and without examining any evidence whatsoever, they **a priori** define "science" as Materialism, Naturalism, Darwinism, Nihilism, and Atheism; and then, these people censor, ban, suppress, hide, and ridicule any subsequent evidence that falsifies Materialism, Naturalism, Darwinism, Nihilism, and Atheism relying on scientific inferences instead in order to make their case.

Despite his claims to the contrary, when it comes to Neuroscience and Biopsychology, John Pinel is at least a century behind contemporary physical theory and the current prevailing principles of physics, as demonstrated by Henry P. Stapp, Mario Beauregard, and Jeffery M. Schwartz.

"Quantum Physics in Neuroscience and Psychology: A Neurophysical Model of Mind-Brain Interaction"

https://www.researchgate.net/publication/7613549_Quantum_physics_in_neuroscie nce_and_psychology_A_neurophysical_model_of_mind-brain_interaction

http://www-physics.lbl.gov/~stapp/PTRS.pdf

http://mypsyche.us/wp-content/uploads/2017/10/PTRS.pdf

https://sites.google.com/a/lbl.gov/stappfiles/

http://www-physics.lbl.gov/~stapp/

I am no longer materialistic, nihilistic, or atheistic. I no longer have blind faith enough to believe in these things.

I have upgraded my science to Science 2.0. Under Science 2.0, observational evidence, experiential evidence, and eye-witness evidence is given precedence over everything else including scientific inferences. Science 2.0 defines science as observation, lived experiences, eye-witness accounts, knowledge, truth, and fact. Under that definition for science, Materialism and Naturalism are actually unscientific, because Materialism and Naturalism and their derivatives have NO observational evidence supporting their primary assumptions or major premises. There is no way to observe that something doesn't exist. Science 2.0 returns to the original definition for Psychology, which was originally defined as "the study of the human psyche." In fact, Science 2.0 is the study of the Lived Experiences of the Human Psyche.

Pinel makes the assumption that *most people* now understand and accept that human behavior and mental illness has ONLY a physiological basis, or physical basis. I guess I'm no longer *most people*. I used to be a Materialist, Nihilist, and Atheist; but, not anymore.

This claim from John Pinel, that mental illnesses are caused ONLY by brain damage, is successfully FALSIFIED by a much better college psychology textbook:

Barlow, D. H. & Durand, V. M. (2015). *Abnormal Psychology: An Integrative Approach* (7th ed.). Stamford, CT: Cengage Learning.

http://mypsyche.us/wp-content/uploads/2017/10/Abnormal-Psychology.pdf

Throughout their whole book, Barlow and Durand successfully demonstrate that most research scientists and the preponderance of the scientific evidence have clearly shown that mental illnesses and abnormal behaviors are the result of multiple different factors – a genetic biological developmental factor, a sociological cultural environmental factor, and a psychological or choice or will factor – an Integrative Model or Integrative Approach which has been called by scientists from all different research disciplines the **BioPsychoSocial Model** of mental illness, psychopathology, and abnormal behavior.

Under the BioPsychoSocial Model, human behavior and mental illnesses are subject to physiological influences, but do not have an exclusive physiological basis or physiological cause. Thanks to psyche or non-local consciousness, we human beings can break our conditioning and override our genetic inheritance at will.

I never fully realized how much I appreciated and enjoyed Barlow and Durand's Integrative Model or BioPsychoSocial Model until I started reading John Pinel's book, which is decades behind the cutting-edge research which is found in *Abnormal Psychology: An Integrative Approach*.

The integrative approach or holistic approach to Abnormal Psychology, Psychopathology, Mental Illness, and Personality Disorders shows us that there is a psychological component (a psyche component) that's completely separate from and has a different function and source than Nature (Bio) and Nurture (Social).

This book, *Abnormal Psychology: An Integrative Approach*, solves the Nature vs. Nurture Debate by adding a third component, the "psycho" portion of the BioPsychoSocial Model which is comprised of the human psyche. It works, and it makes logical sense. The human psyche can be immediately identified any time that some kind of CHOICE, or INTENTION, or WILL is involved in the scenario or case study.

Barlow and Durand also introduced me to the Diathesis-Stress Model. A diathesis is a genetic vulnerability or a genetic susceptibility to a psychiatric disorder, such as the diathesis-stress model of depression. A diathesis is genetic programming. Our genome is indeed programmed into us. Our genome is God's Signature. All throughout their book, Barlow and Durand separated the Bio or genetic portion of a mental illness from the PsychoSocial (psyche/environment) portion of a mental illness.

In contrast, John Pinel dismisses the Mind-Brain Problem and the Nature vs. Nurture Problem by labeling them as "incorrect ways of thinking about the biology of behavior." He erroneously merges everything into biology, and then erroneously states that evolution created all of our biology and therefore all of our psychology. Barlow and Durand are decades, if not centuries, ahead of Pinel.

Continuing with John Pinel:

> **The physiological-or-psychological debate and the nature-or-nurture debate are based upon incorrect ways of thinking about the biology of behavior. What is wrong with the old ways of thinking about the biology of behavior (Cartesian Dualism), and what are the new ways?** (*Biopsychology*, p. 22.)

Can you see anything wrong with this statement or idea? I can. First, he makes the claim that the mind-brain problem or dualism problem is the result of wrong-thinking. He even asks you to think about what's wrong with the old Cartesian Dualism way of thinking. By portraying it and labeling it as old, he is trying to imply that it is outdated and has been proven false. In other words, he is telling us that as scientists we shouldn't be thinking about these kinds of things. Pinel is actually trying to suppress scientific inquiry into this subject. Then, Pinel *jumps straight to the conclusion* that Cartesian Dualism is the result of wrong-thinking. He's implicitly trying to suppress scientific exploration, and scientific thinking, and scientific discovery by defining science from the beginning as Materialism, Naturalism, Darwinism, Nihilism, and Atheism.

Secondly, he implicitly and surreptitiously confounds behavior with biology or the physical brain – a logic fallacy that's called a *category error*. Throughout this chapter of his book, John Pinel is repeatedly *jumping to conclusions*, *affirming the consequent*, and *begging the question*, which are also logic fallacies. He assumes right up front that behavior is exclusively a function of biology; and then, he labels the **physiological-or-psychological debate** or the Mind-Brain Problem as "incorrect thinking" or "wrong think".

Thirdly, John Pinel actually back-pedals on the subsequent pages, because the observational evidence or research evidence makes him do so.

> **Next, it was argued convincingly that behavior always develops under the combined control of both nature and nurture, not under the control of one or the other.** (p. 24.)

Then he proceeds to dismiss it and debunk it, returning to his original claim that the Mind-Brain Problem and the Nature vs. Nurture debate are the result of "incorrect ways of thinking"; and, he tries to explain why. However, throughout the rest of his book, he is forced into an interactive model of sorts by the evidence and the scientific research that has been done. So, all is not lost.

Meanwhile, in his opening salvo, without examining any evidence whatsoever, John Pinel automatically dismisses Cartesian Dualism or the Afterlife, the Nirvana or the Non-Local, and the Psyche or Non-Local Consciousness as being non-existent.

John Pinel continues:

The point here is that although the changes in self-awareness displayed by the patient were very complex, they were clearly the result of brain damage: Indeed, the full range of human experience can be produced by manipulations of the brain. (*Biopsychology*, p. 22.)

The hidden implication here is his claim that every spiritual experience, and out-of-body experience, non-local experience, and Cartesian Dualism experience is the result of brain damage or mental illness. Again, John Pinel is simply *jumping to conclusions* and *begging the question*, which are egregious logic fallacies.

The full range of human experience can be produced by brain stimulation.

THIS STATEMENT OF FACT IS FALSE. Observational evidence and experiential evidence falsifies this claim. Wilder Penfield's book, *The Mystery of the Mind*, falsifies this claim; and, Wilder Penfield was a practicing neuro-surgeon, whereas John Pinel is not.

Whole books have been written on this subject demonstrating that this claim is wrong; and, Pinel deliberately ignores ALL of them, pretending that they do not exist.

Remember, the things that can't be observed with nor detected by our physical instruments have to be experienced in order to KNOW that they are real and true.

Anytime you witness willful actions, intent, decisions, choices, chosen behaviors, purpose, and design, you are in fact witnessing PSYCHE or Non-Local Consciousness in action, because these things have been observed to survive the death of our physical body and our physical brain. Experiential evidence or observational evidence have proven that Psyche or Non-Local Consciousness survives the death of our physical brain, or clinical brain death. Our legal system is based upon intent or choice, which is purely a psyche function or psychological function, because our genes, biology, brain stimulation, and cultural societal environment can't make us believe anything or do anything that we don't want to do. In other words, our genes and our society do NOT make our choices for us – instead our psyche or non-local consciousness does; and, our psyche or non-local consciousness continues to make choices and continues to remember those choices and experiences, long after our physical brain is dead or offline.

Like I said, whole books have been written which FALSIFY John Pinel's claim with actual observational experiential evidence. Let me mention a few of them.

Penfield, W. (1975). *The Mystery of the Mind*. Princeton, New Jersey: Princeton University Press.

Beauregard, M. & O'Leary, D. (2007). *The Spiritual Brain: A Neuroscientist's Case for the Existence of the Soul*. New York: HarperCollins.

Beauregard, M. (2012). *Brain Wars: The Scientific Battle over the Existence of the Mind and the Proof That Will Change the Way We Live Our Lives*. New York: HarperOne.

Barlow, D. H. & Durand, V. M. (2015). *Abnormal Psychology: An Integrative Approach* (7th ed.). Stamford, CT: Cengage Learning.

http://mypsyche.us/wp-content/uploads/2017/10/Abnormal-Psychology.pdf

Do you see how educating yourself on this subject can put you into a position to identify and falsify claims such as the ones that Materialists and Naturalists and Darwinists repeatedly make in their college textbooks?

Remember, *scientific inferences* are easy to debunk and defeat, because they are nothing but science fiction to begin with. Scientific inferences are logic fallacies. We can't touch the observational evidence or empirical evidence, because it is unassailable. But, we can blast and destroy all of the scientific inferences, because they aren't real and because they are nothing more than *ad hoc just-so stories* designed to fit a person's pre-conceived notions. Fiction is easy to debunk and destroy! Remember that!

The theory of evolution is nothing but scientific inferences, or science fiction. That is what I have observed and learned during my research. Obviously, your mileage may vary because you might have been looking in different place than I have been looking. Remember, you won't get any of this kind of "psyche information" from a Materialist, Naturalist, Darwinist, Nihilist, or Atheist because these people have nothing like this to give you. They simply dismiss psyche out-of-hand before doing any science, and then leave it at that for the rest of their lives.

Every now and then, Pinel talks about complexity in his book. Evolution of any kind can't do complexity, because evolution by definition in principle is dumb, blind, and random. Thanks to the second law of thermodynamics or entropy, it's physically impossible for random processes and chaos to do complexity, synergy, design, creation, fine-tuning, teleology, organization, syntropy, and life.

> **The response properties of dorsolateral prefrontal neurons suggest that decisions to initiate voluntary movements may be made in this area of cortex.** (*Biopsychology*, p. 194.)

If you *beg the question*, start with the conclusion that ALL decisions are made in the brain, and then insert that pre-chosen conclusion as the first premise in ALL of your logical arguments, then a case can be made that voluntary movements are being made in every part of the brain. Anytime that you choose to use your pre-chosen conclusion as evidentiary proof that your pre-chosen conclusion is true, you will always end up proving that your pre-chosen conclusion is true. It's cheating and a logic fallacy, but it works every time. That's how they prove that the theory of evolution is true. They use the theory of evolution as evidentiary proof that the theory of evolution is true.

Selective motor control is a function of the Human Psyche.

WE KNOW that voluntary actions and choices are a function of the Human Psyche.

How do we know?

WE KNOW because the Human Psyche continues to make decisions and choices long after its physical body is dead and gone. You have to be willing to allow ALL of the evidence into evidence, if you want to find and know the truth.

WE KNOW from the Orthodox Interpretation of Quantum Mechanics by Henry P. Stapp that the Human Psyche triggers or requests voluntary actions first; and then, Nature's Psyche implements the plan by collapsing the necessary wave functions. Otherwise, physical matter just sits there and does nothing.

It's amazing what we can learn and know, when we finally decide to allow all of the evidence into evidence.

3. Nobody, who understands the evidence, resists Evolution as our Designer and Creator.

Remember, the things that can't be observed with nor detected by our physical instruments have to be experienced in order to KNOW that they are real and true.

Now, John Pinel will tell us what "correct think" is and looks like.

> **So far in this section, you have learned why people tend to think about the biology of behavior in terms of dichotomies** [Mind-Brain Problem, Nature vs. Nurture Debate, Determinism vs Free Will, and any form of Cartesian Dualism]**, and you have learned some of the ways why this way of thinking is inappropriate.** [He wants to suppress and stop this type of inappropriate thought.]

> **Now, let's look at the** [officially sanctioned] **way of thinking about the biology of behavior that has been adopted by many biopsychologists.**

> **Like other powerful ideas, it is simple and logical. This model boils down to the single premise that all behavior is the product of interactions among three factors:**

> **(1) the organism's genetic endowment, which is a product of its evolution;**

> **(2) its experience; and**

> **(3) its perception of the current situation.**

> **Please examine this model carefully, and consider its implications**.

> (*Biopsychology*, p. 24).

I find this statement simply illogical, because it doesn't match with the observational evidence that we have on hand as a race.

Your genome is the product of evolution.

NO! That's an assumption, which cannot be verified nor supported with observational evidence. Experience has taught us that a genetic endowment or genome cannot be the product of evolution, because evolution can't design, create, and produce genomes nor any new genetic programming or genetic information. This claim that "your genome is the product of evolution" is an *ad hoc just-so story*, designed after the fact to fit his pre-chosen conclusions. It's a logic fallacy, not science. It is *wishful thinking*, and nothing more. *Wishful thinking* is also a logic fallacy. By his own admission, everything that John Pinel writes in this chapter is based upon *wishful thinking* (*scientific inferences*), rather than observational evidence.

John Pinel doesn't realize that evolution (genetic drift), natural selection, and random mutations are powerless to provide anyone or anything with a genetic endowment, because evolution and mutation/selection DID NOT EXIST until after God designed and created the genomes in the first place. An organism's genetic endowment cannot be a product of its evolution, because evolution did not exist until after God gave organisms a genetic endowment to begin with.

Furthermore, John Pinel doesn't realize that "perception of the current situation" is exclusively a function of Psyche, because a psyche or non-local consciousness continues to perceive the situation long after the physical body and physical brain are dead and gone, according to the abundant empirical evidence derived from Near-Death Experiences (NDEs), Out-of-Body Experiences (OBEs), and other Non-Local Non-Physical Experiences.

Neurons are defined as cells of the nervous system that are specialized for the reception, conduction, and transmission of electrochemical signals.

Specialized by whom?

It can't be evolution. The Materialists, Naturalists, and Darwinists teach that evolution of any kind can't do specialization and complex specificity, because evolution is the product of random chance. The ONLY logical answer is Nature's Psyche, or God's Psyche, because the Human Psyche isn't consciously aware of its brain's neural specialization.

Migration is the movement of neurons and glial cells from the site of their birth in the ventricular zone of the neural tube to their ultimate location within the brain and nervous system. This was unexpected. They travel! Each neuron and glial cell moves through the brain to its ultimate pre-chosen destination within the brain before final differentiation and synaptogenesis. The information exchange taking place at the quantum level in order to accomplish these tasks could be on the order of petabytes. Not only do neurons and glial cells undergo differentiation and synaptogenesis, but they also miraculously form into functional brain structures such as the thalamus, hippocampus, cerebellum, and cortices. The Hand of God is all over this neurodevelopment process, even if we can't see it or wrap our mind around it.

The Materialists, Naturalists, and Behaviorists define Biological Psychology as the study of how the structures and processes of our nervous system determine our behavior. They get this wrong from the beginning. Biological processes don't control and determine our behavior, the Human Psyche does. However, biological processes can interfere with and even at times prevent the Human Psyche from controlling its physical body and choosing its fate.

Neurogenesis is the growth of new neurons. Who regulates and times and chooses the migration routes and differentiation types of newborn neurons? It can't be evolution, because evolution doesn't have brains enough to do so.

In Psychology, they actually use two different words to describe our senses – "perception" and "sensation". The Materialists, Naturalists, Behaviorists, and Atheists naturally but erroneously conflate or equate the two words.

"Perception is the higher order process of integrating, recognizing, and interpreting complex patterns of sensations." These people say the words, but these people don't really mean what they say, because their Materialism and Naturalism is messing them up.

Sensations are physical input coming in from our five senses and registering on the neurons of our physical brain. Sensation is the process of detecting the presence of physical stimuli.

In contrast, perception, integration, cognition, recognition, and interpretation are ALL functions of the Human Psyche. Recognition and interpretation require Intelligence or Psyche.

Our physical senses end, when our physical body and physical brain die and cease to exist; but, our perceptions, cognitions, and thoughts continue onward into the spirit realm or non-local realm long after our physical body is dead and gone.

Psyche is the ultimate causal agent. Psyche existed long before it brought the first particles of physical matter into existence.

The book, *Irreducible Mind*, adequately corrects Pinel's misconceptions with observational evidence, experiential evidence, and research evidence; so, I won't attempt to do so here.

Kelly, E. F., Kelly, E. W., Crabtree, A., Grosso, M., & Gauld, A. (2007). *Irreducible Mind: Toward a Psychology for the 21st Century*. Plymouth, United Kingdom: Rowman and Littlefield.

However, I find it interesting to point out that John Pinel is correct to employ an "interactive approach" or integrative approach to behavior, abnormal behavior, brain disease, and psychopathology even though he defines and presents these things incorrectly.

If you eliminate the falsehoods from his simple, logical, single premise, John Pinel's "interactive approach" boils down to the **BioPsychoSocial Model** of reality and abnormal psychology. In my own life, I have even had a practicing clinical psychologist tell me that everything in psychology is BioPsychoSocial now – everything. It's integrative, holistic, and interactive.

Yet, John Pinel will persist in his misconceptions, falsehoods, and logic errors because he has officially rejected dualistic thinking and has labeled it as "incorrect thinking".

Nonetheless, the organism's genetic endowment from God does indeed represent the "bio" portion of the BioPsychoSocial Model. The organism's experiences are the result of its psyche interacting with other psyches in its environment, society, and culture. The organism's experiences and developmental flaws (both physical and psychological) are the result of its interaction with its environment. The organism's experience map directly to the "social" portion of the BioPsychoSocial Model. Finally, the organism's perceptions of its current situation, whether locally or non-locally, represents the human psyche in action in the case of human beings. The human psyche is immediately identified any time a decision or a choice is called for. Consequently, the organism's perception of its current situation along with its choices as to how to respond to the current situation successfully maps to and represents the "psycho" portion of the BioPsychoSocial Model.

The BioPsychoSocial Model, if taken at face value and backed by real observational eye-witness evidence from both the physical and the non-physical realms, ends up be a vastly superior representation of Reality than John Pinel's simplified scaled-down model which attributes everything to evolution and makes evolution our designer, creator, and god.

Yet, John Pinel will persist in his misconceptions, falsehoods, and logic errors because he has officially rejected dualistic thinking and has labeled it as "incorrect thinking". His stated goal is to get your thinking "in line" with evolutionary thinking and evolutionary psychology, because he considers it to be simple, parsimonious, powerful, and logical. John Pinel will continue to claim that nobody who understands the evidence rejects evolution as our designer and creator, because they are smarter than that and know better.

Throughout his book, John Pinel repeatedly and deliberately promotes the FALSIFIED Zeitgeist (materialism and the theory of evolution), and deliberately rejects ALL of the observational evidence and experiential evidence that demonstrates that his chosen Zeitgeist is wrong. Throughout his book, he pays homage to the creative genius of evolution with worshipful adoration.

The theory of evolution was at odds with the various dogmatic views embedded in the 19th-century Zeitgeist, so it met with initial resistance. Although resistance still exists, virtually none comes from people who understand the evidence. (*Biopsychology*, p. 25).

This is posturing; and, it's false!

I have observed that the people who understand the observational evidence the most are in fact the people who end up resisting the Theory of Evolution the most, because there is NO observational evidence supporting Macro-Evolution, Spontaneous Generation, Abiogenesis, Materialism, and Naturalism. Macro-Evolution, or Spontaneous Generation, or Creation by Rocks was FALSIFIED by Louis Pasteur in 1859, the very same year that Charles Darwin published "On the Origin of Species".

All they really managed to do with the theory of evolution is to replace one dogmatic view with a different dogmatic view – they replaced one religion with a different religion. Darwinism is a religion. Materialism and Naturalism and Atheism are religions. They have to be taking on blind faith as being true, because there is NO observational evidence supporting them.

Ironically, the people who understand the evidence the most and the best are the ones who resist the theory of evolution the most. Pinel gets this one completely wrong – turning the whole thing upside down in an attempt to shift the burden of proof to the opposition. That's what the Materialists and Naturalists do – they assume that their theories are true, and then demand that everyone else falsify their theories, rather than trying to verify their own theories themselves. In other words, they try to shift their burden of proof to someone else. That's cheating! That's definitely not science! It is religion and dogma.

Stephen Jay Gould understood the evidence as well as anyone, when it comes to evolution. As a dedicated evolutionist and paleontologist, he could see dozens of different things that were wrong with evolution; and, Gould spent his whole career trying to patch them up with concepts like punctuated equilibrium.

Remember, those who understand the evidence and the logic and the science the BEST are the ones who are resisting evolution the most. This truth plays out over and over again in dozens of different books that I have read. The people who resist evolution are the people who are actually thinking about science and trying to do science. In contrast, the Darwinists, Naturalists, and Materialists hide behind the dogma simply parroting their talking points and gushing profusely about the creative genius of evolution.

I have watched Richard Dawkins debate different opponents; and, I was shocked to discover that I understand infinitely more science than he does; and, I actually understand Christianity, Islam, and Buddhism too, whereas he doesn't. I have actually studied this stuff; and, he hasn't. Dawkins hasn't done any science nor learning for fifty years; and, he's still parroting what he was taught fifty years ago in his college classes having made no real progress in his understanding of science since then. I have observed that Dawkins' opponents during his debates understand science as a whole a lot more than he does; and, they understand Christianity infinitely more than he does. Yet, it's Dawkins who always sets himself up as the brilliant expert on these subjects – posturing.

Contrary to Pinel's claim, I have observed that the people who understand science the most are in fact the people who are resisting evolution the most – and I'm talking the whole of science and not just zoology or biology. The truly smart and brilliant people are the ones who understand science well enough that they can actually see the myriad of different things that are wrong with the theory of evolution or Creation by Evolution, especially since random mutations and natural selection have NEVER been caught in the act of designing and creating new genetic information and NEVER will be because they can't. That's what the science is really trying to teach us; but, the Materialists, Naturalists, Darwinists, Nihilists, and Atheists are determined not to listen. Head in the sand where they simply parrot their same talking points that they have been espousing for thousands of years.

Remember, those who understand the evidence against evolution are those who end up resisting evolution the most. Observe for yourself if it is not so!

The very existence of something like **saltation** or **punctuated equilibrium** puts lie to the ongoing claim that "nobody who understands the evidence disagrees with the theory of evolution". Such a claim is an *over-generalization* logic fallacy, or a *posturing* logic fallacy.

There are a few other books from people who understand the evidence and the science, and then use that evidence to falsify Materialism, Naturalism, and the Theory of Evolution.

Meyer, S. C. (2009). *Signature in the Cell: DNA and the Evidence for Intelligent Design*. New York: HarperCollins.

Meyer, S. C. (2013). *Darwin's Doubt: The Explosive Origin of Animal Life and the Case for Intelligent Design*. New York: HarperOne.

Sanford, J. (2014). *Genetic Entropy* (4th ed.). Cornell University: FMS Foundation.

Sanford, J. C., Marks, R. J., Behe, M. J., Dembski, W. A., & Gordon, B. L. (Eds.). (2013). *Biological Information: New Perspectives*. Hackensack, NJ: World Scientific.

Wells, J. (2000). *Icons of Evolution: Science or Myth? Why Much of What We Teach About Evolution Is Wrong*. Washington, DC: Regnary Publishing.

Denton, M. (1986). *Evolution: A Theory in Crisis*. Chevy Chase, MD: Adler and Adler.

The very existence of these books FALSIFIES John Pinel's claim that none of the people who understand the evidence resist the theory of evolution. Eugenie Scott, another proponent of evolution, makes the same exact claim; and, she's wrong. I have observed that the ones who know the most are in fact the ones who resist the theory of evolution the most, because they KNOW why the theory of evolution is false and they KNOW how to falsify it.

This truism is supported by the observation and the fact that Materialism, Naturalism, and Atheism are based upon a refusal to look at evidence. There's no way on earth to know the most, if you always refuse to look at the evidence.

Whenever I have watched Richard Dawkins debate others, I was shocked and amazed at how ignorant and uneducated he really is. There's huge gaping holes in his knowledge; and, he doesn't even know it. Yes, he's rich and notorious, but he got rich and popular by telling the Materialists and Atheists exactly what they want to hear. He got rich by tricking them and deceiving them, just as he was tricked and deceived by those who came before him.

They say that ignorance is bliss; but, Richard Dawkins always comes across to me and my friends as nervous, irascible, neurotic, angry, and afraid. There's no peace there. He seems angry and unhappy; and, it's always clear to me that his opponents know infinitely more about Christianity and science than he does.

Kelly, E. F., Kelly, E. W., Crabtree, A., Grosso, M., & Gauld, A. (2007). *Irreducible Mind: Toward a Psychology for the 21st Century*. Plymouth, United Kingdom: Rowman and Littlefield.

The book, *Irreducible Mind*, is a tour de force when ends up being a vastly superior "study of the human psyche" than John Pinel's book *Biopsychology* even begins to be, because John Pinel denies the existence of psyche or non-local consciousness right from the beginning.

When it comes to my pursuit of the truth, John Pinel's book ended up being a great disappointment, especially whenever he deviated from the observational evidence into personal speculation and scientific inferences. However, when it comes to my obsession with debunking and falsifying the theory of evolution, John Pinel's book ended up being a gold mine of logic fallacies from which to harvest and hone my presentation. I think John Pinel employs, and therefore demonstrates, every logic fallacy that I have ever learned about and studied while trying to make his case for the beauty and simplicity of the theory of evolution.

4. Evolution is simple, beautiful, logical, parsimonious, and true.

John Pinel quotes:

> **Evolution is both a beautiful concept and an important one, more crucial nowadays to human welfare, to medical science, and to our understanding of the world than ever before. It's also deeply persuasive – a theory you can take to the bank ... the supporting evidence is abundant, various, ever increasing, and easily available in museums, popular books, textbooks, and a mountainous accumulation of scientific studies. No one needs to, and no one should, accept evolution merely as a matter of faith.** (*Biopsychology*, p. 25.)

Actually, this is false.

Science is Observation!

There is NO observational evidence and will NEVER be any observational evidence supporting macroevolution, spontaneous generation, or abiogenesis because these things are scientifically impossible.

In fact, in 1859, Louis Pasteur falsified Spontaneous Generation, Creation by Rocks, Macroevolution, and Abiogenesis. Louis Pasteur used a scientific method to falsify Materialism, Darwinism, Naturalism, Creation by Rocks, and the Theory of Evolution the very same year that Charles Darwin published "On the Origin of Species". How ironic is that? The Theory of Evolution was falsified by Science and a Scientific Method the very same year that it was introduced to the world. It was stillborn, yet it still continues to be promoted and preached throughout the world. Darwinism and the Theory of Evolution and Materialism is the most successful deception and lie that has ever been foisted on the human race.

The theory of evolution is a theory that you can take to the bank.

Actually, that's what they do – they take it to the bank. Remember, these people make their living by trying to convince you that the theory of evolution is true. They are trying to sell you a product, a bill of goods; and, millions of people are buying. I had to drop a Benjamin on this book just so that I could get properly indoctrinated into the theory of evolution – so that I could learn some "correct thinking" and "bring my thinking into line". A hundred bucks, just so that I could be brainwashed – conditioned to think the way they think. It's a bit Orwellian if you ask me, with signs of the brave new world that these people are trying to create for us through eugenics, which surfaces from time to time throughout our history.

I have had a great deal of fun debunking John Pinel's book, *Biopsychology*; but, I don't think that was the book's intended purpose. Notice that I do not and cannot attack nor debunk the actual observational evidence in the book, because true observational evidence is unassailable. I can only attack the *scientific inferences*, for which there are NO observations and will never be any observations to support them. Scientific inferences are easily attacked and debunked, because they are logic fallacies to begin with. By John Pinel's own admission, the theory of evolution is a scientific inference from beginning to end; and, scientific inferences are fundamentally flawed and a great deal of fun to attack and debunk.

> **Scientific inference is the fundamental method of biopsychology and of most other sciences [like the theory of evolution] – it is what makes being a scientist fun. This section provides further insight into the nature of biopsychology by defining, illustrating, and discussing scientific inference.**

> **The scientific method is a system for finding things out by careful observation, but many of the processes studied by scientists cannot be observed.**

> **The empirical method that biopsychologists and other scientists use to study the unobservable is called <u>scientific inference</u>. Scientists carefully measure key events they can observe and then use these measures as a basis for logically inferring the nature of events that they cannot observe.**

> **Like a detective carefully gathering clues from which to recreate an unwitnessed crime, a biopsychologist carefully gathers relevant measures of behavior and neural activity from which to infer the nature of the neural processes that regulate behavior.**

> **The fact that the neural mechanisms of behavior [and evolution] cannot be directly observed and must be studied through scientific inference is what makes biopsychological research such a challenge – and as I said before, so much fun.** (p. 13.)

Scientific inference is a logic fallacy.

These people love to tell stories, because it's a lot of fun. I love to speculate as much as anyone, but it's wrong to then turn around and use my speculations or scientific inferences as empirical evidence or observational evidence. Such a process is not only deceptive, but it's also a *category error* logic fallacy – the confounding or the equating of unlike kinds. To confound something means to mix up something with something else so that the individual elements become difficult to distinguish from each other. That's how *category error* logic fallacies are produced, by confounding and conflating and equating dissimilar things.

John Pinel literally and deliberately commits a *category error* by defining and categorizing scientific inference as an empirical method. They are not the same thing at all. He even says that he's making such an equation; but obviously, he doesn't identify the logic fallacy that he is committing, because he doesn't even realize that he's committing a logic fallacy. **"The empirical method that biopsychologists and other scientists use to study the unobservable is called <u>scientific inference</u>."** This sentence is literally a *category error* logic fallacy – the equating and confounding of unlike kinds or unlike categories. The evolutionists do this all the time, most of the time without even realizing it.

Remember, *scientific inference* is a logic fallacy, NOT empirical evidence.

Empirical evidence is by definition evidence that has been verified by observation or experience rather than theory or pure logic. Scientific inference is nothing but *ad hoc just-so story-telling* and therefore nothing but theory and pure logic. Notice how John Pinel employs one logic fallacy after another in order to make his case for the theory of evolution. It's what they all do. It's unavoidable, because the theory of evolution is nothing but a scientific inference to begin with; and, *scientific inferences* are logic fallacies.

I found it absolutely fascinating when John Pinel admitted that the Materialists, Naturalists, Darwinists, Behaviorists, Nihilists, and Atheists treat their scientific inferences as empirical evidence. How fascinating is that? It explains a lot, doesn't it?

Scientific inferences are logic fallacies! Yet these people treat their logic fallacies as empirical evidence, which is also a *category error* logic fallacy – treating scientific inferences as scientific evidence. Amazing, is it not? This discovery and revelation from John Pinel was worth the price of admission, and it made his whole book worth it.

I have upgraded my science to Science 2.0. Science 2.0 is based exclusively on empirical evidence. Under Science 2.0, science is defined as observation and lived experiences, because it is based upon eye-witness evidence. In contrast, evolutionists like John Pinel base their science on scientific inferences and define science as Materialism, Naturalism, Darwinism, Nihilism, and Atheism. There is a difference between the two; and, trying to confound the two is a *category error* logic fallacy.

Science 2.0 gives precedence to Lived Experiences over scientific inferences, which means that when it comes to the things of the mind (the spiritual sciences or non-physical sciences), Non-Local Experiences and Out-of-Body Experiences take precedence over scientific inferences and personal interpretation of scientific evidence every time. Scientific inference is treated as a logic fallacy under Science 2.0 rather than as empirical evidence, as the Materialists and Naturalists typically treat scientific inferences. When it comes to Science 2.0, Materialism, Naturalism, Darwinism, Nihilism, and Atheism are unscientific, because there is NO observational evidence nor empirical evidence supporting their major premises or primary assumptions. Scientific inferences are Bad Science, because they are logic errors.

> **Evolution is both a beautiful concept and an important one, more crucial nowadays to human welfare, to medical science, and to our understanding of the world than ever before. It's also deeply persuasive – a theory you can take to the bank ... the supporting evidence is abundant, various, ever increasing, and easily available in museums, popular books, textbooks, and a mountainous accumulation of scientific studies. No one needs to, and no one should, accept evolution merely as a matter of faith.** (*Biopsychology*, p. 25.)

Ever since the theory of evolution was introduced over a hundred and fifty years ago, they have been pouring in concrete trying to keep it from disintegrating and falling apart. The thing keeps springing leaks, and the Darwinists and Evolutionists keep trying to plug them up before the dam breaks. Unfortunately, the dam broke long ago; but, they still haven't realized it or noticed it.

As already mentioned, Stephen Jay Gould's **punctuated equilibrium** was an attempt to plug some of the leaks caused by the theory of evolution. The very existence of

129

the theory of punctuated equilibrium puts lie to their ongoing claim that the theory of evolution is perfect, simple, beautiful, logical, flawless, parsimonious, and true.

Remember, those who know the most science are those who know most what's scientifically wrong with the theory of evolution! There's a learning curve there. It's really easy to be deceived, if you don't have a broad base in the whole of science. That's actually one of my strengths. I'm a generalist. I'm good at everything and master of nothing. I have a strong broad base in the whole of science, and it's just getting better all the time, because that's what interests me the most.

The theory of evolution is convoluted and messy. There are thousands of exceptions to the rule; and, with the millions of missing links, it only goes downhill from there. Clearly, beauty is in the eye of the beholder, because to me evolution is ugly, if I'm going to be frank about the matter. Many people find that admission severely upsetting and disturbing, because to them Evolution is their religion and their god, and they really do teach their children that they have descended from apes and are actually proud of that fact.

However, I KNOW FOR A FACT that random mutations and natural selection can't design and create anything at all, that mutation/selection can't produce new programming code nor any new information, that mutation/selection would still take trillions of years to convert chimpanzees into humans through natural means even with God's direct intervention preventing all deleterious mutations and keeping those chimpanzees alive for trillions of years so that they can evolve naturally into humans, and that mutation/selection didn't exist before God created the genomes in the first place, which leaves me with nothing to recommend the theory of evolution. It's hard to see beauty in something that can't do the job that has been ascribed to it.

5. The proof for Evolution is so obvious that nobody needs to take it on faith as being true.

I have been a skeptic all of my life; but unlike most, I was an unbiased equal-opportunity skeptic. I was a scientist, or at least I tried to be. I turned my skepticism to the theory of evolution every bit as much as I turned it to religion and God. For me personally, agnosticism was a reasonably logical position to hold – neither in nor out. Even during my Materialism, Nihilism, and Atheism phase, I still continued to have nagging doubts about the theory of evolution, because things didn't add up or make sense.

The past couple of years, I have focused my skepticism on the theory of evolution, because the evidence against it finally hit critical mass. Science itself has proven the theory of evolution false, or falsified the theory of evolution; and, that's all I really needed to know in order to get me started. I KNOW that the theory of evolution is false, because I KNOW why it is false, and I can actually explain why it is false; and, I have been sharpening my claws ever since.

Evolution is both a beautiful concept and an important one. It's also deeply persuasive – a theory you can take to the bank. The supporting evidence is abundant, various, ever increasing, and easily available in museums, popular books, textbooks, and a mountainous accumulation of scientific studies. No one needs to, and no one should, accept evolution merely as a matter of faith. (*Biopsychology*, p. 25.)

He's telling us here not to doubt the veracity of the theory of evolution, because there's no need to doubt it, because it has been proven true beyond a shadow of a doubt.

Well, I doubt that.

When it comes to the organized worship of evolution, John Pinel and his book *Biopsychology* is the ultimate best (worst) of the bunch that I have encountered so far.

It took me a couple of chapters to figure out what he had done, but eventually it became clear to me that he had renamed Comparative Psychology and had called it the Evolutionary Perspective instead. Not only did he want you to know and believe that evolution is your creator and your god; but, he also chose to give evolution ALL the credit for each and every science experiment that we human beings have done throughout the centuries. Obviously, Pinel wasn't the only one doing this kind of stuff – conflation and obfuscation – but he was indeed one of the worst. I decided to dissect and debunk his book. John Pinel received this dubious honor, because I was forced to buy and read and memorize his book as part of a psychology class that I took in college.

I found it instructive to observe everything that he collected under the Evolutionary Perspective within his book; and, I couldn't help commenting from time to time in the margins of my book. Here's some of what I wrote.

It makes me laugh every time he talks about lesioning the brain or cutting on the brain under his "evolutionary perspective". Cutting brains has nothing to do with evolution. Evolution isn't going to be cutting on our brains in a million years. In one of his "evolutionary perspectives", he talks about suctioning out parts of a monkey's brain. Yes, evolution sucks, but evolution isn't going to be sucking people's brains out – not in a million years!

When Pinel put "brain suctioning" into the context of his evolutionary perspective, it made me smile and laugh, because it really is as if evolution has sucked these people's brains out. Not to worry, I too have been the victim of Darwinian lobectomies throughout different parts of my life. Visualizing evolution cutting people's brains out is a fitting metaphor, because that seems to be what happens for real where the theory of evolution is concerned. These people engage in skeptical, questioning, and critical thinking while doing science, except when it comes to the theory of evolution. Whenever these poor people are dealing with Materialism, Naturalism, Classical Physics, and the Theory of Evolution, these people leave their brains at the door and treat these subjects as religious truths that should never be questioned, rather than treating them as scientific disciplines that should be questioned, critiqued, debunked, and falsified.

Observe carefully and see for yourself if it is not so. Materialism, Naturalism, Darwinism, Behaviorism, and Atheism dumb people down in a major way. I watch the PhD Atheists debate online, and I feel embarrassed for them, because I quickly realize that I know way more science than they do, and I also understand religion infinitely more than they do. These people come across as being ignorant and uneducated. There's NO depth nor breadth there. It's as if these poor people have had their brains cut out and are mentally handicapped; and, they are proud of it.

Evolution or random mutations cause all of our genetic diseases, which means that evolution could never have created a perfectly functional genome to begin with. Evolution is entropy or the second law of thermodynamics. Evolution only messes up the genome that God gave us. That's the way entropy works at the physical level – it destroys and degenerates – it NEVER designs and creates. Mutations are faulty genes, which result in faulty proteins.

His comparative studies of humans with other species has nothing to do with random mutations and natural selection. There's NO mention of random mutations and natural selection anywhere within his "evolutionary perspective". He's doing comparative anatomy, and nothing else!

Under his "evolutionary perspective", he states:

If the factors that promote accurate regeneration [of neurons] **can be identified and applied to the human brain, it might be possible to repair human brain damage.** (*Biopsychology*, p. 250.)

And, evolution won't have anything to do with it! Evolution is not going to be repairing damaged brains – not in a million years. Evolution can only produce entropy or random mutations, which result in brain damage, death, and extinction. That's my evolutionary perspective.

He talks about "research" under his evolutionary perspective. He based his "program of recovery" from a brain tumor on cognitive exercise and physical exercise. Evolution has nothing to do with any of this! Human Psyches choose to do research, not evolution. He, his psyche, chose this particular program of recovery. Evolution had nothing to do with it.

Imagine for a second that the theory of evolution is 100% true. What would your brain be like if the theory of evolution were 100% true? What would your eyes and physical body be like if the theory of evolution were 100% true? Evolution is entropy or random chaos. Evolution is the same thing as taking your brain, your eyes, and your physical body, tossing the whole thing into a blender, turning the blender on, and leaving the blender on for billions of years of time. That's the way evolution really works. Evolution is entropy and random chaos. Evolution is a blender. ALL of our heritable diseases and mental illnesses are a gift from evolution or random mutations.

Enter Quantum Mechanics, Psyche, and Quantum Neuroscience.

He talks about evolution developing the human brain and human body much more slowly than any other species. Why would evolution do that? It wouldn't. It can't. What are the evolutionary precursors? There aren't any! The children of God are God's most unique creation, because there are NO evolutionary precursors for our intelligence, our language skills, our mathematical skills, and our ability to read and write. Your ability to read and understand this is evidentiary PROOF that the theory of evolution is false, because there are NO evolutionary precursors for this ability. Don't you feel special?

God slows everything down for us, including our neurodevelopment, so that we can live it, experience it, and remember doing so – and so that we need our parents to survive. That family relationship is important to God, even if it isn't important to us. We tend to learn the most through the pains and joys that come from our family relationships.

Under his evolutionary perspective, he talks about blind-folding an eye. Evolution doesn't have anything to do with blind-folding an eye. Evolution wouldn't do such a thing in a million years.

Pinel doesn't just throw a bone to evolution – he throws the whole animal!

Experience Fine-Tunes Neurodevelopment. (*Biopsychology*, p. 224.)

Notice that it's the Human Psyche's choices and experiences that fine-tune that person's neurodevelopment, and NOT evolution. Evolution or random mutation doesn't have anything further to do with our development, after Genetic Recombination has taken place – the production of our sperm and egg.

He talks about the "complex programs of neurodevelopment going wrong." Evolution of any kind cannot do complex programming, by definition in principle, because evolution is supposed to be completely blind. This means that evolution could never have produced the complex programming code within our genome.

To turn something that is dumb, blind, without hands and a mind into a computer programmer would be a miracle of monumental proportions; but, that's exactly what Pinel does to evolution. He turns evolution into our Designer, Programmer, Engineer, and Creator.

The adult brain displays plasticity. What do you think is the evolutionary significance of this ability? (p. 231.)

NONE!

Evolution doesn't contribute anything of significance! Evolution is entropy and random chaos. Evolution leads us towards death and extinction. That's its only significance.

Discuss the evolutionary significance of the slow development of the human brain. (p. 231.)

NONE!

There are NO evolutionary precursors for such a thing! Our slow development should make us the least fit to survive and should have driven us extinct millions of years ago; but, somehow it didn't. Maybe evolution doesn't work as advertised?

The structure and physiology of each type of somatosensory receptor seems to be specialized for a different function. (*Biopsychology*, p. 172.)

Ya think?

If it weren't, it wouldn't work!

Pinel talks about specialization all throughout his book. Specialized by whom? By definition in principle, evolution can't do specialization or intelligent design! Each structure within the brain seems to be specialized for a different function. In fact, the brain is all about physical functionality. It's as if God or Nature's Psyche has mapped or specialized our brain for physical functionality. Go figure!

How difficult it is to identify and shed conventional approaches even when they clearly haven't been working, and how often solutions to long-standing problems become apparent when approached from a new perspective. (*Biopsychology*, p. xxiii.)

You need to get rid of evolution, because clearly it hasn't been working and it lacks explanatory power. Only by getting rid of Materialism, Naturalism, and Darwinism can long-standing problems be immediately resolved. Of course, the new perspective that he wants you to embrace is the theory of evolution. Eventually, he makes that point clear. It's a bait and switch that he's running here.

In his section on hunger and eating, he wants you to abandon the set-point models, and the positive-incentive models, and adopt his settling-point model instead. He asks his readers, "Can you change your thinking?" I have no problem changing my thinking whenever I'm presented with a better model. In the margins of my book, I asked him in return, "Can you change your thinking about evolution?" It's the Human Psyche who chooses or decides what it considers to be a positive incentive or a negative punishment.

The "Evolutionary Perspective" should have been called Comparative Psychology, because evolution had nothing to do with it. Evolution didn't run the science experiments, we did. I get the feeling that he's trying to trick us and deceive us. He's definitely trying to make evolution infinitely more important than it actually is.

Throughout this chapter, you will need to put aside your preconceptions and base your thinking about hunger and eating entirely on the empirical evidence. (*Biopsychology*, p. 291.)

Yes, and you need to do the same thing with the theory of evolution and realize that there's NO empirical evidence supporting it.

We engage in sexual behavior not because we have an internal deficit, but because we have evolved to crave it. The evolutionary pressures of unexpected food shortages have shaped us and all other warm-blooded animals, who need a continuous supply of energy to maintain their body temperatures, to take advantage of good food and eat it. (*Biopsychology*, p. 293.)

Sorry, but evolution had nothing to do with this!

ALL of the evidence tells us that these "cravings" and "needs" were designed into us and programmed into us from the beginning. Think about it logically. Evolution would NEVER design and create warm-blooded animals in a trillion years, because they require a continuous supply of energy to maintain. That's highly inefficient; and, being warm-blooded puts these animals at a severe evolutionary disadvantage. The warm-blooded animals should all be extinct by now, and many of them already are. Predation is an evolutionary dead-end, and it should end up in extinction, because eventually the predator runs out of prey.

These people use *scientific inferences*, logic fallacies, in order to support their evolutionary inference. I used **converging methods** to demonstrate and prove that Materialism, Naturalism, and Darwinism are false. It's really easy to falsify the theory of evolution by choosing to allow ALL of the evidence into evidence.

Ah, that's how he's using the Evolutionary Perspective, as Comparative Psychology – comparing humans with rats and monkeys – thereby giving evolution much more credit than it deserves. Anytime we humans run a science experiment comparing humans with animals, John Pinel gives evolution all the credit. It's the *shotgun logic* fallacy. He's spitting into the wind hoping to get some of it onto himself. He can't miss, as long as he gives evolution credit for everything.

Ah, here he is using the *most people* logic fallacy in order to provide evidentiary support for the theory of evolution. I have observed that most people are typically wrong when it comes to their pre-chosen conclusions.

In the case of humans, courtship displays and social dominance issues are a product of the Human Psyche, not evolution. God didn't put as much programming into us as He did the animals. The Human Psyche can override its Nature and its Nurture at will.

Nature-nurture issue. The debate about the relative contributions of nature (genes) and nurture (experience) to the behavioral capacities of individuals. (*Biopsychology*, p. 490.)

Notice how the Individual Psyche is hidden within the Nature vs. Nurture Debate. In the end, it's all about the individual – the individual Human Psyche. The individual Human Psyche survives the death of its physical body and physical brain according to the empirical evidence from Near-Death Experiences, Out-of-Body Experiences, and after-death Life Reviews.

When it comes to human evolution, the whole thing is nothing but *a fictional ad hoc just-so story* designed to match pre-chosen conclusions. It's physically impossible for human beings to have evolved from chimp-like ancestors. Macro-evolution of that kind has NEVER been observed and NEVER will be, because it's physically impossible.

The first chordates with spinal bones to protect their dorsal nerve cords evolved about 25 million years later. (*Biopsychology*, p. 27.)

How does he know? Was he there to see it happen? Of course not! It's an inference based completely upon wishful thinking and a blind leap of faith. According to Pinel, we were a bunch of spineless jelly fish until evolution got around to making some spinal bones for us. All of those spineless monkeys in the fossil record tell us that it must be so. They actually slithered across the ground like a snake rather than climbing in trees.

We didn't descend from monkeys and apes. Random mutations and natural selection can't do that kind of genetic programming and genetic restructuring. It's physically impossible. The laws of physics, including the Law of Entropy, tell us clearly and conclusively that it's physically impossible to get order, structure, programming, and abiogenesis out of random mutations or random chaos. It can't be done. It's not possible. Apes cannot breed and produce humans. Chimpanzees always produce chimpanzees. That is what we have observed. It's physically impossible for humans to evolve from chimp-like ancestors.

He bases his whole religion on *scientific inferences*, which by his own definition are things that have NEVER been observed. You can't base a science on things that have NEVER been observed; yet, that's exactly what he is doing when it comes to the theory of evolution. It's fraudulent and deceptive to do so. Amazing! He's basing all of his science on *scientific inferences*, which are logic fallacies. It would be incredible, if I hadn't see it and read it with my own eyes.

All those monkeys that he credits with being our ancestors weren't human – they weren't descendants of Adam and Eve. The scriptures repeatedly tell us that Adam was a son of God. Adam didn't descend from the apes. Adam and Eve descended from the Gods.

The lesson of the scriptures is that the Biblical God has to repeatedly intervene behind the scenes in order to keep some of us alive, because the human race as a whole is determined to destroy itself. Militant Atheists or Mass-Murdering Communists, as well as Darwinian Fascists or Nazis, are the pinnacle of what the theory of evolution has had to offer us during the twentieth century – survival of the fittest used as an excuse for pogroms, killing fields, and genocide.

http://bookrev.allthings.computer/the-true-face-of-the-enemy/

The true face of the enemy.

No one needs to, and no one should, accept evolution merely as a matter of faith, because ALL of the empirical evidence falsifies the theory of evolution and tells us beyond a shadow of a doubt that it is false. There is NO empirical evidence verifying the truthfulness of Darwinism, Materialism, Naturalism, Nihilism, and Atheism. It's physically impossible to provide empirical evidence to support the major premises or primary assumptions of Materialism, Naturalism, and their derivatives.

By His Own Admission, the Theory of Evolution Is Worthless

Theories that cannot be tested have little use. (*Biopsychology*, p. 34.)

Amen!

That truth definitely applies to the theory of evolution. Even with lots of coaxing in a laboratory, when it comes to the theory of evolution, we can't replicate it, test it, nor verify it. We KNOW that it is false, because macro-evolution or abiogenesis or spontaneous generation has NEVER been observed or caught in the act. The theory of evolution can't be tested or verified, so it is of little use to us.

One of the most interesting aspects of this theory is its evolutionary implication. Goodale has suggested that the conscious awareness mediated by the ventral stream is one thing that distinguishes humans and their close relatives from their evolutionary ancestors. (*Biopsychology*, p. 155.)

We don't have evolutionary ancestors. Nobody has evolutionary ancestors, because evolution of any kind cannot create new kinds or new types and classes from existing ones. When it comes to evolutionary ancestors, these people are making things up out of thin air. It's physically impossible for two chimp-like ancestors to magically give birth to two genetically compatible male and female humans at the same time in the same place. It can't be done, which means that it wasn't done. Evolution of any kind from one species into a completely different sexually incompatible species is physically impossible among sexually reproducing species.

Among all the various deceptions and lies that came from his evolutionary perspective, the following sentence from his book is actually true.

The members of a species can produce fertile offspring only by mating with members of the same species. (*Biopsychology*, p. 27.)

This is a TRUTH among sexually reproducing species that completely FALSIFIES the theory of evolution, whether he realizes it or not. Effectively what it means is that NO sexually reproducing species has an evolutionary ancestor!

Periodically, Pinel succeeds in falsifying the theory of evolution without trying to do so and without realizing that he is doing so. Imagine what he could do if he actually tried to falsify the theory of evolution. That thing would go down and go down hard.

Ah, now he's getting into genetics, which is something that has been observed, replicated, tested, and verified in a laboratory. His credibility is going to go up now, because now we are no longer dealing with a convenient fiction, but something that has actually been observed. But still, he's going to find ways to mess it up and give evolution all the credit, even though it's physically impossible for random mutations and natural selection to do genetic engineering and the design and creation of genes and proteins. In fact, random mutations, natural selection, and evolution (genetic drift) didn't even exist, until after God designed and created the proteins and the genomes in the first place. Nevertheless, John Pinel is going to insist that evolution did it all, even though that's physically impossible.

We KNOW that Darwinism is false because its claims are physically impossible.

When it comes to evolution, I don't value it nor worship it anywhere near as much as they do – which annoys them to no end. I've seen through to the truth of the matter, and now there's no going back.

Some people are going to find all of this boring and repetitive. I know that I do. But, we have to eliminate all of the falsehoods, before we can find and know the truth – that's the scientific method! The scientific method was designed to eliminate all the falsehoods, in the hope that someday we will end up with the truth. The scientific method gets very boring and repetitive much of the time, but it's necessary. The discovery of the truth starts by eliminating everything that is false, falsified, and impossible – things such as Naturalism, Materialism, Nihilism, Darwinism, Behaviorism, Determinism, Physical Reductionism, and Atheism. Eliminate these falsified philosophies, and we automatically come a lot closer to the truth. That's just the way things work.

How often have I said to you that when you have eliminated the impossible [and the false], whatever remains, *however improbable*, must be the truth? — Sherlock Holmes.

The best way to find and know the truth is to live it and experience it for yourself, or to choose to trust someone who has. The second-best way to find and know the truth is to eliminate everything that is false and everything that has been falsified. If you successfully eliminate everything that is false, then ONLY the truth will remain. So, let's press onward.

More than a decade after the human genome was described, these medical miracles have yet to be realized. (*Biopsychology*, p. 41.)

Yet, he wants us to believe that evolution produced all of these miracles at will billions of years ago. He truly wants us to believe that random mutations and natural selection are infinitely smarter than our geneticists.

Epigenetics is the study of all mechanisms of inheritance other than the genetic code and its expression. (*Biopsychology*, p. 41.)

He wants you to believe that evolution was and is in control of all the non-genetic means of inheritance, but that's physically impossible by definition in principle. Of course, he doesn't see that, because he doesn't want to see that.

Nevertheless, he wants us to believe that random mutations and evolution (genetic drift) designed and created our genome, our genes, our proteins, and our epigenetic mechanisms from scratch billions of years ago. However, that's physically impossible. Random mutations, natural selection, genetic drift, and evolution didn't even exist until after God designed and created the genes and the proteins in the first place.

Ah, this one is a *correlation error*, *category error*, and *conflation error*. Technically, as I see it, these scientists were more likely breeding for IQ or intelligence and NOT behavior! Intelligence is heritable; whereas, our chosen behaviors are a function of volition and psyche, even within the animals. They are not breeding for behavior, because behavior is chosen by psyche or non-local consciousness. It's IQ or intelligence that has been proven to be heritable. They are going after smarter rats with their breeding experiments, NOT well-behaved rats. There's a difference!

By their own admission, Behaviorists are only interested in observable behaviors – controllable and predictable behaviors. They could care less what's happening with the Human Psyche or the mind, because they can't detect that and measure that with their physical instruments. As long as you call it what it is, Behaviorism has its uses.

Ever since his *scientific inference* admission, wherein he admitted that he is deliberately using scientific inferences as empirical evidence, I have doubted and questioned everything else he has written, especially his interpretation of them. *Scientific inference* is a logic fallacy. By his own admission, he has built ALL of his science within the first couple of chapters of his book upon logic fallacies. I can only hope that things will get better later on in the book, when he is finally forced to switch over to experimental evidence and observational evidence. I was forced to buy this book and read it for a class, and I have been highly disappointed with the all the logic fallacies and scientific inferences within the first couple of chapters of this book, especially his *conflation* of Comparative Psychology with what he has chosen to call the Evolutionary Perspective. He's trying to make evolution our creator and god, because according to him, evolution IS our creator and god.

Periodically throughout this book, he labels belief in psyche or spirit as "wrong-thinking" while at the same time labeling belief in evolution as "correct-thinking" or "right-think". This is an *ad hominem* logic fallacy. It's designed to intimidate and enforce compliance.

In the development of individuals, the effects of genes and experience are inseparable. In the development of differences among individuals, they are separable. (*Biopsychology*, p. 46.)

Choice is a product of Psyche. Anytime we catch choice in action, we have witnessed Psyche in action. It's really simple to understand; but, most people refuse to do so. Even though it is clear and obvious to the open-minded that "experience" is PsychoSocial and therefore a product of both Psyche and the Environment, he deliberately excludes the psyche from consideration. Consequently, all of his models of reality are incomplete.

Psyche is also separable. Our psyche or choices determine most of our differences, especially among identical twins who were raised together in the same environment.

This is the BioPsychoSocial tripod. It's extremely stable and full of explanatory power. However, the Materialists, Darwinists, and Naturalists are determined to eliminate an essential leg of the tripod – the Human Psyche – and the resulting BioSocial model falls over every time. You can't keep it standing, because it's missing an essential ingredient. It's missing one of the legs of the tripod. Pinel never mentions the BioPsychoSocial Model of Reality within his book, because he has denied its existence; however, it's there to be found if he were to have gone looking for it. The BioPsychoSocial Model is my preferred model of reality – it's the ultimate model of reality, in my humble opinion, as long as you don't let the Materialists and Naturalists remove the psycho psyche from it.

Remember, the Human Psyche can override its Nature and Nurture at will, which is an observed and experienced FACT that he never mentions in his book.

He wants you to completely reject an immaterial object or person that we call Psyche or Non-Local Consciousness, and then replace psyche instead with another immaterial object – a non-person – that he calls Evolution. He wants evolution to be your creator and god.

Humans have a completely different origin and pedigree than the primates and other animals. Homology is not proof of common ancestry! Homology is a convenient fiction. In fact, homology is better used as evidence for a Common Designer. Homology makes more sense when used that way. Technically, homology or similarity is NOT proof of a common ancestor nor is it proof of a common designer. These are *scientific inferences* – educated guesses and nothing more.

He relies exclusively on *scientific inferences* to make his case. They all do. With scientific inferences, you can have any result you want simply by *affirming the consequent* and *jumping straight to the conclusion* you want to be true. But, these are logic fallacies. Homology is the greatest and best "scientific proof" of evolution that they have available to them, and it's based totally on a logic fallacy. They start with the pre-chosen conclusion that homology is proof of common ancestry, and then they insert that chosen conclusion into all of their arguments as one of the hidden premises. In other words, they are *begging the question*. They are using their chosen conclusion as evidence and proof that their chosen conclusion is true; but, they can't even see that they are doing so. Nevertheless, they could have just as easily and logically started with the pre-chosen conclusion that homology is proof of a common designer, and then they could have inserted that pre-chosen conclusion into all of their arguments as evidence and proof that homology proves the existence of a Common Designer.

By *begging the question*, you can have any result you want and prove that result to be true by using your pre-chosen conclusion as evidentiary proof that your pre-chosen conclusion is true. That's the way the Materialists, Naturalists, and Darwinists do science; and, they don't even know it.

Oh, I see what he's doing now. His evolutionary perspective is comparative anatomy – comparing humans with other animals. He's trying to give evolution all the credit for the comparisons or the homologies. Tricky!

If he were being fair and open-minded, he should have called it Comparative Psychology throughout his book and NOT the Evolutionary Perspective, because evolution never happened because chemical evolution or spontaneous generation is physically impossible.

He calls it the Evolutionary Perspective; but, evolution isn't giving these rats anxiolytic drugs. Evolution isn't sticking electrodes into rats' brains, zapping their brains, giving them epilepsy or kindling epilepsy, and then calling the whole process an Evolutionary Perspective. Kindling of epilepsy in animals doesn't have anything whatsoever to do with random mutations and natural selection! It's not something that evolution would do to rats. Evolution isn't zapping their brains giving them epilepsy. We are. There's a HUGE difference between evolution and intelligent beings.

Evolution isn't giving MPTP and Parkinson's Disease to nonhuman primates and rats in order to see what happens. His Evolutionary Perspective has nothing whatsoever to do with random mutations and natural selection! It's a confound and a conflation!

This Evolutionary Perspective is a *confound* logic fallacy or *conflation* logic fallacy that he is repeatedly employing throughout his book, in an attempt to make evolution seem infinitely more important and effective than it really is. Random mutations and natural selection don't have anything to do with the things that he is assigning to it. This is the classical *category error* that he's repeatedly using throughout his book to make his case and bolster his chosen belief in the Theory of Evolution.

Since evolution can't design and create proteins and genomes, what can it do?

Evolution can do entropy, chaos, disease, stillbirth, developmental abnormalities, brain damage, cancer, death, and the extinction of species. Evolution has been caught in the act of doing so! Random mutations are an agent of chaos, not an agent of design and creation. Remember that, because it's important to know.

Evolution is the cause of our genetic diseases, NOT the source of our genome and proteins. That's just the way it is, because that's the way it has been observed to be. Under Science 2.0, direct observation trumps wishful thinking and philosophical speculation every time.

You have to eliminate all the falsehoods, before you are motivated to go out and find the truth. If you are satisfied with the falsehoods and the falsified theories – as the

Materialists, Naturalists, and Atheists are – then you have no reason to look for the truth. In fact, the truth is an annoyance to you. I KNOW, because I used to be a Materialist, Nihilist, and Atheist.

The purpose of Quantum Neuroscience is to find out how things truly work. It's a different way of doing science, because it is based upon Science 2.0, which allows ALL the evidence into evidence and then pursues a preponderance of the evidence.

You will have to decide for yourself if I'm making sense, just like I had to decide for myself that Pinel wasn't making sense. This is something that nobody else can do for you.

Pinel's brainwashing is total and complete. He has chosen to see things no other way besides evolution. Over and over again, he states that evolution is a proven fact. It's a *conflation error*, a deliberate deception. YES, evolution (micro-evolution, random mutations, and natural selection) has been observed and caught in the act. NO, evolution (macro-evolution, spontaneous generation, chemical evolution, and abiogenesis) has been repeatedly FALSIFIED due to the fact that macro-evolution is physically impossible and has never been observed. He, like everyone else, *conflates* the two. But, there's a huge difference between micro-evolution and macro-evolution. He wants you to believe that they are the same; but, that's a *category error* logic fallacy if you do. Evolution (genetic drift) functions at the micro scale; but, it can't design and create genes and proteins at the macro scale. Evolution doesn't work at the macro scale.

The following is found in his book:

> **No one needs to, and no one should, accept evolution merely as a matter of faith.** (*Biopsychology*, p. 25.) [His implication of course is that the truth of the theory of evolution has been proven beyond a shadow of a doubt. Technically, that's a lie, especially given the fact that macro-evolution or spontaneous generation has been falsified trillions of times in thousands if different ways.]

Yet, that's exactly what they do – take it on blind faith as being true, because there is no actual evidence for design and creation by evolution. The whole theory of evolution is nothing but a *scientific inference*, which is a logic fallacy. Evolution is genetic change, or genetic drift. There's no way on earth than random change can produce order out of chaos or creation out of nothing.

The truth is that no one needs to, and no one should, accept evolution merely as a matter of faith, because there is NO empirical evidence supporting it. ALL of the empirical evidence we have on hand falsifies the theory of evolution. There's no need to accept evolution on blind-faith as being true, when we KNOW from all the evidence that it is false.

Over and over again, Pinel tells us to engage in "Critical Thinking" – with the exception of evolution, which he wants you to take on blind faith as being true. Oh, he's going to hate my Critical Thinking in relationship to the theory of evolution, if he were ever to actually read it. I'm destroying his religion here, and he's not going to like that. I one time had an atheist tell me to stop making fun of her religion in print and online. She preferred the blissful ignorance rather than the hard, cold scientific facts.

Sorry, folks; but, Macro-Evolution, Materialism, Naturalism, Darwinism, Abiogenesis, and Spontaneous Generation were FALSIFIED in 1859 by Louis Pasteur, the very same year that Charles Darwin published "On the Origin of Species". We KNEW that these things were false before Darwin published his book; and, we've gone downhill ever since.

Okay, I'm not going to apologize for finally finding the truth. I have looked hard and long for the truth. We have to eliminate everything that is false, before we can find and know the truth. That's what the scientific method is all about – eliminating everything that is false! We can't use the scientific method to prove things to be true; but, we definitely can use it to falsify things such as Materialism, Naturalism, Nihilism, Darwinism, and Atheism, which we have done. If you successfully eliminate all of these false philosophies which have already been falsified, then eventually only the truth will remain. It's elementary.

But, complains Pinel, "Most scientists believe in the theory of evolution and teach the theory of evolution." Well, most scientists are wrong. That's the *most-scientists* logic fallacy that he is employing there, and it doesn't work, because it's a logic fallacy. It's an *appeal to authority – credentialism* – which are logic fallacies. These people are demanding that you believe their *ad hoc just-so stories* or logic fallacies because they have PhDs and know the truth; whereas, you don't. Don't be intimidated by their ignorance, blindness, and lack of knowledge just because they have a PhD and have been authorized to write college textbooks for public consumption. Learn to think for yourself, because nobody else will.

Brutal! I'm red in tooth and claw, especially when it comes to the theory of evolution. I will survive!

I actually take genetics much more seriously than macro-evolution, because genetics has been observed and caught in the act; whereas, macro-evolution or spontaneous generation has not. I have much more faith in genetics, because it's an observed science and a practiced science. I have NO faith in macro-evolution because it has never been observed and has never been reproduced. Instead, macro-evolution (spontaneous generation, abiogenesis, chemical evolution) has been falsified trillions of times over.

Our brain and physical body are infinitely more complex than your computer, tablet, or smart phone; but, the Atheists and Darwinists want you to believe that your brain, eyes, body, genome, proteins, computer, tablet, and smart phone self-assembled out of thin air from scratch. The Darwinists want you to believe on blind-faith that evolution designed and created all of these things.

Ask yourself, "Since evolution understands all of these different physical processes perfectly, why don't we? Who taught evolution how to do all of these things?" Evolution knows everything, and evolution can do everything at will. I want evolution's teacher to be my teacher, don't you?

ONLY intelligent beings like humans can do code, programming, language, and writing. A genome is a four-dimensional computer program, or four-dimensional programming code. There's NO way on earth that the rocks or the monkeys or mutation/selection could have written such a thing! We human beings can't even design and write such a thing! Our very best operating systems are one-dimensional – linear – computer code or programming code, and nothing more.

God gave us human beings, their direct descendants both spiritually and physically, the ability to speak, write, code, and read so that the rational and sane among us would automatically KNOW that we human beings have NO evolutionary ancestors.

Why don't nonhuman primates learn language? (*Biopsychology*, p. 639.)

One reason is so that God can demonstrate clearly and conclusively that there are NO evolutionary precursors to language so that we KNOW that human beings have NO evolutionary ancestors. Science, logic, and observation falsify the theory of evolution from every angle that we can possibly imagine. Language is ultimately a function of the Human Psyche, because it survives the death of our physical brain.

141

There is no way in the universe that evolution (genetic drift) or mutation/selection could have designed, programmed, engineered, manufactured, and created our four-dimensional genome given the FACT that we human beings can't even do such a thing. We don't have the brain power or mental fire-power to do such a thing. It's beyond us.

Everything falsifies the theory of evolution; and, there is nothing that verifies it. We'll never be able to verify the theory of evolution because it's physically impossible to get atoms to self-assemble into functional genes and proteins. The theory of evolution fails every test of science that we can possibly imagine. The theory of evolution is dead. Long live Quantum Mechanics!

The Four-Dimensional Genome

https://creation.com/four-dimensional-genome

https://www.youtube.com/watch?v=K3faN5fU6_Y

John Pinel and others like him want you to believe that evolution (genetic drift), natural selection, and random mutations designed and created our Four-Dimensional Genome. There is NO way on earth that random mutations could ever produce a four-dimensional genome! Yet, John Pinel assures us that it did.

If you believe such a thing, then you have been duped, tricked, and deceived because natural selection and random mutations can't do design and creation and because mutation/selection and evolution (genetic change) didn't even exist until AFTER the first genome and the first life form were designed and created.

Come on, let's think logically and rationally about this for once in our lives. We claim to be scientists after all.

Instead, John Pinel writes:

> **In most cases, mutations disappear from the gene pool within a few generations because organisms are less fit. However, in rare instances, mutations increase fitness and in so doing contribute to rapid evolution.** (*Biopsychology*, p. 38).

Rapid evolution! It's magic!

This is an *ad hoc just-so story* designed to fit his pre-conceived chosen conclusions. He's exaggerating. There is NO observational evidence supporting it and there never will be, because the vast majority of mutations are deleterious and contribute to rapid extinction and death, not progressive evolution, construction, edification, and design. Science has observed that random processes cannot design and create. Instead, random processes promote Chaos, Entropy, Disease, Disorder, Disorganization, Loss of Information, Death, Destruction, and Extinction. Random mutations are agents of Chaos and Entropy. This is what the scientists have actually observed, whenever they have gone looking.

The theory of evolution is nothing but *wishful thinking*, which is a logic fallacy. "Rapid evolution" is wishful thinking and posturing. There's no such thing as rapid evolution. Evolution by definition in principle is slow and ponderous taking millions of years to accomplish. It's deceitful to say otherwise.

I'm just telling it as it is. I'm telling you what the scientists have observed and experienced for themselves. Evolution (genetic change) of any kind cannot design and create new and interesting and functional genomes. Evolution, Mutation, and Selection ONLY lead to death and extinction. This is what we have truly observed!

Survival of the fittest is more a function of luck than actual value. If you observe carefully, especially among humans, the fittest and healthiest are often the first to die, while the welfare cases seem to breed like rabbits. Survival of the fittest is just another convenient fiction. It doesn't do anything to change or modify your genes!

Then John Pinel starts talking about Mitochondrial Eve.

There's a massive problem with Mitochondrial Eve – someone had to design her, create her, and make her in the first place, or she wouldn't have existed. The evolutionists just assume her into existence ex nihilo – like magic – without ever once explaining how she could have realistically gotten there. And, the Darwinists aren't able to explain how a compatible Mr. Mutant got there as well, as if by magic. The evolutionists just assume Mitochondrial Eve and Mr. Mutant Adam into existence ex nihilo out of thin air.

That's NOT science! That's wishful thinking and nothing more. It's science fiction – story telling at its best.

WE KNOW for a fact thanks to science experiments, realistic modeling of the mutation/selection process, and observation of reality that Random Mutations and Natural Selection could never have produced a Mitochondrial Eve from a bunch of apes over long periods of time, because mutation/selection can't do design, programming, engineering, manufacturing, and creation. Mutation and selection only effectively and reliably do disease, death, and extinction. Mutations and Natural Selection are agents of Chaos, or Genetic Entropy – not design and creation. The creative powers of random mutations are nothing but science fiction.

It would literally take trillions of years and direct intervention from God preventing ALL deleterious mutations and keeping that population of apes alive for trillions of years in order for Random Mutations and Natural Selection to perform as advertised and actually convert a population of apes into a population of humans naturally, step-by-step generation-by-generation; and, that's with God-given and God-created apes to begin with.

(See: Sanford, J. (2014). *Genetic Entropy* (4th ed.). Cornell University: FMS Foundation.)

Deep time actually works against evolution through mutation and selection resulting in genetic entropy, diseases, cancer, death, and the extinction of species. In order to fit more closely with the truth of the situation, Charles Darwin's book, *On the Origin of Species by Means of Natural Selection*, should have been called, **On the Extinction of Species by Means of Natural Selection**; and then, his book would have ended up matching more closely with reality and actual scientific evidence and scientific observations.

In order for Random Mutations and Natural Selection to function and perform "as advertised" in the design and creation of life, God would have to create the genomes and the life forms to begin with; then God would have to intervene and prevent ALL deleterious mutations from taking place; God would have to force the creation of a compatible Mr. Mutant and Mrs. Mutant at each significant step along the way; and then finally, God would have to intervene and prevent that evolving or changing species from going extinct over trillions of years of deep time just so that population of apes can evolve "naturally" into human beings, through Random Mutations and Natural Selection one mutation and one generation at a time.

(See: Sanford, J. (2014). *Genetic Entropy* (4th ed.). Cornell University: FMS Foundation.)

That's the power of the Theory of Evolution in action – it actually PROVES the need for God, which effectively ends up being Scientific Proof of God's Existence, if you eliminate all the wishful thinking and think about it logically and rationally as some people have done.

The Darwinists and Evolutionists deliberately ignore ALL of these scientific realities – they have to, because Materialism, Naturalism, and Atheism of any kind are based upon a refusal to look at evidence and a refusal to think deeply, critically, and logically about these kinds of things. The Darwinists and Evolutionists simply conclude that evolution can design and create anything that it sets its mind to, and they leave it at that forcing you to fill in the blanks. In other words, they shift the burden of proof to you.

Well, I took the challenge and successfully FALSIFIED Materialism, Naturalism, Darwinism, Nihilism, and Atheism through a preponderance of the evidence.

Remember, Mitochondrial Eve is nothing but a fictional story masquerading as scientific fact. It's fiction – just as good as any science fiction story that you might watch on television or in the movie theater. In fact, in the newer *Battlestar Galactica* television series, Mitochondrial Eve was created by the Cylons and descended from the Cylons. At least they realized that she had to have been created; and conveniently, she was genetically compatible with human males.

The whole theory of evolution from beginning to end is science fiction, because with science fiction you can have any result you want simply by a little *ad hoc just-so story-telling*.

Remember, Mitochondrial Eve is absolutely worthless without a carefully designed genetically compatible Mr. Mutant Adam to go along with her. Sexually reproducing species CANNOT evolve naturally through natural selection and random mutations, because mutation/selection would have to magically and knowingly design and create compatible Mr. and Mrs. Mutants every step of the way throughout the whole process over trillions of years of time. A sexually reproducing species will go extinct trillions of times over, before ever once magically producing a compatible Mr. and Mrs. Mutant simultaneously out of thin air by blind random chance. That's the reality of the science!

John Pinel ends his section on the "evolution of mate bonding" with this statement and question:

> **Critical Thinking: Think about how the information presented in this section on the evolution of mate bonding might relate to events that you have experienced or observed in your daily life. Has your newly acquired evolutionary perspective enabled you to think about these events in new ways?** (*Biopsychology*, p. 34.)

Yes, I guess so.

I realized for a fact that mate bonding could never have evolved, because evolution (mutation/selection) would have had to have magically designed and created a compatible Mr. and Mrs. Mutant every step of the way throughout billions of years of time, in order for evolution (genetic drift) to have had any influence on mate bonding whatsoever. Evolution is science fiction – an ad hoc just-so story and nothing more.

According to the evolutionary perspective, homosexuality shouldn't exist. Homosexuality should have been selected against and eliminated billions of years ago, because homosexuality has no survival value where a species as a whole is concerned. The fact that homosexuality does indeed exist tells me that there is something seriously wrong

144

with the evolutionary perspective. There's more going on there than just evolution and natural selection.

It's clear to me that the theory of evolution is stillborn, because it cannot perform as advertised. The theory of evolution is nothing but science fiction masquerading as fact. This is GOOD SCIENCE and good stuff that I have written here, but only if you are not worshipfully adoring evolution right now. I KNOW that the Theory of Evolution is false, because I KNOW why it is false, and I can explain it to you. That's something I didn't have a few years ago, but now I do. I wish I had all of this fifty years ago, when I first became aware of this subject.

An evolutionary perspective is absolutely worthless, because evolution (genetic drift), natural selection, and random mutations cannot do the things which the Darwinists and Naturalists say that these things can do. Natural selection and random mutations cannot do quantum mechanics, quantum telepathy or quantum telekinesis or quantum tunneling, non-locality, science and scientific experimentation, psyche or intelligence or non-local consciousness, design, programming, genetic engineering, planning, thinking, teleology, genomes, physical brains, physical bodies, eyes, or anything else of value.

If you are going to burn down the house, you are obligated to put something better in its place – enter Quantum Mechanics, Non-Local Consciousness, and Quantum Neuroscience! We're getting there.

6. We observe Evolution in real-time.

Confound: to mix up something with something else so that the individual elements become difficult or impossible to distinguish from one another.

The Premier Confound in science is micro-evolution and macro-evolution. They confound and conflate these two all the time, and I believe that many of these people don't even realize that they are doing so. I have been a scientist all of my life. As a scientist, I went for thirty years of my life or more without knowing that there's a significant difference between micro-evolution and macro-evolution. I have also observed that the Materialists, Naturalist, Darwinists, Nihilists, and Atheists don't want us to discover that there is a difference between micro-evolution and macro-evolution, because such a discovery helps us to discover why the Theory of Evolution is false.

Circumstantial evidence is evidence that relies on an inference to connect it to a conclusion of fact – like a fingerprint at the scene of a crime. By contrast, direct evidence supports the truth of an assertion directly, without need for any additional evidence or inference.

Remember, scientific inferences are circumstantial evidence. Scientific inferences are logic fallacies. Furthermore, there is no direct evidence and will never be any direct evidence supporting the major premises or primary assumptions of Materialism, Naturalism, Darwinism, Nihilism, and Atheism because it's physically impossible to prove that Psyche, or God, or the Non-Local does not exist. These falsified materialistic philosophies are pseudo-sciences at best and downright fraudulent hoaxes at worst.

Darwinism or the Theory of Evolution is based exclusively on *scientific inferences* and *circumstantial evidence*, which are logic fallacies. Charles Darwin put together a large body of circumstantial evidence; and, the Materialists and Naturalists have been treating it as empirical evidence and scientific fact ever since. That's not science – that's blind-faith religious dogmatism instead.

Darwin was not the first to suggest that species evolve (undergo gradual orderly change) from preexisting species, but he was the first to amass a large body of supporting [circumstantial] evidence and the first to suggest how evolution occurs.

Darwin presented three kinds of evidence to support his assertion that species evolve:

1. He documented evolution of fossil records.

2. He described striking structural similarities among living species, which suggested that they had evolved from common ancestors.

3. He pointed to the major changes that had been brought about in domestic plants and animals by programs of selective breeding.

However, the most convincing evidence of evolution comes from direct observations of rapid evolution in progress. For example, Grant observed evolution of the finches of the Galapagos Islands – a population studied by Darwin himself – after only a single season of drought. (*Biopsychology*, p. 25.)

This one from John Pinel is a confound, or a sleight, or a dodge – an attempt at trickery and deception, a bait and switch – because its truthfulness depends upon your chosen definition for evolution.

Remember, NO macro-evolution has taken place, because they are still finches!

This one here is a confound, a *category error* logic fallacy, where he erroneously equates or confounds macro-evolution with micro-evolution.

Macro-evolution is when two finches mate and produce genetically compatible male and female cats. That's macro-evolution, the kind of evolution that John Pinel is suggesting is taking place among the finches of the Galapagos. But, macro-evolution is nothing but a fictional story. It is science fiction, and nothing else, because Macro-Evolution, Spontaneous Generation, Abiogenesis, Materialism, Naturalism, and Darwinism were FALSIFIED in 1859 by Louis Pasteur. Macro-evolution didn't happen, because it can't happen. It's impossible.

These people are deliberately confounding micro-evolution with macro-evolution, presenting them as if they are the same exact thing, which they are not. This is literally the bread and butter of the theory of evolution, and how they successfully trick you into believing in the theory of evolution. It's magic and sleight-of-hand. Now you see it; now you don't. They do it because it works. Their ultimate goal is to trick you into believing that the theory of evolution is true. John Pinel and my college professor fell for their tricks, deceptions, and lies. There was a time when I fell for them too; but, I have since learned how to see through their deceptions to the truths that they are trying to hide from us.

1. The fossil records don't evolve, can't evolve. Fossils don't evolve either, because they are dead already. Only the Materialists and Naturalists believe that dead physical matter can somehow evolve, because these people have chosen to believe in Creation Ex Nihilo. These people actually anthropomorphize and deify evolution and fossil records. John Pinel literally said that Darwin had documented the "evolution of fossil records". After all, evolution can do anything it sets its mind to, including the evolution of fossil records. This argument is fundamentally flawed, because it doesn't make any sense to begin with. Fossil records can't evolve or change without some creative interference from human beings. Yes, Darwin's Tree of Life does exist, and it has indeed evolved or changed over

time; but, Darwin's Tree of Life is creative fiction – science fiction – and has nothing to do with Abiogenesis, Macro-Evolution, or Spontaneous Generation, which have been falsified by science and the collective observations of the human race.

2. Striking structural similarities (homology) is evidence of a Common Designer, not a common ancestor, because evolution of any kind cannot do design and creation and there is NO such thing as macro-evolution or spontaneous generation. This one here from John Pinel is a *bait and switch* and a *category error*, as well as *begging the question* and *affirming the consequent*. Remember, analogous structures and homologous structures are evidence for a Common Designer, because evolution (genetic drift), natural selection, and random mutations cannot do design and creation.

3. NO major changes, NO macro-evolution, ever takes place through selective breeding. The dogs are still dogs hundreds of years later, and millions of years later, and billions of years later, and trillions of years later. This one here is simply a lie.

Those finches are still finches, and they will always be finches. You can go look at them right now or a trillion years from now, and they will still be finches. There is NO such thing as Macro-Evolution. It doesn't exist, and it never will, because it's impossible.

But, evolution has been caught in the act!

This claim is **true** if we define evolution as micro-evolution – random mutations and natural selection. The effects of random mutations, natural selection, and selective breeding are indeed observed and experienced in real-time here and now. Selective breeding actually requires intelligence and intelligent intervention in order to produce results. The ability to adapt to different environments was in fact built into the genomes of animals by God; so of course, the finches adapt to seasons of drought, and then they have been observed to adjust back to seasons of wet and plenty. They are still the same species. They are still finches; and, they have been for the past 150 years ever since Darwin first observed them. There has been NO macro-evolution taking place during that whole time.

But, evolution has been caught in the act!

This claim is **false** if we define evolution as macro-evolution – abiogenesis, creation ex nihilo, chemical evolution, or spontaneous generation. Remember, there is NO such thing as creation ex nihilo, or spontaneous generation, or abiogenesis. God had to design and create and deploy the genomes and the living cells in the first place BEFORE there could be any natural selection, or random mutations, or evolution (genetic change).

Those who make this claim, that we observe evolution in real-time, are in fact confounding or equating micro-evolution with macro-evolution – a logic fallacy which is called a *category error*, because these things are NOT the same thing at all.

Macroevolution, spontaneous generation, chemical evolution, and abiogenesis have NEVER been observed in the wild nor in the lab. These things simply don't exist, because they have never been observed, and they can't be reproduced nor replicated. According to the Materialists and Naturalists, replication, or repeated observations, is the hallmark of TRUE SCIENCE or REAL SCIENCE. Macroevolution FAILS in this regard, just like it fails everywhere else. Macro-evolution has never been observed.

Everywhere else in science, the unobserved, unobservable, non-repeating, non-replicated, and non-replicable is automatically (and at times erroneously) considered FALSE; but, they make a *special exception* where the Theory of Evolution is concerned because it's their baby and they will never do anything to hurt it or harm it. Nevertheless, the Theory of Evolution is a prime example of the *special pleading* logic fallacy in action. These people go out of their way to make a *special exemption* for the Theory of Evolution so that it doesn't have to obey the rules of science like everything else does.

These people are hypocrites. They demand that everything else follow the rules of science, but always make an exception for the Theory of Evolution.

Why do they do this?

It's because they are trying to trick you and deceive you into believing in the Theory of Evolution – it's their religion. I KNOW, because I used to be a Materialists, Nihilist, and Atheist; and at the time, I desperately wanted these things to be true.

When it comes to macro-evolution, you can run the science experiment to test it right here, right now.

I give you any pre-biotic soup of your choosing – preferably one that doesn't kill the organism that you have chosen to test. Create a closed system so that no contaminants can get into your experiment. Select any single celled living organism that you want and introduce it into your specially selected pre-biotic soup. Everything you need for life is already there in that single living cell that you have selected. Next, poke a big hole in that living cell so that the contents of that cell drain out into the pre-biotic soup and so that the contents of the soup enter into that formerly living cell.

Now, sit back and observe how long it takes for that cell to spontaneously regenerate and come back alive again. However long it takes for that cell to regenerate and come alive again is a measurement of how long it takes for abiogenesis or macro-evolution to do its trick.

Are you still waiting? You better stick a camera on it so that you don't miss it, while you are getting lunch.

Actually, this is not a fair test of macro-evolution, because macro-evolution or chemical evolution is in fact Creation Ex Nihilo – creation from nothing by nothing. After all, I have given you the pre-biotic soup of your choosing along with a single cell that is already alive; but, this science experiment will indeed give you a very good feel for how long it takes for things to spontaneously regenerate and how long it takes for abiogenesis or macro-evolution to do its thing.

Are you still waiting?

You will wait for all eternity, and nothing will ever happen. Instead, due to Brownian motion and sunlight, that formerly living cell will decompose over time and become an integral part of the pre-biotic soup that you had chosen to sustain it and keep it alive. Entropy will reign supreme. ONLY psyche or conscious intention can override entropy!

Remember, Louis Pasteur ran similar science experiments in 1859, which successfully falsified Spontaneous Generation, Abiogenesis, Macro-Evolution, Materialism, Naturalism, Darwinism, and Creation Ex Nihilo.

Remember, entropy or the second law of thermodynamics prevents macro-evolution, or chemical evolution, or abiogenesis, or spontaneous generation by stirring the pot and producing chaos instead. The ONLY way to get biological life is to design it and create it in the first place. There is no such thing as macro-evolution. It's nothing but science fiction.

7. Homology proves that Evolution created your brain.

Homology is.

Homology (similarity) exists; therefore, the Theory of Evolution is true.

That's the sum-total of their argument in support of the Theory of Evolution – it is based upon homology or similarities between species. That's it! They base their whole science on this one single scientific inference. Ironically, *scientific inference* is also a logic fallacy, and these people don't even realize it.

Your brain looks like an upgraded version of the chimpanzee brain, so obviously evolution created your brain from something like a chimpanzee's brain. It's self-evident. Nothing else needs to be said.

Actually, *self-evidence* is a logic fallacy, because self-evidence is synonymous with no evidence. It *begs the question*!

By their own admission, the Materialists, Naturalists, and Darwinists treat their scientific inferences as empirical evidence, especially if they can use their scientific inferences **after the fact** to make "predictions" which have already come true. It works, because you can prove any ad hoc story true through scientific inferences. With scientific inferences, you can have tailor-made evidence that fits your chosen conclusions perfectly; and, you don't even have to make any observations nor do any of those tedious science experiments either. Scientific inferences supporting the theory of evolution are a time-saver that's guaranteed to get you the grant money that you deserve. Try it, it works every time; and, it makes doing science a lot of fun!

Deciding whether a structural similarity is analogous or homologous requires careful analysis of the similarity. (*Biopsychology*, p. 31).

Actually, it requires nothing but "scientific inferences" or guesswork, and careful story-telling. Homology, convergent evolution, and analogy are in FACT the pinnacle of *ad hoc just-so story telling*! You can have any conclusion you want simply by making it up out of thin air.

"Analogous" refers to having a similar structure because of convergent evolution – a bird's wings and a bee's wings are analogous.

"Homologous" refers to having as similar structure because of a common evolutionary origin or a common evolutionary ancestor – the human arm and the bird's wing are homologous.

Clear as mud!

Somehow, evolution always magically accounts for all the similarities and all the differences between species. How does it do that? The dice are loaded, and the cards are stacked – methinks! Heads I win, tails you lose. They cheat in order to win.

Spandrels are non-adaptive characteristics that evolve because they are related to evolutionary changes that are adaptive. Spandrels are a fictional attempt to make evolution seem more efficacious than it really is. These people are imbuing evolution with the ability to do non-adaptive tasks. With spandrels, even the things that hinder survival of the fittest are pulled under the umbrella of evolution and given survival value. This is sophistry. Heads I win, tails you lose. Spandrels sounds like scoundrels.

The Darwinists and Evolutionists use homology as proof of a common ancestor and as proof of evolution. Don't fall for the ruse. Homology is not convincing evidence of a common ancestor. Homology could just as easily be used as "evidence" for a common Designer and Creator. Furthermore, analogous structures and convergent evolution actually FALSIFY the claims of homology and a common ancestry; but, they *cheat* and *conflate* the two by saying a priori that everything is done by evolution. They start with the pre-chosen conclusion that evolution did it all, and then they insert their pre-chosen conclusion into all of their arguments as evidentiary proof that their pre-chosen conclusion is true – this is the *begging the question* logic fallacy in action.

> **Structures that are similar because they have a common evolutionary origin are termed homologous; structures that are similar but do not have a common evolutionary origin are termed analogous. The similarities between analogous structures result from convergent evolution, the evolution in unrelated species of similar solutions to the same environmental demands.** (*Biopsychology*, p. 31).

Heads I win, tails you lose.

Notice how they conveniently convert "analogous similarities," which prove that the theory of evolution is false and prove that there is NO common ancestor, into "convergent evolution" which then magically proves that the theory of evolution is true. This is in fact the very best in *ad hoc just-so story-telling* ever created out of thin air, in my humble opinion. It doesn't get any better. Heads I win, tails you lose. Whether there is in fact a common ancestor or not, the theory of evolution still ends up being true, because both analogous structures and homologous structures are defined as proof of evolution, because by definition and fiat homologous structures prove common ancestry and analogous structures prove convergent evolution. Tricky, huh?

Notice how convenient this is. There is no way on earth to determine if a structure is homologous or simply analogous, because there is no way to tell the difference between the two from fossilized bones! So, the Darwinists and Evolutionists simply declare a priori, without any examination of evidence, that both homology and analogy prove that the theory of evolution is true. They design this thing right from the start so that they can't lose. That's the power of *ad hoc just-so story-telling* and *begging the question*. You can have any result you want simply by making up a convincing story. You can do the same thing with Creationism and Theism if you want to; but, it's Bad Science, no matter how it is employed or by whom.

Remember, evolution (genetic drift), chemical evolution, natural selection, and random mutation are limited to the physical level. Materialism, Naturalism, Darwinism, Nihilism, and Atheism can't touch the quantum level or the psyche level. All that these people can do is to say that the psyche level does not exist. That's not very scientific, is it? Neither is homology!

> **Before considering the mechanisms by which sex differences in the brain develop, it is important to understand the nature of the differences. Sex differences develop independently in different parts of the brain at different points in time and by different mechanisms. Different patterns of gene expression exist in the brains of male and female mice before the gonads become functional.** (*Biopsychology*, p. 325.)

Each sexually-reproducing species has two different brain designs – male and female. How did evolution do that, and why?

Evolution didn't do it! Evolution couldn't do it.

It's physically impossible for two chimp-like ancestors to mate and produce genetically compatible male and female humans, in the same place and at the same time. It can't be done, which means that it wasn't done. Evolution can't do complexity, specificity, foresight, design, engineering, manufacturing, and creation. Evolution can't create two different types of brains yet have them be genetically compatible at the same time. Evolution is entropy and random chaos. Evolution is the second law of thermodynamics. Evolution or random mutations destroy and disable, not design and develop. Evolution can't do design and creation. It's physically impossible.

According to the theory of evolution, there should be no homosexuality, which tells me clearly and conclusively that there's something seriously wrong with the theory of evolution. It doesn't work as advertised!

8. The human mind is the same thing as one's physical brain.

Over and over again throughout his book, John Pinel *confounds* and *conflates* the human psyche with the human brain. These are logic fallacies. He is of the opinion that every mental illness is caused by damage to the physical brain. There is no other explanation that he is willing to consider. He has already made up his mind on the subject.

Remember, John Pinel declares evolution to be the foundation of Biopsychology and Neuroscience. By axiom and fiat, John Pinel defines psyche and non-local consciousness out of existence.

When it comes to brain damage, he doesn't realize that the psyche is receiving messed-up input from the physical brain, and instead assumes as a matter of fact that one's psyche is his physical brain.

The point here is that although the changes in self-awareness displayed by the patient were very complex, they were clearly the result of brain damage: Indeed, the full range of human experience can be produced by manipulations of the brain. (*Biopsychology*, p. 22.)

Once again, he is *begging the question* and *jumping to conclusions*. In other words, he is making *scientific inferences*, a logic fallacy. He makes these conclusions in the complete absence of convincing observational evidence. John Pinel simply assumes the evidence into existence. Powerful, isn't it?

Actually, you can manipulate the brain all you want, and it will never produce your after-death Life Review. The brain has never been caught in the act of doing so. Wilder Penfield observed long ago that there's NO brain manipulation that can force you to choose one belief over another. Choice or Psyche doesn't happen within the brain or through brain manipulation. You can force someone to lift their hand against their will using brain manipulation (a physical function); but, you can't force someone to believe in God through brain manipulation (a psychic or spiritual function). I repeatedly get the impression that John Pinel is trying to deceive us. Everything he writes about evolution and the "non-existence of psyche" has been falsified already; but, I guess it's left up to someone like me to do it in print, since nobody else will.

From the beginning, a priori, John Pinel assumes the psyche or the mind out of existence, and then uses that assumption as proof that we humans don't have a psyche or a mind – the very definition of the *begging the question* logic fallacy. In his next proof of concept, John Pinel conflates or equates chimpanzees with humans – another trick in the evolutionists' arsenal. Imagine how much fun you can have and how many different scientific inferences you can generate simply by comparing chimpanzees with humans! The sky's the limit! Yet, *scientific inferences* are logic fallacies.

John Pinel compares chimpanzees with humans and then provides this summary:

The point of this case is that even nonhumans, which are assumed to have no mind, are capable of considerable psychological complexity – in this

case, self-awareness. Although their brains are less complex than the brains of humans, some species are capable of high levels of psychological complexity [without having a mind of their own to rely upon]. (*Biopsychology*, p. 23.)

The point is that this whole thing is an assumption based upon a nest of assumptions. This one here is actually a *straw-man* logic fallacy. He assumes that we assume that chimpanzees don't have a psyche or a mind – thereby constructing the straw man; and, then he uses those assumptions to demonstrate and prove that we humans don't have a psyche or a mind because chimpanzees are self-aware like human beings are. That's his whole point – to use chimpanzees to prove that humans don't have a psyche or a mind. But, it's nothing more than *circular reasoning*.

This whole argument from John Pinel is a *faulty assumption* from beginning to end. And to think, there are actually people in this world who buy into it, or similar things like it. They are quoting him online as if it's Gospel Truth. It's a *pyramid scheme* from beginning to end, and it treated as self-evident despite being logically false.

He even draws pictures on page 32, showing how the human brain evolved, starting with a bass, then a rat, then a cat, then a chimpanzee, and finally a human. It must be so, because the pictures say that it is so; after all, a picture is worth more than a thousand words. Who needs evidence when you can provide a drawing?

Again, this whole thing is based upon the *category error* logic fallacy, the *begging the question logic* fallacy, and the *ad hoc just-so story-telling* logic fallacy – select the conclusion that you want, and then draw the pictures to match.

Category errors take place whenever you try to equate and compare chimpanzees with humans, and then draw conclusions or scientific inferences from that comparison. The whole thing is a logic fallacy from beginning to end. These people are trying to trick you into believing that the theory of evolution is true and that the theory of evolution is God; but, it's a category error from the start, because Creation by Evolution doesn't belong to the category that we call "God" and doesn't even belong to the category that we call "truth", because evolution of any kind cannot design and create. Simple truth!

Creative Thinking: Intelligence is the product of the interaction of genes and experience, and it is dumb to try to find out how much comes from genes and how much comes from experience. The point of this metaphor, in case you have forgotten, is to illustrate why it is nonsensical to try to understand interactions between two factors by asking how much each factor contributes. (*Biopsychology*, p. 24.)

John Pinel is actually trying to suppress scientific inquiry; and, he's wrong to do so. But, that's exactly what the Materialists, Naturalists, Darwinists, Nihilists, and Atheists do – suppress scientific inquiry in an attempt to enforce your belief in the Theory of Evolution. In an act that's extremely unscientific, he is demanding that you not question the Theory of Evolution or how much it contributes to human behavior.

John Pinel and the Materialists also deliberately limit you to only two factors – nature and nurture – genes and environment. They ALL ignore the psychological factor or **choice factor** controlled by Psyche or Non-Local Consciousness, which is in fact the Third Factor in the BioPsychoSocial Model of mental illness and reality.

John Pinel erroneously states that intelligence is the product of genetic and environmental interaction. He is wrong. Actually, intelligence is the product of Psyche or Non-Local Consciousness, because your intelligence, life experiences, and memories go with you into the Non-Local Realm when your physical brain dies or goes offline, according to the empirical evidence from Near-Death Experiences (NDEs) and Out-of-Body Experiences (OBEs). Science 2.0 gives precedence to Observational Evidence over scientific inferences, which are logic fallacies.

Yes, the brain's fire-power is partially due to genetics and environmental influences; but even then, the psyche has to choose to exercise its brain, or the brain will become nothing. It's the Psyche who chooses to study and to practice. Nothing in your genes or environment can force you to practice your piano lessons if you don't want to.

Furthermore, it is NOT dumb to question how much comes from genes, how much comes from experience, and how much comes from psyche. The whole Science of Heritability and Twin Studies are based upon the percentage of a mental illness that can be traced to genetics – as wonderfully demonstrated in the much better college textbook, *Abnormal Psychology: An Integrative Approach*, by Barlow and Durand. Heritability is a statistical approximation, rather than dogma – an average or a mean. Once these scientists know the heritability of a mental illness, then they KNOW that the other part of the mental illness is due to PsychoSocial factors rather than genetics. It's actually very powerful science, because it's based upon observation and measurements of those observations. Heritability is infinitely better science than the Theory of Evolution and Materialism, which have no observational support but are scientific inferences instead.

> **Creative Thinking: Intelligence is the product of the interaction of genes and experience, and it is dumb to try to find out how much comes from genes and how much comes from experience. The point of this metaphor, in case you have forgotten, is to illustrate why it is nonsensical to try to understand interactions between two factors by asking how much each factor contributes.** (*Biopsychology*, p. 24.)

What's dumb is to label his statement as "creative thinking", because it's nothing of the sort. It's censorship. It's an *ad hominem logic* fallacy. My hope after reading the first couple of chapters of his book, which were an attempt to preach the Theory of Evolution, was that he would return to the Observational Science later on in his book. All of his logic fallacies were starting to get to me.

John Pinel is literally trying to suppress scientific inquiry. He doesn't want you to discover anything that might falsify Materialism, Naturalism, and the Theory of Evolution. That's not science – that's religion! He's trying to make you feel stupid for questioning the veracity of the Theory of Evolution; and, he comes up with some weird science as a result.

If you want Real Science, and not *ad-hoc just-so stories*, you are going to have to get rid of Materialism, Naturalism, Darwinism, and Scientific Inferences first; and then pursue some real Observational Evidence and Eye-Witness Evidence instead. You can't successfully build the truth upon something like Materialism and Naturalism and Darwinism, which have been repeatedly falsified by Observational Evidence or Scientific Evidence.

9. Evolutionary Psychology Is the Bee's Knees.

In his book, *Biopsychology*, John Pinel dedicates a few pages towards uniting Evolutionary Psychology with human behavior. After all, evolution can do everything, including psychology.

> **He states that social dominance and courtship displays are the product of evolution and the result of our evolutionary history** (*Biopsychology*, p. 26.)

He's wrong of course. These things are a function of psyche, not evolution, especially in the case of human beings or human psyches. We are not as heavily pre-programmed by instincts as the animals are. We human beings or human psyches can trump nature and nurture at will. I KNOW, because I have done so many times myself.

"Evolutionary Psychology" is absolutely worthless when it comes to the study of Psychology.

Why?

Well, think about it logically.

Psychology was originally defined as "the study of the human psyche."

Can you now see why "evolutionary psychology" is an oxymoron?

Evolutionary Psychology is a contradiction in terms.

Why?

It's because Evolutionists like John Pinel and Behaviorists in particular successfully redefined psychology as "the study of observed behavior and the study of brain functions"; and, the truth is that evolution can't touch nor change the human psyche or any psyche.

As one of the hidden assumptions upon which Materialism and Naturalism are based, Materialists, Naturalists, Darwinists, Nihilists, and Atheists publicly and openly teach that psyche or non-local consciousness does not exist. The primary assumption or major premise of Materialism, Naturalism, Darwinism, Nihilism, and Atheism is the conclusion that psyche, non-local consciousness, quantum non-locality, spirit matter, and the non-physical transdimensional realm DO NOT EXIST.

So, how on earth could evolutionary psychology ever be the "study of the human psyche" when the theory of evolution and Darwinism repeatedly tell us that the human psyche doesn't exist? It's impossible. Evolution (genetic drift), natural selection, and random mutations have absolutely nothing to teach us about the human psyche. They don't even touch the human psyche, because the human psyche was in existence long before genomes, DNA, evolution, natural selection, and random mutations were designed and created by PSYCHE or Non-Local Consciousness; and, your Psyche will be in existence long after your physical body and physical brain are dead and gone, according to the empirical evidence from Near-Death Experiences (NDEs), Out-of-Body Experiences (OBEs), and other types of Non-Local Experiences.

I have upgraded my science to Science 2.0. Under Science 2.0, observational evidence, experiential evidence, and eye-witness evidence is given precedence over everything else including scientific inferences. Science 2.0 defines science as observation, lived experiences, eye-witness accounts, knowledge, truth, and fact. Under that definition for science, Materialism and Naturalism are actually unscientific, because Materialism and Naturalism and their derivatives have NO observational evidence supporting their primary assumptions or major premises. There is no way to observe that something doesn't exist. Science 2.0 returns to the original definition for Psychology, which was originally defined as "the study of the human psyche." In fact, Science 2.0 is the study of the Lived Experiences of the Human Psyche – all of them, including our non-local experiences.

John Pinel is another "scientist" who apparently defines science as Materialism, Naturalism, Darwinism, Scientism, Nihilism, and Atheism.

John Pinel writes:

The evolutionary approach has been embraced by many psychologists. Indeed, a new field of psychology, termed evolutionary psychology, has coalesced around it. Evolutionary psychologists try to understand human behaviors through a consideration of the pressures that led to their evolution. Some of the most interesting and controversial work in this field has focused on questions of sex differences in mate bonding, questions you may be dealing with in your own life. (*Biopsychology*, p. 33.)

John Pinel literally fell for all their trickery, deceptions, and lies.

Evolutionary psychologists START with the conclusion that evolution designed and created us; and then, these people spend the rest of their careers making up *ad hoc just-so stories* to fit their personally chosen pre-conceived conclusion. This whole process is a logic fallacy that's called *begging the question* or *jumping to conclusions*. *Ad hoc just-so stories* and *scientific inferences* (scientific guesswork) are also logic fallacies. When it comes to evolutionary psychology, the whole field is based upon logic fallacies.

I find it highly ironic that they developed a whole field of psychology that's based upon something that has been falsified and something that doesn't even exist – Macroevolution or Spontaneous Generation. They call this highly excellent ultimate psychology, Evolutionary Psychology; and, they give this thing the explanatory powers of a god. Evolution can do anything it sets its mind to – I just know it.

I love Psychology – the study of the human psyche; but, I have learned to despise and hate Evolutionary Psychology, because it's a deceptive hoax and fraud from beginning to end. Harsh, I know; but, I have caught these people lying to me thousands of times in countless numbers of different ways. I hate being lied to. When it comes to Evolutionary Psychology, the whole thing leaves a bad taste in my mouth every time. It makes me want to hurl; but, there's nothing that I can do about it, except to expose the lie.

They tell us that evolution chose where to place our eyes on our head. They tell us that evolution put the eyes of predators on the front of their faces so that they can focus on their prey, and that evolution placed the eyes of prey on the side of their heads so that they can watch out for predators. There's no logical mechanism for evolution (genetic drift), random mutations, and natural selection to choose whether a species is going to be a predator or prey. That's a decision that God has to make, because evolution by definition in principle cannot make decisions and choices. It's a category error logic fallacy to say that evolution makes these kinds of choices for us, because evolution can't make choices. It's physically impossible. Repeatedly, the Materialists, Naturalists, and Behaviorists assure us that choice is physically impossible, which means that every choice is being made at the quantum level by Someone Psyche. Evolution could never produce such a thing as an eye, let alone decide where to place them on our head.

Of course, they want you to believe that evolution has a mechanism for choosing whether a species is going to be active during the day or at night. These people imbue evolution with all the capabilities of a God.

There's no way on earth that evolution (mutation/selection) can produce sexually-reproducing species, because it would have to produce a compatible Mr. Mutant and Mrs. Mutant successfully every single time that it "designs and creates a new genome" in order to keep the process going over long periods of time. Think about it. Think about it critically. Thinking about the evidence, thinking about the reality of the situation, simply lets you KNOW that there's no way on earth that evolution could have successfully created a compatible Mr. and Mrs. Mutant each and every time that it needed to do so. Only a God could do such a thing, and evolution is no god.

While introducing Evolutionary Psychology and after introducing Evolutionary Psychology, John Pinel presents some "ad hoc just-so stories" of his own based upon his pre-conceived conclusions. He's been completely brainwashed into believing in the Theory of Evolution, and he doesn't seem to know it. It's all about keeping up appearances.

The human brain appears to have evolved from the brains of our closest primate relatives. (*Biopsychology*, p. 33.)

This is impossible.

Why?

Think about it logically and critically.

First of all, we KNOW for a fact that evolution, mutation, and selection can't do brains because they have NO brains.

Furthermore, apes cannot breed and produce humans. Chimpanzees always produce chimpanzees with NO exception.

There's no way on earth that this claim can possibly be true, because there's NO observational evidence supporting it. Instead, ALL of the observational evidence, experiential evidence, empirical evidence, and scientific evidence is telling us clearly and conclusively that evolution (genetic drift), mutations, and natural selection cannot design and create brains.

If you read carefully and study the science and their claims, the theory of evolution is self-defeating. The Materialists, Naturalists, and Darwinists can't get their stories straight; and, the Theory of Evolution is a fictional story after all. Occasionally, these people will mess up and slip in KNOWN TRUTHS which completely falsify the Theory of Evolution.

The brain is a finely tuned electrochemical organ. (*Biopsychology*, p. 54.)

Finely tuned by whom? ALL fine-tuning requires an intelligent Fine Tuner! Evolution by definition in principle can't do fine-tuning. It's physically impossible! That reality and truth falsifies the theory of evolution. When it comes to the theory of evolution, these people don't think for themselves. They just parrot what they have been told in school by other Materialist, Naturalists, and Atheists who have never learned how to think about things logically and rationally.

John Pinel actually writes:

The members of a species can produce fertile offspring only by mating with members of the same species. (*Biopsychology*, p. 27.)

Think about what he is saying! He's actually writing the truth here, because what he says here in this sentence actually matches with the observational evidence and experiential evidence.

Have you caught it yet?

This truth by definition in principle completely eliminates Mutants, unless two genetically compatible mutants – Mr. Mutant and Mrs. Mutant – happen to miraculously be born at the same time in the same location over and over again throughout billions of years of history – a process which is both statistically and logically impossible. It would take a God to do such a thing reliably and consistently – to produce a compatible Mr. and Mrs.

Mutant each and every time in order to keep the process of evolution going for billions of years.

The truth is that evolution of new species from pre-existing species is impossible, because **"the members of a species can produce fertile offspring only by mating with members of the same species."** (*Biopsychology*, p. 27.)

Checkmate!

I just falsified the Theory of Evolution, which is the basis of John Pinel's religion. Science itself defeats the Theory of Evolution, as do logic and common sense. Once you falsify the Theory of Evolution, then you simply KNOW that evolutionary psychology is nothing but science fiction – an *ad hoc just-so story*, and nothing more.

But then after making this statement of fact and truth, John Pinel proceeds to **"piece together the evolutionary history of our species"**. There is NO evolutionary history for any species, because **"members of a species can produce fertile offspring only by mating with members of the same species"**. His very statement FALSIFIED the theory of evolution. An evolutionary history is nothing but a *scientific inference*, *a fictional story*, an *ad hoc just-so story*. An evolutionary history is based upon a wide variety of logic fallacies. It's nothing but a fictional story; and, anything derived therefrom, including evolutionary psychology, is fruit from the poisoned tree.

I'm not going to apologize for catching these people in their trickery, deceptions, and lies. I'm a scientist, and not an apologist. The Theory of Evolution is Bad Science, because it's based exclusively on logic fallacies such as *scientific inferences*, and *ad hoc just-so stories*, and outright lies.

Over and over again, the Materialists, Naturalists, Darwinists, Nihilists, and Atheists demand that everything else conform to the rules of science, but they always make a *special pleading* or *demand an exemption* where their theories are concerned. In other words, these people want you thinking critically about everything else except the Theory of Evolution, Materialism, Naturalism, Nihilism, and Atheism; but when it comes to those things, these people want you dumb, blind, gullible, silent, censored, and contained. While interacting with them online, I discovered that these people are a very demanding bunch. These people think critically and skeptically about everything, except for Materialism and the Theory of Evolution, which they eagerly accept on blind faith alone. They believe it because they desperately want it to be true. I KNOW, because I used to be a Materialist, Nihilist, and Atheist; and, I desperately wanted these things to be true at the time. I was wrong.

Remember, genes DO NOT CONTROL our chosen behaviors. Our psyche does. Genes do influence the functionality and the health (or sickness) of the organism, though. People in our environment can try to force us to do the things they want us to do; but, that's force, not choice. Choice and chosen behaviors are a function and a product of the Human Psyche. The Human Psyche can override its Nature and Nurture at will. Psyche is the ultimate causal agent. Before there were genes and an environment, there was Psyche; otherwise, genes and our environment would not exist.

All throughout his book, *Biopsychology*, John Pinel promotes "critical thinking". Over and over again, he tells us to **think critically**. That's precisely what I do when it comes to the Theory of Evolution. I'm brutal and merciless when it comes to this thing, because I KNOW FOR A FACT that collectively these people are trying to trick us, deceive us, and brainwash us into believing in the Theory of Evolution. I won't stand for it; and, I won't fall for it – not anymore, because I know better now. Fool me once, shame on you. Fool me twice, shame on me.

Now all they can do to me is to stir me up enough to write a critical response, as I have done here. I have gotten very good at critiquing and debunking the Theory of Evolution. I did most of this from memory and from my notes in the margins of his book, which provided the talking points. The only thing I had to look up were the quotes from his book.

How did I do? Did I kill a sacred cow or two, or do you still stubbornly believe in the Theory of Evolution hoping beyond all hope that it might someday be proven true? There's no helping some of us. I KNOW because I used to be a Materialist, Nihilist, and Atheist, so I know how it can go. I know that some of you are thinking that I have been woefully deceived.

ALL of the magic is attributed to Evolution instead of God, but they are the same person in the end. They have to be, or we wouldn't exist.

10. The End of Materialism, Naturalism, and Darwinism.

Okay, it's not actually true that Materialism, Darwinism, and Naturalism are at an end, because there are still millions of people who are eager to fall for their trickery, deceptions, and lies. So, the only thing that I can safely and accurately say is that My Materialism, My Nihilism, and My Atheism are at an end because I no longer believe in these things. I know better now, because I know why they are false and how to falsify them.

The exclusion of evidence is unscientific; but, that's precisely the way that the Materialists, Naturalists, Darwinists, Nihilists, and Atheists do science – by deliberately excluding evidence. Naturalism and Atheism of any kind are based upon a refusal to look at evidence. In the summer of 2017, I upgraded my science to Science 2.0, which allows ALL of the evidence into evidence and pursues a preponderance of the evidence, rather than excluding evidence as Materialism and Naturalism deliberately do.

I have had many Darwinists, Materialists, Naturalists, and Atheists tell me that I don't understand the Theory of Evolution – meaning or implying that I simply do not understand the science nor the magical creative powers of genetic change, random mutations, and natural selection. Their complaint is that I can't see the magic nor the beauty of it all, or how perfectly it works. As I see it, my problem is that I now understand a bit too much – especially when it comes to the science and what that science (observations and experiences of the human race) really means and is trying to teach us. I looked behind the curtain and within the black box, and I didn't like what I saw and found.

I'm a bit too educated to believe in the theory of evolution anymore. I can't do it anymore, because I KNOW why it is false and I KNOW how to falsify it. I understand the Philosophy of Science too much, now; so, I easily catch these people in their logic fallacies, deceptions, and lies. They say that ignorance is bliss; but now, I have too much knowledge to draw upon, and my ignorance is gone. You can't maintain a belief in something that you KNOW is false, especially when you know why it is false. Knowledge kills blind faith. That's just the way it is.

I'm no longer sitting in the dark waiting for the Materialists, Naturalists, and Darwinists to explain all of this to me. I decided to take the bull by the horns and figure it out for myself. I'm glad I did. Eventually I realized that the Materialists, Naturalists, and Darwinists are lying to us, and deliberately trying to deceive us. Ouch!

When it comes to neuroscience, the second greatest fiction and myth that we have received from the Materialists, Darwinists, and Naturalists is that our brains are wired together like some kind of CPU or RAM.

The dendrites in our brain transmit neurotransmitters which produce post-synaptic potentials (PSPs). The axons in our brain transmit action potentials. The relative refractory period is a period of time after the absolute refractory period during which a higher than normal amount of stimulation from the dendrites is necessary in order to fire the neuron or trigger another action potential.

Some neurons display Action Potentials that are longer, that have lower amplitude, or that involve multiple spikes. (*Biopsychology*, p. 80.)

They are ALL neurons, so who decides the length, amplitude, and spikiness of an Action Potential within each and every neuron? Granted, this greatly increases the diversity, frequency, versatility, and complexity of Action Potentials; but, who designed and implemented this functionality in the first place and why?

In this case, a "potential" is an electrical current comprised of ions, or electrically charged atoms. The Materialists and Naturalists call dendrites and axons "wires". They do so in order to give you the erroneous impression that your whole brain is wired together into a single functioning whole like a CPU or RAM, which technically isn't true at all. To counteract this deception, I will occasionally make the statement that there are NO wires in your brain. They, of course, will disagree with this statement and point you directly to dendrites and axons in order to counter that statement. However, thanks to the synaptic clefts and synapses that separate every neuron and astrocyte, each neuron is functionally and electrically separated from every other neuron and astrocyte by synapses, which effectively means that there are NO wires in your brain. In other words, your brain is NOT wired together into a single CPU or RAM like the Materialists and Naturalists claim that it is. Technically, your brain is NOT comprised of circuits like these people claim that it is. A wire is a single, usually cylindrical, flexible strand or rod of metal. Wires are used to bear mechanical loads or electricity and telecommunications signals. There's NOTHING like this within a physical brain. There are NO wires in your brain.

There are two types of synapses in our brain – electrical synapses and chemical synapses. The electrical synapses give the erroneous impression that the neurons in our brain are wired together like some kind of CPU or RAM. There are NO wires in our brains. Alas, electrical synapses are just like chemical synapses in their functionality. They both let through ions which can influence a neuron's Excitatory Post-Synaptic Potentials (EPSPs) and Inhibitory Post-Synaptic Potentials (IPSPs). Everything that comes into a neuron through chemical synapses and electrical synapses integrates directly into a single BIT of information, a Post-Synaptic Potential. Even the effects of G-Proteins end up affecting only that single BIT of information – the Post-Synaptic Potential. Whether a neuron is fed by thousands of chemical synapses or hundreds of electrical synapses, ALL of that input gets reduced and integrated into a single BIT of information – the Post-Synaptic Potential or Membrane Potential of the neuron.

With all of these different potentials, it gives Nature's Psyche plenty of locations where it can interact with and interfere with our physical brain on our behalf. Remember, it takes Nature's Psyche to get the FIRST action potential started in a chain of action potentials. Telepathy from the Human Psyche to Nature's Psyche is instantaneous. Nature's Psyche has no physical limitations; but, the Human Psyche seems to be limited by its assigned body and brain.

Pinel and others claim that the graded EPSPs and IPSPs are conducted passively, instantly, and decrementally to the axon initial segment. He says the same thing about conduction in myelinated axons. Technically, the "instantly" part is physically impossible. There's a bit of delay. It's not instantaneous because it decrements or degrades along the way. The only way that instantaneous conduction would be possible is if it is quantum and telepathic – in other words, if it has Non-Local Psyche as its cause. Quantum mechanisms are indeed instantaneous; but, the same reality does not apply to physical mechanisms. In the physical realm, God deliberately slowed everything down for us so that we could choose it, experience it, learn from it, use it, stop it, and actually remember having done so.

Oh, it's happening, alright. Our brains really are processing data, running computer code, crunching numbers, transferring data, building cities, weighing decisions, and storing memories. It's just not happening at the physical level. It's all happening at the quantum level. The physical input of sensory data, and the physical output of motor commands, is the only thing happening at the physical level within the brain. It all happens at the end of an axon, where the switch is either turned ON or OFF. A synapse is a switch. Our brain is comprised of a 100 trillion switches or synapses. Our brain is also comprised of a 100 billion stand-alone switches or neurons, which are either ON or OFF. Furthermore, our brains are comprised of quadrillions of ion channels that are either OPEN or CLOSED. Everything within our brains reduces to a single BIT. There's nothing else there within our brain. Everything else is happening at the quantum level. All the important and interesting stuff is happening at the quantum level, outside our conscious awareness.

Evolution or genetic change can't do the quantum level. Evolution can only function on the physical level, which makes evolution worthless when it comes to Psychology or the study of the Human Psyche.

But, we are going to worship evolution anyway.

John Pinel dedicates a few pages of his book, *Biopsychology*, to human evolution and our evolutionary history.

> **The actual ape ancestor of humans is likely long extinct. It is difficult to reconstruct the events of human evolution because the evidence is so sparse. Only a few partial hominin fossils dating from the critical period have been discovered.** (*Biopsychology*, p. 29).

The actual ape ancestor of humans likely never existed.

But, as proof of concept, he then shows a picture of the skull of Australopithecus girl, one of our putative ancestors. She's a monkey! Any idiot can see that she's a monkey!

> Remember, **"The members of a species can produce fertile offspring only by mating with members of the same species."** (*Biopsychology*, p. 27.)

There's no way that Australopithecus girl could be our ancestor, because she's a completely different species. Any fool can see that she's a monkey, just by looking at her skull.

Nevertheless, he emphatically states:

> **The first Homo species are thought to have evolved from one species of Australopithecus about 2 million years ago.** (p. 30.)

These people extrapolate the whole of evolutionary theory from a few partial fragments of bone – this is the pinnacle of *ad hoc just-so story telling*! It doesn't get any better than this, and it makes science a lot of fun according to John Pinel. This is story telling at its very best. You can have proof for any conclusion you want simply by adjusting your story to fit your pre-chosen conclusions – a process which is called *begging the question*, a logic fallacy. The theory of evolution is the pinnacle of science fiction.

It's all fiction!

Scientific inference. The logical process by which observable events are used to infer the properties of unobservable events. (p. 494.)

Scientific inferences are the SOURCE of all our science fiction! It's a Black Box. You can put anything you want into a Black Box, and it works. He actually states that he has chosen to use scientific inferences as empirical evidence – this is a logic fallacy, that's called *begging the question*. Begging the question is taking your pre-chosen conclusions or scientific inferences and then choosing to use them as empirical evidence. These people conclude a priori that the theory of evolution is true; and then, they use their pre-chosen conclusions or scientific inferences as evidentiary proof that the theory of evolution is true. It's *circular reasoning* or *begging the question*; but, that's how they prove that the theory of evolution is true, because there is no other way to do so. It's nasty, it's deceptive, and it's cheating; but it doesn't matter, so long as you prove that the theory of evolution is true.

The advantages of life on land were so great that natural selection transformed the fins and gills of bony fishes to legs and lungs respectively – and so it was that the first amphibians evolved about 400 million years ago. (*Biopsychology*, p. 28.)

And so it was! Evolution said, "Let there be legs and lungs", and it was so.

So let it be written, so let it be done.

This is a classical demonstration of the *ad hoc just-so story-telling* logic fallacy in action. It doesn't get any better than this!

This claim is impossible and false, because natural selection can't do that kind of design and creation thanks to genetic entropy.

He also gets his facts wrong!

Natural selection cannot transform anything! In fact, random mutations are the "creative genius" behind the theory of evolution, not natural selection. Natural selection is synonymous with luck – it's simply the result of environmental interactions, and NOT an intelligent designer, creator, and transformer. Remember, natural selection cannot transform anything.

The truth is that natural selection and random mutations are a function of genetic entropy and extinction, not a cause of genetic programming and new genomes.

[For the Real Science, see: Sanford, J. (2014). *Genetic Entropy* (4th ed.). Cornell University: FMS Foundation.]

We humans have little reason to claim evolutionary supremacy. (p. 30.)

Assuming facts not in evidence.

Rapid evolutionary changes (i.e., in a few generations) can be triggered by sudden changes in the environment or by adaptive genetic mutations. Whether human evolution occurred gradually or suddenly is still a matter of intense debate among paleontologists. (p. 30.)

This one here is *begging the question* or *jumping to conclusions*. He is assuming a priori that fossils prove evolution, when in fact, the fossil record actually falsifies Darwinism, Darwin's Tree of Life, and the Theory of Evolution.

This one is also a *hidden confound* or *bait-and-switch*, wherein he is ignorantly equating micro-evolution with macro-evolution.

Evolution does not progress to preordained perfection – evolution is a tinkerer, not an architect. (p. 30.)

This one is an *ad hoc just-so story* – he's literally telling evolution what it can and cannot do. It is science fiction, and nothing more.

This is also a *category error* logic fallacy. He's equating or categorizing evolution with tinkerers, architects, designers, manufacturers, engineers, designers, and creators. He's literally anthropomorphizing evolution.

Over and over again, throughout his whole presentation, John Pinel anthropomorphizes, personifies, and deifies evolution – giving it all the powers, capabilities, intelligence, skills, and foreknowledge of a God. Clearly, Evolution is his god. He worships the thing.

This is simply unbelievable!

By his own admission, he gets all of this from *scientific inferences*, which are in fact logic fallacies. He bases his whole religion – his whole belief system – on *scientific inferences*, which are by definition in principle things that have never been observed and can never be observed. The whole thing is extremely unscientific, to say the least, because it's based upon the complete lack of observational evidence.

Ironically, John Pinel sabotages his whole presentation at the very end of his presentation on evolution, without even realizing it.

It is easy to speculate about how particular human behaviors evolved without ever having one's theories disproved because it is not possible to know for sure how an existing behavior evolved. Good theories of behavioral evolution have predictions about current behaviors built into them so that the predictions – and thus the theory – can be tested. Theories that cannot be tested have little use. (*Biopsychology*, p. 34.)

John Pinel then presents some *ad-hoc just-so stories* built to fit his *pre-conceptions* and designed to prove that the theory of evolution makes predictions that have since been proven true. In other words, the evidence is manufactured out of thin air and *built to fit* his pre-conceived conclusions. It is trickery and magic from beginning to end. John Pinel literally puts the cart before the horse in yet another logic fallacy.

Using natural selection and random mutations alone to produce genomes and brains from raw physical matter cannot be tested, because mutation/selection cannot produce brains and genomes from scratch. In fact, evolution (genetic change), random mutations, and natural selection did NOT exist until AFTER God designed, created, and produced the first genomes and the first life forms in the first place. The theory of evolution cannot be tested or verified, which means that it's useless and worthless according to Pinel.

So, there's no way to test **"Creation Ex Nihilo by means of Natural Selection and Random Mutations"**, because mutation and selection didn't exist until after God created the genomes and the genetic drift (evolutionary processes) in the first place. Theories that cannot be tested are simply worthless. In contrast, every single day we test, replicate, and prove Design and Creation by Intelligent Human Beings. The creation of *ad-hoc just-so stories* from thin air PROVES design and creation by intelligent beings.

Everything around us PROVES Design and Creation by Intelligent Beings. It's unavoidable. The existence of the theory of evolution and its associated *just-so ad hoc stories*, along with my rebuttals and debunking, prove the veracity and the existence of Design and Creation by Intelligent Beings. If you can read this, you too are one of those Intelligent Beings – Intelligent Human Psyches – who can design and create practically anything that they set their minds to. It is a *category error* logic fallacy to place evolution (genetic change), random mutations, and natural selection into the same category as Intelligent Human Psyches or Intelligent Beings; but, that's exactly what the Materialists, Naturalists, Darwinists, Nihilists, and Atheists DO ALL THE TIME. These people actually anthropomorphize and deify evolution, natural selection, and random mutations imbuing these lifeless concepts with the powers of a God.

In short, "theories that cannot be tested have little use"; and, this reality applies specifically to "ex nihilo design and creation by mutation/selection", because this is a theory that can never be tested, because there is no such thing as spontaneous generation, and because mutation/selection cannot design and create.

Remember, Materialism, Naturalism, Darwinism, Macro-Evolution, Abiogenesis, and Spontaneous Generation were proven false (falsified) in 1859 by Louis Pasteur, the very same year that Charles Darwin gave us "On the Origin of Species". We had the truth to begin with, until the Materialists and Naturalists tried to hide it from us.

Furthermore, the book *Genetic Entropy* by John Sanford successfully FALSIFIES design and creation by Random Mutations and Natural Selection.

Sanford, J. (2014). *Genetic Entropy* (4th ed.). Cornell University: FMS Foundation.

Remember, the theory of evolution is nothing but science fiction.

John Pinel tells us that theories – like the theory of evolution and creation by mutation/selection – which cannot be tested, verified, replicated, nor observed have little value or use to us. For once, he is right. The theory of evolution, and in particular evolutionary psychology, are absolutely worthless when it comes to the study of the Human Psyche. Evolution or genetic drift can't even touch the Human Psyche! Remember that.

By preaching, teaching, and worshipping the Theory of Evolution, John Pinel seriously damaged his credibility – it called into question all of his other personal interpretations and scientific inferences as well. As a result, I don't trust him; and, I question everything that he says or writes. By relying upon scientific inferences and treating scientific inferences as empirical evidence, it made most everything he wrote in his first two chapters of *Biopsychology* easy to debunk and demolish. Such is the nature of Scientific Inference – it's a logic fallacy and easily defeated.

Anytime the Materialists, Naturalists, Darwinists, and Atheists use the word "evolution" – such as "evolution produced your brain, eyes, proteins, and genome" – you can easily substitute the word "God" for the word "evolution", and the sentence will still make perfect sense.

Why?

It's because the Materialists and Naturalists imbue evolution with all of the foresight, teleology, powers, intelligence, and abilities of a God. Evolution is their God.

This might just end up being the most important chapter in this book for you.

As a former Materialist, Nihilist, and Atheist, I finally realized one day that if I continue to talk myself out of everything, then eventually I'm going to end up with nothing.

That's what the Materialists, Naturalists, Darwinists, Nihilists, and Atheists end up with – NOTHING! Nothing from nothing leaves nothing. Atheism is Creation by Nothing from Nothing. Technically, atheism is belief in nothing; and, it therefore produces nothing.

I can assure you that I know what I'm talking about, until I'm blue in the face; and, it won't make any difference. You, your psyche, is the only one, whom you are going to believe in the end. You are going to have to decide for yourself whether I know what I'm talking about, or not. That's something which nobody else can do for you.

I continue my critique of John Pinel's "evolutionary perspective" in another book that I have written, entitled "BioPsychoSocial: Including Psyche or Light into Our Theoretical Models".

https://www.amazon.com/dp/B0713NDHVW

On page 426 of *Biopsychology*, Pinel introduces the BioPsychoSocial Model without realizing that he is doing so. As long as you refuse to let the Materialists, Naturalists, Darwinists, Behaviorists, and Atheists get their hands on the BioPsychoSocial Model and destroy it, the BioPsychoSocial Model successfully FALSIFIES an evolutionary perspective and the theory of evolution.

Barlow, D. H. & Durand, V. M. (2015). *Abnormal Psychology: An Integrative Approach* (7th ed.). Stamford, CT: Cengage Learning.

The first parts of John Pinel's book, *Biopsychology*, is focused on Neuroscience. Those are the chapters that I covered in this book, *Quantum Neuroscience*. The last part of Pinel's book is focused on Abnormal Psychology and Mental Illness. When it comes to Abnormal Psychology, Barlow and Durand's book is vastly superior to Pinel's book.

Why?

Abnormal Psychology: An Integrative Approach uses the BioPsychoSocial Model to explain Abnormal Psychology.

Biopsychology uses the Evolution Model to explain Abnormal Psychology.

The BioPsychoSocial Model is infinitely superior to the Evolution Model, because the BioPsychoSocial Model has a lot more explanatory power.

My book about the BioPsychoSocial Model seems to be the correct place to continue this critique of Pinel's evolutionary perspective.

So, why do I bother to mention Pinel's book if there's something infinitely superior that we could be learning from?

It's simple. I upgraded my science to Science 2.0, which allows all of the evidence into evidence and pursues a preponderance of the evidence. Pinel's book presents the foil. You have to allow the opposition to present their evidence; and, I allow Pinel to do so.

According to Science 2.0, the best way to find and know the truth is to live it and experience it for yourself, or to choose to trust someone who has. The second-best way to find and know the truth is to eliminate everything that is false so that only the truth remains.

I allow Pinel to present everything that is false about modern-day Neuroscience; then, I proceed to eliminate it, debunk it, critique it, or falsify it so that ONLY Quantum Neuroscience remains. There is method to my madness.

Redeemed by Truth

Pinel, J. (2014). *Biopsychology* (9th ed.). New York: Pearson.

Throughout his book, John Pinel encourages us to "think critically," which is extremely easy to do whenever one is dealing with scientific inferences, because scientific inferences are extremely easy to debunk with observational evidence and experiential evidence. It doesn't require any particular genius to be able to demolish *scientific inferences*, because they are logic fallacies to begin with. I can also attack and destroy his *personal interpretations* of the research evidence, or scientific research. But, I can't attack the observational evidence, or the operational evidence derived from the research, unless they used shoddy methodologies, a process which can be nearly impossible to detect while reading through the summaries of scientific research.

At the end of chapter 2 of *Biopsychology*, John Pinel does redeem himself when he finally abandons scientific inferences and turns to actual scientific research instead. Most of the time, the scientific research speaks for itself; and, most of the time, I find the scientific research compelling and convincing. Scientific research is certainly much better than *scientific inferences* or *ad hoc just-so stories* concocted out of thin air, even though scientific research can definitely be interpreted incorrectly.

Every now and then the Evolutionists and Naturalists slip-up and actually reveal the truth to us.

Species. A group of organisms that is reproductively isolated from other organisms. The members of one species cannot produce fertile offspring by mating with members of another species.

The members of a species can produce fertile offspring only by mating with members of the same species. (*Biopsychology*, pp. 495, 27.)

He can't see it nor understand it; but, that definition for "species" FALSIFIED the theory of evolution! Down it goes in flames! Can you see it? We are trained in our public schools not to see it; but, the evidence is there nonetheless for anyone who is willing to look and see.

As part of his redemption, Pinel also introduces us to Epigenetics.

Our genomes were designed in such a way that our Chosen Behaviors and our Environmental Encounters can actually reprogram our DNA; and, some of these epigenetic changes can be passed on to the next generation.

Epigenetics is the study of stable heritable changes in gene function that do not involve changes in the DNA sequence.

https://en.wikipedia.org/wiki/Epigenetics

The former assumption that monozygotic twins are genetically identical was disproven [by epigenetics]; and, the common practice of referring to monozygotic twins as *identical twins* should be curtailed. (*Biopsychology*, p. 48.)

In his book, *Biopsychology*, John Pinel mentions the Histone Remodeling version of epigenetics, and the DNA Methylation version of epigenetics, which result from our chosen interactions with our environment. The scientists are discovering new types of epigenetics every day. There is no way in this universe that Random Mutations could have ever designed, created, and maintained such a complex system. We human beings can't do such a thing, so why should we ever believe that evolution did it? Evolution is supposed to be dumb and blind, which it is.

The Four-Dimensional Genome:

https://creation.com/four-dimensional-genome

https://www.youtube.com/watch?v=K3faN5fU6_Y

Remember, Experience is the result of our Chosen Behaviors and our Interaction with Our Environment. Experience is PsychoSocial in nature and origin.

> **Studies of the heritability of intelligence were conducted on middle-to-upper class families, and the heritability estimates for intelligence tended to be about .75.**

> **Turkheimer and colleagues assessed heritability of intelligence in two samples of 7-year-old twins: those from families of low socioeconomic status (SES) and those from families of middle-to-high SES.**

> **The heritability estimates for intelligence in the middle-to-high-SES twins was, as expected, about .70. However, the heritability estimate for intelligence in the twins from low-SES families was only .10.**

> **This effect was subsequently replicated and extended to other age groups: babies and adolescents.**

> **One major implication of this finding is that it forces one to think of intelligence as developing from the interaction of inheritance and experience, not from one or the other. It seems that one can inherit the potential to be of superior intelligence, but this potential is rarely realized in a poverty-stricken environment.**

> **Reducing poverty would permit many of the poor to develop their genetic potential.** (*Biopsychology*, p. 48.)

I find the Scientific Research infinitely more convincing than his scientific inferences, although this time around when it comes to this particular set of research studies, I came to essentially the same conclusions or personal interpretations that he did.

Do you understand what these research results mean?

This research clearly demonstrates that heritability estimates tend to run high and tend to be inflated, because heritability estimates can be substantially lowered by negative environmental influences. This is GOOD SCIENCE, because it's based upon evidentiary research or experimental research. We are no longer slogging through scientific inferences or guesswork here. This is good stuff – possibly my most favorite part of his whole book.

Environmental influences and our chosen behaviors have a much greater impact on our intelligence than our genes do. If heritability estimates (BIO) are as low as 10% in the twin studies of low-SES groups, then that means that chosen behaviors (PSYCHO) and environmental influences (SOCIAL) account for up to 90% of these twins' intelligence. Think about it. Up to 90% of these low-SES twins' intelligence is PsychoSocial in nature and origin. Furthermore, it means that evolution (genetic change and random mutations) can only account for up to 10% of these low-SES twins' intelligence, at most. Intelligence isn't a function of evolution (genetic drift), after all.

166

PsychoSocial Factors have a much greater influence on intelligence than the heritability estimates tend to indicate. Your intelligence can actually be increased when you – your psyche – choose to study and learn. Furthermore, your intelligence can be increased by being exposed to an enriched and enriching environment which gives you greater opportunities to learn.

This also means that the genetic component of intelligence can be as low as 10%, because environmental factors and psyche are not being taken into full consideration while developing heritability estimates. Genetics are given more credit than they deserve.

Even if the theory of evolution were absolutely true, which it is not, evolution could only realistically account for up to 10% of our behavior and intelligence, at most. Remember, evolution (genetic change and random mutations) only had a direct impact on you personally at the moment of your conception, and then evolution has been basically worthless and inactive ever since. The rest of your behavior is chosen into existence while you are interacting with your environment; and, even your genes are malleable through epigenetics, environmental influences, and chosen behaviors. You are NOT simply the sum total of your genes, as the Materialists and Naturalists claim – a reality to which the BioPsychoSocial Model and Research so adequately attests.

As a consequence of all of this, I'm probably being extremely generous by giving evolution up to 10% of the credit, because genetic change through random mutations has hopefully had much less influence in your life than that, because the vast majority of mutations are harmful, deleterious, dangerous, and even fatal. Evolution is not your friend; and, evolution is not your creator. Evolution is instead a source of disease, death, and extinction. If you can still read this, you can be thankful that natural selection hasn't found a way to eliminate you, yet. Hopefully, only 1% of your behavior can be attributed to evolution (genetic change and random mutations), because evolution isn't a good thing. Evolution and random mutations are synonymous with disorder, chaos, death, and extinction.

Therefore, whenever the so-called scientists try to tell you that evolution designed and created your genome and your physical brain, YOU KNOW that they are grossly overreaching, severely exaggerating the situation, and even downright lying to you.

Why?

It's because these people are trying to tell you that your genome and your physical brain is 100% determined by evolution and your evolutionary history; whereas, all of the scientific research and common sense logic tells you that NONE of your genome is the result of evolution (natural selection and random mutations) and only a small fraction of your physical brain can be the result of evolution or genetic drift – hopefully a very small part, otherwise, you are going to end up with a lot of brain damage as a result.

If you are a true scientist, you don't want to be caught worshipping genetics and evolution, because these things have nowhere near the amount of power and influence that the Materialists and Naturalists and Darwinists think these things have.

Reducing poverty would permit many of the poor to develop their genetic potential.

This is true, only if their psyche chooses to develop their genetic potential. If they have no desire to develop their genetic potential nor any desire to take advantage of their social opportunities and environmental windfalls, then nothing will come of it.

Genetics give us our physical potential, as well as many of our physical limitations. Environment provides us with opportunities (and obstacles) for developing our genetic potential. However, psyche or choice determines in the end whether a person takes advantage of his genetic potential and the environmental opportunities that come his way, or not. The boy can be black and seven feet tall and raised in a ghetto surrounded by basketball players, but if he chooses not to play ball, he'll never be an NBA star. Psyche trumps Nature and Nurture at will!

Remember, Psyche or Non-Local Consciousness provides and determines our desires, our intentions, our thoughts, our dreams, our hopes, our will-power, our resilience, our decisions, our choices, our beliefs, our personality, our feelings and emotions, and our chosen actions or chosen behaviors.

Psyche or Non-Local Consciousness is identified by chosen behaviors and choices in general; and, we continue to make choices long after our physical brain is dead and gone, according to the empirical evidence from Near-Death Experiences (NDEs) and Out-of-Body Experiences (OBEs). Remember, any time you witness a chosen behavior, you are in fact observing Psyche or Non-Local Consciousness in action. This Psyche or Non-Local Consciousness is the "psycho" portion of the BioPsychoSocial Model; and, it is identified by chosen behaviors.

Automatic reflexes and autonomic functioning are biological in nature and origin; but, chosen behaviors, chosen actions, chosen beliefs, and chosen thoughts are purely a function of Psyche. Reflexes occur too rapidly to be dependent on mental processes, because they are controlled by the spinal cord. Our reflexes and instincts were programmed into us by God through our genome.

Remember, reflexes and autonomic processes are due to design and programming; and thus, they require a Designer and a Programmer. Anything that is caused by chance is unreliable and can't be replicated on demand. That's one of the main reasons why evolution or random mutations FAIL to meet the demand for replication that the Materialists and Naturalists say is the hallmark of real science. Design and Creation by Evolution can't be replicated nor reproduced on demand, because chemical evolution can't design and create proteins, let alone the matching genes to go along with them. The theory of evolution FAILS the very science that the Materialists and Naturalists have created and demand that we believe in, because spontaneous generation, abiogenesis, chemical evolution, and macro-evolution can't be replicated on demand.

Remember, anytime we witness decisions and choices between options, and anytime we witness chosen behaviors, we are witnessing Psyche in action.

How do we know?

WE KNOW because the Materialists, Naturalists, Darwinists, and Behaviorists have told us that choice is physically impossible. Therefore, any time we witness choice and chosen behaviors, then we simply KNOW that those choices are taking place at the quantum level under the command and control of Someone Psyche.

These research studies and empirical realities demonstrate and prove that when it comes to Psychology (the Study of the Human Psyche), the BioPsychoSocial Model is the best model of all, because it matches most closely with the true reality of the situation. Everything else pales in comparison.

Epigenetics

Epigenetics is physically impossible.

Epigenetics is the study of stable heritable changes in gene function that do not involve changes in the DNA sequence.

https://en.wikipedia.org/wiki/Epigenetics

The former assumption that monozygotic twins are genetically identical was disproven [by epigenetics]; and, the common practice of referring to monozygotic twins as *identical twins* should be curtailed. (*Biopsychology*, p. 48.)

Epigenetics means "not of the genes"; and, it refers to non-genetic means by which traits are passed from parents to offspring. Epigenetics is some type of Black Box. Technically, epigenetics is physically impossible.

At times, we do identify physical processes that are producing epigenetics; but, who is controlling and triggering these physical processes? Who designed and implemented the functionality in the first place?

There are physical forms or physical manifestations of epigenetics. The ones that repeatedly show up in our college textbooks are as follows:

DNA Methylation

DNA Methylation is an epigenetic mechanism wherein a methyl group attaches to a DNA molecule.

Who is doing this to the DNA?

Why?

It tends to reduce the expression of adjacent genes, which explains part of the why; but, it still doesn't explain who is doing this to the DNA.

However, according to the Orthodox Interpretation of Quantum Mechanics, by Henry P. Stapp, it is Nature's Psyche who is collapsing the wave function thereby making these various epigenetic mechanisms physically REAL and truly there. It's certainly not the Human Psyche who is doing any of this, because we aren't consciously aware of any of this.

Quantum Neuroscience starts by identifying the physically impossible, and then it continues by trying to find a scientific explanation for the physically impossible.

You need an Interpretation of Quantum Mechanics that explains what the Human Psyche and Nature's Psyche are doing at the quantum level. If your Interpretation of Quantum Mechanics doesn't do that, then it's worthless. You also need an Interpretation of Quantum Mechanics that explains how God's Psyche does His science and His work through quantum mechanics. If your Interpretation of Quantum Mechanics can't do that, then it's worthless.

Histone Remodeling

A histone is a protein around which DNA is coiled. Histone Remodeling is an epigenetic mechanism wherein histones change their shape and in so doing influence the shape of the adjacent DNA. This can either increase or decrease gene expression.

Histone remodeling alters the functionality of the DNA. That explains why it is being done.

But, who is doing this?

How do they know how to do this? How do they know where to do this?

Who designed this capability and functionality into the DNA?

Well, it wasn't evolution, because evolution can't do design and creation.

According to the Orthodox Interpretation of Quantum Mechanics, by Henry P. Stapp, it is Nature's Psyche who is collapsing the wave function thereby making these various epigenetic mechanisms physically REAL and truly there.

As far as the original design and original functionality is concerned, the only one we can realistically turn to is God's Psyche, because evolution, genes, and proteins didn't exist until after God designed and created the genes and the proteins in the first place. It's physically impossible for chemical evolution to design and create proteins, along with the matching genes to go with them.

RNA Editing

Messenger RNA is a strand of RNA that is transcribed from DNA by transcription factors, and the mRNA carries the genetic code out of the cell nucleus to a ribosome to be used in the synthesis of a protein.

The transcription factors are enzymes; and, it's physically impossible for them to KNOW which protein is wanted or needed by the cell and where the associated gene is located that will produce that protein. They are molecules. There's NO brain or database within these things! Yet, the transcription factors seem to be more intelligent and knowledgeable than we are.

MicroRNAs are short strands of RNA that have been found to have major effects on gene expression and protein synthesis.

RNA Editing is an epigenetic mechanism wherein messenger RNA is modified through the actions of small RNA molecules and other proteins.

Who requisitions these things?

Who knows how to use them?

How do these things know what they are doing?

Who makes them do this?

It's a black box of some kind.

If I were to give you the task of using microRNAs or proteins to make a specific gene produce different types of proteins, would you know how to do this? Could you do it? These molecules are smarter and more intelligent than we are, and they allegedly existed millions of years before we did. That's physically impossible! Chemical evolution can't make proteins and genes from scratch.

Who designed this functionality into the DNA?

Well, it isn't evolution!

RNA Editing is a manifestation of Intelligence, and evolution has no psyche or intelligence. Furthermore, evolution can't do design, specificity, complexity, or functionality. It's physically impossible for evolution, random processes, and natural selection to do these kinds of things. Evolution can't do anything at the quantum level or psyche level. Evolution is restricted to the physical level; and to add insult to injury, natural selection doesn't even touch our genes. Natural selection is absolutely worthless where genetic engineering is concerned.

According to the Orthodox Interpretation of Quantum Mechanics, by Henry P. Stapp, it is Nature's Psyche who is collapsing the wave function thereby making these various epigenetic mechanisms physically REAL and truly there.

As far as the original design and original functionality is concerned, the only one we can realistically turn to is God's Psyche, because evolution, genes, and proteins didn't exist until after God designed and created the genes and the proteins in the first place. It's physically impossible for chemical evolution to design and create proteins, along with the matching genes to go with them.

Non-Physical Forms of Epigenetics

There also seems to be non-physical forms of epigenetics that never show up in the college textbooks that are being written by the Materialists, Naturalists, Behaviorists, and Atheists. These people control our school systems, and they are masters at destroying and suppressing any evidence of the non-physical. It's what they do, and they are extremely good at it. Nevertheless, the truth will prevail in the end. It always does. Eugenie Scott, Richard Dawkins, and Dolores Umbridge can't shut it down, no matter how hard they might try to do so.

McTaggart, L. (2002). *The Field: The Quest for the Secret Force of the Universe*. New York: HarperCollins.

In her book, *The Field*, McTaggart discusses lots of different "biological weirdness" that is taking place within biology at the quantum level. Non-physical epigenetics is part of that biological weirdness or spooky action at a distance that has been tied to biology by cutting-edge scientists.

Seek, and ye shall find the truth. Knock, and the truth shall be opened unto you. I can testify that this process works. I KNOW, because I used to be a Materialist, Naturalist, Nihilist, and Atheist until I started allowing the truth and the evidence to work on me and work for me.

Darwinism IS a Form of Materialism

Darwinism is a form of Materialism — the philosophical belief that Physical Matter and/or Physical Processes is all that there is and all that there will ever be.

More specifically, Darwinism IS Creation by Evolution or Creation by Chance. The processes of Mutation and Selection, or Mutation/Selection, IS synonymous with Creation by Chance. In other words, Creation by Mutation/Selection IS the very same thing as Creation by Chance.

Furthermore, the Theory of Evolution IS Creation by Evolution, or Creation by Chance, or Creation by Mutation/Selection. What most people don't realize is that Creation by Chance IS the same thing as Creation by Magic, or Creation by NOTHING, or Creation Ex Nihilo.

Technically, Materialism IS Creation by Physical Matter since the Materialists have chosen to believe that Physical Matter is the ONLY thing that exists and thus the ONLY thing that could have done design and creation. Creation by Physical Matter IS Creation by ROCKS! The Materialists literally believe that the Rocks designed and created everything.

Once we get the correct definition for things, THE TRUTH often becomes OBVIOUS and CLEAR for the majority of people.

Most of us instinctively KNOW that Chance, Mutation/Selection, Evolution, Physical Matter, and Rocks CANNOT design and create and manufacture anything at all. Therefore, most of us instinctively KNOW that Materialism, Darwinism, and the Theory of Evolution ARE FALSE.

According to THE SCIENTIFIC METHOD, a Good Theory produces replicable, reliable, and predictable RESULTS. Creation by Evolution, Creation by Chance, Creation by Physical Matter, and even Design and Creation by Mutation and Selection have produced NO RESULTS whatsoever! Creation by Evolution of any kind has NEVER produced RESULTS and will NEVER produce RESULTS, because evolution of any kind cannot design and create and produce RESULTS! Consequently, Materialism and Darwinism and Naturalism are BAD SCIENCE, because they are not replicable on demand and produce NO predictable RESULTS!

A Scientific Hypothesis is a Testable Prediction! Forming a Hypothesis or a Testable Prediction is the first part of THE SCIENTIFIC METHOD. Ironically, there is NO WAY to TEST Creation by Evolution, Creation by Mutation/Selection, Creation by Chance, Creation by Abiogenesis, Creation by ROCKS, and Creation by NOTHING because these things cannot Design and Create and Manufacture and Produce anything in the first place!

Ironically, most of the Darwinists and Materialists believe that Creation by Evolution or the Theory of Evolution IS their strongest and most convincing theory. But, they are WRONG, and they don't even know it!

Why?

It's because ROCKS have infinitely more Consciousness, Awareness, Intelligence, Spirit, Mind, and Life than evolution or chance or mutation/selection can even begin to have. There is NO Life nor Consciousness in evolution, chance, or mutation and selection.

The ROCKS have infinitely more Consciousness, Awareness, and Intelligence in them than the various forms of evolution do. Therefore, Creation by ROCKS is in fact the Materialists' and Darwinists' strongest and most believable theory. Yet, most of the

Materialists and Darwinists instinctively KNOW that Rocks cannot design and create and manufacture and produce anything at all.

Annoyingly, the Materialists and Darwinists in my College textbooks talk about evolution as if the thing were alive and were a God. They couldn't be more WRONG! Rocks are more alive, conscious, intelligent, aware, and God-like than evolution is!

Once we get the correct definition for these things, then THE TRUTH of the Matter typically becomes OBVIOUS and CLEAR. Materialism, Darwinism, Naturalism, and Atheism are obviously and clearly FALSE, because none of these things have ever been caught in the act of designing and creating and manufacturing and producing anything at all. Rocks have more intelligence, consciousness, and awareness than evolution or mutation/selection does!

For me personally, Creation by ROCKS is a lot more logical, intelligent, credible, and believable than Creation by Evolution. Yet, I also KNOW for a fact that Rocks cannot design and create and manufacture anything at all. So why on earth would I be stupid enough to believe that Evolution can do so? Yet, when I was a Materialist and an Atheist, I was indeed stupid enough and blind enough to believe that Evolution could design and create. I was WRONG! I know better now!

Evolution ain't got no soul! Evolution is dumb and blind without hands and a mind.

THE TRUTH is all that you really want, unless of course you are a Materialist, Darwinist, Naturalist, or Atheist. This reality is obvious also.

The Materialists, Darwinists, Naturalists, and Atheists are actively and deliberately trying to hide THE TRUTH from us and from themselves as well. They use trickery and deception in order to do so. Self-deception works, and it works every time, especially against the PhD's who think that they know everything that there is to know about everything. The PhD's are easy marks for something like Darwinism, Materialism, Naturalism, or Atheism because these things ARE the LIES that the PhD's desperately want to be true and desperately want to believe in.

Materialism has been the most popular and the most favorite LIE of the intellectuals, throughout the whole of human history. Materialism IS the LIE that the most people have wanted to believe in.

By telling it as it really is, I have observed that it PRODUCES RESULTS and gets me closer to THE TRUTH. You want Science that produces RESULTS, not the garbage that the Materialists and Darwinists try to feed to you.

As a Scientist, Philosopher, Logician, and Theoretician, I have learned to love the evaluation of evidence. I have learned to love asking my own questions, evaluating the evidence, and distinguishing between good evidence and weak evidence.

Ironically, I soon discovered that when it comes to the different types of Materialism, there is NO EVIDENCE whatsoever to support the claims of Materialism; and, "No Evidence" is in fact as weak as it can get when it comes time for us to evaluate the evidence. And, when it comes to Science, it is extremely important to see and understand and experience how Scientific Evidence is evaluated.

Materialism IS the claim that the Spiritual does not exist. There is NO EVIDENCE and can be NO EVIDENCE to support that claim. Consequently, when it comes to Materialism there can be NO evaluation of the evidence because there is NO EVIDENCE to support the claims of Materialism and therefore no evidence to evaluate. Instead, ALL of the evidence stands firmly against Materialism telling us clearly and conclusively that Materialism is FALSE. Materialism is simply a Religion that has to be taken completely on blind-faith as being true, because there can be NO EVIDENCE to support the claims of Materialism.

A Materialist has infinitely more blind-faith than a Theist can even begin to muster, because there can be NO SCIENTIFIC EVIDENCE nor any OBSERVATIONAL EVIDENCE to support the claims of Materialism. The lack of supporting Evidence IS the primary reason why Materialism is FALSE. Instead, the preponderance of the evidence tells us clearly and conclusively that Materialism is FALSE. That's what I learned when I tried to apply THE SCIENTIFIC METHOD to Materialism and evaluate the evidence associated with Materialism. I learned that there is NO EVIDENCE to support the claims of Materialism.

Now pay attention!

Despite the FACT that I have debunked and falsified the Theory of Evolution thousands of times in hundreds of different ways, don't ever once let me convince you that there is NO such thing as evolution or random mutations.

Evolution is REAL, very REAL. Evolution or random mutation is entropy or the second law of thermodynamics. It's REAL. Entropy is very REAL at the physical level; and, it works as advertised at the physical level.

Whenever the Materialists, Naturalists, Darwinists, Nihilists, Behaviorists, and Atheists start talking about evolution, they get most everything wrong, because evolution or entropy cannot design and create. However, these people do indeed get one thing perfectly right. Evolution, or random mutation, or entropy is indeed the CAUSE of ALL of our heritable diseases, developmental diseases, and heritable mental illnesses.

Remember, the Theory of Evolution is FALSE because random mutations or entropy cannot design and create genes, proteins, and life forms. However, evolution or random mutation or entropy is very REAL; and, it can indeed destroy genes, proteins, and life forms. Do you see how that works? It's important to understand.

Dark Matter and Dark Energy Are Non-Physical or Immaterial

The more you study Science, the more things you will encounter that PROVE that Materialism is FALSE.

Once you have proven Materialism FALSE and gotten rid of Materialism, then you are finally ready to start studying the True Nature of Reality.

Every Scientific Discipline now has an Immaterial or Non-Physical aspect, or area of study. Materialism has been proven FALSE. We scientists can think of dozens of different Non-Physical things which we KNOW exist but are obviously immaterial, or spiritual, or non-local, or non-physical in nature. Dark Matter and Dark Energy are just a couple of these Non-Physical Realities which exist in our universe.

The information that I have been able to gather on Dark Energy and Dark Matter makes it abundantly clear that these forces are unseen, intangible, and non-physical – like magnetism and gravity. They are in fact spiritual in nature. Any kind of force or field is immaterial, or non-physical, or spiritual. That's the very definition of a force or a field.

I found a documentary that provided an adequate introductory summary to Dark Matter and Dark Energy. I will now quote some of the highlights.

British TV Documentary, "Edge of the Universe", 2008, Season 1 Episode 3, "Final Frontier".

> Space can be twisted out of shape by the massive, unseen force of gravity.

> There's nowhere near enough matter in all these galaxies to pull the universe into a curved shape. But more alarmingly, there is only a fraction of the matter required to keep it flat. To make sense of his unexpected discovery, Brian Boyle has invoked two of the most mysterious forces in the universe: dark matter and dark energy.

> Although the galaxy survey has only found one third of the mass required to make the universe flat, we now believe that the remaining two thirds of the matter energy in the universe is contained in this so-called dark energy.

> Astronomers have long speculated that, in addition to the familiar matter of stars and planets, there is a vast proportion of the universe that is invisible, made up of dark matter and dark energy. But where's the proof? These forces are just theories. Until now.

> In one of the most remarkable discoveries of recent years, the evidence has been found in the microwave background radiation. This radiation, which has travelled through space since the beginning of time, has confirmed the existence of these invisible cosmic forces. In proving that the universe is flat, the microwave background has also confirmed that astronomers were right to assume the existence of dark energy and dark matter. Everything we see is just a small fraction of the total cosmic mass. Most of the universe is invisible.

> Astronomers have discovered that galaxies are not only rushing apart from each other, they are gathering speed. It seems that four billion years ago, the universe changed up a gear and ever since has been expanding faster and faster. But what is the force driving this cosmic acceleration?

> Now if it's true that the rate of expansion is picking up, that suggests that there's something like an antigravity force, something opposing the normal force of gravitation, the pulling force, something pushing galaxies apart faster and faster. This antigravity force is dark energy. Astronomers now believe the universe must be

buzzing with this invisible energy. Space, it seems, is far from empty. We just can't see what's out there. The unseen forces of the universe, gravity and dark energy, are locked in the ultimate battle. At stake is the destiny of the universe. Dark energy will defeat gravity and succeed in stretching the universe into oblivion.

This model we have requires dark energy that we've never seen, and dark matter which also we have never seen. And the atoms that we know about are making up only 5% of the universe.

Anything that we KNOW to exist, which is unseen or unseeable, will in fact be spiritual in nature or non-physical in nature. When Scientists use the words "unseen", or "force", or "field" to describe something, that particular "something" will in fact be immaterial, non-physical, and PROVE beyond a shadow of a doubt that Materialism is FALSE.

The tangible stuff that we can get our hands on makes up only 5% of this universe! Materialism only covers 5% of the universe; and, the Materialists formally reject all the rest. Not only does that make Materialism incomplete and FALSE, but that also makes Materialism BAD SCIENCE. The Materialists should be embarrassed about their claims and what they are trying to force us to believe. The Materialists should be embarrassed and ashamed of the garbage that they have been trying to feed us.

Since 95% of this universe in Non-Physical, Non-Local, Trans-Dimensional, and Spiritual in nature, that leaves plenty of room for God to exist and do His thing. And, those of us who have gotten to know the Biblical God know that He has promised us or covenanted with each one of us to NEVER reveal Himself to us and NEVER interfere in our lives, unless we first ask Him to do so and give Him permission to do so. The Biblical God never comes and never makes an appearance unless He is first asked to do so and given permission by us to intervene in our lives. And even then, He only does just enough to make our righteous desires possible, but we still have to do our part and do the work necessary to make our righteous desires a reality. The Biblical God will never do anything for us that He knows that we can do for ourselves.

Once you have successfully proven to yourself that Materialism, Naturalism, and Atheism are FALSE, and why they are FALSE, then that opens up the possibility that their opposites might in fact be true.

Materialism is the claim that the Spiritual or the Non-Physical does not exist. Once you know that Materialism is FALSE, then you simply know that Spirit Matter or Dark Matter exists. You also know that there are a lot of Non-Physical forces and fields that also exist.

Naturalism is the claim that the supernatural does not exist. Once you know that Naturalism is FALSE and various reasons why it is FALSE, then you know that the supernatural can in fact exist and are free to start looking for signs of the Supernatural. The Biblical God, Jesus Christ, went out of His way to make His presence known as documented in the Bible, the Book of Mormon: Another Testament of Jesus Christ, the Doctrine and Covenants: Yet Another Testament of Jesus Christ, and the Pearl of Great Price. The Biblical God hasn't exactly been hiding from us. But, the Materialists and Naturalists have in fact been deliberately hiding from the Biblical God, which is why they know nothing about Him. You will NEVER learn anything about the Biblical God or the Supernatural from a Materialist, Naturalist, or Atheist because these people have chosen to learn NOTHING about these subjects. When it comes to true and accurate information about the Biblical God, a PhD Materialist or a PhD Naturalist is the dumbest and most uneducated person that you could possibly ask. A child knows more about the Biblical God than a PhD Materialist does.

Atheism is the claim that God does not exist. As a Scientist, it dawned on me that God must exist in order to have done all of the Science, Design, Creation, and Production of Genomes and Life Forms that needed to be done 3.8 billion years ago and again 542 million years ago during the Cambrian Explosion, because as a Scientist I KNOW that the water and the rocks could NEVER have designed and created all of the Genomes and Life Forms on this planet as the Materialists and Atheists claim. Materialism is creation by Rocks. Atheism is Creation by NOTHING or Creation by CHANCE. I'm an intelligent Scientist, and I KNOW for a fact that Rocks, CHANCE, and NOTHING have never even once been caught in the act of designing and creating genomes and life forms, and never will be caught in the act of doing these kinds of Science. I KNOW that Materialism, Naturalism, and Atheism are FALSE, because I KNOW scientifically and logically why they are FALSE.

The only thing that the Materialists and Atheists can provide in rebuttal is mocking and ridicule – their ongoing claim that I'm lying to you, or that I don't know what I'm talking about, or that I don't know anything about Science, or some other such unprovable nonsense. I'm embarrassed for the Materialists and feel sorry for them. Some of them have got it bad and are really making fools of themselves in public. There is NO evidence and can be NO evidence to support the materialistic claim that the Spiritual or the Non-Physical does not exist. Materialism is based upon a complete and total blind-faith in something that can never be proven true, because there will never be any evidence to support Materialism and its claims.

In contrast, I have gathered together an endless array of EVIDENCE from every conceivable Scientific Discipline which demonstrates clearly and conclusively why Materialism is FALSE and which taken together PROVE beyond a shadow of a doubt that Materialism is FALSE. Once I had proven to myself that Materialism is false, then I was finally ready to start doing Science for real.

Whenever the Scientists start talking about unseen things that are "inferred" to exist, then you just simply know that they are talking about something Immaterial, Non-Physical, and Spiritual in origin. The SCIENTIFIC EVIDENCE tells us that these things MUST EXIST, even though we can't see them, touch them, nor get our physical instruments to detect them directly. In the same manner, Science PROVED to me that God must exist in order to have done all of the Science that needed to be done 3.8 billion years ago, which I KNOW for a fact that the Rocks and Chance could NEVER have done. Rocks and Chance can't do Science, and they can't design and create genomes and life-forms from scratch either.

Logic and Science tells me that God must exist in order to have done all of the Science, Design, and Creation that needed to be done 3.8 billion years ago when the water and the rocks were the only other things on this earth. After all, 3.9 billion years ago when all there was on this planet was water and rocks, Darwinian processes, evolution, natural selection, and random mutations didn't exist yet; so, they couldn't have designed and created the first genomes and life forms on this planet. Somebody else had to do the job instead. Materialism IS Creation by Rocks or Creation by Physical Matter, and I just simply know that Rocks and Physical Matter cannot design and create and do Science.

I'm embarrassed to admit that I was an Atheist and a Materialist for a while. I'm proud to say that now I know better. It was Science itself that convinced me that Materialism and Atheism are FALSE, because Science taught me why they are FALSE. Once I knew that Naturalism is FALSE, then I was ready and willing to look for and accept signs of the Supernatural or God; and, I discovered quite clearly that the Biblical God has indeed revealed Himself to us in the books that He had a hand in producing and writing, as documented and recorded in the Bible, the Book of Mormon: Another Testament of Jesus Christ, the Doctrine and Covenants, and the Pearl of Great Price. I have observed, through a preponderance of the evidence, that the Biblical God Jesus Christ has gone out of His way to make His presence and His existence KNOWN to us. Whenever the Being of Love and Light is identified by name during our Near-Death Experiences, He's always Jesus Christ, our Savior and Redeemer.

Getting rid of My Atheism and My Materialism opened doors into everything else. In other words, I'm now FREE to study and learn about the other 95% of this universe which the Materialists and Atheists have formally rejected. I'm finally free now to be and to become a Real Scientist. Nothing is off-limits for me anymore! I'm no longer afraid to study and learn about the Non-Physical, or the Non-Local, or the Trans-Dimensional, or the Spiritual. I can handle it now. Such wasn't the case when I was a Materialist and an Atheist. Back then, I didn't have a clue much of the time.

Science becomes infinitely more interesting once you are willing and ready to take the Non-Physical or the Spiritual into account.

Dark Matter and Tachyons

In his book, *Consciousness Beyond Life: The Science of the Near-Death Experience*, Pim Van Lommel convinced me that in Quantum Non-Locality or the Non-Local Spirit Realm, the Quantum Objects that reside there can theoretically exist at infinite velocities much faster than the speed of light.

Based upon that evidence, I came to the conclusion that spirit matter is the same thing as physical matter, EXCEPT for the fact that spirit matter exists at velocities greater than the speed of light and physical matter exists at velocities less than the speed of light. Then I eventually came to the conclusion that immaterial psyche exists at an infinite velocity or an infinite frequency capable of being instantaneously and simultaneously everywhere in its native original format.

God marries our Psyches or Intelligences to spirit matter and physical matter in order to slow them down, so that we psyches can have Lived Experiences, learn from our experiences, and remember having done so.

I eventually realized that matter existing at different velocities or existing in different phases can actually occupy the same space because they are out-of-phase with each other. Therefore, Psyche, spirit matter, and physical matter can occupy the same space because they have different phases of existence or exist at different velocities in different dimensions.

I eventually was taught and realized that a Quantum Object is a Particle of Matter, and that matter has two main states of existence according to the Quantum Law of Complementarity – a spiritual state and a physical state. Physical matter exists at velocities less than the speed of light, because of all the MASS or resistance to acceleration that physical matter has; whereas, spirit matter with nearly no mass exists at velocities greater than the speed of light.

In time, I came to realize that Dark Matter is in fact Spirit Matter.

For a very long time, I thought I was the ONLY one on the planet who realized that Spirit Matter exists at velocities faster than the speed of light and is therefore undetectable by physical instruments directly.

Last night 29JUL2017, I was reading in my most favorite introduction to Popular Physics or Speculative Physics – a book entitled *Physics of the Impossible* by Michio Kaku – the following description of Tachyons.

> **Tachyons live in a strange world where everything travels faster than light. As tachyons lose energy [mass], they travel faster, which violates common sense. In fact, if they lose all energy [or mass], they travel at infinite velocity. As tachyons gain energy [or mass], however, they slow down until they reach the speed of light. (p. 280).**

That's the exact SAME definition for Spirit Matter that I have been giving in all of my books for the past year or so.

Tachyons ARE Spirit Matter!

Of course, Michio Kaku doesn't realize that Tachyons are Spirit Matter, because as a Materialist he isn't looking for Spirit Matter. But, I instantly recognized the definition for Spirit Matter that I have been providing for the past year in my books within the definition that Michio Kaku provided for Tachyons.

Cool, huh?

Well, at least I thought it was cool.

Tachyons live in a strange world that we have been calling the Non-Local Realm or the Spirit Realm. As tachyons or spirit matter lose energy, their mass or resistance to acceleration decrease, which makes perfect sense to me. In fact, if the tachyons or spirit matter were to lose ALL of their mass, they would become immaterial Psyche and exist at an infinite velocity being instantaneously and simultaneously everywhere in their native original format.

Remember, ZERO MASS = Infinite Velocity. MASS or Inertia is resistance to acceleration. No resistance to acceleration results in an infinite velocity. It's always made perfect sense to me, ever since I learned about it.

Remember, Matter is a Quantum Object, and a Quantum Object has two mutually exclusive states of existence – a spiritual state faster than the speed of light and a physical state slower than the speed of light.

As God inputs more mass or energy into a Tachyon or particle of Spirit Matter, its velocity slows down until it reaches the speed of light. Then, by consciously inserting additional energy or mass into that Tachyon or particle of Spirit Matter, God slows that particle of matter to sub-light velocities thereby converting that particle of Spirit Matter into a particle of Physical Matter.

There in Michio Kaku's paragraph, I found independent confirmation of what I have been trying to teach my readers for the past year of my life.

Non-Local Matter or Spirit Matter exists in the Non-Local Realm at velocities greater than the speed of light. Spirit Matter has a little mass and consists of Tachyons that exist at velocities greater than the speed of light. Should a Quantum Object ever shed ALL of its mass, then it ceases to be matter and becomes Psyche or Non-Local Consciousness instead. Psyche, or Non-Local Consciousness, IS the fundamental unit of reality. This is a Psyche Ontology.

Psyche ACTS; whereas, matter of any kind REACTS to psyche's command. Psyche is the ACTOR, because in its native original format it exists at an infinite velocity which means that it is instantaneously and simultaneously everywhere in its native original format. God marries our psyches or intelligences to spirit matter and physical matter in order to slow us down so that we are subject to the passage of TIME and can therefore have experiences and remember having had them. Each time that God introduces limitations into our lives or our psyches, God also introduces new and interesting opportunities to learn and grow.

Tachyons ARE Spirit Matter, because they exist at velocities greater than the speed of light. I also have reasons for believing that Dark Matter is Spirit Matter. From my perspective, Science itself has PROVEN the existence of Spirit Matter, in the form of Tachyons or Dark Matter.

I found this Science extremely useful and interesting.

Explaining and Defining Dark Matter

As a race, we KNOW that Spirit Matter exists, because Out-of-Body Travelers have seen it, lived it, experienced it, and observed it.

Science IS Observation!

Since we KNOW from the Lived Experiences of the human race that Spirit Matter exists, how do we explain Spirit Matter scientifically? Or, has it already been explained?

In my most favorite introduction to Popular Physics or Speculative Physics – a book entitled *Physics of the Impossible* by Michio Kaku – the following description of Dark Matter was instructive.

> **Gravity, being the warping of space-time, can freely float into the space between universes.**

> **In fact, there is one theory that states that dark matter, an invisible form of matter that surrounds the galaxy, might be ordinary matter floating in a parallel universe. As in H. G. Wells's novel *The Invisible Man*, a person would become invisible if he floated just above us in the fourth dimension. Imagine two parallel sheets of paper, with someone floating on one sheet, just above the other.**

> **In the same way there is speculation that dark matter might be an ordinary galaxy hovering above us in another membrane universe. We could feel the gravity of this galaxy, since gravity can ooze its way between universes, but the other galaxy would be invisible to us because light moves underneath the galaxy. In this way, the galaxy would have gravity but would be invisible, which fits the description of dark matter.**

> **Yet another possibility is that dark matter might consist of the next vibration of the superstring. Everything we see around us, such as atoms and light, is nothing but the lowest vibration of the superstring. Dark matter might be the next higher set of vibrations. (pp. 239- 240.)**

In ALL of his different descriptions of Dark Matter, Michio Kaku is describing Spirit Matter. Of course, since Kaku is a Materialist, he doesn't realize that he's describing Spirit Matter when he defines and describes Dark Matter; but, he is.

The way Michio Kaku described Dark Matter is the way that Out-of-Body Travelers like William Buhlman have been describing Spirit Matter and the Spirit Realm for a couple of decades now.

Spirit Matter is simply ordinary matter that exists in a different phase, or a higher vibration, or a higher membrane, or another dimension besides physical matter. Dark Matter or Spirit Matter IS a higher set of vibrations than physical matter.

Physical matter or localized matter is the lowest vibration, or the lowest level, or the lowest frequency, or the lowest phase, or the lowest dimension of matter. In contrast, Spirit Matter or Dark Matter exists at a higher vibration, or a higher frequency, or a higher membrane, or a higher dimension than physical matter.

Although Dark Matter or Spirit Matter cannot be detected nor seen with our physical instruments, its presence and existence is KNOWN by the fact that it has influence on Gravity and thus the Cosmic Density of our Universe as a whole.

As far as I know, I'm the first person on the planet to realize that Dark Matter IS in fact Spirit Matter; but, it's impossible to read everything, so it's impossible to know for sure.

William Buhlman and other Out-of-Body Travelers have OBSERVED that the lowest level of the Non-Local Realm or Spirit Realm IS a direct mirror image of our physical universe. Every physical object has a spiritual counterpart existing at the lowest vibrational phase of the Non-Local Realm or Spirit Realm. At higher dimensions or higher vibrational states in the Spirit World, things change and there's no longer the one-to-one mapping of the spiritual with the physical.

Higher levels or higher vibrations of the Spirit World or Non-Local World become more ethereal, rarefied, refined, light, and energy-like. The higher levels become more and more defined and controlled by Psyche or Non-Local Consciousness.

In other words, the Spirit World or Dark Matter World is a series of membranes or parallel universes stacked right on top of each other, vibrating at higher and higher frequencies, until the vibrations become infinite and matter ceases to exist and we enter into the Realm of Thought, or Non-Local Consciousness, or Psyche.

Again, I quote from William Buhlman.

Journal Entry, October 2, 1982

I hear the buzzing, engine-like sounds and will myself out-of-body. I step to the bedroom door and automatically request "Clarity now!" My vision improves and I step through the door, into the living room. Still feeling a little out of sync, I verbally repeat my request with more emphasis, "Clarity now!" I feel my awareness and vision snap into place.

My thoughts are clear and I make a verbal demand, "I need to see the form I'm in now!" Instantly I feel an intense sensation of being drawn within myself. I'm suddenly different, weightless as though I'm floating in space. As I look forward I see a sparkling, bluish white form. For some reason, I seem to know that I'm looking at my nonphysical body from a different perspective. I stare in amazement at this form before me that shines and flows with energy and light. It looks like an energy mold created from a million tiny points of light; it radiates a bluish glow but appears to have a defined outer structure. The body of light before me is naked and is identical to my physical form. Even though my body looks firm, there is a noticeable energy motion and radiation present. I can see what appears to be an ocean of blue stars throughout my body. It's difficult to describe because the stars are stable, yet moving at the same time; the light and energy of my body appear to change and flow almost like the waves of an ocean.

As I stare at the body of light, it hits me that I must be in another body. Yet I can't perceive any form or substance; I'm like a viewpoint in space without shape or form of any kind. As I reflect upon my new state of being, I feel a sensation of rapid motion and I'm instantly back within my physical body.

Lying still and reviewing my experience, I'm struck by an inescapable conclusion: I must possess multiple energy-bodies. The form I just experienced was noticeably lighter (less dense) than even my second nonphysical body. I realize that the traditional view of our possessing two bodies — a physical body and a spiritual body — is far too simplistic; we are much more complex than this. Just as there are multiple nonphysical energy dimensions within the universe, each of us must consist of multiple energy-bodies or vehicles of expression.

Now I seriously wonder just how many nonphysical bodies or forms this involves. I suspect that there must be one within each dimension of the universe and that all of these are interrelated and connected, just as the physical body is connected to its first nonphysical (spiritual) body.

Buhlman, William L., "Adventures Beyond the Body: How to Experience Out-of-Body Travel" (pp. 34-35). HarperCollins. Kindle Edition.

William Buhlman suggests that we have multiple different spirit bodies, each residing at higher and higher vibrational levels or frequencies or dimensions or membranes in the Non-Physical Realm. I tend to visualize this as ONE Non-Local Body or ONE Non-Physical Body that can function in ALL dimensions or at ALL vibrations of the superstring; but, that's basically the same thing as having a non-physical body or a spirit body in EACH dimension.

The FIRST or lowest vibrating nonphysical body and realm is a direct mirror of this physical universe and our physical body.

But, as the Psyche and Spirit Body cycle into higher and higher dimensions or vibrational states, you arrive at a situation where the Psyche is separate from the highest vibrating spirit body; and, the Psyche is simply an immaterial viewpoint floating in space vibrating at an infinite frequency. At an infinite frequency and an infinite velocity and NO mass, the superstring transmutes and transforms and comes ALIVE. At infinity, it reaches the limit, achieves unity, integrates, and becomes Psyche or Non-Local Consciousness.

Throughout this whole process, it's Psyche who is doing the experiencing and the remembering – NOT the spirit body and definitely NOT the physical body. It's the Psyche or Non-Local Consciousness who has and remembers experiences and events, NOT the matter. The matter, whether spirit matter or physical matter, is simply OBSERVED and CONTROLLED by the Psyche.

M-Theory or Membrane Theory actually explains how the different membranes of the Spirit Realm or Non-Local Realm relate to our Physical Realm or Local Realm. The local Physical Realm is the lowest vibrational state of the superstring and the lowest membrane. The Non-Local Spirit Realm consists of the higher vibrational states of the superstring and therefore the higher membranes, or higher dimensions, or higher parallel universes that are out-of-phase with our physical universe.

M-Theory explains perfectly the Lived Experiences and the Out-of-Body Non-Local Experiences of the human race.

ALL of this matches perfectly with what Michio Kaku was saying about Dark Matter.

Dark Matter or Spirit Matter IS the "next higher set of vibrations" on the Superstring. Physical Matter IS the lowest vibration of the Superstring.

Dark Matter or Spirit Matter exists in a Higher Membrane, or a Higher Dimension, or a Parallel Universe just above our Physical Membrane or Physical Universe.

Dark Matter IS Spirit Matter or Non-Local Matter.

Explaining and Defining Dark Energy

In my most favorite introduction to Popular Physics or Speculative Physics – a book entitled *Physics of the Impossible* by Michio Kaku – Kaku successfully equates Zero-Point Energy with Dark Energy. He is the first person whom I have encountered to have done so. I have been equating the Zero-Point Field of Light with Dark Energy for over a year now – today's date is 30JUL2017; and, I thought that I was the only one doing so.

From Kaku's book, we have the following:

> At first, "zero-point energy" (or the energy contained in the vacuum) seems to violate the First Law of Thermodynamics. Although zero-point energy defies the laws of Newtonian mechanics [because it is non-local or spiritual in nature and origin], the notion of zero-point energy has reemerged recently from a novel direction.

> When scientists analyze the data from satellites currently orbiting the Earth, such as the WMAP satellite, they have come to the astounding conclusion that fully 73 percent of the universe is made of "dark energy", the energy of a pure vacuum. This means that the greatest reservoir of energy in the entire universe is the vacuum that separates the galaxies in the universe.

> This dark energy is so colossal that it is pushing the galaxies away from each other, and may eventually rip the universe apart in a Big Freeze.

> Dark energy is everywhere in the universe, even in your living room and inside your body. The amount of dark energy in outer space is truly astronomical, outweighing all the energy of the stars and galaxies put together. (Kaku, 2008, p. 270).

Michio Kaku provides the first independent confirmation that I have found so far which equates Zero-Point Energy with Dark Energy. I have been equating the two for over a year now, and it's nice to have a world-class physicist agree with me.

However, I'm probably the first person on the planet to realize that the Biblical God Jesus Christ was talking about and teaching about Dark Energy or the Zero-Point Field of Light over a hundred and fifty years ago in 1832. God called it the Light of Christ.

Doctrine and Covenants 88: 6-7:

> 3 Wherefore, I now send upon you another Comforter, even upon you my friends, that it may abide in your hearts, even the Holy Spirit of promise; which other Comforter is the same that I promised unto my disciples, as is recorded in the testimony of John.

> 4 This Comforter is the promise which I give unto you of eternal life, even the glory of the celestial kingdom;

> 5 Which glory is that of the church of the Firstborn, even of God, the holiest of all, through Jesus Christ his Son —

> 6 He that ascended up on high, as also he descended below all things, in that he comprehended all things, **that he might be in all and through all things, the light of truth**;

7 Which truth shineth. **This is the light of Christ. As also he is in the sun, and the light of the sun, and the power thereof by which it was made.**

8 As also he is in the moon, and is the light of the moon, and the power thereof by which it was made;

9 As also the light of the stars, and the power thereof by which they were made;

10 And the earth also, and the power thereof, even the earth upon which you stand.

11 And the light which shineth, which giveth you light, is through him who enlighteneth your eyes, which is the same light that quickeneth your understandings;

12 **Which light proceedeth forth from the presence of God to fill the immensity of space —**

13 **The light which is in all things, which giveth life to all things, which is the law by which all things are governed, even the power of God who sitteth upon his throne, who is in the bosom of eternity, who is in the midst of all things.**

How did Jesus Christ KNOW about Dark Energy or the Zero-Point Energy, over a hundred and fifty years before our scientists discovered it?

For me personally, the relatively recent discovery of Zero-Point Energy or Dark Energy became Scientific Proof of God's Existence, because I realized that God KNEW about Dark Energy hundreds of years before we did, if not billions of years before we did.

Dark Energy IS the presence of God throughout this universe. As Michio Kaku said, Dark Energy or Zero-Point Energy is in everything, including your living room and your physical body.

In his book, *Why the Universe Is the Way It Is*, Hugh Ross calls Dark Energy the "stretching force in this universe," which is counteracting gravity and causing this universe to expand.

While reading that book over a year ago, it dawned on me that Jesus Christ was telling us about Dark Energy or the Light of Christ back in 1832. Dark Energy or the Zero-Point Energy is God's Presence in this universe – the Light of Christ; and, we wonder how Jesus Christ was able to perform all those different "miracles" while He was on this earth in the flesh. **"This dark energy is so colossal that it is pushing the galaxies away from each other."** Well, that answers a few questions, doesn't it? If you had that kind of power within you or at your command, I imagine that you could do a few things that would seem to counteract or violate the Laws of Thermodynamics. Don't you think?

I have known for over a year now that Dark Energy or the Zero-Point Field of Light IS the Light of Christ and emanates from the presence of God; but, it was nice to have Michio Kaku confirm what I have known for some time now.

SCIENCE 2.0

I have upgraded my science to Science 2.0; and, Science 2.0 defines science as direct observation or lived experiences – not scientific inference or just-so ad hoc stories. Science 2.0 deliberately and knowingly admits ALL of the evidence into evidence, including the non-local evidence or the lived experiences or the direct observations of the human race. Science 2.0 is based upon the Law of Witnesses. A true eye-witness is infinitely more valuable than a million "just-so stories" and a trillion "scientific inferences".

2 Corinthians 13: 1: "In the mouth of two or three witnesses shall every truth be established."

This is the way that the Biblical God operates – through eye-witnesses or lived experiences and direct observations. There's no scientific inferences, circumstantial evidence, or just-so ad hoc stories where God is concerned. He works through eye-witnesses. This is the best way of doing science – through direct observation. **Remember, direct evidence supports the truth of an assertion directly, without need for any additional evidence or inference.** Direct Observation, or Direct Evidence, or Direct Experience IS Science, because Science is Knowledge. Eye-witnesses KNOW the truth, rather than making inferences and concocting ad hoc just-so stories.

Remember, the Gods were there during the organization and the seeding of this earth; consequently, the KNOW how it was done, because they did the job in person. Eye-witness testimony is infinitely more valuable to us than ad hoc just-so stories and scientific inferences or scientific guesswork.

Lived experiences and direct observations such as the following had a powerful influence on my philosophy of life, worldview, and scientific theories. I simply prefer to follow the evidence, rather than making *scientific inferences* and *jumping to conclusions*. Watch some of these videos and decide for yourself what kind of evidence, or observation, or proof they provide. For me personally, these videos successfully falsify Materialism, Naturalism, Nihilism, and Atheism. Your mileage may vary, of course, depending upon what you personally want to believe.

Howard Storm:

https://www.youtube.com/watch?v=Vm647n1360A

https://www.youtube.com/watch?v=UPj4wci_bcI

Dr. Mary Neal:

https://www.youtube.com/watch?v=DX473dF7ChY

Ian McCormack:

https://www.youtube.com/watch?v=HbTAmN4m2lQ

Joe Hadwin:

https://www.youtube.com/watch?v=IOhOynR9Jxg

Another:

https://www.youtube.com/watch?v=N4ut09jDdV0

Science IS Observation. After I made that realization, I now treat Lived Experiences of any kind as Scientific Observations and Scientific Evidence. Lived Experiences or Psyche Experiences ARE the BEST way of finding and knowing the truth in any realm of existence, including the Non-Local Realms. Lived Experiences or Psyche Experiences are a better way of finding and knowing the truth than the Scientific Methods. I never realized that before; but, I understand it and accept it now. It only took me 55 years to figure it out. Lived Experiences, including Out-of-Body Experiences and Near-Death Experiences ARE Scientific Evidence because they are Direct Scientific Observations of the Non-Local Realm or Spirit Realm.

I'm sorry, but I found these people and their lived experiences and observations infinitely more convincing than the posturing and "scientific inferences" which I received from Kalat, Pinel, and Quammen. I have chosen to believe that Observational Evidence is better than circumstantial evidence, which means that I have chosen to accept the belief that Lived Experiences are vastly superior to scientific inferences and just-so ad hoc stories concocted out of thin air.

I can guarantee you, though, that you will see the whole thing differently than the way I now see it, if you have instead chosen to believe on faith that Materialism, Naturalism, Darwinism, Nihilism, and Atheism are true. No amount of evidence will get you to change your mind, unless you actually want to change your mind.

By the time I encountered these videos, I was ready to change my mind, because science itself and scientific observations had started to prove to me that Materialism, Naturalism, and Darwinism are false.

Obviously, you have every right to fall for the "scientific inferences," logic fallacies, and just-so stories of the Materialists, Darwinists, and Naturalists; but, do so knowing that that's what you are doing and that that's what you want to do – and not on blind faith that these people are always telling you the truth like I used to do.

Once I caught the Materialists, Naturalists, and Darwinists in their deceptions, trickery, logic fallacies, and lies, I could no longer go back. I was forced to go forward. I wanted to go forward. I don't like being lied to. Obviously, your mileage may vary. You will do whatever you personally want to do and choose to do. That is as it should be, and that is as it was designed to be.

INTRODUCTION TO SCIENCE 2.0

Most people don't know that the Scientific Method is fundamentally flawed; and, that fact or reality is where all of the confusion and contention comes from where science is concerned.

The Scientific Method is based upon a logic fallacy that is called *affirming the consequent*. The whole of modern-day Science, Materialism, and Scientific Naturalism is based upon this logic fallacy; and, most scientists don't even know it, because they have never studied the Philosophy of Science.

The weaknesses of science and the scientific methods explain why science needs to be upgraded to Science 2.0.

Science is Observation.

If you observe carefully, you will notice that the Materialists, Naturalists, Darwinists, Nihilists, and Atheists define science as Materialism, Naturalism, Darwinism, Behaviorism, Determinism, Scientism, Physicalism, Atomistic Reductionism, Nihilism, and Atheism. These philosophies have been falsified by Observational Evidence, Experiential Evidence, Eye-Witness Evidence, and Empirical Evidence. In other words, Materialism, Naturalism, and their derivatives have been falsified by Observations. Notice carefully, and you will observe that the promoters of these false and falsified philosophies rely upon *affirming the consequent* and the suppression of evidence in order to make their case and meet their burden of proof.

Defining science as Materialism, begging the question, suppressing evidence, and similar practices result in a wide variety of different logic fallacies known variously as *circular reasoning, affirming the consequent, begging the question, jumping to conclusions, scientific inferences, personal interpretations, confirmation biases,* and *ad hoc just-so stories.* These are the bread and butter of Scientific Materialism and the foundation upon which modern-day science is built. It's a house of cards – a pyramid scheme.

Science 2.0 resolves the problem by defining Science as Observation, Experience, Experimentation, Knowledge, Fact, Truth, and Evidence. Science 2.0 is a paradigm shift in our way of thinking and in the way that we do Science. Science 2.0 allows all of the evidence into evidence.

Furthermore, under Science 2.0, there is NO officially sanctified interpretation of the scientific evidence. You are encouraged to make your own interpretation of the evidence. Science 2.0 allows all of the evidence into evidence, and then leaves it up to YOU to interpret the evidence, decide for yourself what to make of the evidence, and choose for yourself how to use that evidence in your own life. You are going to do so anyway, so why not encourage you to do so right from the start?

Science 2.0 is the way that Science should have always been done but wasn't.

DEFINING SCIENCE 2.0

As an upgraded science, Science 2.0 absolutely needs an Account of Origins that matches perfectly with the Observational Evidence and the Lived Experiences of the human race. Consequently, a great deal of time and effort has been placed into finding and/or developing such a thing. You will have to judge for yourself if the process and this project has been a success, or not. All I can do is to present the evidence. You will have to decide for yourself if I have met my burden of proof through a preponderance of the evidence, or not.

The Materialists, Naturalists, Darwinists, Nihilists, and Atheists define "science" as Materialism and Naturalism. Their whole paradigm is based upon a logic fallacy that's been variously called *begging the question*, *affirming the consequent*, or *jumping to conclusions*. In other words, they chose the conclusion that they wanted, long BEFORE they did any actual science. These people jump directly to that conclusion every time they try to do science and interpret science, a process which is very unscientific to say the least. The world is in desperate need of a new and better definition for "science" – a new and better paradigm.

Ironically, Materialism and Naturalism and their derivatives are religions or philosophical worldviews, and NOT science. Materialism and Naturalism are simply a way of interpreting scientific evidence – a way that has actually been falsified by observation, experience, science, scientific evidence, the scientific methods, and the lived experiences of the human race. Materialism, Naturalism, Darwinism, Nihilism, and Atheism are unable to meet their burden of proof because the preponderance of the evidence stands against them and falsifies them; consequently, these false and falsified religions or philosophies have to be taken on blind-faith as being true because they cannot be lived nor experienced for real. Materialism and Naturalism exist only in a person's imagination. They aren't real.

Something is said to be "scientifically accurate" if it matches with reality, meaning that it has been observed and experienced in real life by a real living person or psyche. Materialism, Naturalism, Atheism, and Nihilism are by definition "unscientific" because they don't match with reality, meaning that they have NEVER been observed nor experienced in real life by anyone, nor can they be.

Materialism, Naturalism, Darwinism, Nihilism, and Atheism ARE DEAD. What this means is that they have been falsified trillions of times in thousands of different ways by scientific evidence, or observational evidence, or experiential evidence. We are simply in the process of waiting for the old guard to die, so that a new and better paradigm can have its day in the sun. The world is in desperate need of a new and better definition for "science".

While we are waiting for this new and better paradigm as well as that new and better definition for science to hit critical mass and become generally accepted throughout the world, I have chosen to upgrade my science to Science 2.0. Under Science 2.0, I have chosen to admit ALL of the evidence into evidence. Nothing is excluded now. Everything is taken into consideration, as it should be if we are ever to do Real Science.

Science 2.0 is based exclusively on Observational Evidence and the Lived Experiences of the human race. Under Science 2.0, eye-witness accounts, observational evidence, experiential evidence, direct observations, and lived experiences of any kind trump and take precedence over "scientific inferences," circumstantial evidence, and philosophical speculation. Remember, *scientific inferences* or someone's *personal interpretation of scientific evidence* IS the logic fallacy upon which Materialism, Naturalism,

Darwinism, Nihilism, and Atheism are based. Just present the Observational Evidence or the Lived Experiences, and then let people make up their own mind as to what it all means. Let the people do their own interpretation of the evidence. You can tell people what the evidence means to you, as I repeatedly do; but, don't be dogmatic or religious about your personal interpretations and scientific inferences.

Even though *personal interpretation of the scientific evidence* or *scientific inferences* ARE the last and final step in the Scientific Method, this process of *personal interpretation* or *scientific inference* IS a logic fallacy which introduces ALL of the falsehoods, deceptions, trickery, and lies into science as a whole. Pick any scientific discipline you like, and then observe that it derives and receives ALL of its falsehoods, deceptions, and lies from the *personal interpretations, philosophical speculations, personal guesswork, wishful thinking,* and *scientific inferences* which are given to it by fallible and sometimes-deceptive human beings. Under Science 2.0, just present ALL of the Observational Evidence, Eye-Witness Evidence, and the Lived Experiences of the human race; and then, let people make up their own mind as to what it all means to them.

I'm a scientist; and, I'm not going to apologize for trying to upgrade my science to something better!

While developing Science 2.0, I will present eye-witness accounts, observational evidence, experiential evidence, and lived experiences from every scientific discipline, person, and religion on the planet that I can find; but, you are going to have to make up your own mind as to what that evidence means to you.

Under Science 2.0, there are NO EXPERTS at scientific inferences or scientific interpretations of the data – your personal interpretation is just as good as mine. There is no priesthood under Science 2.0 who will bless and sanctify the religious dogma that's being presented as fact, as we find when it comes to Materialism, Naturalism, Darwinism, Nihilism, and Atheism which have priests and pastors who have given themselves PhDs in their chosen sciences and have thereby made themselves the official interpreters of their scientific disciplines. If you don't agree with their personal interpretation of the scientific evidence, then you are censored, ridiculed, censured, banned, and excommunicated from doing their brand of science.

There's NONE of that under Science 2.0, because under Science 2.0 there are NO high priests of scientific inference and personal interpretation; and therefore, there is NO religion and NO religious dogma where the scientific information and observational evidence are concerned. When it comes to Science 2.0, there is NO official dogma that has to be taken on blind faith as being true, like there is under Materialism, Naturalism, Darwinism, Nihilism, and Atheism. Science 2.0 is a completely open-minded approach to scientific evidence – ALL the evidence. I chose to follow the evidence wherever it might lead me.

Under Science 2.0, as the oracle repeatedly says in the *Matrix*, you make up your own damned mind as to what it is that you want to believe. As Neo and the architect so adequately conclude, it's about choice. It all comes down to choice! Your beliefs are chosen – chosen beliefs and chosen behaviors are the pinnacle and hallmark of free will. It all comes down to choice.

I will present the eye-witness evidence and experiential evidence; and, I will often tell you what that evidence means to me personally. But in the end, you are going to have to make up your own damned mind as to what it is that you personally want to believe. That is as it should be; and, that's the way science should be done.

While developing Science 2.0, I gleaned eye-witness evidence or observational evidence from every scientific discipline and every religion and nationality on the planet that

I possibly could. Neo wanted guns, lots and lots of guns. I want evidence, lots and lots of evidence. The BEST way to find and know the truth is to live it and experience it for yourself, or to choose to trust someone who has. Science 2.0 is based exclusively upon the Lived Experiences of the human race – all of them. Brutal, huh?

Remember, "experiment" and "experience" have a similar etymology – they derive from the same root word. Science is observation, and a science experiment is the same thing as experience.

Science 2.0 is based upon the Lived Experiences of the human race, including all of our science experiments as well as our non-local out-of-body experiences. Under Science 2.0, Lived Experiences of any kind are treated as Scientific Evidence. Whether we are dealing with the spiritual or the physical, the truth of a matter is established through a preponderance of the evidence, as if we were in a court of law – according to the Law of Witnesses. If there are no witnesses, then there is no evidence to be had. The whole purpose of Science 2.0 is to find different ways to account for and explain the observational evidence, the eye-witness evidence, and the experiential evidence provided by the Lived Experiences of the human race.

Under Science 2.0, observational evidence and experiential evidence is given precedence or priority over "scientific inferences" and "personal speculation". Under Science 2.0, Materialism, Naturalism, Darwinism, Nihilism, and Atheism are considered "unscientific" because they have NO observational evidence supporting their major premises or primary assumptions. These pseudo-sciences, speculative philosophies, and dogmatic religions have NO eye-witnesses to sustain them, and therefore they violate the Law of Witnesses.

The Law of Witnesses

2 Corinthians 13: 1: "In the mouth of two or three witnesses shall every truth be established."

Science 2.0 is based upon eye-witness evidence, observational evidence, and the lived experiences of the human race, including both the physical evidence that has been provided by our scientists so far and the non-local evidence which has been provided by the lived experiences of the human race. The best way to find and know the truth is to live it and experience it for yourself, or to choose to trust someone who has.

Science is Observation. Science is Knowledge. Science is Truth. Science is comprised of Lived Experiences.

Under this particular definition for science, Materialism, Naturalism, Darwinism, Nihilism, and Atheism are unscientific because they have NO observational evidence supporting their primary assumptions or major premises. Their lack of eye-witness evidence or lack of observational evidence FALSIFIES them.

Science 2.0 is based upon the Law of Witnesses as established by God himself. The Law of Witnesses is the best way to do science, because the best way to find the truth and know the truth is to live it and experience it for yourself or to choose to trust someone who has.

The Law of Witnesses is technically the ONLY way to find and know the truth. The Law of Witnesses is definitely better than the "scientific inferences," "circumstantial evidence," and "philosophical speculation" which we receive from the Materialists, Naturalists, Darwinists, Nihilists, and Atheists. Why? It's because science IS observation and lived experiences, not scientific inference and personal speculation.

The Law of Witnesses IS Science!

Science 2.0 defines science as the Law of Witnesses, rather than defining science as Materialism and Naturalism as the Materialists and Naturalists do.

The Dedicated Opposition

Every scientific theory and philosophical idea needs a dedicated opposition to give it some teeth and staying power – Science 2.0 is no exception.

Since Materialism, Naturalism, Darwinism, Nihilism, and Atheism have been FALSIFIED trillions of times in thousands of different ways, they really cannot serve as the opposition which Science 2.0 needs, because they have NO observational evidence supporting them. Science 2.0, which is based exclusively on the observational evidence and lived experiences of the human race, can adequately serve as a replacement for Materialism, Naturalism, Darwinism, Nihilism, and Atheism because Science 2.0 automatically subsumes ALL physical observation or physical evidence that has ever been produced; but, Science 2.0 can't actually derive "inspiration" or "opposition" from the falsified philosophies of Materialism and Naturalism because they have been FALSIFIED already.

Instead, Science 2.0 repeatedly turns to Young-Universe Creationism (YUC) and Young-Earth Creationism (YEC) for the dedicated and cunning opposition that it needs to get established. The YUCs and the YECs are extremely intelligent; and, they actually teach science that can be debunked or falsified with observational evidence, which is exactly the purpose of Science 2.0 – getting rid of falsehoods through observational evidence and the lived experiences of the human race. Over and over again, Science 2.0 turns to the YUCs and the YECs in order to get the opposition that it needs.

The YUCs and the YECs teach us directly the flaws and the weaknesses of relying exclusively upon "scientific inferences" or "personal interpretations of scientific evidence and scriptural evidence" while at the same time rejecting or trying to explain away the Observational Evidence – as do the Materialists, Naturalists, Darwinists, Nihilists, and Atheists who also deny, resist, or reject the Observational Evidence and the Lived Experiences of the human race. The people who deny the Observational Evidence and try to explain it away through philosophical logic, scientific inferences, and personal interpretations of the data end up being the dedicated opposition which Science 2.0 needs in order to make its case and demonstrate its superiority.

Under Science 2.0, Lived Experiences and Observation Evidence trump personal interpretations of data, scientific inferences, and philosophical conclusions. The weakest part of the Scientific Method has always been the final step in the process – the interpretation given to the scientific evidence.

FOUNDATIONAL TENETS OF SCIENCE 2.0

The BEST way to find and know the truth is to live it and experience it for yourself, or to choose to trust someone who has. If you are going to choose to trust someone, use your brains and choose to trust someone whom you deem to be absolutely trustworthy. You should do this with any type of evidence.

Although Science 2.0 allows all of the evidence into evidence, since the personal interpretation of the evidence is left up to YOU under Science 2.0, carefully examine that evidence and decide for yourself whether it is trustworthy or not. Under Science 2.0, YOU are called upon to decide for yourself whether evidence is reliable or not.

Science 2.0 is based upon the Burden of Proof Methodology and encourages you to pursue and follow the Preponderance of the Evidence.

Since there is NO officially sanctioned interpretation of the evidence under Science 2.0, you are left free to judge for yourself whether evidence is trustworthy and reliable, or not. This means that you are completely free to reject evidence if you deem it untrustworthy, unreliable, faked, or false – or if it fails to meet its Burden of Proof through a Preponderance of the Evidence.

The SECOND-BEST way to find and know the truth is to eliminate everything that is False and everything that has been falsified, so that ONLY the Truth remains.

We start this process by eliminating Materialism, Naturalism, Darwinism, Nihilism, Behaviorism, Hard Determinism, Reductionistic Atomism, and Atheism because they are False and have been falsified by Science and Observation. If you can successfully eliminate everything that is false and everything that has been falsified, then logically only the truth will remain standing.

Even though the Scientific Methods cannot be used to prove anything true, they can indeed be used to prove things false. When you falsify something, you do in fact prove that it is false. The Scientific Methods along with Direct Observations are extremely powerful tools for falsifying False Philosophies and False Theories such as Materialism, Naturalism, Darwinism, Nihilism, and Atheism.

It works.

How often have I said to you that when you have eliminated the impossible [and the false], whatever remains, *however improbable*, must be the truth? — Sherlock Holmes.

Eliminating the impossible and the false is what Science 2.0 is supposed to do, but only if you make it work for you. Just like in a court of law, Science 2.0 leaves it up to you to decide for yourself if the evidence meets its Burden of Proof through a Preponderance of the Evidence.

Finally, Science 2.0 is based upon Quantum Mechanics. The Materialists, Naturalists, Nihilists, and Atheists don't study and use Quantum Mechanics and Non-Local Consciousness, because these people don't believe in these things. They don't think these things are real and true, so they just ignore these things and/or sweep them under the carpet.

NEUROSCIENCE

Neuroscience is a worthy challenge for any scientist, genius, or intellectual. The complexity of Neuroscience is astronomical. This is a physical science discipline where there's still a lot more to be discovered and understood. Remember, we have to understand physical structures and physical processes before we can learn to identify and come to recognize the things that don't have a physical explanation. We each have to start where we are, and then go on from there.

Neuroscience is the study of how the brain works, from a physical perspective. It helps to study the physical aspects of the brain searching for clues as to how Quantum Mechanics, the Human Psyche, and Nature's Psyche might interface with it and interact with it.

We have to understand how Classical Physics and standard Neuroscience work, before we can see signs of Psyche and try to figure out how Quantum Non-Local Consciousness through Quantum Mechanics interfaces with and controls our physical brain.

There's a lot of interesting free material online that can help us to discover and understand Neuroscience. Some of my favorites are as follows.

Bear, M. F., Connors, B. W., & Paradiso, M. A. (2007). *Neuroscience: Exploring the Brain* (3rd ed.). New York: Lippincott Williams and Wilkins.

This is the old version of their book, which might explain why everyone seems to be handing it out for free online. This is good stuff, if you're into neuroscience!

Neuroscience: Exploring the Brain.

https://canvas.brown.edu/courses/851434/modules

BCP2

https://canvas.brown.edu/courses/851434/files/40331111/

http://mypsyche.us/wp-content/uploads/2017/12/BCP2.pdf

BCP3

https://canvas.brown.edu/courses/851434/files/40426262/

http://mypsyche.us/wp-content/uploads/2017/12/BCP3.pdf

BCP4

https://canvas.brown.edu/courses/851434/files/40426324/

http://mypsyche.us/wp-content/uploads/2017/12/BCP4.pdf

BCP5

https://canvas.brown.edu/courses/851434/files/40426574/

http://mypsyche.us/wp-content/uploads/2017/12/BCP5.pdf

BCP6

http://mypsyche.us/wp-content/uploads/2017/12/BCP6.pdf

BCP7

https://canvas.brown.edu/courses/851434/files/40676532/

http://mypsyche.us/wp-content/uploads/2017/12/BCP7.pdf

BCP20

http://mypsyche.us/wp-content/uploads/2018/01/BCP20.pdf

BCP23

http://mypsyche.us/wp-content/uploads/2017/12/BCP23.pdf

BCP24

http://mypsyche.us/wp-content/uploads/2017/12/BCP24.pdf

BCP25

http://mypsyche.us/wp-content/uploads/2017/12/BCP25.pdf

The Essentials

http://mypsyche.us/wp-content/uploads/2017/12/Transporters.pdf

http://mypsyche.us/wp-content/uploads/2017/12/Neurotransmission.pdf

http://mypsyche.us/wp-content/uploads/2017/12/Shunting-Inhibition.pdf

http://mypsyche.us/wp-content/uploads/2017/12/Equilibrium-Potential.pdf

Neuroscience: Exploring the Brain is the best introductory book on neuroscience that I have found so far; so, I'm going to point to it, quote it, and promote it whenever possible.

Huge chunks of this book are also searchable for free on Google. I bought a hardcover copy of the book, which can be purchased inexpensively on Amazon, since the 3rd edition is an older edition of their book.

I have taken Neuroscience a lot farther than the Materialists and Naturalists want us to take it or will allow us to take it. But, that's the only way to discover things that these people refuse to look at, consider, or know. They wear their ignorance like a bad of honor, and they are proud of it. They actually hand out PhDs in Ignorance. The whole purpose of our public education system is to indoctrinate us into Materialism, Naturalism, Scientism, Darwinism, Nihilism, and Atheism. The goal is to dumb us down, so that we never question Scientific Naturalism – just only believe. I KNOW, because I used to be a Materialist, Nihilist, and Atheist. They successfully brainwashed me into Materialism, Naturalism, and Scientism; and, I was completely clueless as to what they had done to me, because I was taught not to question it.

Quantum Neuroscience and Science 2.0 start with what we already KNOW to be real and true from the Lived Experiences of the human race; and then, these sciences try to find a logical explanation for what we are observing and experiencing, typically by using Quantum Mechanics and Quantum Non-Local Consciousness (Psyche) in order to do so. We each have to start with what we know, and then go on from there.

I see signs of Psyche, or Intelligence, or Quantum Non-Local Consciousness throughout the whole of Neuroscience, but only because I'm looking for it now. I'm searching for it, because I KNOW that it exists thanks to all the wide and varied experiential evidence and observational evidence from Out-of-Body Experiences (OBEs), Near-Death Experiences (NDEs), and other types of Non-Local Experiences or Spiritual Experiences.

NDEs, OBEs, and the Orthodox Interpretation of Quantum Mechanics provide scientific evidence and scientific proof that Psyche and Spirit Matter really do exist; and, Orthodox Quantum Mechanics explains how the Human Psyche negotiates with Nature's Psyche in order to command and control the human brain. Thereby, Quantum Neuroscience is born. Quantum Neuroscience is the scientific study of how the Human Psyche interacts with and controls its physical brain.

By definition, in principle, Materialism, Naturalism, Classical Physics, and Darwinism cannot explain how the Human Psyche interacts with Nature's Psyche in order to use quantum mechanical processes to successfully command and control its physical brain. The Orthodox Interpretation of Quantum Mechanics can and does explain all of this; but, Materialism and Naturalism cannot. Nevertheless, we have to understand the concepts of Classical Physics in order to identify the things that don't have a physical explanation and that require a Non-Local Quantum Mechanical Explanation instead. Some brain processes are impossible and don't make any sense in terms of Classical Physics or Naturalism; but, they become possible and make a lot of logical sense in terms of Quantum Mechanics and Non-Local Consciousness. That's just the way it is.

Remember, you have to study and know the physical explanations for the different brain processes and functions, before you can figure out where they are inadequate or fail.

Notice how we typically have to turn to atoms, molecules, proteins, genes, neurotransmitters, ion channels, ion receptors, peptides, and microbiology in order to find the things that cannot be explained in terms of Materialism, Naturalism, and Classical Physics because the existence of these things is physically impossible yet there they are. Remember, Quantum Mechanics and Non-Local Consciousness (Psyche) can be used to explain all the different things that don't have a physical explanation.

Quantum Neuroscience is a scientific attempt to identify and understand the brain processes which don't have a logical rational physical explanation. Contrary to Materialism and Naturalism which try to suppress the evidence and limit the evidence, Quantum Neuroscience is based upon Science 2.0, which allows ALL of the evidence into evidence and pursues a preponderance of that evidence. Quantum Neuroscience is what I'm going to introduce and study next. There will still be lots of unanswered questions when I am done; but, with Quantum Neuroscience I take neuroscience a lot farther than it can be taken by Classical Physics alone. I hope you enjoy the ride as much as I did.

The world is in desperate need of something like Quantum Neuroscience, and now it exists. Quantum Neuroscience is the bridge between the quantum level and the physical level, where Neuroscience is concerned. It works for me; and when I'm done, I hope that it will make sense to you as well.

Ironically, Quantum Neuroscience was already an observed science and a proven science decades before I coined the term. Quantum Neuroscience has been sitting there for over a century waiting for someone to pick it up and run with it. Now I have.

HYPOTHETICAL MODELS FOR HOW THE BRAIN WORKS

Physical Models

Communication within the Nervous System

I'm a computer scientist, programmer, and certified computer support specialist. I know how computers communicate with each other – through BYTES of information consisting of data and computer code. It's not information or code or data, unless you have at least a BYTE to work with.

Every time that I would see titles such as these – "The Exchange of Information between Neurons" and "Communication within the Nervous System" – at the start of a chapter, I would find myself exclaiming, "Finally, someone is going to explain to me how neurons communicate with each other!" I would get really excited thinking that somebody was finally going to hand me the Holy Grail.

And, what was the result every time?

To paraphrase Shakespeare, "To fire or not to fire, that is the question." And the end result of information exchange between neurons is essentially a yes or no response. (*Biological Psychology*, p. 110.)

That's it!

That's all there is to know about neurons and information exchange between neurons. There's nothing else to discover.

You can't pass even a single BYTE of information through a neuron. It's physically impossible.

So, how are neurons communicating with each other?

Yes or no.

So, how is ALL of that parallel CPU processing, petabytes of RAM memory storage, terabytes of data exchange, and gigabytes image recognition software being processed and stored within our physical brain?

IT'S NOT!

It's physically impossible.

The physical brain can't even process a single BYTE of information at the physical level. If there are any BYTES within our brain they are ALL being mapped, processed, manipulated, and stored at the quantum level, because it's physically impossible for them to exist at the physical level.

ALL of the physical models for brain functionality FAIL because they are physically impossible to implement.

OUCH!

The CPU Model

Klein, S. & Thorne, B. M. (2006). *Biological Psychology*. New York: Worth Publishers.

Biological Psychology mentions the Modular Model of Brain Function.

Rather than acting like a general problem-solving device whose every part is capable of any function, the brain is really a collection of devices that assists the mind's information-processing demands. One of the most intriguing modules of the left hemisphere is the "interpreter" module. (*Biological Psychology*, p. 480.)

These are processors, not memory modules. It's the Human Psyche, or the mind, who does the understanding and the interpretation. Thoughts and memories are quantum waves. They survive the death of our physical brain.

One view of sensorimotor function is that the sensorimotor system comprises a hierarchy of central sensorimotor programs. According to this view, all but the highest levels of the sensorimotor system have certain patterns of activity programmed into them and that complex movements are produced by activating the appropriate combination of these programs.

Practice Can Create Central Sensorimotor Programs

Although central sensorimotor programs for many species-typical behaviors develop without practice, practice can generate or modify them. (*Biopsychology*, pp. 208-209.)

Programmed by whom? Evolution can't do programming. It has never been observed doing so. Programs require Programmers! Programs don't spontaneously generate out of thin air. Activated by whom? Activation requires an Activator! Someone Psyche has to be programming and activating these things.

Practicing behaviors is a function of choice; and, choice is a function of the Human Psyche or Non-Local Consciousness. WE KNOW this is so because the brain and physical body just sit there and do nothing unless some kind of choice is made by the Human Psyche. WE KNOW this is so because the Human Psyche continues to make choices after our physical brain is dead. WE KNOW this is so, because it has been observed in brain scans that the Human Psyche can turn on parts of its brain in the complete absence of physical input, simply by choosing to concentrate on and think about a specific topic or image.

By definition, in principle, Evolution can't do programming, choice, design, and fine-tuning. Evolution is dumb, blind, and random. Random processes cannot do programming. It's physically impossible. Practice is a function of choice or the Human Psyche. Evolution has nothing to do with it. ONLY the Human Psyche can choose to practice.

For me personally, this was the foundational information that I needed to design and create my Cerebral Processing Units Model (CPU Model) for brain functionality. I used the CPU Model for months, until I found something better. We have to go with what we know. You launch out in faith with the hope that you will learn something better as you go along, which I did.

Under this CPU Model, the brain is visualized as being comprised of multiple different programmed Cerebral Processing Units (CPUs); and, programs can be stringed together into sequences of unified functionality. The Materialists and Naturalists call this Memory and

Learning – Procedural Memories; but, it's more accurately described as programmed functionality instead. The individual steps don't have to be remembered – the whole sequence just plays out automatically as practiced. Memory is no longer needed, because the brain and body are running on autopilot.

The Response-Chunking Hypothesis is the idea that practice combines the central sensorimotor programs that control individual actions into a single program that controls all the sequences of action or chunks of behavior.

Instead of having to think of every individual keystroke, a master typist or pianist plays things "whole words" or "whole melodies" at a time, chunking letters and notes into one united whole. All of the individual actions or programs are chunked into a single stream or sequence. The same thing happens to a gymnast or martial artist.

As my research continued over the months, I came to realize that the CPU Model has some serious flaws and weaknesses. The CPUs and RAM can't be implemented at the physical level thanks to the scrambling and randomizing effects of synaptic clefts; and, the CPU Model or Programming Model typically refuses to take the quantum level into consideration.

Functional Segregation is the idea that the brain is organized into different areas, each of which performs a different function. In the sensory systems, for example, the different areas of the primary, secondary, and association cortices analyze different aspects of the same sensory stimulus. This model of brain functionality suggests that the brain is organized into different Maps of Physical Functionality and/or the brain is organized into different functional Cerebral Processing Units (CPUs).

So, who mapped all of this functionality into your physical brain? The Materialists and Naturalists and Darwinists will tell you that evolution did it. However, I discovered that NOTHING survives passage through a synaptic cleft at the physical level – not even programming code or evolution. If we want a realistic explanation for how all of this mapping is being done, operated, and updated, we have to look someplace else besides evolution.

Nature's Psyche is organizing our brain structures and cortices into functional Cerebral Processing Units (CPUs) both at the quantum level and the physical level. The physical level is obvious, because we can observe the different brain structures with our physical instruments. WE KNOW that Nature's Psyche is mapping physical functionality into the CPUs within our brain at the quantum level, because there are NO wires in our brain that can be used to map and program all of that functionality into the brain at the physical level. Nature's Psyche differentiates and creates all the different structures within our brain at the physical level; and then, Nature's Psyche maps functionality onto our different brain structures at the quantum level because it's physically impossible to make CPUs and RAM out of neurons or stand-alone switches at the physical level. Even if it were possible to make functional CPUs and RAM and Memory Systems out of a 100 billion stand-alone switches or neurons, those CPUs and RAM would still have to be programmed and operated at the quantum level by Nature's Psyche. Intervention from Nature's Psyche is essential if you want to have a functional brain. It can't be avoided.

Remember, there's NO physical entity in the synaptic cleft called "evolution," who is moving the neurotransmitters and degradation enzymes around with tractor beams at the quantum level. Evolution doesn't work at the quantum level.

Along the same lines, neuroscientists have observed that our brains seem to have some sort of hierarchical organization into a series of levels that can be ranked with respect to one another – such as the primary sensory cortex, secondary sensory cortex, and the association cortex performing more detailed analyses along the way. Our brains seem to be organized into ever-increasing levels of functionality. Our physical brain is all about physical functionality; and, Nature's Psyche seems to have designed it to be so.

CPU Models for brain functionality have a few major flaws, because they are materialistic and naturalistic in nature and origin. The CPU Model doesn't work, without a lot of help from Nature's Psyche at the quantum level or the psyche level. As a purely physical model at the physical level, the CPU Models and RAM Models fail due to the randomizing and integrating effects of our synaptic clefts. You can't even pass a single BYTE of information through a synaptic cleft and have it survive intact at the other end – not at the physical level at least.

Remember, there are NO wires, no CPUs, no RAM, and no Memory Systems within a physical brain. They don't exist at the physical level, thanks to the synaptic clefts which scramble or randomize everything that comes their way reducing all of that information into a single Post-Synaptic Potential, which determines whether the post-synaptic neuron fires or doesn't fire.

A neuron is a stand-alone switch, which is either ON or OFF. It either fires or it doesn't fire. With no wires to connect them, you can't make functional CPUs, RAM, and Memory Systems out of a 100 billion stand-alone switches or neurons. It's physically impossible.

But, who knows what's happening at the quantum level or the psyche level?

I guess Nature's Psyche knows; but, the Human Psyche certainly doesn't.

Diffuse Modulatory Systems

Diffuse Modulatory Systems involve dumping neurotransmitters diffusely into the extracellular fluid of the brain, involving something called an indirect synapse, undirected synapse, or non-directed synapse. These are the neurotransmitters (typically monoamines) that have to be maintained at a specific balanced level within the brain, or mental illness and abnormal behavior is the result.

Anatomy and Functions of the Diffuse Modulatory Systems

The diffuse modulatory systems differ in structure and function, yet they have certain principles in common:

1. Typically, the core of each system has a small set of neurons (several thousand).

2. Neurons of the diffuse systems arise from the central core of the brain, most of them from the brain stem.

3. Each neuron can influence many others, because each one has an axon that may contact more than 100,000 postsynaptic neurons spread widely across the brain.

4. The synapses made by many of these systems release transmitter molecules into the extracellular fluid, so they can diffuse to many neurons rather than be confined to the vicinity of the synaptic cleft.

We focus on the modulatory systems of the brain that use either NE, serotonin (5-HT), dopamine (DA), or ACh as a neurotransmitter. Recall from Chapter 6 that all of these transmitters activate specific metabotropic (G-protein-coupled) receptors and that these receptors mediate most of their effects; for example, the brain has 10–100 times more metabotropic ACh receptors than ionotropic nicotinic ACh receptors.

Because neuroscientists are still working hard to determine the exact functions of these systems in behavior, our explanations here will necessarily be vague. It is clear, however, that the functions of the diffuse modulatory systems depend on how electrically active they are, individually and in combination, and on how much neurotransmitter is available for release. (*Neuroscience*, p. 498.)

These Diffuse Modulatory Systems don't work right if there's too little neurotransmitter in the system or too much. This is the physical model that the drug manufacturers rely upon the most, because the diffuse modulatory systems are where their drugs can have the most impact, for good or ill.

It has been observed that drugs that block the synthesis or release of serotonin increase aggressive behavior.

Too much serotonin, and heart disease seems to be the result. Too little serotonin, and depression, anxiety, OCD, and aggression seem to be the result.

Too much dopamine seems to feed schizophrenia and psychosis. Too little dopamine causes Parkinson's Disease.

Memories Distributed Among Synapses

There are many distributed memory systems and distributed processing systems that have been suggested over the years. These models typically have our memories and CPU processing being done within our synapses; but, our synapses scramble and randomize everything that comes their way. It's physically impossible to store memories as BYTES within a synaptic cleft. It's physically impossible to store megabytes of memories within an ion receptor. It's also physically impossible to store even a single BYTE of information within a neuron, because a neuron is a hardware BIT. You can't store a BYTE within a hardware BIT – not at the physical level at least.

If all that distributed memory and distributed processing is happening for real, and it seems as if it is, then WE KNOW that it's ALL happening at the quantum level or the psyche level, because it's physically impossible at the physical level.

I keep waiting for them to explain to me how information (plural) or BYTES are being exchanged between neurons. THEY ARE NOT! Not at the physical level at least. Who knows what's happening at the quantum level through quantum waves or telepathy?

Mind-Over-Matter Models

Pay Attention!

It has been observed in brain scans that different parts of the brain turn on or fire up depending upon what the Human Psyche chooses to pay attention to – even in the complete absence of physical input to the area of the brain that's being turned on by the Human Psyche's choice to pay attention. The Human Psyche can turn on its brain, even in the complete absence of physical stimuli or physical input.

Psyche is the ultimate causal agent!

In contrast, evolution can't choose to pay attention to anything. Evolution can't do Quantum Mechanics, because evolution isn't a psyche or a person. Evolution can't do anything at the quantum level or psyche level.

"Chapter 10: Attention Must Be Paid" was particularly important as Proof of Concept for Quantum Neuroscience.

http://publicism.info/psychology/mind/11.html

http://mypsyche.us/wp-content/uploads/2018/01/The-Mind-and-the-Brain.pdf

This is good stuff that everyone should have access to, in my humble opinion

The state of selectively processing simultaneous sources of information is known as attention. In the visual system, attention enables us to concentrate on one object over many others in our visual field. Attention has to do with preferential processing of sensory information.

Neural Reflections of Attention

Human brain imaging studies and recordings from individual neurons show that cortical activity is significantly altered by the attention we exert. (*Neuroscience*, p. 644-645.)

Attention is a function of the Human Psyche. The Human Psyche chooses what part of the sensory input to pay attention to.

There are neural reflections from the Human Psyche's attention. This is mind-over-matter caught in the act.

The Human Psyche can choose to pay attention to specific ideas and mental images, and thereby get Nature's Psyche turn on specific parts of its brain, in the complete absence of any physical input from its sensory systems. With eyes closed and taped shut, the Human Psyche can heat up or turn on the occipital lobes simply by choosing to visualize a beautiful picture in its mind's eye. That's how WE KNOW that it is the Human Psyche who is paying attention and NOT something in the brain.

Selective attention is the ability of the Human Psyche to focus on a small selective subset of the multitude of physical stimuli that are being received at any one time.

Attention of any type is a function of the Human Psyche. It involves choice.

We consciously perceive only a small subset of the many stimuli that excite our sensory organs at any one time and largely ignore the rest. The process by which this occurs is selective attention.

Selective attention has two characteristics: It improves the perception of stimuli that are its focus, and it interferes with the perception of the stimuli that are not its focus.

Attention can be focused in two different ways: by internal cognitive processes (endogenous attention) or by external events (exogenous attention).

Endogenous attention is thought to be mediated by top-down (from higher to lower level) neural mechanisms; whereas, exogenous attention is thought to be mediated by bottom-up (from lower to higher) neural mechanisms.

Where do top-down attentional influences on sensory systems originate? There is a general consensus that both prefrontal cortex and posterior parietal cortex play major roles in directing top-down attention. (*Biopsychology*, pp. 184-185.)

Pinel is a Materialist, Naturalist, Darwinist, and Atheist. They almost always get things wrong. These people can be extremely fascinating to study – to see what they get wrong and then to try to figure out why they got it wrong. That's why I quote them extensively in this book. I'm trying to figure out how things really work; and, it starts by figuring out what these people are getting wrong.

Can you sense where he's getting a few things wrong?

Stimuli, sensation, or physical input is a function of the brain; and, so is physical output or movement of the physical body. The physical brain was designed by God and implemented by Nature's Psyche for physical functionality. In contrast, perception is a function of the Human Psyche and Nature's Psyche. The Materialists and Naturalists deliberately conflate sensation with perception, because these people are trying to trick you and deceive you.

Consciousness is a function of the Human Psyche.

How do we know?

WE KNOW from the fact that Near-Death Experiencers (NDErs) say that they are super-conscious or hyper-conscious when they are dead and in the spirit world. Death is consciousness on steroids.

Attention is chosen into existence by the Human Psyche. WE KNOW this is so because we choose to go places and choose to pay attention to things after our physical brain is dead and gone.

In contrast, evolution doesn't have senses and sensory information to draw upon, so it can't concentrate and pay attention to what it is doing and where it is going. Evolution really is dumb and blind, which makes it worthless.

Selection of any kind involves choice; and, choice is the primary function of the Human Psyche according to the Orthodox Interpretation of Quantum Mechanics. Stimuli are input from the physical body; but, perception is a function of the Human Psyche. Again, WE KNOW this is so, because dead people (NDErs) are sentient and perceive their environment long after their physical body is dead and gone.

Attention and intention are a function of the Human Psyche. WE KNOW this is so, because it involves choice. Your psyche chooses what it's going to pay attention to and what it's going to ignore. Your psyche chooses what it is going to do with its physical body. The Materialists, Naturalists, and Behaviorists assure us that choice is physically impossible, which means that ALL of our choices are taking place at the quantum level or the psyche level.

Cognitive processes are a function of the Human Psyche. We continue to have cognitive processes long after our brain is dead and gone. Thoughts, memories, choices, and cognitions are quantum waves. They survive the death of our physical brain.

In his book, Pinel repeatedly talks about a hierarchy – different levels of brain functionality. By choosing to allow ALL of the evidence into evidence, we simply KNOW that the Human Psyche is at the TOP of all of these different top-down processes.

There is a general consensus among everyone who chooses to allow all of the evidence into evidence that the Human Psyche plays the major role in directing top-down attention. Furthermore, all physical stimuli and sensory input feed in from the bottom-up and converge on the Human Psyche and Nature's Psyche at the top.

The Human Psyche is at the top of the hierarchy. WE KNOW this is so, because the Human Psyche survives the death of its physical brain. It comes out on top.

In his book, he compares the brain and motor output to a finely-tuned corporation. Fine-tuning is a function of Psyche. WE KNOW this is so, because physical matter cannot fine-tune itself. Physical matter has never been caught in the act of abiogenesis, spontaneous generation, chemical evolution, or macro-evolution of any kind. Physical matter can't act in this manner, because it's physically impossible. God prevents the atoms from spontaneously generating and coming alive. WE KNOW this is so, because it has never been observed and has never been replicated in any kind of lab. It's physically impossible. Since it obviously it happened, WE KNOW that it happened at the quantum level or the psyche level under the command and control of Someone Psyche.

The Human Psyche has direct access to ALL of the sensory input that registers on the neurons of its physical brain. Then, the Human Psyche learns to pay attention to the information that interests it, while ignoring the other information that's coming in. The Human Psyche chooses what to do with all of that physical input; and, Nature's Psyche implements that choice both at the quantum level and the physical level, because it's Nature's Psyche who collapses the wave function.

By paying attention, the Human Psyche chooses what it wants to remember most; and then, Nature's Psyche collapses the necessary wave functions at both the quantum level and the physical level in order to make it so. This is how the physical brain really works, both at the quantum level and the physical level.

With Science 2.0 and Quantum Neuroscience, I chose to take everything into consideration. In other words, I chose to pay attention to everything. I wanted to know how everything works, and now I do.

Selective attention and attention really can't be explained at the physical level. The only part of this that can be explained at the physical level is ALL of the sensory input that's registering on the different neurons of the physical brain. There is NO physical mechanism for deciding which neurons to pay attention to and which neurons to ignore. Where attention and choice are concerned, choosing what parts of the human body to pay attention to is a function of the Human Psyche who is in control of that particular human body. That is the BEST and most parsimonious explanation that can be given, because it takes both the quantum level and the physical level into consideration.

The best that the Materialists and Naturalists can give you is to tell you that they don't know how selective attention is done, but it's got to be some kind of brain function because there is nothing else. They just as well cut out their brains for all the good that it does them. I think that would get their attention and thought that would get their attention.

Attention and selective attention are Quantum Neuroscience in action.

Looking for Love in All the Wrong Places

The Materialists, Naturalists, Darwinists, and Behaviorists mess up attention, just like they mess up everything else. I figured that would get their attention.

Remember, the Human Psyche is slaved to the physical brain; whereas, Nature's Psyche is not. The Human Psyche can't pay attention to things in the external world that the brain can no longer sense or process. If the brain lacks the machinery or the visual registers, then the Human Psyche can't perceive these things, and neither can the brain of course.

What's happening within the brain when we shift our attention to something?

First, we look at human brain imaging studies that show activity changes associated with the allocation of attention and then turn to animal studies that reveal the effects of attention on individual neurons. These experiments show the consequences of giving attention to a location of feature. Only a few recent experiments have explored brain areas that might be involved in guiding attention. (*Neuroscience*, p. 649.)

Again, these are Materialists, Naturalists, and Behaviorists.

How can you tell?

They talk about the brain guiding attention rather than responding to attention. Psyche acts or guides or chooses. Physical matter responds. That's the way things really work.

The "WE" who is doing the shifting of the attention is the Human Psyche, not the physical brain. The Human Psyche continues to shift attention long after its brain is dead and gone. Of course, being Naturalists, they conflate "WE" with the physical brain.

Raw physical matter can't guide brain areas nor pay attention to a specific location or specific features – at least not at the physical level. Who knows what the rocks can do at the quantum level?

Functional MRI Imaging of Attention to Location

A key observation made in behavioral studies of attention is that enhancements in detection and reaction time are selective for spatial location. When we know where an important stimulus is more likely to appear, we move our attention to it and process the sensory information with greater sensitivity and speed. A common analogy is that attention works like a spotlight, moving to illuminate objects of particular interest or significance.

Recent experiments using functional magnetic resonance imaging (fMRI) imaging of the human brain suggest that there may be selective changes in brain activity associated with spatial shifts in attention. (*Neuroscience*, p. 649.)

Who is moving the spotlight? Someone Intelligent has to move the spotlight, or it's just going to stay where it currently is.

Through fMRI and PET scans, we can actually observe Psyche priming the pump or turning on different brain areas in real-time.

Finding Neural Correlates of Attention

Everything still comes down to the right place in the brain and the right behavioral question. (p. 653.)

Actually, everything comes down to getting the Human Psyche to pay attention. The human body just sits there and vegetates, and the physical brain idles in default mode, unless the Human Psyche makes them do something.

Remember, correlation is NOT proof of causation. Once again, these Naturalists are deliberately breaking this cardinal rule of science by attributing attention to physical causes or neural causes. Attributing attention to physical causes is a *category error* logic fallacy, because the Human Psyche continues to pay attention and continues to make choices long after its physical body is dead and gone, according to the empirical observational evidence from Near-Death Experiences (NDEs) and Shared-Death Experiences (SDEs.)

A close association appears to exist between eye movement and attention. (p. 657.)

Ya think?

But then we get this in the next sentence.

Recent experiments suggest that the brain circuitry responsible for directing the eyes to objects of interest might also play a critical role in guiding attention. (p. 657.)

We get a correct observation, followed by a faulty interpretation or a faulty correlation.

Physical matter can't guide attention. It's physically impossible! According to the Materialists, Naturalists, and Behaviorists, choice and attention are physically impossible, which means that our choices of what to pay attention to are taking place at the quantum level or the psyche level.

You (your psyche) can't pay attention to something physical, if you can't sense it or see it with your physical machinery. If the physical brain can't see it, then the Human Psyche can't pay attention to it. However, there's nothing physical in the brain that's guiding attention. It's the Human Psyche who chooses to pay attention. The Human Psyche can choose to visualize a picture or a scene in its mind's eye and actually turn on the associated visual systems within its brain, even with its eyes closed and even in the complete absence of visual input.

Attention is a function of the Human Psyche, and not the physical brain. They always get this one wrong.

Studies of attention provide a powerful reminder of the flexibility of the human brain. By allocating more mental energy to one location or feature, we can enhance our sensitivity or reaction time. At the same time, we ignore competing stimuli that are of less interest. Reflections of our attentional bias can be clearly seen in brain imaging studies. Perhaps most surprising is the finding that changes in attention are reflected in the receptive field properties of individual neurons spread across the brain.

Perhaps you have wondered why we even need attention. After all, if sensory information successfully makes its way into the brain, why not process all of it? One possibility is that the brain simply cannot handle all the incoming sensory information simultaneously. For example, striate cortex in a macaque monkey occupies about 10% of all cerebral cortex. By current estimates, there are more than twenty other visual areas, but many are much smaller than striate cortex. It is likely that these other areas cannot process all of the detailed information represented in area VI. If this is true, attention plays a key role in selecting what information should receive the limited processing resources of the brain. For this reason, some scientists believe there may a close relationship between attention and the brain mechanisms of conscious awareness. (*Neuroscience*, pp. 658-659.)

Actually, it is not the physical brain that is flexible. It's the Human Psyche. The physical brain simply does the same thing every time a stimulus registers on the brain, until the Human Psyche chooses to pay attention to something else or chooses to do something else. There's NO flexibility in the human brain. It simply does whatever the Human Psyche and Nature's Psyche tells it to do.

Whenever the Human Psyche changes attention, Nature's Psyche turns on the parts of the brain needed to accomplish the new assigned task.

Allocation is a function of choice; and, choice is a function of the Human Psyche. According to the Orthodox Interpretation of Quantum Mechanics by Henry P. Stapp, the Human Psyche makes the choices and pays the attention, and Nature's Psyche collapses the necessary wave functions in order to make those choices physical and real. That's the way things really work within our brain, both at the quantum level and the physical level.

Mental energy is Psyche. Quantum waves are Psyche. Memories and thoughts are quantum waves.

It's the flexibility of Psyche that we are studying here, whenever we start to study attention, concentration, intention, decisions, and choices. As long as the brain is functioning properly at the physical level, it has NO choice but to do what the Human Psyche and Nature's Psyche tells it to do. When the brain is damaged or malfunctioning, lots of glitches happen over which Nature's Psyche and the Human Psyche don't have complete control. There's nothing that Nature's Psyche can do about broken machinery, broken genes, and broken proteins. We have to turn to God in order to have that type of physical functionality fixed. Whether God chooses to fix things for us or not, the Human Psyche still has a choice in what to do with the physical functionality that remains. Sometimes the stubbornness of the Human Psyche can actually force Nature's Psyche to remap the lost functionality onto other parts of the brain – as in the case of stroke victims.

Remember, the brain's mechanisms for conscious awareness are Quantum Mechanics and Psyche, which are the same thing as the mechanisms behind awareness, consciousness, attention, selectivity, and choice. The Human Psyche provides the "conscious awareness" or the "attention"; and, Nature's Psyche provides the "brain mechanisms" both at the quantum level and the physical level, through Quantum Maps of Physical Functionality and by collapsing the necessary wave functions to produce that physical functionality.

One thing they do get right is the FACT that our physical brain can't store ALL of our memories simultaneously, and therefore Nature's Psyche has to offload a huge chunk of our memories into the cloud for later recall and use.

It's also TRUE that our physical brain cannot process all of the detailed information that comes its way at the physical level. Our brain has to offload huge chunks of that quantum parallel processing to Nature's Psyche in the cloud in order to get it done in a timely fashion.

Attention from the Human Psyche does indeed SELECT what information should receive the limited processing resources of the brain. Neurons associated with sensory input are registers, and the Human Psyche does indeed choose which input and therefore which neurons to pay attention to, since it seems to be impossible to pay attention to them all.

See what can be done by choosing to allow ALL of the evidence into evidence!

Other Mind-Over-Matter Models

Gene knockout techniques are a form of genetic engineering for creating organisms that lack a specific gene. The amazing thing is that Nature's Psyche compensates for the missing gene. Let's say for example that you have genes A, B, and C in that order. Now, let's use genetic engineering to knockout gene B. Nature's Psyche KNOWS that gene B is missing. Nature's Psyche continues to treat gene C as gene C rather than treating gene C as the missing gene B. Using the map for our genome contained in God's Database, Nature's Psyche knows when a gene is missing; and, Nature's Psyche goes along business-as-usual despite the missing gene. Some functionality is lost due to the missing gene, but the whole system doesn't go down just because that gene is missing. The non-local map or index of our genome is an ISAM (indexed sequential access method) or a random-access method, and NOT a sequential-access method for indexing our genome. If one Christmas light burns out, the whole string or chromosome doesn't go out as a result. This reality suggests Intelligence or Psyche – mind-over-matter.

Purpose and meaning are a function of the Human Psyche.

Semantic analysis is an analysis of the meaning of language. Only Psyche can do meaning and analysis; and, only the Human Psyche can do the full range of language on both sides of the veil.

Explicit memories are conscious memories.

Implicit memories are memories that can't be recalled by the Human Psyche and physical brain – unconscious memories.

Episodic memories are explicit memories for the particular events and experiences of one's life.

Semantic memories are explicit memories for general facts and knowledge.

All of these semantic memories and episodic memories seem to survive the death of our physical brain, according to the after-death Life Reviews that are on record. This seems to suggest that our after-death Life Review Memories are being recorded by Nature's Psyche or the Angels in Heaven holographically as quantum waves at the quantum level, and then being stored in God's Database.

Quantum waves are thoughts; and, they are being produced, broadcast, retrieved, and interpreted by some kind of Psyche or Intelligence at the quantum level.

Out-of-Body Travelers have observed that the spirit body doesn't have thoughts and doesn't make memories while its Psyche or Intelligence is separated from it. Consciousness and sentience reside within the Psyche, and not the spirit body. Psyche is thought. Psyche is memory. Psyche is intelligence. Psyche is a quantum wave or a vibrating thought. In string theory, a string is a thought or a quantum wave. That's why I have chosen to separate the Psyche from the spirit body and treat them as two separate entities. I noticed in the scriptures, which the Biblical God Jesus Christ had a hand in writing and producing, that He has done the same thing; and/or, His chosen prophets have.

The exteroceptive sensory system is the five sensory systems that register stimuli from outside the body – vision, hearing, touch, smell, and taste. The Materialists and Naturalists tell us that these five sensory systems or Maps of Physical Functionality are interpreting this data and assigning meaning to this data. The Darwinists will tell you that evolution is doing it. But, that's physically impossible. NO type of meaning, memory, purpose, analysis, or interpretation can get through a synaptic cleft at the physical level. A synaptic cleft scrambles and randomizes everything that comes its way, reducing it ALL to a single post-synaptic potential. It's the Human Psyche who is interpreting this sensory data and deciding what to do with it.

This is a whole different way of looking at memories and brain functionality than what you will get from the Materialists and Naturalists who deliberately restrict themselves to the physical level and choose to ignore the quantum level or psyche level.

An electroencephalogram (EEG) measures the gross electrical activity of the brain. It's using physical instruments to record brain waves. Imagine all the quantum waves that are taking place at the quantum level, that can't be detected with our physical instruments.

Electron microscopy is used to study the fine details of cellular structure. Functionality doesn't seem to be disturbed, which means that functionality and thoughts are not taking place at the physical level, but they are instead taking place at the quantum level through the mediation of the Human Psyche and Nature's Psyche. If thoughts and functionality were taking place at the physical level, then something like fMRI or electron microscopy should scramble a person's brain and thoughts, and it should destroy the person's functionality and personality.

We see signs of Psyche all throughout Neuroscience, but the Materialists and Naturalists choose to ignore it.

Throughout this book, I try to cover the full range of Neuroscience from a quantum perspective or a non-local perspective. You will have to decide for yourself if any of it has value to you or not.

Quantum Maps of Physical Functionality

When I first started writing this book, I developed the (Cerebral Processing Unit) CPU Model, because it's clear to everyone that each structure in the brain has been programmed to perform a specific function. I even got to the point where I had finally concluded, based on all the evidence, that it is Nature's Psyche who has programmed this functionality into our brains at the quantum level; and then, Nature's Psyche uses this CPU programming in order to get things done for us at the physical level. It's a useful model and easy to relate to.

However, as the months of research went by, it finally became clear to me that there are NO wires in our brain and that a synaptic cleft will scramble and integrate ALL programming code, memories, and data that comes its way into a single BIT of information – a Post-Synaptic Potential in the Post-Synaptic Neuron. When I made these two discoveries, I realized that the CPU Model is not the best model for describing brain functionality at the physical level. There are no CPUs, no RAM, and no Memory Systems within our brains at the physical level. This type of brain functionality only exists at the quantum level. It clearly exists, but only at the quantum level, because it's physically impossible at the physical level.

Further research led me to observe that the 100 billion stand-alone switches or neurons within our brain, rather than being organized into CPUs and RAM, have instead been organized by Nature's Psyche into Maps of Physical Functionality at the quantum level. I understand maps and how they work, almost as well as I understand CPUs and RAM. It's just as easy to say that physical functionality has been mapped into our brains by Nature's Psyche at the quantum level, as it is to say that physical functionality has been programmed into our brains by Nature's Psyche at the quantum level. Mapping and programming are interchangeable – basically the same thing, especially at the quantum level; but, the Maps of Physical Functionality Model comes closer to telling it as it really is, in my humble opinion.

When you look in a mirror, you are not seeing yourself. You are seeing a map or a reflection of yourself comprised of quantum waves – in this case, visible light waves. But, it's a map any way you choose to look at it, NOT programming. The design and programming for this physical feature was done by God someplace else besides the map. The Maps of Physical Functionality Model moves us one more step away from the CPU and RAM Models, which is appropriate, because there are no CPUs and no RAM within a physical brain; but, there are indeed Maps of Physical Functionality within the physical brain organizing the 100 billion stand-alone switches into different and useful collections of switches, which turn ON and turn OFF specific functions for us at the physical level under the command and control of Nature's Psyche.

Rather than going through and changing everything in this book over to the Maps of Physical Functionality Model, I intend to leave the CPU Model in place, because it has its uses and a lot of other people have chosen to use it as well. Programming and mapping at the quantum level are basically the same thing anyway. For all I know, there might in fact be CPUs, RAM, and Memory Systems mapped into our brains at the quantum level by Nature's Psyche; so, the CPU Model might be valid after all.

I just KNOW that at the physical level, the Maps of Physical Functionality Model is a better and more accurate model than the CPU Model (or the RAM and Memory Systems Model which the Materialists and Naturalists have chosen to use), because there are NO wires, no CPUs, no RAM, and no Memory Systems within a physical brain at the physical level thanks to the terminating effects of synapses and synaptic clefts at the physical level.

Obviously, you will have to decide for yourself which model you prefer; but, I'm leaning towards the Maps of Physical Functionality Model more and more as each day goes by, because it's the one the matches best with reality and because it's the one that has the most explanatory power. Quantum Mechanics has infinitely more explanatory power than Materialism, Naturalism, Scientism, Darwinism, Nihilism, Behaviorism, Atheism, and Classical Physics combined.

Quantum Maps of Physical Functionality is infinitely more versatile and powerful than the CPU Models and the RAM Models for brain functionality, which we typically get from the Materialists and Atheists, because Quantum Maps of Physical Functionality work at both the quantum level and the physical level. Anytime you choose to involve the quantum level or the psyche level, you are going to end up with a more powerful model in the end – a model that actually matches with the whole of reality, and not just the physical sub-set of reality.

The maps are made and exist at the quantum level; and, there's no limit placed on them at the quantum level. The functionality is observed and experienced at the physical level. The physical level always has its limitations.

Quantum Maps of Physical Functionality is inherently more powerful than anything we can get from the Materialists, Naturalists, and Classical Physicists.

With Quantum Maps of Physical Functionality, any number of different CPUs existing at the quantum level can be mapped onto brain structures at the physical level by Nature's Psyche. Any number of different RAM modules or memory systems existing at the quantum level can be mapped onto a single specific brain structure at the physical level, for different types of physical input and for different types of quantum memories, quantum waves, and quantum thoughts manifesting at the physical level. Cut out a specific brain structure, and consolidated quantum memories can be instantly remapped to a different brain structure on the fly. But, cut out a part of the brain that has been mapped for physical functionality – physical input or physical output – and that physical functionality may or may not return depending upon a lot of different factors.

With Quantum Maps of Physical Functionality, different programs and different memory modules existing at the quantum level can actually be mapped onto the same brain structure (hippocampus) at the physical level, at will, under the command and control of Nature's Psyche. One map can be swapped with a different map at the quantum level, thereby producing different physical functionality, different physical input, and different physical effects at the physical level within the same physical brain structure.

Due to the fact that Nature's Psyche created and organized both the physical brain and these various quantum maps, the physical brain and these quantum maps or quantum memories exist in symbiosis with each other. The physical functionality from all of these different quantum maps works adequately well at the physical level as long as the physical structure remains intact. The quantum maps are typically mapped to a specific physical structure within a specific physical brain at both the quantum level and the physical level. If the physical structure gets damaged, the custom quantum map(s) that was made for that specific physical structure by Nature's Psyche ceases to be useful. It either needs to be redone or discarded. A young child's brain can be remapped. An adult's brain was actually designed to sacrifice neurons and to discard the functionality once it is lost or no longer working right.

Quantum memories or quantum thoughts are different though. They can travel, and they can be mapped onto any physical brain structure that Nature's Psyche sees fit.

With Quantum Maps of Physical Functionality, different programs along with their associated data can be loaded into the same brain structure one after another at the quantum level and then used by Nature's Psyche to produce physical input and physical output at the physical level. Maps of Physical Functionality at the quantum level gives our brain multi-use functionality at the physical level, which is something that is physically impossible through physical matter alone within a physical brain thanks to the randomizing and integrating effects of our synaptic clefts.

Maps of Physical Functionality, being produced at the quantum level by Nature's Psyche and then being used by Nature's Psyche to do things for us at the physical level, ends up as the BEST Model for explaining brain functionality at both the quantum level and the physical level; and therefore, it ends up being the BEST Model for Quantum Neuroscience, in my humble opinion.

Remember, if there is any CPU functionality or RAM storage taking place within our brains, it's taking place at the quantum level and not the physical level, because it's physically impossible at the physical level. At the physical level, neurons are stand-alone switches with NO wires connecting them. We have to include the quantum level or the psyche level, if we want to know what's really happening within our brain. This is the ultimate discovery of Quantum Neuroscience.

This discovery is sustainable, replicable, demonstrable, and verifiable once a person realizes and accepts the fact that there are NO wires within a brain. Neurons are switches, which are either ON or OFF. It's physically impossible to create functional CPUs and RAM and Memory Systems out of 100 billion stand-alone switches without some kind of wiring to connect them. Remember, the "wiring" connecting our neurons and astrocytes is implemented at the quantum level, because it doesn't exist and can't exist at the physical level.

I needed something that would bridge the quantum level and the physical level; and, I believe that I have found what I need in Quantum Maps of Physical Functionality, which Nature's Psyche maps onto our brains at the quantum level and then uses in order to get things accomplished for us at the physical level.

The Orthodox Interpretation of Quantum Mechanics from Henry P. Stapp explains how most of this works at the quantum level; and, we can observe how much of this works at the physical level. The Quantum Maps of Physical Functionality are the bridge or interface between the quantum level and the physical level.

If the physical brain structure gets damaged or removed, then the associated Maps also cease to exist at the physical level, unless Nature's Psyche maps the functionality onto some other brain structure at the physical level, which may or may not happen depending upon the extent of the physical damage, the type of map or functionality, the age of the individual, and the level of brain development that has already taken place. We can't perform these physical functions at the physical level if the mapped functionality gets destroyed at the physical level, even though these functions are being mapped onto our brains at the quantum level by Nature's Psyche. If the part of our brain that has been assigned by Nature's Psyche to experience and process physical events gets destroyed, then the Human Psyche is no longer able to experience and remember those types of events. The Human Psyche is slaved to its physical brain; but, Nature's Psyche is not.

Physical functionality learned and practiced by the Human Psyche and mapped onto our brain at the physical level by Nature's Psyche doesn't always come back if the physical structure is destroyed. However, consolidated declarative memories mapped for us at the quantum level by Nature's Psyche seem to be able to "travel" anywhere in the brain no matter what part of the brain gets destroyed. Thoughts and memories are quantum waves; and, quantum waves are something completely different than Physical Functionality.

Physical functionality can be destroyed by destroying parts of the brain; but, thoughts, memories, psyche, and quantum waves cannot be destroyed by destroying parts of the physical brain. That's why our long-term consolidated declarative memories seem to travel or seem to be distributed throughout our brain. The thoughts, memories, and personality traits that survive the death of our brain exist and are stored at the quantum level or the psyche level; and, they can be mapped to any part of our brain. This is what all the observational evidence and experimental evidence are telling us is actually happening for real.

Have you ever strained really hard to recall a memory for a specific fact, only to find it physically impossible to do so; but then hours later when you are relaxed and no longer thinking about the subject, the memory pops into your mind whole, unbidden, but totally intact? Quantum Maps of Physical Functionality EXPLAIN this phenomenon infinitely better than Materialism, Naturalism, and Classical Physics can. You see, if ALL of our memories were really stored within our physical brain as the Materialists and Naturalists claim, then ALL of those memories should be instantaneously and simultaneously available at all times and in every situation, because a physical realm is the ultimate best consensus reality that God has ever created. If those memories really existed at the physical level within our brain as the Materialists and Naturalists claim, then just like the hard drive in your computer, those memories would be instantly available at all times thanks to the physical reliability of the physical realm. The very fact that the vast majority of our memories are NOT available to us on demand – but have to be "called up" or "recalled" – is proof positive that those memories are not being stored in a physical format within our physical brain. These observations are proof of concept for Quantum Neuroscience.

From what I can tell, Quantum Memories and Quantum Maps of Physical Functionality seem to explain everything that's going on within our brains, both at the quantum level and the physical level. I chose to go with this model for brain functionality, until I find something better.

Technically, it took me over 55 years to discover Quantum Maps of Physical Functionality, because I had to **overcome** my brainwashing or conditioning in Materialism and Naturalism, which wasn't easy to do. It took me four or five months to discover and develop Quantum Maps of Physical Functionality, once I was actively looking for it. Again, the major obstacle was that I had to **overcome** my brainwashing and conditioning in the CPU Models, RAM Models, and Wired Brain Models that my neuroscience teachers were feeding me in order to discover the Quantum Maps of Physical Functionality Model, which wasn't easy to do. As a former Materialist, Nihilist, Skeptic, and Atheist, it hasn't been easy for me to **overcome** these things; but, it was necessary if I wanted to discover how our brain really works at both the psyche level and the physical level.

Quantum Maps of Physical Functionality is an extremely powerful concept.

It's theoretically possible to map the output from zillions of quantum computers crunching out results at the quantum level into a single atom or neuron at the physical level. It's theoretically possible to map exabytes of quantum level RAM or quantum memories into a single atom or neuron that exists at the physical level. Maps of Physical Functionality are also bi-directional. Therefore, through Quantum Maps of Physical Functionality, it's theoretically possible at the quantum level for an atom to know everything that God knows, if that knowledge has been mapped onto that atom by God or Nature's Psyche; and, since we are functioning at the quantum level where all of this mapping is taking place, it's possible for God to know everything that that specific atom or neuron knows.

Powerful, is it not?

Conclusion

Do you see what I'm doing here? The principle is the important thing to learn, not necessarily the specifics.

I'm trying to adjust my theories in order to make them match with and explain ALL the observational evidence and ALL the proven FACTS that we have on hand as a race. Science is observation and experience, or it should be. I'm trying to adjust my theories and make them match with all the observations.

In contrast, the Materialists, Naturalists, and Darwinists deny the evidence, reject the evidence, ridicule and ban and destroy the evidence; and then, these people make up an *ad hoc just-so story* out of thin air to explain what they think is happening or want to be happening.

I learn best through comparison and contrast.

One of these days, I hope to find a theory that matches with ALL of the evidence and explains everything that has ever been observed or experienced. Maybe I already have. Quantum Maps of Physical Functionality, existing at both the quantum level and the physical level, seems like a good place to start. I intend to go with it, until I find something better. That's what science is all about.

Now pay attention!

Despite the FACT that I have debunked and falsified the Theory of Evolution thousands of times in hundreds of different ways, don't ever once let me convince you that there is NO such thing as evolution or random mutations.

Evolution is REAL, very REAL. Evolution or random mutation is entropy or the second law of thermodynamics. It's REAL. Entropy is very REAL at the physical level; and, it works as advertised at the physical level.

Whenever the Materialists, Naturalists, Darwinists, Nihilists, Behaviorists, and Atheists start talking about evolution, they get most everything wrong, because evolution or entropy cannot design and create. However, these people do indeed get one thing perfectly right. Evolution, or random mutation, or entropy is indeed the CAUSE of ALL of our heritable diseases, developmental diseases, and heritable mental illnesses.

Remember, the Theory of Evolution is FALSE because random mutations or entropy cannot design and create genes, proteins, and life forms. However, evolution or random mutation or entropy is very REAL; and, it can indeed destroy genes, proteins, and life forms. Do you see how that works? It's important to understand.

COMPUTER SCIENCE AND QUANTUM NEUROSCIENCE

While doing research for Quantum Neuroscience, I observed that our brains seem to be partitioned into Cerebral Processing Units, and our brains seem to be functioning as a collection of CPUs.

Each structure within our brain seems to be an intricately designed, exquisitely fine-tuned, carefully engineered, precision-wired, and marvelously manufactured piece of machinery that computer scientists would call a CPU – with cache for working memory. There's one serious flaw in this CPU Model, though. Do you know what it is?

For most of the duration while writing this book, I visualized each structure in our brain as a Cerebral Processing Unit, which was created by Nature's Psyche at the quantum level and is now being operated by Nature's Psyche through quantum waves at the quantum level in order to get things done for us at the physical level.

Why did I choose to move brain functionality to the quantum level, instead of leaving it on the physical level as the Materialists and Naturalists do?

The efficiency of man-made CPUs comes from the fact that everything is wired together directly. There's NONE of that observable within our physical brain, however! There are NO wires in our brain, at least not at the physical level. Our brain is like having a hundred billion stand-alone switches just sitting there, all separated from each other by synaptic clefts. Our brain is comprised of 100 billion switches with NO wires connecting them.

Imagine it!

How are you going to run a program through that or on that?

You can't!

It's physically impossible.

You'd just as well have a hundred billion marbles or rocks, just sitting there. Neurons are individual BITS – ON or OFF. Neurons are switches.

I'm a computer scientist. I KNOW how these things work. I've studied them and used them for decades of my life.

WE KNOW precisely what's happening within the synaptic cleft at the physical level. It's called INTEGRATION. ALL of the synaptic input into a neuron is being integrated into a single Post-Synaptic Potential – a single BIT of information. That's it! Neurons are switches. They turn on and turn off physical functionality. So, who is assigning a specific physical function to each one of these switches, and who is throwing the switches at will? Well, it isn't the synapses, and it isn't evolution. That's physically impossible. It has to be Nature's Psyche, because the Human Psyche isn't consciously aware of these things. Remember, Psyche can do the physically impossible at will.

According to the Orthodox Interpretation of Quantum Mechanics, it is Nature's Psyche who collapses the wave function, thereby making the Human Psyche's choices actual, physical, and real. It has to be Nature's Psyche who is mapping all of this different physical functionality onto each of the different neurons in our brain, because the Human Psyche isn't aware of these things.

No matter how I try to come at it, I just can't visualize running a program through a hundred billion marbles, even if half of them are white representing ON and half of them are black representing OFF. I'm a computer scientist, and I can't see how it can be done. It's physically impossible. I also can't visualize how we could ever turn a hundred billion switches into functional CPUs and RAM, without some wiring to make the connections.

Even an abacus requires an intelligent being to push the beads around, in order to add things up! Our brains need the same thing!

If someone has indeed gone throughout our brains assigning eight neurons to function together as a single byte of data, or programming code, or distributed memory, then clearly all of that invisible wiring, organization, programming, functionality, and the assignment or mapping of data or code to the different bytes is taking place outside of our brains in the non-local realm or quantum realm because there are NO wires in our brains connecting neurons together into bytes of physical functionality.

If such a thing has happened, and Nature's Psyche has programmed and/or mapped our brains to do processes or functions or memory storage at the quantum level resulting in Maps of Physical Functionality, then suddenly we are looking at Quantum Neuroscience and not just plain-old neuroscience.

There's overlap between CPU programming and Maps of Physical Functionality. Our autonomic nervous system, brain stem, and reflexes seem to be programmed or hard-wired into us. They either work, or they don't. Nature's Psyche triggers these things as needed. The Map of our Physical Output, the primary motor cortex, seems to be programmed or hard-wired by Nature's Psyche, even though it is subject to some malleability or neuroplasticity.

However, the Maps of our Physical Input seem to function as working memory, cache, registers, or buffers. The locations of these registers or neurons are selected and mapped at the quantum level by Nature's Psyche, but their day-to-day functionality seems to be ephemeral and they function as registers and buffers rather than as CPUs and RAM. Our physical senses are Maps of Physical Functionality, and not CPUs and RAM. The information or the input consolidates at it travels through the thalamic relays, to the primary cortex, onward to the secondary cortex, and then finally to the association cortex. But, it's all a Map of Physical Functionality and not a CPU or RAM. It can't be a CPU or RAM, because there are NO wires in our brain.

There are NO visible wires connecting the neurons and synapses in our brain; yet, the different structures in our brain are functioning as if each one of them is a carefully designed CPU. That's physically impossible. In order to explain this invisible functionality scientifically, we are forced to turn to Quantum Mechanics for an explanation of the invisible wiring, invisible hardware structures, and the invisible organization of our brain into bytes of data and bytes of programming code. And, we are forced to turn to Nature's Psyche as the programmer and operator who is making the whole thing work as Cerebral Processing Units (CPUs) invisibly behind the scenes.

Nature's Psyche has constructed and mapped each structure in our physical brain to function as a unique Cerebral Processing Unit (CPU). Neurons register and map sensory input from our physical environment, which the Human Psyche and Nature's Psyche can then access and use. Nature's Psyche also maps physical functionality or motor skills – physical output – directly onto specific neurons in our brain. However, the neurons in our brain are NOT wired together like the transistors in our computers. There are NO wires in our brain.

The synaptic clefts terminate, randomize, and scramble every message, memory, or programming code that comes their way, at the physical level at least. Nothing makes its way through a chemical synapse, except ON or OFF. This means that if our brains are wired together by Nature's Psyche to function as CPUs and RAM, then that wiring only exists in the quantum realm or the psyche realm, because there is NO physical wiring within a physical brain.

It seems to me that Nature's Psyche is mapping our brain at the quantum level onto different and distinct Maps of Physical Functionality. Our brains are all about physical functionality. This is a one-to-one map controlled by Nature's Psyche. According to the Orthodox Interpretation of Quantum Mechanics, by Henry P. Stapp, it is Nature's Psyche who collapses the wave function thereby making the Human Psyche's choices ACTUAL, PHYSICAL, and REAL. We know that Nature's Psyche is in control of the physical processes within our brain, because we (the Human Psyche) are not consciously aware of these things.

Maps of Physical Functionality under the control of Nature's Psyche at the quantum level is a completely different way of looking at brain functionality than the one that the Materialists, Naturalists, Darwinists, Nihilists, Behaviorists, and Atheists give us. Their whole purpose in life is to remove Psyche, God, Non-Locality, Choice, and Quantum Mechanisms from the equation. Their maps of physical functionality are defined as being physical maps within the physical brain, but that's physically impossible thanks to the synaptic clefts that would scramble and destroy any type of physical map, CPU, or RAM that we might try to put into our brain. Synaptic clefts and neurons reduce everything to a single BIT of information – ON or OFF. That's it. You can't shove a CPU or RAM or a map into a single hardware BIT or neuron.

Executive function is defined as a collection of cognitive abilities – planning, insightful thinking, and reference memory – that appear to depend upon the prefrontal cortex. Executive function is also a Map of Physical Functionality. The prefrontal cortex, in the third eye position, seems to be the location within the human body where the Human Psyche resides. Therefore, it is fitting that the executive function is the Map of Physical Functionality through which the Human Psyche chooses to get things done.

Produced by God's Psyche, your genome is also a Map of Physical Functionality, but it exists at the physical level rather than the quantum level; therefore, a genome can be physically flawed by evolution or random mutations. It has been proven many different times in many different ways that evolution of any kind can't make functional genes and proteins and life forms; but, evolution (genetic change) or random mutations can definitely destroy a genome. Nature's Psyche functions flawlessly; but, when the physical genes and physical proteins are damaged or missing thanks to evolution, then there's nothing that Nature's Psyche can do to compensate. It requires an act of God to fix your damaged genes and proteins as well as the developmental damage that came along with them.

https://www.youtube.com/watch?v=FYt8sv4vzQs&t=14s

The Materialists and Naturalists call our dendrites and axons "wiring", but that's a misnomer and is actually misleading. Remember, the Materialists and Naturalists and Darwinists always choose the wrong interpretation for everything. It's what they do. That's how they maintain their self-deception. Self-deception works, and it works every time. Our brains are not CPUs in the traditional sense of a computer CPU. There are no wires in our brains.

Nevertheless, all throughout my research and my essays in this book, I talk about our brain being organized into unique and different Cerebral Processing Units (CPUs), that were made by Nature's Psyche and then mapped by Nature's Psyche at the quantum level to perform specific physical functions. I found the CPU metaphor or analogy to be a useful and powerful concept, so I kept it. And, it works perfectly fine, as long as you fully realize that there are NO physical wires within our brain, and that the "wiring" or the mapping of our Cerebral Processing Units is being done at the quantum level by Nature's Psyche instead.

Remove the map, and the physical functionality is lost, even though the consolidated memories remain.

In this physical realm, good things don't just spontaneously generate out of thin air. You (your psyche) have to put some work into it, or it will never exist. The same reality applies to Nature's Psyche. Nature's Psyche has to map physical functionality into each of our brain structures, or it won't exist for Nature's Psyche and the Human Psyche to use.

Whether we choose to model the physical brain as CPUs, RAM, and/or Maps of Physical Functionality, one thing WE KNOW for sure is that chemical evolution, random mutations, and natural selection DID NOT map our brains with physical functionality at the quantum level like Nature's Psyche did, because evolution doesn't have any intelligence, psyche, or brain. There's no way in the universe that evolution could be the person or the entity who is mapping out our different brain structures at the quantum level, collapsing the wave functions, and thereby imbuing our neurons with purpose and physical functionality. Evolution can't do the quantum level, and there are NO wires in our brains.

Based upon all the evidence at my disposal, I have concluded that Nature's Psyche has mapped each structure or cerebral processing unit within our brain with physical functionality of some kind. Our brain is all about mapped physical functionality, and NOT wires or computer processing units or random-access memory.

I fought this and resisted this for months on end; but, you've got to let the evidence teach you what it's trying to tell you, or you will never find and know the truth. I realized in the end that all I really wanted is the truth. What I have come up with based upon the observational evidence, experiential evidence, and empirical evidence is vastly superior to anything that the Materialists and Naturalists have manufactured out of thin air. Good enough!

NEUROTRANSMITTERS AND SYNAPSES

Neuroscientists, Classical Physicists, Materialists, and Naturalists have been telling us for decades now that there are NO direct connections between the neurons within our physical brain. Instead, signals are transmitted chemically from one neuron to another by dumping a bunch of neurotransmitters into the synaptic cleft between neurons.

This never made any sense to me, and it has bugged me to no end. According to the Laws of Classical Physics, those neurotransmitters should diffuse randomly and evenly throughout the brain fluid in the synaptic cleft, and then just sit there vibrating under the effects of Brownian motion and Entropy for the rest of eternity. According to the Laws of Classical Physics, we can visualize the synaptic cleft as a tiny garbage can. What happens when you dump junk into a garbage can? That's what should be happening between the synapses in our physical brain according to Classical Physics. According to Classical Physics, our neurotransmitter systems shouldn't work at all, because random diffusion is not a reliable mechanism for getting anything done. The random nature of neurotransmitter release tells us that our brains should be firing off randomly and uncontrollably all the time, because that's the natural result of random processes.

Neurotransmitters, also known as chemical messengers, are endogenous chemicals that enable neurotransmission. They transmit signals across a chemical synapse, such as a neuromuscular junction, from one neuron (nerve cell) to another "target" neuron, muscle cell, or gland cell. Neurotransmitters are released from synaptic vesicles in synapses into the synaptic cleft, where they are received by receptors on the target cells. Many neurotransmitters are synthesized from simple and plentiful precursors such as amino acids, which are readily available from the diet and only require a small number of biosynthetic steps for conversion. Neurotransmitters play a major role in shaping everyday life and functions. Their exact numbers are unknown, but more than 100 chemical messengers have been uniquely identified.

– Wikipedia

One day, I finally realized that it's physically impossible to transfer a message through a synaptic cleft, because a synaptic cleft is a randomizer, scrambler, or jamming device. A synaptic cleft will scramble any message that you try to send through it. Well now, that's a problem, isn't it? And this problem kept resurfacing over and over again, anytime that I started to study neuroscience. I went into neuroscience with my eyes wide open, and I have encountered one enigma after another ever since. I quickly encountered the physically impossible!

My bachelor's degree was in psychology. Over and over again, I was taught Neuroscience, Brain Anatomy, and Brain Functioning from one class to the next.

For me personally, as I studied the intricacies of neurotransmitters and synapses, it became obvious and clear to me that there is NO WAY in the universe that the blind and dumb processes of evolution (mutation/selection) could have ever designed, created, produced, fine-tuned, and field-tested the neurotransmitter and neuropeptide systems. If one of these is just slightly off, then you end up with Parkinson's Disease or Schizophrenia or something even worse – death.

The neurotransmitter and synapse system is scientific proof of God's necessity, which ends up becoming Scientific Proof of God's Existence.

Nevertheless, over and over again, I found myself asking, "Why did God design it this way?"

It seemed highly inefficient to me, purposefully inefficient! What's to keep one of these things from randomly misfiring a target neuron? As I see it, according to the laws of Classical Physics, there should be a lot of echoing and false positives going on, with all those neurotransmitters floating randomly between synapses, or floating between the pre-synaptic terminal buttons and post-synaptic receptors.

Think about it.

Because of synapses and the neurotransmitter system, there is NO direct contact between neurons. The circuits in our computers and the wires in our homes are directly connected to each other; consequently, the electrical transmission in our computers and our homes is much faster and more reliable, without any of that nasty quantum interference or quantum indeterminacy going on. Only when the circuitry in our computer chips gets too small and/or the heat rises does quantum bleed, quantum tunneling, and unreliability start to take place, causing the computer to miscalculate and/or crash.

Computers are much faster and much more reliable and efficient than the human brain, because the electric circuits are connected directly rather than being stopped at synapses.

So, why did God design and create our physical brain with such inefficiencies as neurotransmitters and synapses? It didn't make sense to me. I was still a Materialist at my core; and, I liked the efficiency of computers. I did not like the element of chance that was being introduced by all of those different neurotransmitters, and their random diffusion throughout the synaptic cleft. It seemed like a nasty situation to me. Random diffusion and entropy are an integral part of Classical Physics. According to Classical Physics, our brains shouldn't work.

The situation only got worse as I gained additional knowledge.

The Monoamine Neurotransmitters – dopamine, epinephrine, norepinephrine, and serotonin – are often dumped en masse by neurons that have highly branched axons with many varicosities (synapses beaded all along the axon) that release these Monoamines diffusely into the extracellular fluid between neurons; and then, the scientists observe glial cells called Astrocytes monitoring the Monoamine levels between synapses making sure that there's not too many and not too few. Apparently, the Monoamines have to be maintained at the correct level of concentration between the synapses and the cells for proper brain functioning. If you have too much dopamine floating between the synapses, then you develop Psychotic and Schizophrenic symptoms. If you have too little dopamine, then you develop Parkinson's Disease. If you have too little serotonin floating between the synapses, you get depression, anxiety, and OCD. The Monoamines have to be maintained at a certain balanced finely-tuned level between neurons at all times, or brain malfunctions occur.

Clearly, this random diffusion of neurotransmitters willy-nilly between the synapses is functioning differently than what's being presented to us in our college textbooks, especially in the case of the Monoamines. The Monoamines seem to be functioning as lubricant or facilitators, rather than as transmitters of actual information. I didn't like it – the randomness, uncertainty, and unpredictability of it all. Even worse, ALL of the Neuroscience chapters talk about Serotonin Reuptake. Why would there need to be any reuptake of serotonin, if the brain is trying to maintain a constant level of serotonin between the synapses that use serotonin? I was getting conflicting information that didn't make sense to me. Is it Serotonin Reuptake that's actually taking place between the neurons that use serotonin, or is it balancing and maintaining a certain level of serotonin between these synapses that's taking place? I still haven't answered that question, and nobody else seems to have figured it out either.

I KNOW for a fact that random chance seldom turns out well. I was confused. I couldn't visualize how the random Brownian motion of neurotransmitters between synapses could ever result in anything but random chaos, diffusion, entropy, and signal noise. I wasn't seeing it, and I wasn't liking it. It was bugging me severely. I KNOW how diffusion works – there should be billions of false-starts and false-positives and signal noise happening every microsecond with all of those neurotransmitters floating randomly between the synapses all the time. How could anything bring order out of all that chaos and randomness?

I like everything to make sense, and this never made any sense to me.

So why did God design our brains with such obvious inefficiencies? Why didn't He just wire everything directly like any sensible person would do?

Aside from the obvious – God designed it that way in order to make His presence and influence KNOWN because there's no way that mutation/selection could have or would have designed, fine-tuned, constructed, and created our neurotransmitter system and physical brain in such an inefficient and complex manner – I still found myself asking, "Why did God design it that way?"

I kept returning to it again and again over the years.

From my point of view, synapses and neurotransmitters seem to be highly inefficient, slowing everything down and introducing chance or quantum indeterminacy into whole the system. With all those neurotransmitters floating around randomly, we should be nothing but one spastic mess and never be able to pass on information from one neuron to the next in a timely fashion. It didn't make sense to me, because we seem to be functioning just fine despite all those random neurotransmitters haphazardly making connections to receptors just whenever they get lucky enough to do so.

Why did God do this? Why didn't He make it efficient and reliable like our computer chips, or the wiring in our cars and homes?

The answer wasn't immediately forthcoming. It took months, or even years, of study and research and thought. I'm still not sure that I completely know or understand; but at least now, I have some ideas.

My questioning and research continued.

Who determines what type of neurotransmitter and neuropeptide to produce within a presynaptic neuron, and who decides what sub-type of neurotransmitter receptor to build and deploy across the synaptic cleft on the postsynaptic neuron?

Anytime we observe decisions and choices being made, we have caught Someone Psyche in the act. How do we know? WE KNOW because the Materialists, Naturalists, Behaviorists, Darwinists, and Atheists assure that choice is impossible at the physical level. Since choice is physically impossible, anytime we observe chosen behaviors and decisions being made, we simply KNOW that it's taking place at the quantum level or the psyche level under the command and control of Someone Psyche. Simple. Parsimonious. True.

Neurotransmitters and synapses?

They do slow things down, and maybe that's one of their purposes – to give us the chance to change one of our chosen actions should we change our mind. When it comes to computers, cars, and homes, there's no way to intervene and stop the thing once the switch has been thrown. All you can do is to hope that the software and/or wetware running the thing has recovery capabilities.

There are also practical reasons to consider. Neuroscientists have suggested that if all of our neurons were directly connected with each other like the circuitry in our computer chips, then our brain would get too hot and burn up just like what happens when our computer chips get too hot from running at too high of a frequency or without adequate cooling.

But, I now believe it goes way beyond these practical issues.

God made it physically impossible for our brains to function as CPUs and RAM.

Our brains are comprised of a hundred billion stand-alone switches that we call neurons. Neurons are either ON or OFF – just like a switch. What makes it really bad is that there are NO wires connecting those switches. How are you going to create a CPU or RAM out of a bag full of switches with NO wires to connect them? It ain't gonna happen! It can't be done. It's physically impossible.

Furthermore, ALL of the synapses on that neuron REDUCE to that single BIT. Even if those synapses on the neuron contained some kind of message, like "Hello World", that message is reduced to a single bit – ON or OFF – when that message gets to the neuron. Everything in your brain reduces to a single BIT at the physical level – a switch, which is either ON or OFF.

Finally, NO physical message or memory can survive the synaptic cleft. Let's say you manage to stick a message, like "Hello World", into a neurotransmitter vesicle – which is physically impossible because neurotransmitters function as single bits of information, either ON or OFF. A synapse has only two neurotransmitters at most – a regular neurotransmitter, and sometimes a neuropeptide. But, let's say that you did have "Hello World" lined up and stored in a presynaptic vesicle, comprised of seven different neurotransmitter types. Once that vesicle is dumped into the synaptic cleft, that message is automatically scrambled, into "W l d l e H l o r o" or some such; and then, all those neurotransmitters arrive randomly at the post-synaptic membrane in a completely different order; and, some of them NEVER arrive at the post-synaptic membrane.

So, the received message ends up being "roll," or "oro," or "Hollow," or "led" or some such; and, the received message will end up being different every time, even though you might start with "Hello World" every time. NO message or memory can survive a synaptic cleft. It's physically impossible. The synaptic cleft is a randomizer, scrambler, or jamming device. God is making it abundantly clear and obvious that it's physically impossible to transfer messages and memories through a physical brain, and it's physically impossible to store memories and messages and programming code within a physical brain.

The ONLY thing you can get through a synaptic cleft is a message for the next neuron(s) in line to turn on and fire – one BIT of information at most. That's it! A single BIT does not a memory make! All of the Materialists, Naturalists, and Darwinists are telling us that our memories are stored within our synapses as synaptic weights; but, that's physically impossible. It can't be done, which means that it isn't being done.

There's no getting past this. It's physically impossible to transmit messages from one neuron to the next within our brains; and, it's physically impossible to store memories and messages within our brains, because NOTHING can get through a synaptic cleft and survive except for a single BIT of information telling the next neuron in line to turn on and fire.

The brain is extremely disappointing once you figure out what it really is – a bag full of switches with NO wires connecting any of those switches.

So, how is the magic being done, if it's not being done physically with stand-alone switches within our brains?

The answers finally started to come to me when I started to study and understand the **Orthodox Interpretation of Quantum Mechanics** and how the Human Psyche uses quantum processes to interact with Nature's Psyche and the physical brain, through the Quantum Zeno Effect or Quantum Telepathy. Remember, telepathy is WiFi at the quantum level. With Quantum Telepathy or the Quantum Zeno Effect, suddenly we have a scientific mechanism that allows the Human Psyche or Non-Local Consciousness to interact with and control its physical brain. This is extremely important due to its vastly superior explanatory power.

Without spirit or psyche, our brains aren't going to work! It's physically impossible.

However, if you are willing to allow Quantum Mechanics, Nature's Psyche, and the Human Psyche in to play, then instantly everything can be explained in a logical and scientific manner.

I spent months adjusting and adapting this thing as further information and ideas became available to me. I had to learn to understand both the physical processes and the quantum mechanisms, before I could understand how the physical brain really works. It bugged me to no end for a few months NOT to have a physical explanation for how the brain works; but slowly, I got used to it, because that's the way it is.

Speaking of the physical processes, even out-of-body travelers have noticed that something keeps our brains idling in default mode, while our psyche and spirit body are exploring the astral plane. That something is Nature's Psyche or God's Psyche. On the physical side of the equation, about 80% of the energy that a person uses each day goes to maintain his or her resting metabolic processes. 80% of our energy reserves are used to keep our brains and bodies idling or running in default mode, while the Human Psyche is exploring dreamland or the astral plane. Our default mode network is the network of brain structures that tend to be active when the brain is in default mode or simply idling. The default mode is the command and control for our autonomic functions; and, our default mode consists of all our built-in programming and reflexes both of a physical nature and a quantum nature.

> **If all your fat calories were stored as glycogen, you would likely weigh well over 225 kilograms (600 pounds).**

> **Physical exercise normally accounts for only a small proportion of total energy expenditure. About 80% of the energy you expend is used to maintain the resting physiological processes of your body and to digest your food. Our bodies are efficient machines, burning only a small number of calories during a typical workout.** (*Biopsychology*, pp. 289, 307.)

Well, that took some planning on God's part, didn't it? Complex physical functionality doesn't spontaneously generate out of thin air.

Most of our energy is used just sitting there.

Up to 25% of a neuron's ATP energy is being used by the sodium potassium pumps, just to keep the battery charged. A neuron is a battery, it's either charging or discharging. God designed these things to idle while the Human Psyche is away from them; but, the physical body and spirit body don't form nor retain any new memories while the Human

Psyche is outside of them and away from them. It's interesting and instructive to study and observe how things really work while the Human Psyche is separated from its spirit body and physical body.

The explanatory power, which we get from Quantum Mechanics and Quantum Non-Local Consciousness (Psyche), is infinitely superior to anything that we can get from Materialism, Naturalism, and Classical Physics.

All I was ever looking for was a physical explanation for why our brain and neurotransmitter systems work. All I ever found was physical explanations why they shouldn't work.

Remember, NONE of the functionality within the human brain can be explained in terms of Classical Physics, because it's physically impossible for our brains to function as CPUs or RAM, because there are NO wires connecting the neurons or switches in our brains. I'm 100% certain, though, that the different structures within our brain are functioning as highly organized and highly efficient Cerebral Processing Units (CPUs), even though that's physically impossible.

So, how is this done, and who is doing it?

The only logical scientific answer that we have is that it's being done by quantum mechanisms, and Nature's Psyche or God's Psyche is the person who is doing it – commanding and controlling it. It can't be the Human Psyche, because we aren't consciously aware of any of these things. Our brains are 100% action at a distance; and, action at a distance or nonlocality is a quantum mechanism, and not a physical mechanism. It's physically impossible for our brains to work

Quantum Mechanics goes way beyond the limitations of Classical Physics or Materialism; and, ALL of the Psychology textbooks explain Neuroscience from the perspective of Classical Physics. A Neuroscience built upon Evolutionary Psychology, Materialism, and Naturalism is completely worthless when it comes to the Quantum Mechanical processes that control the neurotransmitters in our brains. It's clear to me that evolution couldn't and wouldn't have designed, created, and controlled our physical brain through such an inefficient neurotransmitter system, with neurotransmitters floating around randomly between bunches of synapses following the laws of Classical Physics. If you want the full picture and a complete understanding of how our brains really work, you have to abandon Materialism and Naturalism and Darwinism, and turn to Quantum Mechanics instead. There is no other way.

Remember: as scientists, we need an interpretation of Quantum Mechanics that matches with and EXPLAINS the Quantum Zeno Effect and the Observational Evidence from Near-Death Experiences (NDEs), Out-of-Body Experiences (OBEs), and other Non-Local Experiences. The **Non-Local Consciousness Interpretation** from Pim van Lommel, and the **Orthodox Interpretation** from Henry P. Stapp do just that – they EXPLAIN the Quantum Zeno Effect (Telepathy), NDEs, OBEs, and other types of Non-Local Non-Physical Experiences and Events. We need an interpretation of Quantum Mechanics that matches with the Observational Evidence and actually explains the Observational Evidence! There's no other way to get at the truth of the matter.

Quantum Zeno Effect Verified – atoms won't move while you watch them.

https://phys.org/news/2015-10-zeno-effect-verifiedatoms-wont.html

http://mypsyche.us/wp-content/uploads/2017/10/Zeno-effect-verified—atoms-wont-move-while-you-watch.pdf

One verified experimental observation of the Quantum Zeno Effect is how atoms freeze into position whenever you, your psyche or non-local consciousness, observes them or looks at them or uses them. The Quantum Zeno Effect only makes sense from a non-local perspective or a spiritual perspective, because there is telepathic "action at a distance" taking place between your psyche and the psyche within nature or physical matter, according to the proven and observed Quantum Zeno Effect.

Verification of the Quantum Zeno Effect provides PROOF of **action at a distance**, or telepathy between you and the atoms, because there is NO physical contact between you and the atoms being observed, yet the atoms freeze or stop moving whenever you look at them. They KNOW that you are looking at them. That's physically impossible; but, Quantum Mechanics and Psyche do the physically impossible.

Remember, we need an Interpretation of Quantum Mechanics that explains what's really happening in our brains. I found two. It was hard to do, because the Materialists and Naturalists and Atheists are trying to hide this information from us, and these people are very good at what they do.

The Non-Local Consciousness Interpretation of Quantum Mechanics.

Van Lommel, P. (2010). *Consciousness Beyond Life: The Science of the Near-Death Experience*. New York: HarperCollins.

This book taught me what I call the Non-Local Consciousness Interpretation of Quantum Mechanics, which is the science behind Near-Death Experiences.

The Orthodox Interpretation of Quantum Mechanics.

This article introduces the Orthodox Interpretation of Quantum Mechanics.

Schwartz, J. M., Stapp, H. P., & Beauregard, M. (2004). *Quantum Physics in Neuroscience and Psychology: A Neurophysical Model of Mind-Brain Interaction*. Published Online: Phil. Trans. R. Soc. B.

http://www-physics.lbl.gov/~stapp/PTRS.pdf

http://mypsyche.us/wp-content/uploads/2017/10/PTRS.pdf

https://www.researchgate.net/publication/7613549_Quantum_physics_in_neuroscience_and_psychology_A_neurophysical_model_of_mind-brain_interaction

http://escholarship.org/uc/item/4w8665vk

This paper summarizes everything I have been looking for during the past fifty-five years of my life. That's a long time to wander in darkness looking for the truth; but luckily, our generation finally has the truth, if we know where to find it and recognize it as true when we do find it.

This paper and Henry P. Stapp's papers and books taught me what Stapp calls the **Orthodox Interpretation of Quantum Mechanics**, which explains the interplay or the scientific interface between mind and matter.

It gets very complex!

Obviously, the complexity of the whole system establishes the NEED for God's direct intervention in the design and construction of the physical brain; but, God had other reasons for designing the brain and the neurotransmitter system this way, besides just providing proof of His existence, which it does.

The answer comes by studying the different Processes involved in the Orthodox Interpretation of Quantum Mechanics, as developed by John von Neumann and Henry P. Stapp.

http://mypsyche.us/wp-content/uploads/2017/10/Orthodox-Interpretation.pdf

http://mypsyche.us/wp-content/uploads/2017/10/Henry-P-Stapp-Articles.pdf

Science 2.0: https://www.amazon.com/dp/B0771K6WTX

The Orthodox Interpretation of Quantum Mechanics:

Process 1 is the Human Psyche – the Actor – Command and Control. The thing that makes the requests, decisions, and choices.

Process 2 is the chosen physical reality, which is the result of the collapse of the quantum wave function.

Process 3 is Nature's Psyche, which can only respond with a yes or a no, depending upon what the physical laws or God's Laws allow. Process 3 is the thing that collapses the wave function in Process 2. Nature's Psyche simply responds to the requests of the Human Psyche to the best of its ability.

Quantum Zeno Effect: Process 1 (the Human Psyche) interacts with Process 3 (Nature's Psyche) telepathically and telekinetically through the Quantum Zeno Effect, or Quantum Telepathy. Telepathy is WiFi at the quantum level.

The Orthodox Interpretation answers all our questions by explaining what's really going on at the quantum level.

All of the physical complexity of the physical brain and the neurotransmitter system proves God's necessity, which successfully proves God's existence; but, that's not the main reason why God designed it that way with synapses and neurotransmitters.

Neurotransmitters and Synapses: Why did God design it this way?

To give Psyche or Non-Local Consciousness something to do, and a chance or a way to intervene in the whole process.

There are two types of synapses in our brain – electrical synapses and chemical synapses. The electrical synapses give the erroneous impression that the neurons in our brain are wired together like some kind of CPU or RAM. There are NO wires in our brains. Alas, electrical synapses are just like chemical synapses in their functionality. They both let through ions which can influence a neuron's Excitatory Post-Synaptic Potentials (EPSPs) and Inhibitory Post-Synaptic Potentials (IPSPs). Everything that comes into a neuron through chemical synapses and electrical synapses integrates directly into a single BIT of information, a Post-Synaptic Potential. Even the effects of G-Proteins end up affecting only that single BIT of information – the Post-Synaptic Potential. Whether a neuron is fed by thousands of chemical synapses or hundreds of electrical synapses, ALL of that input gets reduced and integrated into a single BIT of information – the Post-Synaptic Potential or Membrane Potential of the neuron.

In their papers and their books, these scientists (Schwartz, Stapp, and Beauregard) repeatedly state that Process 1 (the Human Psyche) interacts with Process 3 (nature's psyche or consciousness) which controls Process 2 (the physical matter), through the **ions** or electrolytes that flood into the neuron (are fuel injected into the neuron) at the quantum level and trigger the action potential or the electrical charge within the neuron.

Ions are positively and negatively charged particles. Ion channels are pores in neural membranes through which specific ions pass.

If I gave you the task of forming atoms into the first functional gene or protein, how would you accomplish that task? You would have to use quantum mechanisms at the quantum level in order to form all of these stand-alone atoms into a functional gene or protein. There is NO other way. Left to their own devices, atoms just sit there and vibrate under the effects of Brownian motion, random diffusion, and entropy. You need some kind of Syntropy or Psyche in order to make them organize or assemble into something useful and worthwhile, because the spontaneous generation of proteins and genes out of thin air is physically impossible.

The quantum level is where everything starts to get interesting. Classical Physics or Materialism is bland in comparison.

The Human Psyche (Process 1) interacts with and controls Process 2 and Process 3 at the quantum level, through the ion exchange which triggers the action potential within a neuron – a brain cell.

The Human Psyche also interacts with and controls Process 2 and Process 3 at the quantum level, through all the different neurotransmitters floating around between the synapses of the physical brain.

The smaller can reside within and control the larger. The telepathic and telekinetic can also control the larger. Psyche or Non-Local Consciousness is telepathic and telekinetic, according to the Quantum Zeno Effect. The Psyche is smaller than the strings, electrons, quarks, atoms, molecules, ions, neurotransmitters, and neuropeptides; consequently, both the Human Psyche and Nature's Psyche can reside within and control telepathically all of these different Quantum Objects or Physical Objects through the Quantum Zeno Effect.

According to the Quantum Zeno Effect, Psyche or Non-Local Consciousness has to intervene with focused attention in order to successfully flood the synapses with the right type neurotransmitters, which kind of suggests that the neurons don't have a mind of their own or a life of their own. Instead, the neurons respond to Psyche's queries and commands. The neurons react or respond to the Human Psyche, rather than being actors in their own right. Only the autonomic systems and reflexes have been programmed to react independently of the Human Psyche's commands. ALL chosen behaviors are a function of the Human Psyche.

Process 1, or the Human Psyche, is the Actor – Command and Control. It's active. It chooses and decides. It queries and probes and tries.

Process 3, or Nature's Psyche, is passive. It simply responds with a Yes or No answer, in large part depending upon what's physically possible. Nature's Psyche or Process 3 collapses the wave function (process 2), thereby converting an infinite number of possibilities into a single chosen reality. Nature's Psyche or Nature's Non-Local Consciousness simply follows God's Laws. If the human's request is physically lawful or physically possible, then Nature's Psyche answers with a Yes; if not, then Nature's Psyche answers with a No.

Nature's Psyche has covenanted to follow God's Laws – the physical laws of nature. Nature's Psyche is rather constrained; but, this does explain why the Biblical God Jesus Christ was able to do the things that He did while here in mortality. Nature's Psyche was simply obeying God's Laws or God's Commands.

Since the neurotransmitter system is a quantum system, God designed it that way so that the human psyche could intervene, command, and control the physical brain directly and efficiently, through the Quantum Zeno Effect or Process 1. The Human Psyche is also a quantum system. The neurotransmitter system is the way that your psyche controls its physical brain. This means that there are trillions of places (synapses) within your physical brain where your psyche can intervene at the quantum level and have a direct impact on the outcome of the events.

In contrast, there is NO active psyche or consciousness within a computer system. A computer system is pure Process 2 – the collapse of the wave function – with the associated hope that there's no quantum bleed, or quantum indeterminacy, or quantum tunneling taking place. You want your computer to be reliable and rock solid, after all. You don't want your computer to develop a mind of its own, because chaos ensues. The way you tame that savage beast is to make the circuits and transistors big enough (buffered enough) and/or keep the computer chip cool enough, so that there is no quantum bleed or quantum indeterminacy or quantum tunneling taking place within the computer chip.

I'm a computer scientist and a certified computer repair technician – I have stress tested CPUs with Prime 95 and have observed the results of what happens when the computer chips get too hot or are run at too high of a frequency – they produce errors in the calculations, which means that the electric current tunneled or teleported on you at the quantum level and/or the Process 2 wave function didn't collapse properly. Computer chips and CPUs are designed and rated to meet a certain specification. If you exceed those specifications with too high of a frequency and/or too much heat, the electric current tunnels on you and escapes or jumps the track.

Imagine it!

God actually designed and created a piece of machinery – your physical brain – that actually interacts with and processes the telepathic exchange between Process 1 (the Human Psyche) and Process 3 (Nature's Psyche) at the quantum level. Now that's some radically advanced technology! I'd like to see the Materialists and Naturalists do that. There's no way in the universe that evolution (random mutations and natural selection) could have designed and created such a thing. We human beings or human psyches can't even design and create such a thing, so why on earth would anyone believe or expect evolution to have been able to design and create such a machine?

We are infinitely more intelligent and versatile and reliable than random mutations and natural selection; yet, even we human beings wouldn't know where to begin if we were trying to design and create a physical machine that was capable of interacting with and processing the telepathic exchange between the Human Psyche and Nature's Psyche.

Are you starting to see why God designed and created our physical brains with synapses and neurotransmitters? All of those neurotransmitters seem to be how our Psyche interacts with and controls its physical brain, telepathically through the Quantum Zeno Effect. In contrast, computers have no neurotransmitters and synapses, because a computer has no psyche or non-local consciousness that it needs to interact with.

The Human Psyche needs a mechanism through which to interact with and control its physical brain – the brain is a mechanical object after all. The neurotransmitter system is that mechanism by which the Human Psyche controls its physical brain. The synaptic cleft and the ion exchange gives the Psyche plenty of time to intervene, which wouldn't be possible if the brain were one continuous wired circuit like a computer or a car.

I find it fascinating how God hides in plain sight where the Atheists, Materialists, and Naturalists refuse to look for him, and therefore can't see Him nor find Him. Depending upon who is doing the poll, I still think it's possible that the majority of the scientists don't believe in God and don't believe in the existence of the psyche or soul. These scientists wouldn't even dream of trying to create a physical machine that can interact with, monitor, and respond to the telepathic and telekinetic exchanges which are taking place between the Human Psyche and Nature's Psyche, through the Quantum Zeno Effect. Such an idea is completely beyond their ken. Instead, they ridicule and mock, censor and block, these kinds of Science. They take pride in their ignorance.

Even if they were to try to create such a machine, how would they go about connecting or marrying a specific Controlling Psyche to their physical machine? We humans wouldn't even know where to begin; yet, the Materialists and Naturalists and Darwinists want you to believe that evolution (random mutations and natural selection) mastered these details and this science at the beginning of time; and, they have been merrily controlling the whole process ever since. These people simply don't know what they are talking about, because they have never studied Near-Death Experiences (NDEs) and have never studied nor understood the Orthodox Interpretation of Quantum Mechanics and the telepathic Quantum Zeno Effect.

Ignorance is bliss – unless you are trying to figure out what's really going on.

The fact that electrical synapses or direct connections exist in the physical brain, yet God chose not to use them for neurotransmission, is trying to tell us something extremely important – I do believe.

First of all, chemical synapses are telling us that evolution (genetic drift), random mutations, and natural selection didn't design and build our physical brain, because chemical evolution cannot design and create a single functional protein let alone the matching gene to go with it; and, many of our neurotransmitters are proteins which are built from genes. The existence of genes, proteins, neuropeptides, and carefully placed and carefully differentiated chemical synapses is Scientific Proof of God's Existence.

Second of all, God used chemical synapses in order to slow everything down, so that the Human Psyche can intervene in the process and stop a process should it change its mind and decide to do so. If our brain were wired like a computer with electrical synapses, then there would be no way for us to stop an action once it has started. Computers are all-or-nothing machines. Chemical synapses allow the Human Psyche to have and experience a trillion different variations between all and nothing depending upon what it decides that it wants to get done. Chemical synapses increase our agency as well as our ability to repent and change our minds. With electrical synapses, chances are good that we would be nothing but automatons, a box of reflexes, or preprogrammed machines, and then all of our creativity and versatility would be lost.

Think about it logically. With over 100 different types of neurotransmitters in existence, that reality adds up to at least a hundred times more versatility than what's possible from electrical synapses or transistors by themselves. Agency, free will, versatility, and creativity are what we get from chemical synapses, instead of being pre-programmed with electrical synapses or transistors like our computers are.

This is cool stuff to think about; but, you are going to have to decide for yourself if any of it is real and true. It makes sense to me, but does it make sense to you?

Reuptake is the drawing back into the pre-synaptic terminal button of neurotransmitter molecules after their release into the synaptic cleft. Reuptake is the more common of the two mechanisms that God has designed and commissioned for the deactivation of a released neurotransmitter. How is this done? Why is it done? When is it done? Who controls the timing of all of this? What's their motivation? Who or what makes the neurotransmitters move?

Reuptake transporters or reuptake pumps are described as "drawing in" neurotransmitters as if they have some kind of tractor beam that pulls in the neurotransmitter molecule to the reuptake pump. If you are going to use quantum waves to pull in neurotransmitters, then you are also going to have to use quantum waves to locate these specific molecules within the synaptic cleft, because tractor beams require some kind of targeting mechanism. Who is controlling all of this quantum level machinery? It's as if the whole process is alive. With quantum waves, tractor beams, quantum telekinesis, quantum telepathy, and action at a distance, we are no longer in Kansas anymore. Classical Physics no longer applies, and we have entered the Twilight Zone.

When it comes to neurotransmitters and neurotransmission, this is something that you have to get a feel for at the gut level, or quantum level, or psyche level because at the physical level it should be nothing more than random diffusion, Brownian motion, entropy, and chaos. Neurotransmission seems to work too well and too reliably for it to be purely restricted to Classical Physics and the physical realm. Also, things mess up way too badly whenever there are too many (or too few) neurotransmitters in the synaptic cleft over a prolonged period of time, for it to be nothing more than just a random accident. Remember, random accidents can't do anything deliberate.

The average number of neurotransmitters in the synaptic cleft determines the future functionality and efficiency of the synapse and the post-synaptic neuron. Furthermore, when you use drugs (SSRIs) to change the average number of neurotransmitters within a synaptic cleft, it can take six weeks to six months for those changes to have a noticeable effect on the physical functionality of the brain. There's a delay, which means that there's new mapping being put into place at the quantum level by Nature's Psyche and there's new or different physical functionality being put into place at the physical level by Nature's Psyche or God's genetic programming.

The metabotropic receptors connected to G-proteins were designed by God to alter neurotransmission efficiency by up-regulating or down-regulating the synapse; and, they were also designed by God to alter the internal functionality of the neuron at the genetic or molecular level; but, it can sometimes take a while for them to do so, possibly because of the random nature of the process. Also, there is plenty of observational evidence suggesting that it takes a while for Nature's Psyche to map physical functionality onto neurons and glial cells, especially during neurodevelopment. Remember, a Map of Physical Functionality is being built into the neurons or switches within our brain BOTH at the quantum level and the physical level. There's a lot required to get synapses and neurotransmission to work right.

Furthermore, if there are too few neurotransmitters in the synaptic cleft, then the physical functionality that Nature's Psyche has mapped to that particular synapse is reduced or lost. Consequently, Nature's Psyche has a choice to make – it can eliminate the synapse completely, or it can up-regulate the synapse at the post-synaptic membrane in an attempt to improve synaptic functionality for that particular synapse. Nature's Psyche decides whether to eliminate a synapse or save a synapse, because the Human Psyche isn't consciously aware of any of these things.

If there are too many neurotransmitters in a synaptic cleft on average, Nature's Psyche can add reuptake transporters (pumps) and autoreceptors to the presynaptic neuron in order to decrease the average number of neurotransmitters in a synaptic cleft.

The advantage of presynaptic facilitation and inhibition (compared to EPSPs and IPSPs) is that they can selectively influence one particular synapse rather than the entire pre-synaptic neuron. (*Biopsychology*, p. 86.)

The natural question is to ask, "Who wired these things this way?" Who decided how these axons and dendrites should be wired? Okay, it's not actually "wiring". There are NO wires within a physical brain. It's a physical mapping of some sort; but, whenever we start talking about maps, programming, blueprints, layouts, and design, we are talking about planning, teleology, foresight, intention, and choice which always requires Someone Psyche in order to get the job done.

There's a balancing act taking place within a synaptic cleft BOTH at the quantum level and the physical level, under the command and control of Nature's Psyche.

Evolution doesn't have anything to do with any of this, because by definition in principle evolution doesn't have a psyche and evolution can't do anything at the quantum level. In fact, the Materialists, Naturalists, Darwinists, Nihilists, Behaviorists, and Atheists teach and preach that the psyche level or the quantum level does not exist. Evolution is restricted to the physical level.

The ONLY thing that evolution can do to neurotransmissions and synaptic clefts is to mess up and destroy our genes and proteins at the physical level. Evolution can destroy physical functionality; but, evolution can't design and create functionality – BOTH at the quantum level and the physical level.

Evolution of any kind is impotent, when it comes time to design and create something new, such as a gene or a protein. Remember, the chemical evolution of proteins from scratch is physically impossible – not to mention the matching genes to go along with them. Chemical evolution can't design and create proteins and genes. It's physically impossible. Proteins and genes have to be designed and created at the quantum level, or they won't exist. Evolution can't do anything at the quantum level, because evolution is restricted to the physical level.

God did NOT use evolution to design and create our proteins and genes, because that's physically impossible. God used something infinitely superior – quantum mechanisms or psyche mechanisms. God used Quantum Mechanics (the Priesthood Power of God) to design and create our genes and proteins, because that's the ONLY way it can be done!

My working theory is that God gave us neurotransmitters instead of wires to slow everything down for us so that we can live it, experience it, choose it, stop it, learn from it, and actually remember having done so.

Of course God also gave us neurotransmitters and synaptic clefts that scramble and randomize every memory and piece of information that comes their way so that any thinking person will KNOW that it's physically impossible to construct functional CPUs, RAM, and Memory Systems out of the stand-alone neurons or switches or hardware bits within our brains, because there are NO wires in the first place within our brain to connect those hypothetical CPUs and RAM together from the neural bits of information into bytes and into one functional whole.

Neurodevelopment and Synaptogenesis

It's during beginnings and endings that God's presence, influence, and necessity becomes obvious and known.

The timing of Neurogenesis, Neural Migration, Logistics, Selected Neuron Destinations, Neural Differentiation, Dendritic Growth, Axonal Growth, Synaptogenesis, Synaptic Targeting by Axons, Synaptic Pruning, and Neuroplasticity ARE the pinnacle of command and control. The information exchange required to accomplish these tasks borders on the infinite, because some of these processes are still ongoing within an adult human brain. There's no way in the universe that all of this information can be stored within our genome. It's physically impossible. You can't store the infinite within something as limited as the human genome.

The neurotransmitters and synaptogenesis are where Neuroscience, Non-Local Consciousness, Nature's Psyche, the Quantum Zeno Effect, and Quantum Mechanics start to get really interesting. The neurotransmitters are how the Human Psyche commands and controls its physical brain.

Current studies estimate that the average adult male human brain contains approximately 86 billion neurons. As a single neuron has hundreds to thousands of synapses, the estimated number of these functional contacts is much higher, in the trillions (estimated at 0.15 quadrillion).

The total number of neurotransmitter types is not known, but is well over 100.

The human brain has a huge number of synapses. Each of the 10^{11} (one hundred billion) neurons has on average 7,000 synaptic connections to other neurons. It has been estimated that the brain of a three-year-old child has about 10^{15} synapses (1 quadrillion), where the exchange of neurotransmitters takes place. There are quadrillions of neurotransmitters floating between your synapses right now.

The longest axons in the human body are those of the sciatic nerve, which run from the base of the spinal cord to the big toe of each foot.

Betz cells are large pyramidal neurons in the primary motor cortex that synapse directly on the motor neurons in the lower regions of the spinal cord. The Betz neurons are the largest in the central nervous system, sometimes reaching 100 µm in diameter. Their axons run down the spinal cord.

These are the FACTS – the proven, observed, and experienced Science. I put them in bold lettering with the hope that you will remember them.

Source: Uncle Google and Wikipedia

During synaptogenesis, the amount of intelligence and information exchange being employed by Nature's Psyche is simply off the charts! If you missed God's handiwork elsewhere, you can definitely see it here. Where the human brain is concerned, Nature's Psyche has at least 86 billion neurons to differentiate – to choose what type of neurotransmitter these neurons will make and what type of neuron they will be. Furthermore, Nature's Psyche has one quadrillion synapses (comprised of over 100 different types of neurotransmitters) to map-out and construct during neurodevelopment. There's no way on earth that all of that intelligence and information for one quadrillion synapses – their type and their precise target location on the dendrites or neuron – could ever be contained within the 3.2 billion base pairs of our genome. The neurons have to be getting and reading this 3D logistical map for the synaptogenesis of one quadrillion synapses from someplace else besides our genes. There's no way in the universe that random mutations and natural selection could have ever designed and created all of that.

Chemical evolution can't even design and create a single functional protein, meaning that chemical evolution can't even design and create a single functional gene, let alone the 3D wiring map and neurotransmitter typing for one quadrillion synapses. Nature's Psyche has to get these one quadrillion synapses right and correctly placed, with their associated transmitter types and directed targeting of axon growth and dendrite growth, or your brain isn't going to work right. The quadrillions of pieces of information for this 3D Synaptogenesis Axonal-Dendritic Growth Map cannot be successfully stored in our physical genome – it's physically impossible. Nature's Psyche has to be getting its information from someplace else besides our genome, during synaptogenesis. So, where is Nature's Psyche getting all of this information from?

It's during beginnings and endings that God's presence, influence, and necessity becomes obvious and known.

Back in the summer of 2015, I started looking for and praying for Scientific Proof of God's Existence; and, this is it – one quadrillion synapses specifically typed and specifically targeted at a specific location on a specific dendrite on a specific axon on a specific set of neurons. Somehow while using God's Database, Nature's Psyche weaves together a hundred billion neurons and a hundred billion glial cells into functional and unique Cerebral Processing Units (CPUs) that actually work. God's necessity, and therefore God's existence, doesn't get more obvious than this. Each act of precision fine-tuning IS scientific proof of God's existence. There's no way that all of this information can be stored within our genome – it's physically impossible. God's blueprint for the growth logistics, typing, and the positional layout of our quadrillion synapses and their associated axons and dendrites has to be stored someplace else besides our genome. That's just logical common sense.

The other piece of common sense is that there's NO CHANCE whatsoever that random mutations and natural selection could have designed and created and overseen any of this. We human beings couldn't do such a thing, so why on earth would anyone believe that evolution did? The existence of each of those quadrillion synapses along with their precise typing and their perfect precision positioning on the targeted dendrite or targeted neuron cell body falsifies Materialism, Naturalism, Darwinism, and the Theory of Evolution. Any fool KNOWS that targeting requires intelligence, precision, and skill. There's NO WAY in the universe that random mutations or natural selection could do such a thing, because evolution of any kind is dumb and blind with no intelligence whatsoever. Random chance simply doesn't cut it, where 3D-targeting and precision fine-tuning are concerned. Any fool knows that random chance can't do 3D targeting and precision fine-tuning! Evolution, by definition in principle, cannot do command and control; so, who did?

When it comes to our quadrillion synapses, and the quadrillions of neurotransmitters that flow through them, God is bragging and showing off, because it's obvious to any rational person that random mutations and natural selection could never have designed and created and constructed such a thing. God's necessity and God's existence doesn't get more obvious than this.

Yet, the authors of ALL of my neuroscience and biopsychology textbooks want you to know and believe that evolution (random mutations and natural selection) designed and created your physical brain including the positioning and the typing your one quadrillion synapses, even though that's physically impossible. It's very important to these people that you know and believe that evolution designed and created it all.

Evolution Designed and Created Our Physical Brain

Bear, M. F., Connors, B. W., & Paradiso, M. A. (2007). *Neuroscience: Exploring the Brain* (3rd ed.). New York: Lippincott Williams and Wilkins.

Pinel, J. (2014). *Biopsychology* (9th ed.). New York: Pearson.

Klein, S. & Thorne, B. M. (2006). *Biological Psychology*. New York: Worth Publishers.

Wickens, A. (2009). *Introduction to Biopsychology* (3rd ed.). Essex, England: Pearson.

Kalat, J. W. (2008). *Introduction to Psychology* (9th ed.). Belmont, CA: Wadsworth, Cengage Learning. **"The brain is the product of Evolution."** (p. 12.)

Pinel, J. (2014). *Biopsychology* (9th ed.). New York: Pearson. **"This is a particularly important chapter for you."** (p. 21.) **"Genetic endowment is a product of its evolution."** (p. 24.) **"The human brain appears to have evolved from the brains of our closest primate relatives."** (p. 33.) **"The human eye, a product of 600 million years of evolution."** (p. 132.)

In the first year of medical school, I was introduced to the human brain, the crowning achievement of evolution. (*Neuroscience*, p. 150.)

Evolution: The process by which succeeding generations of organisms change in physical appearance, function, and behavior through a process of natural selection.

As we noted earlier, variations in genetic makeup are responsible for both individual differences within a species and differences among species. Have you ever asked yourself why we have a nervous system? Some species survive perfectly well without one and have done so for hundreds of millions of years. Primitive forerunners of animals such as single-celled protozoa or primitive multicellular organisms such as sponges do not have specialized nerve cells or an organized nervous system to respond to environmental stimulation. So how did we come to possess a complex system that enables us to laugh, cry, throw a ball, enjoy a great meal, and appreciate a painting? We can find the answer by studying the evolution of the nervous system.

Evolution is the process by which succeeding generations of organisms change in physical appearance, function, and behavior, according to Darwin's theory of evolution – the origin of species by natural selection. These changes begin with an increase or decrease in the frequency with which specific genes are represented in the population of a species over successive generations, and the changes occur as species adapt to changing environmental conditions. We can trace the evolutionary development of the nervous system by looking at present-day animals, from the most primitive

to the most advanced, highly complex species. (*Biological Psychology*, pp. 78-79.)

https://books.google.com/books?id=p4fNxxuuOgwC&pg=PA78&lpg=PA78#v=onepage&q&f=false

If we take the emergence of primates as the starting point, then the evolution of the human brain has taken place over a period of at least 70 million years. This is a long time, especially as human civilization has existed only for around 3,000 years. The gradual process of evolution has resulted in new structures emerging and taking over the roles of older ones. However, this does not mean that the more primitive regions of the brain have become redundant. Rather, they remain incorporated into the neural circuits of the brain and still have vital roles to play. In short, the brain always functions as a collective entity, although it also exhibits a hierarchy of function where newer structures are more likely to be involved in complex behaviors. Another feature of evolutionary development is cephalization, that is, the massive increase in size of the brain in relation to the body. This trend is most noticeable in the cerebral cortex, which has become so large and complex in humans that it has developed ridges and fissures in order to increase its surface area. In fact, the cerebral cortex is not dissimilar to a screwed-up sheet of newspaper, and it is this adaption that gives the external surface of the forebrain its distinctive wrinkly appearance. (*Introduction to Biopsychology*, p. 36.)

https://catalogue.pearsoned.co.uk/assets/hip/gb/hip_gb_pearsonhighered/samplechapter/NEW%20Wickens%20Ch1.pdf

Since this is indisputably true, this means that our evolutionary ancestors didn't have brains for at least 70 million years, while evolution was making the human brain. Just so! Since we are making up a story here out of thin air, we can jump to any conclusion that we want and assume that evolution designed and created everything; and then, we can adjust everything else to make it fit our pre-chosen conclusion.

We know that evolution designed and created our physical brains, eyes, neurotransmitter systems, proteins, and genomes because these people were there and saw how it was done. Therefore, it's extremely important for you to know and believe that evolution is your designer, creator, and God. This is a particularly important chapter for you and your life.

The authors of ALL of my neuroscience and biopsychology textbooks desperately want you to know and believe that evolution (random mutations and natural selection) designed and created your physical brain, including the positioning and the typing your original one quadrillion synapses, even though that's physically impossible. It's very important to these people that you know and believe that evolution designed and created it all.

However, natural selection is completely worthless. The only time it comes into play is when something dies – gets selected against. There's absolutely nothing that your death can do to influence the mapping, typing, construction, layout, and expression of your genes! Natural selection is over-rated – way off the charts. It has to be, if natural selection is going to take the place of God. Natural selection is an idol – a false god.

Natural selection does absolutely nothing to modify our genes, so natural selection can have no part in our design and creation. We can simply ignore it until we die. Natural selection does nothing whatsoever, except kill you dead! Just ignore it. Natural selection is a sleight used to trick us and deceive us, in order to take our minds off of what's really important in all of this. (See: Sanford, J. (2014). *Genetic Entropy* (4th ed.). Cornell University: FMS Foundation.)

Genetic recombination during the creation of our gametes and random mutations are the only things that actually change our genes and the positioning and typing of our genes; and, these things do NOT create any new genes nor any new information. Genetic recombination and random mutations just shuffle or modify what God has already given us. Remember, evolution, random mutations, and natural selection did not exist until AFTER God designed and created the genomes, and the genes, and the associated proteins in the first place. This is logical common sense.

Random mutations cannot design and create. It's physically impossible! Random chance doesn't know what to do with itself. Random chance is not an all-seeing, all-knowing, prescient God.

Chemical evolution of genes and proteins from scratch is physically impossible. It can't be done.

During neural migration and neurodevelopment, the neurons will use cell-adhesion molecules (CAMs) to get to where they are going. Neurons worm their way through the brain in order to get to where they are assigned to be. CAMs are molecules on the surface of cells that have the ability to recognize specific molecules on the surface of other cells and bind with them.

How did evolution make these CAMs?

It didn't! It's physically impossible for random mutations or chemical evolution to design, field-test, and fine-tune these types of things, because evolution didn't even exist until after God designed and created the genomes and the living cells in the first place.

Scientists have proven that chemical evolution cannot create proteins, which means that chemical evolution cannot create the matching genes to go with them. Design and creation by evolution is physically impossible.

Since science and scientific observations have established that random mutations and random chance cannot design and create anything whatsoever, what is it that random mutations can do?

Random mutations can give us developmental diseases, heritable diseases, random diseases, fatal diseases, genetic entropy, cancer, death, entropy, chaos, and extinction; and, that's exactly what we observe them doing! Random mutations can destroy what God has given us. It's nasty, and it's real!

Random Mutations Designed and Created Our Genome and Physical Brain

Let's observe and document some of the blessings of random mutations.

Alzheimer's Disease, Schizophrenia, and Autism are heritable and have a sizable genetic component. Huntington's disease is caused by a mutated dominant gene. Phenylketonuria is by the genetic mutation of two recessive genes. Our worst, most debilitating, and most lethal mental illnesses and physical illnesses are the result of a mutant gene. Thank you evolution! We couldn't have done it without you. ALL of our heritable diseases are caused by evolution or random mutations. Thank you evolution and random mutations for these wonderful gifts!

Shunting Inhibitions

Now that we have established that random chance or random mutations could never have been our Designer and Creator, let's look at what really happened during our neurodevelopment, neurogenesis, and synaptogenesis.

In my humble opinion, the best way to KNOW that God had a hand in our neurodevelopment, neurogenesis, and synaptogenesis is to carefully observe some of the results of this process.

Principles of Synaptic Integration

Most CNS neurons receive thousands of synaptic inputs that activate different combinations of transmitter-gated ion channels and G-protein-coupled receptors. The postsynaptic neuron integrates all these complex ionic and chemical signals and gives rise to a simple form of output: action potentials. The transformation of many synaptic inputs to a single neuronal output constitutes a neural computation. The brain performs billions of neural computations every second we are alive. As a first step toward understanding how neural computations are performed, let's explore some basic principles of synaptic integration. Synaptic integration is the process by which multiple synaptic potentials combine within one postsynaptic neuron.

Inhibition

So far, we've seen that whether or not an EPSP contributes to the action potential output of a neuron depends on several factors, including the number of coactive excitatory synapses, the distance the synapse is from the spike-initiation zone, and the properties of the dendritic membrane. Of course, not all synapses in the brain are excitatory. The action of some synapses is to take the membrane potential away from action potential threshold; these are called inhibitory synapses. Inhibitory synapses exert a powerful control over a neuron's output (Box 5.6).

[Editor's Note:

V_m = membrane potential or the Voltage of the membrane.

mV = millivolts.

E_{ion} = an equilibrium potential for an ion is the membrane potential that results if a membrane is selectively permeable to that ion alone.

EPSP = excitatory post-synaptic potential.

IPSP = inhibitory post-synaptic potential.

λ = Lambda, the dendritic length constant.

r_m = membrane resistance.]

IPSPs and Shunting Inhibition.

The postsynaptic receptors under most inhibitory synapses are very similar to those under excitatory synapses; they're transmitter-gated ion channels. The only important differences are that they bind different neurotransmitters (either GABA or glycine) and that they allow different ions to pass through their channels. The transmitter-gated channels of most inhibitory synapses are permeable to only one natural ion, Cl^-. Opening of the chloride channel allows Cl^- ions to cross the membrane in a direction that brings the membrane potential toward the chloride equilibrium potential, E_{Cl}, about -65 mV. If the membrane potential were less negative than -65 mV when the transmitter was released, activation of these channels would cause a hyperpolarizing IPSP [making the membrane potential more negative].

Notice that if the resting membrane potential were already -65 mV, no IPSP would be visible after chloride channel activation because the value of the membrane potential would already equal E_{Cl} (i.e., the reversal potential for that synapse; see Box 5.4). If there is no visible IPSP, is the neuron really inhibited? The answer is yes. Consider the situation illustrated in Figure 5.20, with an excitatory synapse on a distal segment of dendrite and an inhibitory synapse on a proximal segment of dendrite, near the soma.

Activation of the excitatory synapse leads to the influx of positive charge into the dendrite. This current depolarizes the membrane [increases the membrane potential or makes it more positive] as it flows toward the soma [neuron cell body]. At the site of the active inhibitory synapse, however, the membrane potential is approximately equal to E_{Cl}, -65 mV. Positive current therefore flows outward across the membrane at this site to bring Vm [the membrane potential back] to -65 mV. This synapse acts as an electrical shunt, preventing the current from flowing through the soma to the axon hillock [where the next action potential will be triggered].

This type of inhibition is called shunting inhibition. The actual physical basis of shunting inhibition is the inward movement of negatively charged chloride ions, which is formally equivalent to outward positive current flow. Shunting inhibition is like cutting a big hole in the leaky garden hose – all the water flows down this path of least resistance, out of the hose, before it gets to the nozzle where it can "activate" the flowers in your garden.

Thus, you can see that the action of inhibitory synapses also contributes to synaptic integration. The IPSPs can be subtracted from EPSPs, making the postsynaptic neuron less likely to fire action potentials. In addition, shunting inhibition acts to drastically reduce r_m [membrane resistance] and consequently λ [dendritic length constant], thus allowing positive current to flow out across the membrane instead of internally down the dendrite toward the spike-initiation zone.

The Geometry of Excitatory and Inhibitory Synapses.

Inhibitory synapses in the brain that use GABA or glycine as a neurotransmitter always have a morphology characteristic of Gray's type II (see Figure 5.7b). This structure contrasts with excitatory synapses that use glutamate, which always have a Gray's type I morphology. This correlation between structure and function has been useful for working out the

geometric relationships among excitatory and inhibitory synapses on individual neurons. In addition to being spread over the dendrites, inhibitory synapses on many neurons are found clustered on the soma and near the axon hillock, where they are in an especially powerful position to influence the activity of the postsynaptic neuron. (*Neuroscience*, pp. 122, 126-129).

[This is what it takes to STOP an action potential that has already been triggered someplace else! IPSPs are used to bleed-off EPSPs. For over 100 pages, the authors of this college textbook have been building towards Shunting Inhibition. They want you to know and believe that evolution designed, created, and now oversees this whole process. Notice that based upon mathematical computations, these authors chose -65 mV as the membrane's resting potential, instead of the -70 mV that is typically chosen in the other science books on this subject. It's a bit confusing at first and it bugged me, but you get used to it. Different types of neurons have different resting potentials anyway, so it really doesn't matter.]

http://faculty.fiu.edu/~fasoto/courses/COMPNEURO/papers/bear-ch5.pdf

http://mypsyche.us/wp-content/uploads/2017/12/bear-ch5.pdf

http://mypsyche.us/wp-content/uploads/2017/12/Shunting-Inhibition.pdf

[This is the old version of their book, and one of these guys put this online on one of their websites, so I take this to mean that they want us to find it and read it. It's excellent science, in my humble opinion. Neurogenesis and synaptogenesis are the pinnacle of command and control – logistics and planning doesn't get any better or more impressive than this.]

The precise positioning of the inhibitory synapse in relationship to the excitatory synapse is essential for proper functioning of a shunting inhibition! God's handiwork is all over this! Random mutations and natural selection can't do precision fine-tuning of this sort – not in a quadrillion years. There's no way in the universe that evolution could have designed and created this! When it comes to shunting inhibitions, God is bragging and making His presence and His necessity obvious and known. There's no way that properly functioning shunting inhibitions could have come about by random luck and random chance, because precision, intelligence, targeting, differentiation, specificity, and fine-tuning are essential in this process. Both the excitatory synapse and the inhibitory synapse have to be placed perfectly in relationship to each other in order for them to work right.

Think about it logically and rationally. Evolution (genetic drift), random mutations, and natural selection by definition, in principle, cannot do "correlation between structure and function" nor the "working out the geometric relationships among excitatory and inhibitory synapses on individual neurons". There's nowhere near enough information storage capacity in our 3.2 billion base-pair genomes to work out the 3D geometry and the logistical, timing, migrational, axonic-dendritic growth instructions, neuron differentiation, synapse differentiation, and placement relationships among the one quadrillion synapses within a single human brain.

Scientists have observed that each axon grows towards a specific synaptic receptor on a specific dendrite or neuron. How does an axon choose which synapse to grow towards? It only has a quadrillion to choose from. Some axons are a meter long. That's precision targeting there! How does each dendrite or post-synaptic neuron choose what type of neurotransmitter receptors to build and where to build them? There can be as many as 7,000 neurotransmitter receptors or synapses on each neuron and its associated dendritic tree. How do neurons choose where to build our synapses and what type of synapse each one of them should be? Choice is a function of Psyche or Intelligence! Random chance, natural selection, and random mutations cannot do choice! By definition, in principle, evolution can't do Psyche, Intelligence, Intention, or Choice. Where the Materialists, Naturalists, Darwinists, Nihilists, and Atheists are concerned, Psyche, Choice, and Intelligence are the enemy – their very existence falsifies Materialism, Naturalism, and their derivatives.

The axons grow towards a specific synapse at a specific location on a specific neuron. How do they do that? Well, evolution isn't doing it, because according to the evolutionists evolution can't do targeting and teleology. Likewise, each neuron KNOWS precisely how to grow its dendritic tree, and precisely where to place each and every synapse on its surface, and specifically what type of synapse each carefully positioned synapse should be. By definition, in principle, evolution of any kind cannot do specificity! Evolution is dumb and blind. Therefore, WE KNOW that evolution isn't doing any of this! So, who is?

Precisely fine-tuned neurodevelopmental necessities like this are found all throughout the human brain; and, observational science has repeatedly proven that evolution of any kind cannot do fine-tuning, specificity, geometric relationships, correlation, and precision targeting. By definition, in principle, evolution, random mutations, and natural selection are dumb and blind; therefore, evolution could never have designed and created our shunting inhibition networks and made them work. The writing is on the wall for everyone to see. Logical common sense tells us that evolution can't do "the geometry of excitatory and inhibitory synapses". It requires way too much intelligence to get it right.

Imagine what you would do if you were assigned the task of placing one quadrillion synapses in a physical brain, selecting their precise location on a dendrite or a neuron, telling each axon which synapse they should grow towards, and choosing what type of neurotransmitters and/or neuropeptides that each one of those synapses will transmit and receive. Could you do it? Would it work when you are done? Either Nature's Psyche is infinitely more intelligent than we are, or Nature's Psyche has access to infinitely more information than we do – it's probably a little bit of both, actually. I'm not going to apologize for discovering these things. I worked really hard to do so.

While studying all of this, I got the distinct impression that God used chemical synapses instead of electrical synapses in order to prove to us beyond a shadow of a doubt that He exists. It worked on me. I'm a believer now.

God is hiding in plain sight where the Materialists, Naturalists, and Atheists will never find Him and never see Him. I KNOW, because I used to be a Materialist, Nihilist, and Atheist; and, it was impossible for me to see and find God at that point in my life. Like everyone else, I just assumed that random chance designed and created all of this, and I didn't give it any further thought at the time. Well, I'm finally thinking and pondering now; and, you can witness the results if you have any desire to do so. I'm no longer hiding from evidence like I used to do.

The Materialists, Naturalists, Darwinists, Nihilists, and Atheists are trying to use science to deceive us and trick us into believing that evolution can do science, technology, correlation, specificity, teleology, design, creation, thought, planning, shunting inhibition, mapping, 3D geometric relationships, command, and control; but, you can learn to see through the deception and the ruse if you choose to do so. If I can do it, anyone can. I'm not all that bright – certainly not as bright as the New Atheists or the Brights claim to be. Just ask them, and they will tell you that it is so.

There's a reason why they are rich, and I am not. The New Atheists are telling the Materialists, Naturalists, and Atheists precisely what they want to hear, and I am not.

Nobody is willing to pay for good news. They expect it to be given to them for free. But, millions are eager to pay handsomely for bad news, and they gobble it up like candy. Being told that they completely cease to exist when they die, and that God does not exist, is precisely what they want to be told and want to hear. They will pay good money for that.

The good news is that God knows precisely what He is doing, even if we don't. The bad news, for me at least, is that I will never make any money from my chosen profession. My books were written for my benefit, education, and learning alone; and so far, for nobody else. It is what it is. I learned a lot once I finally decided to go looking for it.

Unanswered Questions about Neurodevelopment

Kalat, J. W. (2008). *Introduction to Psychology* (9th ed.). Belmont, CA: Wadsworth, Cengage Learning. **"The brain is the product of Evolution."** (p. 12.)

Well, that covers everything, doesn't it! What more needs to be said? Simple. Parsimonious. Final.

With this statement of FACT, we don't have to do anymore science, now do we? It explains everything, doesn't it?

According to the Materialists, Naturalists, Darwinists, and Atheists the brain is a dynamic network of interconnected structures that has been wired together by evolution into a massive 3D array of networked circuits, or CPUs and RAM. The whole thing happened by accident or random chance. I'm not exaggerating here. This is what these people really truly believe and are preaching to the world every day in our public schools.

> **Most of us tend to think of the brain as a three-dimensional array of neural elements "wired" together in a massive network of circuits. However, the brain is not a static network of interconnected structures. It is a plastic (changeable), living organ that continuously changes in response to genetic programs and environment.** (*Biopsychology*, p. 214.)

The brain is not a static network. The brain is a dynamic network that has been wired together by evolution and continuously evolves in response to genetic programs and our environment.

Who wired the brain together in the first place? Who designed and manufactured the individual brain structures? Who made the physical brain dynamic rather than static? Who interconnected the structures into one functional whole? Who gave it life and who is now giving it life? Who is making our brain respond to our genetic programs and our environment? Who wrote the genetic programs, and who provided the environment in the first place?

Evolution, of course, even though that's physically impossible.

Nature (genetic programs) vs. Nurture (environment) is an incomplete paradigm or methodology; and, this is the paradigm that the Materialists and Naturalists have limited themselves to or have restricted themselves to following and using.

The truth? Our brain is a plastic living organ that continuously changes in response to our genetic programs, in response to our interaction with our environment, in response to what the Human Psyche chooses to learn and choose to do, and in response what Nature's Psyche chooses to do to the physical brain and chooses to map onto the physical brain.

241

According to the Orthodox Interpretation of Quantum Mechanics by Henry P. Stapp, it is Nature's Psyche who collapses the wave function, thereby making the Human Psyche's choices physical and real.

God's presence, influence, and necessity are most obvious when it comes to beginnings and endings. In the case of Neuroscience and Quantum Neuroscience, we are talking about neurodevelopment, neurogenesis, neural migration, neural differentiation, and synaptogenesis whenever we are talking about beginnings and the end product of brain development.

The development of a human body from a zygote and the neurodevelopment of the brain and nervous system are two of the greatest mysteries and supernatural miracles that we have ever encountered.

We realize that the whole thing is physically impossible, once we start to tally all the data and information that would be needed to accomplish the task if you or I were in control of the whole process.

THIS!

First, cells must *differentiate*; some must become muscle cells, some must become multipolar neurons, some must become glial cells, and so on. Second, cells must make their way to appropriate sites and align themselves with the cells around them to form particular structures. And third, cells must establish appropriate functional relations with other cells. (*Biopsychology*, p. 215).

There are around 200 major cell types in the human body. With a BYTE of information per cell in the human body, we can assign a number between 0 and 255 to each cell representing the 200 different cell types known to exist; but, that doesn't handle any sub-typing, nor does it handle migration and synaptogenesis.

An estimated 37.2-trillion-cells is the lowest number that I have seen for the number of cells in the human body. With a BYTE per cell, that means that at least 37.2 terabytes or 37.2 trillion bytes would be needed just to store the differentiation information for each cell in the physical body. If you were to store all of that information in sequential order, then the same information could be used to control the birth order of each cell in the human body. Good enough.

But, how is that information being passed from the quantum level where it is being stored to the physical level where it is being used to accomplish cell differentiation? Who put all of this information into the quantum level in the first place?

The Materialists, Naturalists, and Darwinists tell us that it's evolution who is doing all of this for us. Do you believe them?

The human genome only holds 750 megabytes; so, where's the other 37 terabytes being stored, so that evolution can build our body from the ground up; and, how is evolution accessing and using all of that information? 37-terabytes is 37 different terabyte-size hard drives that evolution has put someplace within our body and is now accessing to build and control the cell differentiation processes within our physical body. Evolution is amazing, is it not? When it comes to idols and false gods, evolution is the BIG ONE!

The 37 terabytes handle just cell differentiation. How many petabytes of information are needed for cell migration, axon targeting, and synaptogenesis within the brain itself? Who is handling all of this information, and where is it being stored? It can't exist in the genome nor can it be stored in the genome; so, where does it exist and how is it being stored? How is it being accessed and used? How is all of this being done? There's not enough information in DNA nor our genome to do all of this! Clearly, it's being done; but, it's not being done at the physical level, because it's physically impossible to do it at the physical level.

How is evolution getting 37 hard drives slammed into our genome or our brain, just for cell differentiation alone?

It's NOT. It's physically impossible.

The physically impossible is where we see Quantum Mechanics, God's Psyche, and Nature's Psyche start to come into play. Someone Psyche can do the physically impossible at the quantum level in order to get things done for us at the physical level. Remember, evolution can't do the physically impossible, because evolution by definition in principle is restricted to the physical level.

Each stem cell, which divides into a stem cell and a daughter cell, somehow KNOWS which of the 200 cell types that daughter cell should be. How does it know? Where's the 37 terabytes of cell differentiation information being stored; and, how is it being accessed and used?

Who or what tells each cell in the physical body how and when to differentiate into two different types of cells? Who or what tells each neuron and astrocyte in the brain what type of structure it's supposed to form and what type of functionality it's supposed to implement? Random evolution or genetic drift simply will not do!

There are nearly 10,000 neuron sub-types within the human body. Who tells each neuron which of the 10,000 sub-types it's supposed to become? Who tells each neuron the number of dendrites and axon branches that it's supposed to produce? Who controls, directs, and targets axon growth? Who tells two adjacent neurons where to place their synapses so that they can interact with each other?

Evolution, of course, even though that's physically impossible.

Each neuron and glial cell, which is born from neural stem cells, migrates to a pre-chosen destination within the brain before taking up shop. How do they KNOW where to go? How do they KNOW what method of migration to use? How do they KNOW what they should become when they get there, because they differentiate again after reaching their pre-chosen destination? After they have arrived at their final destination, how do they choose the targets for their axons? How do they choose the location, types, and sub-types for each of the quadrillion synapses within a child's brain? How do they organize themselves into distinctive and different brain structures? Where's all of this knowledge and information coming from? Where is it being stored? It's going to take a few petabytes of data or information just for synaptogenesis. Where are all those petabytes being stored, and how are they being accessed and used?

Migration of neurons and glial cells – who taught them and tells them how to do this? These are brain cells that have NO synaptic connections yet! How are they communicating with each other? They aren't even connected to each other, and never will be. Evolution is no longer in play here!

The Neuroscientists talk about physical cues that have been put into place for use during the migration of neurons. Neurons slither along radial glial cells to the destination that God or Nature's Psyche has chosen for them. Glial cells worm their way through the brain to the final destination that God or Nature's Psyche chose for them. Who put these physical cues into place in the first place? Who designed this whole process? Who is implementing it?

The Neuroscientists talk about guidance molecules that brain cells can use to sniff out their destination. Who is making all the right guidance molecules in all the right places at the right time? Who is telling each migrating brain cell which specific guidance molecule to worm its way towards, and which specific guidance molecules to avoid?

With these physical cues and guidance molecules, there's communication taking place here. Who taught these brain cells this language, by which they can communicate with each other during migration and neurodevelopment? They seem to be born KNOWING this unique language. How's that possible? It's not at the physical level, because there is no such thing as genetic memory at the physical level. If we have any kind of genetic memory or genetic learning, it's being stored at the quantum level and then being deployed at the quantum level when it's needed.

Most cells engage in both radial and tangential migration to get from their point of origin in the ventricular zone to their target destination.

The guidance molecules play a central role in neurodevelopment because a brain cannot function normally unless each class of developing neurons arrive at the correct location. (*Biopsychology*, pp. 216-217.)

Amen!

Every neuron and glial cell migrates from its birth place in the center of the brain to its final destination in the outer portions of the brain. Who chooses their target destination? How are the neurons and glial cells told where to go? How do these brain cells know their target destination? Who or what motivates the brain cells to move? How do these brain cells KNOW that they have reached their target destination? How do the brain cells sense and implement all of this information?

By definition in principle, targeting and fine-tuning require help, intervention, and some kind of choice from Someone Psyche. According to the Materialists, Naturalists, and Behaviorists, choice and psyche are physically impossible, which means that anytime we catch Someone Psyche making choices, we are observing quantum mechanisms taking place at the quantum level, because choice and psyche are impossible at the physical level.

The Materialists, Naturalists, Darwinists, Behaviorists, and Atheists stick neurodevelopment into a huge BLACK BOX and simply state that evolution is handling it all. Yeah, right. By definition in principle, evolution only has access to the 750 megabytes within our genome; and, most of that is being used to manufacture proteins. Evolution doesn't have access to the petabytes of data needed to properly construct a functional human brain. Evolution doesn't have access to the petabytes of information being stored in God's Database non-locally in the cloud. Evolution can't transmit that petabytes of quantum information to the different brain cells when and where they need it.

Any time we catch the Materialists and Naturalists using BLACK BOXES such as evolution to explain what's going on, WE KNOW automatically that they are actually talking about Nature's Psyche and God's Psyche, even though these people don't believe in the existence of psyche or non-local consciousness. Only Someone Psyche could access the exabytes of information being stored within God's Database in the cloud at the quantum level, and then use that information to get things done for us at the physical level. You can't shove petabytes of data into the 750 megabytes of our genome. It's physically impossible! Since this information clearly exists, it's obviously being stored and used and standardized and shared somehow, some way, just not at the physical level. It's impossible to store and access petabytes of information at the physical level within our 750-megabyte genome.

The physically impossible points us directly to Quantum Mechanics, Nature's Psyche, God's Psyche, and God's Database, which exist at the quantum level or the psyche level. The physically impossible is where science really starts to get interesting. Don't you agree? The observed and proven existence of the physically impossible falsifies Materialism, Naturalism, and their derivatives. I find that fascinating as well.

All you really want is the truth, unless of course you are a Materialist, Naturalist, Darwinist, Nihilist, Behaviorist, or Atheist. These people are trying desperately to hide truth from us where we will never see it, never find it, never hear anything about it, and never think about it. It's what they do. Atheism and Naturalism of any kind are based upon a refusal to look at evidence, a refusal to look at the truth. In contrast, Science 2.0 and Quantum Neuroscience allow ALL of the evidence and then pursue a preponderance of the evidence. It's a new and better way of doing science.

Aggregation

Once developing neurons have migrated, they must align themselves with other developing neurons that have migrated to the same area to form structures of the nervous system. This process is called aggregation.

Both migration and aggregation are thought to be mediated by cell-adhesion molecules (CAMs), which are located on the surfaces of neurons and other cells. Cell-adhesion molecules have the ability to recognize molecules on other cells adhere to them.

Elimination of just one type of CAM in knockout mice has been shown to have a devastating effect on brain development.

Gap junctions between adjacent cells have been found to be particularly prevalent during brain development. (*Biopsychology*, p. 217.)

Who designed, holds, and stores the master blueprint that neurons can use to align themselves into functional brain structures? Who or what is holding back aggregation until the necessary cells have migrated to the correct location? Who does all the timing, logistics, and information exchange associated with aggregation? Who gave these brain cells this aggregation ability?

Who designed the CAM molecules?

Who taught the brain cells how to use these CAM molecules? Who taught the CAM molecules to recognize other molecules and adhere to them? More importantly, who is telling the brain cells where to go, how to get there, and when they have successfully arrived at their pre-chosen destination during brain cell migration? Who tells the different brain cells how to aggregate?

Who put the gap junctions between adjacent cells? The placing of gap junctions and ionotropic receptors is a part of synaptogenesis. Who chooses the location of these synapses? Who oversees their differentiation, sub-typing, and construction?

It has been estimated that there are one quadrillion synapses within a child's brain, which would require at least 3 petabytes of RAM for the neuro-transmitter typing, neuro-transmitter sub-typing, and the neuropeptide typing – just to get synaptogenesis done properly at the physical level. I'm guessing that we need at least another 7 petabytes of RAM or information exchange for the location logistics and timing logistics of synaptogenesis. Where are the 10 petabytes of RAM being stored, and how are the brain cells accessing this information when they need it? Ten-petabytes ends up being 10,000 terabyte-size hard drives. How are you going to shove that into a 750-megabyte genome? You can't! It's physically impossible!

Evolution ONLY has access to our 750-megabyte genome, and nothing else! Evolution ONLY functions at the physical level, and nowhere else! Evolution by definition in principle can't do the physically impossible. Isn't it amazing what evolution has done for us so far? Actually, evolution hasn't done anything for us so far, except maybe mess up the proteins that Nature's Psyche has been using to build or construct our physical cells, including our brain cells.

Since it's physically impossible to shove 10 petabytes of physical RAM into our 750-megabyte genome, we simply KNOW that that 10 petabytes for synaptic typing and synaptogenesis is being stored in God's Database at the quantum level; and, it is being accessed and used by Nature's psyche at the quantum level in order to get things done for us at the physical level.

This is where we see Quantum Neuroscience and the hand of God in action! There's nowhere near enough information and programming code within our 750-megabyte genome to handle brain cell migration, aggregation, axon growth, and synaptogenesis! Synaptogenesis alone requires petabytes of information exchange at the quantum level in order to get the job done right. That's physically impossible at the physical level. You can't slam or shove 1,024 terabyte-size hard drives (a petabyte of physical data storage) into a 750-megabyte genome or a physical brain. It can't be done at the physical level, which means that it isn't being done at the physical level.

Clearly, ALL of this functionality exists, and ALL of this information exchange is taking place, or our brains wouldn't work, which means that since it's physically impossible, ALL of this information exchange, information storage, and the implementation of this functionality is taking place at the quantum level because it can't be done at the physical level.

This is Quantum Neuroscience in action! Quantum Neuroscience begins where our physical genome ends!

The ONLY thing evolution ever does for us is to mess up our God-given genome and give us messed-up non-functional proteins as a result. Thank you, evolution! Oh, evolution is real all right! Evolution (genetic change) and random mutations have given us ALL of our heritable diseases and heritable mental illnesses. That's what we have gotten from evolution, and NOTHING ELSE. Evolution can't do anything else for us except mess us up at the physical level, because evolution doesn't have access to the quantum level. Evolution is restricted to the physical level and our physical genome. Evolution can't touch anything else.

Kalat, J. W. (2008). *Introduction to Psychology* (9th ed.). Belmont, CA: Wadsworth, Cengage Learning. **"The brain is the product of Evolution."** (p. 12.)

Now, we have finally got our brain cells migrated to the correct, pre-chosen, targeted destinations, differentiated and aggregated into the correct brain structures; and NOW, it's time to send out the axons and dendrites so that we can do synaptogenesis. Thank God evolution knows how to do all of this, because the Human Psyche isn't consciously aware of any of these physical processes that are taking place within our brain.

There's no way in the universe that you or I would know how to do any of this. How would you go about accessing and using the petabytes of data that are being stored non-locally at the quantum level in God's Database, so that you can successfully accomplish neurodevelopment, neural migration, neural aggregation, and synaptogenesis without messing anything up any further than evolution has already messed it up?

How would you do neurodevelopment if I were to assign the task to you? I know how I'd do it.

I'd get evolution go do it! Evolution can do anything it sets its mind to.

Kalat, J. W. (2008). *Introduction to Psychology* (9th ed.). Belmont, CA: Wadsworth, Cengage Learning. **"The brain is the product of Evolution."** (p. 12.)

Thank God evolution knows how to do all of this, or we wouldn't have a brain.

If I only had a brain . . .

Once again, much like neural migration, the Neuroscientists and Psychologists have the growing axons and axonal growth cones groping towards their pre-chosen destination, sniffing their way to their pre-chosen destination, and searching for the correct route to their pre-chosen destination.

The chemoaffinity hypothesis [has the axon growth cones sniffing out their target destination.]

The topographic gradient hypothesis has been proposed to explain accurate axonal growth involving topographic mapping in the developing brain.

According to this hypothesis, axons growing from one topographic surface to another are guided to specific targets.

Remarkably, most growth cones reach their correct targets. (Pinel, *Biopsychology*, pp. 218-219.)

Most growth cones reach their correct targets. That is kind of remarkable, amazing, miraculous, and supernatural, isn't it? Remember, these growing axons don't have any synapses yet that they can use to communicate with other neurons. So, how are they communicating with each other and their target neurons?

This is probably my most favorite chapter in Pinel's book, *Biopsychology*, due to ALL of the sub-text that points us directly to Nature's Psyche, Quantum Mechanics, the Quantum Level, and Quantum Neuroscience; yet, Pinel isn't even aware that he is pointing us to these things.

Again, Someone Psyche has to be doing ALL of this targeting and fine-tuning, because it's physically impossible for evolution or random mutations to do all of this on the fly for us. Evolution can't have anything to do with this, because it's physically impossible. Targeting, teleology, logistics, timing, and fine-tuning by definition in principle require the intervention of Someone Intelligent or Someone Psyche in order to get them done in the first place, and to get them done correctly.

Remember, a growing axon can literally go ANYWHERE that it wants to go! Think about it! So, who is telling that axon growth cone where to go? Someone has to be telling each axon growth cone where to go, or we are NOT going to end up with a functional brain when we are done! There are NO wires connecting the neurons in our brain. So, how do the growth cones know where to go? How are the growth cones told where to go? How do the growth cones know when they have reached their target destination and that it's now time to shut down and form a synapse instead?

This, once again, is where we see God's Signature or the Hand of God in action!

A growing axon can literally go ANYWHERE; so, Someone Psyche or Someone Intelligent has to be telling it where to go, or it's going to go nowhere useful or right.

This is Quantum Neuroscience in action!

The more "macro" you get, the easier it becomes to find a physical explanation for what is being observed. The more "macro" you get, the easier it becomes to find a materialistic and naturalistic explanation for what is being observed.

But, we are NOT there yet!

Once again, the Neuroscientists talk about guidance molecules, chemical signals, the chemoaffinity hypothesis, topographic gradients, topographic maps, neural maps, and the topographic gradient hypothesis which the axon growth cones are using to find their specific target destinations.

Who is controlling these guidance molecules? Who is telling the growth cone which guidance molecule to pursue, and which ones to reject? Who is dropping these breadcrumbs all along the route that it has chosen for the growing axon to follow?

He talks about **topographic mapping in the developing brain.**

Who put this neural map or topographic map into our brain? These topographic maps or Quantum Maps of Physical Functionality seem to be some sort of GPS system that the growth cones use to select their eventual target destination. An invisible quantum 3D GPS system and/or a highly detailed and complex quantum 3D map within our physical brain requires remote communication (Action at a Distance) and telepathic communication (the Quantum Zeno Effect or Quantum Waves) between the growth cones and the neurons at the quantum level through Nature's Psyche, in order to implement it successfully and correctly at the physical level.

And, Someone Psyche had to put that map there in the first place! Maps require Mapmakers! Maps don't just spontaneously generate out of thin air.

The beauty of a 3D Quantum Map is that it can be stored within a single atom, and it can be mapped to any physical structure that Nature's Psyche sees fit.

This is where the tires hit the pavement, where Quantum Neuroscience is concerned.

Think about it logically.

We can study the physical effects of this Quantum Telepathic Communication System or Quantum 3D Topographic Map with classical physics and our physical instruments; but, we cannot detect the language nor the precise messages that are being communicated telepathically and chemically between the neurons and growth cones. Nor can we detect the language and the precise messages that are taking place at the quantum level or the psyche level between the individual molecules within a neuron or cell. We can't read a neuron's thoughts. WE KNOW, that this information is being exchanged at the quantum level, by the functional physical results that we are observing and recording at the physical level. The quantum information has to be getting through to these neurons and growth cones, or we wouldn't have a functional physical brain when Nature's Psyche is done making our brain.

WE KNOW that evolution didn't design and create any of this, because evolution can't do complexity, and because evolution can't design, create, and access 3D quantum topographic maps that are being built and stored at the quantum level within God's

Database. WE KNOW that evolution isn't doing any of this because axon growth and synaptogenesis require petabytes of information exchange at the quantum level in order to get the job done right; and, evolution only has access to our 750-megabyte genome, and nothing else!

A physical genome is needed for making physical proteins at the physical level, and for making these protein maps or genes heritable at the physical level; otherwise, we wouldn't even have a genome. Everything could be done at the quantum level by Nature's Psyche instead.

Nevertheless, there has to be some place in the process where we switch from the quantum level to the physical level; and, our physical genome and physical proteins seem to be it. Our genes and proteins are precisely where evolution can mess us up, destroy us, make us ill, make us insane, hurt us, kill us, and make us extinct. God has given evolution (genetic drift) or random mutations – entropy or the second law of thermodynamics – some influence over our genes and our proteins.

Thank God evolution doesn't have any access to the quantum processes that are in charge of our neurodevelopment and physical development, and under the command and control of Nature's Psyche. Imagine the horrors that evolution could produce if it did have access to these 3D quantum topographic maps that are being stored in God's Database and being used by Nature's Psyche to do neurodevelopment, brain cell migration, cell differentiation, aggregation, and synaptogenesis. Evolution is entropy – evolution scrambles and randomizes everything that it "touches", influences, or gets near at the physical level.

Kalat, J. W. (2008). *Introduction to Psychology* (9th ed.). Belmont, CA: Wadsworth, Cengage Learning. **"The brain is the product of Evolution."** (p. 12.)

Well, evolution has now successfully performed all of our neurogenesis, brain cell migration, brain cell differentiation, brain structures, axon growth, and dendrite growth. Now, it's time for evolution to do synaptogenesis. Now it's time to actually connect the neurons or astrocytes together at the physical level. Before now, we were doing all of this brain cell communication, molecular guidance, and brain development from the quantum level. It's finally time to get something in place at the physical level, don't you think?

Thank God evolution knows what it's doing, or who knows what we would end up with! We would be mindless and brainless, wouldn't we?

Synapse Formation.

Once axons have reached their intended sites, they must establish an appropriate pattern of synapses. A single neuron can grow an axon on its own, but it takes coordinated activity in at least two neurons to create a synapse between them.

Perhaps the most exciting recent discovery about synaptogenesis (the formation of new synapses) is that it depends on the presence of glial cells, particularly astrocytes.

Current evidence suggests that astrocytes play a much more extensive role in synaptogenesis by processing, transferring, and storing information supplied by neurons.

The brain must be "wired" according to a specific plan in order to function. (*Biopsychology*, p. 220.)

Wiring requires a Wirer – Someone Intelligent or Someone Psyche to do the wiring!

A plan requires a Planner – Someone Intelligent or Someone Psyche to do the planning, and then to implement the plan.

Wiring and planning don't spontaneously generate out of thin air, despite what the Materialists and Naturalists might claim.

Pinel puts "wired" in quotes, because even he knows that there are NO wires in our brain. He knows on some level that the Materialists and Naturalists and Darwinists are lying to us, whenever they suggest that there are wires in our brains.

The brain ONLY has synapses, and synapses TERMINATE messages and memories by integrating the whole thing into a single Postsynaptic Potential.

He talks about the axons reaching an intended site.

Who tells the axons what their intended site is, or who gives the growing axons their intentions?

Intention is a function or product of Psyche, NOT physical matter. It requires some kind of choice; and according to the Materialists, Naturalists, and Behaviorists, choice is physically impossible and so is psyche.

He talks about axons forming the appropriate pattern of synapses.

How?

Patterning or mapping requires a Mapmaker.

How do two neurons coordinate their mutual synapses?

This requires some kind of language and communication system that is fully understood by both of the neurons involved. Alas, the ONLY information that can be stored within a neuron at the physical level is YES or NO. That's it! ALL of the rest of the communication between neurons is taking place at the quantum level or the psyche level.

Of course, since he has limited himself exclusively to the physical level, he absolutely needs something at the physical level to be coordinating all of these different physical processes. He gets excited and assigns it all to the astrocytes.

He talks about information (plural) being supplied by the neurons to the astrocytes.

How is this being done at the physical level?

Remember, the ONLY type of information that neurons can supply at the physical level is YES or NO! And, the synapses haven't even been built yet; so, the neurons can't even supply YES or NO to the astrocytes at the physical level.

Nevertheless, since he needs a physical culprit to do all of this synaptogenesis, processing, transferring, and storing information supplied by neurons, he chooses the astrocytes to get the job done, even though that's physically impossible.

Who's telling the astrocytes what type of synapse to make?

Evolution, of course! Who else could it be?

Kalat, J. W. (2008). *Introduction to Psychology* (9th ed.). Belmont, CA: Wadsworth, Cengage Learning. **"The brain is the product of Evolution."** (p. 12.)

With one quadrillion synapses in a child's brain, and 3 BYTES per synapse just for neurotransmitter typing, neurotransmitter sub-typing, and potentially neuropeptide typing, that means that evolution has to be providing God or Nature's Psyche with 3 petabytes of RAM or information just to select the type of synapse that evolution or the astrocyte is going to build for us.

Where is evolution storing all of this information?

Evolution can't be storing it within our genome, because our 750-megabyte genome can't hold 3 petabytes of information. Apparently, evolution is storing all of this information in the cloud, non-locally, at the quantum level within God's Database. If we start bringing 3D location addressing and timing into play, evolution would probably need another 7 petabytes of information in order to successfully do synaptogenesis.

Imagine it!

Somewhere within your body or your genome, evolution has 10 petabytes of data set aside, which it uses for doing the synaptogenesis within your brain. That's just the 171 billion brain cells, or 86 billion neurons that we are talking about. Where's evolution storing the petabytes of information that it needs to properly map, differentiate, and construct your physical body, and the 37.2 trillion cells that it contains?

Pinel talks about a Master Plan that evolution or the astrocytes are using to do synaptogenesis.

The brain must be "wired" according to a specific plan in order to function.

YES, the brain has to be organized according to a specific Master Plan in order to function properly. So, how did evolution create this Master Plan? Where is evolution storing this Master Plan right now? And, how is evolution accessing this Master Plan and then communicating this Master Plan to the neurons and the astrocytes?

Where is this Master Plan or 3D Blueprint being stored, because there's NOT enough DNA in our genome to store it there? Our genome only holds 750 megabytes at most.

> **The truth is that long before the nervous system is fully developed, neurons begin to fire spontaneously and begin to interact with the environment. The resulting patterns of neural activity fine-tune subsequent stages of neurodevelopment as the animal grows. This fine-tuning constitutes the critical final phase of normal development.** (*Biopsychology*, p. 224.)

How? Why? Who or what motivates the neurons to fire spontaneously?

It can't be random mutations and natural selection, because these things can't do telepathy at the quantum level.

There is NO such thing as spontaneous generation at the physical level. An influx of Excitatory Postsynaptic Potentials (EPSPs) is NEEDED to fire a neuron! Everything we know about neurons at the physical level tells us that they can't fire spontaneously; yet somehow, they do. How are they doing that?

It can't be genetic programming nor fine-tuning as he suggests, because design, programming, and fine-tuning REQUIRE the intervention of Someone Psyche at the quantum level in order to come into existence in the first place! So, how are these neuron's firing spontaneously at the physical level? It's physically impossible as far as I know; yet, the neurons retain this ability to fire spontaneously throughout the remainder of their lives. It's not just a feature that's limited to neurodevelopment.

There's only ONE logical answer that I can think of, and it requires the Orthodox Interpretation of Quantum Mechanics to provide that answer. Nature's Psyche can trigger the first action potential in a chain, by collapsing the necessary wave functions.

Simple. Parsimonious.

Of course, he blames it all on genes and the environment, which is physically impossible. There's nothing in our physical environment nor within our genes that can reach into a neuron telepathically and make that neuron fire out-of-the-blue. Firing a neuron spontaneously requires some kind of decision or choice; and, choice is physically impossible according to the Materialists, Naturalists, and Behaviorists. The CHOICE to spontaneously fire a neuron has to be done at the quantum level or the psyche level, because it's physically impossible at the physical level.

Neuroplasticity in Adults

The accumulation of evidence has made clear that mature brains are continually changing and adapting.

In adult mammals, substantial neurogenesis seems to be restricted the olfactory bulbs and hippocampuses — although low levels have been observed in the cortex under certain conditions. (*Biopsychology*, p. 225.)

Adults can repent and change for the better!

For centuries the Neuroscientists preached that there is NO change taking place within the adult brain. They were wrong once again. The Materialists and Naturalists are almost always wrong. It would be impossible for learning and growth to take place if the adult brain were static and frozen in time. Of course, now the Materialists and Naturalists are telling us that evolution gave our brains the ability to change during adulthood. Isn't that nice of evolution to give us this gift?

Still, NOTHING is going to change within the adult human brain, unless the Human Psyche living within that brain decides to make a change! My brain and body just sit there doing NOTHING, until I decide to make them do something. This is how things really work. The Human Psyche makes the decisions and the choices, and then Nature's Psyche collapses the necessary wave functions in order to make those decisions and choices actual and physically real.

Pinel mentions cell death, or apoptosis. The cell goes pop!

Genetic programs inside neurons are triggered and cause them to commit suicide.

Apoptosis removes excess neurons in a safe, neat, and orderly way. But, apoptosis has a dark side as well. If genetic programs for apoptotic cell death are blocked, the consequence can be cancer; if the programs are inappropriately activated, the consequence can be neurodegenerative disease. (*Biopsychology*, p. 220.)

We started with beginnings, and now we are at endings.

Who wrote these genetic programs inside the neurons? Who triggers these programs causing the cells to commit suicide? How did evolution manage to create these genetic programs in the first place; and then subsequently, how did evolution manage to mess them up?

Well, WE KNOW for a FACT that evolution and random mutations can mess up our genes and proteins, because it has been caught in the ACT of doing so. Evolution is entropy, or the second law of thermodynamics. Entropy or evolution is very much active at the physical level.

However, WE ALSO KNOW for a FACT that evolution can't do anything at the quantum level. By definition, in principle, evolution is restricted to the physical level, and there it stays. Evolution doesn't exist at the quantum level.

Evolution isn't doing anything for us, or against us, at the quantum level. Nature's Psyche performs flawlessly at the quantum level or the psyche level. So, that means that we ONLY have to worry about evolution messing things up for us at the physical level.

Interesting, is it not?

Kalat, J. W. (2008). *Introduction to Psychology* (9th ed.). Belmont, CA: Wadsworth, Cengage Learning. **"The brain is the product of Evolution."** (p. 12.)

Imagine for just a second what your brain would be like if this statement were 100% true, and your brain really is the product of evolution.

Evolution or random mutation is the very pinnacle of entropy or chaos!

What would your brain really be like if it were the produce of evolution?

You would be deaf, blind, and unable to touch, smell, or taste; but, it would be much worse than that.

Take your brain and your eyes, put them into a blender, and turn the blender on. Leave the blender on. That's what your brain and eyes would really be like if they were the product of evolution.

Thank God the theory of evolution is FALSE, because now the blind can see, assuming of course that evolution or random chaos hasn't gotten its "hands" on our genome and our proteins.

SEEING THE PHYSICALLY IMPOSSIBLE

Have you ever wondered why Classical Physicists, such as Albert Einstein, vigorously opposed Quantum Mechanics all of their lives? It's because Quantum Mechanics is supernatural in nature and origin. Quantum Mechanisms involve faster-than-light Action at a Distance, which is physically impossible. Quantum Mechanics is supernatural; and therefore, Quantum Mechanics falsifies Materialism, Naturalism, and their derivatives.

Each structure within our brain seems to be an intricately designed, exquisitely fine-tuned, carefully engineered, precision-wired, and marvelously manufactured piece of machinery that computer scientists would call a CPU – with cache for working memory. There's one serious flaw in this CPU Model, though. Do you know what it is?

I don't know whether you can see it or not, because most of us have trained ourselves not to be able to see it or understand it. Huge numbers of us don't want to see it or understand it.

Nevertheless, when it comes to Neurodevelopment and Quantum Neuroscience, WE SEE God hiding in plain sight where only those who are looking for Him will be able to see Him and find Him.

https://en.wikipedia.org/wiki/Neural_stem_cell

https://en.wikipedia.org/wiki/Neuroblast

https://en.wikipedia.org/wiki/Glioblast

https://en.wikipedia.org/wiki/Development_of_the_nervous_system

https://en.wikipedia.org/wiki/Synaptogenesis

Neuroblasts are formed when a neural stem cell, which can differentiate into any type of mature neural cell (i.e. neurons, oligodendrocytes, astrocytes, etc.), divides and becomes a transit amplifying cell. Transit amplifying cells are slightly more differentiated than neural stem cells and can divide asymmetrically to produce postmitotic neuroblasts or glioblasts, as well as other transit amplifying cells. A neuroblast, a daughter cell of a transit amplifying cell, is initially a neural stem cell that has reached the "point of no return." A neuroblast has differentiated such that it will mature into a neuron and not any other neural cell type. — Wikipedia

While there are as many as 10,000 specific types of neurons in the human brain, generally speaking, there are three kinds of neurons: motor neurons (for conveying motor information), sensory neurons (for conveying sensory information), and interneurons (which convey information between different types of neurons). — Uncle Google

1. When stem cells divide, they divide into another stem cell and a daughter cell. Somewhere along the way, at an appropriate time of God's choosing, an embryonic stem cell eventually differentiates into another embryonic stem cell and a daughter neural stem cell.

2. That neural stem cell differentiates into a neural stem cell and a daughter "transit amplifying cell".

3. A transit amplifying cell differentiates either into a neuroblast or a glioblast. A transit amplifying cell is also a type of stem cell, in that it can produce another transit amplifying cell as well as daughter neuroblasts or daughter glioblasts.

4. After they are created, neuroblasts and glioblasts MIGRATE from the location of their birth in the ventricular zone in the center of the brain, through the different layers of cortex within the brain, to the current upper-most layer of the brain to a preselected destination that ONLY God or Nature's Psyche could have given them. There are four major types of migration that the brain cells can choose to use.

5. Neuroblasts are told to slither along the radial glial cells to the current highest level within the brain. Radial glial cells radiate like spokes on a bike's wheel. The neuroblasts do as they are told by God or Nature's Psyche. They migrate from the center or axis of the wheel to the outermost rim. The neuroblasts travel radially or outward from the center of the brain. The last ones born have the furthest to migrate.

6. In contrast, the glioblasts launch out on their own into the sea horizontally or tangentially, and the glioblasts can theoretically go anywhere they want to go. Ain't got no strings on them. Who chooses their eventual destination? Who tells them when they have arrived and to stop migrating?

7. When the glioblasts reach the destination that God or Nature's Psyche has chosen for them, they differentiate once again into oligodendrocytes and astrocytes. Who tells the glioblasts what type of glial cell to become? They have a choice to make. How do they make that choice?

8. When the neuroblasts reach the destination that God or Nature's Psyche has chosen for them, they differentiate once again into as many as 10,000 different specific types of neurons in the human brain. Who tells the neuroblasts what type of neuron to become? They have a choice to make. How do they make that choice? A neuroblast can become one of any 10,000 different types of neurons.

9. God or Nature's Psyche also forms the neurons and glial cells into organized and functional cortices, amygdalae, hippocampus, thalamus, hypothalamus, striatum, cerebellum, and a bunch of other brain structures – as well as the central nervous system and peripheral nervous system within the physical body.

Remember, evolution is telling the neuroblasts and glioblasts how to do all of this and when to do all of this at the physical level BEFORE the neuroblasts and glioblasts have developed synapses and are able to communicate with each other at the physical level. How is evolution doing all of this for us, so that we end up with a properly functioning brain when evolution is done with us? How is evolution communicating all of this information at the quantum level telepathically to the individual neurons and glial cells telling them what to do and when to do it?

Evolution is amazing isn't it? It's almost as if evolution is a God.

How Much RAM Does Evolution Have to Provide in Order to Do Neurodevelopment?

Each structure within our brain seems to be an intricately designed, exquisitely fine-tuned, carefully engineered, precision-wired, and marvelously manufactured piece of machinery that computer scientists would call a CPU – with cache for working memory. There's one serious flaw in this CPU Model, though. Do you know what it is?

How much information and physical RAM does evolution have to provide in order to do neurodevelopment of the physical brain at the physical level?

When does information become a memory, or organized data, or programming code at the physical level?

According to what we have observed and learned from computer science, information becomes a memory or data when those BITS of information are organized into a single BYTE of data, or a BYTE of memories, or a BYTE of programming code. A BYTE is typically where BITS start to become useful at the physical level.

The physical BITS have to be assigned some sort of order or organization by Someone Psyche or Someone Intelligent in order for those BITS to become BYTES, memories, data, and programming code at the physical level. That's when memory is initially stored at the physical level, when a collection of BITS is organized by Someone Psyche or Someone Intelligent into a functional unit called a BYTE.

On the low end, I have seen it estimated that there are 86 billion neurons in the adult brain, and 85 billion glial cells in the human brain, for a total of 171 billion brain cells. Each one of them is born, differentiates, migrates, and then differentiates again when it reaches the destination that God has assigned it.

Neurons and glial cells differentiate at least four different times. Assuming that the differentiation information for each differentiation requires at least a BYTE of data at the physical level, we are looking at four BYTES per 171 billion brain cells just to do cell differentiation within the brain, and nothing else. 171 X 4 = 684. That means that at the physical level, evolution has to provide at least 684 gigabytes of information to these different stem cells, neuroblasts, and glioblasts just to get them to differentiate in the correct sequence, into the correct type of cells, at the correct time, and the correct location – and this assumes that evolution is overseeing, coordinating, and timing the whole process all the way along.

Technically, evolution has to provide a WORD (2 BYTES) of physical RAM for each of the 86 billion neurons in order to successfully differentiate the neurons at the physical level. That's 172 gigabytes of RAM, which evolution has to provide, just to differentiate the neurons properly into their 10,000 different types. Evolution has to decide which of the 10,000 different types of neurons that each neuroblast has to become – where and when – in order to provide us with a functional brain when evolution is done making us. That works out to be 770 gigabytes of physical RAM at the physical level that evolution has to provide in order to properly differentiate the human brain at the physical level.

The human genome is capable of storing 750 megabytes of information at the physical level. That's nearly one gigabyte; and, most of the genome is being used to code for proteins and NOT cell differentiation. So, where is evolution storing the 770 gigabytes of data at the physical level needed to do the cell differentiation within our brains?

Of course, there's another petabyte that evolution is storing for us at the physical level, which evolution is using to make sure that all of the 37.2 trillion cells in the human body differentiate at the correct time, in the correct sequence, and end up in the correct location with the correct functionality, after any type of migration or differentiation or connections are made.

There's also possibly a petabyte of 3D information that needs to be transmitted or given to the neuroblasts and glioblasts during migration, as well as 171 gigabytes for the migration typing information; and then, petabytes of information have to be given to the neurons and glial cells during the growth of their axons and dendrites and during synaptogenesis. Some estimates say that a child is born with one quadrillion synapses. Synapses have to be located within the brain, possibly requiring at least 6 petabytes of 3D data from evolution at the physical level, just for the 3D mapping of synapses within the brain. Synapses also differentiate or do what's called typing. They are assigned by God or Nature's Psyche a neurotransmitter type, a neurotransmitter sub-type, and at times a neuropeptide type. That means that evolution is going to have to provide God or Nature's Psyche with at least 3 petabytes of RAM at the physical level, so that a BYTE is available for each synaptic type that a synapse can have.

Ironically, the ONLY thing that evolution has to work with is the 750 megabytes of our genome that God gave to evolution in the first place to work with. So, where is evolution storing the petabytes of data and physical RAM needed to construct your brain and physical body?

Well, it's NOT happening at the physical level, because it's physically impossible. You can't slam thousands of terabyte-size hard drives into a physical body, physical brain, or physical genome, without them being noticed by someone.

On the conservative side, we are talking at least 10 petabytes at the physical level that evolution needs to provide and then transmit to the 37.2 trillion cells in the human body and the 171 billion cells in the human brain just to accomplish cell differentiation, neurogenesis, neural migration, and synaptogenesis. That's physically impossible! Can you imagine trying to slam 10,000 hard drives into your genome or even into your brain? It can't be done, which means that it isn't being done – not at the physical level at least.

Evolution can't transmit quantum waves or information between neurons, or to neurons and body cells. And, evolution can't store 10 petabytes of information within our 750-megabyte genome. It's physically impossible!

Clearly, this type of information is being stored somehow, somewhere, and then transmitted to and received by the cells within our physical body – even though it's physically impossible. Clearly, it's being done, because we have functional brains when evolution gets done making us; but, since it's physically impossible, this means that it's being done by Someone Psyche at the quantum level through quantum mechanisms.

Remember, when it comes to physical BITS, Someone Psyche has to purposefully and deliberately allocate them as BYTES and WORDS, and then Someone Psyche or Someone Intelligent has to use them as BYTES and WORDS to do computations and to issue instructions at the physical level; otherwise, those BITS just sit there and do nothing useful. BITS at the physical level are absolutely worthless unless Someone Psyche deliberately organizes them or maps them into BYTES and WORDS at the physical level and then uses them as such at the physical level. Computer hardware and computer software don't just spontaneously generate out of thin air. It's physically impossible.

Anytime we encounter the physically impossible, we have encountered the Signature of God and are witnessing God's Handiwork in action! God's Psyche and Nature's Psyche use quantum mechanisms to do the physically impossible at the quantum level or the psyche level. By definition, in principle, evolution can't do the physically impossible, because evolution is limited or restricted to the physical level. Evolution can't do anything at the quantum level or the psyche level.

Finally, evolution is done making your brain. So, what is it that you actually have at the physical level when evolution is done making your brain at the physical level? Well, what's a neuron at the physical level?

A neuron is a switch at the physical level – a single hardware BIT – it is either ON or OFF.

That's it! There's nothing more there to see or experience at the physical level.

To paraphrase Shakespeare, "To fire or not to fire, that is the question." And the end result of information exchange between neurons is essentially a yes or no response. (*Biological Psychology*, p. 110.)

What is this complex code that the neurons are communicating to each other at the physical level?

YES or NO.

That's it!

At the physical level, a neuron is a single hardware BIT – a switch or a register. But, unlike the BITS in a computer, the neuron bits in our brain can't turn ON and stay ON. Neurons are OFF most of the time, until an action potential is triggered, and then they turn ON for a couple microseconds and then turn OFF again. At best, a neuron is a BIT at the physical level.

What did we just learn about physical BITS?

Do you remember?

When it comes to physical BITS, Someone Psyche has to purposefully and deliberately allocate them as BYTES and WORDS, and then Someone Psyche or Someone Intelligent has to use them as BYTES and WORDS to do computations and to issue instructions at the physical level; otherwise, those BITS just sit there and do nothing useful. BITS at the physical level are absolutely worthless unless Someone Psyche deliberately organizes them or maps them into BYTES and WORDS at the physical level and then uses them as such at the physical level. Computer hardware and computer software don't just spontaneously generate out of thin air. It's physically impossible.

We just learned that physical functionality at the physical level is physically impossible where the physical brain is concerned! It requires Someone Psyche or Someone Intelligent to organize those 86 billion stand-alone BITS into functional BYTES and WORDS of programming code and RAM.

YIKES!

If our brain really is organized into functional BYTES and WORDS containing memories, data, and programming code, then ALL of that functionality is being implemented at the quantum level or the psyche level by God or Nature's Psyche, because it's physically impossible at the physical level without the intervention of Someone Psyche or Someone Intelligent who is deliberately mapping those BITS as BYTES and then deliberately using them as BYTES to transmit and process data and programming code.

This is Quantum Neuroscience in action. Only Someone Psyche can successfully use Quantum Mechanics in order to get things done both at the quantum level and the physical level. Through Quantum Mechanics, it's possible for Someone Psyche to do the physically impossible, both at the quantum level and the physical level.

I go through this very same process a number of different times from many different angles – so, this isn't the last time that you will see these types of computations and ideas being presented in this book. I like running the numbers, so I will do so periodically. Running the numbers gives me a solid sense as to where the physical level ends and the quantum level beings; or in this case, I get a sense as to where Neuroscience ends, and Quantum Neuroscience begins.

Evolution ENDS with our 750-megabyte genome that God gave to evolution to mess up for us through random mutations. Nature's Psyche and Quantum Neuroscience BEGIN with God's Database and all the information, memories, data, blueprints, sequential cell birth logistics, differentiation information, migration logistics, synaptogenesis logistics, timing information, and programming code that can't be stored within our genome and isn't being stored within our genome. Where our physical development is concerned, Nature's Psyche and Quantum Neuroscience begin where evolution ends; and, evolution ends with our 750-megabyte genome.

Thank God our genes and proteins are the only thing that evolution can mess up for us – that's bad enough as it is. Evolution or random mutation is synonymous with physical damage. The genomes that God gave Adam and Eve were flawless; but, evolution has been messing them up ever since. Thank you evolution for giving us ALL of our inherited diseases and mental illnesses! Nature's Psyche performs flawlessly for us; but, there's nothing that Nature's Psyche can do with evolved genes and proteins – or damaged genes and proteins – although Nature's Psyche does try to compensate as best as possible.

This is fascinating stuff to discuss and think about – or at least I think it is; otherwise, I wouldn't be researching it and writing about it.

Each structure within our brain seems to be an intricately designed, exquisitely fine-tuned, carefully engineered, precision-wired, and marvelously manufactured piece of machinery that computer scientists would call a CPU – with cache for working memory. There's one serious flaw in this CPU Model, though. Do you know what it is?

The Neuroscientists repeatedly tell us that our brains are wired together into functional structures at the physical level. "Wired" and "wiring" are actually the words that they use. But, there's a major problem with this description of brain functionality. IT'S FALSE. There are NO wires in our brains! It's deceptively fraudulent to say that there is.

Since there are NO wires in our brain at the physical level, that means that there are no CPUs, no RAM, and no Memory Systems within our brains at the physical level. They don't exist at the physical level because they can't exist at the physical level. There are NO wires in our brain.

OUCH!

I used to be a Materialist, Naturalist, Nihilist, and Atheist. This reality and truth was painful for me to accept. I still have a hard time accepting it and believing it. I was fixated for months and even decades on trying to find a physical explanation for brain functionality; but, there are NO wires in our brain at the physical level, which means that there are no CPUs, no RAM, and no Memory Systems within our brain at the physical level. They don't exist at the physical level because they can't exist at the physical level. There is NO physiological explanation for brain functionality! There's nothing there within our brains for evolution to work with at the physical level.

DOUBLE OUCH!

ALL of this marvelous and OBVIOUS supernatural wiring, CPU functionality, RAM memory storage, caching or working memory, and all the other Memory Systems within our brain were implemented, mapped, or put into place at the quantum level by Nature's Psyche; and, they are now being operated and controlled at the quantum level by Nature's Psyche in order to get things done for us at the physical level. WE KNOW it is Nature's Psyche (or God) who is doing all of this brain structuring, brain mapping, and brain functionality for us at the quantum level, because the Human Psyche isn't consciously aware of any of these things.

ANYTHING at the physical level is a highly INEFFICIENT way to map and store data, numbers, instructions, memories, and programming code – even a genome. A genome is a highly inefficient way to store memories or programming code, because a genome exists at the physical level.

We have hypothesized that Psyche is quantum waves if we are talking Quantum Mechanics, or the vibrations within a string if we are talking String Theory. However, it's highly probable that Psyche is in fact the living entity or the living light who broadcasts and receives the quantum waves, and the living entity or living light who sets the frequency of the vibration within a string. Quantum waves and vibrations in strings seem to be thoughts and memories, and not Psyche itself.

If there is a point particle, Psyche is it. At times, I have hypothesized that Psyche or Non-Local Consciousness is an infinite singularity. If it is, then Psyche is the smallest object in existence. If it is, then Psyche or Intelligence is a point particle. If Psyche is an infinite singularity, then that means that Psyche can theoretically hold an infinite amount of information within a point particle that has NO physical size and takes up NO physical space whatsoever. Now, that's EFFICIENT data storage! You can't beat it! It's physically impossible.

The explanatory power of Quantum Mechanics and Quantum Neuroscience is astronomical – out the ying-yang and through the roof as the Materialists and Atheists would say! Quantum Mechanics and Quantum Neuroscience are Supernatural – they take over where Materialism, Naturalism, Darwinism, Nihilism, Behaviorism, and Atheism END.

As you can tell, I have learned to love Quantum Neuroscience.

But, it wasn't easy for me at first. I resisted it for months, before finally giving in to the evidence and going in all the way. I'm there now, but I wasn't there before.

Have you ever wondered why Classical Physicists and Materialists, such as myself, vigorously opposed Quantum Mechanics all of their lives? It's because Quantum Mechanics is supernatural in nature and origin. Quantum Mechanisms involve faster-than-light Action at a Distance, which is physically impossible. Quantum Mechanics is supernatural; and therefore, Quantum Mechanics falsifies Materialism, Naturalism, and their derivatives.

Cell Differentiation

Cell Differentiation is where we see the Hand of God. It's a sequential process, and it has to be performed perfectly every time, or we are going to be messed up even more than we already are.

The differentiation of neurons and glial cells into different functional brain structures – consistently and reliably – suggests the existence of some kind of non-local Master Plan or 3D Blueprint, especially when the differentiation, logistics, and synaptogenesis of one quadrillion synapses is taken into account. It's physically impossible to store petabytes of differentiation in our 750-megabyte genome. It can't be done, which means that it isn't being done.

However, it is hard to tell where the local DNA blueprint ends, and the much larger non-local blueprint begins.

Considering that it's a physical device, the genome of any species is a highly sophisticated, very efficient, and radically advanced piece of hard-coded programming, even in something as simple as a bacterium. Your genome is God's Signature. The intelligence behind your genome is off the charts!

Remember, elegant code requires an Intelligent Programmer. The very existence of our genome and the fact that functional proteins are encoded in our genome is Scientific Proof of God's Existence. Evolution (genetic drift), chemical evolution, random mutations, natural selection, abiogenesis, spontaneous generation, and macro-evolution have NEVER been caught in the act of designing and creating genomes and proteins because they can't. It's physically impossible.

Despite the miraculous nature of our genome, it's still physically impossible to store petabytes of differentiation information within our 750-megabyte genome.

Depending upon the calculations that are done and the estimates that are given, the human body has anywhere from 37.2 trillion cells up to 100 trillion cells; and, each one of them had to be born in the right sequence, come into existence at the correct location, migrate to the correct destination if they migrated after their birth, uniquely differentiate or sub-type when they get to their chosen destination, and then make connections with the other cells around them – especially in the case of neurons and glial cells. It's proper and efficient cell differentiation of neurons and glial cells that constructs and builds the different structures of our physical brain and our nervous system. If neurons and glial cells don't follow their instructions and the correct sequence of birth, migration, and differentiation, then the different structures of our physical brain and nervous system will not form.

There is NO way that all of this logistical information – about the timing of cell birth, the initial typing of a cell, the migration coordinates after cell birth, the subsequent differentiation or sub-typing of each cell, as well as dendritic growth and synaptogenesis in the case of neurons – for 37.2 trillion different cells can be stored within the 3.2 billion base-pairs of our genome. All of those quadrillions of variables and information have to be stored someplace else besides our genome – non-locally or transdimensionally in God's Database – because it's physically impossible to store that amount of information within our genome.

The observed fact that human bodies develop the same way every time tells us that God's 3D Database for Cell Differentiation and Cell Migration does in fact exist; and, the sheer monumental astronomical size of all that quadrillions of bytes of information tells us that God's Database is NOT stored in our genome. Since God's Database can't be stored in our genome, then where is it stored and how is it accessed?

Nature's Psyche accesses God's Database telepathically. Quantum Telepathy is WiFi at the quantum level. It works just as well as WiFi at our macro-level, and most likely even better.

Psyche or Intelligence or Quantum Non-Local Consciousness seems to be an infinite singularity. As an infinite singularity, Psyche is therefore theoretically capable of storing an infinite amount of information in something that has NO physical size and takes up NO physical space, because Psyche is a non-physical, non-local, transdimensional entity or object. As an infinite singularity, Psyche is the ultimate in information storage density and data storage capacity.

So, where are God's 3D Databases for the construction of physical bodies stored, since they can't be stored in the physical genomes that God creates? God's 3D Databases are stored non-locally or transdimensionally within God's Psyche or God's Intelligence. I can think of no other logical explanation for the storage of quadrillions of bytes of information (or possibly an infinite amount of information) within a single physical atom. You see, Nature's Psyche within each physical atom seems to have access to God's Databases to one extent or another, because that's how the physical atoms know what the Physical Laws or God's Laws are. That's why we have order, instead of chaos in this physical universe.

It is obvious and clear, to anyone who is willing to think logically and rationally about this subject of Information Storage Capacity and Cell Differentiation, that Nature's Psyche within the differentiated molecules of our physical body is either infinitely more intelligent than we are or has access to infinitely more knowledge and information than we do. This is logical common sense once we start to realize what's involved in Cell Differentiation. God is in the details.

QUANTUM NEUROSCIENCE: THE OBSERVED AND EXPERIENCED FACTS

Quantum Neuroscience is the study of how the physical brain functions at the quantum level, including the study of how the Human Psyche interacts with and controls its physical brain at the quantum level. Quantum Neuroscience is an attempt to understand and explain the physically impossible. The quantum level is the foundation of the physical level; therefore, Quantum Neuroscience subsumes Neuroscience.

For a couple of months, while studying Neuroscience (brain anatomy and brain functioning) and also studying Quantum Mechanics at the same time, I realized that we needed a new science discipline called Quantum Neuroscience, which is the science of how the Human Psyche interacts with and controls its physical brain at the quantum level.

Quantum Neuroscience is the kind of neuroscience that I wanted to study and learn about, because the quantum level is where the Human Psyche interacts with and controls its physical brain.

But, you will never find anything like this in our public schools, because the Materialists, Naturalists, and Atheists who run and control our public schools won't allow anything like this into evidence, because it falsifies Materialism and Naturalism. These people don't allow anything into our public schools that falsifies Materialism, Naturalism, Darwinism, Scientism, Determinism, Atomism, Behaviorism, Materialistic Reductionism, Nihilism, and Atheism; and, they have the US Supreme Court on their side too. Ignorance is permitted, encouraged, fostered, and promoted in our public schools and government agencies.

Over and over again while working on my bachelor's in psychology, I was re-introduced to Neuroscience, Brain Anatomy, Brain Function, and the Neurotransmitter Systems from the perspective of Classic Physics, Materialism, Darwinism, and Naturalism. It got kind of boring. It got to the point where I wanted to learn and understand how the physical brain is functioning at the Quantum Level. I wanted to study Quantum Neuroscience, but there wasn't any such thing. Instead, I started to study Quantum Mechanics and tried to piece together Quantum Neuroscience for myself, since nobody else had ever tried to do so.

The quantum level is where everything starts to get interesting, because it does indeed falsify Materialism, Naturalism, and their derivatives.

I wish I could have studied Quantum Neuroscience in college; but instead, I was left to develop it for myself. Nevertheless, I learned a lot while doing so; and, I continue to discover new and interesting ideas every day.

Quantum Neuroscience

I'm interested in Neuroscience at the Quantum Level – which is something that I have labeled and called "**Quantum Neuroscience**".

Quantum Neuroscience is the scientific study, the experiential study, and the observational study of how the Human Psyche interacts with and controls its physical brain.

Quantum Neuroscience is based upon the principles and ideas associated with Quantum Mechanics.

https://en.wikipedia.org/wiki/Quantum_mechanics

http://mypsyche.us/wp-content/uploads/2017/11/Quantum_mechanics.pdf

Ironically, when it comes to Quantum Neuroscience, it really isn't theory, because it has all been verified and proven to exist already! Quantum Mechanics is the most-verified, best-proven, and most-used science that we have! Quantum Neuroscience was already proven science before I coined the term.

The only thing that was really needed was to get the correct Interpretation of Quantum Mechanics, which can be found in the Orthodox Interpretation of Quantum Mechanics as developed and presented by Henry P. Stapp. Then instantly, you find yourself looking at Quantum Neuroscience, whenever you apply the Orthodox Interpretation to Neuroscience itself.

I recognized the correct and true Interpretation of Quantum Mechanics immediately when I first saw it. The correct Interpretation of Q. M. is the one that has the most explanatory power! The correct Interpretation of Q. M. is also easy to identify, because it is the one that the Materialists, Naturalists, and Atheists have been trained to fight against and reject. The Materialists, Naturalists, Darwinists, Nihilists, and Atheists are the Masters of identifying the truth and then formally rejecting it. If there is anything that these people are fighting against, ridiculing, and rejecting, then that's precisely what you should be finding, learning, and embracing if you want to know the truth. I recommend that you start with Quantum Non-Local Consciousness (Psyche) and the Orthodox Interpretation of Quantum Mechanics. These are the things that the Materialists and Naturalists fight against the most.

https://sites.google.com/a/lbl.gov/stappfiles/

http://www-physics.lbl.gov/~stapp/

As you read through this book, observe how I adapt Neuroscience to fit with the Orthodox Interpretation of Quantum Mechanics. Even if I get some of my adaptations wrong, the overall practice or principle is solid, reliable, replicable, and sound; and, the explanatory power is through the roof to infinity and beyond! You get better at it, the more you do it!

Quantum Neuroscience is based upon Science 2.0. Science 2.0 allows ALL of the evidence into evidence, and then pursues a preponderance of the evidence.

Quantum Neuroscience is also based upon the observational evidence and experiential evidence obtained from Near-Death Experiences (NDEs), Observational Experiences, Out-of-Body Experiences (OBEs), Shared-Death Experiences (SDEs), after-death Life Reviews, and other types of Non-Local Experiences or Spiritual Experiences.

We start with what WE KNOW to be true from NDEs, OBEs, and Life Reviews; and then, we try to explain it logically and scientifically; and, we try to use Science and Quantum Mechanics to figure out how the Human Psyche controls its physical brain.

Although it is safe to say that WE KNOW that the Human Psyche controls its physical brain through telepathy or the Quantum Zeno Effect, how does the Human Psyche KNOW or learn which of the 100 billion neurons to trigger first in order to get a specific desired-action started?

I slowly develop an answer to this question by pursuing all the evidence that I could find. It helped me to realize that Process 1 of the Orthodox Interpretation of Quantum Mechanics is performed by the Human Psyche – this is the conscious request or the choice process. Process 2 is the collapse of the wave function, thereby making one of the infinite possibilities become actual and real. And, Process 3 is Nature's Psyche (or God's Psyche), who has the power to collapse the wave function. The Human Psyche makes a request or a choice, and Nature's Psyche collapses the wave function making that request or choice actual and real. It's really simple to understand, once you decide to do so.

So, how does the Human Psyche know which of the 100 billion neurons it needs to trigger in order to raise its finger? It doesn't know! The Human Psyche isn't consciously aware of these things. It's Nature's Psyche who knows which neuron it needs to fire in order to get your finger to rise. Nature's Psyche collapses the wave functions and grants your requests. It all happens automatically under the control of Nature's Psyche.

Nature's Psyche is like a genie – it grants all of your wishes and makes them come true – as long as your wishes are actually possible in a physical realm and aren't a violation of the Physical Laws or God's Restrictions and God's Laws. God has placed some limitations on us and Nature's Psyche, which Nature's Psyche will not help us get around, because Nature's Psyche has covenanted to obey God's commands. That's what brings order and life to this physical universe. Nature's covenant with God to obey God's commands is also what keeps your physical body from quantum tunneling away on you one atom at a time, until you disintegrate and dissolve into nothing. God's Psyche is the One who brought order out of chaos where this physical universe is concerned. Nature's Psyche within the atoms does what God commands.

You, the Human Psyche, make the requests and the choices; and, Nature's Psyche does all the heavy lifting, thereby making your choices actual and real. God's Psyche provides the order or the physical laws – the construct; and, God holds all the KEYS to the different quantum mechanisms. God provides the standards, structures, cosmological constants, and limitations. God is the one who reins in Nature's Psyche preventing it from running amok. This is what the Orthodox Interpretation of Quantum Mechanics tries to teach us. This is Quantum Neuroscience in a nutshell.

This explains how the Human Psyche is able to interface with and control its physical body and physical brain. It's all based upon waves or quantum mechanisms. Remember, waves are WiFi at the quantum level, which means that telepathy is WiFi at the quantum level. The Human Psyche communicates with Nature's Psyche through waves or quantum mechanisms; and, Nature's Psyche collapses the wave functions making your physical body and physical brain do the things that you want done. This is how the Human Psyche interfaces with, commands, and controls its physical body and physical brain. It's done through Waves, or WiFi, or Quantum Mechanics, with Nature's Psyche as the mediator.

Remember, it's physically impossible to develop, study, and understand Quantum Neuroscience with Materialism, Naturalism, and Classical Physics. Something has to give! We have to up our game if we want to figure out how things really work.

Out-of-Body Travelers have observed that Quantum Telepathy (the Quantum Zeno Effect or Mind-to-Mind WiFi Connections) is the standard practice or the modus operandi in the Non-Local Realm or the Spirit Realm. Remember, telepathy is WiFi at the quantum level. Everybody communicates telepathically in the Spirit World – Psyche-to-Psyche. The same Out-of-Body Travelers have also observed that Quantum Teleportation (Quantum Tunneling) is the standard practice or the modus operandi in the Non-Local Realm. They can teleport. They can also levitate and fly. Out-of-Body Travelers have observed that Non-Consensus Realms or Unorganized Realms or Unorganized Spirit Matter in the Spirit

World automatically conform themselves to the demands of their Psyche or Non-Local Consciousness. Spirit matter reorganizing itself to meet the demands or expectations of their Psyche is a type of Quantum Telekinesis which seems to be common in the Non-Local Realms.

Our Physical Realm is completely different, because with our physical realm God deliberately slowed everything down so that we can live it, experience it, participated in it, learn from it, and remember having done so. Pain is our greatest teacher. It motivates us to learn. There are real and lasting consequences for our chosen acts here in this physical realm.

Our Physical Realm is also different, because God designed it to be and intended it to be the Ultimate Consensus Reality, where I can expect this essay to be on my computer tomorrow and next week and next month whenever I go looking for it. I can also expect my house, my car, my dog, and my computer to be there tomorrow in some form or another. These kinds of blessings and expectations aren't fully available in the Non-Local Realms or the Spirit World, because most of it is completely unorganized spirit matter and exists as non-consensus realities awaiting the arrival of some psyche or non-local consciousness to give it order, meaning, structure, purpose, and life.

The following book was the first book about Quantum Mechanics that actually made logical sense to me. All sorts of lights went on when I first read this book.

Van Lommel, P. (2010). *Consciousness Beyond Life: The Science of the Near-Death Experience*. New York: HarperCollins.

This book, from Pim van Lommel, taught me what I call the **Non-Local Consciousness Interpretation of Quantum Mechanics**, which is the science behind Near-Death Experiences.

Lommel's book provides scientific evidence, experiential evidence, observational evidence, empirical evidence, and eye-witness evidence that Non-Local Consciousness or Psyche does in fact exist; and then, his book tries to develop an explanation for it all through Quantum Mechanics.

The following paper eventually surfaced after years of studying Quantum Mechanics. The following article is a great introduction to **Quantum Neuroscience**.

Schwartz, J. M., Stapp, H. P., & Beauregard, M. (2004). *Quantum Physics in Neuroscience and Psychology: A Neurophysical Model of Mind-Brain Interaction*. Published Online: Phil. Trans. R. Soc. B.

http://www-physics.lbl.gov/~stapp/PTRS.pdf

http://mypsyche.us/wp-content/uploads/2017/10/PTRS.pdf

https://www.researchgate.net/publication/7613549_Quantum_physics_in_neuroscie nce_and_psychology_A_neurophysical_model_of_mind-brain_interaction

http://escholarship.org/uc/item/4w8665vk

This paper summarizes everything I have been looking for during the past fifty-five years of my life. That's a long time to wander in darkness looking for the truth; but luckily, our generation finally has the truth, if we know where to find it and recognize it as true when we do find it.

This paper and Henry P. Stapp's articles and books taught me what Stapp calls the **Orthodox Interpretation of Quantum Mechanics**, which explains the interplay or the scientific interface between mind and matter at the quantum level.

https://sites.google.com/a/lbl.gov/stappfiles/

http://www-physics.lbl.gov/~stapp/

Henry P. Stapp's free articles repeatedly led me to the following book:

Schwartz, J. M. (2002). *The Mind and the Brain: Neuroplasticity and the Power of Mental Force*. New York: HarperCollins.

> **By applying a new and scientifically testable method that would empower OCD patients actively and willfully to change the focus of their attention, I just might help them learn to overcome their disease. But I had a hunch that I might achieve something else, too: demonstrating, with the new brain imaging technology, that patients could systematically alter their own brain function. The will, I was starting to believe, generates a force. If that force could be harnessed to improve the lives of people with OCD, it might also teach them how to control the very brain chemistry underlying their disease.** (p. 7.)

Schwartz, J. M. (2002). *The Mind and the Brain: Neuroplasticity and the Power of Mental Force*. New York: HarperCollins.

(HarperCollins generously permits quoting from their books for review and promotional purposes. For me personally, this book from Jeffrey Schwartz is the textbook for Quantum Neuroscience. It doesn't get better than this.)

I chose this book, *The Mind and the Brain*, to be the foundational textbook for **Quantum Neuroscience**. I have been looking for this book all of my life, and I finally found it in October 2017. This book represents real, efficacious Neuroscience; and, it takes everything into account including Quantum Mechanics and Psyche or Mind.

Instead of using Neuroscience, Biopsychology, and Brain Scans to preach and promote Darwinism, Determinism, Materialistic Reductionism, and the Theory of Evolution as most of the Neuroscientists and Biopsychologists do, Jeffrey Schwartz actually used Neuroscience and Brain Scans to develop an effective treatment and potential cure for Obsessive-Compulsive Disorder (OCD). Jeffrey Schwartz does REAL Psyche-Therapy, by using Psyche or the Mind to desensitize the OCD Circuit and rewire the human brain, through a process known as Cognitive-Behavior Therapy. It works. The Brain Scans show a change and a normalization of brain functioning, after the Psyche-Therapy or Cognitive Therapy has taken hold. Jeffrey Schwartz and team really do use Psyche or Mind to rewire and normalize the human brain. Since the 1980's when Cognitive Therapy saved my life, I have had a great deal of admiration for Cognitive Therapy and what it can do. I always had a feeling that Cognitive Therapy, Friendship Therapy, or Psyche-Therapy could rewire and normalize the human brain, but now I KNOW. Science is KNOWLEDGE – or at least it should be.

During my research, I discovered that the different books from Jeffrey Schwartz, Henry Stapp, and Mario Beauregard are excellent foundational material for **Quantum Neuroscience**. These three scientists set out to discover how the Human Psyche interacts with and controls its physical brain at the quantum level; and, the evidence and information within their books tell us that these authors succeeded in doing so.

I submit their books into evidence as Proof of Quantum Neuroscience and Non-Local Consciousness.

Schwartz, J. M. (2002). *The Mind and the Brain: Neuroplasticity and the Power of Mental Force*. New York: HarperCollins.

Beauregard, M. & O'Leary, D. (2007). *The Spiritual Brain: A Neuroscientist's Case for the Existence of the Soul*. New York: HarperCollins.

Stapp, H. P. (2017). *Quantum Theory and Free Will: How Mental Intentions Translate into Bodily Actions*. Cham, Switzerland: Springer International Publishing.

Van Lommel, P. (2010). *Consciousness Beyond Life: The Science of the Near-Death Experience*. New York: HarperCollins.

Beauregard, M. (2012). *Brain Wars: The Scientific Battle over the Existence of the Mind and the Proof That Will Change the Way We Live Our Lives*. New York: HarperOne.

Stapp, H. P. (2007, 2011). *Mindful Universe: Quantum Mechanics and the Participating Observer* (2nd ed.). New York: Springer-Verlag.

Stapp, H. P. (2004). *Mind, Matter and Quantum Mechanics* (2nd ed.). New York: Springer-Verlag.

Stapp, H. P. (2009). *Mind, Matter and Quantum Mechanics* (3rd ed.). New York: Springer-Verlag.

Stapp, Henry. *Quantum Physics of Consciousness*. Quantum Reality and Mind (Kindle Locations 461-518). Cosmology Science Publishers. Kindle Edition.

Penfield, W. (1975). *The Mystery of the Mind*. Princeton, New Jersey: Princeton University Press.

Marshall, P. D., Kelly, E. F., & Crabtree A. (Eds.). (2015). *Beyond Physicalism: Toward Reconciliation of Science and Spirituality*. London, United Kingdom: Rowman and Littlefield.

Kelly, E. F., Kelly, E. W., Crabtree, A., Grosso, M., & Gauld, A. (2007). *Irreducible Mind: Toward a Psychology for the 21st Century*. Plymouth, United Kingdom: Rowman and Littlefield.

Obviously, we have to understand and master Quantum Mechanics if we want to understand **Quantum Neuroscience** and how the physical brain really functions at the quantum level.

The following is the Science. It has been observed and experienced; and therefore, is KNOWN to be true. The problem comes in trying to explain it and interpret it. In this next section, I place in **bold** everything that has been experimentally verified as being real and true. You are going to see a lot of **bolding** where Quantum Mechanics is concerned.

How Did Evolution Learn How to Design and Control Your Physical Brain?

Kalat, J. W. (2008). *Introduction to Psychology* (9th ed.). Belmont, CA: Wadsworth, Cengage Learning. **"The brain is the product of Evolution."** (p. 12.)

Pinel, J. (2014). *Biopsychology* (9th ed.). New York: Pearson. **"Genetic endowment is a product of its evolution."** (p. 24.)

Natural selection is the idea that heritable traits, which are associated with high rates of survival and reproduction, are preferentially passed on to future generations. Natural selection doesn't change our genes! Natural selection doesn't touch our genes, because that's physically impossible. Natural selection is absolutely worthless for evolution or genetic change!

Furthermore, evolution (genetic change) and random mutations can't do design and creation at the quantum level. By definition in principle, evolution is restricted to the physical level. Evolution can't do anything at the quantum level or the psyche level. Isn't that fascinating? Evolution can't even do what they say that it does. Can you feel the burn?

> **The multiplier effect is a mechanism by which the behavioral effects of a gene are increased because the gene promotes the choice of experiences that have the same behavioral effects.** (*Biopsychology*, p. 489.)

How?

This multiplier effect is *begging the question*. It's *circular reasoning*. Chosen behaviors are a function of the Human Psyche. According to the Materialists, Darwinists, Behaviorists, and Naturalists, choice is physically impossible. According to these people, genes can't do anything at the quantum level or psyche level, because according to these people the psyche level doesn't exist. So, how are these genes running our lives and making choices for us? It's physically impossible! Choice implies some kind of Intelligence or Psyche at the quantum level, because choice is physically impossible. It's amazing what you end up with when you finally choose to allow all of the evidence into evidence.

How did Evolution, Random Mutations, and Natural Selection learn how to design, create, and control your genome, your proteins, and your physical brain?

They didn't!

In fact, our physical functionality isn't even controlled by our physical brain. Instead, Nature's Psyche maps physical functionality onto the different structures of our brain; and then, Nature's Psyche uses these Maps of Physical Functionality to control our brain and physical body for us. However, our decisions and choices are controlled by our psyche – the Human Psyche. The Human Psyche makes the decisions and the choices, and then Nature's Psyche collapses the wave functions and/or turns ON the switches needed to make those decisions and choices physically real. This is what's really happening within the physical brain, both at the physical level and the quantum level. It explains everything that we need to know about the physical brain.

Quantum Telepathy: The proven and observed Quantum Zeno Effect. Remember telepathy is WiFi at the quantum level. Quantum Telepathy often involves Quantum Telekinesis.

Quantum Teleportation: The proven and observed Quantum Tunneling Effect.

Quantum Telekinesis: Using psyche to move molecules with our mind. Many molecules "swim" through the cytoplasm of a physical cell and go specifically where they need to go. This is often some kind of mind-over-matter phenomenon, as each molecule in a living cell seems to have a mind or a psyche of its own. Quantum Scientists have called this thing Nature's Psyche. Any type of

mind-over-matter telekinesis or propulsion is also what the quantum physicists call Action at a Distance.

How did evolution, random mutations, and natural selection learn how to master and control quantum telepathy, quantum teleportation, and quantum telekinesis? Well, they didn't, because that's physically impossible! Evolution of any kind by definition in principle is limited exclusively to the physical level and the physical realm. Evolution can't do anything at the quantum level.

Ribosome: A structure in the cell's cytoplasm that translates the genetic code from strands of messenger RNA (mRNA) into proteins.

It has been proven scientifically that it is physically impossible for chemical evolution to produce functional proteins, let alone the matching genes to go with them. Your proteins and genes are not a product of evolution, because that's physically impossible. So, there's no way in the universe that your eyes, brain, and genome can be a product of evolution. It's physically impossible.

Ribosomes make proteins. Ribosomes are made of proteins. Proteins are used to make ribosomes. How were ribosomes made in the first place, if there were no ribosomes to make the proteins that are used to make the ribosomes? It's physically impossible to make something that doesn't exist yet out of something that doesn't exist yet.

Ribosomal protein synthesis – the production of new ribosomes – is done by pre-existing ribosomes. They never do describe how the very first ribosome was produced! It just magically popped into existence out of thin air, I guess, just like everything else that was made by evolution.

The physically impossible is where science really starts to get interesting!

Transduction: The conversion of one form of energy to another. Photosynthesis and the conversion of light waves into electrochemical signals in the retina of our eye. How did evolution learn how transduction works and master that technology?

Replication: Cell division. Replication is the process by which the DNA molecule duplicates itself.

Replication is overseen by Nature's Psyche at the quantum level.

How do we know?

During replication, the correct nucleotides move into the correct locations as if they know what they are doing and why they are doing it. Each chromosome strand attracts free-floating complementary bases.

Why is this so significant?

Nucleotides have NO special affinity for one another, or no special attraction to one another. That's how they are able to go anywhere onto the genome and act as programming code, because there's nothing there that evolution could have used to arrange them into a specific pre-ordained order. Evolutionary necessity does not apply and cannot apply to nucleotides; but, somebody or someone is forcing them to come together during replication, at the quantum level.

These nucleotides are free-floating under the effects of random diffusion, and then suddenly someone turns on the tractor beam and draws the correct ones in to the correct position. There's targeting, purpose, intention, selection, choice, and tractor beams here –

all quantum processes that are physically impossible at the physical level. These things seem to line up in advance and are there waiting for their turn as if they are on some kind of schedule. It's as if these molecules are alive!

How does evolution move these nucleotides into the correct position? How did evolution, random mutations, and natural selection get the incentive and the desire to master cell replication; and, how did evolution learn and master genetics and the human genome? Evolution didn't. That's physically impossible. Evolution can't act on atoms and molecules at the quantum level. Evolution only works on genes at the physical level. Think people, think! The theory of evolution is a fraudulent hoax.

Gene Expression or Protein Synthesis: This is the physical process of converting the information contained within a gene into a functional protein at the physical level.

How does evolution know how to do all of this? How is evolution controlling all of these things? How does evolution design and manufacture a functional protein out of atoms from scratch at the quantum level? How does evolution manufacture and deploy the matching gene to go along with that newly created protein, at the quantum level? You can't manipulate atoms with physical instruments. You have to use some kind of wave at the quantum level in order to move atoms around and make them do things. We can use different kinds of waves to move atoms around and make them do things. How did evolution learn how to do so billions of years ago? Who taught evolution how to do all of these different things?

Activation: The prescient binding of the correct amino acid and the correct codon to each transfer RNA (tRNA) molecule.

Transfer RNA are molecules of RNA that carry amino acids to ribosomes during protein synthesis. Each kind of amino acid is carried by a different kind of transfer RNA molecule. There are 20 amino acids and 64 possible codons. Some amino acids are bound to more than one codon during activation.

During activation, how does the tRNA molecule know in advance which codon and amino acid to grab? That's physically impossible, because it requires foresight, telepathy, and action at a distance.

Somehow, the activation factors that bind the amino acid and codon to the tRNA molecule KNOW in advance what type of codon is going to be needed in the future at the ribosome. The activation factors or the proteins associated with activation are prescient! They can either see the future, or they can read the incoming mRNA molecule telepathically at a distance, because the activation factors KNOW precisely what amino acid and what codon to attach to the tRNA molecule that they are working on. Transfer RNA molecules also move, or they quantum tunnel to their destination. All of this is physically impossible. We are looking at Quantum Neuroscience here, not Materialism, Naturalism, and Classical Physics. Activation factors, amino acids, codons, and tRNA molecules are controlled by Nature's Psyche at the quantum level and NOT classical physics at the physical level.

How do evolution, natural selection, and random mutations do prescience or quantum telepathic communication between molecules? How do evolution, natural selection, and random mutations do quantum tunneling and quantum telekinesis? How does evolution make molecules move? Who taught evolution how to activate tRNA molecules? There's no way in the universe that evolution could be controlling our amino acids, codons, and tRNA molecules at the quantum level. By definition, in principle, evolution doesn't do foreknowledge or prescience. Evolution can't because evolution of any kind doesn't have access to the quantum level. Evolution is trapped at the physical level, and there it stays.

Transcription: The location of the correct gene, and the production of the needed messenger RNA (mRNA) molecule for synthesis or construction of the desired protein.

How do evolution, natural selection, and random mutations know and remember where the correct genes are located, each time a living cells requests the production of a certain protein? How evolution controlling all of these things? How does evolution choose the target of the mRNA molecules, and then make the mRNA molecules move to the target that evolution has selected and not to someplace else? How does evolution tell the mRNA molecules where it wants them to go?

Translation: The ribosome's assembly of proteins from mRNA and the correct amino acid and codon combinations found on tRNA.

How do evolution, natural selection, and random mutations get the tRNA molecules to line up perfectly with the mRNA sequence long before reaching the ribosome for protein synthesis or the construction of a protein? How does evolution choose the target of the tRNA molecules, and then make the tRNA molecules move to the target that evolution has selected and not to someplace else? How does evolution tell the tRNA molecules where it wants them to go? Clearly, evolution, random mutations, and natural selection can do the physically impossible at the quantum level or psyche level.

Activation, transcription, translation, gene expression, and protein synthesis are the smoking gun where Quantum Neuroscience is concerned, because these molecules are NOT being controlled by our genes at the physical level. Manipulations of our atoms, molecules, amino acids, codons, transcription factors, genes, and proteins are NOT being caused nor controlled by our genes and proteins. Manipulation of our genes, proteins, molecules, and atoms IS being done at the quantum level by Nature's Psyche completely outside the conscious awareness of the Human Psyche. Activation, transcription, translation, gene expression, and protein synthesis are not being done by our genes, contrary to what the Materialists, Naturalists, and Darwinists claim. Evolution can't do anything at the quantum level, because it doesn't exist at the quantum level.

Who or what propels the mRNA and tRNA molecules?

There are indications that some kind of ionic propulsion is attached to these things to make them move through the cytoplasm; however, specific targeting requires Intelligence or Psyche to get that job done. Somebody has to steer these things! Propulsion is worthless if you don't have a way to steer the thing and a place where you want it to go. According to the Orthodox Interpretation of Quantum Mechanics, Nature's Psyche collapses the wave function, which means to me that Nature's Psyche chooses the target, the route, and the destination of the various different mRNA and tRNA molecules. These things also have to rotate into the proper orientation so that they can dock with their target molecules. Targeting and docking are functions of Psyche or Intelligence, not random luck or random mutations.

If I were to drop you into space within a ship that has ionic propulsion that actually works, would you know how to steer the thing to a space station orbiting the moon, and then successfully dock your ship with that space station when you get there?

According to the Materialists and Naturalists, evolution knows how to do targeting, steering, and docking, so why don't you? You're smarter than evolution, aren't you? Apparently not!

These are molecules that we are talking about here; and, they seem to be alive! That's physically impossible according to the Materialists and Naturalists; but, totally possible at the quantum level or psyche level.

The tRNA and mRNA are highly intelligent, motile, and know precisely where they are going and how to get there. They are imbued with a purpose or a mission. Who taught the tRNA, mRNA, and the ribosomes what the codons mean, and which amino acids should be attached to a specific codon? Who designed and manufactured this whole process in the first place? It's physically impossible for these types of things to spontaneously generate out of thin air. Combining atoms into functional molecules, messengers, proteins, and genes requires a vast amount of planning, intelligence, and foresight.

Oh, there's a lot more going on here than meets the eye. The mRNA and tRNA molecules are like little vessels or submarines moving through the cytoplasm; and, somebody onboard has to be telling them where to go and what to do when they get there. Otherwise, they are just going to sit there, vibrate, and do nothing – or move around randomly if they have some kind of ionic propulsion nailed to their tail. This somebody who is doing all of this targeting and motivation is Nature's Psyche or God's Psyche, because all of this is taking place completely outside the conscious awareness of the Human Psyche. Evolution can't do any of these things, because they are physically impossible. Evolution of any kind can't do the physically impossible; but, Psyche certainly can. Psyche IS physically impossible after all.

Protein Biosynthesis:

https://en.wikipedia.org/wiki/Protein_biosynthesis

http://mypsyche.us/wp-content/uploads/2017/12/Protein_biosynthesis.pdf

Genetic Code and Gene Expression

Structural genes contain the information necessary for the synthesis of proteins. Proteins are long chains of amino acids; they control the physiological activities of cells and are important components of cellular structure.

Enhancers are stretches of DNA whose function is to determine whether particular structural genes initiate the synthesis of proteins and at what rate. The control of gene expression by enhancers is an important process because it determines how a cell will develop and how it will function once it reaches maturity. Enhancers are like switches because they can be regulated in two ways: They can be turned up, or they can be turned down.

Proteins that bind to DNA and influence the extent to which genes are expressed are called transcription factors. Many of the transcription factors that control enhancers are influenced by signals received by the cell from its environment. (*Biopsychology*, p. 39.)

How did evolution, random mutations, and natural selection learn how to do protein synthesis?

The standard answer is that all of these things are physical machines, and they were programmed by evolution to do what they do; but, evolution can't do programming. Programming requires an intelligent Programmer; and, machines require an intelligent Designer and Creator. Evolution, random mutations, and natural selection cannot do programming, planning, design, engineering, fine-tuning, field-testing, teleology, manufacturing, and creation.

How did Evolution learn how to master and control quantum mechanics, teleportation or quantum tunneling, quantum telekinesis, quantum WiFi or quantum telepathy, signaling between the environment and transcription factors, quantum non-locality, genetics, ion channels, neurotransmitters and neurotransmitter receptors, neuropeptides, transduction, replication, transcription, translation, activation, and protein synthesis?

It didn't.

It's physically impossible. There's no psyche or intelligence in evolution (genetic drift), random mutations, and natural selection. Evolution can't do psyche or intelligence. By definition, in principle, evolution can't do complexity and science. Random mutations and natural selection are physical reactions at best – not creative geniuses. There's no intelligence there in evolution. Design and creation by evolution is nothing but a fictional construct, based upon blind faith and wishful thinking.

The very existence of transduction, replication, activation, transcription, translation, genomes, neurotransmitter systems, ion channels, physical brains, fine-tuning, teleology, quantum mechanics, and protein synthesis is Scientific Proof of God's Necessity, which ends up being Scientific Proof of God's Existence. There's no other logical explanation for these things. These are things that random mutations and natural selection cannot do and cannot master and learn. It's so obvious, that I sometimes wonder how I missed it for the first fifty years of my life; but, I missed it because I wasn't looking for it. No seeking, then no finding. I used to be a Materialist, Naturalist, Nihilist, Skeptic, and Atheist. We don't look for these types of things, because we don't want them to be true and don't want them to exist. I KNOW because I have been there and done that. I have lived it and experienced it.

The BEST way to find and know the truth is to live it and experience it for yourself, or to choose to trust someone who has; and, I learn best through comparison and contrast – by comparing evolution with everything else.

As they say, God is in the details. Evolution (genetic drift), random mutations, and natural selection can't do details, because these things are by definition in principle dumb and blind. Evolution and its derivatives are a function of luck; and, blind luck cannot do design and creation. That is what we have experienced and observed.

Our very best scientists and medical doctors can't figure out how to fine-tune and properly rebalance our neurotransmitter systems so as to alleviate or eliminate our worst mental illnesses such as depression and schizophrenia; so, why on earth would anyone in their right mind believe that random mutations and natural selection did the job blindly, one generation at a time, billions of years ago? Such a belief is illogical and irrational.

Transduction, replication, activation, transcription, translation, genomes, neurotransmitter systems, ion channels, physical brains, physical bodies, fine-tuning, teleology, quantum mechanics, and protein synthesis are a function of Psyche, Intelligence, or Non-Local Consciousness. There are signs of intelligence there, which required an Intelligent Source. There's no other logical explanation for their origins or effectiveness.

Quantum Non-Locality

Most books that I have read about Quantum Mechanics define Non-Locality as "Action at a Distance" or "Quantum Entanglement" – the idea being that at the Local Level or the Physical Level or the Level of Classical Physics, actions are directly connected through atoms and molecules, and there is NO action at a distance taking place in the Physical Realm; whereas, at the Quantum Level or the Non-Physical Level or the Non-Local Level, the effects of Action at a Distance and Quantum Entanglement have been observed and proven to exist.

In other words, Quantum Mechanics and Quantum Effects do NOT follow the rules of Classical Physics, and therefore do NOT make sense from a materialistic, naturalistic, and physicalistic perspective. Quantum Mechanics only makes sense from a Non-Local or Non-Physical perspective; and, Quantum Entanglement is only a small part of Quantum Non-Locality.

However, in order to satisfy Einstein and match with Einstein's theory of special relativity, the scientists erroneously defined quantum nonlocality as not allowing for faster-than-light communication, thus forcing quantum nonlocality to be compatible with special relativity and limiting it to the speed-of-light and Classical Physics.

As a result of all this, I modified and enhanced the definition of Quantum Non-Locality in the following manner.

Local means physical. Local means Classical Physics. Local means located in our 3D Physical Space-Time Realm. Local means having a location in physical space and physical time. Local means being subject to the rules and limitations of Classical Physics. Local means being slower than the speed-of-light. Remember, physical matter slows everything down and introduces entropy, aging, time, speed limits, and other physical limitations into the equation in a major way. Physical matter slows everything down so that we can live it, experience it, learn from it, and remember having done so. A Physical Realm is the Ultimate Consensus Reality – it's highly reliable. In this Physical Realm, I can trust that this paper will be on my computer and on the cloud tomorrow when I go looking for it.

In contrast, Non-Local means non-physical. Non-Local means Quantum Mechanics. Non-Local means not being subject to any physical limitations, including the speed-of-light. Non-local means being able to exceed the speed-of-light and travel at an infinite velocity (teleport) at will. Non-Local means that the Quantum Object (particle of matter) under consideration is spirit matter and is located in the Non-Local Realm, or Transdimensional Realm, or the Spirit World. Non-Local means being subject to the rules and limitless capacity of Quantum Mechanics. Non-Local means that the Quantum Object (particle of matter), Quantum Psyche (non-local consciousness), and the Quantum Effects (telepathy, teleportation, multi-tasking, and transdimensionality) EXIST and FUNCTION in the Non-Local Realm or Transdimensional Realm, which many people call the Spirit World. In the Non-Local Realm, everything happens instantaneously at the speed of thought.

Action at a Distance or Quantum Entanglement have been proven experimentally to exist at the Quantum Level. It's real proven Science. It has been experienced and observed.

https://en.wikipedia.org/wiki/Quantum_entanglement

http://mypsyche.us/wp-content/uploads/2017/11/Quantum_entanglement.pdf

Bell's Theorem Verified, and Quantum Entanglement Proven True:

https://en.wikipedia.org/wiki/Bell%27s_theorem

http://mypsyche.us/wp-content/uploads/2017/11/Bells_theorem.pdf

https://en.wikipedia.org/wiki/Bell_test_experiments

http://mypsyche.us/wp-content/uploads/2017/11/Bell_test_experiments.pdf

But, Action at a Distance or Quantum Entanglement is NOT limited to the speed-of-light, as the Materialists and Naturalists claim. Psyche and Quantum Objects (spirit matter) in the Non-Local Realm have no speed-limit, and therefore are capable of infinite velocities and teleportation or Quantum Tunneling.

Quantum Tunneling is teleportation. It's not limited to sub-light speeds.

Psyche (non-local consciousness), Quantum Objects (particles of matter), and Quantum Zeno Effects (telepathy and thoughts) are theoretically capable of infinite velocities, which means that they can be instantaneously and simultaneously anywhere and everywhere. They can teleport and multi-task in the Quantum Non-Local Realm.

The materialistic and naturalistic scientists box-in quantum non-locality and force it to fit into the limitations of Classical Physics. These people refuse to look at the other side of the equation – the spiritual side or the non-local side. Consequently, quantum non-locality as it is typically presented in all of the science literature has been castrated and neutered so that it matches with Classical Physics and Einstein's Theory of Special Relativity limiting everything to sub-light speeds, which means that they make Non-Locality completely worthless from a Quantum Mechanical perspective.

Remember, there are NO physical limitations in the Quantum Realm or the Non-Local Realm. To suggest that there are results in a confound where the scientists forcefully equate Quantum Non-Locality with Classical Physics thereby defeating the whole purpose of Non-Locality before we even start to look at it and study it.

See if it is not so:

Whilst quantum nonlocality improves the efficiency of various computational tasks, it does not allow for faster-than-light communication, and hence is compatible with special relativity.

https://en.wikipedia.org/wiki/Quantum_nonlocality

http://mypsyche.us/wp-content/uploads/2017/11/Quantum_nonlocality.pdf

Notice how they erroneously try to limit Non-Locality to the speed-of-light, in order to make it compatible with Classical Physics and Einstein's Special Theory of Relativity. They are trying to define Non-Locality in terms of Classical Physics rather than defining it in terms of Quantum Mechanics. It's a *category error* or a *confound*, that they are employing here in order to make their case.

When it comes to Quantum Mechanics and Quantum Non-Locality, it can be fascinating to observe how the Materialists and Naturalists mess it up and try to distort it, in order to make it seem more like Classical Physics.

These people try to define Quantum Non-Locality in terms of Classical Physics and Einstein's Special Relativity.

Why do they do this? Why do the materialistic scientists give Quantum Non-Locality physical limitations?

It's because the very existence of Quantum Non-Locality or the Transdimensional Spirit Realm FALSIFIES Materialism, Naturalism, and Classical Physics; and, they can't have that, now can they?

Quantum Mechanics has to make sense to them from a physical perspective or a naturalistic perspective, so they force it to do so, and keep forcing it to do so until Quantum Mechanics becomes Classical Physics; and then, they call the theory completed and done.

The truth is always hidden in the fine-print and the sub-text when it comes to the Materialists and Naturalists who write this stuff.

The Wikipedia states:

According to the formalism of quantum theory, the effect of measurement happens instantly. It is not possible, however, to use this effect to transmit classical information at faster-than-light speeds.

https://en.wikipedia.org/wiki/Quantum_entanglement

http://mypsyche.us/wp-content/uploads/2017/11/Quantum_entanglement.pdf

In other words, in the Non-Local Realm or Quantum Non-Locality, every effect happens instantly. There are NO speed-limits and NO physical limits in the Non-Local Realm, Quantum Non-Locality, or the Spirit World. Yet, he's still trying to imply that quantum effects and quantum entanglement cannot propagate faster than the speed-of-light; but, that's not exactly what he says either. He's trying to straddle the fence and satisfy both camps – or both competing factions – yet also in this case, he is making sure that everything he writes is technically true.

He is right, throughout. It is impossible to transmit physical information or classical information at faster-than-light speeds, because physical matter and this physical realm are subject to physical limitations, including entropy, time, and the speed-of-light. Physical matter is subject to physical limitations.

In contrast, in the Non-Local Realm or the Spirit World or Quantum Non-Locality, every effect happens instantly, just as he says. There are NO speed limits and there is no aging nor entropy in the Non-Local Realm. That's what has been experienced and observed by Out-of-Body Travelers and Near-Death Experiencers.

Consequently, it is inaccurate, incomplete, and even deceptive to try to define Quantum Non-Locality in physical terms or in terms of Classical Physics as the Materialists and Naturalists always try to do. There are NO physical limitations and NO speed limits in the Quantum Realm, or Non-Local Realm, or the Spirit World. Everything in the Quantum Realm at the Quantum Level happens instantly, just as the Wikipedia says.

They are trying to split a very fine line here, when it comes to Quantum Non-Locality and Quantum Entanglement, because the Quantum is intricately entangled with the Physical. The Quantum is the foundation upon which the Physical is based and built. The Physical is the Quantum, except for the fact that God has introduced physical limitations into the Physical Realm in order to slow everything down and turn that Physical Realm into the Ultimate Consensus Reality.

These people have observed that the effects of Quantum Entanglement or Quantum Non-Locality are instantaneous over vast distances; but, they have also observed that it is impossible to transmit physical information or classical information faster than the speed of light. The scientists have detected both the instantaneous effects of the Quantum Realm as well as all the different physical limitations associated with this Physical Realm. In order to reconcile what they are observing in their experiments, the Materialists and Naturalists try to forcefully equate Quantum Mechanics with Classical Physics, and it can't be done; but, that doesn't stop them from trying, as all of their distorted definitions and explanations of quantum nonlocality and quantum mechanics attest.

The confusion comes along whenever they try to confound or equate the two – Quantum Mechanics and Classical Physics. They are two completely separate things – two completely different things – and should be treated as such. Quantum Mechanics is Spiritual Mechanics, or the way that spirit matter and psyche work – instantaneously and simultaneously at any speed one desires. In contrast, Classical Physics is a list of all the physical limitations that God has placed upon our physical universe, in order to slow things down and create the Ultimate Consensus Reality, a Physical Reality.

Remember, whenever you encounter Quantum Non-Locality, you should automatically remove ALL the physical limitations that the Materialists and Naturalists try to apply to it or force upon it. When you do, you will get closer to the truth of the situation.

In fact, if you want to find and know the truth, you start by eliminating everything that is false, such as Materialism and Naturalism and their derivatives. Remember, the very existence of Quantum Mechanics and the Quantum Non-Local Realm (or Spirit World) FALSIFIES Materialism and Naturalism; so of course, the Materialists and Naturalists are going to try to distort and change the situation if they can, by defining Quantum Non-Locality as something other than what it really is.

That's why I had to expand upon and improve the definition of Quantum Non-Locality, so that it ends up being better and more truthful than what is typically presented to the world.

Quantum Uncertainty

Allegedly, Quantum Indeterminacy and the Uncertainty Principle introduce Choice and Free Will back into the equation by including Psyche, or Non-Local Consciousness, or the Observer into the mix. While experimenting with the Quantum Uncertainty Principle, scientists observed that the elementary particles respond to the scientist's chosen observations and chosen actions. This observation eventually led to the discovery of the Quantum Zeno Effect, which will be mentioned later as another experimentally verified quantum phenomenon.

If we choose to look at an atom, it does one thing; but, if we choose to look someplace else or choose to focus our attention someplace else, then the atom responds differently. This observed reality is indirectly related to the Quantum Uncertainty Principle; and, clearly demonstrates that elementary particles are conscious, can read your mind, and can make choices.

The Materialists, Naturalists, Behaviorists, and Determinists have tried to define Psyche, Free Will, and Choice out of existence.

Notice all the different materialistic and naturalistic interpretations of Quantum Mechanics that have been developed – most of them are worthless because they don't match with the observational evidence, experimental evidence, and experiential eye-witness evidence. Some of these have NO observational evidence supporting them. If it has never been experienced nor observed by anyone, then it really does not exist.

https://en.wikipedia.org/wiki/Interpretations_of_quantum_mechanics

The Measurement Problem is intimately associated with the Uncertainty Principle as well as the different interpretations that are given to Quantum Mechanics.

> **The measurement problem in quantum mechanics is the problem of how (or whether) wave function collapse occurs. The inability to observe this process directly has given rise to different interpretations of quantum mechanics, and poses a key set of questions that each interpretation must answer.**

The Uncertainty Principle and Quantum Indeterminacy put choice or free will back into Science and Physics by falsifying materialistic determinism. Of course, the Materialists and Naturalists never present it this way, now do they? The Quantum Uncertainty Principle has been experimentally verified.

Of course, they are right in that the Uncertainty Principle all by itself does not demonstrate and prove self-determination and free will. However, all the different science experiments that were run trying to verify and prove the existence of the Uncertainty Principle ended up showing us that each atom, electron, photon, and elementary particle has a Psyche or Non-Local Consciousness of its own; and, these experiments ended up demonstrating to us that each elementary particle can read your mind and KNOWS whether you are looking at it or not, and then that particle CHOOSES to respond accordingly.

The Heisenberg Uncertainty Principle has been experimentally verified in many different science experiments.

The famous Double-Slit Experiments are an offshoot of their attempt to understand the Uncertainty Principle and the other effects of Quantum Mechanics. I couldn't determine which came first. The Uncertainty Principle seemed to inspire a bunch of the different Double-Slit Experiments where they tried to measure everything; but, the Double-Slit Experiments were also used to establish the veracity of Heisenberg's Uncertainty Principle and Wave-Particle Duality. My research indicated that both the Double-Slit Experiments and Heisenberg's Uncertainty Principle came into being at the same time in 1927.

The Double-Slit Experiments do an excellent job demonstrating how Quantum Objects (particles of matter) choose to act, while they are observed and while they are unobserved. The Uncertainty Principle has been verified experimentally many times; and, both the Double-Slit Experiments and the Uncertainty Experiments have demonstrated and proven the existence of Particle-Wave Duality.

The Materialists and Naturalists try to explain this stuff out of existence. A materialistic and naturalistic explanation or interpretation ruins all of these experiments and makes them nonsensical; but, I have observed that they all make perfect sense if interpreted from a Non-Local Perspective or from a Spiritual Perspective.

Furthermore, versions of the experiment that include detectors at the slits find that each detected photon passes through one slit (as would a classical particle), and not through both slits (as would a wave). However, such experiments demonstrate that particles do not form the interference pattern if one detects which slit they pass through. These results demonstrate the principle of wave–particle duality.

https://en.wikipedia.org/wiki/Double-slit_experiment

http://mypsyche.us/wp-content/uploads/2017/11/Double-slit_experiment.pdf

Wave Particle Duality

https://en.wikipedia.org/wiki/Wave%E2%80%93particle_duality

http://mypsyche.us/wp-content/uploads/2017/11/Wave–particle_duality.pdf

The following article discusses the Double-Slit Experiments.

One of the greatest puzzles of the double-slit experiment – and quantum physics in general – is why electrons seem to act differently when being observed. While electrons traveling through a barrier with two slits create interference patterns when unobserved, these interference patterns disappear when scientists detect which slit each electron travels through.

As one of the most famous experiments in quantum physics, the double-slit experiment demonstrates how the quantum world is very different from the classical world.

When macroscale objects are shot at a barrier with two slits, the objects travel straight through the slits and leave two straight lines on the wall behind the barrier [they follow the laws of Classical Physics]. But when electrons are used instead of macroscale objects, they do not leave two straight lines on the wall but an interference pattern of many lines [they follow the laws of Quantum Mechanics]. Because the interference pattern remains even when the electrons are shot one at a time, the experiment seems to suggest that each electron somehow travels through both slits at the same time and interferes with itself, like a wave instead of a particle.

The second unusual part of the double-slit experiment is that the electrons stop creating an interference pattern when scientists set up a detector near one of the slits to determine which slit(s) an electron is passing through. Under these circumstances, the electrons simply create two straight lines, the same as classical particles.

Throughout the years, scientists have demonstrated different versions of the two-slit experiment.

Read more at: https://phys.org/news/2011-01-which-way-detector-mystery-double-slit.html#jCp

The atoms, photons, and electrons really are sentient and aware of their environment, and they really do read your mind and KNOW whether you are looking at them or not, and then choose to respond accordingly.

A quantum object or electron while traveling as a wave is functioning telepathically, sensing its environment as it goes. Telepathy is like radar or a radio wave but at a much smaller scale. In contrast, a macroscale object such as a bullet functions according to Physical Laws and simply does what the Physical Laws tell it to do.

Many scientists use the Uncertainty Principle and Quantum Mechanics to establish the existence of Free Will and Choice.

"Why Quantum Physics Ends the Free Will Debate"

https://www.youtube.com/watch?v=IFLR5vNKiSw

But then, the materialistic and naturalistic philosophers successfully punch holes into their arguments and using sophistry "prove" that free will does not exist. Round and around they go. Who is right? Nobody knows!

One thing I am sure of is that Philosophy is worthless if it is used to prove a lie to be true, which is very easy to do using Sophistry or Philosophy.

In order to resolve this whole issue, we have to observe that Out-of-Body Travelers and Near-Death Experiencers continue to make decisions and choices even while their physical brain is clinically dead and offline. Choice or Free Will is a function of Psyche; and, Psyche or Non-Local Consciousness is the fundamental unit of reality and the creator of reality and existence.

The whole of Quantum Mechanics when taken together as a united front does indeed falsify Determinism and does indeed seem to establish the existence of Psyche and Free Will instead.

Remember: The Physical Realm and physical brain is determined by Psyche or Non-Local Consciousness, which means that the Physical Realm is controlled by and governed by Psyche. Psyche is a Causal Agent. Psyche, by definition in principle, makes decisions and choices. In contrast, physical matter or a macroscale object simply reacts to Psyche's commands – it's determined to do so.

Each elementary particle has some kind of Psyche or Non-Local Consciousness; and, Nature's Psyche knows what God's Laws are or what the Physical Laws are, can read your mind and KNOWS whether you are looking at it or not, and can CHOOSE to respond accordingly.

Remember, Nature's Psyche within each elementary particle is determined to obey God's Laws or the Physical Laws and Quantum Laws first; and then, Nature's Psyche is determined to obey the Human Psyche's commands so long as they don't violate God's Laws.

While traveling spiritually as a wave, a Quantum Object or Particle of Matter is traveling telepathically sensing its environment as it goes. It's functioning according to the rules of Quantum Mechanics. It can read your mind and KNOWS whether you are looking at it, or not.

While traveling as a macroscale object, a Quantum Object or Particle of Matter is traveling like a bullet and is simply obeying the Laws of Classical Mechanics. It KNOWS its size, and it knows whether it is traveling as a group or not, and it responds accordingly.

These experiments demonstrate that each Quantum Object has a bit of Nature's Psyche within it, which is conscious, sentient, telepathic, and aware of its environment; and, the Psyche driving these Particles of Matter is determined to obey God's Laws first, and then our Human Psyche's requests next, followed by the sentient animals of course. The elementary particles choose to respond according to what they KNOW to be true. They have covenanted with God to do so; in other words, they obey God's Laws, or the laws associated with Quantum Mechanics. Quantum Mechanics is the Priesthood of God.

Quantum Law of Complementarity

Different types of Quantum Complementarity have also been verified experimentally.

In physics, complementarity is both a theoretical and an experimental result of quantum mechanics, also referred to as principle of complementarity, closely associated with the Copenhagen interpretation. It holds that objects have complementary properties which cannot all be observed or measured simultaneously.

The complementarity principle was formulated by Niels Bohr, a leading founder of quantum mechanics. Examples of complementary properties that Bohr considered:

Position and momentum

Energy and duration

Spin on different axes

Wave and particle

Value of a field and its change (at a certain position)

Entanglement and coherence

https://en.wikipedia.org/wiki/Complementarity_(physics)

http://mypsyche.us/wp-content/uploads/2017/11/Complementarity_physics.pdf

There are many different types of complementarity that have been observed and experienced in Quantum Mechanics. Complementarity has been verified experimentally. Most everything in Quantum Mechanics has been verified experimentally by now. Quantum Mechanics is REAL SCIENCE.

The only thing in Quantum Mechanics that has been repeatedly FALSIFIED is the materialistic and naturalistic interpretations of Quantum Mechanics – a classical physics interpretation or naturalistic interpretation of Quantum Mechanics never works, because Quantum Mechanics FALSIFIES Materialism, Naturalism, and their Derivatives such as Darwinism, Nihilism, and even Atheism.

As far as I know, the following has not been experimentally verified; and, I don't think it can be. It has been experientially verified, though, through Out-of-Body Experiences and Near-Death Experiences, but not with physical experiments.

After reading Pim van Lommel's book, *Consciousness Beyond Life*, I enhanced the Quantum Law of Complementarity to include Non-Local Consciousness and Quantum Non-Locality into the mix. I wanted a comprehensive theory that covered it all. I needed something that made sense to me, as a whole.

I started visualizing the wave aspect of a Quantum Object as the non-local or spiritual part of a Quantum Object. Whereas, I considered the particle aspect of a Quantum Object to be the localized or physical part of a Quantum Object.

According to this Quantum Law of Complementarity, a Quantum Object or Particle of Matter can manifest as either a wave or as a particle but not as both simultaneously. This means that a Quantum Object can either be spiritual (non-local) or physical (local), but not both simultaneously. In other words, some kind of phase shift or dimensional change takes place whenever a particle of spirit matter is converted into a particle of physical matter by God's Laws or God's Psyche, which is the only Psyche we know of that has the ability to convert spirit matter into physical matter at will.

I took my inspiration for all of this from:

Bishop, B. G. (1998). *The LDS Gospel of Light*. USA: Ponce de Leon.

Van Lommel, P. (2010). *Consciousness Beyond Life: The Science of the Near-Death Experience*. New York: HarperCollins.

Taking all of this into consideration, I eventually developed what I call the Quantum Law of Complementarity, which states that a Quantum Object (elementary particle of matter) can exist in either a Non-Local Wave-Like Spiritual State or a Local Particlized Physical State; but, a Quantum Object or Particle of Matter cannot exist in both states simultaneously. Furthermore, it takes Psyche, or Non-Local Consciousness, or God's Word of Command to convert a particle of spirit matter into a particle of physical matter. I also concluded that Dark Matter and Tachyons are in fact Spirit Matter.

I developed this theory in another book. I now quote from that book:

The Quantum Law of Complementarity

When it comes to "wave-particle duality", the wave is the Spiritual Non-Local Non-Physical Trans-Dimensional Universal aspect of a Quantum Object, and the particle is the physical or solid or material or located aspect of the Spiritual Wave, when OBSERVATION causes the Spiritual wave function to collapse and become local or Physical in nature.

Nevertheless, Particle-Wave Duality is a misnomer that we scientists inherited from the Materialists, because there is NO Duality involved when it comes to Quantum Objects.

According to the Quantum Law of Complementarity, Quantum Objects are either a wave or a particle, but they can't be both at the same time. While a Quantum Object is a wave, it is omnipresent simultaneously and instantaneously everywhere. When the Probability-Wave collapses, the Quantum Object becomes physical in nature and becomes fixed or localized in our Physical Space-Time.

After the wave function collapses, the Quantum Object is no longer Spiritual and no longer an omnipresent Probability-Wave; but, it is instead physical in nature and has been placed or localized in Space-Time.

Heisenberg's uncertainty principle indicates that a Quantum Object is fundamentally changed from a Spiritual state of existence to a Physical state of existence when that Quantum Object is consciously observed. That Quantum Object ceases to be a Spiritual Omnipresent Probability-Wave and becomes a Physical Particle localized in Space-Time instead, when it is consciously observed.

Conscious observation literally creates physical reality.

It is the physical that is created and NOT the Spiritual.

Consciousness and the Spiritual have always existed and cannot be created. Consciousness and the Spiritual can ONLY be organized and given order and enlivened; these things cannot be designed and created and manufactured.

The physical NEEDS an Intelligent Consciousness to organize it, "create it", and give it order and life. The physical cannot design and create and self-assemble and enliven itself. The physical was designed to be acted upon and to react, NOT to ACT. Only an Intelligent Living Consciousness has the innate ability to ACT, design, create, manufacture, do science, and collapse Quantum Wave functions into physicality.

The Physical derives from the Spiritual; and, the Spiritual has always existed and will always exist. That's what we learn from "Particle-Wave Duality" and the Heisenberg uncertainty principle, when our minds are not being darkened and blinded by Materialism. We learn that "Particle-Wave Duality" is a materialistic misnomer and that according to the Quantum Law of Complementarity, a Quantum Object is either in a wave-state of existence or in a particle-state of existence; but, it cannot exist in BOTH states at the same time.

In fact, what we call "the speed of light" seems to be the demarcation line between the two. If a Quantum Object is traveling or existing slower than the speed of light, then it is a particle and is Physical in nature. When a Quantum Object is traveling and existing faster than the speed of light, then it is a Probability-Wave and Non-Physical and Non-Local in nature with the innate ability for infinite velocity or instantaneous velocity.

A Quantum Object has two states of being, as a Non-Local Probability-Wave or what we sometimes call a Spirit, and as a localized Physical Particle in our physical 3D Space-Time; but, the Quantum Law of Complementarity states that a Quantum Object cannot be in both states of existence at the exact same time.

When the Quantum Wave function collapses, the Physical Particle materializes into our 3D Physical Space-Time. Should Physical Matter be dematerialized by Conscious will, then that Physical Matter will cease to exist here in our 3D Space-Time and will instead convert back into Probability-Waves in Non-Local Space or what we often call the Spirit Realm. The Physical is limited or localized to our 3D Space-Time. The Spiritual or the Quantum or the Probability-Wave is unlimited and simultaneously present everywhere in Non-Local Space or the Spirit Realm. There is NO Physical Matter and NO distance and NO space in the Non-Local Spirit Realm. In the Non-Local Quantum Realm, everything is instantaneously and simultaneously present in a medium which takes up NO Physical Space whatsoever.

I have observed that Quantum Mechanics makes perfect sense, once a person abandons Materialism and realizes that the local-part of a Quantum Object or particle is Physical in nature and that the Non-Local or "wave" part of a Quantum Object and Quantum Mechanics is Spiritual in nature.

Understanding Spirit or Nonlocality brings Quantum Mechanics into the realm of logic and common sense. The Quantum Law of Complementarity explains it all. A Quantum Object can either be in a Spiritual State of existence as a Probability-Wave, or it can be in a Physical State of existence as a Physical Particle; but, it can be in both states of existence at the same time. This realization covers everything.

According to Quantum Mechanics, a quantum object has two different complementary states of existence, a spirit-like wave-like non-local infinite-velocity simultaneously-everywhere quantum wave state and a localized sub-light-speed finite physical particle state, after the quantum object has been observed by a living consciousness. Quantum objects can be either in spirit matter format or in physical matter format, but NOT in both formats at the same time according to the Quantum Law of Complementarity. A quantum object manifests as either a spiritual non-local wave or as localized space-time particle, but not as both at the same time. Most scientists get this one wrong thinking that a quantum object is simultaneously wave-like and particle-like, which is not the case. A quantum object is in either one state or the other at any given point in time. (Van Lommel, 2010; & Bishop, 1998).

Mark My Words. (2016). *Quantum Mechanics: From a Non-Physical Spiritual Perspective*. https://www.amazon.com/dp/B01J023TGU

As stated, I took my inspiration for all of this from:

Bishop, B. G. (1998). *The LDS Gospel of Light*. USA: Ponce de Leon.

Van Lommel, P. (2010). *Consciousness Beyond Life: The Science of the Near-Death Experience*. New York: HarperCollins.

As you can tell, though, if you were to make a comparison, I adjusted and enhanced things quite a bit. I'm a theoretician, so it's what I do; but, you are going to have to decide for yourself if any of this is of value to you.

I needed it to make sense to me; so, I kept thinking about it and studying it and adjusting it until I finally had something that made sense to me and matched with the Observational Evidence, Experimental Evidence, and Experiential Evidence. The Quantum Law of Complementarity was the result – a Quantum Object or Particle of Matter can be in either a physical state or a non-local spiritual state, but not in both states simultaneously.

In other words, whatever we encounter in the Non-Local Realm or Spirit World is in a different phase or dimension than what we encounter here in the physical world. Consequently, a Quantum Object that is in a spiritual dimension can occupy the same exact space as a Quantum Object that is in a physical dimension, because they are out of phase with each other. This concept is called Phase-Shifting. If you were to phase-shift your physical body, you would then be able to walk through the walls; and, there would be two Quantum Objects occupying the same space simultaneously.

The Spirit World is right here where you are now occupying the same space, but out of phase or in a different dimension. When God converts a particle of spirit matter into a particle of physical matter, one of the things He does is to phase-shift it or move it into our physical dimension.

The concept of Phase-Shifting is based in large part on Quantum Non-Locality and the Quantum Law of Complementarity.

My best friend and his truck phase-shifted, and an elk passed right through his truck; so, I KNOW the phenomenon is real and can truly happen if God chooses to intervene and make it so. Of course, it was nothing that my best friend did. It was something that God did for him in order to save him.

Just because I have never experienced phase-shifting doesn't mean that nobody else has. I don't have to experience Phase-Shifting in order to know that the phenomenon is real and truly happens, from time to time.

Remember, the BEST way to find and know the truth is to live it and experience it for yourself, or to choose to trust someone who has.

Out-of-Body Travelers like William Buhlman have observed and testified that there are many different phases, dimensions, or levels in the Astral Plane or the Non-Local Transdimensional Realm.

I have experienced Quantum Tunneling or Teleportation, though. God stopped the passage of time and teleported me and my car to safety. I KNOW for a fact that God can teleport physical matter instantaneously anywhere He wants it to go, because I have lived it and experienced it.

God holds ALL the KEYS for the teleportation and phase-shifting of physical matter, so we will NEVER be able to verify and replicate these Quantum Phenomena at will. The fact that they have been lived and experienced by real human beings will have to suffice for now. Science 2.0 gives precedence and preferential value to Lived Experiences over scientific inferences and philosophical wishful thinking; and, Lived Experience is the only way we can effectively discuss and deal with these types of Non-Local Quantum Phenomena.

I simply wanted to understand the Science behind it all, and now I do. It's based upon the Quantum Law of Complementarity, the Quantum Law of Non-Locality, Quantum Tunneling, and Phase-Shifting through different dimensions which occupy the same space. Good enough!

As usual, I will close this section by mentioning my observation that you will NEVER get anything like this from a Materialist, Naturalist, Nihilist, or Atheist. Instead, these people will simply tell you that these things do not exist and that they are hallucinations and figments of the imagination. Ironically, Materialism and Naturalism end up being the delusion and the figment of their imagination, once you understand Quantum Mechanics and how things really work. Quantum Mechanics FALSIFIES Materialism, Naturalism, and ALL their derivatives such as Darwinism.

These Quantum Phenomena are REAL, and they have been verified experimentally and observationally; whereas, Materialism and Naturalism are the delusions and wishful thinking, being nothing more than philosophical concepts that have NO observational evidence whatsoever supporting them. Materialism, Naturalism, Darwinism, Nihilism, Behaviorism, Scientism, and Atheism are based exclusively on blind faith, self-deception, delusions, a refusal to look at evidence, wishful thinking, scientific inference, and philosophical speculation. They are philosophical concepts and nothing more.

Remember, philosophy is absolutely worthless if it is used to prove a lie to be true, as the philosophers and sophists and scientific naturalists so often try to do.

The Orthodox Interpretation of Quantum Mechanics

There's NO physical explanation for Action at a Distance (telekinesis), Psyche (non-local consciousness), Quantum Tunneling (teleportation), the Quantum Zeno Effect (telepathy), and Syntropy (the opposite of entropy).

Henry P. Stapp has spent his career developing the Orthodox Interpretation of Quantum Mechanics. It's a model that explains how the Human Psyche or Non-Local

Consciousness interacts with and controls its physical brain, through the Quantum Zeno Effect or Process 1.

http://mypsyche.us/wp-content/uploads/2017/10/Henry-P-Stapp-Articles.pdf

http://mypsyche.us/wp-content/uploads/2017/10/PTRS.pdf

http://mypsyche.us/wp-content/uploads/2017/10/Orthodox-Interpretation.pdf

http://mypsyche.us/wp-content/uploads/2017/10/A-Quantum-Mechanical-Theory-of-the-Mind-Brain-Connection.pdf

Process 1 – the Human Psyche – the Actor, Chooser, and Agent. The Human Psyche is the Ultimate Causal Agent.

Process 2 – the physical object – the collapse of the wave function into a single unique reality from among the infinite number that are possible.

Process 3 – Nature's Psyche – the Reactor or Responder. The mediator of the physical laws or God's Laws. Nature's Psyche decides what is physically possible and/or physically lawful, and then chooses to respond either Yes or No to the Human Psyche's commands. ONLY Nature's Psyche can collapse the wave function, thereby making the Human Psyche's requests REAL and ACTUAL. There is a part of Nature's Psyche within each physical atom.

The Psyches within rocks and glass are passive receivers and recording devices (Process 3), not actors and agents (Process 1).

Quantum Zeno Effect – the telepathic connection between the Human Psyche and Nature's Psyche. Telepathic signals are like radio signals, but on a much smaller scale and at a much higher frequency. The Human Psyche is the transmitter, and Nature's Psyche is the receiver. The Quantum Zeno Effect is the telepathic signal or "radio wave" between the two.

In a very real sense, Psyche is the Quantum Zeno Effect, or Action at a Distance, or Telepathy. Telepathy is radio waves at the quantum level. Telepathy is WiFi at the quantum level. Each of these things is KNOWN to exist by the effect that they have on individual atoms. The Quantum Zeno Effect has been verified experimentally many different ways.

Quantum Neuroscience starts by identifying the physically impossible, and then it continues by trying to find a scientific explanation for the physically impossible.

You need an Interpretation of Quantum Mechanics that explains what the Human Psyche and Nature's Psyche are doing at the quantum level. If your Interpretation of Quantum Mechanics doesn't do that, then it's worthless. You also need an Interpretation of Quantum Mechanics that explains how God's Psyche does His science and His work through quantum mechanisms. If your Interpretation of Quantum Mechanics can't do that, then it's worthless.

The Quantum Zeno Effect Is the KEY

The Quantum Zeno Effect has been verified experimentally many different ways.

The Materialists and Naturalists who write of the "Official Interpretations" of Quantum Mechanics never seem to take the Quantum Zeno Effect far enough for it to be of any use to us; but, it's there to be found none-the-less, even though they try to water it down and turn it into classical physics as much as possible.

https://en.wikipedia.org/wiki/Quantum_Zeno_effect

http://mypsyche.us/wp-content/uploads/2017/11/Quantum_Zeno_effect.pdf

The true power of the Quantum Zeno Effect comes into play only after one reads and understands some of the following books and articles:

Penfield, W. (1975). *The Mystery of the Mind*. Princeton, New Jersey: Princeton University Press.

Beauregard, M. & O'Leary, D. (2007). *The Spiritual Brain: A Neuroscientist's Case for the Existence of the Soul*. New York: HarperCollins.

Beauregard, M. (2012). *Brain Wars: The Scientific Battle over the Existence of the Mind and the Proof That Will Change the Way We Live Our Lives*. New York: HarperOne.

Stapp, H. P. (2017). *Quantum Theory and Free Will: How Mental Intentions Translate into Bodily Actions*. Cham, Switzerland: Springer International Publishing.

J. M. Schwartz, H. Stapp, and M. Beauregard, "Quantum Theory in Neuroscience and Psychology: A Neurophysical Model of Mind/Brain Interaction," *Philosophical Transactions of the Royal Society B: Biological Sciences 360* (2005): 1309–27.

http://www-physics.lbl.gov/~stapp/PTRS.pdf

http://mypsyche.us/wp-content/uploads/2017/10/PTRS.pdf

https://www.researchgate.net/publication/7613549_Quantum_physics_in_neuroscience_and_psychology_A_neurophysical_model_of_mind-brain_interaction

http://escholarship.org/uc/item/4w8665vk

Schwartz, Jeffrey M. *The Mind and the Brain: Neuroplasticity and the Power of Mental Force*. New York: HarperCollins.

I chose this book, *The Mind and the Brain*, to be the foundational textbook for Quantum Neuroscience. I have been looking for this book all of my life, and I finally found it in October 2017. This book represents real, efficacious Neuroscience; and, it takes everything into account including Quantum Mechanics and Psyche or Mind.

Instead of using Neuroscience, Biopsychology, and Brain Scans to preach and promote Darwinism, Determinism, Materialistic Reductionism, and the Theory of Evolution as most of the Neuroscientists and Biopsychologists do, Jeffrey Schwartz actually used Neuroscience and Brain Scans to develop an effective treatment and potential cure for Obsessive-Compulsive Disorder (OCD). Jeffrey Schwartz does REAL Psyche-Therapy, by using Psyche or the Mind to desensitize the OCD Circuit and rewire the human brain, through a process known as Cognitive-Behavior Therapy. It works. The Brain Scans show a change and a normalization of brain functioning, after the Psyche-Therapy or Cognitive Therapy has taken hold. Jeffrey Schwartz and team really do use Psyche or Mind to rewire and normalize the human brain. Since the 1980's when Cognitive Therapy saved my life, I have had a great deal of admiration for Cognitive Therapy and what it can do. I always had a feeling that Cognitive Therapy, Friendship Therapy, or Psyche-Therapy could rewire and normalize the human brain, but now I KNOW. Science is KNOWLEDGE – or at least it should be.

Jeffrey Schwartz mentions and uses the Quantum Zeno Effect a lot in his papers and books.

The Quantum Zeno Effect

There is a mechanism in place, whereby the Human Psyche can and does control its physical brain at the quantum level. This mechanism is called the "Quantum Zeno Effect" – or "Process 1" in the Orthodox Interpretation of Quantum Mechanics.

The Quantum Zeno Effect is telepathy and telekinesis. The Quantum Zeno Effect is Mind-Over-Matter or Psyche-Over-Matter. The verified and proven Placebo Effect is also a part of the Quantum Zeno Effect, as is Action at a Distance or Telepathy. ALL of these things have been verified and replicated quantitatively through scientific experiments and scientific observations.

The Quantum Zeno Effect Has Been Verified – atoms won't move while you watch them.

https://phys.org/news/2015-10-zeno-effect-verifiedatoms-wont.html

http://mypsyche.us/wp-content/uploads/2017/10/Zeno-effect-verified—atoms-wont-move-while-you-watch.pdf

Those are the atoms out there. They freeze whenever you look at them, because you are telepathically connected with them; but, what happens to the atoms inside your brain whenever your Psyche needs them or commands them to do something?

Answer: A Process 3 response takes place, and those atoms and neurotransmitters en masse within your physical brain respond to your Psyche's request with a Yes or a No, depending upon what's physically and realistically possible. Your Psyche can actually make those atoms and neurotransmitters within your physical brain move and do things – telekinetically. Your Psyche can trigger an action potential anywhere within your physical brain, at the quantum level, through the Quantum Zeno Effect or Telepathy.

It's called an action potential in part because it results in actions. Someone has to make a decision or a choice in order to fire the first action potential in a chain of action potentials; or, nothing would ever happen. The very first action potential in a chain of action potentials needs a reason fire, or it never will. Raw physical matter has never been caught in the act of driving itself. Someone has to drive the bus, or the bus is going to crash and go nowhere in the end.

The Psyche's thoughts and feelings and choices have to transfer into one's physical brain somewhere along the way; and, it's through the neurotransmitters and ions and action potentials that the Human Psyche does so at the quantum level. In other words, someone or something has to trigger the first action potential in a chain of action potentials, or no thoughts and no ideas and no actions would ever take place within our physical brain. The Human Psyche has to get the process going, or nothing would ever happen within the human brain. Imagine a physical brain where no action potentials are taking place and no neurotransmitters are being released. The thing would be dead, would it not?

The Quantum Zeno Effect – Telepathy and Telekinesis – is proven and observed science. The Quantum Zeno Effect is how the Human Psyche commands and controls its physical brain. Telepathy or the Quantum Zeno Effect is Action at a Distance. Action at a Distance has also been verified experimentally dozens of different times by different scientists. Action at a Distance is usually called Entanglement, and it becomes scientific proof of Nonlocality. Action at a Distance is also scientific proof of Telepathy.

Telepathy is like radio waves, but on a much smaller scale. You and all those atoms are telepathically connected. Their psyche knows when your psyche is looking at them or needing them. The Human Psyche acts (Process 1), and Nature's Psyche reacts (Process 3).

The atoms out there in the physical world react while you are watching them – they freeze into place, just like the Quantum Zeno Effect predicts. This reality has been observed and experienced and proven to be true. Telepathy is real. The freezing of those atoms is a telekinetic effect. It truly happens, when the atoms are observed.

Likewise, all of those neurotransmitters and ions within our physical brain have a Process 3 or a Psyche – Nature's Psyche. The neurotransmitters and ions were designed to react to the Human Psyche's commands through Process 1 or the Quantum Zeno Effect. Each neurotransmitter and ion is a physical object (Process 2); but, each ion and neurotransmitter and atom in the physical brain has a Process 3 Psyche which is obligated by physical law, or God's laws, or God's commands to obey its Human Psyche's commands whenever physically possible or realistically possible. It's the influx of ions that triggers the action potential. Someone or something has to get the first action potential going somewhere within the physical brain, or no thoughts and computations would ever take place within the physical brain.

The Quantum Zeno Effect is the remote control or radio control. The Human Psyche is the transmitter. Nature's Psyche is the receiver of the radio signal or the telepathic signal. And, the various different ions, neuropeptides, molecules, atoms, and neurotransmitters are the remote-control cars or the remote-control vehicles which are being controlled by the Human Psyche through remote control or telepathy or "radio waves", which the scientists have called the Quantum Zeno Effect or Process 1.

I promise you, if you are a Materialist, Naturalist, Nihilist, or Atheist, the Quantum Zeno Effect will seem like nothing but science fiction to you. I KNOW, because I used to be a Materialist, Nihilist, and Atheist; and, I still have some of that programmed within me from my fifty years of Scientism. Scientism is the philosophical belief that only the scientific method and the physical sciences can be used to find and know the truth. I have fifty years of conditioning and brainwashing into Materialism and Scientism that I have to deal with. There's still a part of me that rebels against everything that Quantum Mechanics is trying to tell us and teach us. It seems foreign and unreal to me. The Quantum Zeno Effect seems fictional or fake to me. Some part of me resists it. I assume it's all that conditioning and programming into Scientism and Materialism, which has taken place throughout my life, that's the source of my resistance to the Quantum Zeno Effect and Telepathy.

What is truly ironic, though, is that it's Materialism, Naturalism, Darwinism, Behaviorism, Determinism, Physicalism, Nihilism, Scientism, and Atheism that are in fact the fiction and the ruse. There's no observational evidence supporting the major premises or primary assumptions any of these falsified philosophies, and there never will be. They all have to be taken on blind faith as being true.

Instead, it's the Quantum Zeno Effect, or Mind-over-Matter, or Psyche-over-Matter, or the Placebo Effect which has been observed, experienced, verified, and proven to be true trillions of times in dozens of different ways. The Quantum Zeno Effect, Action at a Distance, Telepathy, and the Placebo Effect are the REAL SCIENCE; whereas, it's the

Materialism and Naturalism and their derivatives that are in fact the pseudo-science and the fakes.

How ironic is that?

The fiction is found in the Materialism and the Naturalism, upon which Darwinism, Scientism, Nihilism, and Atheism are based. Whereas, ALL of the observational evidence, experimental evidence, eye-witness evidence, experiential evidence, and empirical evidence stand firmly behind the Quantum Zeno Effect or Psyche-over-Matter as being real and true.

Spooky, is it not?

The Quantum Zeno Effect is the KEY to understanding how the Human Psyche controls its physical brain. The Human Psyche is the remote-control device; whereas, the physical brain is the remotely controlled "car" that is being driven and controlled and steered by its assigned psyche. It has been observed that when the physical brain dies or goes offline, the Human Psyche goes on.

Remember, Quantum Mechanics and the Quantum Zeno Effect are the verified and proven Science; whereas, Materialism and Naturalism and their derivatives are the falsified "science". ALL of the evidence stands firmly behind Quantum Mechanics; and, verification of Quantum Mechanics and the Quantum Zeno Effect falsifies the claims of Materialism and Naturalism. When Materialism and Naturalism go down, Darwinism, Nihilism, Atheism, Behaviorism, Determinism, Physical Reductionism, and Scientism go down with them as fruit from the poisoned tree.

If we successfully eliminate everything that is false and everything that has been falsified, then ONLY the truth will remain. We start this process by eliminating Materialism and Naturalism and Darwinism, which is why I spend so much of my time doing so, whenever I start doing Science. The falsehoods have to be eliminated, or we will never find the truth. There is no other way.

Figure 8: Quantum Effects of Attention

The rules of quantum mechanics allow attention to influence brain function. The release of neurotransmitters requires calcium ions to pass through ion channels in a neuron. Because these channels are extremely narrow, quantum rules and the Uncertainty Principle apply. Since calcium ions trigger vesicles to release neurotransmitters, the release of neurotransmitter is only probabilistic, not certain. In quantum language, the wave function that represents "release neurotransmitter" exists in a superposition with the wave function that represents "don't release neurotransmitter"; each has a probability between 0% and 100% of becoming real. Neurotransmitter release is required to keep a thought going; as a result, whether the "wash hands" or "garden" thought prevails is also a matter of probability. Attention can change the odds on which wave function, and hence which thought, wins.

Today, as we derive a scientific worldview from quantum mechanics, we cannot be sure that this theory, too, will not be superseded. For now, however, we are left with the fact that the laws of nature, as Wigner elegantly stated in the epigraph at the beginning of this book, cannot be written without appeal to consciousness. The human mind is at work in the physical universe. The effect of attention on the brain offers a rational, coherent, and intuitively satisfying explanation for the interaction between mind and brain, and for the causal efficacy of mental force. It describes the

action of the mind as we actually experience it. Consciousness acts on, and acts to create out of an endless universe of predetermined possibilities, the material world — including the brain. Mental effort can speed up the rate at which attention is focused and questions are posed. This speeding up, through the Quantum Zeno Effect, tends to sustain a unified focus on one aspect of reality — which prevents the selected stream of consciousness from losing focus and diffusing. Quantum theory, with the Quantum Zeno Effect, seems to explain how human volition acts in our lives.

For scientifically minded people seeking a rational basis for the belief that truly ethical action is possible, James's epigram — "Volitional effort is effort of attention" — must replace *Cogito ergo sum* as the essential description of the way we experience ourselves and our inner lives. The mind creates the brain. We have the ability to bring will and thus attention to bear on a single nascent possibility struggling to be born in the brain, and thus to turn that possibility into actuality and action. The causal efficacy of attention and will offers the hope of healing the rift . . . between science and moral philosophy. It is time to explore the closing of this divide, and the meaning that has for the way we live our lives.

Schwartz, Jeffrey M. *The Mind and the Brain: Neuroplasticity and the Power of Mental Force* (pp. 361-364). New York: HarperCollins.

In summary, the Quantum Zeno Effect or Telepathy is radio waves at the quantum level, with one very important distinction. Everything at the quantum level, or the non-local level, or the spiritual level has NO physical limitations. This means that spirit matter, quantum telepathy, quantum tunneling, and psyche have no inherent speed limits and are therefore capable of traveling or moving at an infinite velocity, which means that they can teleport from one location to another instantaneously, while in the Non-Local Realm and separated from one's physical body.

Out-of-Body Travelers and Near-Death Experiencers have also at times reported being able to hear everyone's thoughts and prayers simultaneously (multitasking and telepathy) and being in two places at once on opposite sides of our earth simultaneously (multitasking and teleportation and time-sharing). Quantum Objects or Particles of Spirit Matter in their spiritual wave-like state can exist in multiple different locations simultaneously – this is known as Quantum Superposition. Of course, the Materialists and Naturalists and Atheists never explain nor define Quantum Superposition in this manner, because they don't believe in these kinds of things. So remember, Quantum Superposition is being in two places at the same time.

Quantum Superposition is the "physical" or spirit matter counterpart of the telepathic Quantum Zeno Effect, which is a function of Psyche or Non-Local Consciousness.

The BEST explanation for Quantum Superposition that I have found so far is in *The Mind and the Brain: Neuroplasticity and the Power of Mental Force* (pp. 268-293).

Just as the Quantum Zeno Effect is telepathy, Quantum Tunneling is teleportation. All of these things are possible in the Non-Local Realm or Spirit World without speed limits, because there are NO physical limitations in the Transdimensional Realms. There is no aging and no entropy in the Non-Local Realms, either. Everything there is immortal.

I have a section entitled, "Quantum Zeno Effect", in my book "Science 2.0: I Upgraded My Science", which you might find interesting, if you want to know more about the Quantum Zeno Effect, and some of the books I mentioned above.

The Quantum Zeno Effect is the KEY to understanding how the Human Psyche controls its physical brain and starts action potentials from scratch, through the calcium ion channels in the synapses and the neurotransmitter system as a whole.

The Quantum Zeno Effect or the telepathic connection between the Human Psyche and Nature's Psyche has been repeatedly observed, experienced, and demonstrated experimentally; therefore, WE KNOW that it is real and truly exists.

Quantum Telepathy

Quantum Telepathy is synonymous with Quantum Waves. Telepathy is WiFi at the quantum level. Thoughts and memories are quantum waves or the vibrations within strings.

Fourier analysis is a mathematical procedure for breaking down a complex wave form (an EEG signal) into component waves of varying frequency. There are tons of waves being produced by Nature's Psyche and ions within our physical brain. Some of these waves can be detected by a physical instrument that we call an EEG. However, some of these waves cannot be detected with our physical instruments, because they are quantum waves, and they originate and propagate at the quantum level.

Neurons are switches – ON or OFF. The brain is comprised of switches, except they aren't wired together. The neurons are just sitting there, emitting chemicals.

If any of the neurons are "wired together" or connected with each other, it's being done wirelessly through WiFi or Quantum Telepathy. Telepathy is WiFi or radio waves at the quantum level. Quantum waves are thoughts; and, thoughts are produced by Psyche.

If the neurons are indeed segregated into separate bytes of data or programming, that functionality is being done wirelessly at the quantum level, because NONE of that is visible or detectable at the physical level.

At the physical level, the switches are either ON or OFF, and NONE of it means anything to us, except when turning on a specific switch in your brain makes your finger rise off the table.

Nature's Psyche KNOWS which switch will make your finger rise, but the Human Psyche isn't consciously aware of these things.

Brains: I don't know how they do what they do, but they do.

A neuron is a battery, not a CPU. There's no detectable CPU in a neuron. A neuron is either charging or discharging, just like a battery. That's it! A neuron has NO detectable RAM for storing memories. A neuron is a switch – ON or OFF. It's as if all of our long-term consolidated memories are being stored someplace else besides our brain. That's what the observational evidence and experimental evidence repeatedly suggests.

If I were to give you 100 billion switches but NO wires, how would you turn them into functional CPUs or functional RAM? You can't without wires! It's physically impossible. But, let's suppose that you have WiFi or Quantum Telepathy built into each one of those switches, then how would you proceed to turn those switches into functional CPUs or RAM?

Well, for every neuron in the brain, you would have to go through the brain and connect eight switches (neurons) together wirelessly to form a functional unit called a byte, assign some sort of arbitrary but standardized meaning to each of the 256 different types of bytes that are possible, set the eight switches in each of the bytes to the correct meaning that you want that byte to represent (and keep doing so for over eighty years of time), and then force all of these different switches and bytes to function together as a united whole – either as a CPU or as RAM.

If brains are made up of 100 billion stand-alone switches, with no wires or connections between them, then why do brains work? There's NO physical reason why they should work, so why do brains work? The only answer I can give is that Nature's Psyche is assigning functionality and purpose to each of the neurons, or structures, or CPUs in the brain; and, Nature's Psyche is connecting those stand-alone switches together wirelessly into a functional whole. The mechanism that Nature's Psyche uses to get the job done is Quantum Telepathy or WiFi at the quantum level.

If you hook up an EEG to your scalp, what's the only detectable thing coming from your brain as a whole? Waves! Waves are WiFi at the quantum level! If the neurons in your brain are communicating with each other and sending complex messages to each other, they are doing so wirelessly through different types of waves. They are using WiFi at the quantum level or Quantum Telepathy. They definitely are NOT using synapses to send messages to each other, because synapses are randomizers. A synaptic cleft will scramble any message that you try to send through it! It's physically impossible to send messages of any kind through 100 billion stand-alone switches that have NO wires connecting them. It can't be done, so that means that it isn't being done.

When it comes to the brain, God gave us 100 billion switches to look at. 100 billion switches with NO wires connecting them. God made it extremely clear to every Computer Scientist, Materialist, Naturalist, Darwinist, and Atheist that our brains shouldn't work, because it's physically impossible to send messages of any kind from one neuron to the next through a synaptic cleft. It can't be done. A synaptic cleft is a randomizer, scrambler, and jamming device. A neurotransmitter receptor is also a bit – OPEN or CLOSED. The receptor receives the ON signal letting it know that the message has started, and then it receives an OFF signal letting it know that the message has ended; but, there's NO message! You can't send a message through a single stand-alone bit or switch. It's physically impossible.

So, how is the magic being done? Well, the only thing we KNOW for sure is that it isn't being done through the synaptic clefts, because it's physically impossible to send messages through synaptic clefts!

The Materialists and Naturalists refuse to think about the physically impossible, because these types of things falsify Materialism and Naturalism; but, the physically impossible is precisely where science starts to get interesting.

We have observed that it's Nature's Psyche who reaches out and fires the appropriate neuron whenever the Human Psyche finds something interesting. According to the Orthodox Interpretation of Quantum Mechanics, from Henry P. Stapp, it is Nature's Psyche who collapses the wave function. The Human Psyche simply chooses what it wants its brain to pay attention to. Nature's Psyche accomplishes the request or the task by collapsing the appropriate wave functions, thereby bringing the Human Psyche's desires into physical reality. That's how the Human Psyche interacts with and controls its physical brain – through Nature's Psyche and Quantum Mechanics. This is Quantum Neuroscience. It's really simple to understand, once a person chooses to do so. It's a lot simpler than trying to find a physical explanation for everything. That's physically impossible!

Reality Check

Telepathy and Telekinesis: Ask yourself why God could prevent fallen mortal beings such as us from reaching out with our Human Psyche, seeing the individual atoms with our mind, and then rearranging those atoms with our mind at will. If you were a God, is that the kind of power and ability that you would want to give to a Militant Atheist, a Muslim terrorist, the Christian Inquisition, Hitler, James Bond, or the devil living next door to you?

Action at a Distance: Ask yourself why God would prevent fallen mortal beings like us from reaching out with our Human Psyche, steering buses into gas stations, triggering the detonation of nuclear missiles at will, choking people to death with our mind, stopping their hearts at will, calling up earthquakes, steering hurricanes, starting fires, and transforming viruses into lethal forms. If you were a God, is that the kind of power and ability that you would want to give to a Militant Atheist, a Muslim terrorist, the Christian Inquisition, Hitler, James Bond, or the devil living next door to you?

Quantum Tunneling or Teleportation: Ask yourself why God would prevent fallen mortal beings like us from reaching out with our Human Psyche, teleporting ourselves and a bomb into parliament, and then teleporting away to safety before the bomb goes off? If you were a God, is that the kind of power and ability that you would want to give to a Militant Atheist, a Muslim terrorist, the Christian Inquisition, Hitler, James Bond, or the devil living next door to you?

Action Potential

"The greatest potential for control tends to exist at the point where action takes place." — Louis A. Allen

Your physical brain is comprised of over 100 billion neurons. Our physical brain functions through action potentials. Our brain, our neurons, and our action potentials have been observed experimentally and are known to exist.

The action potential is the transmission of a nerve impulse and takes place when the neuron fires. An action potential is the change in electrical potential that occurs between the inside and outside of a nerve or muscle fiber when it is stimulated, serving to transmit nerve signals. The action potential allows nerve cells to transmit a signal over a distance. An action potential results in some kind of action, experience, calculation, or felt emotion.

All neurons are electrically excitable, maintaining voltage gradients across their membranes by means of metabolically driven ion pumps, which combine with ion channels embedded in the membrane to generate intracellular-versus-extracellular concentration differences of ions such as sodium, potassium, chloride, and calcium. Changes in the cross-membrane voltage can alter the function of voltage-dependent ion channels. If the voltage changes by a large enough amount, an all-or-none electrochemical pulse called an action potential is generated, which travels rapidly along the cell's axon, and activates synaptic connections with other cells when it arrives, through the release of neurotransmitters into the synaptic cleft. Source – Uncle Google and Wikipedia

The dendrites in our brain transmit post-synaptic potentials. The axons in our brain transmit action potentials. In this case, a "potential" is an electrical current comprised of ions, or electrically charged atoms. The Materialists and Naturalists call dendrites and axons "wires". They do so in order to give you the erroneous impression that your whole brain is wired together into a single functioning whole like a CPU or RAM, which technically isn't true at all. To counteract this deception, I will occasionally make the statement that there are NO wires in your brain. They, of course, will disagree with this statement and point you directly to dendrites and axons in order to counter that statement. However, thanks to the synaptic clefts and synapses that separate every neuron and astrocyte, each neuron is functionally and electrically separated from every other neuron and astrocyte by synapses, which effectively means that there are NO wires in your brain. In other words, your brain is NOT wired together into a single CPU or RAM like the Materialists and Naturalists claim that it is. Technically, your brain is NOT comprised of circuits like these people claim that it is. A wire is a single, usually cylindrical, flexible strand or rod of metal. Wires are used to bear mechanical loads or electricity and telecommunications signals. There's NOTHING like this within a physical brain. There are NO wires in your brain.

Getting an Action Started

So, what triggers the first action potential in a chain of action potentials? Who or what decides which of the 100 billion neurons in your physical brain to fire first, whenever you have a thought or an idea or a desire to do something? Someone has to get the process or sequence started, or nothing would ever happen. Even in complete sensory deprivation, someone or something continues to fire off action potentials within the physical brain. Who's doing that?

It has to happen somehow. Someone or something has to be giving those action potentials their origin, organization, purpose, and ultimate goal. This is logical common sense. Imagine what would happen if the neurons in your brain were all firing randomly by chance as the Materialists and Naturalists claim is happening. It would be chaos. You really would be dancing to your DNA and look like a total idiot.

So, which came first, the action potential or the neurotransmitter release? Actually, the Psyche came first, and then it either triggered an action potential or it triggered the release of neurotransmitters into the synaptic cleft. Either one would get a chosen action started in the first place.

The following authors believe that the Human Psyche controls its physical brain through the ion channels in the neuron's membrane, at the quantum level through the Quantum Zeno Effect.

Figure 8: Quantum Effects of Attention

The rules of quantum mechanics allow attention to influence brain function. The release of neurotransmitters requires calcium ions to pass through ion channels in a neuron. Because these channels are extremely narrow, quantum rules and the Uncertainty Principle apply. Since calcium ions trigger vesicles to release neurotransmitters, the release of neurotransmitter is only probabilistic, not certain. In quantum language, the wave function that represents "release neurotransmitter" exists in a superposition with the wave function that represents "don't release neurotransmitter"; each has a probability between 0% and 100% of becoming real. Neurotransmitter release is required to keep a thought going; as a result, whether the "wash hands" or "garden" thought prevails

is also a matter of probability. Attention can change the odds on which wave function, and hence which thought, wins.

Schwartz, Jeffrey M. *The Mind and the Brain: Neuroplasticity and the Power of Mental Force* (p. 364). HarperCollins. Kindle Edition.

This area of physics, quantum physics, is the study of the behavior of matter and energy at the subatomic level of our universe. Briefly, the synapses, the spaces between the neurons of the brain, conduct signals using parts of atoms called ions. [The influx of calcium ions into the presynaptic membrane triggers the release of neurotransmitters into the synaptic cleft.] The ions function according to the rules of quantum physics, not of classical physics. What difference does it make if quantum physics governs the brain? Well, one thing we can dispose of right away is determinism, the idea that everything in the universe has been or can be predetermined. The basic level of our universe is a cloud of probabilities, not of laws. In the human brain, this means that our brains are not driven to process a given decision; what we really experience is a "smear" of possibilities. But how do we decide between them?

One quantum mechanics discovery that may help us understand how we decide is the quantum Zeno effect. Physicists have found that if they observe an unstable elementary particle continuously, it never decays — even though it would almost certainly decay if it were not observed. In quantum physics, it is not possible to separate the observer entirely from the thing observed. They are part of the same system. The physicists are, essentially, holding the unstable particle in a given state by the act of continuing to measure it. In the same way, experiments have shown that, because your brain is a quantum system, if you focus on a given idea, you hold its pattern of connecting neurons in place. The idea does not decay, as it would if it were ignored. But the action of holding an idea in place truly is a decision you make, in the same way that the physicists hold a particle in place by deciding to continue to observe it.

Beauregard, Mario. *The Spiritual Brain: A Neuroscientist's Case for the Existence of the Soul* (Kindle Locations 808-831). HarperCollins. Kindle Edition.

These two physical theories, classical and quantal, are contradictory. Yet orthodox quantum mechanics combines them to produce a rationally coherent understanding of the connection between mind and brain. This quantum approach constitutes a way of comprehending that connection that is far more reasonable than what is attainable within the materialistic framework, which is fatally flawed by the omission of our causal minds from the theory of the mind-brain connection.

The oft-heard claim that "quantum mechanics is not relevant to the mind-brain problem because quantum theory is only about tiny things", is absolutely contrary to the basic quantum principles. Being 'big' does not tend to make a quantum system truly classical! Quantum mechanics is explicitly designed to cover 'big' systems, but by becoming 'big' a quantum system does not become classical!

Indeed, the fact that quantum mechanics is explicitly designed to cover big things is important to the solution of the mind-brain problem. For the quantum mechanical dynamics leads to the evolution of the brain, via

Process 2, into a mixture of many different brain states that correspond to many different potential experiences, and hence to the need for the added Process 1 that selects for consideration some perceivable small part of the existing mixture, which nature will then promptly either actualize or reject.

This proliferation in the brain of representations of many different alternative possible immediate courses of action is assured by the structure of ion channels. Ion channels are large brain molecules, each having a small tube (a channel) through which ions of a particular kind — say calcium ions — can flow single file, under specific brain conditions, into the interior of a neuron, where they tend to cause that neuron to release, in due course, a "vesicle" of neuro-transmitter molecules into the gap that separates that neuron from a neighboring neuron. The narrowness of the ion tube ensures that the ion that enters the interior of the neuron has a large uncertainty in its direction of motion. Hence each ion channel in the brain is a source of dynamical uncertainly in the Process-2-generated evolution of the quantum state of the brain. The resulting macroscopic state of the brain will thus tend to evolve into a quantum "mixture" of many different classically describable brain states, each with a different perceptual correlate, between which the mind-dependent quantum Process 1 is free to choose. Thus the pertinent-for-us essence of quantum mechanics is the causal dynamical linkage that QM specifies between our conscious thoughts and our atomic-particle-based brains.

The quantum mechanically entailed causal effects of our mental intentions upon our material brains is in complete harmony with our normal intuition, which is based on our lifetimes of first-hand empirical evidence. Our minds are promoted by quantum dynamics from the absurd role of impotent witnesses of events they cannot affect to causally effective instigators of intended bodily actions. Our minds thus have a natural reason to exist, which is to help us to achieve what we value, not to deceive us into believing we are something we are not!

The conclusion here, and in what follows, is that the realistically interpreted orthodox quantum conception of reality provides not only dynamical explanations of all well-established ordinary empirical data, but, automatically, also the foundation of a rationally coherent dynamical understanding of how our conscious minds can affect our material brains, and hence our material bodies, in ways concordant with both our conscious intentions and the empirical data of everyday life. Those ubiquitous first-hand data, which seem to confirm the causal power of our mental intentions, need not be interpreted as "illusions" or "delusions", as the Newtonian-particle-based materialistic physics appears to demand. Likewise, the problem of the seeming incompatibility of "free will" and "determinism" is resolved by noting that the QM law of evolution incorporates the inputs from our "free" (not-materially-coerced) choices into the causal dynamics! Hence there is natural causal evolution, and thus no causal gap or incompatibility that needs to be explained.

Stapp, Henry P. *Quantum Theory and Free Will: How Mental Intentions Translate into Bodily Actions* (Kindle Locations 470-500). Springer International Publishing. Kindle Edition.

Pim van Lommel believes that Non-Local Consciousness controls its physical brain and physical body through many different quantum and non-local processes.

DNA as a Source of Information for Each Cell

As the only person-specific and permanent cell component in our body, DNA plays an essential role as interface for the body's design, the continuity of all bodily functions, and the interaction between nonlocal consciousness (and memories) and the body. This approach retains the already-discussed interface model based on nuclear spin resonance (quantum spin correlation).

All matter, including all of our body's cells, molecules, and atoms, is made up of 99.999 percent emptiness or vacuum, and this vacuum is filled with energy and information that originates in nonlocal space, just as the universe around us is saturated with information and energy. As a result, our DNA is always in contact with all possible forms of information from nonlocal space. DNA transmits information directly and nonlocally via coherent systems in remote molecules, cells, and organ systems. But information is also communicated indirectly via signal proteins, messenger proteins, and antibodies produced by DNA and transported in the bloodstream to the cell membrane. Information is also exchanged via the DNA-induced electromagnetic fields in neurons, which find their way to the body via the autonomous nervous system (the sympathetic and the parasympathetic systems) and the central nervous system. Finally, there is indirect information exchange from the brain (hypophysis, epiphysis or pineal gland, brain stem) via hormones and neuropeptides, which are also produced by DNA in certain cells. DNA appears to be the direct and indirect personal coordinator of all information required for the optimum function of our body. And for this our individual DNA receives the necessary information from nonlocal space.

This conclusion is reinforced by our immune system, which protects our individual organism from foreign invaders such as viruses, bacteria, and foreign cells that enter the body with blood transfusions and tissue and organ donations. Governed by DNA, the immune system must identify self and nonself antigens, coordinate the generation of necessary antibodies from an unprecedented number of options, and provide an immunological memory storage that remains directly accessible and up-to-date throughout life. People enjoy lifelong immunity from the infectious diseases they suffered in childhood.

Where in the body could this immunological memory be stored while the body's composition changes every second? And how could this ever-changing immunological information be stored in DNA? In my opinion, this immunological information could also be stored in nonlocal space and directly accessible to the individual DNA in each cell via nonlocal information exchange. This view seems to be corroborated by an article in Nature that provides evidence of resistance to certain antibiotics among strains of bacteria in animals living in the wild in extremely remote areas, thus ruling out any contact with the antibiotic in question. We can only assume that the bacterial DNA received information via nonlocal space from strains that developed resistance as a result of irresponsible and careless use of antibiotics elsewhere in the world.

Nonlocal Information Transfer via DNA

In view of these facts and arguments, DNA probably plays a central role in the reciprocal information transfer between nonlocal space and the field of resonant and coherent cell structures. I would compare DNA to the high-speed processor in my computer. This processor, which consists of a tiny oscillating quartz crystal and a couple of million transistors on a few dozen square millimeters, is constantly moving, switching, and copying data at a speed of four hundred million bits per second. The processor with its oscillating crystal does not contain any information itself, but it transmits information that enters encoded in the form of electromagnetic waves with certain frequencies. For living systems too, the fact that all organisms possess rhythmic oscillation, vibration, or periodic movement is essential for (nonlocal) information exchange. Each living cell is composed of countless vibrating molecular structures. All molecules (including DNA) and atoms in a human cell are part of a living organism with an oscillating activity, each with its own characteristic frequency between 100 and 1,000 Gigahertz.

The oscillating activity of cells and the propagation of waves in and between cells are nonlinear processes (quantum processes), which have been demonstrated in calcium ions in genes, in proteins, and in cellular networks of neurons and cardiac muscle cells. There is evidence for transitions from simple to complex oscillating behavior, for instance during the development of action potentials (in the heart or neurons) or of chaotic patterns and complex feedback mechanisms in living systems. This means that in living organisms many processes, like the development of the electrical signals in the heart or brain, or feedback mechanisms in and between cells, are considered to be typical for quantum processes. Scientists have even found evidence of this oscillating property in activated white blood cells. The oscillating activity triggers resonance between molecules with identical frequencies, thus producing a cohesive whole of vibrating molecules. Resonance refers to vibration with the same frequency. The coherence of vibrating molecules creates powerful interference patterns that, in an ordered state, not only behave like a whole but actually become a whole. The parts lose their individual identity. Recent experiments with epithelial cells in the intestine appear to prove this. When a group of cells was damaged by toxic substances and significantly changed as a result, a group of identical detector cells, which are mechanically separated and which cannot communicate via chemical or electrical mechanisms, underwent an identical significant change despite a lack of exposure to any toxic substances. Their synchronous (coherent) response without direct contact suggests nonchemical and nonelectric remote communication between these cells. The authors of the research paper do not exclude biophotons as a possible source of this remote information exchange.

According to developmental biologist Brian Goodwin, the differentiation of cell functions during the embryonic phase cannot be accounted for solely by the genetic code contained in DNA structure. Goodwin too proposes self-organizing fields in and among cells to explain the differentiation and coordination of cells and cell systems. What else could explain the hundreds of thousands of well-coordinated chemical reactions in each cell every second, coupled with a reciprocal feedback mechanism in cells, organs, and entire organisms (systems biology)? Besides, chemical processes sometimes happen a million times faster in living organisms than under the most auspicious laboratory conditions. How can the living cell, governed by DNA, achieve this? It is highly likely that nonlocal information exchange between cells and cell systems plays a role. Another possible explanation is the fact that all cells are interconnected because they originate from a single source, namely the fertilized egg cell with the person-specific DNA. As we saw, Alain Aspect's experiment, which provided definitive proof of nonlocal information exchange, also made use of two particles originating from one and the same source.

The electrocardiogram (ECG), which shows the heart's electrical activity, can be registered on the skin of people's arms, legs, and chest because this electrical activity can be found in each of the body's cells. Presumably all of the body's quintillion cells are interconnected via the heart's rhythmically changing electromagnetic field. Also the registration of the brain's electrical activity, the EEG, reveals the heart's electrical activity. It is possible that self-organization enables the heart, with its intensive electromagnetic fields with coherent patterns, to create reception potential (an interface) for certain aspects of our consciousness and to transmit this information via its electromagnetic fields to the body as a whole. However, this supposition requires a great deal of additional research.

To understand correctly the evident effect of nonlocal information I would like to mention the effective functioning of groups of thousands and sometimes millions of living organisms, such as bees, wasps, ants, and termites. These colonies are examples of living and self-organizing systems composed of animals with different tasks but with a collective consciousness coordinated by the queen. If the queen is isolated from her colony but alive, everything continues as normal, but if the queen is killed away from her colony, chaos ensues and all work stops. The queen coordinates at a distance (nonlocally) — and probably on the basis of her DNA function — all of the colony's activities by creating and maintaining a collective consciousness.

Van Lommel, Pim. *Consciousness Beyond Life: The Science of the Near-Death Experience* (pp. 291-295). HarperCollins. Kindle Edition.

Remember, Science is KNOWLEDGE. Science is observation and experience. Science is NOT Materialism and Naturalism, because the major premises or primary assumptions of Materialism and Naturalism have NEVER been observed nor experienced, nor can they be. It's impossible to observe that something does not exist. It's impossible to prove that Materialism and Naturalism are true. Materialism, Naturalism, Darwinism, Nihilism, and Atheism have to be taking on blind faith as being real and true, because they can never be experienced nor observed. They can never be caught in the act; therefore, they can never become knowledge.

How do you catch something that – by fiat, declaration, definition, principle, premise, and axiom – "Does Not Exist", in the act of not existing and in the act of doing nothing? You can't! Contrary to what the Materialists and Naturalists claim, Materialism, Naturalism, Darwinism, Nihilism, and Atheism are NOT self-evident nor axiomatic. The Materialists, Naturalists, and Atheists are asking us to take a massive leap of faith into the unknowable; and, millions are willing to take the leap.

I upgraded my Science to Science 2.0 by redefining Science as Observation, Experience, Eye-Witness Evidence, Experimental Evidence, and KNOWLEDGE.

WE KNOW from the Quantum Zeno Effect that Psyches, and Telepathy between Psyches, do in fact exist. The Quantum Zeno Effect has been demonstrated and proven to be true experimentally. WE KNOW from Near-Death Experiencers and Out-of-Body Travelers that the Human Psyche and its assigned Spirit Body do in fact exist and do in fact continue to function and live after the physical brain is offline or dead.

Based upon what we already KNOW from Science and Observation, what I and the others are trying to do here is to EXPLAIN SCIENTIFICALLY how a quantum entity like the Human Psyche interfaces with and controls its assigned physical brain at the quantum level, by using what we already KNOW to be real and true in order to do so – the Quantum Zeno Effect. We are trying to do Science or apply KNOWLEDGE.

In contrast, the Materialists and Naturalists are trying to suppress and hide this type of KNOWLEDGE, because it falsifies Materialism and Naturalism. So, ask yourself which camp is really trying to do Science here, and which camp is trying to censor, ban, block, and suppress KNOWLEDGE? The answer is obvious, once we start thinking about these things logically and rationally. The writing is on the wall, so to speak. We simply KNOW from observation and experience which camp is really trying to do Science and develop and use Knowledge, and which camp is trying to suppress and hide the Knowledge, Observations, Experiences, and Science that they personally don't like.

The Quantum Zeno Effect is remote control or telepathy. The Human Psyche controls its physical brain through the ion channels in the neuron's membrane, at the quantum level through the Quantum Zeno Effect. The Non-Local Consciousness coordinates everything within its assigned physical body at a distance, nonlocally, through the Quantum Zeno Effect. Do you have a better explanation? I don't; so, I'm going to run with it. It's the simplest explanation and the best explanation as far as I'm concerned, because WE KNOW that the rocks or raw physical matter or an ocean full of RNA and amino acids cannot control and animate themselves, nor can they spontaneously generate and come alive on us.

The Quantum Zeno Effect is a remote control telepathic connection between the Human Psyche and Nature's Psyche – which in this case would reside within the atoms of the ion channels within the cell membrane of the brain's neurons. The Human Psyche would also have some level of control over the ions and the DNA within its assigned physical body. The proven and observed Placebo Effect is Mind-Over-Matter, and it functions at the quantum level on individual atoms, when it is working properly and is at its most effective. Nature's Psyche resides within and controls the various different elementary particles, vibrating strings, quarks, and leptons. The Placebo Effect is also telepathy between the Human Psyche and Nature's Psyche.

Telepathy or the Quantum Zeno Effect is like radio waves, but on a much smaller scale and a much higher frequency. The Human Psyche or Process 1 is the transmitter of the "radio waves" or thoughts or commands. Nature's Psyche or Process 3 is the receiver of the "radio waves". The Quantum Zeno Effect is the "radio waves" or the telepathic connection between Process 1 and Process 3. The ions, DNA, ribosomes, proteins, ion channels, gap synapses, and neurotransmitters are the remotely controlled cars in this scenario.

I put together a remote-control car. It works. I control the thing "at a distance" through radio waves, using a transmitter in my hands and a radio receiver located on the remote-control car. The Human Psyche does the same exact thing with its physical brain and physical body, remotely or telepathically or nonlocally, at the quantum level through the Quantum Zeno Effect. The Quantum Zeno Effect is remote control, or action at a distance, or telepathy. It works. This is the foundation of Quantum Neuroscience.

This was all really easy to understand and accept, when I was finally introduced to it. Of course, it took me over 55 years to find it, because the Materialists and Naturalists are trying to hide this type of information from us, and they have been very successful at doing so. These people don't want you to discover any of this stuff, because it FALSIFIES Materialism and Naturalism. These people have a vested interest in making sure that you don't discover anything about how the Human Psyche controls its physical brain.

Remember, the physical brain is a machine or a mechanism. Someone has to throw the switch and turn it on, or nothing would ever happen. Someone has to fire the first action potential in a chain of action potentials, or there would never be any actions. The physical brain is a computer – hardware. And, WE KNOW that all hardware has to be designed, engineered, fine-tuned, manufactured, and produced. We also KNOW that computers need a Programmer and a User or an Operator; otherwise, computers won't do anything. Computers like the physical brain don't design and create and program themselves from scratch as the Materialists, Naturalists, and Darwinists claim. So, who was your brain's Designer, Organizer, Creator, Originator, and Programmer? Who is your physical brain's current Operator or User?

Psyche is known to be an Actor and a Chooser and a Decider. It has been experienced as such. Psyche has the innate ability to be a Designer, Organizer, Creator, Originator, and Programmer. Computer hardware like a physical brain also needs an Operator, or it ain't gonna do nothing. Cutting edge scientists have determined that the Human Psyche can operate and control its physical brain remotely and telepathically through the Quantum Zeno Effect at the quantum level. It makes sense to me. Someone has to be driving the bus. Buses don't drive themselves.

Only Psyche can do choice, or decision-making. Physical matter simply reacts or responds to Psyche's decisions and choices. Raw physical matter, or the rocks, have never been observed making choices and acting on their own. The rocks and glass are passive receivers and recording devices, not actors and agents.

Psyche is the Chooser and the Actor, whether we are talking about the Human Psyche or Nature's Psyche. A rock, or raw physical matter, or a physical brain simply responds or reacts to Psyche's decisions, choices, desires, and commands. This is what we have observed. Furthermore, the Human Psyche continues to make decisions and choices long after the physical brain is dead and gone, according to the empirical evidence from Near-Death Experiences (NDEs) and Out-of-Body Experiences (OBEs) and other types of Non-Local Experiences.

Science is KNOWLEDGE. Science is Observation and Experience. WE KNOW from NDErs and OBErs that the Human Psyche and its Spirit Body do in fact exist and are real; therefore, it logically follows that the Human Psyche or Human Intelligence has to have some way of controlling its physical brain and physical body, while the Human Psyche is located and residing within its physical body.

The way that the Human Psyche controls its physical brain is by triggering action potentials and/or releasing neurotransmitters telepathically through the Quantum Zeno Effect. This is the best explanation, for this observed and experienced phenomenon, that I have found so far.

Quantum Tunneling

Quantum tunneling is not a good thing when you're trying to shrink transistors even smaller. Transistors need barriers. When electrons start tunneling through these barriers, you get problems. Big problems. In fact, quantum tunneling sets a fundamental limit on how small transistors can get to be. If any internal barriers get thinner than a nanometer, too much current will tunnel through when the transistor is off. That's not a good thing.

https://hackaday.com/2015/07/31/quantum-mechanics-in-your-processor-tunneling-and-transistors/

Quantum tunneling is teleportation. Once again, this quantum phenomenon has been observed and caught in the act by our scientists. It's real and it truly happens.

Furthermore, Astral Travelers or Out-of-Body Travelers have observed that their spirit body can teleport anywhere in this universe instantaneously at psyche's command. There are no speed limits in the Non-Local Realm, because there are no physical limitations in the Non-Local Realm. Typically, during Near-Death Experiences, God has these people flying through a tunnel so that they know that they are no longer in Kansas; and, they can also sense whether they are going up, or going down. However, these same people usually teleport back into their physical body or snap back into their physical body instantaneously when they are done with their Out-of-Body Experience, no matter where they might have been in this universe.

If spirit matter can teleport, and it can because it has been experienced doing so, then technically there's no reason why physical matter can't teleport. Physical matter and spirit matter the same and are Quantum Objects after all – just at a different phase or different density. So, why can't you teleport your physical body anywhere in the universe at will, like we can do with our spirit body? We have observed that Quantum Mechanics supports the teleportation of physical objects like atoms; so, why can't we simply teleport our whole physical body anywhere we want it to go?

The answer is that each particle of physical matter has covenanted with God to obey God's laws and God's commands; and, God has commanded the physical matter within your physical body not to teleport away on you. Physical matter is the ultimate consensus reality – it's supposed to be stable and reliable, not tunneling away on you willy-nilly. In a physical reality, I can depend upon this paper being on the cloud and on my computer tomorrow when I go looking for it. Such a thing wouldn't be possible if physical matter could tunnel or teleport away at will. Yes, electrons have a bit more flexibility and can teleport at will, but God still makes them stay with the atoms and molecules that they have been assigned to.

God teleported me and my car to safety, so I KNOW that God can teleport cars and physical bodies at will instantaneously to any destination in this universe that He desires, simply by stopping the passage of time for that particular physical object. It was nothing I did. It was something that God did for me to save me. God holds all the KEYS for the teleportation of physical matter.

Each particle of physical matter has covenanted to obey God's Laws and God's command. Your physical body isn't going to teleport away from you, atom-by-atom or en masse, because God has commanded it not to. Thank God, that He knows what He's doing!

When it comes to a physical consensus reality, it would be bad if the individual atoms could teleport away anywhere in the universe at will – it would be chaos. God negotiated with chaos and covenanted to bring order to chaos and physical matter, in exchange for a covenant from physical matter to obey God's will and God's Command. God's command and control is why there's order in this physical universe, rather than chaos. When it comes to physical objects and physical consensus realities, Quantum Tunneling is bad – it has to be contained and controlled, or chaos ensues.

Remember, the Quantum Zeno Effect is telepathy and Quantum Tunneling is teleportation. These phenomena have been experienced and observed, and thereby proven to be real.

Science 2.0 and Quantum Neuroscience are based upon Observations and Experiences, not scientific inferences, philosophical speculation, and wishful thinking.

God deliberately designed physical matter so that it cannot be pushed or accelerated faster than the speed of light. David Bohm defined physical matter as condensed light or frozen light, so we are never going to be able to accelerate it faster than the speed of light no matter what our science fiction stories try to tell us, because physical matter has by definition in principle been permanently slowed down to velocities below the speed of light so that we can live it, experience it, learn from it, and remember having done so.

However, God does have a workaround, though; but, only He has the KEYS to this workaround. It's called Quantum Tunneling. When it comes to Quantum Tunneling or teleportation of physical objects, God knows how to transform the improbabilities into certainties, or the uncertainties into guarantees.

As stated, I have experienced this myself. God teleported me and my car to safety. I was right on top of the car that had pulled out in front of me, and I going 40 mph or more. I felt time stop. It felt as if the whole of eternity had folded into me. I only had time enough to hit the brakes. When the passage of time resumed, I was someplace else in the middle of a lawn with no memory of how I got there. My car was stopped and turned off. I was no longer going 40 mph; and, instead of pointing north, my car was pointing east. That's Quantum Tunneling or teleportation that I experienced there. I can think of no other scientific explanation for it. Of course, it wasn't anything that I did. It was something that God did for me in order to save me, and in order to show me that it can indeed be done.

God can teleport or Quantum Tunnel physical objects by stopping the passage of time for those particular physical objects. From their perspective at the speed of light, photons Quantum Tunnel or teleport to their destination, because they experience no passage of time while doing so. According to the Theory of Relativity, time and entropy stop at the speed of light, and objects traveling at the speed of light from our perspective experience no passage of time from their perspective. Time itself comes into existence by slowing down to velocities below the speed of light. If we were living at the speed of light or beyond, we would experience no passage of time, no entropy, and no aging. We would be immortal and eternal. We could teleport instantaneously to anywhere we desire to go, subject to any limitations that God might impose on us, because He does hold all the KEYS to teleportation or Quantum Tunneling after all.

As fallen mortal physical beings, we are never going to be able to figure out how to teleport or Quantum Tunnel anything, either, because God won't let us. This is a power or a skill that He is keeping to Himself.

Elementary Particles

In particle physics, an elementary particle or fundamental particle is a particle whose substructure is unknown; thus, it is unknown whether it is composed of other particles. Known elementary particles include the fundamental

fermions (quarks, leptons, antiquarks, and antileptons), which generally are "matter particles" and "antimatter particles", as well as the fundamental bosons (gauge bosons and the Higgs boson) and gluons (strong nuclear force), which generally are "force particles" that mediate interactions among fermions. A particle containing two or more elementary particles is a composite particle. – Wikipedia.

If there is such a thing as a point particle – something smaller than a vibrating string – that point particle would be Psyche or Non-Local Consciousness. Either way, point particle or not, a Psyche resides within and controls vibrating strings telepathically; yet, Psyche also seems to be infinite in range, and scope, and sphere of influence. Psyche IS the vibration or the music within the vibrating strings. Psyche is the thing which CHOOSES which frequency the strings will vibrate at, and consequently the type of physical object that the strings will manifest as.

The fundamental unit and substructure of an elementary particle or fundamental particle would be Psyche or Non-Local Consciousness. This reality is a Psyche Ontology, wherein Psyche is the fundamental unit of reality. Psyche is the Ultimate Cause, or the Ultimate Causal Agent. This is the Ultimate Model of Reality. It doesn't get more elementary than this.

As far as I know Psyche has never been observed. Consensus Realities and Non-Consensus Realities have been observed and experienced in the Non-Local Realm. Spirit matter and spirit bodies have been observed and experienced in the Spirit World; but, Psyche has never been observed.

Why?

It's because Out-of-Body Travelers always describe their Psyche, Intelligence, Personality, or Non-Local Consciousness as a "spark" or an "immaterial viewpoint in space," which is looking at, perceiving, and experiencing its spirit body from an immaterial third-person perspective floating in space – literally a viewpoint in space. When they go looking for their Psyche or Intelligence, there's nothing there to be seen. These people experience Me-ness, or Identity, or Individuality, or Intentionality; but, there's nothing there to observe whenever they go looking for their Psyche or their Spark.

If there is such a thing as a point particle, Psyche or Non-Local Consciousness is definitely it. It's as if Psyche IS the radio waves, or the force fields, or the dark energy, or the vibrations in the strings, or the telepathy within the Non-Local Realm – completely immaterial – comprised of neither spirit matter nor physical matter. Psyche is immaterial non-physical Light, Life, and Intelligence. This is what has been observed and experienced. Psyche appears to be an infinite singularity, literally an immaterial viewpoint in space.

Psyche IS the Quantum Zeno Effect. The Human Psyche exercises quantum control of its neurotransmitters through the Quantum Zeno Effect. The Quantum Zeno Effect is telepathy and telekinesis.

Psyche is the Placebo Effect – Mind-Over-Matter.

Psyche also seems to work with or work through the gluons, bosons, gravitons, photons, and dark energy. Remember, the smaller can reside within and control the larger. Psyche is the smallest particle of them all, which makes Psyche or Non-Local Consciousness the fundamental unit of reality – the most elementary of the elementary particles.

Psyche is the Ultimate Cause, meaning the Ultimate Causal Force and the Ultimate Causal Agent, in the multiverse and in every dimension that we can possibly imagine. We are talking about a Psyche Ontology here; wherein, Psyche or Non-Local Consciousness is the fundamental unit of reality and the ultimate causal force behind all reality and existence. This is the Ultimate Model of Reality. It doesn't get more fundamental than this.

Quantum Mechanics Falsifies Materialism, Naturalism, and Darwinism

It's November 2017 as I write this. It can be fascinating to observe that books about Quantum Mechanics, which were written twenty or thirty years ago, actually try to support the Theory of Evolution and make accommodations for Darwinism.

The authors of those books never once realize that the very existence of Quantum Non-Locality (the spirit world and spirit matter), the Quantum Zeno Effect (telepathy and psyche), Quantum Entanglement (action at a distance), Quantum Uncertainty (choice), and Quantum Tunneling (teleportation) FALSIFIES all the claims of Materialism and Naturalism and all their derivatives. Darwinism, Nihilism, Determinism, Behaviorism, Atomistic Physicalism, Materialistic Reductionism, and Scientism are all derivatives of Materialism and Naturalism. When Materialism and Naturalism go down, Darwinism, or Spontaneous Generation, or Abiogenesis, or the Theory of Evolution goes down with them, as fruit from the poisoned tree.

Remember, Quantum Mechanics and Non-Local Consciousness FALSIFIES Materialism, Naturalism, and ALL of their derivatives such as Darwinism, Nihilism, Atheism, Behaviorism, Determinism, Atomistic Reduction, and even Scientism. When Materialism and Naturalism go down, then all of the other falsehoods go down with them. The truth or Quantum Mechanics is what remains.

Try to explain how random mutations and natural selection get at and control the ions in your physical brain, the neurotransmitters and neuropeptides in your physical brain, the ion channels in your physical brain, the transcription and translation and replication and activation processes in the cells of your physical body, or even the programming in your physical genome. You can't, because it's physically impossible. The only time that random genetic change has any real influence in your life is during the construction of your gametes, the sperm and the egg that eventually became your own personal zygote or fertilized egg. Genetic recombination, a random shuffling of your mother's two chromosomes and your father's two chromosomes, went into the production of your egg and your sperm respectively. That's the only time that random processes and chance had anything real and productive to do with your genome and your life. Thereafter, random mutations have been the cause of diseases, cancer cells, death, and the eventual extinction of our species through a process called Genetic Entropy. Random mutations do absolutely nothing to oversee, design, and control your neurotransmitters, neuro-receptors, ions, and ion channels. They can't. The only thing random mutations could do is muck things up and bring it all to a screeching halt. Scientists have observed that deleterious mutations outnumber beneficial mutations a million to one. If evolution really were our designer and creator and god, all life forms on this planet would be extinct by now, because evolution (genetic drift) of any kind is a source of entropy, death, and extinction – not progress, syntropy, growth, and life.

Likewise, the ONLY time that natural selection has anything to do with your life and your genome is when you get selected against and die. Natural selection is not an all-seeing God monitoring, regulating, and controlling every molecule and atom in your physical body as the Materialists, Naturalists, and Darwinists claim. Instead, natural selection is absolutely worthless for you, because it always results in your death whenever and however you get selected against. Think about it logically and rationally. Evolution (genetic drift) of any kind had nothing whatsoever to do with your physical brain – the only thing evolution will do for you is to cause the death of your physical brain when you finally get selected against and are killed or die.

Evolution of any kind is a source of entropy, death, and extinction – and not a source of design and creation. Evolution can't do design and creation. That is what we have actually observed and know to be true. Evolution isn't good for you, because it's leading us inexorably to the extinction of our species. The Theory of Evolution is the greatest hoax ever perpetrated against mankind – the lie that millions of people desperately want to be true and desperately want to believe. Alas, the truth can never successfully be built upon a lie; and, the claim that evolution designed and created your genome, your physical brain, your neurotransmitter system, protein synthesis, and your eyes is the greatest lie that has ever been spread among humankind. All of the observational evidence tells us that it is false.

Ions

In physiology, the primary ions of electrolytes are sodium (Na+), potassium (K+), calcium (Ca2+), magnesium (Mg2+), chloride (Cl−), hydrogen phosphate (HPO42−), and hydrogen carbonate (HCO3−). The electric charge symbols of plus (+) and minus (−) indicate that the substance is ionic in nature and has an imbalanced distribution of electrons, the result of chemical dissociation. Sodium is the main electrolyte found in extracellular fluid and potassium is the main intracellular electrolyte; both are involved in fluid balance and blood pressure control.

The cell membrane of the axon and soma contain voltage-gated ion channels that allow the neuron to generate and propagate an electrical signal (an action potential). These signals are generated and propagated by charge-carrying ions including sodium (Na^+), potassium (K^+), chloride (Cl^-), and calcium (Ca^{2+}).

There are several stimuli that can activate a neuron leading to electrical activity, including pressure, stretch, chemical transmitters, and changes of the electric potential across the cell membrane. Stimuli cause specific ion-channels within the cell membrane to open, leading to a flow of ions through the cell membrane, changing the membrane potential.

If you are alert, you notice that both the sodium and the potassium ions are positive. Neurons actually have a pretty strong negative charge inside them, in contrast to a positive charge outside. This is due to other molecules called anions. They are negatively charged, but are way too big to leave through any channel. They stay put and give the cell a negative charge inside.

So, when an axon is at rest, the anions give it a negative charge, the sodium pumps keep sodium out and potassium in, and the sodium gates and potassium gates are all closed. Because of the positive-negative difference between the inside and outside, this resting state is called a resting potential – a steady membrane potential. The word potential refers to the fact that there is a potential for change here. We use the same term to refer to a battery that is just sitting there, not connected to anything: It, too, has a resting potential. Some neurons have a resting potential of -70 mV and others have -65 mV.

When the action potential reaches the axon ending, it causes another ion (calcium, Ca^{++}) to enter the cell, which in turn causes the vesicles – the tiny bubbles full of neurotransmitters – to release their contents into the synaptic gap ... amazing, isn't it? Source: Uncle Google and the Wikipedia.

See also:

http://webspace.ship.edu/cgboer/actionpot.html

https://faculty.washington.edu/chudler/ap.html

Neuropeptides

Neuropeptides are small protein-like molecules (peptides) used by neurons to communicate with each other. They are neuronal signaling molecules that influence the activity of the brain and the body in specific ways. Different neuropeptides are involved in a wide range of brain functions, including analgesia, reward, food intake, metabolism, reproduction, social behaviors, learning and memory. Neuropeptides are related to peptide hormones, and in some cases peptides that function in the periphery as hormones also have neuronal functions as neuropeptides. – Wikipedia.

Tropic hormones are hormones that trigger the release of other hormones.

Our brains are NOT wired together. There are NO wires, CPUs, or RAM within our brain. Our brain is a collection of stand-alone switches or hardware BITs that we call neurons.

However, our blood vessels and circulatory are "wired together" or plumbed together. Our hormone system or endocrine system is an infinitely better Communication System than the neurons within our brain. There's actual signaling and functionality that has been programmed into our hormones.

The endocrine and circulatory systems are the real "wiring" within our body – NOT the brain – but, it's plumbing not wiring. Technically, there are no wires within our body.

Neurotransmitters

The most interesting and complex chapter in any Neuroscience textbook is the chapter discussing neurotransmitters, neuropeptides, action potentials, synapses, and gap junctions or electrical synapses. Of course, most of the PhDs who write these textbooks want you to believe that evolution (random mutations and natural selection) designed, created, and maintains this whole system. These people want you to take a blind leap of faith and conclude that evolution designed and created it all. But, the more I studied the physical brain and the neurotransmitter system, the more I could see God's handiwork and Signs of Psyche (or Psychic Influence) throughout the whole physical brain. This time around, I could see it, because I was prepared to see it and was actually looking for it. The writing was on the wall.

Someone has to be driving the bus. Buses don't just drive themselves, as the Materialists, Naturalists, and Darwinists claim. The physical brain is computer hardware. Someone has to program the computer. Computers don't just program themselves, as the Materialists, Naturalists, and Darwinists claim. Evolution isn't a programmer. Evolution isn't alive. Evolution has no brain and no foresight. Evolution is dumber than rocks.

There's no way in the universe that evolution could have designed, created, programmed, engineered, manufactured, fine-tuned, balanced, and produced our physical brain, genome, and neurotransmitter system. We humans can't do such a thing, so why on earth would anyone ever believe that evolution created our physical brain and physical genome through blind random processes from scratch? They believe simply because they want to believe; but, there's no logical reason for doing so. These people are emotionally vested in the theory of evolution, and they will defend it with their very lives, if called upon to do so.

It doesn't matter what it is – Materialism, Naturalism, Darwinism, Physicalism, Materialistic Reductionism, Scientism, Behaviorism, Nihilism, or Atheism -- they take the whole thing on blind faith as being real and true, because they desperately don't want its opposite to be true. It's their religion. But they are wrong. Evolution cannot design and create. It has NEVER been observed doing so; and, it never will be, because evolution can't design and create and fine-tune anything at all, especially something as complex as a neurotransmitter system.

Adapted from the Wikipedia and Uncle Google:

Neurotransmitters, also known as chemical messengers, are endogenous chemicals that enable neurotransmission. They transmit signals across a chemical synapse, such as a neuromuscular junction, from one neuron (nerve cell) to another "target" neuron, muscle cell, or gland cell. Neurotransmitters are released from synaptic vesicles in synapses into the synaptic cleft, where they are received by receptors on the target cells. Many neurotransmitters are synthesized from simple and plentiful precursors such as amino acids, which are readily available from the diet and only require a small number of biosynthetic steps for conversion. Neurotransmitters play a major role in shaping everyday life and functions. Their exact numbers are unknown, but more than 100 chemical messengers have been uniquely identified.

The neurotransmitters floating between the synapses are where the Human Psyche converts chance into choice, through the Quantum Zeno Effect. The neurotransmitters are command and control. The neurotransmitters are where the Human Psyche interfaces with its physical brain and controls its physical brain.

The neurotransmitters are crucial for proper brain function. Too little serotonin in the synapses results in depression and OCD. Some types of schizophrenia are caused by too much dopamine in the neurotransmitter system; yet, too little dopamine, and the person ends up with Parkinson's Disease. The number of neurotransmitters floating between the synapses has to be properly and exquisitely balanced, or the brain doesn't function correctly, and certain mental illnesses are the result.

The neurotransmitters are where Neuroscience, Non-Local Consciousness, the Quantum Zeno Effect, and Quantum Mechanics start to

get really interesting. **The neurotransmitters are how the Human Psyche commands and controls its physical brain.**

Current studies estimate that the average adult male human brain contains approximately 86 billion neurons. As a single neuron has hundreds to thousands of synapses, the estimated number of these functional contacts is much higher, in the trillions (estimated at 0.15 quadrillion).

The total number of neurotransmitter types is not known, but is well over 100.

The human brain has a huge number of synapses. Each of the 10^{11} (one hundred billion) neurons has on average 7,000 synaptic connections to other neurons. It has been estimated that the brain of a three-year-old child has about 10^{15} synapses (1 quadrillion), where the exchange of neurotransmitters takes place. There are quadrillions of neurotransmitters floating between your synapses right now.

These are the FACTS – the proven, observed, and experienced Science. I put them in bold lettering with the hope that you will remember them. Source: Uncle Google and the Wikipedia.

See also:

http://webspace.ship.edu/cgboer/actionpot.html

https://faculty.washington.edu/chudler/ap.html

Notice that Materialism, Naturalism, and their derivatives have NEVER been observed, never been experienced, and don't really in fact exist, because they are philosophical concepts and not science. Materialism and Naturalism have NO observational evidence supporting their major premises or primary assumptions, which claim that Quantum Mechanics and Quantum Non-Locality do not exist and isn't real.

Scientific Naturalism and Quantum Mechanics are mutually exclusive. If one of them is true, then the other one is false. The observation of Quantum Mechanics in action, the experience of Quantum Non-Locality, and the experimental verification of the Quantum Zeno Effect and Quantum Tunneling and Quantum Entanglement FALSIFY Materialism and Naturalism.

Synaptic Regulation and Control vs. Random Diffusion and Entropy

Neuroscientists, Classical Physicists, Materialists, and Naturalists have been telling us for decades now that there are NO direct connections between the neurons within our physical brain. Instead, signals are transmitted chemically from one neuron to another by dumping a bunch of neurotransmitters into the synaptic cleft between neurons.

This never made any sense to me. According to the Laws of Classical Physics, those neurotransmitters should diffuse randomly and evenly throughout the brain fluid in the synaptic cleft, and then just sit there vibrating under the effects of Brownian motion and Entropy for the rest of eternity. According to the Laws of Classical Physics, we can visualize the synaptic cleft as a tiny garbage can. What happens when you dump junk into a garbage can? That's what should be happening between the synapses in our physical brain according to Classical Physics.

As mortal physical beings, we ALL instinctively understand how Classical Physics works. We also instinctively understand how Random Diffusion works.

When that perfume bomb walks by you, you are hoping that the room is big enough that the shock and awe will diffuse quickly so that the effect is no longer above the sensory threshold within the room. If you drop and break a bottle of perfume in the corner of a room, before long the smell has diffused throughout the room. There's no way to put that genie back in the bottle according to Entropy and the Second Law of Thermodynamics. Given enough time, that whole bottle of perfume will evaporate and diffuse randomly and evenly throughout the whole room, and eventually the whole house. That's the way that Entropy and Random Diffusion works according to the Laws of Classical Physics.

The very same thing should happen between the synapses within the synaptic cleft according to the Laws of Classical Physics.

From what we know about diffusion, it always results in randomness, chaos, and entropy. Diffusion never results in order, organization, and purpose. The goal of random diffusion is an equal distribution of neurotransmitter molecules throughout the synaptic cleft, NOT the painting of a post-synaptic membrane with neurotransmitter molecules.

Although it is safe to say that WE KNOW that Psyche controls a physical brain through quantum telepathy or quantum waves or the Quantum Zeno Effect, how does the Psyche KNOW or learn which of the 100 billion neurons to trigger first in order to get a desired action started? There's more going on here than random diffusion!

Who knows what's going on at the quantum level under the control of Nature's Psyche within a synaptic cleft? God knows – I guess.

There's one thing that I do know for sure. God made our neurotransmitter systems so complex, variable, random, uncipherable, and confusing that it should be obvious and clear to anyone who is paying attention and thinking rationally that evolution and random mutations could never have designed, created, and fine-tuned our neurotransmitter systems in a trillion years. Evolution can't do fine-tuning and field-testing. It's physically impossible. Likewise, for some of the very same reasons, evolution can't convert spirit matter into physical matter.

Someone or something has to be forcing order and organization and purpose onto our neurotransmitter systems and the physical brain.

I get the impression that we still have no idea what's really going on in the synaptic cleft between synapses. The apparent results suggest that it is infinitely more organized and controlled than the random diffusion of neurotransmitters would ever allow or suggest.

Something has to be forcing the right neurotransmitters in the synaptic cleft to connect with and engage the correct receptors on the post-synaptic membrane in a timely and orderly fashion; otherwise, chaos and entropy would ensue. Random diffusion does not produce order, sequentiality, and organization. Random diffusion produces chaos and entropy.

Now, let's discuss some of the observed evidence that suggests that way more is going on between the synaptic cleft than just random diffusion, chaos, and entropy.

The more you study this, the more you realize that there's a lot more going on here than the Materialists and Naturalists want you to know or would ever be willing to accept as "science".

Autobots and Autoreceptors

Neuroscientists, Classical Physicists, Materialists, and Naturalists have been telling us for decades now that there are NO direct connections between the neurons within our physical brain. Instead, signals are transmitted chemically from one neuron to another by dumping a bunch of neurotransmitters into the synaptic cleft between neurons.

They have been backtracking ever since.

After the neurotransmitters dump into the synaptic cleft, lots of different things happen. The Materialists and Naturalists and Atheists who write our college textbooks can't tell us how these things happen. These people simply say that they do and leave it at that. There is NO explanation for any of this according to the rules of Classical Physics.

John Pinel is a Darwinist, Materialist, and Naturalist. Let's observe how he describes these neurotransmitter processes.

Pinel, J. (2014). *Biopsychology* (9th ed.). New York: Pearson.

The unconventional neurotransmitters act in ways that are different from those that the neuroscientists have come to think of as typical for such substances. One class of unconventional neurotransmitters, the <u>soluble-gas neurotransmitters</u>, include <u>nitric oxide</u> and <u>carbon monoxide</u>.

These neurotransmitters are produced in the neural cytoplasm and immediately diffuse through the cell membrane into the extracellular fluid and then into nearby cells. They easily pass through cell membranes because they are soluble in lipids. Once inside another cell, they stimulate the production of a second messenger and in a few seconds are deactivated by being converted into other molecules. They are difficult to study because they exist for only a few seconds.

Soluble-gas neurotransmitters have been shown to be involved in <u>retrograde transmission</u>. At some synapses, they transmit feedback signals from the postsynaptic neuron back to the presynaptic neuron. The function of retrograde transmission seems to be to regulate the activity of the presynaptic neurons.

Over 100 neuropeptides have been identified. The actions of each neuropeptide depend on its amino acid sequence.

You will encounter discussions of the <u>putative</u> (hypothetical) behavioral functions of various neurotransmitters in subsequent chapters. However, this chapter ends with descriptions of three particularly influential lines of research on neurotransmitters and behavior. (*Biopsychology*, pp. 93-94, 96.)

Yikes!

Like I said, it only gets more and more indescribable, the more that we try to understand it and describe it.

Notice how he anthropomorphizes and enlivens these things. He has them acting, behaving, and even controlling our behavior. That's a major error in critical thinking for a Darwinist.

According to the Materialists, Naturalists, and Classical Physicists, neurotransmitters and neuropeptides cannot "ACT". They are molecules – physical particles – they can't ACT! They should just diffuse randomly throughout the synaptic fluid and then just sit there and do nothing.

So, how on earth are these neurotransmitters and neuropeptides ACTING and BEHAVING unconventional ways? And, how are they controlling our behavior?

They can't, unless they have some sort of psyche or non-local consciousness of their own, that we aren't aware of.

At least he gets the diffusion part right, because diffusion is indeed a part of Classical Physics and Darwinism, and it should work as explained.

And here in these quotes, we suddenly have neurons communicating with each other through soluble gases, random diffusion, and osmosis – completely overthrowing decades of Neuroscience!

How in the world do these gas molecules and the postsynaptic neuron KNOW that the synaptic cleft and the presynaptic neuron needs regulating? How do these little buggers go directly to the correct presynaptic neuron and regulate it, without interfering with and messing up the other presynaptic neurons that don't need any interference or regulation? Remember, the gas molecules have got only a couple of seconds to do so! They've got to get moving now! Yet, the process is described as random diffusion, which can NEVER produce the kinds of "actions" and "behaviors" that he's talking about here, according to the Laws of Classical Physics.

Pinel, J. (2014). *Biopsychology* (9th ed.). New York: Pearson.

Neurotransmitter molecules that leak from their vesicles are destroyed by enzymes.

Action potentials cause vesicles to fuse with the presynaptic membrane and release their neurotransmitter molecules into the synapse.

Released neurotransmitter molecules bind with autoreceptors [on the presynaptic terminal button] **and inhibit subsequent neurotransmitter release. Neurotransmitters produce inhibitory feedback via autoreceptors.**

Released neurotransmitter molecules bind to postsynaptic receptors.

Released neurotransmitter molecules are deactivated by either reuptake or enzymatic degradation. (*Biopsychology*, p. 95.)

These Materialists, Darwinists, Naturalists, Neuroscientists, and Classical Physicists describe the actions, behavior, and function of these neurotransmitters as if they are Autobots.

I have watched the Transformer Movies – well the last one was a dog and I couldn't get interested in it – but the other two were cool.

The Autobots need some kind of Psyche or Spark in order to come alive. Without the All-Spark, they are simply inanimate metal.

Here he has neurotransmitters leaking from their vesicles.

Why would they leak?

The word leak is code for Quantum Tunneling or Teleportation; but, he is a Materialist and Naturalist, so he has to use the word "leak" instead, because these people don't believe in telepathy and teleportation. Apparently, you don't want these things inside the presynaptic neuron mucking things up, so the synapse has to clean them up if they leak. So, who's controlling the cleanup process, and who knows that the vesicle has leaked? How do these little enzymes know that there has been a spill on aisle one, or is the whole neuron filled with enzymes just in case a neurotransmitter escapes?

The presynaptic neuron should "know" that it has released neurotransmitters and not to release anymore, so why on earth does that presynaptic neuron need autoreceptors to prevent it from releasing additional neurotransmitters?

How do these neurotransmitters KNOW that they need to produce inhibitory feedback at the autoreceptors? What makes them do so?

Who or what tells these inhibiting neurotransmitters to disengage from the autoreceptors when it is finally time for the presynaptic terminal button to release additional neurotransmitters? In other words, now the autoreceptor has been blocked, and the presynaptic neuron can no longer release neurotransmitters. Who or what detects that additional neurotransmitters are now needed and clears or primes the autoreceptors so that the presynaptic neuron is now free to release additional neurotransmitters?

How does this overlord know that the synaptic cleft needs additional neurotransmitters?

How do the autoreceptors in the presynaptic neuron KNOW that the transmission has been received at the postsynaptic neuron, and that it is now time to clear and prime the autoreceptors?

Remember, Neuroscientists, Classical Physicists, Materialists, and Naturalists have been telling us for decades now that there are NO direct connections between the neurons within our physical brain. There can't be any communication between neurons according to the Laws of Classical Physics, because there is NO direct physical connection between neurons. Yet, the neurons seem to be communicating with each other nonetheless.

Now, the postsynaptic neuron has finally received the transmission.

Who or what tells the degradation enzymes and the presynaptic neuron that it's now time to start the reuptake process and clear the synaptic cleft in preparation for the next action potential?

Who tells the degradation enzymes when the postsynaptic neuron has received the transmission and it is time to start the degradation process? Those things can't be eating on the neurotransmitters the whole time; otherwise, most of the transmissions would be blocked from ever reaching the postsynaptic neurons.

How does the presynaptic neuron KNOW that the synaptic cleft has been completely cleared of the excess neurotransmitters and the garbage from the degradation process, so that the synaptic cleft doesn't get clogged with gunk and cease to function?

What happens to the neurotransmitters that have bound to the postsynaptic receptors? You can't just dump them back into the synaptic cleft where they can keep reattaching to the postsynaptic receptor creating signal noise, or can you?

Who or what is controlling all of this?

There should be NO command and control according to Materialism, Naturalism, and the Laws of Classical Physics! It should all be happening randomly, but it isn't! Houston, we have a problem! This whole neurotransmitter system is under way more control than it should be!

Who's controlling it all, and how can we make him stop so that Materialism, Naturalism, and Classical Physics will continue to remain real and true?

Drugs Are the Pinnacle of Materialism and Naturalism and Psychiatry

The Materialists, Naturalists, and Psychiatrists teach and believe that drugs are the ONLY solution to our mental illnesses, because physical matter is the only thing that exists.

Some agonists of a particular neurotransmitter bind to postsynaptic receptors and activate them, whereas some antagonistic drugs, called <u>receptor blockers</u>, bind to postsynaptic receptors without activating them and, in so doing, block access to the usual neurotransmitter. (*Biopsychology*, p. 96.)

So, why doesn't the physical brain just shut down and turn off with all of those receptor blockers binding to postsynaptic receptors without activating them? The whole thing should come to a screeching halt, should it not? The brain is just a machine after all. If you block all of its transmissions, the brain should just turn off and die. Receptor blockers should kill us, if they are in fact functioning the way that the Materialists and Naturalists claim.

Or, with all of those agonistic drugs binding to the postsynaptic receptors and activating them randomly, why isn't complete chaos the result?

Who or what's preventing the drugs from messing everything up, throwing the whole thing out of balance, and killing us dead? Why do the drugs work at all? They shouldn't, according to the laws of Classical Physics. The drugs should produce random synaptic noise at best, and complete physical death at worst.

When it comes to our neurotransmitter system, there is more than meets the eye.

Reuptake Transporters

I used to be a Materialist, Nihilist, and Atheist because I considered myself to be a Scientist – and for other personal reasons as well. I understand the mind-set, because I have lived it and experienced it for myself.

It should be nothing but random chaos and entropy between the synapses according to the Laws of Classical Physics. Yet, everything seems to function perfectly and reliably.

How is that possible?

It's NOT possible according to Materialism, Naturalism, Random Diffusion, Entropy, and the Laws of Classical Physics.

It ONLY becomes possible through the Laws of Quantum Mechanics and the influence of Nature's Psyche. There is no other logical explanation. Psyche produces Order out of Chaos.

Pinel, J. (2014). *Biopsychology* (9th ed.). New York: Pearson.

If nothing intervened, a neurotransmitter molecule would remain active in the synapse, in effect clogging that channel of communication. However, two mechanisms terminate synaptic messages and keep that from happening. These two message-terminating mechanisms are <u>reuptake</u> by transporters and <u>enzymatic degradation</u>.

Reuptake is the more common of the two deactivating mechanisms. The majority of neurotransmitters, once released, are almost immediately draw back into the presynaptic buttons by transporter mechanisms.

In contrast, other neurotransmitters are degraded (broken apart) in the synapse [synaptic cleft] **by the action of <u>enzymes</u> – proteins that stimulate or inhibit biochemical reactions without being affected by them.**

Terminal buttons are models of efficiency. Once released, neurotransmitter molecules or their breakdown products are drawn back into the button and recycled, regardless of the mechanism of their deactivation. Even the [neurotransmitter] **vesicles, once they have done their job, are drawn back into the neuron from the presynaptic membrane and are used to create** [and fill] **new vesicles.** (*Biopsychology*, p. 90.)

Models of efficiency?

Why?

They shouldn't be according to the Laws of Classical Physics, Materialism, Naturalism, and Darwinism! It should be maximum diffusion, random chaos, and entropy within the terminal buttons and between the synapses, according to the Laws of Classical Physics – these are physical particles we are talking about, after all!

But, the Neuroscientists and Biopsychologists talk about the synapses as if they were alive, sentient, aware, and in control of the whole process. They can't be, according to Materialism, Naturalism, Darwinism, and the Laws of Classical Physics!

Nevertheless, after the signal has been successfully transmitted to the post-synaptic neuron, someone or something turns on the Reuptake Transporters, and they start sucking neurotransmitters out of synaptic cleft and actively return them to the pre-synaptic neuron where they will be recycled and reused. Or, someone tells the degradation enzymes that it's now time for them to do their work and break apart the neurotransmitters that remain in the synaptic cleft; and, then the Transporters turn on their tractor beams and suck up the debris.

Transporters are stationary mechanisms in the membrane of a cell that actively transports ions or molecules or neurotransmitters across the membrane. This is important to understand, because I was repeatedly given the impression that transporters were little shuttles that went out into the synaptic cleft collecting neurotransmitters shuttling them back to the pre-synaptic neuron. Instead, transporters are described as tractor beams that pull in or draw in neurotransmitters when you turn these transporters on.

How do these Reuptake Transporters know that the post-synaptic neuron has received the signal and that it's now time to collect up all the remaining neurotransmitters and return them to the pre-synaptic neuron? Well, the logical conclusion according to Classical Physics is that the Reuptake Transporters must be getting told by pony express that the post-synaptic neuron has received transmission and now it's time to suck in all the excess neurotransmitters and return them to the pre-synaptic neuron. Turn on the vacuum cleaner, and let's suck these things up.

However, the truth is that we are talking about atoms and molecules here, and there are no physical mechanisms that you can attach to an atom and make it do things. Ionizing the thing is the best you can do. And, if there were physical ionic propulsion units attached to the neurotransmitters in the first place to make them swim, then those things would interfere with neurotransmission, unless they have someone onboard providing targeting and steering. If these neurotransmitters have ionic propulsion units, targeting capabilities, and rudders for steering, all of this functionality is being produced at the quantum level and not the physical level. From the other perspective, the only way you can tractor beam in a molecule is through quantum waves at the quantum level, a process that's called Action at a Distance or Telekinesis.

Oh, it's a model of efficiency alright; but, not at the physical level. ALL of the efficiency and functionality associated with these molecules is taking place at the quantum level under the control of Nature's Psyche.

Now, the REAL questions start to come into play.

John Pinel stated: "If nothing intervened, a neurotransmitter molecule would remain active in the synapse, in effect clogging that channel of communication."

So, who intervenes and oversees the whole process making sure that the synaptic clefts remain clear of garbage? What happens to the garbage when nobody intervenes to clean it up?

John Pinel wants you to believe that evolution designed and controls this whole neurotransmitter process; but, that's impossible.

Why?

Well, think about it logically. Evolution, random mutations, and natural selection didn't even exist until AFTER God designed and created the genomes and the physical brains in the first place. Evolution doesn't have a psyche or a brain, so evolution didn't have anything to do with our genomes, our physical bodies, our brains, or our lives. The only time evolution came into play for you was when you were conceived, and when you die (get selected against). Evolution didn't have anything to do with our brains, our neurotransmitter system, or our psychology and behavior. Evolution didn't even exist until after God designed and created the brains and genomes in the first place. Remember that.

Why is this important?

It's because the Darwinists, Materialists, and Naturalists really truly want you to believe that evolution designed and now controls your Neurotransmitters, Neuropeptides, Degradation Enzymes, and Reuptake Transporters. Yet, the REALITY of what we experience and observe tells us clearly and conclusively that the Darwinists, Materialists, and Naturalists are wrong.

Those Reuptake Transporters KNOW telepathically where all the different excess neurotransmitters are and use tractor beams to suck them in.

OR, someone transmits a signal to all of those degradation enzymes and tells them that it is now time to do their work and break apart the remaining neurotransmitters remaining in the synaptic cleft. They can't do their cleanup work until they KNOW that the synaptic transmission has been successfully received at the post-synaptic neuron. The degradation enzymes know precisely where all the different neurotransmitters are and proceed to do their work efficiently and effectively. And then: magically, telepathically, and telekinetically all of that leftover garbage from the breakdown of neurotransmitters within the synaptic cleft is drawn back into the pre-synaptic neuron where it is recycled and reused – otherwise, the synaptic cleft would fill up with that garbage and clog shut, never to function again.

All of this happens at "light-speed" because the pre-synaptic neuron has to recycle and reset in milliseconds so that it's ready for the next action potential to come down the pipeline. The whole action potential, transmission process, and cleanup process takes about 3 milliseconds; and by then, the whole synaptic cleft has to be cleared and the pre-synaptic neuron readied to process the next action potential that comes its way.

"Neurotransmitter molecules or their breakdown products are drawn back into the pre-synaptic terminal button and recycled".

HOW?

Again, he is describing this in terms of Classical Physics. They ALL describe these processes in terms of Classical Physics.

Is there a big Hoover vacuum cleaner in each presynaptic terminal button sucking in all of this garbage (breakdown products) back into the synapse, and keeps doing so until it KNOWS that the synaptic cleft is completely clean and pure?

This whole thing doesn't make sense to me from the perspective of Materialism, Naturalism, and Classical Physics – and it NEVER did! That's my problem with all of this. I used to be a Materialist, Nihilist, and Atheist; and, NONE of this makes any sense from the perspective of Materialism and Naturalism and Classical Physics.

It is indeed the very Model of Efficiency in practice, but not in theory. In theory, according to Classical Physics, it should be nothing more than random diffusion, entropy, and chaos.

The REALITY doesn't match at all with Classical Physics, Materialism, and Naturalism.

If you see a truck bearing down on you, you can't afford to wait for one of these neurotransmitters to leisurely and randomly make its way to the appropriate receptor on the post-synaptic membrane, according to the Laws of Classical Physics. Someone or something has to pick this thing up and make it go to the receptor immediately, or you are going to get run over. And, this reality and need applies to each and every synapse along the way. Some kind of FORCE is needed to make these neurotransmitters engage immediately, whenever they are desperately needed; otherwise, you are going to die.

So, who taught these things how to do all of these functions in a timely and efficient manner, rather than just floating around randomly as the Materialists and Naturalists claim happens? These people want you to believe that evolution designed, created, and now controls this whole process. John Pinel, the author of *Biopsychology*, is one of those who wants you to believe that evolution designed, created, and now controls this whole neurotransmitter process; but, it's NOT possible according to the Laws of Classical Physics. There is NO command and control where Classical Physics is concerned – there's just random diffusion and entropy and chaos, where the Second Law of Thermodynamics reigns supreme.

Do you suddenly see where Classical Physics, Materialism, Naturalism, and Darwinism completely FAIL to explain what's really going on within the synaptic clefts between all the synapses?

Well, I can see it and sense it; and, it's certainly a different feeling than what I was feeling before I discovered how the Human Psyche controls its physical brain.

According to the Laws of Classical Physics, it should be maximum entropy, maximum diffusion, and maximum random chaos between the synapses within the synaptic cleft; but, it is not.

Instead, it's the very model of efficiency, organization, power, elegance, control, and grace.

How's that possible?

There's only one logical explanation.

Everything that's happening within the synaptic cleft is happening telepathically and telekinetically according to the Laws of Quantum Mechanics. The ONLY way that the synapses and terminal buttons could ever be "models of efficiency" is if they are being controlled at the quantum level telepathically by some sort of Psyche or Intelligence. There is NO other way. Classical Physics doesn't cut it; and, Classical Physics certainly doesn't explain it. You can't do Quantum Neuroscience with Classical Physics! It doesn't work.

This is an observed and experienced example of Nature's Psyche interacting with and communicating with Nature's Psyche. ALL of the atoms and molecules within the synaptic cleft are talking with each other, communicating with each other, and coordinating their actions with each other telepathically through the Laws of Quantum Mechanics which allows such a thing to happen for real. Quantum Telepathy or Quantum Telekinesis is a force that actually makes these individual particles move to pre-selected targets and destinations within the synaptic cleft. There is no other logical explanation for what we are witnessing within the synaptic cleft.

Remember, according to the Laws of Classical Physics, it should be maximum entropy, maximum diffusion, maximum randomness, and maximum chaos between the synapses within the synaptic cleft; but, it is not.

Quantum Telepathy and Nature's Psyche have the power and the ability to override entropy and bring order, purpose, communication, direction, assignments, location sensing, motion, and organization out of the chaos and random diffusion that should reign supreme between the synapses within the synaptic cleft according to the Laws of Classical Physics.

Can you understand now why the neurotransmitter system never made any sense to me, and why it bugged me so much? I was interpreting it and trying to understand it from the perspective of Materialism, Naturalism, and Classical Physics as everyone else tries to do; but, it can't be done that way. It makes NO sense from the perspective of Classical Physics.

Classical Physics, Materialism, Naturalism, Nihilism, and Atheism have NO answers for what we are truly witnessing taking place within neurons and between the synapses within the synaptic cleft, because the whole thing is being organized, moderated, motivated, and controlled telepathically and telekinetically by Nature's Psyche as a whole through the Laws of Quantum Mechanics. It's a Quantum Process, and NOT a physical process. It's the result of Quantum Telepathy and Quantum Telekinetics, and NOT the result of the maximum diffusion, entropy, and chaos which are typically associated with Classical Physics.

It is organization – command and control and orchestration – taking place within the synaptic cleft; the very Model of Efficiency, NOT random diffusion, entropy, and chaos like I used to visualize it when I was visualizing it from a materialistic and naturalistic Classical Physics perspective.

Sorry, Dorothy, but we are not in Kansas anymore. We have entered the Twilight Zone of Quantum Mechanics where everything happens telepathically and telekinetically at a distance, according to the observed and proven Laws of Quantum Mechanics. The Quantum Zeno Effect or Telepathy IS Psyche, or at least the way that Psyche works and acts. Psyche or Intelligence is a force – a telepathic force – a mental force – a telekinetic force – a nonlocal nonphysical force; and, each elementary particle has a Psyche of its own that the Orthodox Interpretation of Quantum Mechanics calls collectively "Nature's Psyche". These things are coordinating with each other and interacting with each other telepathically and telekinetically at the quantum level between the synapses of each and every neuron within your physical brain. There's no other way to explain the marvelous efficiency of what we are witnessing and experiencing, because it shouldn't be that way according to the Laws of Classical Physics.

The Laws of Quantum Mechanics ARE the Laws of Psyche or the Laws of Non-Local Consciousness. It's the very Model of Efficiency – the very Model of Command and Control – the very Model of Coordinated and Motivated Effort. In fact, though, we still continue to describe these neurotransmitters in terms of Classical Physics. They are "drawn back into the pre-synaptic neuron", or they are "transported back to the pre-synaptic neuron". The neurotransmitters are somehow drawn to the correct receptors on the post-synaptic neuron. We completely forget that Quantum Tunneling (teleportation) is an observed and proven phenomenon at the quantum level. When that truck is bearing down on you, and you need to move right now, either the ion channels open immediately in the specific neurons that are needed to start you running and/or the necessary neurotransmitters are teleported directly to the appropriate receptors so that you start running immediately – instead of standing there like a dear in caught in the headlights waiting for some neurotransmitter to randomly make its way to the appropriate receptor to get the whole process moving.

When a specific neurotransmitter is needed immediately, then Psyche reaches out and teleports it directly to where it needs to go; and, an action like running is started immediately. This is what we experience and observe in REAL LIFE. There's NO waiting for a neurotransmitter – the right neurotransmitter – to randomly and blindly find its way to the correct receptor, like I used to believe and like what's typically taught to us in our Public College textbooks. Our neurotransmitter system does NOT function according to the Laws of Classical Mechanics. It can't! Our neurotransmitter system has to function according to the Laws of Quantum Mechanics, or it wouldn't function at all in the first place.

The Darwinists, Materialists, Naturalists, and Classical Physicists want you to believe that evolution (random mutations and natural selection) designed and controls all of this. It can't, because evolution cannot function as a mental force or a psychic force like Psyche can. Evolution isn't intelligent, and it isn't alive. Evolution is not a person or a mind. There's no way that Evolution and Classical Physics could ever be in control of our neurotransmitter system. I wouldn't work. It would be random diffusion, chaos, and entropy instead.

The writing was on the wall, so to speak.

All of this is still extremely foreign to me, because I am a scientist and theoretician; and, I was raised on Materialism, Naturalism, Darwinism, Scientism, and Classical Physics – I think naturally and automatically in terms of Classical Physics, and it's holding me back. I had to let all of these go, if I truly wanted to understand what's really happening within the synaptic clefts between the neurons of our physical brain, because the only way to explain logically and rationally what we are witnessing and experiencing is through Quantum Telepathy, Nature's Psyche, Quantum Telekinesis, Quantum Tunneling, and the Laws of Quantum Mechanics. There is no other way.

All of this tells us that Psyche must exist, and that God's Psyche must exist; otherwise, this whole physical universe would be nothing but a randomly diffused homogeneous cloud of hydrogen gas and nothing more. These observations tell us that Psyche or Non-Local Consciousness has the power to override Entropy and the Laws of Classical Physics and thereby bring Order and Organization out of random chaos.

God is in the details; but, the only way we will ever find Him is if He reveals himself to us; and, He has indeed revealed himself to us within the Laws of Quantum Mechanics. We find God by going and looking for Him. God has promised to leave us alone if we don't want to have anything to do with Him.

Here once again, we have an example of God hiding in plain sight where the Materialists, Nihilists, and Atheists like me will never see Him and never find Him, because He is functioning telepathically and telekinetically through the Non-Local Laws of Quantum Mechanics, which the Materialists and Naturalists and Classical Physicists do NOT accept as being real and true.

Sodium-Potassium Pumps, Ion Channels, and Neurotransmitters

The more you study the neuron potentials and the neurotransmitter system, the more the unanswered and unanswerable questions begin to multiply. When the neuroscientists try to describe these processes and machines in terms of Classical Physics, Materialism, and Naturalism, unexplainable logic contradictions are automatically introduced because the whole process doesn't add up and doesn't make any sense from a materialistic and naturalistic perspective.

The whole thing unravels and falls apart with the sodium-potassium ion pumps. Let me try to explain.

Some Na^+ ions do manage to enter resting neurons despite the closed sodium channels and K^+ ions do exit, then why does the resting membrane potential stay fixed?

At the same rate that Na^+ ions leaked into resting neurons other Na^+ ions were actively transported out and at the same rate that K^+ are leaked out of resting neurons, other K^+ ions were actively transported in.

Such ion transport is performed by mechanisms in the cell membrane that continually exchange three Na^+ ions inside the neuron for 2 K^+ ions outside. These transporters are commonly referred to as sodium-potassium pumps.

Since the discovery of sodium-potassium pumps, several other classes of transporters (mechanisms in the membrane of a cell that actively transport ions or molecules across the membrane) have been discovered. (*Biopsychology*, p. 78).

This is all fine and good until you start adding excitatory postsynaptic potentials (EPSPs) and inhibitory postsynaptic potentials (IPSPs) into the picture.

EPSPs are graded post-synaptic depolarizations, which increase the likelihood that an action potential will be generated. An EPSP takes the membrane potential of a neuron away from the resting potential's natural negative pole and closer to the positive pole.

IPSPs are graded post-synaptic hyperpolarizations which decrease the likelihood that an action potential will be generated. An IPSP makes the membrane potential of a neuron more negative, taking the neuron further away from the -55 mV of the threshold potential that's needed to trigger an action potential.

The membrane potential is the difference in electrical charge between the inside and the outside of a neuron or cell.

An action potential is a massive momentary reversal of a neuron's membrane potential from about -70 mV to about +50 mV. An action potential is when a neuron turns ON and FIRES. A neuron is a single hardware BIT. It is either ON or OFF. That's all the information that ever passes through a neuron at the physical level – ON or OFF, YES or NO. A neuron is a switch or a gate.

The thing that NOBODY seems to realize is that all of the information, memories, and data that allegedly comes in through the synapses on a single neuron get merged or integrated into a SINGLE post-synaptic potential. Everything coming into a neuron gets reduced to a single BIT of information, which is called a post-synaptic potential. Nothing else survives passage through a synaptic cleft and synapse, except IPSPs and EPSPs.

It's physically impossible to send a message or a byte of information through a synapse and have that message or memory survive the transmission, because a synaptic cleft scrambles, randomizes, and then integrates everything that comes its way into single BITs of information – ON or OFF, OPEN or CLOSED. Single hardware BITs can't function as CPUs, RAM, and Distributed Memory without some kind of external organization, command, control, and programming. It's physically impossible for the brain to function as CPUs, RAM, Maps, and Distributed Memory Storage at the physical level. If any of that functionality exists, it's being done at the quantum level by Nature's Psyche, because it's physically impossible for it to be happening at the physical level.

Fascinating, is it not?

A ligand is a molecule that binds to another molecule. Neurotransmitters are ligands to their receptors. The question I immediately have with neurotransmitters is, "Who or what draws them in? What makes them move? Who docks them?" It requires some kind of force, field, syntropy, thought, or quantum wave to move, steer, and navigate an atom or a molecule to a specific target or ligand in a timely fashion. We know from studying Classical Physics that unguided and unmotivated atoms and molecules could sit there under the effects of Brownian motion, entropy, and random diffusion for all eternity and never do anything useful.

Ionotropic [neurotransmitter] receptors are associated with ligand activated ion channels.

When a neurotransmitter molecule binds to an ionotropic receptor, the associated ion channel usually opens or closes immediately inducing an immediate postsynaptic potential. For example, in some neurons, EPSPs (depolarizations) occur because the neurotransmitter opens sodium, thereby increasing the flow of Na^+ ions into the neuron. In contrast, IPSPs (hyperpolarizations) often occur because the neurotransmitter opens potassium channels or chloride channels, thereby increasing the flow of K^+

ions out of the neuron or the flow of Cl⁻ into it, respectively. (*Biopsychology*, pp. 88-89).

That sodium-potassium pump KNOWS precisely what the internal voltage of the associated neuron is, because it has the necessary machinery needed to read the internal voltage of a neuron cell. Its primary goal is to keep the resting potential within the neuron at -70 mV.

But now, it gets really funky and freaky.

The sodium-potassium pump is tasked with maintaining the internal voltage of the neuron at -70 mV; yet, the ionotropic receptors are tasked with changing the internal voltage of the neuron either through depolarization (changing the resting membrane potential from -70 to -67 mV for example) or through hyperpolarization (changing the resting membrane potential from -70 to -72 mV for example).

If left to their own devices, the sodium-potassium pumps and the ionotropic receptors are working at cross-purposes and should cancel each other out or interfere with each other; but, they don't.

Why?

One theory, rather weak in my humble opinion, is that the sodium-potassium pump KNOWS telepathically whether a change in voltage is being caused by quantum leakage or is being caused by an IPSP or an EPSP; and, the sodium-potassium pump can adjust its performance accordingly. There are no known wires within a neuron keeping the sodium-potassium pumps apprised of what's the true source of the voltage changes within the cell; but, it KNOWS anyway, and responds accordingly either by eliminating the effects of the quantum leakage or by allowing the changes in the resting potential if caused by IPSPs and EPSPs.

However, I have read that the sodium-potassium pumps continue to pump during an action potential, which means that the sodium-potassium pumps were actually designed by God to eliminate the random signal noise that is being produced by stray neurotransmitters and the occasional EPSP. According to this model, the sodium-potassium pumps were designed to cancel out the effects of the ionotropic receptors; so, if there were too many sodium-potassium pumps on the neuron, then there would be no action potentials. That wouldn't be good. We could stop breathing or some such if there were no action potentials.

Either way, this whole system requires balancing, fine-tuning, command, control, and intelligence in order to get it to work right.

Remember, ALL of the synapses on a single neuron feed into a single Postsynaptic Potential (PSP). A neuron functions as a brake, dampener, and integrator. A neuron combines ALL of the input into a single output, an action potential; and, ALL of that input is simply erased by the sodium-potassium pumps if that input isn't sufficient enough to trigger an action potential. A neuron is a switch, a single hardware BIT. It is either ON or OFF. That's it. There's NO memory storage and NO computation taking place within a neuron. It's physically impossible.

The sodium-potassium pumps are powered by ATP molecules; and, the flow of ATP to the pump determines how quickly and efficiently it works – limited by the flow of course.

ATP molecules are where we see the hand of God in action, which is why the Materialists and Naturalists avoid talking about them if they can. The ATP are free-roaming molecules. They would just sit there and vibrate under Brownian motion and random diffusion if they were in a jar of Gatorade. However, when ATP are within a living cell, they come alive with purpose, activity, intention, and motion. They move directly to their assigned target, discharge their energy, and the resulting ADP moves directly to the ATP Synthase molecular machine for recharging.

It's as if each sodium-potassium pump has some sort of ongoing wireless WiFi connection with each and every ion and ATP molecule within the neuron and simply KNOWS that ion's purpose or assignment. It's as if there is a person in each and every neuron driving the thing, regulating the thing, and making it do whatever this person or psyche wants it to do.

We see signs of Psyche or Non-Local Consciousness all throughout the functioning of a neuron. The signs are there that this thing must exist, because somebody has to be controlling that sodium-potassium pump and feeding it information telling it what to do and how to respond to the changes of the internal voltage within a neuron – wirelessly or telepathically. Remember, telepathy IS WiFi at the quantum level; but, somebody has to be sending and monitoring those wireless signals, or they wouldn't do any good. The target molecules also have to know what the messages or quantum waves mean.

We see signs of Psyche or Non-Local Consciousness all throughout the functioning of a neuron.

Ionotropic receptors are neuron receptors that are associated with ligand-activated or neurotransmitter-activated ion channels. Metabotropic receptors are neuron receptors that are associated with signal proteins and G-proteins.

Who decided which ion channel a particular neurotransmitter receptor will open or close? Who decided the amount of voltage change in the resting potential that will take place each time an ionotropic receptor is triggered or activated? Who is preventing the sodium-potassium pumps from cancelling out the EPSPs and the IPSPs? How does the presynaptic neuron KNOW when the postsynaptic neuron no longer has any need for the neurotransmitters in the synaptic cleft so that the presynaptic neuron can start the reuptake process? Who holds back the degradation enzymes and prevents them from destroying the neurotransmitters in the synaptic cleft, until the neurotransmitters have successfully done their job? How are the degradation enzymes told that it is now time for them to move into action and clear the synaptic cleft? How do the degradation enzymes know where their target neurotransmitters are located, when these enzymes are finally turned loose to do their job? Who or what clears or empties the post-synaptic ionotropic and metabotropic receptors? Who decides how long to keep a neurotransmitter bound to a receptor? What prevents other neurotransmitters from randomly connecting to those ionotropic and metabotropic receptors once they have been cleared and readied for the next action potential? Who decided what receptor subtypes to place on the post-synaptic membranes? Who decided what type of neurotransmitter and neuropeptides to produce in the pre-synaptic terminal buttons? Who taught these neurotransmitters, ions, pumps, receptors, messengers, and ion channels how to do all these different functions rather just floating around randomly and firing-off randomly as the Materialists and Naturalists claim is happening?

The only logical answer to these questions is Psyche or God – Nature's Psyche or God's Psyche.

If nothing or nobody intervened in this process, then neurotransmitters would remain active in the synaptic cleft, in effect clogging up that channel of communication and keeping the post-synaptic neuron in a constant state of firing-off one action potential after another until the brain burns out. But apparently, Nature's Psyche intervenes and keeps everything running like a finely-tuned machine, where the neurons and neurotransmitters are concerned. WE KNOW it is Nature's Psyche doing these quantum tasks, because the Human Psyche isn't consciously aware of any of these processes.

We see signs of Psyche or Non-Local Consciousness all throughout the functioning of a neuron. The metabotropic receptors on the post-synaptic neuron receive input from neuropeptides functioning as neurotransmitters, thereby releasing a G protein into the cytoplasm of the post-synaptic neuron. That G protein then goes directly and binds either to an ion channel opening it up, or that G protein goes and creates a second messenger RNA which can actually enter into the nucleus of the neuron and bind directly to the DNA, thereby altering the gene expression of that particular neuron. Can you imagine what would happen if all of this was happening randomly, willy-nilly, according to the laws of Classical Physics and Random Diffusion as it's typically explained as happening? These G proteins have the power to reprogram the DNA. What would be the result if Materialism and Naturalism were true, and these G proteins were just randomly altering the genes and the gene expression through blind luck or random happenstance alone? Thank God that someone or something is overseeing the whole process and actually KNOWS when a specific post-synaptic neuron needs to have its programming or DNA modified, and when it does not – and the Materialists and Naturalists actually want you to believe that Evolution designed, created, and now controls this whole process through Classical Physics. Give me a break!

We see signs of Psyche or Non-Local Consciousness all throughout the functioning of a neuron. For example, selective serotonin reuptake inhibitors (SSRIs) are typically used to treat depression, anxiety, and OCD. Their method of action is to block the reuptake of serotonin by the presynaptic neuron, thus resulting in a lot more serotonin neurotransmitter floating around within the synaptic clefts.

Why is it helpful to have more serotonin in the synaptic cleft? Why isn't increased signal noise, false positives, random action potentials, and chaos the result of having more serotonin floating around in the synaptic cleft? How is extra serotonin a good thing? What prevents those extra neurotransmitters from causing the post-synaptic neuron to misfire randomly? Blocking the reuptake of serotonin should result in random chaos, according to Materialism, Naturalism, and Classical Physics; but, somehow magically it does not. Instead, contrary to all logic, having extra serotonin floating around randomly and diffusely in the synaptic clefts actually improves the situation and reduces the depression, anxiety, and OCD. Why? How? It doesn't make any sense!

I could only come up with one logical explanation for what we are observing, and it involved getting rid of Materialism, Naturalism, and Classical Physics; and then, switching to Quantum Mechanics instead.

Where serotonin seems to be concerned, these neurotransmitters float out there in the synaptic cleft until the human psyche reaches out telepathically and telekinetically and uses some of them, and then plugs them into or teleports them into the post-synaptic receptors precisely when and where they are needed. ONLY when there are not enough of those serotonin neurotransmitters out there in the synaptic cleft for the human psyche or nature's psyche to reach out and grab and use, do we start to have problems like depression, anxiety, and OCD as a result.

Directed synapses are synapses at which the site of neurotransmitter release and the site of neurotransmitter reception are in close proximity. Sometimes the Astrocytes will even build a wall around directed synapses in order to prevent the neurotransmitters from escaping into the cerebral spinal fluid.

If we study this long enough, we eventually run into non-directed synapses, which are synapses that have multiple release sites called "varicosities" all along the axon, and these particular non-directed neurotransmitters are released some distance from their intended site of reception. The neurotransmitters from varicosities are widely dispersed diffusely throughout the cerebral spinal fluid just floating there waiting to be used.

Why?

The only logical answer that I could come up with is so that there will always be a neurotransmitter in the system for the human psyche or nature's psyche to reach out, grab, and use whenever it needs one to start up an action potential that hasn't already been initiated. In other words, when the human psyche makes a decision or a choice, it literally reaches out telepathically to the right neurotransmitters and the correct specific post-synaptic neuron and teleports (quantum tunnels) nearby neurotransmitters directly to the necessary post-synaptic receptors to get the desired action and action potential started immediately. Only when there aren't any neurotransmitters nearby does the human psyche and nature's psyche start to run into problems and difficulties; and, glitches result.

Someone or something is there in the machine, reaching out telepathically and telekinetically grabbing neurotransmitters precisely when it needs them; yet, completely ignoring them and completely blocking reception at the post-synaptic neurons when it doesn't need these diffusely-spread non-directed neurotransmitters that are just floating around out there in the cerebral fluid between the neurons. Contrary to the way that it's typically described by the Materialists and Naturalists, serotonin and dopamine seem to be non-directed neurotransmitters that have to be maintained at a certain level within the cerebral spinal fluid, in order to prevent glitches and malfunctions from happening.

Psychosis consists of hallucinations and delusions – faulty and fake sensory input. Delirium tremens (DTs) is the phase of alcohol withdrawal syndrome that's characterized by hallucinations, delusions, agitation, confusion, hyperthermia, and tachycardia.

There's a ton of faulty input and faulty output associated with psychosis, which means that there's a lot of faulty and random neurotransmission going during psychosis. I KNOW, because I experienced a couple of years of substance-induced psychosis due to all the different brain-altering prescription drugs that they had me on, at the time. It's a hellish way to exist. I pray to God every day that I never have to go back there ever again. I separated from reality, lost all sense of time, was delusional, hallucinating, paranoid, couldn't sleep, had bugs crawling under my skin, and wired so extreme that I had to kill myself. Oh, that was a lot of fun, let me tell you. When I finally chose to go cold-turkey, the withdrawal lasted six months. The dude was crazy nuts.

Before going cold-turkey, I attempted suicide by overdosing on the sleeping pills. I had to get out of there; and, the people on the radio and television knew me by name and were telling me that I had failed, that I was going to hell, and that I had to kill myself. My suicide attempt landed me in the looney bin for another month, where they promptly got me addicted to everything all over again. Towards the end, I was even misdiagnosed as being schizophrenic. But, my psychosis was iatrogenic – physician-created. The medical doctors and psychiatrists gave me my mental illness; and after six months of withdrawal, my mental illness or psychosis came to an end when I finally got all their drugs out of my system and my brain finally re-balanced and normalized. Substance-induced psychosis is often iatrogenic. Death by medical doctor is the third most common way to die. They turn you into an addict and kill you slowly, but they get the job done in the end if you let them.

The Human Psyche is slaved to the physical machinery, so when the physical machinery is spontaneously producing faulty input and faulty output, there's not a whole lot that the Human Psyche or Nature's Psyche can do about it. Our brains were programmed by Nature's Psyche to idle in default mode, when the Human Psyche is not in command; and, lots of weird things happen when our default mode is faulty and malfunctioning.

It has been observed that too much dopamine contributes to psychosis and schizophrenia; whereas, too little dopamine causes Parkinson's Disease. In a healthy, normal, undamaged, and properly developed physical brain, someone or something is able to keep that physical brain perfectly balanced where dopamine levels are concerned; and, that very same someone KNOWS when the system is out-of-balance and actually takes steps to rectify the situation. ONLY when the brain is messed up, broken, damaged, or not functioning properly is this someone unable to make extra dopamine or unable to eliminate too much dopamine in order to keep the whole neurotransmitter system perfectly in balance; and, random glitches are the result.

The Materialists, Naturalists, and Darwinists want you to believe that evolution (mutation and selection) designed, created, fine-tuned, and balanced the dopamine system. If that were true, then we should be able to cure Parkinson's Disease and schizophrenia at will, because we human psyches are infinitely more intelligent than natural selection and random mutations.

It finally dawned on me while studying all of this that the Human Psyche can literally reach out telepathically and telekinetically to any neuron it wants, open the Ca^{2+} ion channels in the presynaptic membrane of that particular neuron, and thereby trigger the release of neurotransmitters into the synaptic cleft thereby starting the neurons firing or increasing the rate of neuron firing. This is called exocytosis, and it is triggered either by action potentials or by the Human Psyche, which cause the Ca^{2+} ion channels to open in the presynaptic membrane resulting in the release of neurotransmitters into the synaptic cleft.

The Human Psyche makes a decision or a choice or a demand, and Nature's Psyche responds by grabbing the necessary neurotransmitters or opening the correct ion channels, thereby getting the whole action potential process rolling in the first place.

It's the completely on-or-off nature of action potentials and sodium-potassium pumps that makes Psyche's presence so obvious and so necessary; and, it's the completely open-or-closed nature of ion channels that makes Psyche's presence and influence so necessary and essential. The machinery won't work right without Psyche, especially if the machinery were actually functioning according to the rules of Classical Physics, Materialism, and Naturalism as some of these people claim is happening.

A few of the neuronal processes can be successfully explained in terms of classical physics and electrical conduction. Excitatory post-synaptic potentials (EPSPs), and inhibitory post-synaptic potentials (IPSPs) are graded potentials. Graded responses are responses whose magnitude is determined by the magnitude of the stimuli that induce them. In contrast, Action Potentials are all-or-none responses. Ironically, ALL of the synaptic input that feeds into a single neuron gets integrated or reduced into a single Post-Synaptic Potential, or a single BIT of information. This is crucial to understand and get right, if you truly want to know how the physical brain works.

Different types of neurons have different membrane resting potentials, typically anywhere from -70 mV to -65 mV. When the depolarization reaches about -55 mV, a neuron will fire an action potential. This is the threshold potential. In neuroscience, the threshold potential is the critical level to which a membrane potential must be depolarized to initiate an action potential. IPSPs hyperpolarize a neuron; and, EPSPs depolarize a neuron.

A neuron is effectively a switch – it either FIRES or it doesn't. It's all-or-none – ON or OFF. A neuron isn't a CPU or RAM. A neuron is a switch, gate, register, or single hardware BIT. This is crucial to understand and get right, if you truly want to know how the physical brain works. Remember, there are NO wires in your brain. Your brain is comprised of a 100 billion switches or neurons, with NO wires connecting them. How can you build a CPU or RAM without any wires? You can't! It's physically impossible. This is crucial to understand and get right, if you truly want to know how the physical brain works.

Electrical conduction along the myelinated segments of the axon and electrical conduction through the neuronal cytoplasm of EPSPs and IPSPs is passive, which means that it happens at the speed of electricity instantly and decrementally. When an excitatory neurotransmitter activates or triggers an ionotropic receptor, that ion channel opens briefly raising the resting potential of the neuron at the location of the ion channel by a couple of millivolts. That local change in voltage diffuses instantly and decrementally throughout the rest of the neuron which typically means that there is little detectable change in the voltage level by the time it diffuses completely throughout the neuron and reaches the axon hillock and the axon itself. This effectively means that potentially dozens of these excitatory neurotransmitters have to trigger dozens of ionotropic receptors in order to actually raise the voltage enough throughout the whole neuron so as to trigger an action potential. This means that the triggering of an action potential is deliberate because it requires potentially dozens of integrated EPSPs in order to raise the voltage enough in the whole neuron to actually trigger an action potential. Electrical conduction, passive conduction, decremental conduction, saltatory conduction, action potentials, excitatory post-synaptic potentials (EPSPs), and inhibitory post-synaptic potentials (IPSPs) actually make logical sense in terms of classical physics and random diffusion because that's how electrical currents work in the physical world.

However, whenever we switch to the neurotransmitter side of the issue, we suddenly find ourselves having to deal with Psyche and Quantum Mechanics. An example: for as long as the neurotransmitter remains bound to the receptor-channels or the ionotropic receptors, the alteration in membrane permeability responsible for the EPSP or IPSP continues. Who or what decides how long to keep a neurotransmitter bound to the ionotropic receptor, thereby increasing or decreasing the effect and duration of an EPSP or IPSP? Who or what decides to keep the ion channel open for a prolonged period of time? Raw physical matter can't do it, because it is inert and reactive, not active. There's only one logical answer. Psyche can intervene telepathically and telekinetically forcing a single neurotransmitter to stay bound to a single receptor-channel long enough for that single open ion channel to raise the voltage enough throughout the whole neuron so as to trigger an action potential. Psyche or Non-Local Consciousness seems to have dozens of different ways to trigger an action potential when it needs one. There's nothing random about it, as the Materialists and Naturalists claim. It's deliberate and purposeful. The Human Psyche literally reaches out telepathically and telekinetically locking these neurotransmitters into the appropriate receptors until the Human Psyche gets what it wants done. This is the essence of Quantum Neuroscience, which is the science of how the Human Psyche interacts with and controls its physical brain.

Even when we do finally figure out a Classical Physics explanation for some of the cerebral processes, it still doesn't change the fact that someone or something has to start the first action potential in order to get the ball rolling. My butt simply sits there in the chair, and it would sit there until I die, unless I (my psyche) chooses and decides to get it moving. It's my decision to move it that reaches into my brain telepathically and telekinetically starting the first action potential and getting the thing moving. So, even if we do figure out the physical mechanisms for some of these things, that still doesn't change the fact that nothing happens until your psyche decides to make it happen. We are not billiard balls as the Materialists, Naturalists, Darwinists, and Determinists claim. We are the person behind the pool cue who gets the billiard balls rolling in the first place.

The Materialists, Naturalists, and Atheists are simply focusing on the obvious; and, they are completely and deliberately ignoring and rejecting the most interesting stuff because it involves Psyche, Quantum Non-Locality or Quantum Non-Physicality, Quantum Mechanics, Quantum Telepathy, Quantum Telekinesis, Quantum Tunneling, and Quantum Teleportation. Psyche or Non-Local Consciousness is reaching out telepathically and telekinetically and making these things move. These people don't like that! They continue to insist that Psyche, telepathy, telekinesis, mind-over-matter, the non-local, and the non-physical DO NOT EXIST. The Materialists and Naturalists completely reject Quantum Neuroscience. Instead, these people say that we don't know how these cerebral processes work – they just do.

John Pinel in his book, *Biopsychology*, repeatedly states that we don't know how these things work – they just do. He's right. Most of this stuff is physically impossible according to the Laws of Classical Physics, Materialism, Darwinism, and Naturalism. I used to be a Materialist, Nihilist, and Atheist. We materialists and naturalists have NO logical explanation for half the things that are happening in the physical brain because we deny and reject Quantum Non-Locality, Quantum Telepathy, Quantum Teleportation, Quantum Telekinesis, Quantum Tunneling, and Quantum Non-Local Consciousness by stating officially that these things DO NOT EXIST. We define them out of existence before we even start to do science, which means that most of the time we are unable to explain how brain mechanisms really work.

In his book, *Biopsychology*, evolutionist, naturalist, materialist, nihilist, and atheist John Pinel states that the pre-synaptic neuron is the very model of efficiency; but, the only way that that is possible is if Materialism, Naturalism, and Darwinism are false and instead the neurons, neurotransmitters, resting potentials, pumps, ions, ion channels, receptors, neuropeptides, and synaptic clefts are all being monitored and controlled telepathically and telekinetically by the Human Psyche and Nature's Psyche dynamically and actively through the Laws of Quantum Mechanics.

There's NO randomness or Classical Physics where the functioning of the neurotransmitter system, neuropeptide system, ion system, and synaptic clefts are concerned. It's all being controlled by something or someone, or it wouldn't work right; and, it would be the very model of chaos, entropy, and random diffusion instead.

Quantum Neuroscience is the scientific study of how the Human Psyche interacts with and controls its physical brain, by negotiating with Nature's Psyche telepathically. Telepathy is nothing but WiFi at the quantum level. Quantum Neuroscience is a function of Psyche or Non-Local Consciousness. There is no other logical explanation for what we are observing. Quantum Mechanics is also a function of Psyche or Non-Local Consciousness; otherwise, telepathy (the Quantum Zeno Effect) and teleportation or telekinesis (Quantum Tunneling) would not be observed, would not exist, and would not be scientifically verifiable through our science experiments. WE KNOW that these things exist, because they have been experienced and observed and caught in the act.

Someone or something is out there forcing the neurotransmitter system and the ion system of the physical brain to work right in an orderly and efficient manner. ONLY when the neural pathways and the physical brain are damaged, does this "thing" fail to do what it's supposed to do. In their books and articles, John von Neumann, Henry P. Stapp, Mario Beauregard, and Jeffery M. Schwartz call this thing **Nature's Psyche**.

"Quantum Physics in Neuroscience and Psychology: A Neurophysical Model of Mind-Brain Interaction"

Quantum Neuroscience is a function of Psyche, whether we are talking about the Human Psyche or Nature's Psyche.

The physicists call Nature's Psyche the physical laws and the physical constants of the universe. The more creative among the bunch call it the Zero-Point Field of Light. The Astrophysicists seem to call this thing Dark Energy.

The New Agers call this thing the Quantum Sea of Light.

The theologians call Nature's Psyche the Mind of God, or the Light of Christ, or the Holy Spirit, or God's Psyche.

WE KNOW it exists because Out-of-Body Travelers have experienced it; and, so have the prophets of God.

It's way too orderly and way too perfect for everything within a neuron and a synaptic cleft to be happening according to the laws of Classical Physics, random diffusion, blind luck, and random Brownian motion. Someone or something is driving the neuron and controlling the neuron and KNOWS precisely what is happening within the neuron and outside the neuron instantaneously and simultaneously. There is no other logical explanation besides a Quantum Mechanics explanation or a Psyche explanation.

It really is as if someone is in the neurotransmitter system of the physical brain and within the ion system of the physical brain forcing them to do what it wants the neurotransmitters and ions to do and forcing them to work right and work efficiently.

In my Neuroscience class, there were a couple of questions that came up; and, the answers I chose are pertinent to this discussion. Luckily, these questions weren't actually read and graded, or I would have failed the course because my answers aren't acceptable to the Materialists, Naturalists, Darwinists, and Atheists to design and run these types of college courses.

1. All mammals share the same basic divisions of the central nervous system; i.e., spinal cord, myelencephalon, metencephalon, mesencephalon, diencephalon, and telencephalon. The relative sizes of these divisions differ greatly between species. Humans, for example, have a much larger telencephalon relative to the rest of the brain than other mammalian species. What behavioral differences do you think arise from this anatomical difference?

WE KNOW from the Lived Experiences of the human race, particularly Near-Death Experiences (NDEs) and Out-of-Body Experiences (OBEs), that the human psyche or our non-local consciousness continues to think, feel, perceive, choose, behave, and do things long after our physical body and physical brain are dead and gone. The thought processes of the human psyche are much more complex and detailed than the thought processes observed in any other species. Human psyches are able to design and create complex programming and complex machines; and, we also have innate language communication abilities which the other species do not

331

have, apparently on both sides of the veil. Human psyches communicate telepathically, mind-to-mind, in the non-local spirit realm; and, human psyches communicate verbally, through feelings, through body language or gestures, and through writing here in this physical realm.

We human beings need a larger telencephalon to interface successfully the more complex thoughts, ideas, theories, feelings, emotions, teleology, and intelligence of our human psyche with the physical realm surrounding us.

The much larger telencephalon in humans successfully falsifies the theory of evolution, because there are NO evolutionary precursors for such a thing. The human telencephalon is way over-built, far more than fitness and survival demand. External intervention from God was needed to design and build such a thing, because NO evolutionary process would ever have done so.

2. The nervous system has groups of neurons that perform specialized tasks. For example, in the peripheral nervous system the somatic nervous system and the autonomic nervous system do different (and in a sense opposite) things. What advantages come from having the nervous system set up in this manner?

It's nice to have the autonomic nervous system running things for us in the background, because then we don't have to think about these things in order to get them done. We don't have to constantly concentrate in order to keep breathing and to keep our heart beating. This is highly advantageous.

Out-of-Body travelers have experienced and observed that their physical body keeps itself alive while their psyche or non-local consciousness is separated from their physical body and physical brain. It's nice that our physical body has some automatic programming built into it, so that our spirit and psyche can leave it behind for a while and our body won't die on us while our spirit and mind are exploring the astral plane.

While studying Quantum Neuroscience, we learn that the human psyche controls its physical brain directly through telepathy (the Quantum Zeno Effect) and telekinesis or teleportation (Quantum Tunneling). We know that the human psyche also controls its somatic nervous system (voluntary nervous system) through the very same Quantum processes.

Volition

Will is a function of Psyche or Non-Local Consciousness, because CHOICE is involved. The somatic nervous system is often defined as our voluntary nervous system – it's under our volitional control. Voluntary behavior or chosen behavior are controlled by the Human Psyche. Autonomic functions and reflexes were programmed into us by God.

If Psyche, choice, and free will did not exist, we wouldn't have a somatic nervous system. The whole thing would be autonomic. According to the Materialists, Naturalists, and Behaviorists, choice is physically impossible, which means that choice is taking place at the quantum level or the psyche level.

The neurotransmitter system and the ion system within the physical brain has been described as the very model of efficiency. That's impossible according to Materialism, Naturalism, and Classical Physics. It should be nothing but random diffusion, chaos, and entropy within the synaptic cleft; but, it's not, thanks to Psyche and Quantum Mechanics.

The sodium-potassium pump and the ionotropic receptors should cancel each other out and there should be no IPSPs and EPSPs as a result; but, there are, thanks to Nature's Psyche or God's Psyche and Quantum Mechanics. Somebody is driving the bus where the neurons are concerned.

Quantum Neuroscience is the scientific study of how the Human Psyche interacts with and controls its physical brain.

While studying Quantum Neuroscience, we learn that the human psyche controls its physical brain directly through telepathy (the Quantum Zeno Effect) and telekinesis or teleportation (Quantum Tunneling). We know that the Human Psyche also controls its somatic nervous system (voluntary nervous system) through the very same Quantum processes, by negotiating telepathically with Nature's Psyche or God's Psyche in order to do so.

Classical Physics, Materialism, and Naturalism cannot explain these processes logically and rationally, because these people are constantly stating that Nature's Psyche, God's Psyche, and the Human Psyche DO NOT EXIST.

Instead, these people want you to believe that some magical invisible entity or force, which they call Evolution, designed, controls, oversees, and manages this whole process perfectly and flawlessly. These people are willing to take that blind leap of faith and choose to believe that Evolution is controlling the sodium-potassium pumps, ion channels, neurotransmitters, and the 10^{14} synaptic clefts within the human brain through Classical Physics, or Materialism, or Naturalism. If they are willing to take that kind of blind leap of faith – and they are and they do – then they shouldn't be mocking and ridiculing the people who choose to believe that Nature's Psyche or God's Psyche is controlling and overseeing the whole shebang, because the Biblical God Jesus Christ has been observed and experienced during our Near-Death Experiences (NDEs), Out-of-Body Experiences (OBEs), and other types of Non-Local Experiences or Spiritual Experiences, whereas Evolution's Psyche has not.

Evolution's Psyche or God's Psyche? One of these has been experienced and observed in action, and one of these has not.

Someone is in the neurotransmitter system, synaptic clefts, and the ion system of the physical brain forcing them to do what it wants them to do and forcing them to work right; and, I don't think that it's Evolution's Psyche, because the Materialists and Naturalists keep assuring us that Psyches, including Evolution's Psyche, do not exist. Materialism and Naturalism and Darwinism are self-defeating, because they say that psyche does not exist; yet, they treat evolution, random mutations, and natural selection as if these things have a psyche, are intelligent, are teleological, are sentient, are prescient, are omniscient, are telepathic, are telekinetic, and are alive. These people treat Evolution as if it were a God, while at the same time claiming that it is completely random and blind. Their claims are illogical and irrational – nothing more than convenient ad hoc just-so fictional stories.

WE KNOW from the collective Lived Experiences of the human race that God's Psyche does in fact exist because it has been lived and experienced by human beings or human psyches; and, we also KNOW through a process of elimination and through a complete lack of observation that Evolution's Psyche does not exist because it has never been lived nor experienced. Remember, the best way to find and know the truth is to live it and experience it for yourself, or to choose to trust someone who has.

If you successfully eliminate everything that is false, everything that has been falsified, everything that is impossible, and everything that has never been observed nor experienced, then ONLY the Truth will remain. We start by eliminating Materialism,

Naturalism, Darwinism, Abiogenesis, Spontaneous Generation, Macro-Evolution, Creation by Rocks, Creation by Blind Luck, Creation by Nothing, Nihilism, Determinism, Physicalism, Scientism, and Atheism because they have NEVER been lived nor experienced, nor can they be lived and experienced. They simply don't exist as an entity or a force that can be experienced and observed and interacted with.

Then we choose instead to embrace what has been experienced, observed, and caught in the act – namely, Nature's Psyche or God's Psyche, because WE KNOW from observation and experience that such a thing exists. Remember, the BEST way to find and know the truth is to live it and experience it for yourself, or to choose to trust someone who has.

Through Quantum Mechanics and Psyche the Impossible Becomes Possible

My best friend is a scientist; and, I was explaining all of this to him last night. He basically asked me if I have ever experienced situations where the neurons aren't fast enough to get you out of harm's way, everything slows down, and your Psyche or Spirit takes complete control totally bypassing all the neurons and controlling your whole physical body directly as a whole in order to get you to safety faster than your physical body is capable of accomplishing the task?

My best friend has told me that he has experienced that particular phenomenon, every now and then – two times as a teenager when he crashed his friend's dune-buggy and when he crashed his car, and once later on when he or the elk phase-shifted and the elk passed through his truck unscathed. As a teenager, he experienced the "bubble of protection" phenomenon where suddenly you become an indestructible and untouchable superhero for a few seconds until the emergency is over.

His question actually triggered the memory of four times when such a thing happened in my own life – something took control of my physical body and instantaneously and automatically my physical body literally did the impossible. Until now, I never had a scientific explanation for what happened to me, and just considered it a fluke or random serendipity. All of these things happened when I was a teenager and doing the stupid things that teenagers do.

1. One time, the Scoutmaster took the troop, and we went skeet shooting. I had never used a shotgun before, and I had never taken a shot at clay pigeons; but, most of the other boys had, and they were very good at it. The Scoutmaster slung a clay pigeon out fast and at an odd angle. One expert took the shot and missed. Another expert took, the shot and missed. Instinctively automatically, I lifted my shotgun up, aimed from my right hip, pulled the trigger, and took out the clay pigeon a couple of feet off the ground. The experts were all excited and asking, who did that, who did that? I told them that I did. They didn't believe me. I had never demonstrated that type of skill or capability before or since. It just happened automatically.

2. I was never any good at sports. I was a jogger and learned how to juggle; but, I could never learn how to be functional and useful at basketball, baseball, football, and soccer. One time we were playing tag football. They kicked the football to me, and I caught it – something that I had never done before. Then something just took over, time slowed down, my situational awareness amped up to the max, and I started to run. The big experienced kids and the adults felt like putty in my hands. It was almost as if I could read their minds – I automatically jinked and jagged in response to what they were thinking and

doing. I literally watched these kids and adults slipping, sliding, and falling onto the ground as they tried to adjust while trying to catch me; and, I just zig-zagged my way past them all in complete control of the situation. I almost got the touchdown – but one of the adults caught up to me from behind and tagged me out before I got to the end-zone. I'd never done anything like that before or since.

3. In high school, I caught a ride home with one of the seniors who was nice to me. He had this big old clunker with a V8 in it and no seat-belts; and, he liked to punch the hammer at the stop lights so that we could feel the G Forces. He was turning left and punched the hammer. The G Forces pressed me against the door to my right, the door snapped open, and I launched into the street head-first. I was airborne and ballistic. Then something just took over. I was no longer in control. It felt as if I was a passive first-person witness of what was happening to me. I was still inside my body but no longer in control of my body. I was being controlled by something at the quantum level. I automatically tucked and rolled. I somersaulted on the asphalt and/or the dirt along-side the road, rolling over two or three times to bleed off the speed, and came up running. I ran and climbed into the back seat where the door was still open, and I closed the door.

I felt nothing, so I can't tell for sure how many times I rolled, nor could I sense whether I was rolling on asphalt, or the dirt to the side of the road, or down the middle of both. I remember running on the dirt and back into the asphalt in order to get into the car that had stopped ahead of me ten or twenty feet.

In high school, I never used a backpack. I carried ALL of my books in my left hand. When I got back into the car and shut the door, ALL of my books were still there in my left hand, where they were before I was jettisoned out the door onto the asphalt. There wasn't a scratch or bruise on me, and no damaged clothing. I came through the whole thing completely unscathed. There's no way on earth that I could have done that on my own.

My friend sitting next to me on the left saw the whole thing as I launched out of the door and then through the back window of the car as I did the somersaults and came up running. He still says to this day that it's the coolest thing he has ever seen in his life.

I was in some sort of bubble of protection, because I didn't have a scratch on me.

Within a bubble of protection, you temporarily become an untouchable indestructible super hero. My best friend experienced this phenomenon. A careful study of Near-Death Experiences will reveal that some others have experienced a bubble of protection as well. It's not anything that I did. It's something that God did for me in order to save me.

I had a scab an inch and a half in diameter the time that God didn't intervene and save me. Imagine what I would have had by being launched from a car onto the asphalt, if God hadn't intervened and saved me.

4. The strangest and most miraculous and most-quantum thing that has ever happened to me took place while I was driving to work at age 18. I was driving down the road at 40 mph, the speed limit. From nowhere, a car pulled out in front of me just a few feet from me and there was nowhere for me to go but into that car. I remember pressing the brakes and that was it; there wasn't time for anything else.

Suddenly, I felt the whole of eternity fold into me; and, time literally stopped. I could sense everything around me 360 degrees. I could sense that there was no place for me to go, with cars to my left and that other car forward of me towards the right.

Then I kind of sensed where I wanted to be or wished that I could be.

Then snap.

Instantly, time resumed, and I was some twenty feet away in the middle of a lawn with my car completely stopped and my car turned off. I had gone from 40 mph to zero instantly; and, both I and my car had teleported twenty or thirty feet forward into the center of the lawn to my right.

From my perspective, I had experienced the physically impossible. I had experienced Quantum Tunneling or Teleportation. The observer in the shop came running out and asked if I was okay. He kept looking at me and my car. He told me that there's no way on earth that I could have missed that other car, but I did. I don't know what it all looked like from his perspective, but he said that that's the coolest thing he had ever seen in his life.

I was as cool as a cucumber – completely totally calm. There hadn't been enough time to get my adrenaline going. It all happened automatically. I really didn't have anything to do with it. It was something that God did for me in order to save me.

Conclusions: There is no logical explanation for any of this according to the Laws of Classical Physics. In fact, when I went through my materialistic, nihilistic, and atheistic phase, I literally forgot about these things as if they had never happened to me.

It's interesting how the mind works, especially when we are determined to keep it limited to Classical Physics, Materialism, and Naturalism. Materialism and Naturalism and Atheism produce a blinding force that actually prevents us from remembering and understanding Quantum Events, or Spiritual Events, or Psychic Events. This is truly a mind-over-matter phenomenon. We literally forget the things that we don't want to be true.

Whenever you start sharing experiences like these, you soon discover that everybody has had them, and then over time dismissed them as luck, flukes, or serendipity – if they remember them at all. We tend to forget these types of things, especially if we are determined to pursue Materialism, Naturalism, Darwinism, Nihilism, and Atheism instead.

I was addicted to prescription drugs during my Atheistic phase, and experienced Substance-Induced Psychotic Disorder. I had separated from Reality, and I was hallucinating and delusional. I had literally forgotten how to turn on the television and play a DVD in the machine. Others had to do that for me while I was high on all that crap and going through withdrawal. My memories were gone; and, I had nothing going on at the time that I wanted to remember. It was six months of hellish withdrawal. I had no sense of time, so they had to force me to eat because I never got hungry. I could no longer add or subtract. I had to start over from scratch after I got sober, although I could still read.

God and/or Nature stared to slowly return my memories to me during the year after I got sober; and, the process continues to this day. According to the Laws of Classical Physics, NONE of my memories should have returned, because they were ALL gone during the withdrawal period which lasted six months.

Long-term memories are defined as memories for experiences and facts that endure after the experiences are no longer the focus of attention. Long-term memories are often called consolidated memories. Attention is a product of the Human Psyche. Long-term memory storage is done for us by Nature's Psyche – whether within the physical brain or non-locally in the cloud, it doesn't really matter, because we don't have any control over it anyway. Nature's Psyche stores our memories for us completely outside our conscious awareness of where those memories are actually going and being stored. I have adjusted the definition of long-term memory to more accurately match with the empirical evidence. If it survives the death of your physical body and physical brain, then it's a long-term memory in every sense of the phrase. Thoughts are psyche; and, psyche is memory.

Due to the fact that my long-term memories or access to my long-term memories did indeed return slowly over time after I had gotten through the withdrawal period, it is my current belief that all of our long-term memories are stored in the Transdimensional Realm or Non-Local Realm or Quantum Realm, where they are accessed during our Life Reviews after we are dead. It was an interesting feeling when I first realized that I finally had access to the "encyclopedia" of my long-term memories once again. Wow! I'm back!

WE KNOW from Near-Death Experiences that ALL of our memories are stored Non-Locally in the Spirit World, because they are retrieved and used during our Life Reviews after our physical brain is dead or offline. This KNOWLEDGE has ramifications and even practical applications.

A sensory evoked potential is a change in the electrical activity of the brain, such as a cortical EEG that is elicited by the momentary presentation of a sensory stimulus; but, these sensory evoked potentials are a product of Psyche when they happen in the absence of a physical stimulus.

It has been observed and measured scientifically through PET and fMRI, that whenever the Human Psyche chooses to pay attention to something, Nature's Psyche turns on the parts of the brain that are needed for that type of physical functionality.

The following book mentions and discusses this observed reality in many of its chapters.

Schwartz, J. M., & Begley, S. (2002). *The Mind and the Brain: Neuroplasticity and the Power of Mental Force*. New York: HarperCollins.

"Chapter 10: Attention Must Be Paid" was particularly important as Proof of Concept for Quantum Neuroscience.

http://publicism.info/psychology/mind/11.html

http://mypsyche.us/wp-content/uploads/2018/01/The-Mind-and-the-Brain.pdf

This is good stuff that everyone should have access to, in my humble opinion.

Short-term memory are memories that are stored only until a person stops focusing on them. Stored where? It's physically impossible to store memories within a synaptic cleft! Furthermore, thoughts and memories are quantum waves. WE KNOW that this is so, because our thoughts and memories and personalities survive the death of our physical brain.

Attention consolidates memories, in a normal functioning brain. The Human Psyche has to pay attention to an object or concept for a while before Nature's Psyche will consolidate those memories in long-term storage for later retrieval by the Human Psyche. However, if the parts of the brain – which Nature's Psyche has mapped at the quantum level for the purpose of experiencing, remembering, and consolidating physical events and memories at the physical level – get destroyed or removed, then it's no longer possible for the Human Psyche to experience and remember those types of events or facts or memories. The Human Psyche is slaved to its physical brain. If information, events, and memories don't make their way into the physical brain, then they don't make their way to the Human Psyche.

Short-term memory acts like a buffer – first-in then first-out. If the machinery that Nature's Psyche has mapped to the brain for use in experiencing, remembering, and consolidating physical events and memories at the physical level get destroyed at the physical level, then the Human Psyche will lose all access to that type of functionality, even if Nature's Psyche continues to experience and record those types of events and memories for later use during a person's after-death Life Review. Nature's Psyche isn't required to share what it knows with the Human Psyche; in fact, there's plenty of evidence demonstrating that God has commanded Nature's Psyche NOT to share what it knows with the Human Psyche.

Studying memory and memory storage is the most interesting thing to study, BOTH at the quantum level and the physical level, when it comes to Neuroscience and Quantum Neuroscience.

When it comes to my friend who has Alzheimer's Disease, despite all of the physical damage that has taken place within his brain, I have observed that he can retrieve all of his long-term memories from Non-Local Storage at will – all the ones that were stored there before he got Alzheimer's Disease. However, he can't remember many of the things that happened seven minutes ago. In other words, the Alzheimer's Disease and neurological damage are eliminating his ability to create new long-term memories and store them in Non-Local Storage; and, his short-term memory or working memory is diminishing; but, he continues to have full access to all of his long-term memories that were stored Non-Locally before he got Alzheimer's Disease despite all the neurological damage. My friend will never know nor remember my name, because he met me and got to know me after he got Alzheimer's. He still seems to sense that he has met me and that he knows me, because that apparently took place before the Alzheimer's got too bad for him to be able to remember my name.

I believe that our physical brain is a transceiver. When it is functioning properly, it transfers or transmits working memory or short-term memory into long-term Non-Local Storage, and then forgets it locally. As the physical brain deteriorates, this ability to store short-term memories into long-term Non-Local Storage is slowly lost. There's no other logical explanation for what I am witnessing – he's losing all ability to create and transmit new long-term memories to Non-Local Storage; but, he still retains the ability to retrieve all of his old long-term memories from Non-Local Storage even though his brain is getting to the point where it is so damaged that it should no longer have any memories of anything.

This is amazing stuff when you finally start thinking about it logically, rationally, and from the perspective of Quantum Mechanics, Quantum Tunneling, Quantum Entanglement, Quantum Non-Locality, and the Quantum Zeno Effect.

Astrocyte Glial Cells Regulate Synaptic Transmission through Gap Junctions

I have upgraded my science to Science 2.0

Science 2.0 allows ALL of the evidence into evidence, and then tries to make sense of it as a whole.

Now, I'm going to take everything that we have discussed so far and just throw it into the garbage can, because scientists have made a NEW discovery in recent decades.

It has been discovered that Glial Cells called Astrocytes regulate Synaptic Transmissions through Gap Junctions or Electrical Synapses.

What?

The materialistic and naturalistic Neuroscientists have been telling us for over a century that the physical brain controls itself through neurotransmitters, neurons, and synapses.

What's all this about Glial Cells, Astrocytes, Gap Junctions, and Electrical Synapses? You mean our physical brain has been wired-together directly like a computer chip or an electrical circuit or a computer network, all the way along, and we didn't even know it?

Yep, that's what it means!

Gap junctions function as electrical synapses – direct electrical connections between Neurons and Glial Astrocytes. In other words, most of our nerves and neurons are already wired together directly through Gap Junctions or Electrical Synapses; so, why on earth would there ever be any need for neurotransmitters and synapses? Aren't the neurotransmitters a bit redundant, unnecessary, and inefficient? It seems so, doesn't it?

https://en.wikipedia.org/wiki/Gap_junction

Regulation of Synaptic Transmissions by Astrocytes is called a Tripartite Synapse:

Tripartite synapse refers to the functional integration and physical proximity of the presynaptic membrane, postsynaptic membrane, and their intimate association with surrounding glia as well as the combined contributions of these three synaptic components to the production of activity at the chemical synapse. – Wikipedia.

https://en.wikipedia.org/wiki/Tripartite_synapse

This tripartite synapse seems to function like a transistor or a switch – three terminals and not two.

Astrocytes form Glial Networks – computer networks. Many have suggested that it should have been called Glial-science and NOT Neuroscience.

There's really NO need for the Neurons, Synapses, and Neurotransmitters. The Astrocytes can do it all, because they seem to be wired-together directly through Gap Junctions or Electrical Synapses. So, why on earth would God put Neurons, Synapses, and Neurotransmitters into this system, and make it so inefficient?

At first, I thought that I had found the smoking gun, when I discovered astrocytes and gap junctions; but, such proved not to be the case, as I will now demonstrate.

Although glia often envelop synapses, they have traditionally been viewed as passive participants in synaptic function. Recent evidence has demonstrated, however, that there is a dynamic two-way communication between glia and neurons at the synapse. Neurotransmitters released from presynaptic neurons evoke Ca^{2+} concentration increases in adjacent glia. Activated glia, in turn, release transmitters, including glutamate and ATP. These gliotransmitters feed back onto the presynaptic terminal either to enhance or to depress further release of neurotransmitter. Transmitters released from glia can also directly stimulate postsynaptic neurons, producing either excitatory or inhibitory responses. Based on these new findings, glia should be considered an active partner at the synapse, dynamically regulating synaptic transmission.

http://ltc-ead.nutes.ufrj.br/constructore/objetos/astrocytes%20synapse.pdf

Astrocytes form Computer Networks

Recent research has clearly established that glial cells (particularly astrocytes) and gap junctions play major roles in brain function.

One aspect of astrocytic organization suggests that they too play a role of synchronizing activities of like cells in a particular area. Unlike neurons, astrocytes are distributed evenly throughout a particular area, with only one astrocyte per location and little overlap between the projections of adjacent astrocytes. This suggests that each astrocyte coordinates the activity of neurons in its domain, and with as many as 40,000 processes, each astrocyte has a great potential to coordinate activity. Gap junctions on astrocytes tend to occur at the end of each process, where it comes in contact with processes from adjacent astrocytes.

What could astrocytes be coordinating?

The fact that many astrocytic processes wrap around synapses and are connected to both presynaptic and postsynaptic neurons by gap junctions suggests that each astrocyte may coordinate the activity of synapses in its domain. The hypothesis that synaptic transmission depends on communication among three cells (presynaptic neuron, postsynaptic neuron, and astrocyte) via gap junctions is referred to as the tripartite synapse. (*Biopsychology*, p. 91.)

The Glial Astrocyte makes sure that the post-synaptic neuron actually got the message from the pre-synaptic neuron – if not, the Astrocyte takes steps to rectify the situation. Each Astrocyte coordinates the activity of synapses and neurons under its domain.

These Astrocytes also maintain a certain level of Monoamine Neurotransmitters between synapses for proper brain functioning. In a very real sense, it seems as if it's the Glial Astrocytes that are running the show, and NOT the neurons. The neurons and synapses only seem to be there to slow the process down so that the Human Psyche has time enough to intervene and change its course of action if it desires to do so.

Really, truly, this whole Astrocyte Oversight situation and direct electrical connections (electrical synapses or gap junctions) suggested to me that one of the main purposes of neurotransmitters is to slow everything down and moderate it so that it doesn't all happen at the speed of light. The synapses and neurotransmitters and action potentials give the Human Psyche time enough to intervene and change the outcome and the course of action if it desires to do so.

Once again, nagging questions arise, though. The more you study this thing, the more confusing and complicated it gets, especially from a Classical Physics perspective.

How do these "gliotransmitters" feed back onto the presynaptic terminal either to enhance or to depress further release of neurotransmitters? What makes these gliotransmitters go to the presynaptic terminal and NOT the post-synaptic terminal? What makes them move towards their target rather than diffusing randomly willy-nilly throughout the synaptic cleft and the rest of the brain according to the Laws of Classical Physics and the Second Law of Thermodynamics? How do the Glia and the Astrocytes KNOW that there's too much neurotransmitter in the synaptic cleft, or too little? Apparently, the Astrocytes also KNOW whether the desired transmission has reached the targeted neuro-receptor on the post-synaptic membrane, or not. How is all of this KNOWLEDGE and Information transmitted, coordinated, and controlled? Who or what keeps it all straight and in mind?

The Astrocyte KNOWS how many neurotransmitters are in the synaptic clefts under its control; and, the Astrocyte KNOWS what types they are. The Astrocyte has to KNOW so that it can regulate and control and coordinate synaptic circuits.

Furthermore, there's evidence that the Monoamine Neurotransmitters – dopamine, epinephrine, norepinephrine, and serotonin – have to been maintained in perfect balance between the synaptic clefts where they are used. If you have too much dopamine in the neurotransmitter system, then Psychosis and Schizophrenia – hallucinations and delusions – are the result. If you have too little dopamine in the neurotransmitter system, then Parkinson's Disease is the result. If you have too little serotonin in your brain, then Depression, Anxiety, and OCD are the result. Apparently, these Monoamine Neurotransmitters have to be maintained in perfect balance within the synaptic clefts where they are used, which means that the Astrocyte has to KNOW how many different neurotransmitters are floating out there in the synaptic clefts, which are under its command. How does an Astrocyte KNOW all of this information?

Again, there's only one logical answer. All of this is happening telepathically and telekinetically at the Quantum Level according to the Laws of Quantum Mechanics. There's NO visible detectable telephone wires and guidelines running from the Astrocytes to each and every molecule and neurotransmitter currently in the synaptic cleft. It's all done through remote control or radio control, telepathically and telekinetically, at the Quantum Level, or nothing would ever get done. There is no other way, because Classical Physics, Random Diffusion, Materialism, Naturalism, and the Second Law of Thermodynamics tell us that ALL of this is impossible; yet, all of this is happening despite the Laws of Classical Physics. Apparently, the Laws of Classical Physics do not apply within the synaptic clefts of our physical brain.

The ONLY logical answer is that all of this information exchange, knowledge, command, and control is taking place telepathically and telekinetically at the quantum level within each synaptic cleft in our physical brain, according to the observed and proven Laws of Quantum Mechanics. There really is no other way that we know of by which all of this can be done.

Furthermore, even electrical synapses, gap junctions, or direct connections between neurons and astrocytes only do OPEN and CLOSED. They really aren't wires after all! It's deceptive to claim that they are. You can send ions and the occasional errant small molecule through an "electrical synapse"; but, you can't send gigabytes of video and memories through the thing, because the thing functions as a gate, either OPEN or CLOSED. OPEN would start the message or the memory; and CLOSED would be the signal to end the message; but then, there would be NO message. It's physically impossible to send a coherent message or memory through a gate, or a gap junction, a synapse, or a single switch. You would need to signal when the message starts and when the message ends, as well as sending an actual message, which is physically impossible through a gate or a switch like a synapse.

Single ions and the very smallest of molecules are the only thing that get through these gates or gap junctions; and, if any kind of message or memory is being organized and sent through these gates, it's not being done at the physical level. The formation of ions and molecules into memories and messages is taking place at the quantum level, if it's taking place at all. The transmission of messages and memories through gap junctions and synaptic clefts is taking place at the quantum level, if it's taking place at all. The storage of petabytes of memories is taking place at the quantum level, if it's taking place at all.

Memories are a product and function of Psyche. The memories that survive the death of our physical body and physical brain are taking place at the quantum level, if they are taking place at all.

This is what I have been getting from my study and research into Quantum Neuroscience. Each atom and molecule in the physical brain has a Psyche; and, that little Psyche or Intelligence can be commanded and controlled telepathically and telekinetically by the Human Psyche (and Nature's Psyche within the Astrocytes) through the Quantum Zeno Effect. I have found no other logical explanation for what we are witnessing and experiencing, when it comes to synapses and our neurotransmitter systems, because ALL of this is impossible according to Classical Physics, Materialism, and Naturalism. It ALL keeps pointing back to Quantum Mechanics, the Quantum Zeno Effect, the Human Psyche, Nature's Psyche, Telepathy, and Telekinetics.

Yet, the Evolutionists and Naturalists want you to believe that evolution knows how to design, create, fine-tune, balance, organize, and control all of these things. That's impossible, because evolution (random mutation and natural selection) didn't even exist, until AFTER God designed, programmed, engineered, fine-tuned, created, and manufactured the genomes, the physical bodies, and the physical brains in the first place.

Think about it logically and rationally. That's what I finally chose to do.

When it comes to the transmission and storage of memories, especially the memories that survive the death of our physical brain, there's more than meets the eye and it's not happening at the physical level.

Information Storage and Cell Differentiation

It has been observed that each neuron in the physical brain produces one type of neurotransmitter and one type of neuropeptide, although many of the neuroscientists seem to claim differently.

Now, I have some questions that have been bugging me for a while.

Who decides what type of neuron each neuron will become?

There are interneurons, such as internuncial neurons, relay neurons, association neurons, connector neurons, intermediate neurons, or local circuit neurons.

There are many different types of multipolar neurons, such as Purkinje cells, Dogiel cells, and five different types of ganglion cells.

Who decides where and what type of neuron a neuron will be? Who guides the direction and the growth of its axon and dendrites?

Multipolar neurons have more than two processes extending from its cell body – multiple dendrites and a branching axon.

Why?

It's the same genes in every neuron, so who decides how many dendrites a neuron will produce and how many branches its axon will have? Choices are being made within these neurons; and, choice is a function and a product of Psyche. According to the Materialists and Naturalists and Behaviorists, choice is physically impossible, which means that choice is taking place at the quantum level or psyche level. Someone Psyche is making these choices for each one of these multipolar neurons.

Who determines what type of neurotransmitter and what type of neuropeptide to produce within each neuron, and who decides what type of neurotransmitter receptors to build across the synaptic cleft on each post-synaptic membrane? Who controls cell differentiation and sequential task assignments during the construction of the physical body and the physical brain? Who holds the 3D blueprints for this thing? Who reports the defects in the different cells of the physical body and to whom do they report those defects? Who hands out the assignments and tells all the different molecules where to go during the routine maintenance of the physical body? Who is driving the bus, operating the computer, and running the show?

During my studies of Neuroscience, I discovered that neurons and glial cells migrate.

What?

Yes, neurons and glial cells migrate and aggregate! They also differentiate and become different types of neurons and glial cells with different types of functions or missions somewhere along the way. They also magically organize or aggregate into the appropriate Cerebral Processing Units (CPUs), when they reach their assigned destination.

They migrate! How do they do that?

The official answer is that evolution does it. Evolution does everything, including neural migration. Any fool knows that!

A fertilized egg [zygote] is totipotent, that is, the cell has the ability to develop into any class of cell in the body. However, after about 4 days of embryological development, newly created cells lose their totipotency and begin to specialize. At this stage, developing cells have the ability to develop into many, but not all, classes of body cells and are said to be pluripotent. As the embryo develops, new cells become more and more specialized. Eventually, new cells can develop into different cells of only one class. These new cells are said to be multipotent. Most developing cells will eventually become unipotent; they can develop into only one type of cell.

Why do stem cells have an almost unlimited capacity for self-renewal? This capacity results from the fact that when a stem cell divides, two different daughter cells are created: one that eventually develops into some type of body cell, and one that develops into another stem cell.

Migration

Once [neuron and glial] cells have been created through cell division in the ventricular zone of the neural tube, they migrate to the appropriate target location. During this period of migration, the cells are still in an immature form. Two major factors govern migration: time and location. In a given region of the tube, subtypes of neurons [and glial cells] arise on a precise and predictable schedule and then migrate together to their prescribed destinations.

This neural proliferation does not occur simultaneously or equally in all parts of the [neural] tube. Most cell division in the neural tube occurs in the ventricular zone – the region adjacent to the ventricle (the fluid-filled center of the tube).

Glia-mediated migration. This is one of two major modes of neural migration during development, by which immature neurons move out from the central canal along radial glial cells.

Neural tube. The tube that is formed in the vertebrate embryo when the edges of the neural groove fuse and that develops into the central nervous system.

Neural proliferation. The rapid increase in the number of neurons that follows the formation of the neural tube.

Radial glial cells. Glial cells that exist in the neural tube only during the period of neural migration, and that form a network along which radial migration takes place.

Radial migration. Movement of immature neural and glial cells in the developing neural tube in a straight line outward toward the tube's outer wall.** (*Biopsychology*, pp. 215-216, 485, 490, 493.)

This is probably the best physical explanation of neurodevelopment that I have found so far. Of course, it doesn't do anything to explain what's happening at the quantum level.

The neural grove folds and fuses to make the neural tube. To make a tube, I would create an opening in something that already exists, and then go from there. Someone Psyche CHOSE this method! We know, because it's always repeated. Who makes the edges of the neural grove move and fuse? What's their motivation? It's as if these cells are alive, know what they are doing, and why they are doing it. If this neural tube isn't done right, the brain is missing or spina bifida is the result. Genetic mutations or evolution can mess up this process severely.

It's my theory that Nature's Psyche always performs flawlessly; but, Nature's Psyche can't do anything about bad genetics and bad proteins. ONLY God can correct those kinds of problems at the quantum level.

Proper and correct differentiation and the sequential timing of that differentiation is what's truly miraculous in all of this – the same genes, but thousands of differences and trillions of different destinations. How is all that timing, differentiation, and migration information programmed into the genes? Well, it isn't. It's physically impossible to store petabytes of timing, differentiation, and migration logistics information within a 750-megabyte genome.

The ventricular zone is the region adjacent to the ventricle in the developing neural tube, and the zone where neural proliferation takes place. The ventricular zone is the place where neural cells and glial cells are born.

Differentiating stem cells into neurons and glial cells, then migrating those cells to a completely different location than their origin or birthplace, differentiating them again into different types of neurons and glial cells depending upon their chosen destination and purpose, and then the synaptogenesis that "wires" all these cells together into functional Cerebral Processing Units (CPUs) shows us the Hand of God in action. There's intelligence, sensors, telepathy, communication, action at a distance, breadcrumbs, pathways, purpose, invisible blueprints, schedules, telekinesis, growth, motion, planning, and organization behind this process – which is physically impossible!

I eventually started calling the different structures of the brain "Cerebral Processing Units", because all of my many different neuroscience books talked about there being "neural circuits" within these different structures or CPUs. Eventually I realized that our brain is comprised CPUs or Modules of Mapped Physical Functionality, especially when I finally learned that our brain has no memory engrams, RAM, or consolidated memory storage whatsoever.

The Materialists and Naturalists are satisfied with this explanation of neural development, but I wanted more.

Who motivates the glial cells to stretch out a wire from the ventricular zone to the tube's outer wall? Who tells them to do so and makes them do so? What's their motivation?

Who motivates the newly born neurons and glial cells to slither along the radial cells until they reach their assigned destination? Who chooses the specific radial cell that they are supposed to climb? Who decides what type of brain structure they are supposed to form when they reach their pre-chosen destination? Why don't the cells just sit there and vegetate under the effects of Brownian motion, like everything else in the physical world does?

Motivation requires command and control, which requires Psyche! Only God or Nature's Psyche could motivate these newborn cells to do things beside just sitting there and vegetating. Differentiation requires specificity and specialization. By definition, in principle, evolution (random mutations) doesn't do complex specificity and specialization. It can't. Differentiation and complex specificity and specialization are also functions of Intelligence or Psyche. Using God's Database, Nature's Psyche decides where these newly born brain cells are supposed to migrate and what these newly born brain cells are supposed to become and form when they get to their assigned destination.

If I were to give you thousands of newly born neurons and glial cells, could you form them into a fully functional thalamus or a hippocampus? Why not? Evolution can do it, so why can't you?

Columnar organization is the functional organization of the neocortex into vertical columns. The cells in each column form a mini-circuit that performs a single function. (*Biopsychology*, p. 481.)

He's talking about Cerebral Processing Units (CPUs) and Neural Circuits. He's NOT talking about Random Access Memory (RAM) or memory storage. He's talking about functionality, and NOT memory storage. The Materialists and Naturalists deliberately conflate the two, because they are trying to trick and deceive you; but, Mapped Functionality or CPUs are not the same thing as memory storage or RAM. I'm a computer scientist, so I KNOW how these things work. I can tell the difference between the two.

It's fascinating to leave the physical level behind and try to study the brain at the quantum level or atomic level. But, we need a good handle on the physical level, before we can start to explore and understand the quantum level.

According to Uncle Google: On the low end, it is estimated that there are 37.2 trillion cells in the human body. It is estimated that the human genome stores 6 billion bits or the 750 million bytes. The number of cell types within the human body is 200. It would take at least a byte of information storage capacity to store the differentiation information for each one of our cells. Can you see the problems multiplying here? I certainly can. I'm a computer scientist after all. I know how these things work!

Your body has many different kinds of cells. Though they might look different under a microscope, most cells have chemical and structural features in common. In humans, there are about 200 different types of

cells, and within these cells there are about 20 different types of structures or organelles. — Uncle Google.

There are 200 different types of cells in the human body. There's definitely enough information storage capacity in our genome to code for each unique type. But, there's definitely not enough information storage capacity in the human genome to code for the sequential typing information or differentiation information, birth addressing and timing, migration instructions, and eventual final location addressing, which means that there are times when the cells either have to fly blind, rely on chemical lures and guidance cells, or receive destination instructions and logistics from Nature's Psyche or God's Blueprint, and then take instructions telepathically and physically from neighboring cells after the migrating cells know that they have reached their final destination.

Who chooses what type of daughter cell a dividing stem cell is going to produce, and when to produce it – timing and location? A stem cell can become anything! Nature's Psyche or God's Blueprint is the only logical and realistic answer. There's not enough information storage capacity within our genome to handle all of this differentiation information, migration information, and synaptogenesis for the whole human body. Each daughter cell has to be assigned to one of the 200 different cell types. It's physically impossible to store 37.2 trillion bytes of differentiation information within the 6 billion bits or the 750 million bytes that our genome holds. Remember, it's physically impossible, which means that we are looking for something other than a physical explanation!

Whenever there is a physical explanation for any of these things, KNOW that it required Someone Psyche to design and create and map that physical functionality in the first place.

The problem with stem cell treatments is that the newly added stem cell doesn't always integrate and differentiate properly when implanted; and other times, the stem cells differentiate into something other than the desired type of cell. It's as if these stem cells have a mind of their own. They don't always do what we want them to do.

It's impossible to know where the genetic coding ends and Nature's Psyche and God's Blueprint begin.

Furthermore, the Materialists, Naturalists, Darwinists, Behaviorists, and Atheists are hoping that there's no such thing as God's Non-Local Blueprint or Nature's Psyche, because these people want to be able to control everything, which they won't be able to do if God is keeping some of these things to Himself.

Who chooses the appropriate target location for the migration of glial cells and neurons? They can theoretically go anywhere! Who organizes these migrating brain cells into a functional thalamus, striatum, hippocampus, cerebellum, amygdala, and cortices? The only logical answer is Nature's Psyche or God's Non-Local Blueprint. This requires some intelligence, command, control, coordination, and communication to get the job done! It is estimated that there are 86 billion neurons and 85 billion non-neuronal cells within the human brain. 3D location addressing takes up a ton of storage space, and so would the CPU processing needed to get the job done. At a conservative two bytes per dimension of the 3D address, we are looking at 6 bytes * 171 billion cells just to store the 3D addressing for the human brain, assuming that that's the way it is done. That's 1,026 billion bytes or a terabyte of information storage capacity needed to store the 3D migration addressing information within our physical genome. But, that's physically impossible, because our genome only stores 750 million bytes at most. So, where is all this migration data being stored? The only logical answer is that it's being stored non-locally in the cloud within God's Blueprint for the development and construction of the human body.

When the cells reach their assigned destination, they differentiate once again. I think this requires another byte of information per brain cell, or an additional 171 gigabytes of information shoved into our 750-million-byte genome, which is physically impossible.

Then synaptogenesis happens. YIKES!

It is estimated that there might be up to a quadrillion synapses within a new-born child's brain, and during that person's lifetime, they get pruned back to an estimated 0.15 quadrillion, or 150 trillion. At one quadrillion synapses and 6 bytes per synapse for location addressing with at least 2 more bytes for synapse typing and possibly 2 more bytes for timing information, we are looking for at least 10 petabytes of RAM in order to store all of that synaptic differentiation information within our 750-million-byte genome – assuming that God hasn't found a better way to do this thing and store this information. Obviously, it's physically impossible to store 10 petabytes within 750 million bytes. It can't be done.

The construction of our physical brain and its unique Cerebral Processing Units (CPUs) happens on a "precise and predictable schedule". It's physically impossible to store this precise and predictable schedule or blueprint within our genome, so where is it being stored? The only thing that makes logical sense is that it's being stored non-locally in the cloud within God's Database, and Nature's Psyche has direct access to this database. All of this happens automatically outside of our conscious awareness, so it has to be happening someplace else besides the Human Psyche and our physical brain.

So, let's recap.

The neurons are manufactured near the central neural tube, and then they migrate to their eventual destination. They can go anywhere after they are manufactured. Who tells the neurons where and when to proliferate along the neural tube, because neural proliferation does not occur simultaneously nor equally along the neural tube? Who tells each neuron where to go and what to become after it has been manufactured in the ventricular zone of the neural tube? Who controls the timing and the location of the totipotency, pluripotency, multipotency, and unipotency of cells? In other words, who controls the timing and the locations of cell differentiation, including the differentiation and logistics of neurons? Who or what keeps track of all this information and the logistics of this whole thing? There's nowhere near enough information storage capacity within our genome to control and decide such a thing!

These are legitimate scientific questions that need to be answered.

Evolution can't be doing it! The only time evolution (random mutations and natural selection) has any serious, potentially productive, and lasting effect on your physical body is when you are conceived (random genetic recombination of your gametes) and when you die (get selected against). The rest of your life, evolution really doesn't apply nor come into play for you. Evolution isn't Psyche, and Evolution is not psychic!

DNA isn't doing it either! We can't blame everything on our genes, although the Materialists, Naturalists, Atheists, and Darwinists certainly try to do so, simply telling us that we are dancing to our DNA. That's physically impossible; but, they don't know that, nor do they want to know it. Why is it physically impossible for us to be puppets simply dancing to our DNA? Let's go where the Materialists and Naturalists refuse to go.

What is the data storage capacity of the human genome?

The 2.9 billion base pairs of the haploid human genome correspond to a maximum of about 725 megabytes of data, since every base pair can be coded by 2 bits. – Wikipedia.

Of course, they keep adjusting the estimates for these things, so we have to go with what we currently have. I have seen estimates as high as 3.2 billion base-pairs, and I have tended to vacillate between the two extremes. I'll go with the conservative estimate for this round of calculations, since I went with a higher estimate during the previous round.

Many intelligent and observant scientists have done the math and concluded that there is nowhere near enough data storage capacity within the human genome to store ALL of the cell differentiation information, the different cell types, the task assignment and process information for the interaction between molecules and cells, the neurotransmitter assignments, the neuropeptide assignments, the 3D sequential differentiation blueprint for development and production of the physical body, and the 3D location addressing for the estimated 37.2 trillion to 100 trillion cells in the human body. There's nowhere near enough data storage capacity in DNA to coordinate and maintain all of this information nor to develop it all from scratch; yet, the Materialists, Naturalists, and Darwinists want you to believe that evolution (random mutations and natural selection) can handle all of this information on the fly at will, presciently, and all-knowingly. These people literally turn evolution into a God.

Again, what is the data storage capacity of the human genome?

The 2.9 billion base pairs of the haploid human genome correspond to a maximum of about 725 megabytes of data, since every base pair can be coded by 2 bits. – Wikipedia.

Even though DNA is the ultimate physical data-storage-device ever designed and developed in this physical universe, because DNA is comprised of physical matter and is digital code stored on large molecules, DNA is still a highly inefficient and limited means of storing data. 725 megabytes is nowhere near enough data storage capacity to store the 3D blueprint, the 3D location addressing, neurotransmitter types and locations, along with the sequential cell differentiation process that "evolution" needs to KNOW in order to produce an adult human body and physical brain from a zygote.

Neural proliferation and neural migration alone cannot be handled nor controlled by our genome. Imagine the logistics. Imagine 100 billion soldiers (neurons) landing on the beaches and the shores of the ventricular zone along the neural tube and then spreading out from there like an army to design and build our physical brain. Who or what is controlling the logistics of all of this? Each individual soldier has a mission and a position in the overall battle-plan. He knows where he is going, and he knows what he is going to be doing when he gets there.

Scientists estimate that the human nervous system has as many as 1,000 billion neurons – one trillion neurons. Imagine all of these soldiers landing on the shores and beaches of the ventricular zone of the central canal of your spinal cord and migrating from there to your fingers and toes and guts creating your central nervous system and peripheral nervous system along the way. How do they know how to do that? If you were one of these little soldiers, would you know how to play your part in the overall plan? And, who on earth is over-seeing and coordinating the actions, missions, locations, and purposes of these 1,000 billion or these one trillion soldiers who make up your nervous system? There's nobody on earth who could do such a thing – tell each neuron in your physical body where to go and what to do when it gets there. Would you know what to tell each neuron in your physical body – where to go and what to become when it gets there? Think about it logically. 2.9 billion base pairs in your genome is nowhere near enough information storage capacity to command and control the one trillion neurons that make up your nervous system – to tell each one where to go, what to become, how to migrate, how to grow, and what to do when it finally gets to the location it has been assigned.

The lowest number I have seen for the number of cells in the human body is 37.2 trillion. Imagine that you were given control of 37.2 trillion soldiers. Would you know what to do with them? Could you tell them where to go, what to become, and what to do when they finally get to the location that you have chosen for them? The Materialists, Naturalists, and Darwinists want you to believe that Evolution (random mutations and natural selection) knows what each of these 37.2 trillion soldiers are supposed to be doing and where they are supposed to be going. These people want you to believe that Evolution planned, designed, and now controls this whole thing. But, that's physically impossible. There's nowhere near enough information storage capacity within our genome to handle such a thing properly and efficiently.

The whole naturalistic process is further complicated by the fact that Evolution of any kind has NO psyche, NO intelligence, and NO brain. Evolution is dumb and blind by definition in principle. The final nail in the coffin comes from the fact that Evolution, Random Mutations, and Natural Selection didn't even exist until AFTER God designed and created the genomes, physical cells to house and protect those genomes, physical bodies, physical brains, and physical eyes in the first place.

If we truly want to KNOW where all of this information, knowledge, logistics, command, and control are being stored, we have to move beyond our paltry 2.9 billion base pairs towards something capable of storing the information, knowledge, and logistics needed to control 37.2 trillion soldiers to ensure that each one does his part, obeys orders, goes to or migrates to where it is assigned, becomes what it is told to become along the way, and doesn't mess up when he gets there. We need a few variables to tell the cell or soldier where and when to come into existence (at least 5: including 3D location, time of origination, and original cell type), where and when to migrate to its chosen destination (4), what to become when it gets there (1), where it currently exists in the potency hierarchy or differentiation hierarchy and when to change levels (2), and how to function when it finally differentiates, reaches unipotency, and becomes all that it can become (10, 15, 20, 100, 1000, potentially infinite because it's ongoing and dynamic and alive until the cell dies).

Low-balling it, we are easily looking at 1,000 trillion pieces of information or data that need to be handled and coordinated dynamically, intelligently, reliably, and efficiently in order to command and control the development and the ongoing operations of the human body. That's a lot of logistics! We need some kind of information storage platform capable of storing, implementing, controlling, organizing, discriminating, differentiating, and accessing this level of information. There's nothing like that within our physical body! There can't be. It's physically impossible. Our physical brain can't do it. Our physical brain needs its synapses for operations, not data storage. Our genome can't do it. You can't encode 1,000 trillion pieces of information in 2.9 billion base pairs. It's physically impossible. Our computers can't do it, because there's no way to interface our computers with our physical bodies and have our computers design and control our physical bodies; plus, our computers didn't exist billions or trillions of years ago when God designed and created our physical bodies.

Since the Human Psyche seems to be an infinite singularity, it's theoretically capable of an infinite amount of information storage capacity while at the same time having no physical size and taking up no physical space whatsoever. Physical matter severely limits information storage density; whereas, there are NO physical limitations in the Non-Local Spirit Realm, where our Psyche or Non-Local Consciousness originated and resides.

Many different scientists have concluded that the 3D Blueprint, the Cell Differentiation Sequence for producing a physical body, the neurotransmitter receptor locations and receptor typing, transcription control, translation control, activation control, replication control, protein synthesis, migration instructions, timing, and the 3D Location

Addressing for all the cells in the human body are stored non-locally in the Spirit Realm or Transdimensional Realm, because there is nowhere near enough storage capacity in the human genome to store that kind of information in a realistic manner. Or, there's an intelligent architect, engineer, and coordinator there in the non-local realm running the show for us.

There are many different theories and ideas that have been developed to explain where all of that information is being stored, since it can't be stored in DNA within the human genome. There's no way to store 1,000 trillion pieces of variable data in the 2.9 billion base-pairs of our genome! It's physically impossible. Imagine trying to command and control 1,000 trillion soldiers telling them when to act, where to act, how to interact with their environment, how to function internally, and what to do and become at the end of their migration and growth process. Nobody in this physical realm can do that! How would you communicate your orders to these 1,000 trillion soldiers or molecules in your physical body telling them where to go and what to do when they get there?

Uncle Google states:

In summary, for a typical human of 70 kg, there are almost $7*10^{27}$ atoms (that's a 7 followed by 27 zeros!) Another way of saying this is "seven billion billion billion." Of this, almost 2/3 is hydrogen, 1/4 is oxygen, and about 1/10 is carbon. These three atoms add up to 99% of the total!

The number of molecules per cell also has high variability but is estimated to be 23 trillion per cell.

23 trillion molecules per cell with a minimum of 37.2 trillion cells equals 855.6 trillion different variables, just to give each one of these molecules an assignment. This doesn't include any 3D addressing instructions needed to tell them where to go, nor does it include the information needed to explain to these molecules how to perform their assigned tasks when they get to where they have been assigned to go. We are easily looking at petabytes of RAM just to command and control the molecules in our physical body; but, there's no way to store that amount of information within our physical body! That's just data storage. Where's the programming stored, and how does it work? Where is all of this data, programming, and CPU functionality being stored? We might be looking at an exabyte of memory storage capacity needed to command and control the different molecules within our physical body. You can't shove it all into our genome. It's physically impossible.

An exabyte seems to be a magic number. You can theoretically command and control all off the molecules within our physical body with an exabyte of RAM and CPU functionality. Likewise, you should be able to store a lifetime's worth of ALL the memories and experiences and video that a human being typically acquires during the normal lifetime within an exabyte of RAM. You should be able to store all of our cell differentiation and developmental information within an exabyte of RAM and Blueprints. It looks like God has at least 3 exabytes of RAM and CPU functionality dedicated to each one of us, non-locally in the cloud. It's physically impossible to shove 3 exabytes of data into 750 million bytes of genome, as the Materialists and Naturalists are claiming is being done. Since it's physically impossible to store this amount of information within our genome, it's obvious and clear that we are looking for some other explanation besides a physical explanation.

Anytime we find something that's physically impossible, well that's when science really starts to get interesting!

I find all of this fascinating; but, the Materialists, Naturalists, Darwinists, and Atheists are going to find it frustrating, because it FALSIFIES Materialism and Naturalism.

These realities and truths effectively become Scientific Proof of God's Necessity, and therefore, Scientific Proof of God's Existence. They ain't gonna like that!

When it comes to this conundrum, Rupert Sheldrake's Morphogenetic Fields seems to be the theory that's receiving the most flak, ridicule, and suppression from the Materialists, Naturalists, Nihilists, and Atheists. They are trying to shut him down, but Sheldrake is a force of nature. He won't go into that dark night where the Materialists and Naturalists are trying to send him.

Rupert Sheldrake Amazon Page:

https://www.amazon.com/Rupert-Sheldrake/e/B000AQ3F38/

Forces and Fields by nature, in principle, are immaterial and non-local. Bosons, Gluons, Dark Energy or the Zero-Point Field, Gravitons, Photons, Magnetism, Light, the Quantum Zeno Effect, Quantum Tunneling, and Psyche are forces and fields. These things are NON-PHYSICAL, yet they affect and control the physical.

I have observed that if you want to find and KNOW the truth, then the best place to start is by reading and studying whatever the Materialists and Naturalists are fighting against the most. These people have a talent for finding, rejecting, and suppressing the Truth, Knowledge, Observational Evidence, Experiential Evidence, Eye-Witness Evidence, and Experimental Evidence that successfully falsifies Materialism and Naturalism. That's what these people do. Consequently, we do the opposite if we want to find and KNOW the truth. Elementary, my dear friend.

The "scientists" behind TED Talks tend to be Materialists and Naturalists. Study what these people are trying to ban and block.

https://www.youtube.com/watch?time_continue=1&v=4gFi285OhrQ

https://www.youtube.com/watch?v=s42vuf0ahU8

http://highexistence.com/teds-controversy-3-threatening-talks-they-tried-to-censor/

These people tend to ban, censor, and block anything that falsifies Materialism, Naturalism, and Atheism calling it unscientific or creationist. That's what these people do. That's how they do science – form a committee, vote it out of existence, and censor it into extinction. We do the opposite, if we want to find and KNOW the truth.

McTaggart, L. (2002). *The Field: The Quest for the Secret Force of the Universe*. New York: HarperCollins.

Lynne McTaggart interviewed over 70 different scientists whom the Materialists, Naturalists, and Darwinists have been ridiculing, censoring, banning, and blocking – many of them are Quantum Scientists. If you want an introduction to all the different things that the Materialists and Naturalists are banning and blocking, *The Field* is a good place to start. After all, if you truly want to find and KNOW the truth, then you start by reading and studying the science and the knowledge and the evidence, which the Materialists and Naturalists are trying to hide and eliminate.

It works. Try it and see if I'm right, or not.

Memories and thoughts are quantum waves. In her book, *The Field*, McTaggart called quantum waves the "zero-point field". It's the same thing. Quantum waves are vibrating strings. Quantum waves of one type or another are the zero-point field of light, the quantum sea of light, the Holy Spirit, the Light of Christ, Nature's Psyche, God's Psyche, and the Human Psyche. They each tend to operate on a different frequency and they each have different abilities and assignments, but they are all quantum waves of some kind. Quantum waves can be instantaneously and simultaneously everywhere all at once.

If the Materialists, Naturalists, Darwinists, Scientists, Behaviorists, Determinists, Materialistic Reductionists, Evolutionary Psychologists, Nihilists, and Atheists are trying to BURN something down, I want to take a look and see what they are trying to censor, ban, and suppress; and then, make sure I get a copy of it saved onto my hard drive in case they succeed.

I want evidence. Lots and lots of evidence. I'm no longer hiding from the evidence. Been there and done that, when I was a Materialist, Nihilist, and Atheist. It's boring and depressing, and I don't want to go back.

Richard Dawkins: "Darwin made it possible to be an intellectually fulfilled atheist."

Actually, NO.

While I was a Materialist and Atheist, I still had lots of doubts; and, I could sense all the time that there were lots of intellectually stimulating theories and ideas which I was refusing to look at and explore.

Richard Dawkins is satisfied with his ignorance. I was not. Consequently, I changed, and Richard Dawkins did not. Nevertheless, Richard Dawkins is wealthy and famous, and I am not, because Richard Dawkins is telling the Materialists and Atheists what they want to hear, and I am not. It is what it is.

I'm no longer hiding from evidence, and I no longer have any doubts. I KNOW that I have found the truth, because I went looking for it. Sow, and ye shall reap. Seek and ye shall find. Knock, and it shall be opened unto you. Read, study, and learn; and, you will no longer be an intellectually fulfilled atheist. That's just the way it is.

Supernatural Intelligence

We have observed that there is NO randomness where our neurotransmitters are concerned, contrary to the claims of Naturalism and Classical Physics.

When it comes to the undirected neurotransmitters that are randomly diffused throughout the cerebral spinal fluid and have to be maintained at a certain level for proper functioning of our physical brain, it soon becomes clear to us that the ionotropic channel receptors KNOW when they need a neurotransmitter to start an action potential or to perpetuate an action potential; and, they KNOW when they don't.

Whenever the ionotropic receptors and metabotropic receptors KNOW or DECIDE that they need one of the randomly diffused neurotransmitters floating in the synaptic cleft, the receptor reaches out telepathically and telekinetically and draws one in or even potentially quantum tunnels and teleports one in; and, when one of these receptors KNOWS that it doesn't need a neurotransmitter, then it literally blocks and prevents one of these randomly diffused neurotransmitters from randomly and accidentally binding with its receptor and thus accidentally triggering an unwanted action potential.

Furthermore, when one of these ion channels reaches out and grabs the necessary neurotransmitter floating nearby, the ionotropic receptors controlling the ion channels actually choose and decide how long to keep that neurotransmitter bound to its receptor and thus how long to keep the ion channel open.

The level of intelligence and telepathic information exchange within those ionotropic and metabotropic receptors is astounding, but it's nothing in comparison to the little machines and invisible forces that create the messenger RNA from a specific gene (transcription of the mRNA), that transport the messenger RNA to the ribosomes (telekinesis of the mRNA from the nucleus to the ribosome), that assemble the protein (the ribosomes), and that transport the amino acids to the ribosomes for protein synthesis (telepathically and telekinetically controlled transfer RNA or tRNA) .

Proteins are assembled from amino acids using information encoded in genes. Each protein has its own unique amino acid sequence that is specified by the nucleotide sequence of the gene encoding this protein. The genetic code is a set of three-nucleotide sets called codons and each three-nucleotide combination designates an amino acid, for example AUG (adenine-uracil-guanine) is the code for methionine. Because DNA contains four nucleotides, the total number of possible codons is 64; hence, there is some redundancy in the genetic code, with some [of the 20] amino acids specified by more than one codon.

Genes encoded in DNA are first transcribed into pre-messenger RNA (mRNA) by proteins such as RNA polymerase. Most organisms then process the pre-mRNA (also known as a primary transcript) using various forms of Post-transcriptional modification to form the mature mRNA, which is then used as a template for protein synthesis by the ribosome. In prokaryotes the mRNA may either be used as soon as it is produced, or be bound by a ribosome after having moved away from the nucleoid. In contrast, eukaryotes make mRNA in the cell nucleus and then translocate it across the nuclear membrane into the cytoplasm, where protein synthesis then takes place. The rate of protein synthesis is higher in prokaryotes than eukaryotes and can reach up to 20 amino acids per second.

The process of synthesizing a protein from an mRNA template is known as translation. The mRNA is loaded onto the ribosome and is read three nucleotides at a time by matching each codon to its base pairing anticodon located on a transfer RNA molecule, which carries the amino acid corresponding to the codon it recognizes. The enzyme aminoacyl tRNA synthetase "charges" the tRNA molecules with the correct amino acids. The growing polypeptide is often termed the nascent chain. Proteins are always biosynthesized from N-terminus to C-terminus.

https://en.wikipedia.org/wiki/Protein#Synthesis

http://mypsyche.us/wp-content/uploads/2017/12/Protein.pdf

The amount of intelligence, knowledge, telekinesis, and telepathic control that's needed for Protein Synthesis is infinite, especially when one starts computing the odds of Protein Synthesis happening randomly through blind luck alone. Protein Synthesis through random chance is impossible. It can't be done. These molecules know precisely what they are doing and why they are doing it.

The amount of invisible telepathic information exchange taking place for the transcription of mRNA from DNA is simply staggering; and, all the texts simply gloss over it, mechanize it, and effectively dismiss it as if it doesn't exist.

Transcription

The DNA is "unzipped" (disruption of hydrogen bonds between different single strands) by the enzyme helicase, leaving the single nucleotide chain open to be copied. RNA polymerase reads the DNA strand from the 3-prime (3') end to the 5-prime (5') end, while it synthesizes a single strand of messenger RNA in the 5'-to-3' direction. The general RNA structure is very similar to the DNA structure, but in RNA the nucleotide uracil takes the place that thymine occupies in DNA. The single strand of mRNA leaves the nucleus through nuclear pores, and migrates into the cytoplasm.

https://en.wikipedia.org/wiki/Protein_biosynthesis

http://mypsyche.us/wp-content/uploads/2017/12/Protein_biosynthesis.pdf

Genetic Code and Gene Expression

Structural genes contain the information necessary for the synthesis of proteins. Proteins are long chains of amino acids; they control the physiological activities of cells and are important components of cellular structure.

Enhancers are stretches of DNA whose function is to determine whether particular structural genes initiate the synthesis of proteins and at what rate. The control of gene expression by enhancers is an important process because it determines how a cell will develop and how it will function once it reaches maturity. Enhancers are like switches because they can be regulated in two ways: They can be turned up, or they can be turned down.

Proteins that bind to DNA and influence the extent to which genes are expressed are called transcription factors. Many of the transcription factors that control enhancers are influenced by signals received by the cell from its environment.

The small section of the chromosome that contains the [desired] gene unravels. The DNA molecule partially unravels, exposing the structural gene that is to be expressed. A strand of messenger RNA is transcribed from one of the exposed DNA strands and carries the genetic code from the nucleus into the cytoplasm of the cell. In the cytoplasm, the strand of messenger RNA attaches itself to a ribosome. The ribosome moves along the strand translating each successive codon into the appropriate amino acid, which is added to the lengthening protein by a molecule of transfer RNA. When the ribosome reaches the end of the messenger RNA, a codon instructs it to release the completed protein. (*Biopsychology*, p. 39.)

How did the transcription factors learn to read and understand the signals received by the cell from its environment? How do the enhancers know what stretch of DNA they are supposed to oversee, control, turn up, or turn down? How do the transcription enzymes know where the desired structural gene is located? How does the DNA know what part of itself it is supposed to unravel? How is all of this information being exchanged wirelessly at a distance? How are signals between the transcription factors and the environment transmitted and received? Who taught these transcription factors how to read and understand these signals? How does the messenger RNA know where it's supposed to go? What motivates the mRNA to bind with the ribosome? How does each amino acid in the sequence KNOW that its attached codon is the next one to be processed by the ribosome? The amino acids seem to line up perfectly before they ever reach the ribosome. How do they do that? How do they know which ribosome they are supposed to go to and interact with? After the protein has been successfully synthesized and the gene successfully expressed, how does that protein know where it's supposed to go and what it's supposed to do when it gets there? How does the protein get there? How do all of these things know where they are going and what they are supposed to do when they get there?

There's massive amounts of intelligence, information, and telepathic communication taking place between these molecules wirelessly at a distance throughout this whole process of protein synthesis. Remember, telepathy is WiFi at the quantum level.

The DNA is unzipped by the enzyme helicase or transcription enzyme. It's direct, purposeful, and intelligent. The enzyme helicase doesn't parse the whole genome sequentially looking for the gene it needs, unzipping the whole genome as it goes along. Those who do computer programming will recognize that it's not "sequential access method" that's used, but what's called the random-access method or more accurately an indexed sequential access method (ISAM) that this enzyme helicase or transcription enzyme uses to locate the specific gene it needs.

This enzyme helicase or transcription enzyme, which unzips the DNA, KNOWS that the cell needs a specific protein and KNOWS precisely what part of the DNA needs to be unzipped. How does it KNOW? Someone is telling this enzyme helicase that the cells need a specific protein, and this Someone Psyche tells the enzyme helicase precisely where the needed gene is located. There's a great deal of invisible telepathy, intelligence, and information exchange taking place here.

How great?

Well, there's a one-to-one correlation with the Human Genome Project.

The Human Genome Project (HGP) was an international scientific research project with the goal of determining the sequence of nucleotide base pairs that make up human DNA, and of identifying and mapping all of the genes of the human genome from both a physical and a functional standpoint. It remains the world's largest collaborative biological project. After the idea was picked up in 1984 by the US government when the planning started, the project formally launched in 1990 and was declared complete in 2000.

https://en.wikipedia.org/wiki/Human_Genome_Project

It took an international scientific research team ten years to map the human genome; and, we still don't know what most of the individual genes actually do. Think about it!

That little enzyme helicase KNOWS precisely where every gene is located in the human genome, and it KNOWS precisely what each and every gene does; or, that enzyme helicase has access to God's database and God's ISAM file mapping the entire human genome. That little enzyme helicase is billions if not trillions of years more intelligent and technologically advanced than we human beings are, with our paltry human genome project compared to its direct access to the Mind of God.

So, these proteins unzip and construct a messenger RNA (mRNA).

Then that "single strand of mRNA leaves the nucleus through nuclear pores, and it migrates into the cytoplasm".

What motivates it, or moves it, or migrates it?

355

How does the mRNA know where it's going? It doesn't just sit there floating around randomly as the Materialists and Naturalists claim. It KNOWS precisely where it is going, and it proceeds to do so. It KNOWS where those ribosomes are, and it KNOWS that it has to get there, and it what? It apparently proceeds to swim there telekinetically. There are many different ribosomes. How does it know which one to go to? Or, how does it decide? How does the mRNA molecule know where the ribosome is located?

"The mRNA is loaded onto the ribosome."

How?

By whom?

Once the mRNA docks with its chosen ribosome, then additional magic takes place.

Activation and Translation

The synthesis of proteins from RNA is known as translation. In eukaryotes, translation occurs in the cytoplasm, where the ribosomes are located. Ribosomes are made of a small and large subunit that surround the mRNA. In translation, messenger RNA (mRNA) is decoded to produce a specific polypeptide according to the rules specified by the trinucleotide genetic code. This uses an mRNA sequence as a template to guide the synthesis of a chain of amino acids that form a protein. Translation proceeds in four phases: activation, initiation, elongation, and termination (all describing the growth of the amino acid chain, or polypeptide that is the product of translation).

In activation, the correct amino acid (AA) is joined to the correct transfer RNA (tRNA). While this is not, in the technical sense, a step in translation, it is required for translation to proceed. The AA is joined by its carboxyl group to the 3' OH of the tRNA by an ester bond. When the tRNA has an amino acid linked to it, it is termed "charged". Initiation involves the small subunit of the ribosome binding to 5' end of mRNA with the help of initiation factors (IF), other proteins that assist the process. Elongation occurs when the next aminoacyl-tRNA (charged tRNA) in line binds to the ribosome along with GTP and an elongation factor. Termination of the polypeptide happens when the A site of the ribosome faces a stop codon (UAA, UAG, or UGA). When this happens, no tRNA can recognize it, but releasing factor can recognize nonsense codons and causes the release of the polypeptide chain.

https://en.wikipedia.org/wiki/Protein_biosynthesis

http://mypsyche.us/wp-content/uploads/2017/12/Protein_biosynthesis.pdf

Once again, they just gloss over what's really happening.

There are many different ribosomes. How does the tRNA molecule know which one to go to? Or, how does it decide? How does the tRNA molecule know where the ribosome is located? How does this information get exchanged?

If you watch animations of Protein Synthesis, you observe that the tRNA molecules KNOW that they have an appointment with some distant ribosome, and they actually line up in the correct order BEFORE they arrive at the ribosome for the translation process.

How do they do that?

"The enzyme aminoacyl tRNA synthetase charges the tRNA molecules with the correct amino acids."

How does this enzyme KNOW what the correct amino acids are? It has twenty amino acids to select from, and it has 64 possible codons on the upcoming mRNA molecule that it has to match the amino acids to. This process is called activation. This enzyme has to KNOW precisely which amino acid to match with which codon long before the tRNA molecule ever arrives at the ribosome for matching with the mRNA molecule that is docked with the ribosome.

Who or what activates the tRNA molecules, and gets them moving in the correct sequence towards the correct ribosome? How do the tRNA molecules know to which ribosome they should go?

How is that information exchanged?

It's all done through WiFi, telepathically. There is no other logical explanation. We are dealing with psyche, intelligence, quantum mechanics, foresight, quantum telekinesis, and quantum telepathy here – not classical physics, naturalism, and materialism. There's nothing blind or random about any of this. The molecules involved in protein synthesis are intelligent, sentient, and fully aware of what they are supposed to be doing, where, and when.

Somehow, those aminoacyl tRNA synthetase enzymes and the resulting tRNA molecules actually read the mind or the psyche of mRNA molecule; and then, the tRNA molecules and the activation enzymes KNOW IN ADVANCE that they have to grab specific amino acids which match with specific codons on the mRNA molecule, and then after activation make their way to the specific ribosome where the mRNA molecule will eventually dock. Those customized tRNA molecules carrying the correct amino acids and the correct codons arrive at the correct ribosome in the correct sequential order, and the protein is assembled quickly and efficiently. There's no random sampling and Brownian motion going on, according to the animations of this process. Each tRNA molecule KNOWS precisely where it is going, and they all line up in the correct sequence and arrive at the ribosome accordingly, like clockwork.

Imagine the amount of information exchange that would be needed to get 27,000 college students to line up perfectly in alphabetical order; and suddenly, you have some semblance of what it takes to get 27,000 amino acids and tRNA molecules to line up perfectly with the correct codons on the target mRNA molecule in the right order, for the assembly of one of the largest proteins on record. What are the odds of all that happening randomly by pure luck and blind chance, as the Materialists and Naturalists claim that it was done? It's impossible. The odds against it happening by random mutations and natural selection alone are so astronomical as to be literally infinitely impossible. It can't be done. Just like you can't get 27,000 college students to line up in alphabetical order by random mutations, natural selection, blind luck, happenstance, Brownian motion, random diffusion, genetic drift, macro-evolution, abiogenesis, spontaneous generation, and NO information exchange whatsoever!

Yet, the Materialists, Naturalists, Darwinists, Nihilists, and Atheists really truly want you to believe that Evolution (random mutations and natural selection) understands all of this, planned all of this, designed all of this, engineered all of this, created all of this, now controls all of this, and continues to upgrade and improve all of this. It's impossible; but, these people desperately want you to believe in the impossible; and, millions of people do.

I used to be a Materialist, Naturalist, Nihilist, and Atheist. We Materialists and Naturalists refuse to think about these kinds of things; and, when we finally do think about these kinds of things, we quickly realize that we are no longer Materialists, Naturalists, Darwinists, Nihilists, and Atheists. It took me fifty years, but I got there in the end.

Imagine what it would take to get 27,000 college students to line up perfectly in alphabetical order and suddenly you realize what it takes to get 27,000 amino acids and tRNA molecules to line up in perfect order as they approach the ribosome, the mRNA molecule, and the codons during Protein Synthesis. It can't happen randomly through blind luck, random mutations, natural selection, and trillions of years of evolution. It happens dynamically, presciently, telepathically, telekinetically, and perfectly in real-time instead.

Can you imagine how long it would take for the 27,000 tRNA molecules to put together a protein that is 27,000 amino acids long if Materialism and Naturalism were true and this whole process was happening blindly and randomly as the Materialists and Naturalists claim that it is? It would NEVER happen! It's physically impossible.

So, let's give them a few things. Can you imagine how long it would take for the 27,000 tRNA molecules to assemble a protein, if they had to try each of the possible 64 codons sequentially until they found the right codon, for each and every amino acid in the sequence? I take 64 – 4 eliminating some of the starting and ending codons that they technically don't have to try, then divide by two to bring averaging and luck of the draw into the equation; and, I end up with 30 X 27,000 which is 810,000 attempts on average. If it took a tenth of a second each attempt, it would take on average 81,000 seconds for Materialism and Naturalism to try all 64 codons sequentially looking for the correct codon and the right amino acid as the Materialists and Naturalists typically claim that the process is done. That's 1,350 minutes per protein on average, or 22.5 hours.

The reality is much faster because the tRNA molecules know precisely what they are doing, even though it still takes quite a bit of processing time for the ribosome to assemble the protein.

According to Uncle Google:

Many ribosomes can be working on a single strand of mRNA at once. Protein synthesis isn't a slow process, either. A protein chain 400 amino acids long can be assembled in 20 seconds!

The rate of protein synthesis is higher in prokaryotes than eukaryotes and can reach up to 20 amino acids per second.

Consequently, at best speed, a single protein 27,000-amino-acids-long would take 1,350 seconds to build (27,000 divided by 400 times 20 seconds). That's 22.5 minutes.

The only way that ribosomes could be working together on a single strand of mRNA is if they are communicating with each other telepathically sharing with each other their starting and end points in the chain.

I'm not the first person to run the numbers and to come to the conclusion that it is impossible to create a functional protein through chemical evolution and random chance.

Origin: Probability of a Single Protein Forming by Chance:

https://www.youtube.com/watch?v=W1_KEVaCyaA

https://www.youtube.com/watch?v=cQoQgTqj3pU

https://www.youtube.com/watch?v=X7VqomfivC0

I, too, have run the numbers, and have concluded that the Theory of Evolution is impossible. There's no way in this physical universe that evolution (genetic drift), chemical evolution, random mutations, natural selection, abiogenesis, and spontaneous generation could have designed and created life from scratch. It's impossible. These things can't even create a single functional protein by chance, so there's no way that they can create a functional genome and a functional life form. It's impossible. The odds against are so astronomically high that it becomes synonymous with impossible – the limit moves to zero, and NO CHANCE WHATSOEVER is the ultimate result. That's what the math is telling us, and I believe the math infinitely more than I believe the wishful thinking of the Materialists and Naturalists and Darwinists.

Random Mutations and Natural Selection cannot design and create genomes. In fact, evolution (genetic drift), random mutations, and natural selection didn't even exist until AFTER God designed and created the genomes in the first place. Furthermore, the chemical evolution of genomes by blind random chance is even more impossible than protein synthesis by blind random chance, because ALL of the proteins used in the human body are encoded in our genome.

As it is, Protein Synthesis is radically advanced – far beyond anything that the Materialists and Naturalists are willing to imagine, understand, or accept. All the activation enzymes, who are working together binding amino acids to transfer RNA and attaching the correct reverse codons to transfer RNA, KNOW precisely which mRNA molecule they are building for and KNOW precisely which amino acid and codon is next in the sequence. The activation enzymes have a hive-mind and literally KNOW in advance at a distance which of the 20 different amino acids is next in the sequence and which of the 64 possible codons is next in the sequence, for the specific mRNA molecule and protein that they are building for and preparing for. The 27,000 completely different tRNA molecules literally line up in alphabetical order like 27,000 college students lining up in alphabetical order in preparation to receive their diplomas; and, the transfer RNA are processed through the ribosome in an orderly clockwork fashion as if they KNOW precisely what they are doing.

This is science which the Materialists, Darwinists, and Naturalists formally and officially reject; yet, these very same people assure us that evolution, random mutations, and natural selection understand these things, planned and organized these things, and can do these things even though Nature's Psyche, the Human Psyche, and God's Psyche cannot – because they don't exist. According to these people, evolution is infinitely smarter and infinitely more capable than human beings, and evolution has complete control over all of these physical processes. Evolution is their god – they imbue it with the powers and prescience of a god, and they worship it and idealize it and idolize it as if it were God. But, WE KNOW for a fact that evolution of any kind cannot do design, creation, genomes, programming, organization, and protein synthesis because evolution is dumb and blind. There's no intelligence there.

Materialism, Naturalism, and Darwinism are irrational, contradictory, illogical, inconsistent, inefficient, and self-defeating. I used to be a Materialist, Naturalist, Nihilist, and Atheist. I could see it and be it, but that doesn't mean that it is real and true and functional. How long do you think it would take 27,000 college students to line themselves up alphabetically, after being given the assignment to do so? Could they do it in 22 minutes, or would it take them 22 hours? How many errors would they make in the process? There's supposed to be some intelligence there, after all; so theoretically, it can be done – unlike the Theory of Evolution which by definition in principle would make this kind of process impossible to accomplish.

Could 400 college students align themselves alphabetically in 20 seconds, while trying to model the efficiency of protein synthesis? How long would it take evolution, random mutations, and natural selection to align these 400 students alphabetically? How long would it take a single college student to find the correct gene in the human genome matching the precise protein that you want him to make? The transcription enzyme or transcription factor goes directly to the gene it wants, unzips it, and produces the mRNA molecule needed to get the job done. How does it do that? This is really cool stuff to think about; but, only when you are ready for it and are ready to take off the blinders which the Materialists and Naturalists have given you to wear.

This is where the tires hit the pavement.

The amount of intelligence and telepathic information exchange required in order to accomplish protein synthesis is simply staggering. There's nothing random and willy-nilly about it. In fact, it has been proven scientifically that random chance and chemical evolution cannot do protein synthesis, even if given an infinite amount of time to do so. Again, how long would it take evolution, random mutations, and natural selection to align 400 college students alphabetically? It ain't never gonna happen! It's impossible!

Uncle Google states:

The nuclear genome comprises approximately 3,200,000,000 nucleotides of DNA, divided into 24 linear molecules, the shortest 50,000,000 nucleotides in length and the longest 260,000,000 nucleotides, each contained in a different chromosome.

3 billion base pairs equal 6 billion nucleotides.

How long would it take evolution, random mutations, and natural selection to align 3,200,000,000 nucleotides in the correct order? Face reality! It ain't never gonna happen, because it's impossible! How long would it take 3.2 billion college students to line up alphabetically in the correct order? It's theoretically possible, because college students are supposed to be intelligent and semi-efficient; but, for all practical purposes, such a thing is physically impossible due to the logistics. Imagine evolution, random mutations, and natural selection trying to line up 3.2 billion college students alphabetically in the correct order; then, you have some idea of what we are really dealing with here.

What are the odds of a single college student, or even a group of college students, going directly to the gene which encodes the protein that I want them to build? Would they know how to build that protein after they found the right gene? How long could this group of college students keep going straight to the correct genes in the human genome building one correct protein after another whenever asked to do so? How long would it take evolution, random mutations, and natural selection to do such a thing – select the right gene and produce the correct protein that is needed?

There are now recently discovered indications that the "junk DNA" contains backup copies of some essential genes for essential proteins just in case the original genes are damaged or go missing. How do these transcription enzymes know where the backup copies are located, when we human beings don't have a clue that they even exist? How do the transcription enzymes know that a gene is broken and that they need to acquire a better copy if they can? There's way more knowledge and intelligence within transcription enzymes, neurotransmitter receptors, and allocation enzymes than what we human beings have access to.

The transcription enzymes, which go directly to the correct gene and produce the needed mRNA molecule, are either infinitely more intelligent than we human beings are or have access to infinitely more knowledge and information and intelligence than we human beings do. Furthermore, the transcription enzymes also KNOW which gene is dominant and which gene is recessive, whenever they are dealing with heterozygous alleles. They KNOW which gene or allele is supposed to be the best and most reliable gene. How do they do that? There's nothing random, slow, gradual, hunt-and-peck, or step-by-step about it as the Materialists, Naturalists, and Darwinists claim it should be. The very efficiency and reliability of transcription enzymes and allocation enzymes falsifies Materialism, Naturalism, and Darwinism. Transcription enzymes, allocation enzymes, and neurotransmitter receptors are OBSERVED SCIENCE which successfully falsify the claims of Materialism and Naturalism, and actually prove that Materialism, Naturalism, and Darwinism are false.

I once had an atheist tell me to stop what I am doing – poking fun of her religion in public. Am I guilty as charged? Am I successfully making fun of and successfully ridiculing my former religion? I used to be a Materialist, Naturalist, Nihilist, Skeptic, Critic, Doubter, and Atheist after all – I was rude, self-centered, disrespectful, ungrateful, offensive, mean-spirited, and crude just like most of these people typically are.

Have I successfully debunked these things? I have to my satisfaction; but, you are a completely different member of the jury. What do you think? Do I have a point?

Others have told me that I'm brutal, genius, with an incisive mind that cuts straight to the core of the issue. Am I? Or am I just simply blowing smoke up your tailpipe like the Materialists and Naturalists and Darwinists are doing to you? Only you can decide for yourself what you truly want to believe, because you are the only one that you are really truly going to believe after all. You are going to have to decide for yourself if I know what I'm talking about, or not. You are going to have to decide for yourself if I'm trying to trick you and deceive you like the Materialists, Naturalists, and Darwinists are trying to trick you and deceive you.

Our beliefs are chosen; and, only you can choose to believe what it is that you want to believe. The very best things in life such as friendship, charity, compassion, mercy, forgiveness, and love are chosen into existence and believed into existence. These things don't exist until the human psyche wills them into existence, or chooses them into existence, or believes them into existence. You, your psyche, decides whether you are going to love someone or hate that person.

Repeatedly, protein synthesis is portrayed as the very model of efficiency – there's choice and intelligence involved in protein synthesis; and, the only way that is possible is if Nature's Psyche is communicating directly with Nature's Psyche, and with God's Psyche or God's Intelligence or God's Database as well. This is radical science far beyond the fringe. This is trillions of years ahead of anything that the Materialists and Naturalists think about.

The transcription enzymes, activation enzymes, ribosomes and/or translation enzymes, mRNA molecules, tRNA molecules, and neurotransmitter receptors in our physical brain KNOW precisely what they are doing and why they are doing it and how to do it. They are infinitely more intelligent than we are and have access to infinitely more information and knowledge than we do. It all happens through WiFi, telepathically and telekinetically, like precision clockwork. Over and over again, it is described as the very model of efficiency; and it is, because these enzymes and receptors are intelligent, telepathic, sentient, alive, knowledgeable, and aware. These things function in synchronicity like a hive-mind. Have you ever seen murmurations? Now imagine a murmuration at the quantum level throughout the whole of protein synthesis from beginning to end. Now you are starting to see how things really work.

https://www.wired.com/2011/11/starling-flock/

What are the odds of evolution, random mutations, and natural selection doing what these starlings are doing? What are the odds of evolution, random mutations, and natural selection doing protein synthesis? There no chance at all. It's impossible for evolution, random mutations, and natural selection to do protein synthesis, because there's no intelligence there.

These little machines or enzymes are infinitely more intelligent than we human beings are, and/or these little machines have direct access to God's Intelligence and God's Database. There's no other logical explanation for what we observe them doing. There's nothing random about it. Once again, Classical Physics, Materialism, and Naturalism FAIL to explain what's really happening where transcription, activation, translation, neurotransmission, genomes, design, creation, teleology, programming, and protein synthesis are concerned. These processes only make sense in terms of Psyche, Intelligence, Non-Local Consciousness, Non-Locality, Quantum Telepathy, Quantum Telekinesis, Quantum Tunneling, and Quantum Mechanics. Classical Physics completely FAILS to explain what's really going on.

Yes, we can explain what's happening physically in terms of classical physics – these things move here and there, and they bind with this and that. But, we can't explain the telepathic information exchange of knowledge that is taking place at a distance – between the helicase transcription enzyme and the genome, or between the tRNA activation enzymes and the future mRNA codons which the tRNA molecules will be docking with at the ribosome – in terms of Classical Physics, Materialism, and Naturalism. Protein synthesis is way too efficient and way too perfect for it to be happening through random diffusion, Brownian motion, and hunt-and-peck mechanisms as the Materialists and Naturalists assure that it's happening.

When it comes to the fore-knowledge of these transcription, activation, and translation enzymes and the telepathic information exchange of knowledge taking place invisibly and wirelessly between them, Classical Physics, Materialism, and Naturalism completely FAIL to explain what's really going on between these things – despite the claims of the Materialists and Naturalists to the contrary.

We have to switch over to Psyche, Intelligence, WiFi, God's Database, and Quantum Telepathy in order to successfully explain what we are observing – especially when it comes to ionotropic neurotransmitter receptors, ion channels, neurotransmitters, neuropeptides, transcription enzymes, activation enzymes, translation enzymes, mRNA molecules, tRNA molecules, and ribosomes. These enzymes and neurotransmitter receptors function too perfectly and too reliably and too quickly to be operating exclusively according to the limited rules of Classical Physics, Entropy, Random Diffusion, Genetic Drift, Random Mutations, Natural Selection, Brownian motion, Materialism, Darwinism, and Naturalism.

Remember, telepathy is WiFi at the quantum level.

When it comes to these enzymes, neurotransmitter receptors, messenger molecules, and transfer molecules, there is syntropy and telepathy and telekinesis going on here, not entropy and not the second law of thermodynamics. These enzymes, messenger molecules, and neurotransmitter receptors are alive, intelligent, prescient, sentient, aware, and KNOW precisely what they are doing, why they are doing it, and when they should do it. They function way too perfectly and way too reliably and way too presciently to be otherwise than sentient, intelligent, alive, conscious, aware, and psychic. The spirit or psyche within these things KNOWS precisely what it's doing and why it's doing it, even if we don't.

Observe that Classical Physics, Materialism, and Naturalism can only explain the physical results of these unseen forces and unseen processes, not the actual processes themselves.

In terms of explanatory power, the Quantum Physicists and Spiritualists are trillions of years ahead of the Materialists, Naturalists, and Atheists, because Quantum Physicists and Spiritualists have infinitely more to work with. Neuro-development, or the construction of our nervous system, makes no sense whatsoever in terms of Materialism and Naturalism, because there is way too much information exchange, command, and control taking place at the quantum level to be explained away successfully and effectively by Materialism and Naturalism.

Neuro-Development

When it comes to neural development and the construction of our nervous system, God is hiding in plain sight where nobody can seem to see Him.

The development of the neural plate is induced by chemical signals from an area of the underlying mesoderm layer – an area consequently referred to as an organizer.

Once the lips of the neural groove have fused to create the neural tube, the cells of the tube begin to proliferate (increase in number). This neural proliferation does not occur simultaneously or equally in all parts of the tube. Most cell division in the neural tube occurs in the ventricular zone – the region adjacent to the ventricle (the fluid-filled center of the tube).

In each species, the cells in different parts of the neural tube proliferate in a particular sequence that is responsible for the pattern of swelling and folding that gives the brain of each member of that species its characteristic shape.

The complex pattern of proliferation is in part controlled by chemical signals from two organizer areas in the neural tube. (*Biopsychology*, p. 215.)

Notice the anthropomorphization and deification of all these different processes and his description of them. He imbues them with intelligence. He has no choice but to do so, because these individual cells KNOW precisely what they are doing and why they are doing it.

An area of cell development called an Organizer. Only Psyche can do organization, intelligence, planning, and fine-tuning. Whose psyche did all of the organization and fine-tuning BEFORE the first genomes and first life forms and first brains were designed and created? Who chose what chemical signals to emit and when to emit them? Who taught the developing cells how to recognize those chemical signals and grow towards them, rather than growing randomly as the Materialists and Naturalists claim that everything is done?

Each species has a different sequence of cell differentiation and cell growth. Where is the information for this sequence stored? A 3 billion base-pair genome has nowhere near enough information storage capacity to store the organizational sequence for 37.2 trillion cells. Someone is telling these proliferating cells specifically where to start growing, because they do not grow simultaneously nor equally in all parts of the neural tube. Someone has to choose which chemical signals to emit and when to emit them. Furthermore, someone has to teach or tell the newborn cells what these distant chemical signals mean – which ones to grow towards and which ones to avoid.

This is too far out there for Materialism, Naturalism, Darwinism, Random Mutations, and Natural Selection to adequately explain – there's too much invisible information exchange going on here. The newborn cells are born with the ability to sniff-out and then migrate towards specific chemical signals. That's something that we can't do when we are newly born. Either these individual cells are more intelligent than we are, or they have Someone Intelligent guiding and directing their growth and migration. Someone had to design, program, and organize all of this. Evolution, random mutations, and natural selection simply will not do! Someone had to develop and store this invisible sequence-of-organization for each and every species. Evolution couldn't have done this, because it doesn't have access to any of this information. It's physically impossible for evolution to have designed and created all of this.

> **Once cells have been created through cell division in the ventricular zone of the neural tube, they migrate to the appropriate target location. During this period of migration, the cells are still in an immature form, lacking the processes (axons and dendrites) that characterize mature neurons. Two major factors govern migration in the developing neural tube: time and location. In a given region of the tube, subtypes of neurons arise on a precise and predictable schedule and then migrate together to their prescribed destinations.**
>
> **There are two methods by which developing cells migrate. One is somal translocation. In somal translocation, an extension grows from the developing cell in the direction of the migration; the extension seems to explore the immediate environment for attractive and repulsive cues as it grows. Then, the cell body itself moves into and along the extending process, and trailing processes are retracted.**
>
> **The second method of migration is glia-mediated migration. Once the period of neural proliferation is underway and the walls of the neural tube are thickening, a temporary network of glial cells, called radial glial cells, appears in the developing neural tube. At this point, most cells engaging in radial migration do so by moving along the radial glial network.** (*Biopsychology*, p. 216.)

He literally wants you to believe that evolution, random mutations, and natural selection designed and created all of this, and is now overseeing and controlling all of this. It's physically impossible. Evolution, random mutations, and natural selection don't have access to any of this!

Who is choosing the time and location of a cell's birth and what type of cell it will be when it is born? In other words, who is choosing the subtype for each neuron and glial cell, as well as the precise and predictable schedule of their birth? Who is choosing the direction of migration and target of the migration? Who prescribes their destinations? Who chooses and guides the birth, direction, and the growth of the radial glial cells? Who taught each cell how to identify the attractive and repulsive cues as it grows? Who taught the neuron cells how to climb the radial glial cells, and who tells these newly born neurons that the radial glial cells are there and should be climbed. Who decides which neuron will climb which radial glial cell? Who is deciding what each cell will become when it finally reaches its chosen destination? These are legitimate scientific questions. Evolution (genetic drift) can't be doing any of this, because there isn't enough information storage capacity in the genome to handle all of this.

According to Uncle Google:

The nuclear genome comprises approximately 3,200,000,000 nucleotides of DNA, or 3.2 billion base-pairs.

100 billion neurons are interconnected via trillions of synapses. What's more, for each neuron there are some 10 to 50 glial cells providing structural support, protection, resources and more.

The human brain has a huge number of synapses. Each of the 10^{11} (one hundred billion) neurons has on average 7,000 synaptic connections to other neurons. It has been estimated that the brain of a three-year-old child has about 10^{15} synapses (1 quadrillion).

The Materialists, Naturalists, and Darwinists want you to believe that genetic drift (evolution), random mutations, and natural selection planned all of this and is now controlling all of this; but, it is physically impossible. A quadrillion bits of data, information, programming, command, and control cannot be stored within our genome. It's physically impossible. We literally have to get rid of Materialism, Naturalism, Classical Physics, Darwinism, and the Theory of Evolution if we truly want to know what's happening within the molecules and between the molecules of our physical body.

There is no way on earth that the 3.2 billion base-pairs in our genome could store enough information to guide and control the precise and predictable sequential growth of the 37.2 trillion cells in the human body or the precise and predictable sequential construction of the 10^{15} synapses in the human brain. It's physically impossible to store that amount of information in our physical genome! You can't even store precise and predictable sequential construction, sub-typing, and migration information for 100 billion neurons within the 3.2 billion base-pairs of our human genome, let alone the information guiding and controlling the trillions of glial cells along with their sequential construction logistics, sub-typing, and migration.

Someone is also storing and supplying the information for sub-typing of cells or cell differentiation, the location and timing for the birth of each type of cell, and the location and method of each cell's migration and eventual destination, as well as the type of cell it will eventually become when it matures. Someone or something is also telling each neuron how to interact with nearby neurons and how to map out its axon, dendrites, and synapses so as to properly coordinate will all the nearby neurons. Someone is telling each axon what type of neurotransmitters and neuropeptides to make and emit, and where to do so. Someone is also telling teach post-synaptic dendrite what type of neuro-receptors to build, and where to build them. There is no way on earth to store enough information for the synaptic sub-typing of 10^{15} synapses, within the 3.2 billion base-pairs of the human genome. It's physically impossible. But, that information exists and is being stored someplace; and, each neuron has direct access to this information and/or is being fed this information by God's Psyche.

There's nothing random about any of this! Even according to the Materialists, Naturalists, and Darwinists, "it all arises on a precise and predictable schedule".

How?

These people assure us that Evolution is controlling all of this; but, that's physically impossible. Evolution, random mutations, and natural selection don't have access to any of this information; and, these things wouldn't know what to do with this information even if they did have access to it.

The only logical explanation is that each cell in your physical body has direct access to God's database, which is being stored non-locally in the Quantum Realm, Non-Local

Realm, Transdimensional Realm, or Spirit Realm; and/or, each cell in your physical body is conscious, sentient, alive, and fully aware of its purpose, environment, eventual destination, and how to become all that it has been commanded to become. Each cell in your physical body has a Psyche and also has direct access to God's Psyche and God's Database. All of this origination, sub-typing, and destination information is being exchanged telepathically at the quantum level between God's Psyche and Nature's Psyche. There is no other logical explanation, because it's impossible to store all of that information physically within our genome.

Neurons at a specified destination are emitting specifically chosen chemical signals which the newly born neurons are able to sense telepathically and grow towards or move towards during the migration process. These things are alive, conscious, sentient, and aware; and, these newly-born neurons seem to be infinitely more intelligent than we human beings are, or seem to have access to infinitely more information than we human beings do.

Once neurons have migrated to their appropriate positions and aggregated into neural structures, axons and dendrites begin to grow from them. For the nervous system to function, these projections [or processes] must grow to the appropriate targets. At each tip of an axon or dendrite is an amoeba-like structure called a growth cone, which extends and retracts fingerlike cytoplasmic extensions called filopodia, as if searching for the correct route.

Remarkably, most growth cones reach their correct targets. A series of studies of neural regeneration by Roger Sperry in the early 1940s first demonstrated that axons are capable of precise growth.

In one study, Sperry cut the optic nerve of frogs, rotated their eyeballs 180 degrees, and waited for the axons of the retinal ganglion cells, which compose the optic nerve, to regenerate (grow again).

Each retinal ganglion cell had grown back to the same point of the optic tectum to which it had originally been connected. Neuroanatomical investigations have confirmed that this is exactly what happens.

Brains have the ability to reorganize themselves in response to experience. (*Biopsychology*, pp. 218, 250.)

Who decided or chose what type of neural structure for these aggregating neurons to become? They could have become anything, so why did they choose to become the thalamus, or the hypothalamus, or the hippocampus, or the prefrontal cortex? How did they know how to form a thalamus or a hippocampus? Who taught them how to coordinate their efforts? Who or what makes them coordinate with each other and form unique structures within the physical brain? During neural regeneration, who makes the neurons regenerate or who decides that the neurons are going to stop trying to regenerate? Who decides whether to save a neuron or to terminate a neuron? Each neuron is following some kind of invisible 3D blueprint telling it how to grow, how to interact with other neurons, and what to become when it gets done migrating. Axons and dendrites are the physical manifestation of some sort of invisible quantum level schematic, diagram, or map. We KNOW this is so because there are NO wires within a physical brain at the physical level connecting the neurons together.

It's the Human Psyche who chooses whether to have experiences or not. The brain has nothing to do with the choice. The brain just sits there and idles in default mode unless the Human Psyche makes it to do something. A Human Psyche chooses whether to practice or not to practice. The brain has nothing to do with it. Nature's Psyche reorganizes the brain in response to the Human Psyche's choices. This is how things really work, both at the physical level and the quantum level. All I ever wanted to know is how everything works, and now I do – thanks to the Orthodox Interpretation of Quantum Mechanics from Henry P. Stapp.

In Sperry's experiment, each severed axon grew back to the precise location that it was originally connected to. How did it do that? Each axon is given a specific location within the physical brain to grow towards and to "attach to" or synapse upon when the axon gets to where it has been assigned to go. These axons KNOW where they are supposed to go or grow, and they KNOW what to do when they get there. These axons KNOW which of the 10^{15} synaptic connections belongs to them, and no one else.

In order for your brain and nervous system to function properly, each axon and dendrite has to grow to a precise and specific location, and then sub-type and synapse properly when they get there. The only way that's possible is if each axon has access to supernatural information from a Supernatural Source, because there's no way possible to store that monumental amount of information within the human genome.

Our neural map doesn't reside in our DNA. It can't. It's physically impossible. Our neural map is being broadcast from every neuron to every other neuron, and they each KNOW precisely what they are doing and why. The neurons KNOW how to aggregate and form brain structures precisely and predictably. There's only enough information storage capacity in our DNA to make the requisite proteins. There's nowhere near enough information storage capacity in our genome to store our neural map or 3D brain blueprint, let alone the rest of our physical body's 3D blueprint, the timing and location of cell birth, the logistics of migration and growth, as well as sub-typing information and information as to how they should function and interact with nearby cells. The amount of telepathic information exchange taking place at the quantum level between neurons, glial cells, molecules, and atoms during neuro-development is simply off the charts. It's infinite, because it's still ongoing!

This doesn't even take into consideration the amount of information exchange taking place between the atoms and molecules within our physical bodies, right here, right now.

The chemo-affinity hypothesis fails to account for the discovery that some growing axons follow the same circuitous route to reach their target in every member of a species, rather than growing directly to it.

Growth cones seem to be influenced by a series of chemical signals along the route. These guidance molecules are similar to those that guide neural migration in the sense that some attract and other repel the growing axons. Guidance molecules are not the only signals that guide growing axons to their targets. Other signals come from adjacent growing axons.

Pioneer growth cones – the first growth cones to travel along a particular route in a developing nervous system – are presumed to follow the correct trail by interacting with guidance molecules along the route.

At first, it was assumed that the integrity of topographical relations in the developing nervous system was maintained by point-to-point chemo-affinity, with each retinal ganglion cell growing toward a specific chemical label. However, evidence indicates that the mechanism must be more complex.

In most species, the synaptic connections between retina and optic tectum are established long before either reaches full size. Then, as the

retinas and optic tectum grow at different rates, the initial synaptic connections shift to other tectal neurons so that each retina is precisely mapped onto the tectum [topographically], regardless of their relative sizes.

The topographic gradient hypothesis has been proposed to explain the accurate axonal growth involving topographic mapping in the developing brain. Growing axons are guided to their destination by two intersecting signal gradients. (*Biopsychology*, p. 219.)

Now, he's down to the level of individual molecules, which he calls Guidance Molecules. Those Guidance Molecules are communicating at a distance, telepathically or wirelessly, with the molecules in the Growth Cones. Remember, telepathy is WiFi at the quantum level.

Pioneer growth cones are the first growth cones to travel along a particular route in the developing nervous system. They can go anywhere, so who tells them where to go? This is where we see Nature's Psyche come into play. Nature's Psyche is the one who maps physical functionality into our brains by collapsing the wave functions thereby making things physically real.

The target cell is told telepathically what type of Guidance Molecule to emit, and the traveling growth cone is told telepathically what type of Guidance Molecule to sniff out and grow towards.

Who laid down this trail of breadcrumbs, Guidance Molecules, telling each one what it should be or what chemical signal it should emit? Evolution couldn't have done it, because random mutations and natural selection don't work and don't exist at the Quantum Level. By definition, in principle, Evolution can't do telepathy. Evolution, natural selection, and random mutations can't do planning and mapping and complexity, either. So, who is controlling these Guidance Molecules; and, who is telling the individual Growth Cones which collection of Guidance Molecules to follow and which collection of Guidance Molecules to avoid?

The topographic gradient hypothesis is the idea that axonal growth is guided by the relative position of the cell bodies on intersecting gradients, rather than point-to-point coding of neural connections. In other words, it's a Map of Physical Functionality and a non-local 3D Blueprint that's functioning through quantum waves as a GPS system. Someone in control of the axonal growth cone KNOWS precisely where the different cell bodies are located, and it's navigating or weaving through them towards a pre-selected destination. Navigation, targeting, weaving, dodging, and pre-selected destinations are ALL functions of Nature's Psyche that are taking place at the quantum level.

He concluded by suggesting that the growing axons are following some kind of topographical map. He makes this topographical map a 2D map like the ones the US Geological survey provides; but, it has to be at least a 4D map – and possibly an 11D map – including the three dimensions of space and the one dimension of timing.

Then there has to be other dimensions in this topographical map storing the initial layout of the guidance molecules or guidance cells, along with the 3D blueprints for each neural sub-type and neural-structure, along with all the different types of neurotransmitters, synaptic sub-types, ion channels, neuro-receptors, neuropeptides, locations, along with 3D target destination and specialization information when it gets there, and other things needed to make specific structures in the physical brain. There might be enough information in the DNA to tell a cell whether to be a neuron or a bone cell; but, there's nowhere near enough information storage capacity in the genome to tell a neuron what it's supposed to do, where it's supposed to go, what it's supposed to become, and how it's supposed to connect-up with other neurons after it has been assigned to be a neuron, told where to go, and reached its ultimate destination. The intelligence or psyche needed to accomplish all of this is off the scale; and, it all takes place completely outside the conscious awareness of the Human Psyche.

Can you imagine what you would get if natural selection and random mutations really were in control of all these things and if the theory of evolution were really true? It would be complete and utter chaos; and, we wouldn't exist.

When it comes to neuro-development, God is hiding in plain sight where the Materialists, Naturalists, and Atheists are never going to see Him or find Him, because these people refuse to go there and refuse to think about these kinds of things logically and rationally. The most that you will ever get out of these people is that Evolution did it, which is physically impossible. The only thing that evolution ever does for us is entropy, disease, cancer cells, death, and extinction. Evolution is never going to design and create anything, because it can't. It's physically impossible.

Instead, WE KNOW from a careful examination of the evidence that neuro-development for each species is based upon a unique multidimensional blueprint which the individual molecules are accessing telepathically from the Non-Local Realm or the Quantum Realm, and/or is based upon some kind of Non-Local Consciousness and Highly Advanced Intelligence who is in control of the whole process from beginning to end. There's no other logical and rational explanation for what we are observing. There's just way too much information exchange taking place between the molecules at the quantum level for the whole thing to be designed, created, and controlled by natural selection and random mutations which don't even work as advertised at the genetic level let alone at the quantum level.

I KNOW from direct personal experiences that The Truth always wins out in the end, because the Lord is willing and able to back up The Truth with revelatory evidence and confirmatory evidence because it is The Truth that we are dealing with. When it comes to many of these things, I have asked God to reveal to me The Truth; and, he has done so. What's interesting is that He typically picks different ways to answer your questions, in order to demonstrate His versatility and to make His presence and influence more obvious and known.

Once I started asking for Scientific Proof of God's Existence a couple of years ago in 2015, the Lord gave me a handful up front; and, He has given me dozens more as I go along, when I'm ready for them, looking for them, and asking for them. There are many more out there that I still haven't discovered because I don't yet have the necessary foundation for them and I don't yet know that I should be looking for them and asking for them. The Lord doesn't give you revelations and insights if He knows that you aren't going to be doing anything with them, or knows that you have no intention to take advantage of them, or knows that you aren't going to understand them. Sometimes I have to ask God what questions I need to be asking next; and, He finds ways to tell me what questions I should be asking next.

The truly interesting thing about all of this is that if you were to start asking God for Scientific Proof of His Necessity or Scientific Proof of His Existence, He will give you a completely different set than the ones He has given to me, because you are a completely different person with a completely different set of interests, a completely different preparation, a completely different skill-set, and a completely different set of spiritual gifts. I believe that God gives each one of His children a unique piece to the puzzle; and, when we start comparing and sharing the different pieces that we have received, the overall picture begins to emerge.

I Know There Is Something I'm Still Missing

I have been a computer scientist most of my life. I like to think that I know how computers work and how memory is stored at the physical level within a computer. As a result, I found the physical brain severely disturbing. For months, if not years, I couldn't put my finger on it; but, the physical brain has caused me a lot of cognitive dissonance. Something wasn't quite right. In fact, as the months passed, I found many things that aren't quite right about a physical brain.

Bear, M. F., Connors, B. W., & Paradiso, M. A. (2007). *Neuroscience: Exploring the Brain* (3rd ed.). New York: Lippincott Williams and Wilkins.

Pinel, J. (2014). *Biopsychology* (9th ed.). New York: Pearson.

Clearly, our physical brains function and work and do their jobs. So, I start with that as the given. In every hypothesis, one has to start with a given. Then I try to use the Orthodox Interpretation of Quantum Mechanics and Psyche or Non-Local Consciousness in order to explain the aspects and functions of our physical brain that don't make sense in terms of Classical Physics, Materialism, and Naturalism. Quantum Mechanics is a proven science. Psyche or Non-Local Consciousness is an observed science and an experiential science based upon Out-of-Body Experiences and Near-Death Experiences for a start.

Our neurotransmitter systems work too perfectly for them to have a basis in classical physics, random diffusion, Brownian motion, and entropy. Consequently, I know that there's something that I'm still missing when it comes to neurotransmitters; and, I sense that it is important. Those things disturb me greatly. Their functioning is completely counter-intuitive most of the time. Here are some of my initial unanswered questions to show you what I'm getting at.

As I see it, reuptake of neurotransmitters is worthless and counter-productive for non-directed synapses and the neurotransmitters that have to be maintained at a certain level for proper functioning of the physical brain. Why does reuptake exist for serotonin, if blocking its reuptake with SSRIs helps to alleviate anxiety, OCD, and depression? Do serotonin neurotransmitters have a shelf life? Does a malfunctioning brain lose the ability to manufacture adequate levels of serotonin? I don't get it; and, a satisfactory answer wasn't initially forthcoming.

There's more going on here than meets the eye.

Why do SSRIs (Selective Serotonin Reuptake Inhibitors) work? Why would they be needed, and why would they be helpful? How do they improve the situation? Why isn't there lots of false positives and signal noise from having all those extra serotonin neurotransmitters floating willy-nilly in the synaptic clefts? Why would reuptake transporters on the presynaptic neuron be needed for serotonin if having extra serotonin floating in the cerebrospinal fluid helps to alleviate anxiety, OCD, and depression?

The main answer that I got initially is that having enough serotonin floating in the synaptic clefts makes it possible for the post-synaptic neurons to reach out and grab one at will, when it knows that it needs one. If there aren't enough of these serotonin neurotransmitters available to reach out and grab telekinetically or through quantum tunneling, then that's when the various different problems start to arise in respect to serotonin.

The same realities apply to dopamine. If the brain doesn't have enough of the stuff, then Parkinson's Disease and glitches are the result.

But, how does having too much dopamine contribute to schizophrenia? Too little dopamine, and we have problems; too much dopamine, and we have problems. There was something here that I was missing; and, I suspect that my brain-washing in Materialism, Naturalism, and Classical Physics was interfering with my understanding somehow. I needed to up my game and dive into Quantum Mechanics, particularly the Orthodox Interpretation of Quantum Mechanics that is based upon the interplay between the Human Psyche and Nature's Psyche.

The idea that eventually came to mind is that each neurotransmitter has a variety of different functions, and different types of post-synaptic receptors that do different things. It's called "selective" serotonin reuptake inhibitor, because it doesn't block ALL serotonin reuptake ports, only specific ones. But, why does blocking specific serotonin reuptake transporters while leaving others open and functional help to alleviate anxiety, OCD, and depression? I didn't get it. It was still too complex for me.

A partial answer came while reading, *Neuroscience: Exploring the Brain*.

> **The most numerous glia in the brain are called astrocytes. These cells fill the spaces between neurons. Consequently, astrocytes probably influence whether a neurite [axon or dendrite] can grow or retract.**
>
> **An essential role of astrocytes is regulating the chemical content of this extracellular space. For example, astrocytes envelop synaptic junctions in the brain, thereby restricting the spread of neurotransmitter molecules that have been released. Astrocytes also have special [transporter] proteins in their membranes that actively remove many neurotransmitters from the synaptic cleft. A recent and unexpected discovery is that astrocytic membranes also possess neurotransmitter receptors that, like the receptors on neurons, can trigger electrical and biochemical events inside the glial cell. Besides regulating neurotransmitters, astrocytes also tightly control the extracellular concentration of several substances that have the potential to interfere with proper neuronal function. For example, astrocytes regulate the concentration of potassium ions in the extracellular fluid.** (*Neuroscience*, pp. 46-47).

The primary intelligence associated with Nature's Psyche apparently resides within the astrocytes. The astrocytes enclose neurons of a certain type and regulate the neurotransmitters and synaptic clefts between them. Astrocytes shepherd the neurons under their domain. Astrocytes direct and control axon and dendrite growth as well as synaptogenesis. The astrocytes apparently know when the presynaptic membrane has released neurotransmitters, and when the postsynaptic membrane has received the message; and then, the astrocytes trigger and monitor the cleanup process of the directed synapses and/or monitor the neurotransmitter levels associated with the non-directed synapses, which are synapses that have multiple release sites called "varicosities" all along the axon, releasing non-directed neurotransmitters some distance from their intended site of reception. A single astrocyte has been known to have up to 40,000 different processes or electrical synapses. Astrocytes have direct connections to neurons and other astrocytes through gap junctions or electrical synapses. That's a lot of command and control for a single astrocyte. Astrocytes are also implicated in synaptogenesis, the positioning and sub-typing of synapses, including the selection of the neurotransmitters that will be conveyed between the synapses under their command.

The astrocytes seem to be the brains of the operation. Astrocytes appear to be infinitely more intelligent than we are. So, the natural question becomes, "How did the astrocytes learn to do all of these different things, including the construction of our physical brain and the differentiation and synaptogenesis of neurons?" Apparently, the astrocytes know how to create a thalamus, hypothalamus, hippocampus, amygdala, and all the other structures in our physical brain. Where did the astrocytes get that information and knowledge, and how do the different astrocytes coordinate their activities so that functional brain structures are the result?

Of course, the Materialists and Naturalists will tell you that evolution (genetic drift), random mutations, and natural selection taught the astrocytes how to create brains, differentiate neurons, and oversee vast clusters of synapses; but, is that even realistically possible? How would natural selection and random mutations communicate with astrocytes and teach astrocytes how to do their job?

And, how many of us would really be willing to turn our brains and our bodies over to random mutations in the hope of evolving into some higher life form? We have already run this experiment, and we know how it turned out. We have irradiated large groups of people, mutated them severely, and either watched them die of radiation poisoning or watched them die of cancer some years later. Some of my ancestors were part of that experiment. Random mutations don't design and create. They debilitate and destroy. Random mutations and natural selection are by definition in principle dumb and blind, not intelligent and creative.

The astrocytes are highly intelligent and very skilled – much more intelligent and skilled than we human beings are. So, who created the astrocytes and taught them how to do their job? Random mutations couldn't have done the job, and neither could natural selection. There's no intelligence there; and, there's no way for random mutations and natural selection to communicate with the astrocytes and teach the astrocytes how to coordinate their activities and do their jobs! This question really can't be answered in terms of Materialism, Naturalism, Darwinism, and Classical Physics. Instead, we are forced to turn to the Orthodox Interpretation of Quantum Mechanics, Nature's Psyche, the Human Psyche, and God's Psyche if we want an answer that actually makes logical sense and is realistically possible.

There are at least 37.2 trillion cells in the human body each with a different mission, birth location, target location, differentiation assignment, hierarchy, mode of functioning, and set of logistics. There are only 3.2 billion base pairs in the human genome to control them all and differentiate them all. At some point during the development of the human body and the human brain, the astrocytes, neurons, and other cells will run out of the information that they can get from our genome and will be forced to switch to the non-local 3D blueprint of the human body for their commands, knowledge, and instructions. These little cells have direct access to God's blueprint for the human body; whereas, we don't. They are alive, sentient, conscious, intelligent, and aware of what they are doing and why they are doing it. They know and understand their mission and their surroundings. That's how they are able to do what they do. They are more intelligent than we are and/or have access to greater knowledge than we do.

It's difficult to know for sure when a developing physical body and developing physical brain are forced to switch from the Genome's DNA over to God's Non-Local Blueprint; but, we are given clues that do make sense to me.

After observing carefully, repeatedly running the numbers, and thinking about it logically, it was obvious and clear to me that all the information needed to construct or make proteins during protein synthesis is there to be found in our genes and our genome; but, when the ribosome lets go of the mRNA molecule and the completed polypeptide protein strand, that's the END of the genetic instructions for that particular protein molecule. Thereafter, the newly born protein has to get its knowledge and information from some other source besides our genome, in order to discover and know where it is supposed to go and what it's supposed to do and become when it gets there.

That newly born protein can go anywhere; and, someone or something has to tell it where to go and what to do when it gets there. This is logical common sense to me, and came to me from following the evidence. Genes are used to make proteins; but, after the proteins have been constructed or made, then our genes are no longer part of the process and our genes no longer decide what that newly made protein is going to do with itself thereafter. At some point in the process, these newly made proteins and molecules simply take on a life of their own and go about their business as if the know precisely what they are doing and why they are doing it. They are no longer connected to our genome, so they have to be getting their information and knowledge from someplace else besides our genome.

Each protein is requisitioned by someone, and then it is told where to go and what to do after it has been made. The genes come into play primarily while the protein is being constructed and folded. After the protein has been made, then it is turned loose, and it can go anywhere and do anything; and, that's when God's Non-Local Blueprint and Logistics takes over and tells that newly born protein where to go, what to do, and what to become.

Proteins need physical blueprints and instructions that are stored in our genome and our genes. However, the individual lone atoms and molecules within our physical bodies, cell structures, cytoplasm, proteins, and even genomes have to get their operating instructions and blueprints from someplace else non-physical in order to know what they are supposed to do, supposed to become, where to go, when to go, and how to do their mission when they get there, because they are no longer connected to our genes. That's logical common sense.

It seems logical to me that embryonic stem cells are a natural product of the associated genome, because the stem cells can become any type of cell that we can possibly imagine. However, when the cells start to differentiate and migrate, at some point in the process, they are going to exceed the informational capacity of our genome and be forced to switch over to God's Non-Local Blueprint for their knowledge and instructions and logistics. If the cells fail to follow this Non-Local Blueprint or fail to follow the tasks that have been assigned to them, then birth defects will be the result. Someone is telling the individual cells how to interact with each other; and, not all of that information can be stored within our genes because the newly-born or newly-differentiated cells are no longer using their genome directly at that point in their lives. The genome is used for repair and maintenance – the construction of additional proteins – not for telling each cell where to go and what to become when it gets there. That's the way I see it.

After a cell is born, it can go anywhere. Someone is telling each new-born cell where to go and what to do when it gets there – the migration of newly-born neurons from the ventricular zone to the outer part of the neural tube where they differentiate, specialize, coordinate, and engage in synaptogenesis taught me this.

Lots of genetics goes into making the proteins that comprise a physical cell; but, after that particular living cell has been made, something other than genetics takes over and tells that cell where to go and what to do and what to become when it gets there. That newly born cell could go anywhere; but, someone tells it where to go, and how to get there, and how to act when it gets there. Someone also determines the location and the timing of its birth. Cell migration, cell location, cell differentiation, and coordination between individual cells is a miracle of Divine Design, not just a coincidence. There's intelligence behind all of this, not just blind random luck. There's no way that all of that information can be stored within our genome. It's physically impossible. At some point in the process, each individual cell is forced to switch over from its internal genome to God's Blueprint and God's Divine Design, or it's going to fail to do what it was born to do and meant to do.

In the first year of medical school, I was introduced to the human brain, the crowning achievement of evolution. (*Neuroscience*, p. 150.)

I wish I had evolution's press manager and book publisher – I would be radically rich by now, if I did.

NO! The human brain is NOT the crowning achievement of evolution. Cancer is the crowning achievement of random mutations. The extinction of species is the crowning achievement of natural selection, when they get selected against and cease to exist. This person has been duped, just like everyone else. Random chance can't design and create anything; but, the majority of neuroscientists still haven't figured this out for themselves. There's no such thing as spontaneous generation, abiogenesis, macro-evolution, or creation ex nihilo. It doesn't exist, so it can't have any crowning achievements. Sometimes I wish these people would use their brains. It would have saved me a lot of trouble and grief if they would have. Oh well, it is what it is. We each have to start where we are, and then go on from there. At least I'm no longer a Materialist, Nihilist, and Atheist. That's a start.

I learn best through comparison and contrast. When it comes to our neurotransmitter systems, neuropeptides, ion channels, ion pumps, reuptake transporters, and postsynaptic receptors, it's clear to me that there's no way on earth that evolution (genetic drift), random mutations, and natural selection could have figured all this out billions of years ago. We can't always figure it out, so why on earth would anyone believe that evolution figured it out billions of years ago and got it right every time? That's physically impossible, illogical, irrational, and counter-intuitive – to believe that evolution designed, created, and now controls our physical brain and our neurotransmitter systems.

The next question I had was, "Why don't post-synaptic receptor blockers shut-down the physical brain and bring it to a standstill?" The brain functions anyway. We have created such things; but, how could they ever be any good for us? How on earth could blocking transmissions and action potentials be good for us? There's more here to study and learn than we can possibly begin to imagine; and, the answer to this one has been a long time in coming.

It's clear to me that when it comes to our neurotransmitter systems, there's no way on earth that evolution (mutation/selection) could have figured all this out and got it perfectly balanced and correct, when our best minds can't even figure it out. Most of us are still mired in Materialism and Naturalism, and we aren't even asking these questions in the first place. I didn't start asking these kinds of questions until two years ago, back in 2015. I didn't even know that these questions existed. I used to be a Materialist, Nihilist, and Atheist. Materialism and Naturalism make you dumb and blind – just like evolution, natural selection, and random mutations are supposed to be.

The answers only started to come when I got rid of My Materialism and My Naturalism and fully embraced Psyche and Quantum Mechanics instead.

G-proteins are located inside neurons and some other cells, and they are attached to metabotropic receptors in the cell membrane. Most of the time, the G-proteins are used either to enhance synaptic transmission or decrease synaptic transmission – increase or decrease the strength of the Post-Synaptic Potential that is generated by that particular synapse.

This quote explains some of the different complexities that are involved in getting the balance of neurotransmitters perfectly correct, as well as some of the problems that ensue when we don't.

Here you can see the entire G-protein receptor process, from the activation of the receptor by the neurotransmitter dopamine to the opening of the sodium ion channel. Now an anti-psychotic drug comes in from the left and one binds to the receptor. It binds better than dopamine, but can't activate the G-protein. Dopamine tries to bind, but its binding site is occupied by the drug. This is how antipsychotic drugs prevent sodium ions from entering the postsynaptic cell.

Excessive neurotransmission of dopamine is associated with schizophrenia, a clinical condition marked by seriously disordered thought. Antipsychotics, also called neuroleptics, are a class of compounds with a high affinity for several subtypes of dopamine receptors. The chemical structure of the various antipsychotics allows them to bind to dopamine receptors without triggering the postsynaptic response that the binding of dopamine normally would. Because of their ability to block dopamine receptors without causing the opening of ion channels and setting off an action potential, neuroleptics can be administered to schizophrenic patients to help reduce excess levels of dopamine, and to thus help alleviate the positive symptoms of the disorder.

There are approximately thirty antipsychotic drugs presently available in North America. Developed in the 1950's, the "typical antipsychotic drugs" include the phenotiazines (e.g. chloropromazine), the thioanthenes (e.g. chlorprothixene), the butyrophenones (e.g. haloperidol), the diphenypbutylpiperidines (e.g. pimozide), and the dihydroindolones (e.g. molindone). These drugs illustrate a high affinity for the D2 family of dopamine receptors and it is at these sites that they are thought to exert their therapeutic action. Because D2 dopamine receptors are present not only on the post-synaptic membrane, but on the cell bodies, dendrites and nerve terminals of presynaptic cells as well, antipsychotic compounds can interfere with dopaminergic neurotransmission at various sites in both the pre- and postsynaptic cell. Some neuroleptics interfere with the release of DA at the presynaptic terminal, while others block postsynaptic dopamine receptors so that postsynaptic cells cannot recognize dopamine.

Using typical antipsychotic drugs to interfere with dopaminergic neurotransmission in the limbic system and in the cerebral cortex (areas involved in the control of motivated and emotional behavior and in facilitating organized thought) can be helpful in alleviating the positive symptoms of schizophrenia. The inhibition of dopamine transmission in other pathways, however, can result in a wide range of highly undesirable side effects. Unfortunately, when typical antipsychotics are administered, all D2 receptors are blocked, including those in areas of the brain involved in the fine tuning of motor movement (namely, the basal ganglia and cerebellum). As a result, when these drugs are administered to schizophrenic patients, many experience the motor dysfunction characteristic of Parkinson's disease: tremors, akinesia (a slowing of voluntary movements), spasticity and rigidity, and a kathesia (discomfort and a feeling of restlessness in the legs). The use of typical antipsychotics may also interfere with normal endocrine function, and can exert anticholinergic, antiadrenergic, antihistaminic and antiserotonergic actions.

A second class of neuroleptics, known as "atypical antipsychotics," exert the same therapeutic effects of the traditional antipsychotics without producing the undesirable side effects of the older, typical antipsychotics. The atypical antipsychotics illustrate a lower affinity for D2 receptors, yet readily bind with D3 and D4 dopamine receptors. Since the expression of D3 and D4 is restricted to the neurons of the limbic system and cerebral cortex only, the action of these newer antipsychotics is confined to areas involved in the pathology of schizophrenia.

https://web.williams.edu/imput/synapse/pages/IIIB5.htm

See also:

There are five different subtypes of dopamine receptors.

https://www.youtube.com/watch?v=7KLDneIdKk8

You can sense the intelligence in these things and the intelligence behind these things. It's an intelligence that natural selection and random mutations could never have successfully and reliably produced.

While reading this quote, did you finally realize that random mutations are the CAUSE of all our heritable diseases or genetic diseases, including our neurotransmitter imbalances and malfunctioning? Random mutations don't design and build genomes and physical brains. Random mutations destroy these things. There's no way in the universe that random mutations and natural selection could have designed and created our genomes, our physical brains, and neurotransmitter systems. Random mutations and "getting selected against" destroy our genomes and our physical brains – not create them.

"Excessive neurotransmission of dopamine is associated with schizophrenia." Schizophrenia is highly heritable, meaning that it has a strong genetic component. Where schizophrenia is concerned, some kind of genetic mutation and/or resulting brain damage results in excessive production and/or excessive neurotransmission of dopamine at certain receptors resulting in hallucinations and delusions. Psyche, Non-Local Consciousness, and Will-Power cannot override developmental damage caused by random genetic mutations. If the bus is broken, the human psyche is going to have a hard time driving it down the street. The human psyche and quantum mechanics can't always override broken machinery and the results of random genetic mutations. God knows how to do so; but, we human beings or human psyches can't always learn how to do so. God can do genetic engineering; but, we human beings are just learning to do this science.

God Does Genetic Engineering

https://www.youtube.com/watch?v=78SMfWw2xXY

Blocking all of the D2 receptors isn't good for us – Tardive Dyskinesia or Parkinsonian symptoms are the result. If we are going to use neuroleptics or antipsychotics to block dopamine receptors, we need these neuroleptics to be very selective about what they block while at the same time leaving the essential dopamine pathways open and functional.

Serotonin is the neurotransmitter that excites and inhibits the other neurotransmitters.

The 7 general serotonin receptor classes include a total of 14 known serotonin receptors.

https://en.wikipedia.org/wiki/5-HT_receptor

The amino acids Glutamate, Aspartate, Glycine, and GABA are the neurotransmitters in the vast majority of fast-acting directed synapses. Directed synapses have short gaps; and therefore, the cleanup of the synaptic cleft through reuptake is extremely important for proper neurotransmitter functioning of the fast-acting directed neurotransmitters.

In contrast to the amino acid neurotransmitters, the monoamines are small-molecule neurotransmitters with slower and more diffuse effects. The monoamines are non-directed neurotransmitters. They are just dumped into the synaptic clefts at varicosities, diffuse throughout the cerebrospinal fluid, and are maintained at a certain optimal level so that when the post-synaptic receptors need one, they can reach out and grab one. Only when there's functional imbalances and developmental damage caused by random genetic mutations do these monoamines get out-of-balance and/or the monoamine receptors stop functioning properly; and then, the malfunctions need to be treated by medications in an attempt to restore the balance. That's why serotonergic imbalances and malfunctions such as anxiety, OCD, and depression run in families, because they are the result of unfortunate detrimental genetic mutations which are being passed from one generation to the next.

Schizophrenia is the result of too much dopanergic activity at dopamine receptors, resulting in hallucinations and delusions – extra input and thoughts that aren't real. Tardive Dyskinesia is the result of using neuroleptics or antipsychotics to block and damage dopamine receptors that shouldn't be blocked and damaged. Tardive Dyskinesia (TD) is a motor control disorder that results from chronic use of certain antipsychotic drugs. Functionality is lost. Parkinson's Disease is the result of damage to nigrostriatal pathway, which manufactures and distributes dopamine to where it needs to go.

When the physical machinery is damaged, broken, imbalanced, or malfunctioning, the Human Psyche and Nature's Psyche can't always override the side effects. The brain is all about physical functionality, and Nature's Psyche has organized the different structures in our brain into Maps of Physical Functionality or Cerebral Processing Units at the quantum level. Damage or destroy the map, and the physical functionality is lost.

Because the monoamines are randomly diffused throughout the cerebrospinal fluid where they are being used, there has to be enough of them on-hand when the post-synaptic receptors need them to start an action potential; and, the post-synaptic receptors have to be properly designed and therefore functioning properly, or the post-synaptic receptors are going to end up not getting the neurotransmitters they need when they try to grab them, or they will end up grabbing neurotransmitters when they shouldn't be grabbing them resulting in signal noise, false starts, tremors, and other malfunctions. Only when the monoamine neurotransmitters systems are out-of-balance and malfunctioning, due to genetic mutations and brain damage, will we start to experience glitches and mental illnesses caused by these random genetic mutations which evolution or genetic drift gives to us.

Monoamines belong to two categories, catecholamines and indolamines. Dopamine, epinephrine, and norepinephrine are catecholamines. Serotonin is an indolamine. These are the slow-acting diffuse neurotransmitters that can cause chemical imbalances in the brain whenever their receptors are subjected to random genetic mutations which modify their function and result in excitatory and inhibitory imbalances in our neurotransmitter systems. The monoamines have to be perfectly balanced for proper brain functioning, because they are diffusely distributed throughout the cerebrospinal fluid between the synaptic clefts where they are used. They float out there in the cerebrospinal fluid waiting for the post-synaptic neurons to reach out and grab them when these post-synaptic neurons know that they need them. Only when there aren't enough of them – because of an overactive reuptake transporter or manufacturing malfunctions, or when their post-synaptic receptors are malfunctioning due to genetic mutations – do we start to run into the mental illnesses which are caused by monoamine imbalances and/or monoamine receptor malfunctions.

We have the genes for anxiety, OCD, and depression on my mother's side, and SSRIs have proven to be beneficial for me and some of my children, because I inherited some of those mutated and inefficient genes. Evolution (genetic drift), random mutations, and natural selection have not been my friends, during my life. Instead, they have been the source of most of my physical problems, including the chemical imbalances and malfunctioning in my physical brain. Evolution doesn't design and create, it destroys and incapacitates. Evolution, random mutations, and natural selection don't do origin of species – they do extinction of species. That has been my observation and my experience. Evolution or chance cannot do fine-tuning; but, it certainly can do entropy, chaos, and destruction.

Acetylcholine is the small-molecule neurotransmitter at neuromuscular junctions. Acetylcholine is the neurotransmitter for which enzyme degradation is the main mechanism of synaptic deactivation.

There are unconventional neurotransmitters such as anandamide, nitric oxide, and carbon monoxide. Furthermore, over 100 neuropeptides have been identified, and these large-molecules function as neurotransmitters.

Source for some of this information: Pinel, J. (2014). *Biopsychology* (9th ed.). New York: Pearson.

So far, my focus has been on post-synaptic receptors. These are needed to pass a signal from one neuron to the next. If they are messed up by random mutations, the process is going to glitch. But, this still doesn't answer my question about the reuptake transporters. Why do serotonin reuptake transporters exist if blocking them successfully alleviates anxiety, OCD, and depression? Are they there simply to give us one more thing to go wrong?

The gene that encodes the serotonin transporter is called solute carrier family 6 (neurotransmitter transporter, serotonin), member 4 (SLC6A4, see Solute carrier family). In humans the gene is found on chromosome 17.

Mutations associated with the gene may result in changes in serotonin transporter function, and experiments with mice have identified more than 50 different phenotypic changes as a result of genetic variation. These phenotypic changes may, e.g., be increased anxiety and gut dysfunction.

The serotonin transporter removes serotonin from the synaptic cleft back into the synaptic boutons. Thus, it terminates the effects of serotonin and simultaneously enables its reuse by the presynaptic neuron.

Neurons communicate by using chemical messengers like serotonin between cells. The transporter protein, by recycling serotonin, regulates its

concentration in a gap, or synapse, and thus its effects on a receiving neuron's receptors.

Medical studies have shown that changes in serotonin transporter metabolism appear to be associated with many different phenomena, including alcoholism, clinical depression, obsessive-compulsive disorder (OCD), romantic love, hypertension and generalized social phobia

https://en.wikipedia.org/wiki/Serotonin_transporter

How does modifying the presynaptic serotonin reuptake transporters cause or alleviate all of these diseases and mental illnesses? Isn't the post-synaptic receptors where all the action is supposed to be taking place?

A careful study of all of this shows that there's lots of contradictory information where the monoamines are concerned.

Sometimes the monoamines are portrayed as non-directed, diffusely spread, and finely balanced neurotransmitters that are always floating in the synaptic clefts waiting to be used at the very moment the Human Psyche needs them and Nature's Psyche reaches out and grabs them.

Other times, the monoamines are portrayed as normal neurotransmitters which are cleaned up with monoamine degradation enzymes and/or monoamine reuptake transporters after each action potential and exocytosis of the monoamine neurotransmitters from the pre-synaptic neuron into the synaptic cleft.

So, which is it? Monoamines floating at a certain level or concentration in the cerebrospinal fluid, or monoamines being cleaned up from the synaptic clefts after each action potential?

The problem is that it can theoretically be both depending upon which part of the brain one chooses to study, because there are at least 14 different types of serotonin post-synaptic receptors and 5 different dopamine post-synaptic receptors that have been identified; and, it's possible that some of them have reuptake transporters associated with them and some of them don't. The last quote indicated that there may be as many as 50 different serotonin reuptake transporters, with each one contributing to a specific disease or mental illness. I get the impression that there's no way to know for sure what's really going on here, because there are way too many variables and it's way too complex to keep track of it all. When it comes to all of this, I have observed that the neuroscientists repeatedly state that we still don't know how these various processes work; and, I believe them.

The only thing that the neuroscientists seem to be in agreement on is that evolution, random mutations, and natural selection understood perfectly all of these neurotransmitters, reuptake transporters, ion channels, and postsynaptic receptors and their functions billions of years ago, and that evolution still does and is now overseeing the whole process making sure that it will continue to work perfectly for us for the rest of eternity. Yeah, right! The theory of evolution is worthless, because it doesn't function as advertised and can't do what they say it does. It's better to go with Quantum Mechanics on this, because Quantum Mechanics has been repeatedly observed and verified. Quantum Mechanics and Non-Local Consciousness makes a lot more sense than the Theory of Evolution.

As I see it, God has commanded and taught Nature's Psyche within each neurotransmitter receptor, within each ion channel, within each ion pump, and within each reuptake transporter to perform specific functions or tasks at certain times and certain rates whenever the Human Psyche calls upon them to do so. It's only when evolution and random mutations have gotten ahold of these neuro-receptors, ion channels, pumps, and reuptake transporters, that we observe these machines failing to perform their God-given tasks and glitches result instead. We have observed that each neurotransmitter is a multipurpose tool; and, their associated post-synaptic receptors and their associated pre-synaptic reuptake transporters are of different types and each have a different function, purpose, task, timing sequence, and command protocol. The variables multiply and quickly go beyond anything that the typical human brain can comprehend. We really don't understand how these things work; but, we pretend that we do.

Actually, the problem comes from the fact that we scientists typically try to simplify the neurotransmitter systems in an attempt to make it possible for evolution (genetic drift), random mutations, and natural selection to have designed and created these things. Even though I have tried to include Quantum Mechanics, Nature's Psyche, the Human Psyche, and God's Psyche into the mix where neurotransmitter systems are concerned (which nobody has done before), I too have fallen for the desire to simplify all of this into one set of simple rules so that I can see it and comprehend it; but, it can't be done!

Do as I have done and start searching the Wikipedia for these things, and quickly the information overload goes off the charts in an exponential way. Yet, the Materialists, Naturalists, Darwinists, Nihilists, and Atheists want you to believe that evolution, natural selection, and random mutations understood all of this stuff, designed and created all this stuff, and continue to tweak it all to this very day. That's physically impossible, but hope springs eternal with these people.

Neuroscience

https://en.wikipedia.org/wiki/Neuroscience

https://en.wikipedia.org/wiki/Anaxonic_neuron

https://en.wikipedia.org/wiki/Interneuron

https://en.wikipedia.org/wiki/Unipolar_neuron

https://en.wikipedia.org/wiki/Multipolar_neuron

https://en.wikipedia.org/wiki/Neuroglia

https://en.wikipedia.org/wiki/Astrocyte

https://en.wikipedia.org/wiki/Ependyma

https://en.wikipedia.org/wiki/Microglia

https://en.wikipedia.org/wiki/Oligodendrocyte

https://en.wikipedia.org/wiki/Schwann_cell

https://en.wikipedia.org/wiki/Satellite_glial_cell

https://en.wikipedia.org/wiki/Electrical_synapse

https://en.wikipedia.org/wiki/Gap_junction

https://en.wikipedia.org/wiki/Gap_junction_protein

https://en.wikipedia.org/wiki/Ion_channel

https://en.wikipedia.org/wiki/Na%2B/K%2B-ATPase

Monoamines and Transporters

https://en.wikipedia.org/wiki/Monoamine_transporter

https://en.wikipedia.org/wiki/Vesicular_monoamine_transporter

https://en.wikipedia.org/wiki/Plasma_membrane_monoamine_transporter

https://en.wikipedia.org/wiki/SLC22A3

https://en.wikipedia.org/wiki/Neurotransmitter_transporter

Serotonin

https://en.wikipedia.org/wiki/Serotonin

https://en.wikipedia.org/wiki/5-HT_receptor

https://en.wikipedia.org/wiki/Serotonin_transporter

Dopamine

https://en.wikipedia.org/wiki/Dopamine

https://en.wikipedia.org/wiki/Dopamine_receptor

https://en.wikipedia.org/wiki/Dopamine_transporter

Epinephrine

https://en.wikipedia.org/wiki/Epinephrine

https://en.wikipedia.org/wiki/Adrenergic_receptor

http://macromoleculeinsights.com/adrenergic.php

Norepinephrine

https://en.wikipedia.org/wiki/Norepinephrine

https://en.wikipedia.org/wiki/Adrenergic_receptor

https://en.wikipedia.org/wiki/Norepinephrine_transporter

Glutamate

https://en.wikipedia.org/wiki/Glutamate_(neurotransmitter)

https://en.wikipedia.org/wiki/Glutamate_receptor

https://en.wikipedia.org/wiki/Glutamate_transporter

Acetylcholine

Acetylcholine is synthesized in certain neurons by the enzyme choline acetyltransferase from the compounds choline and acetyl-CoA. Cholinergic neurons are capable of producing ACh. An example of a central cholinergic area is the nucleus basalis of Meynert in the basal forebrain.

The enzyme acetylcholinesterase converts acetylcholine into the inactive metabolites choline and acetate. This enzyme is abundant in the synaptic cleft, and its role in rapidly clearing free acetylcholine from the synapse is essential for proper muscle function. Certain neurotoxins work by inhibiting acetylcholinesterase, thus leading to excess acetylcholine at the neuromuscular junction, causing paralysis of the muscles needed for breathing and stopping the beating of the heart.

https://en.wikipedia.org/wiki/Acetylcholine

https://en.wikipedia.org/wiki/Acetylcholine_receptor

https://en.wikipedia.org/wiki/Vesicular_acetylcholine_transporter

Acetylcholinesterase has a very high catalytic activity – each molecule of AChE degrades about 25000 molecules of acetylcholine (ACh) per second, approaching the limit allowed by diffusion of the substrate.

During neurotransmission, ACh is released from the presynaptic neuron into the synaptic cleft and binds to Ach receptors on the post-synaptic membrane, relaying the signal from the nerve. AChE, also located on the post-synaptic membrane, terminates the signal transmission by hydrolyzing ACh. The liberated choline is taken up again by the pre-synaptic neuron and ACh is synthesized by combining with acetyl-CoA through the action of choline acetyltransferase.

https://en.wikipedia.org/wiki/Acetylcholinesterase

Since acetylcholinesterase is abundant in the synaptic cleft and radically efficient, my first question is what prevents it from acting prematurely and disintegrating the acetylcholine before the acetylcholine gets the chance to do its job and bind with the post-synaptic receptors? My second question is how does the acetylcholinesterase know when it's time for it to stop the signal transmission at the post-synaptic receptor? Someone is deciding how long each neurotransmitter stays bound to the post-synaptic receptor site. My third question is how does the liberated choline know to make its way back to the pre-synaptic neuron, and how does it get itself there when it's dumped back into the synaptic cleft? The telekinetic powers of the pre-synaptic reuptake transporters have never been adequately explained to me in a way that actually made sense to me. They say that the correct molecules are simply drawn into the transporters. How? Is it like some kind of magnet or vacuum cleaner; or, are these presynaptic transporters quantum tunneling in the liberated choline? They simply say that it's done, but don't ever seem to explain how. Who's handling all the logistics?

Anyway, it's obvious to me that there's intelligence or psyche there in the acetylcholinesterase molecules or enzymes. They seem to know when to act, and when to refrain from acting. They are so prolific and so efficient that somebody has to be holding them back; otherwise, there would be no acetylcholine neurotransmission.

There's more going on here than can be explained by Classical Physics, Materialism, Naturalism, and the Theory of Evolution. The neurotransmitter receptors and reuptake transporters are highly intelligent, know precisely what they are doing, know when they should do it, know how long they are supposed to do it, and know when they are supposed to stop the process. Psyche or Intelligence is a function of Quantum Mechanics and Non-Local Consciousness, not Classical Physics.

As I see it, someone is running the show behind the scenes. Nature's Psyche knows precisely what it is doing and when it's supposed to do it. As I see it, someone is holding the hounds back until the Acetylcholine reaches the appropriate postsynaptic receptors and does its job.

The Materialists and Naturalists will tell us that evolution and natural selection designed and programmed these things to wait to do their job before clearing the synaptic cleft; but, where is all of that programming, information, knowledge, and data stored within acetylcholinesterase if the Materialists and Naturalists are correct in their assumptions? And, how do the acetylcholinesterase know where they are in the cycle between action potentials? How do they do their timing and self-regulation? How are the acetylcholinesterase enzymes communicating with the post-synaptic receptors, so that they know when it's time to do their job or time to refrain from doing their job? How do the liberated choline make their way back to the pre-synaptic neuron rather than congregating and clogging up the synaptic cleft? Who's handling all the logistics, because the presynaptic reuptake transporters and the post-synaptic receptors are working at cross-purposes after all? This neurotransmission process shouldn't work; but, it does. Why?

The most thoroughly studied transmitter-gated ion channel is the nicotinic ACh receptor at the neuromuscular junction in skeletal muscle. It is a pentamer, an amalgam of five protein subunits arranged like the staves of a barrel to form a single pore through the membrane. Four different types of polypeptides are used as subunits for the nicotinic receptor, and they are designated α, β, γ, and δ (abbreviated $\alpha_2\beta\gamma\delta$).

There is one Ach binding site on each of the 'α' subunits; the simultaneous binding of Ach to both sites is required for the channel to open.

The nicotinic Ach receptor on neurons is also a pentamer, but, unlike the muscle receptor, most of them are comprised of α and β subunits only (in a ratio of $\alpha_3\beta_2$). (*Neuroscience*, p. 152.)

Simultaneous binding is required to prevent or mitigate false starts. How ingenious!

Would you know how to design and build one of these Neurotransmitter-Gated Channels? Would it work right when you are done? Could you design and build a complete neurotransmitter system from scratch? Would it work right when you are done? The Materialists, Naturalists, and Darwinists assure us that evolution (random mutations and natural selection) knew how to do all of this billions of years ago when evolution designed and created our neurotransmitter systems. Do you buy it? If you do, then you are as gullible as I was when I was a Materialist, Nihilist, and Atheist.

Materialism and Naturalism Have All the Answers

As I study and do research, I do find physical explanations for some of this, but not all of it. While looking for physical explanations, the book *Neuroscience: Exploring the Brain* ended up being noticeably superior to *Biopsychology*. *Biopsychology* from Pinel generated more questions than it answered.

We are told that there are physical explanations for all of these questions, and that we just have to wait for Materialism and Naturalism to explain it all to us. I no longer buy into Promissory Materialism. It's a crock and a dodge. Promissory Materialism and Darwinism simply sweep everything under the rug and leave it up to others to meet their burden of proof. Most of the time, Materialism, Naturalism, and Atheism are based upon a refusal to look at evidence and a refusal to take things at face-value. Many of these people actively engage in the suppression of evidence. I have changed. I want evidence, lots and lots of evidence. Neo wanted guns. I want evidence.

https://en.wikipedia.org/wiki/Materialism

http://mypsyche.us/wp-content/uploads/2017/12/Materialism-Wikipedia.pdf

Some people have complained that I'm overly brutal against Materialism, Naturalism, Darwinism, and Classical Physics. They do have a point to a certain extent. We each have to start where we are, and then go on from there. It's obvious that we are physical beings, and we naturally start with physical matter and classical physics and master those before we move on to trying to figure out how spirit matter and psyche work.

The reason I learned to hate Materialism, Naturalism, Darwinism, Nihilism, and Atheism is because I fell for these things and because I have caught these people lying to me and trying to trick us and deceive us dozens of different times in dozens of different ways. I don't like being lied to; so, I'm going to resist it.

This Wikipedia article about Materialism mentions Joseph Smith:

> **Joseph Smith, the founder of the Latter Day Saint movement, taught: "There is no such thing as immaterial matter. All spirit is matter, but it is more fine or pure, and can only be discerned by purer eyes; we cannot see it; but when our bodies are purified we shall see that it is all matter." This spirit element** [spirit matter and psyche] **has always existed; it is co-eternal with God. It is also called "intelligence" or "the light of truth", which like all observable matter "was not created or made, neither indeed can be". Members of the Church of Jesus Christ of Latter-day Saints view the revelations of Joseph Smith as a restoration of original Christian doctrine, which they believe post-apostolic theologians began to corrupt in the centuries after Christ.**

I have studied many of Joseph Smith's other writings, productions, and alleged revelations from God; and after my research, I have concluded that Joseph Smith was either the greatest savant genius that has ever lived on this planet or he really was a prophet of God getting revelations from the Biblical God Jesus Christ just as he claimed to be. It's all too perfect and self-consistent to be otherwise. When it comes to science, quantum mechanics, quantum non-locality, and human psychology, Joseph Smith was clearly centuries ahead of the rest of us.

Intelligence or the Light of Truth is what we have been calling Psyche or Non-Local Consciousness. Spirit matter really is the same thing as physical matter, it's just that they each exist in a different phase or dimension. Multiple physical objects can occupy the same space if each one of them is slightly out of phase with the others. If your physical body were slightly out of phase with your physical surroundings, then you would be able to walk through the walls.

Your spirit body is out of phase with your physical body, so they are able to occupy the same space. Your psyche, intelligence, or non-local consciousness is out of phase with your spirit body and your physical body, so your psyche is able to occupy the same space as your spirit body and physical body.

Joseph Smith got all of this right back in the 1830's; and, the rest of us are starting to discover that he was right, a couple of centuries later. That kind of tells me that he really was getting his revelations and his insights from God. Of course, your mileage may vary depending upon what you are ready and willing to accept and believe. You are a different member of the jury after all; and, the evidence will naturally influence you differently than the way it influenced me. But, I'm no longer hiding from evidence, like I used to do when I was a Materialist, Nihilist, and Atheist.

Other Neurotransmitters

Neurotransmitters are insane; and, they bug me to no end. They are so inefficient and unreliable, because they are based upon random diffusion and random chance. I don't like it. The exotic neurotransmitters can walk through walls and go anywhere.

https://en.wikipedia.org/wiki/Neurotransmitter

https://en.wikipedia.org/wiki/Neurotransmission

https://en.wikipedia.org/wiki/Amino_acid_neurotransmitter

https://en.wikipedia.org/wiki/Glycine

https://en.wikipedia.org/wiki/Aspartic_acid

https://en.wikipedia.org/wiki/Serine

https://en.wikipedia.org/wiki/Gamma-Aminobutyric_acid

https://en.wikipedia.org/wiki/Gaseous_signaling_molecules

https://en.wikipedia.org/wiki/Trace_amine

https://en.wikipedia.org/wiki/Purinergic_signalling

https://en.wikipedia.org/wiki/Anandamide

https://en.wikipedia.org/wiki/Cannabinoid

There are several unusual qualities about endocannabinoids:

1. They are not packaged in vesicles like most other neurotransmitters; instead, they are manufactured rapidly and on-demand.

2. They are small and membrane permeable; once synthesized, they can diffuse rapidly across the membrane of their cell of origin to contact neighboring cells.

3. They bind selectively to the CB1 type of cannabinoid receptor, which is mainly located on certain presynaptic terminals.

CB1 receptors are G-protein-coupled receptors, and their main effect is often to reduce the opening of presynaptic calcium channels. With its calcium channels inhibited, the ability of the presynaptic terminal to release its neurotransmitter (usually GABA or glutamate) is impaired. Thus, when a postsynaptic neuron is very active, it releases endocannabinoids, which suppress either the inhibitory or excitatory drive onto the neuron (depending on which presynaptic terminals have the CB1 receptors).

This general endocannabinoid mechanism is used throughout the CNS, for a wide range of functions that we don't completely understand.

https://books.google.com/books?id=75NgwLzueikC&pg=PA150#v=onepage&q&f=false

We don't completely understand this? Ya think?

This whole process implies that the post-synaptic neuron is sentient, conscious, and aware. It doesn't like being over-activated, so it takes steps to solve the problem. But, that's physically impossible, because there's no identifiable brain within a neuron.

Endocannabinoids are manufactured on demand. By whom? Why? How? Demands require someone who is making the demands, intelligence, as well as a way to communicate those demands.

Endocannabinoids can walk through walls and go anywhere. Who tells them where to go? They can go anywhere they want – or just sit there and go nowhere. So who tells them where to go? What motivates them to move?

Why do they interact with other cells, and not their own cell? Or do they interact with their own cell?

What makes the endocannabinoids move towards a specific receptor on a specific presynaptic terminal? How do they know where to go? How do they know which presynaptic terminal to head towards? Endocannabinoids have potentially 10^{15} synapses (1 quadrillion) to select from. How do they select the right one remotely at a distance? If you were running the show, how would you tell an endocannabinoid where to go? How would you like to have that assignment? One quadrillion is a lot to select from in a matter of micro-seconds. Could you do it? Could you guide and move an endocannabinoid to the destination of your choosing? Or would it just sit there doing nothing?

Endocannabinoids are like little remote-control cars. Someone is clearly controlling them wirelessly at a distance, since the little buggers can walk through walls and eventually lose contact with their originating neuron. They can go anywhere, yet they go to precisely where they are needed. How do they do that? This is Action at a Distance. It is physically impossible!

Endocannabinoids would be completely worthless unless someone is commanding these things, driving these things, and telling them where to go. Random diffusion or random chance is NOT a reliable platform upon which to build a neurotransmitter system. It doesn't work. It can't work! The fact that these endocannabinoids know precisely where they are going and proceed to do so efficiently and reliably is proof to me that there is something going on here besides Classical Physics, Materialism, and Naturalism. The very existence, functionality, and reliability of endocannabinoids falsifies Materialism, Naturalism, and Darwinism.

Yet on the very same page, these people state:

In the first year of medical school, I was introduced to the human brain, the crowning achievement of evolution. (*Neuroscience*, p. 150.)

These people want you to know, remember, and believe that evolution designed, created, and now controls your endocannabinoid neurotransmitters and endocannabinoid systems, even though that's physically impossible. My questions are, "How did evolution (genetic drift), random mutations, and natural selection figure out how our endocannabinoid systems work billions of years ago, when we can't even figure them out here and now?" "How do evolution, random mutations, and natural selection currently command and control our endocannabinoid systems?" It's obvious that Classical Physics isn't driving our endocannabinoid systems, so my last question is, "How did evolution, random mutations, and natural selection learn to do Psyche and Quantum Mechanics?"

It's magic!

Whenever we encounter something that's physically impossible, like the precision targeting and logistics of endocannabinoids and gaseous neurotransmitters, we are forced to turn to Quantum Mechanics and Quantum Non-Local Consciousness (Psyche) for a logical and workable scientific explanation for what we are observing and experiencing. Random diffusion and random chance isn't going to cut it as an explanation for what we are witnessing.

ATP Is Where We See the Hand of God in Action

ATP synthase is an enzyme that creates the energy storage molecule adenosine triphosphate (ATP). ATP is the most commonly used "energy currency" of cells for most organisms. It is formed from adenosine diphosphate (ADP) and inorganic phosphate (P_i). — Wikipedia

https://en.wikipedia.org/wiki/ATP_synthase

I have seen animations of ATP and ATP synthesis from ADP and P_i; and, I still can't figure out who chooses the target for an ADP and ATP molecule and what makes these molecules move towards their chosen target – at the physical level. After ATP has been manufactured, it simply goes to wherever it is needed. How does it know where to go; and, who motivates it to get there? When ATP is used, the ADP waste product knows that it has to go back to the ATP Synthase machine for recharging. How does the ADP know where the machine is located and how does the ADP get there? Both the ATP and ADP molecules know that they have someplace that they need to be; and, they find some way to get there. There are way too many black boxes where ADP, ATP, and ATP synthase are concerned. Somebody is clearly running the show behind the scenes.

ATP is made in the mitochondria.

Then it has to migrate from the mitochondria to the outer walls of a neuron where it is used by the sodium-potassium pumps to maintain the cell's membrane potential.

If we were to flood the neuron with ATP, we could probably depend upon Brownian motion and random diffusion to get the ATP to the sodium-potassium pumps. But, there's an ADP "waste product" being produced by the chemical reaction that uses the ATP for energy; and, this is where we would see Brownian motion, random diffusion, and entropy working against us all the way back! You see, the ADP and AMP waste products have to make their way back to the mitochondria for recharging; and, the same Brownian motion, entropy, and random diffusion that we used to get the sodium-potassium pumps painted with ATP will now prevent the ADP and AMP from making their way back to the mitochondria for recharging. It could take a million years for an ADP or AMP molecule to make its way back to the mitochondria naturally through Classical Physics, especially with all of those incoming ATP blocking the way.

In truth, though, we don't see any of this happening within a living cell.

Each ATP molecule seems to go precisely where it is needed, and the ADP molecule waste product seems to find the nearest mitochondria and returns there directly for recharging. Someone Psyche is motivating these things, making them move, giving them assignments, and telling them where to go after they have completed their assignments. ALL of the animations portray ATP Synthase as the very model of efficiency, as if the hand of God were in it.

When ATP molecules are "made" or "recharged" they are dumped back into the cytoplasm. They are free-roaming. They can go anywhere they want to go. How do they know where to go and what to do? Whenever an ATP molecule interacts with a G-protein and produces a second messenger, that second messenger is also dumped into the cytoplasm and is free-roaming. How does the second messenger know where to go and what to do when it gets there? How is all this communication taking place between the molecules and the living cell? What makes these molecules move? Even if we were to hook up ionic propulsion to these molecules, Someone Psyche still has to steer them in the correct direction and make them dock with their pre-chosen target when they get there.

The second messengers being produced by the interaction of ATP and G-Proteins go to precisely where they are supposed to go. How do they do that? That's physically impossible!

I don't know if you can see and understand any of this or not; but, I sure learned how to do so. I didn't stop at the physical level. I decided and chose to take everything to the quantum level. I decided to do Quantum Neuroscience in addition to regular Neuroscience. They're both interesting and useful, in my humble opinion.

Imagine if I were to give you the task of commanding and controlling the quadrillions of ATP molecules and ADP molecules within your physical body. How would you do so?

Well, you would have to use quantum waves or some kind of telepathy to communicate with them so that you can tell them where to go. And, you would have to pick some sort of quantum mechanism to make them move towards their assigned destination. These are molecules after all. Molecules just sit there and vibrate under Brownian motion when under the control of Classical Physics, Materialism, and Naturalism. You are going to have to use some sort of quantum mechanism in order to make them move – Action at a Distance, Telekinesis, or Quantum Tunneling. You are also going to have to rotate them into the correct orientation and dock them when they reach their destination. These are molecules after all. They are not going to do anything unless you make them do something with quantum mechanisms at the quantum level.

Let's say I give you the task of creating and producing the first functional protein. How would you do so?

Well, first of all, you would have to design the thing atom-by-atom and be relatively confident that it's going to work right when you are done, or it isn't going to be worth your time. You are going to have to gather together the correct number and types of atoms.

Next, you will have to make the correct atoms come together into the right amino acids that you need for the task. Then you would have to make the amino acids line up in a specific order, connect or dock in a specific way, and then fold up into a 3D structure in a specific manner. How would you communicate all of this information to these atoms and molecules? The only way you could do so is at the quantum level with quantum waves or telepathy. How would you make these individual atoms and molecules come together and line up properly? Again, the only way would be to use some kind of Action at a Distance, or Telekinesis, or Quantum Tunneling at the quantum level. You would have to use quantum mechanisms to make these atoms and molecules come together into the proper sequence and structure. Remember, it's physically impossible to do any of this at the physical level, because you don't have any proteins or enzymes yet because you are making the first one.

These same realities apply directly to the ATP Synthase machine and the ATP molecules. Someone Psyche had to make the first one!

These same realities apply directly to genes, proteins, and genomes. Someone Psyche had to make the first one.

It has been proven beyond a shadow of a doubt that chemical evolution of proteins and matching genes is physically impossible. We are talking about atoms and molecules here. They don't do anything unless you make them do something. Remember, Nature's Psyche can use quantum mechanisms to do the physically impossible, to make things from atoms, and to make atoms and molecules do the things that it wants them to do; and, it does. Nature's Psyche and God's Psyche can do the physically impossible at will.

WE KNOW for a fact that machines require designers, engineers, field-testers, fine-tuning, repairmen, and manufacturers to make them and keep them working. The very existence of a machine required Someone Psyche to make it and get it going in the first place.

In contrast, we have this claim:

The evolution of ATP Synthase is thought to have been modular whereby two functionally independent subunits became associated and gained new functionality. — Wikipedia

This claim of evolving ATP Synthase machines is science fiction.

How do we KNOW?

WE KNOW because any organism, that doesn't have a fully functional ATP Synthase machine to begin with, will die immediately and never have progeny or descendants.

Evolution of any kind is nothing more than ad hoc just-so story-telling. It is science fiction, and nothing more.

It's physically impossible for chemical evolution to make ATP Synthase machines from atoms and molecules, because evolution of any kind by definition in principle has NO access to quantum mechanisms at the quantum level. In fact, the very existence of Quantum Mechanics falsifies Materialism, Naturalism, and the theory of evolution.

And, it's physically impossible for an ATP Synthase machine to evolve into existence one gene and one protein at a time, because all of those organisms that don't have an ATP Synthase machine functioning within them from the beginning will immediately go extinct.

Remember, it's physically impossible for random mutations to produce a genetically compatible Mr. and Mrs. Mutant each and every time that a new species, sexually incompatible with the previous species, is produced by luck or random chance. Design, programming, engineering, field-testing, fine-tuning, manufacturing, and creation are physically impossible through abiogenesis, spontaneous generation, chemical evolution, macro-evolution, natural selection, and random mutations.

Evolution is an evolutionary dead-end. Evolution produces the extinction of species, not the origin of species.

The ATP-ADP Cycle should be extremely disturbing to Materialists, Naturalists, and Darwinists because these molecules KNOW precisely where they are going, what they are supposed to do when they get there, and how to do it – something that random diffusion and Brownian motion and entropy should make and would make impossible.

It's at the molecular level, the quantum level, where we see the Hand of God in action.

Amino acids, mRNA, tRNA, rRNA, AMP, ADP, ATP, and proteins KNOW precisely where they are supposed to go and how to get there. They move directly to their pre-chosen coordinates. Amino acids in the wild are incapable of self-assembly into complex functional proteins; yet, within a living cell, these same amino acids KNOW precisely where they are supposed to go and proceed to do so.

Messenger RNA (mRNA) has to travel through the nuclear cytosol, the nuclear pores of the nuclear membrane, and the cytoplasm of the cell to reach a specific ribosome. It would take ALL eternity for it to do so if it were under the control of evolution, Brownian motion, random diffusion, and Classical Physics.

Transfer RNA (tRNA) knows precisely where it's going – to a specific ribosome and a specific Codon on the MRNA molecule that's being read by that ribosome.

When the different proteins, ATP molecules, ADP molecules, mRNA molecules, tRNA molecules, and amino acids are simply dumped into the cytoplasm of a living cell, how do they KNOW where to go and how do they get there? It's as if these molecules are intelligent, sentient, and actively choosing their destination; and, it's as if invisible forces and fields are driving each one of them to a specific pre-chosen destination. You could theoretically ionize these molecules in order to make them move – use some sort of invisible field; but, you still need Someone Psyche to steer them and get them to their pre-selected destination.

What do you think Psyche or Quantum Non-Local Consciousness really is?

Psyche is some kind of conscious intelligence – a chooser and causal agent – as well as some type of invisible force or field like microwaves, radio waves, magnetism, gravity, dark energy, or quantum waves.

Psyche does Action at a Distance just like magnetism, gravity, dark energy, gluons, and bosons. There's no direct physical contact between two magnets, yet they draw themselves together through some kind of invisible force or quantum wave.

The very existence of magnetism falsifies Materialism and Naturalism. Magnetic forces and fields are not physical. They are spiritual and non-local in nature and origin. They are based upon some type of quantum wave and Action at a Distance.

How many other things can you think of that falsify Materialism and Naturalism?

Now that I have started looking, I can think of dozens of different things, including quantum waves, Action at a Distance, and Quantum Mechanics; and, there are probably thousands of different things that falsify Materialism and Naturalism. Once you successfully use observation and science to falsify Materialism and Naturalism, then Darwinism or the theory of evolution goes down with them as fruit from the poisoned tree.

When it comes to atoms and molecules, moving directly and precisely implies knowledge of coordinates and specific destinations, sentience, motivation, intelligence, teleology, purpose, intention, command, control, quantum mechanisms, quantum communication, quantum waves, psyche, and conscious awareness. These atoms and molecules literally come alive when they find themselves in the presence of a living cell and Nature's Psyche.

Remember, pre-chosen coordinates, targeting, and steering imply Psyche, Intelligence, Law, and Will.

Where do these atoms and molecules get brains enough to know how to do all of these things?

Evolution is telling them what to do, according to the Materialists, Darwinists, and Naturalists; but, that's physically impossible!

Proteins KNOW precisely where they are supposed to go and what they are supposed to do after they are born – or their transporters do, if they have transporters.

Once it is created, ATP knows precisely where it is supposed to go. Once the ATP is used, its ADP waste product knows that it is supposed to go back to the ATP Synthase nano-machine for recharging.

There's way too much intelligence happening here for evolution to be the cause. Evolution can't function at the quantum level because evolution has no Psyche or Intelligence. Evolution is limited to the physical level.

Remember, Psyche is the ultimate causal agent, at any level of existence. Psyche can use quantum mechanisms to do the physically impossible.

The interesting thing about Quantum Neuroscience is that it was already an observed science, verified science, and proven science before I coined the term. Quantum Neuroscience has just been sitting here in plain sight for decades waiting for someone to pick it up, identify it, and then use it.

You are welcome to it, because technically I'm not the one who discovered it. The Council of the Gods has been using Quantum Neuroscience, Quantum Mechanics, and Action at a Distance for quadrillions of years – one physical universe after another. It's what they do, and they are very good at it.

Sodium-Potassium Pumps and ATP

Thank God evolution, random mutations, and natural selection knew how to make and balance miniature batteries, potentials, membranes, and pumps billions of years ago, or our skulls would be nothing but a jar of Gatorade. Would you know how to design and create and manufacture these types of things from scratch from individual atoms? No? Well don't worry about it. Evolution knew how to design and manufacture these things billions of years ago, so none of us really need to figure out how any of these things work anyway.

https://en.wikipedia.org/wiki/Sodium-potassium_pump

https://en.wikipedia.org/wiki/Na%2B/K%2B-ATPase

https://en.wikipedia.org/wiki/Action_potential

https://en.wikipedia.org/wiki/Membrane_potential

https://en.wikipedia.org/w/index.php?title=Equilibrium_potential&redirect=no

https://en.wikipedia.org/wiki/Reversal_potential

https://en.wikipedia.org/wiki/Excitatory_postsynaptic_potential

https://en.wikipedia.org/wiki/Inhibitory_postsynaptic_potential

THIS is the other one that bothers me a lot – Sodium-Potassium Pumps and ATP.

Sodium-Potassium Pumps are powered by ATP. I have seen animations of ATP and ATP synthesis from ADP and P_i; and, I still can't figure out who chooses the target for an ADP and ATP molecule and what makes these molecules move towards their chosen target. After ATP has been manufactured, it simply goes to wherever it is needed. How does it know where to go; and, who motivates it to get there? When ATP is used, the ADP waste product knows that it has to go back to the ATP synthase machine for recharging. How does the ADP know where the machine is located and how does the ADP get there? Both the ATP and ADP molecules know that they have someplace that they need to be; and, they find some way to get there. There are way too many black boxes where ADP, ATP, and ATP synthase are concerned. Somebody is clearly running the show behind the scenes.

https://www.youtube.com/results?search_query=atp+synthase

ATP is used to power many different functions within living cells.

The sodium-potassium pump is an enzyme that breaks down ATP in the presence of internal Na^+. The chemical energy released by this reaction drives the pump, which exchanges internal Na^+ for external K^+. The actions of this pump ensure that K^+ is concentrated inside the neuron and that Na^+ is concentrated outside. Notice that the pump pushes these ions across the membrane against their concentration gradients (Figure 3.16). This work requires the expenditure of metabolic energy. Indeed, it has been estimated that the sodium-potassium pump expends as much as 70% of the total amount of ATP utilized by the brain. (*Neuroscience*, p. 66.)

ATP is also a precursor to RNA and DNA. ATP is involved in intracellular signaling, DNA and RNA synthesis, amino acid activation in protein synthesis, and as a binding cassette transporter. ATP seems to be intelligent, conscious, sentient, aware, mobile, and alive. They say that it carries signals. How does the ATP receive these signals, and how does it know where to carry these signals when it receives them? ATP is always precisely where it needs to be, as if it is working on some kind of schedule and using some kind of 3D map.

Even worse, ATP also functions as a neurotransmitter.

Adenosine Triphosphate (ATP) as a Neurotransmitter. Adenosine triphosphate (ATP) is an important extracellular signaling molecule. ATP acts as a neurotransmitter in both peripheral and central nervous systems. In the peripheral nervous system, ATP is involved in chemical transmission in sensory and autonomic ganglia.

Other processes regenerate ATP such that the human body recycles its own body weight equivalent in ATP each day. Source: Uncle Google and Wikipedia.

Who keeps track of all these ATP molecules and tells them where to go and what to do when they get there? How do all the different processes draw in the ATP molecules? Animations of these things show the ATP immediately doing whatever they need to do as if they know precisely where they are going and precisely what they are supposed to do when they get there. Who controls the requisition and replenishment of ATP? The manufacture of the stuff requires energy.

The authors of all the different neuroscience and biopsychology books imply that evolution designed, created, and now controls ATP; but, I don't buy it. I can't visualize how random mutations and natural selection could ever get their hands on ATP in order to tell ATP what to do and where to go; and, evolution by definition in principle cannot do complexity, specificity, logistics, teleology, command and control, design, programming, and creation. Evolution is supposed to be dumb and blind.

The book, *Neuroscience*, states that the Sodium-Potassium Pumps continue to run even during action potentials. Running during action potentials isn't the problem, because the action potentials are so extreme that nothing can override them. Action potentials have to be interrupted and blocked at synapses, which is probably why God gave us chemical synapses – so that we could interrupt and stop action potentials.

The thing that bugs me so much about Sodium-Potassium Pumps is that they are efficient enough to completely eliminate Excitatory Post-Synaptic Potentials (EPSPs) and Inhibitory Post-Synaptic Potentials (IPSPs); so, why don't they? Some of the EPSPs and IPSPs are extremely subtle. Maybe they are getting eliminated as if they never existed. I didn't like that. What if that message was really important and enough neurotransmitter receptors weren't lucky enough to get triggered, in order to keep that message alive? Is Psyche going to have to send the message again? I guess so.

Once again, I'm getting conflicting information here, and I couldn't figure out what parts of it are true and what parts of it are false, nor could I figure out how to reconcile the differences to my satisfaction in terms of Materialism, Naturalism, and Classical Physics.

It ends up being Proof of Psyche.

Concentration and attention are a function of the Human Psyche. Under the Orthodox Interpretation of Quantum Mechanics, the Human Psyche chooses and decides what to pay attention to, what to think about, and what to concentrate on. This is the way things actually work in real life! The Orthodox Interpretation of Quantum Mechanics actually matches with reality and the Lived Experiences of the human race.

According to the Orthodox Interpretation of Quantum Mechanics, the Human Psyche has to keep issuing the request long enough to keep the request alive and make it important (concentration and attention); and thereby, Nature's Psyche gets into the habit of answering "yes" to that question until a whole neural pathway is developed and born. We human beings or human psyches learn to concentrate and pay attention; and, Nature's Psyche ends up habituating and developing neural pathways simply because both of us are fighting against the resistance that's being provided by Sodium-Potassium Pumps all throughout our brain, which are trying to maintain equilibrium and status quo.

The reason why these Sodium-Potassium Pumps bugged me so much is because I have been schooled, conditioned, and brainwashed in Materialism, Naturalism, Darwinism, and Scientism; and, there's no logical explanation for these things and what they do (eliminating EPSPs and IPSPs and action potentials) in terms of Materialism, Naturalism, and Classical Physics. I was forced to turn to Psyche or Non-Local Consciousness as well as the Orthodox Interpretation of Quantum Mechanics in order to find an explanation that I actually found reasonable and satisfactory. When the Human Psyche learns to concentrate and pay attention, it can override the equalizing actions of the Sodium-Potassium Pumps and trigger Action Potentials instead.

The explanation was simple and logical, once I brought Psyche and Quantum Mechanics into the equation. Quantum Non-Local Consciousness (Psyche) has explanatory power that Materialism, Naturalism, Darwinism, and Classical Physics completely lack.

Neurotransmitter Reuptake Transporters

We have a pump for that!

They called these things Transporters. Two things came immediately to mind, both of which were inaccurate and wrong. I have watched the Transporter movies, Star Trek, and The Fifth Element; and, the first thing that comes to mind is a car or a taxi or a shuttlecraft. The other thing that comes to mind is transmats in Dr. Who or the transporters in Star Trek or the blink drive in the Dark Matter series – devices which Quantum Tunnel physical matter to its destination through sub-space.

https://en.wikipedia.org/wiki/Neurotransmitter_transporter

https://en.wikipedia.org/wiki/Reuptake_inhibitor

https://en.wikipedia.org/wiki/Glutamate_transporter

https://en.wikipedia.org/wiki/GABA_transporter

https://en.wikipedia.org/wiki/Glycine_transporter

https://en.wikipedia.org/wiki/Monoamine_transporter

https://en.wikipedia.org/wiki/Equilibrative_nucleoside_transporter

https://en.wikipedia.org/wiki/Vesicular_acetylcholine_transporter

https://en.wikipedia.org/wiki/Enzyme_inhibitor

https://en.wikipedia.org/wiki/Acetylcholinesterase

Eventually, it became clear to me that I wanted to know how Receptors and Transporters really work. That's where the tires hit the pavement.

The book, *Neuroscience*, taught me that Transporters are pumps. Once I was taught that, then the physical dynamics of these things made a bit more sense to me. Initially, due to some drawings in *Biopsychology*, I visualized the Transporters as little ships or vessels floating around in the synaptic cleft collecting up the remnant neurotransmitters and then shipping them back to the presynaptic neuron. I was trying to visualize how that might work. Then other parts of *Biopsychology* eventually made the Transporters seem to be vacuum cleaners sucking up the neurotransmitters from the synaptic cleft. I was confused. The book, *Neuroscience*, did a much better job of explaining how Transporters really work.

Transporters are ACTIVE rather than passive, which makes it much easier to provide a physical explanation for how they work. Transporters are carefully designed pumps. Each Transporter is designed to pump in a specific neurotransmitter. This reality doesn't explain how Transporters draw in distant neurotransmitters and completely clear a synaptic cleft (as I understand it, gradients diffuse and diminish with distance just like magnets get weaker at a distance), but the transmembrane gradients that Transporters use do explain how Transporters are able to draw in nearby neurotransmitters.

Pumping Ions and Transmitters

Neurotransmitters may lead an exciting life, but the most mundane part of it would seem to be the steps that recycle them back from the synaptic cleft and eventually into a vesicle. Where synapses are concerned, the exotic proteins of exocytosis and the innumerable transmitter receptors get most of the publicity. Yet the neurotransmitter transporters are very interesting for at least two reasons: They succeed at an extraordinarily difficult job, and they are the molecular site at which many important psychoactive drugs act.

The hard job of transporters is to pump transmitter molecules across membranes so effectively that they become highly concentrated in very small places. There are two general types of neurotransmitter transporter. One type, the *neuronal membrane transporter*, shuttles transmitter from the extracellular fluid, including the synaptic cleft, and concentrates it up to 10,000 times higher within the cytosol of the presynaptic terminal. A second type, the *vesicular transporter*, then crams transmitter into vesicles at concentrations that may be 100,000 times higher than in the cytosol. Inside cholinergic vesicles, for example, ACh may reach the incredible concentration of 1000 mM, or 1 molar — in other words, about twice the concentration of salt in seawater!

How do transporters achieve such dramatic feats of concentration? Concentrating a chemical is like carrying a weight uphill; both are extremely unlikely to occur unless energy is applied to the task. Recall from Chapter 3 that ion pumps in the plasma membrane use ATP as their source of energy to transport Na^+, K^+, and Ca^{2+} against their concentration gradients. These ion gradients are essential for setting the resting potential, and for powering the ionic currents that underlie action and synaptic potentials. Notice, however, that once ionic gradients are established across a membrane, they can themselves be tapped as sources of energy. Just as the energy spent in pulling up the weights on a cuckoo clock can be reclaimed to turn the gears and hands of the clock (as the weights slowly fall down again), transporters use transmembrane gradients of Na^+ or H^+ as an energy source for moving transmitter molecules up steep concentration gradients. The transporter lets one transmembrane gradient, that of Na^+ or

H^+, run down a bit in order to build up another gradient, that of the transmitter.

The transporters themselves are large proteins that span membranes. There can be several transporters for one transmitter (e.g., at least four subtypes are known for GABA). Figure A shows how they work. Plasma membrane transporters use a *cotransport* mechanism, carrying two Na^+ ions along with one transmitter molecule. By contrast, vesicular membrane transporters use a *countertransport* mechanism that trades a transmitter molecule from the cytosol for an H^+ from inside the vesicle. Vesicle membranes have ATP-driven H^+ pumps that keep their contents very acidic, or high in protons (i.e., H^+ ions).

What is the relevance of all this to drugs and disease? Many psychoactive drugs, such as amphetamines and cocaine, potently block certain transporters. By altering the normal recycling process of various transmitters, the drugs lead to chemical imbalances in the brain that can have profound effects on mood and behavior. It is also possible that defects in transporters can lead to psychiatric or neurological disease; certainly some of the drugs that are therapeutically useful in psychiatry work by blocking transporters. The numerous links between transmitters, drugs, disease, and treatment are tantalizing but complex, and will be discussed further in Chapters 15 and 22. (*Neuroscience*, p. 144.)

http://mypsyche.us/wp-content/uploads/2017/12/Transporters.pdf

This is the best explanation of Transporters that I have found so far; but, I'm sure that there are many others once a person knows what he or she is looking for.

https://en.wikipedia.org/wiki/Cotransporter

https://en.wikipedia.org/wiki/Active_transport

https://en.wikipedia.org/wiki/Passive_transport

https://en.wikipedia.org/wiki/Countercurrent_exchange

https://en.wikipedia.org/wiki/Transport_phenomena

The significance of Transporters comes from their design and creation. It should be clear to any rational person that chemical evolution, evolution (genetic drift over generations), random mutations, and natural selection could never have designed and created a Transporter, which is comprised of multiple different proteins. Chemical evolution can't create a single protein, let alone a machine comprised of multiple different proteins. The Transporters were designed and created by God. There is no other rational explanation for their existence.

Cellular Transport Machinery

https://en.wikipedia.org/wiki/Protein_targeting

Who tells the cellular transport machinery where to take a particular protein? They could take the thing anywhere. This whole process is so recursive that eventually we find ourselves staring into the face of God. Once we find a physical explanation for a process like protein targeting or protein transportation, then we find ourselves asking how the cellular transport machinery select their targets and know where they are going. Eventually, we make our way back to Nature's Psyche, which seems to be in direct contact with God's Psyche and God's Database.

I have asked many times how the proteins get to their destinations after the ribosomes turn them loose. Well, the cellular transport machinery do it. But now I'm asking how the cellular transport machinery select their destinations and know where they are going. On and on and on it goes, where it stops nobody knows. It's like a Matryoshka doll, and eventually the buck stops with God.

Neuropeptides

https://en.wikipedia.org/wiki/Neuropeptide

https://en.wikipedia.org/wiki/Peptide

https://en.wikipedia.org/wiki/G_protein%E2%80%93coupled_receptor

https://en.wikipedia.org/wiki/Metabotropic_receptor

Like everyone else, I didn't even scratch the surface of the 100+ neuropeptides. I know they are there and I know that they tend to modify the function of neurotransmitters and reprogram our DNA, but that's about it when it comes to the neuropeptides.

When it comes to neuropeptides, evolution, natural selection, and random mutations know how all of this works, so there's no logical reason for any of us to try to figure out how they work. We'll just let evolution do it. Evolution can do anything it sets its mind to. That's been my observation.

Autoreceptors

https://en.wikipedia.org/wiki/Autoreceptor

Why are autoreceptors needed, if the reuptake transporters already know how many of their assigned neurotransmitters remain within the synaptic cleft? Can't the reuptake transporters just pass that information on to the neuron rather than leaving it up to the autoreceptors to monitor the synaptic cleft and the post-synaptic receptors waiting for that information to arrive? How do the autoreceptors know when it's time for them to release the neurotransmitter that is bound to them? How do the autoreceptors, reuptake transporters, and post-synaptic receptors know what they are supposed to do and when they are supposed to do it? Who taught them how to do these things, and time these things? How do they coordinate with each other so that they don't interfere with each other, because they are working at cross-purposes after all? How is all this information getting exchanged wirelessly at a distance across the synaptic cleft and throughout the synaptic cleft?

Neuromodulation

https://en.wikipedia.org/wiki/Neuromodulation

Synapses

https://en.wikipedia.org/wiki/Synapse

https://en.wikipedia.org/wiki/Tripartite_synapse

https://en.wikipedia.org/wiki/Gap_junction

https://en.wikipedia.org/wiki/Electrical_synapse

What kind of information is being exchanged between neurons at gap junctions or electrical synapses? How did the neurons learn this language so that they could communicate this information? Information has to be in some kind of code or language, and both the transmitter and the receiver have to understand that language or code; otherwise, they won't be able to communicate with each other.

The answer is much simpler than they let on.

The only signal or message being sent through a gap junction is ON and OFF; or, MORE OPEN and MORE CLOSED.

Gap junctions are narrow spaces or ion channels between adjacent neurons and astrocytes that are bridged by fine tubular channels containing cytoplasm through which ions and small molecules can pass readily.

Who is controlling these things and how are they controlling these things? Who is opening them and closing them? A gap junction or electrical synapse is another Black Box!

The Materialists and Naturalists exaggerate and say that signals or messages are being passed through these things; but, a gap junction or an ion channel is a single hardware BIT. The thing is either OPEN or CLOSED. You can't send memories, bytes, signals, and messages through a single BIT. It's physically impossible. Ions passing through a gap junction will either increase or decrease the Post-Synaptic Potential. That's it! And, a Post-Synaptic Potential represents a single BIT of information – it either FIRES and Action Potential, or it doesn't.

The only other signal that I can visualize being sent through a gap junction is a G-protein which would then be used to either increase or decrease the effectiveness of the post-synaptic membrane by either adding or removing neurotransmitter receptors to the post-synaptic membrane.

https://en.wikipedia.org/wiki/Chemical_synapse

https://en.wikipedia.org/wiki/Excitatory_synapse

https://en.wikipedia.org/wiki/Synaptogenesis

https://en.wikipedia.org/wiki/Axon_terminal

https://en.wikipedia.org/wiki/Immunological_synapse

Potentials

The threshold of excitation is the level of depolarization necessary to generate an action potential, usually about -65 mV.

The Action Potential

https://en.wikipedia.org/wiki/Membrane_potential

https://en.wikipedia.org/wiki/Resting_potential

https://en.wikipedia.org/wiki/Action_potential

Action potentials made no logical sense to me at first!

How does replacing one positive ion K+ with another positive ion Na+ successfully trigger an action potential and the complete reversal of the resting potential and membrane potential? At first glance, it doesn't make sense, and it bugged me for months.

My answer didn't come until months later while I was reading the book *Neuroscience: Exploring the Brain*. I'm glad I bought that book, or I still wouldn't know how these things really work.

The Distribution of Ions across the Membrane

It should now be clear that the neuronal membrane potential depends on the ionic concentrations on either side of the membrane. Estimates of these concentrations appear in Figure 3.15. The important point is that K+ is more concentrated on the inside, and Na+ and Ca^{2+} are more concentrated on the outside.

How do these concentration gradients arise? Ionic concentration gradients are established by the actions of ion pumps in the neuronal membrane. Two ion pumps are especially important in cellular neurophysiology: the sodium-potassium pump and the calcium pump. The sodium-potassium pump is an enzyme that breaks down ATP in the presence of internal Na+. The chemical energy released by this reaction drives the pump, which exchanges internal Na+ for external K+. The actions of this pump ensure that K+ is concentrated inside the neuron and that Na+ is concentrated outside. Notice that the pump pushes these ions across the membrane against their concentration gradients. This work requires the expenditure of metabolic energy. Indeed, it has been estimated that the sodium-potassium pump expends as much as 70% of the total amount of ATP utilized by the brain.

The calcium pump is also an enzyme that actively transports Ca^{2+} out of the cytosol across the cell membrane. Additional mechanisms decrease intracellular Ca^{2+} to a very low level (0.0002 mM); these include intracellular calcium-binding proteins and organelles, such as mitochondria and types of endoplasmic reticulum, that sequester cytosolic calcium ions.

Ion pumps are the unsung heroes of cellular neurophysiology. They work in the background to ensure that the ionic concentration gradients are established and maintained. These proteins may lack the glamour of a gated ion channel, but without ion pumps, the resting membrane potential would not exist, and the brain would not function. (*Neuroscience*, pp. 65- 67.)

https://canvas.brown.edu/courses/851434/files/40426262/

The KEY to the successful triggering of an action potential using only positive ions lies in the equilibrium potentials for the different ions under consideration. The equilibrium potential for potassium (K^+) is -80 mV; and, the equilibrium potential for sodium (Na^+) is +62 mV, thanks to the sodium-potassium pumps, which maintain the membrane potential of the neuron around -65 mV to -70 mV, depending upon which book we choose to read. There is a huge driving force on the sodium, when the sodium ion channels open, thus driving the membrane potential towards the equilibrium potential for sodium, which is 62 mV. That's how opening the sodium ion channels flips the membrane potential from negative to positive so quickly and thereby triggers an action potential.

Consider the ideal neuron. The membrane of this cell has three types of protein molecules: sodium-potassium pumps, potassium channels, and sodium channels. The pumps work continuously to establish and maintain concentration gradients. As in all our previous examples, we'll assume that K^+ is concentrated twentyfold inside the cell and that Na^+ is concentrated tenfold outside the cell. According to the Nernst equation, at 37°C, E_K = -80 mV and E_{Na} = 62 mV. Let's use this cell to explore the factors that govern the movement of ions across the membrane.

We begin by assuming that both the potassium channels and the sodium channels are closed, and that the membrane potential, V_m, is equal to 0 mV. Now let's open the potassium channels only. As we learned in Chapter 3, K^+ ions will flow out of the cell, down their concentration gradient, until the inside becomes negatively charged, and V_m = E_K. Here we want to focus our attention on the movement of K^+ that took the membrane potential from 0 mV to -80 mV.

The Ins and Outs of an Action Potential

Let's pick up the action where we left off in the last section. The membrane of our ideal neuron is permeable only to K^+, and V_m = E_K = -80 mV. What's happening with the Na^+ ions concentrated outside the cell? Because the membrane potential is so negative with respect to the sodium equilibrium potential, there is a very large driving force on $Na+$ ([V_m - E_{Na}] = [-80 mV - 62 mV] = -142 mV). Nonetheless, there can be no net $Na+$ current as long as the membrane is impermeable to $Na+$. But now let's open the sodium channels and see what happens to the membrane potential.

At the instant we change the ionic permeability of the membrane, g_{Na} is high, and, as we said above, there is a large driving force pushing on Na^+. Thus, we have what it takes to generate a large sodium current, I_{Na}, across the membrane. Na^+ ions pass through the membrane sodium channels in the direction that takes V_m toward E_{Na}; in this case, the sodium current, I_{Na}, is inward across the membrane. Assuming the membrane permeability is now far greater to sodium than it is to potassium, this influx of Na^+ depolarizes the neuron until V_m approaches E_{Na}, 62 mV.

Notice that something remarkable happened here. Simply by switching the dominant membrane permeability from K^+ to Na^+, we were able to rapidly reverse the membrane potential. In theory, then, the rising phase of the action potential could be explained if, in response to depolarization of the membrane beyond threshold, membrane sodium channels opened. This would allow Na^+ to enter the neuron, causing a massive depolarization until the membrane potential approached E_{Na}, 62 mV.

How could we account for the falling phase of the action potential? Simply assume that sodium channels quickly close and the potassium channels remain open, so the dominant membrane ion permeability switches back from Na$^+$ to K$^+$. Then K$^+$ would flow out of the cell until the membrane potential again equals E$_K$, -80 mV. Notice that if g$_K$ increased during the falling phase, the action potential would be even briefer.

Our model for the ins and outs, ups and downs of the action potential in an ideal neuron. The rising phase of the action potential is explained by an inward sodium current, and the falling phase is explained by an outward potassium current. The action potential therefore could be accounted for simply by the movement of ions through channels that are gated by changes in the membrane potential. If you understand this concept, you understand a lot about the ionic basis of the action potential. What's left now is to see how this actually happens – in a real neuron.

https://books.google.com/books?id=75NgwLzueikC&pg=PA80#v=onepage&q&f=false

https://books.google.com/books?id=75NgwLzueikC&pg=PA82#v=onepage&q&f=false

https://books.google.com/books?id=75NgwLzueikC&pg=PA83#v=onepage&q&f=false

https://canvas.brown.edu/courses/851434/files/40426324/

This is ingenious! Thank God evolution, random mutations, and natural selection figured out how to do all of this billions of years ago, or we would be in one heck of a fix. Wouldn't we?

The Resting Potential

https://en.wikipedia.org/wiki/Resting_potential

Notice that each different type of cell has a different Resting Potential. It seems to me that if evolution designed and created these cells, then their resting potentials would all be the same, due to evolution's law of conservation. The fact that each type of cell has a different Resting Potential tells me that something is going on here besides evolution. We are looking at design and customization, not evolution, where the Resting Potentials are concerned. Even the resting potential of neurons varies from one type of neuron to another. Astrocytes have a completely different resting potential than neurons; yet, they exist side-by-side in the same extracellular cerebrospinal fluid. This means customized and carefully designed ion channels and pumps for each type of cell. That's really radical, if one thinks about it long enough. God is in the details.

https://en.wikipedia.org/wiki/Depolarization

https://en.wikipedia.org/wiki/Hyperpolarization_(biology)

https://en.wikipedia.org/wiki/Excitatory_postsynaptic_potential

https://en.wikipedia.org/wiki/Inhibitory_postsynaptic_potential

https://en.wikipedia.org/wiki/Non-spiking_neuron

https://en.wikipedia.org/wiki/Transduction_(biophysics)

It can be extremely interesting to study the potentials, or batteries, associated with the nervous system. Materialists, Naturalists, and Darwinists want you to believe that evolution, random mutations, and natural selection designed and created all of these things; but, that's physically impossible. Evolution doesn't have hands and a brain. Evolution can't do protein synthesis, nor design and creation. Evolution (random mutations and natural selection) certainly wouldn't be able to figure out how to make a potential or a battery. It has never been observed doing so.

Neurotransmitter Receptors

Receptors are cells that are specialized to receive chemical, mechanical, or radiant signals from the environment. Receptors are also proteins that contain binding sites for particular neurotransmitters. Receptor sub-types are the different types of receptors to which a particular neurotransmitter can bind. Who decides what kind of receptor sub-type to place into a post-synaptic membrane? This Someone Psyche has a number of different sub-types that it can choose to use. Specialized by whom? Anytime they say that cells have been specialized or differentiated, they are talking about Someone Psyche who either did the job on the fly or programmed that functionality into our genome to begin with. It always reduces to Someone Psyche, because evolution of any kind can't do psyche at the quantum level according to the Materialists, Naturalists, and Darwinists.

Many different polypeptides could serve as subunits of functional receptors. Consider as an example the GABA$_A$ receptor, a transmitter-gated chloride channel. Each channel requires five subunits, and there are five major class of subunit proteins, designated α, β, γ, δ, and ρ.

At least six different polypeptides (designated α1–6) can substitute for one another as an 'α' subunit. Four different polypeptides (designated β1–4) can substitute as a β subunit, and four different polypeptides (γ1–4) can be used as a γ subunit. Although this is not the complete tally, let's make an interesting calculation.

If it takes five subunits to form a GABA$_A$ receptor-gated channel and there are 15 possible subunits to choose from, then there are 151,887 possible combinations and arrangements of subunits. This means there are at least 151,887 potential subtypes of GABA$_A$ receptors!

It is important to recognize that the vast majority of the possible subunit combinations are never manufactured by neurons and, even if they were, they would not work properly. Nonetheless, it is clear that receptor classifications like those appearing in Table 6.1,

while still useful, underestimate the diversity of receptor subtypes in the brain. (*Neuroscience*, p. 141.)

Of course, we all know that evolution, random mutations, and natural selection designed and created all of this billions of years ago, so the obvious question is, "How did they do it?" How did evolution, natural selection, and random mutations pick and choose which of the 115,887 possible subunits to employ and which ones to ignore, when evolution was designing and creating our GABA$_A$ neurotransmitter system? How did evolution get these transmitter-gated ion channels built and make their subunits come together in the right spot? How did evolution get the post-synaptic neurons to choose which type of post-synaptic receptors to build, or how did evolution choose which type of post-synaptic receptors to build on a specific neuron? Inquiring minds want to know. Do you have the answers? I don't. There's no way on earth that evolution could have designed and created all these things. Chemical evolution can't even design and create a single functional protein, let alone 115,887 of them. It's not like evolution has any brains to work with. By definition, in principle, evolution is supposed to be dumb and blind, which it is. The human brain is one seriously complex computer; and, we all know that computers cannot design, create, and program themselves. It's physically impossible. The theory of evolution is the pinnacle of self-deception and wishful thinking.

https://en.wikipedia.org/wiki/Histamine_receptor

https://en.wikipedia.org/wiki/Opioid_receptor

https://en.wikipedia.org/wiki/Glycine_receptor

https://en.wikipedia.org/wiki/Ligand-gated_ion_channel

https://en.wikipedia.org/wiki/Metabotropic_receptor

https://en.wikipedia.org/wiki/G_protein%E2%80%93coupled_receptor

https://en.wikipedia.org/wiki/Voltage-gated_ion_channel

In my humble opinion, neurotransmitter receptors are the enigma. They are typically portrayed as being passive – they just sit there waiting for a neurotransmitter to come and bind with them. They function by molecular diffusion. Ouch! Some of the drawings that I have seen of these transmitter receptors portray them as a narrow cave, and it made me wonder how the neurotransmitters ever found their way into the narrow cave and then into the correct slot. But, I have learned that many of the pictures in the book, *Biopsychology*, are inaccurate and misleading, so I won't link to any of them in order to show you what I mean.

Instead, I eventually realized that additional research was needed to determine if the neurotransmitter receptors really are passive, or not. It seems that they are. They aren't even passive transporters. They truly seem to be passive receptors, in every sense of the word "passive". They just sit there and rely upon random diffusion or random chance to get their job done. That bugs me to no end. I can't even begin to explain how much that bugs me.

Passive systems, random chance, and random diffusion are notoriously unreliable, yet God seems to make these things work, especially where our physical brain is concerned. In the hands of God, a passive system is no longer passive, because He makes it work.

Is that really the best explanation that we have for neurotransmitter receptors, or can some physical explanation be employed to push God one step further back when it comes to the passive receptors in our physical brain? I had to know.

. . . Months later, and I'm still searching for an answer.

Meanwhile, I have to go with what I have.

The signaling network within a single neuron resembles in some ways the neural networks of the brain itself. It receives a variety of inputs, in the form of transmitters bombarding it at different times and places. These inputs cause an increased drive through some signal pathways and a decreased drive through others, and the information is recombined to yield a particular output that is more than a simple summation of the inputs. Signals regulate signals, chemical changes can leave lasting traces of their history, and drugs can shift the balance of signaling power.

Neurons integrate divergent and convergent signaling systems, resulting in a complex map of chemical effects. The wonder is that it ever works; the challenge is to understand how. (*Neuroscience*, pp. 164-165.)

That, too, has been my personal assessment. It's amazing that it even works. It's not supposed to.

Random chance or random diffusion is NOT a reliable platform upon which to build whole systems, such as our neurotransmitter systems. It shouldn't work. It can't work. According to classical physics and the laws of random diffusion, our neurotransmitter receptors shouldn't work, but they do. That tells me that they aren't being controlled by Classical Physics (Materialism and Naturalism), but by Nature's Psyche instead. Our neurotransmitter systems and transmitter receptors work way too reliably and efficiently for it to be otherwise. There's actual intelligence driving them. I see no other logical explanation.

After all of my research, it is still my ultimate conclusion that the various neurotransmitter systems and our neurotransmitter receptors, their efficiency and reliability, can only be explained in terms of Quantum Mechanics, Quantum Telepathy, Quantum Telekinesis, Quantum Tunneling, and the direct intelligent intervention of Nature's Psyche – because according to Classical Physics and the Laws of Random Diffusion or Osmosis they shouldn't work at all. The wonder is that it ever works; and, the miracle is that it actually works reliably and efficiently.

I'm just scratching the surface here. Books like *Neuroscience* and *Biopsychology* are full of unexplainable phenomena from beginning to end – at least they are unexplainable in terms of Classical Physics, Materialism, Darwinism, and Naturalism. These phenomena do make some sense, though, in terms of Intelligence, Psyche, and Quantum Mechanics; or at least, that's the way I see it. Your mileage may vary, however, depending upon what you are willing to accept, study, and believe. The best I can do is to explain what I'm seeing and observing within the scientific evidence, and then leave it up to you to decide for yourself what you personally want to believe.

The Materialists, Darwinists, and Naturalists continue to promise us an explanation for these things that will eliminate the need for Psyche and God; but in the case of our neurotransmitter receptors, these people have failed to deliver. Meanwhile, better and more believable explanations have come to us from the Orthodox Interpretation of Quantum Mechanics, in order to make up for the failure of these people to meet their burden of proof.

Despite their claims to the contrary, when it comes to Neuroscience and an explanation of our Neurotransmitter Systems, the Materialists and Naturalists are at least a century behind contemporary physical theory and the current prevailing principles of quantum physics, as demonstrated by Henry P. Stapp, Mario Beauregard, and Jeffery M. Schwartz.

"Quantum Physics in Neuroscience and Psychology: A Neurophysical Model of Mind-Brain Interaction"

https://www.researchgate.net/publication/7613549_Quantum_physics_in_neuroscience_and_psychology_A_neurophysical_model_of_mind-brain_interaction

http://www-physics.lbl.gov/~stapp/PTRS.pdf

http://mypsyche.us/wp-content/uploads/2017/10/PTRS.pdf

https://sites.google.com/a/lbl.gov/stappfiles/

http://www-physics.lbl.gov/~stapp/

Go with the best, and get rid of all the rest. That's what I decided to do. I'm not going to apologize for finding better science than what the Materialists, Naturalists, and Darwinists have been able to provide us. The truth is infinitely more interesting than their fictional ad hoc just-so stories, anyway. So, why not go with the truth instead?

Neurotransmission

. . . Months later, and I'm still searching for an answer.

Most scientists initially refused to believe that our physical brain is based upon chemical synapses, and not electrical synapses, because we all instinctively know how unreliable and unpredictable random chance really is. There's nothing more unreliable and unpredictable than a chemical synapse. A chemical synapse is based primarily on random chance; and, random chance generates a monumental amount of entropy, resistance, and inertia to have to overcome. According to Classical Physics, chemical synapses shouldn't work reliably and efficiently at all; yet, they do. Why? It's as if Materialism, Naturalism, and Classical Physics are incomplete and lack explanatory power. That can't be, can it?

https://en.wikipedia.org/wiki/Neurotransmission

https://en.wikipedia.org/wiki/Chemical_synapse

https://en.wikipedia.org/wiki/Acetylcholine

https://en.wikipedia.org/wiki/Acetylcholine_receptor

https://en.wikipedia.org/wiki/Vesicular_acetylcholine_transporter

https://en.wikipedia.org/wiki/Acetylcholinesterase

I finally found an explanation for neurotransmission that actually made sense to me in terms of Classical Physics. The Materialists, Naturalists, Darwinists, and Atheists are going to love this one – and they do – at least until we start thinking about the true ramifications of what's really happening here.

The immense chemical complexity of synaptic transmission makes it especially susceptible to the medical corollary of Murphy's Law, which states that if a physiological process can go wrong, it will go wrong. When chemical synaptic transmission goes wrong, the nervous system

malfunctions. Defective neurotransmission is believed to be the root cause of a large number of neurological and psychiatric disorders. The good news is that, thanks to our growing knowledge of the neuropharmacology of synaptic transmission, clinicians have new and increasingly effective therapeutic drugs for treating these disorders. We'll discuss the synaptic basis of some psychiatric disorders, and their neuropharmacological treatment, in Chapter 22.

PRINCIPLES OF SYNAPTIC INTEGRATION

Most CNS neurons receive thousands of synaptic inputs that activate different combinations of transmitter-gated ion channels and G-protein-coupled receptors. The postsynaptic neuron integrates all these complex ionic and chemical signals and gives rise to a simple form of output: action potentials. The transformation of many synaptic inputs to a single neuronal output constitutes a neural computation. The brain performs billions of neural computations every second we are alive. As a first step toward understanding how neural computations are performed, let's explore some basic principles of synaptic integration. Synaptic integration is the process by which multiple synaptic potentials combine within one postsynaptic neuron.

The Integration of EPSPs

The most elementary postsynaptic response is the opening of a single transmitter-gated channel (Figure 5.17). Inward current through these channels depolarizes the postsynaptic membrane, causing the EPSP. The postsynaptic membrane of one synapse may have from a few tens to several thousands of transmitter-gated channels; how many of these are activated during synaptic transmission depends mainly on how much neurotransmitter is released.

Quantal Analysis of EPSPs. The elementary unit of neurotransmitter release is the contents of a single synaptic vesicle. Vesicles each contain about the same number of transmitter molecules (several thousand); the total amount of transmitter released is some multiple of this number. Consequently, the amplitude of the postsynaptic EPSP is some multiple of the response to the contents of a single vesicle. Stated another way, postsynaptic EPSPs at a given synapse are quantized; they are multiples of an indivisible unit, the *quantum*, that reflects the number of transmitter molecules in a single synaptic vesicle and the number of postsynaptic receptors available at the synapse.

At many synapses, exocytosis of vesicles occurs at some very low rate in the absence of presynaptic stimulation. The size of the postsynaptic response to this spontaneously released neurotransmitter can be measured electrophysiologically. This tiny response is a miniature postsynaptic potential, often called simply a "mini." Each mini is generated by the transmitter contents of one vesicle. The amplitude of the postsynaptic EPSP evoked by a presynaptic action potential, then, is simply an integer multiple (i.e., 1X, 2X, 3X, etc.) of the mini amplitude.

Quantal analysis, a method of comparing the amplitudes of miniature and evoked postsynaptic potentials, can be used to determine how many vesicles release neurotransmitter during normal synaptic transmission.

Quantal analysis of transmission at the neuromuscular junction reveals that a single action potential in the presynaptic terminal triggers the exocytosis of about 200 synaptic vesicles, causing an EPSP of 40 mV or more. At many CNS synapses, in striking contrast, the contents of only a single vesicle are released in response to a presynaptic action potential, causing an EPSP of only a few tenths of a millivolt.

EPSP Summation.

The difference between excitatory transmission at neuromuscular junctions and CNS synapses is not surprising. The neuromuscular junction has evolved to be fail-safe; it needs to work every time, and the best way to ensure this is to generate an EPSP of a huge size. On the other hand, if every CNS synapse were, by itself, capable of triggering an action potential in its postsynaptic cell (as the neuromuscular junction can), then a neuron would be little more than a simple relay station. Instead, most neurons perform more sophisticated computations, requiring that many EPSPs add together to produce a significant postsynaptic depolarization. This is what is meant by integration of EPSPs.

EPSP summation represents the simplest form of synaptic integration in the CNS. There are two types of summation: spatial and temporal.

http://mypsyche.us/wp-content/uploads/2017/12/Neurotransmission.pdf

https://canvas.brown.edu/courses/851434/files/40426574/

http://faculty.fiu.edu/~fasoto/courses/COMPNEURO/papers/bear-ch5.pdf

http://mypsyche.us/wp-content/uploads/2017/12/bear-ch5.pdf

Yikes!

The goal in neuromuscular synapses is to slam the postsynaptic neuron with neurotransmitters. Several thousand neurotransmitters per vesicle and 200 vesicles dumped per action potential results in a million neurotransmitters dumped into each neuromuscular junction or synaptic cleft each time an action potential comes along. Think about it. That's insane! What a waste of time and energy and resources, just to transmit a single signal!

Neurotransmission is so inefficient and unreliable that in order to guarantee neurotransmission and guarantee that our muscles actually move, the presynaptic motor junctions dump an average of a million neurotransmitters into the synaptic cleft after each action potential, in order to guarantee that our muscles actually move.

I was right to be worried that neurotransmission through random diffusion – it is a horrible, weak, and inefficient way to get things done. Neurotransmission has bugged me and annoyed me all of my life, because I can sense how inefficient and unreliable it truly is – to rely upon pure chance alone in order to get something done.

According to these people, evolution realized and knew how inefficient and unreliable our neurotransmitter systems are, so when it came to our muscles and our survival, evolution, random mutations, and natural selection took the shotgun approach in order to guarantee that our muscles move. The shotgun approach is fine and dandy, but it typically leaves us with a lot of wasted meat!

A million neurotransmitters per synapse per Acetylcholine action potential just to transmit a single signal – what a waste of energy and resources! That's just insane!

For EACH action potential, our neuromuscular junctions or terminals open 200 vesicles and dump a million neurotransmitters into the synaptic cleft just to guarantee that our muscles actually move. Think about it! One carefully designed gap junction or electrical synapse would have completely eliminated the need for opening 200 vesicles of neurotransmitter and the need to make and dump a million neurotransmitters into the synaptic cleft for each action potential that comes along.

A million neurotransmitters just to transmit a single signal. Does that bug you anywhere near as much as it does me? The sheer inefficiency annoys me to no end. I have never heard of anything so inefficient! We think government is bad. This one takes the cake. Can you imagine sending out a million employees just to deliver a single memo to a single person? Can you say overkill? Of course, it's necessary to do so where random chance is involved. It takes a monumental amount of time, effort, concentration, and work to overcome the effects of entropy and random chance. Think about it. It's radically insane!

If evolution (genetic drift), natural selection, and random mutations were truly as intelligent and efficient as the Materialists and Naturalists and Darwinists claim, then evolution would have used electrical synapses instead chemical synapses. One carefully designed gap junction and electrical synapse would have eliminated the need for a million neurotransmitters each time an action potential comes down the pipeline.

The inefficiency of our neurotransmitters systems and random chance is simply staggering! In fact, it's because of random chance that WE KNOW that evolution and natural selection could never have done the things that the Darwinists and Naturalists say that evolution did. We human beings literally have to load the dice and stack the deck in order to successfully overcome random mutations and random chance. And, evolution would have had to have done the same; and, there's no way that it could, since evolution is based upon random chance or random mutations in the first place.

This is bad. With a million neurotransmitters in the synaptic cleft bathing and painting the postsynaptic neuron, we now have other monumental problems that we have to deal with and overcome. Can you sense what they are?

Cleanup!

Here, we are looking at the reverse problem. Instead of running the risk that the post-synaptic neuron will never receive the signal, we are now looking at signal overload instead. Now we have to try to figure out how to clear the synaptic cleft in a timely fashion in preparation for the next action potential, and how to prevent ongoing false starts, echoes, glitches, and signal noise.

Since we are now talking about our muscles, we are talking about acetylcholine and acetylcholinesterase here.

Now that we have a million neurotransmitters in the synaptic cleft, the degradation enzymes have to get them all cleaned up BEFORE the next action potential arrives. What a waste of time and energy! And think about it logically – until those one million acetylcholine neurotransmitters are cleaned up or removed, the postsynaptic membrane is still subjected to random firing of its neurotransmitter receptors. What a mess! We should be spazzing all the time. Why aren't we?

With one million acetylcholine neurotransmitters, the presynaptic neuron and postsynaptic neuron literally get painted and caked with neurotransmitter. In my read, I discovered that some of the synapses have walls or membranes around them, thereby preventing the escape of neurotransmitters from the synaptic cleft. That, too, would greatly enhance the reception, as well as the cleanup process. They didn't identify which synapses have walls around them, but the acetylcholine synapses would be my choice if I were designing the system.

Someone has to turn the degradation enzymes (acetylcholinesterase) loose at the right time after the postsynaptic neuron has triggered an action potential of its own; and, the degradation enzymes have to engage in a quick and efficient target and destroy mission. Then someone has to remove these highly efficient degradation enzymes from the synaptic cleft, so that they aren't there preventing the next action potential that comes along. Furthermore, the presynaptic neurons have to draw in or suck up the degradation wreckage so that it doesn't clog the synaptic cleft. What a mess – simply because we are dealing with random chance, chemical synapses, and a million neurotransmitters instead of once carefully designed gap junction and electrical synapse. If the axons would have just connected to the dendrites and/or the neurons directly through gap junctions or electrical synapses, all of this wasted energy and time could have been avoided.

It requires intelligence and skill to overcome random chance and to get the job done in the end! It's unavoidable. It's just that with our muscles and neuromuscular synapses, the intelligence and skill takes place at the end of the process instead of at the beginning of the process, because "evolution" wants to guarantee that the signal gets delivered and then leave it up to intelligent design and intelligent intervention to clean up the mess afterwards. What a waste of time and energy and resources! It bugs me to no end! I can't help it. These neurotransmitters irritate me to the extreme, due to their random nature caused by random diffusion and random chance. Neurotransmitter systems and chemical synapses are monumentally inefficient and unreliable.

A gap junction or electrical synapse is infinitely more reliable, predictable, dependable, controllable, and logical than a chemical synapse based upon neurotransmitters, synaptic clefts, and random luck could ever be. Why didn't all-seeing and all-knowing evolution and natural selection use electrical synapses instead of chemical synapses? It would definitely have increased our survival value a million-fold and would have made infinitely better use of our limited resources. Alas, the only logical answer is that evolution, random mutations, and natural selection didn't have anything to do with this and are not controlling the whole show after all. Apparently, God had something else in mind other than efficiency, when it came time for Him to design and create our neurotransmitter systems and chemical synapses.

I think God deliberately slowed everything down giving Him and the Human Psyche the chance to intervene in the action potential process and neurotransmitter systems if they desire to intervene or choose to intervene. Psyche is a function of choice. I think God had something else in mind besides just our survival. Our physical survival isn't all that important to God, because He knows that our Psyche and our memories continue to live and continue to function long after our physical body and physical brain are dead and gone. Instead, God uses the inefficiency of our neurotransmitter systems to slow everything down for us so that we can live it, experience it, choose it, intervene in it, stop it, learn from it, and remember having done so.

According to the Materialists and Naturalists and Darwinists in our quote, the good news is that we now have drugs to solve all of our problems. The BAD NEWS is that neurotransmitter drugs also work by random chance alone, unless of course we are willing to allow some Psyche or Intelligence or God into the system.

Too many neurotransmitters, and we have problems. Too few, and we also have problems. And, all of these various problems arise simply because our neurotransmitter systems are based upon random chance, and are a highly inefficient use of resources. The only way to overcome the natural failures of random chance, and random mutations, is through concentrated and purposeful and deliberate skill and intelligence. The only way to overcome entropy in a closed system is through skill and intelligence!

The degradation enzymes (acetylcholinesterase) that clean up our neuromuscular synapses are the most efficient degradation enzymes that God has ever designed and developed and deployed. That's how God, skill, and intelligence are able to overcome a million neurotransmitters dumped randomly into a synaptic cleft. There has to be some kind of carefully designed force driving acetylcholine and acetylcholinesterase together for Armageddon. It's WAR. It's destruction on a massive scale, only for the presynaptic neuron to have to turn around and build a million new neurotransmitters all over again. What a waste of time and energy. Imagine if you were tasked with wiping out a million soldiers in milliseconds, only to have a million more sent your way a second later. What a mess! It's frustrating for me. I don't like inefficiency. I hate it with a passion. I have always been that way, except when I was high and didn't give a damn.

Of course, the reason why we Materialists and Naturalists hate Psyche, God, and Quantum Mechanics so much is because we can't predict what these things will do and we can't control them. I KNOW, because I used to be a Materialist, Nihilist, and Atheist. We can't control what God chooses to do to us. I didn't like that! But, we can't control what our neurotransmitters do to us, either. Any way we look at it, we lose. We don't like Psyche, God, and Quantum Mechanics, because these things seem to command us and control us, instead of the other way around. But, our neurotransmitters seem to command us and control us as well. It's nasty, especially if prediction, command, and control is your primary goal in life like it used to be for me.

I still see signs of Psyche or signs of Intelligence, even when "evolution" is using the shot-gun approach to get what it wants. It's just that the intelligence takes place at the end of the process where our neuromuscular synapses are concerned, instead of at the beginning. When you are looking at a million neurotransmitters after each action potential, "cleanup is a bitch", as they say! I'd hate to have to clean that up. I hate having to cleanup my brushes after putting on a coat of polyurethane, so imagine what the degradation enzymes are looking at, after each action potential. Why didn't evolution use its brains and create us with electrical synapses instead? I guess evolution doesn't have any brains after all.

I see God's signature and intervention all over this. There's no way on earth that evolution, random mutations, and natural selection could have designed and created any of this; and, we Materialists and Naturalists KNOW that if evolution had designed and created any of this with our personal survival in mind, then evolution would NEVER have used chemical synapses. Evolution would have gone directly for electrical synapses instead, and would have never looked back. As I see it, the very existence of our neurotransmitter systems and chemical synapses is Scientific Proof of God's Existence. Only God would deliberately design and create something so complex, confusing, unreliable, and inefficient, just to show us what He can do and to show us that He exists.

Evolution can't do fine-tuning; and, the overwhelming complexity of our neurotransmitter systems required a monumental amount of fine-tuning and field-testing in order to get them balanced and right. Even then, our neurotransmitter systems are still messed up and highly inefficient, as if they were actually designed to be that way. When it comes to chemical synapses and our neurotransmitter systems, I see God's signature and intervention all over it. God deliberately made us inefficient and deliberately gave us

410

physical limitations to slow things down for us, so that we could actually learn something from the experience. But, it's a monumental mess, and highly inefficient!

As I see it, God is in the details. Evolution could never and would never have designed and created our neurotransmitter systems and chemical synapses. Evolution would have gone directly for electrical synapses instead. That's the way I see it now. Of course, your mileage may vary, because you are a different member of the jury after all, with completely different goals and desires than mine. You may still choose to believe that evolution, random mutations, and natural selection designed and created all of this, if that's what you truly want to believe. That's the beauty of Psyche or Intelligence – it makes decisions and choices, even ones that are faulty or wrong. Our beliefs are chosen into existence.

There's nothing that I can do about your chosen beliefs and chosen behaviors. The only thing I can do is to choose to share with other people what my newly chosen beliefs are; and then according to the Oracle in the Matrix, let them make up their own damned mind what they want to believe and what they want to do with their beliefs.

Choice – it all comes down to choice; and, evolution, random mutations, and natural selection can't do choice. Evolution has no choice in the matter whatsoever. That's the way I see it now. I'm no longer a Materialist, Nihilist, Addict, and Atheist. I chose something better instead; and, my life has gotten progressively better ever since. We can't do anything about our neurotransmitters – not directly anyway; but, we can do anything we want with our choices and our beliefs. God has given us a way to get even, if we want to.

We can do anything we want, no matter how badly we feel. There was a time in my life when I wanted to commit suicide; and, that's precisely what I tried to do. I was suffering from substance-induced psychosis disorder; and, the people on the television and radio were telling me that I was worthless and that I had to kill myself. That's what happens to us when our neurotransmitters get completely out of balance. The whole thing seems to take on a life of its own. The Psychiatrists had prescribed one of everything. I had pills, lots and lots of pills; and, I took bottles of them trying to end it all. I guess I should have taken more than I did, or taken them all, but I wasn't thinking straight at the time.

Conclusions Regarding Neuroscience

When it comes to neurotransmitters and your physical brain, can you imagine what it would be like if you had to micromanage all of this – telling each molecule where to go, what to do, when to do it, how to do it, how to interact with other molecules, and when to stop doing it – like evolution and natural selection do? Thank God we have evolution, random mutations, and natural selection to make all of these decisions for us and to micromanage all of this for us! We would go insane if we had to do it for ourselves.

This is way too much information for my simple brain to encompass and fully understand; yet, the Materialists, Naturalists, Darwinists, Nihilists, and Atheists want you to believe that evolution (genetic drift), natural selection, and random mutations understood all of this billions of years ago when evolution and random mutations designed and created all of these different neurotransmitter systems, neurotransmitters, autoreceptors, neurotransmitter receptors, reuptake transporters, neuropeptides, synapses, astrocytes, gap junctions, potentials, and physical brains at the beginning of time here on this earth.

411

These people really truly want you to believe that evolution (genetic drift), random mutations, and natural selection designed and created your genome, your physical brain, your eyes, your physical body, your thoughts, and your neurotransmitter systems. But, that's physically impossible, because evolution, random mutations, and natural selection didn't even exist until after God designed and created the genomes and the living cells to contain them, in the first place.

Do you really truly believe that evolution and random mutations could have designed and created all of this from scratch? I no longer have blind-faith enough to believe in such a thing. I used to be a Materialist, Nihilist, and Atheist; but, all the evidence got to me eventually and convinced me that I was wrong. It wasn't the first time I was wrong, and it won't be my last. All of this data and observational evidence and scientific evidence about neurotransmitters doesn't allow me to be an intellectually fulfilled atheist anymore.

I see signs of Psyche, Intelligence, Complexity, Sentience, Consciousness, Awareness, Telepathy, Telekinesis, Teleology, Quantum Tunneling, Transdimensional Non-Locality, Quantum Mechanics, Fine-Tuning, Design, Engineering, Specified Complexity, Irreducible Complexity, Creation, Planning, Organization, Cooperation, and God at the molecular level throughout the whole of it, even if others can't. I no longer have blind-faith enough to be an Atheist. After the different molecules and proteins are constructed from our genes, then it is clear to me that something else takes over and takes control of these newly born molecules and proteins and tells them what to do, where to go, and what to become. It's at the level of newly made independently born molecules and proteins that we see God's handiwork and God's blueprint in action.

When it comes to neurotransmitters, there's more than meets the eye – too few, and we have problems – too many, and we have problems. And, there doesn't seem to be anything random about it. If all of this isn't precisely balanced and finely-tuned, we are going to have diseases and mental illnesses as a result. By definition in principle, evolution, random mutations, and natural selection can't do fine-tuning; so, who did? If evolution really did design and create all of this, then evolution is another name for God. It can't be any other way.

Given enough time and resources and coordinated effort, I believe we human beings could eventually design and created a new species. In contrast, given an infinite amount of time, chemical evolution would still be unable to design and create a single functional protein, nor would chemical evolution know what to do with that single protein if it were lucky enough to actually create one.

The *Origin* video from Illustra Media does some mathematical modeling demonstrating that the odds of a functional protein developing by chance are 1 chance in 10^{164}; and, that's with God taking the time to fill an ocean completely full of the right 20 amino acids with the right chirality, and with God forcing 150 of those amino acid sequences to line up every second so that they can see if they will fold up into the right usable protein.

Technically, there's no logical reason for that ocean full of the right 20 amino acids to deliberately line up 150 amino acids in sequence every second in order to try out all of the 10^{164} possible combinations in a timely fashion – naturally, or through undirected processes. God is going to have to provide the ocean full of amino acids in the first place, sustain that ocean full of amino acids in existence for trillions of trillions of trillions of trillions of years, and force 150 of those amino acids to line up in sequence every second for those trillions of trillions of years JUST TO PRODUCE a single protein "naturally". The odds of that whole process happening by random chance, without God's intervention, is ZERO. There is NO CHANCE!

The odds of a functional cell developing by natural processes are 1 chance in $10^{340,000,000}$.

The current estimated age of this universe is 13.8^9, or 13.8 billion years. In comparison to 13.8 billion years, something like $10^{340,000,000}$ is synonymous with IMPOSSIBLE or INFINITY, because there is NO CHANCE that any living cell could ever develop by chance through natural processes in 13.8 billion years of time. In fact, there is technically NO CHANCE that a functional protein could develop by chance in 13.8 billion years, even if we were to start with a God-given ocean full of the right 20 amino acids with the right chirality, no destructive oxygen, and no sunshine to destroy that particle of meat once it comes together.

Our neurotransmitter systems function too perfectly, too well, and too reliably to a product of blind random chance. Think about it logically. I'm fifty-six years old, and I just barely figured out how to tell the difference between a presynaptic reuptake transporter and a postsynaptic neurotransmitter receptor – a major accomplishment for me. There's no way on earth that evolution could have figured all of this out billions of years ago, because you and I are by definition in principle infinitely smarter and more skilled than any type of evolution could ever be; and, we can't even figure out how all of these things work.

In a properly functioning brain, the post-synaptic neurons know when they need a neurotransmitter, and they seem to reach out and grab one, while at the same time blocking reception when they know that they don't need a neurotransmitter. At times, the presynaptic terminal buttons seem to be communicating telepathically with the postsynaptic dendritic receptors, unless there is an astrocyte there to convey the messages between them. How do they do that? Who taught these cells how to communicate with each other and coordinate with each other to form a thalamus or some other brain structure? Apparently they are born knowing how to communicate telepathically with other molecules and neurons, or through electronic gap junctions after they have made the gap junctions.

Other times, it seems as if the presynaptic reuptake transporters and autoreceptors are communicating telepathically and interacting telekinetically with each other and with the different neurotransmitters and molecules in the synaptic clefts. How do the reuptake transporters know that the synaptic clefts need cleaning, when to start the cleanup process, and when the synaptic clefts have been properly cleaned? The laws of random diffusion or random chance would make it impossible to completely clear a synaptic cleft of a specific molecule or neurotransmitter; yet, the transporters seem to do just that. How are the astrocytes able to monitor the contents of the synaptic clefts under their control wirelessly and at a distance? Somehow, those messages are getting transmitted from the post-synaptic receptors to the pre-synaptic uptake transporters and glial astrocytes! There's tripartite communication going on within a tripartite synapse. How do they do that? How do these individual cells know each other's language, and how do they communicate wirelessly at a distance with each other before they have formed gap junctions with each other?

Postsynaptic Receptors seem to have the ability to reach out telekinetically or through quantum tunneling and immediately grab the neurotransmitters that they need when the Human Psyche asks them to do something. Nature's Psyche has to have some way of triggering or starting that first action potential whenever the Human Psyche makes a decision or a choice. Nature's Psyche within the post-synaptic receptors knows how to reach out and grab what it needs when it needs it, in order to get those first action potentials started after the Human Psyche decides that it wants to raise its arm or go for a walk. Only when the brain isn't functioning properly thanks to evolution and genetic mutations does it lose the ability to follow Psyche's Commands; and, glitches, false starts, tremors, faulty bogus input, hallucinations, delusions, depression, anxiety, OCD, and malfunctions are the result.

According to Orthodox Quantum Mechanics, the **Human Psyche** negotiates with or interacts with **Nature's Psyche** in order to trigger and control our physical responses and our physical brain. The Human Psyche queries or commands, and Nature's Psyche responds as well as it can. However, if evolution or random mutations have gotten in the way and damaged or crippled our neurotransmitter receptors, then they won't be able to function properly. The neuro-receptors KNOW what they are supposed to do because God has taught them how to do it and commanded them to do it; but, they can't do it if they haven't been constructed properly in the first place. Evolution and random mutations are the source of all our physical diseases that have a developmental cause. Random mutations and evolution are not our designers and creators. They are our judge, jury, and executioners. Natural selection doesn't create – it causes extinction.

When the Human Psyche decides to move its arm, Nature's Psyche within the appropriate neurons that have been assigned to that task reaches out telepathically and telekinetically and grabs the necessary neurotransmitters to get an action potential started; and, the arm moves. However, if there are no neurotransmitters to grab or if the post-synaptic receptors are mutated and damaged, then Nature's Psyche is going to fail to follow the Human Psyche's commands because it can't do the job that it has been asked to perform and knows how to perform.

I still find myself asking why God chose to do it this way, because there's so many different things that can go wrong. Random mutations messing with our genes cause our neuro-receptors to malfunction, resulting in too few neurotransmitters or too many neurotransmitters engaging the system resulting in glitches, malfunctions, and false starts. Why did God do it this way? If you increase complexity, you also increase the number of things that can go wrong.

Did God do it this way simply to make His necessity and His presence known to us? Did God do it this way to make it clear to us that evolution, random mutations, and natural selection could never have designed, fine-tuned, field-tested, and created our neurotransmitter systems while also making it clear to us that it's evolution (genetic drift) and random mutations which are causing all the damage to our neuro-receptors, reuptake transporters, ion channels, and neurotransmitter systems?

It's obvious to most people who are thinking logically and rationally that evolution, random mutations, and natural selection can't do complexity. The very existence of our neurotransmitter systems, reuptake transporters, ion channels, and our post-synaptic receptors is Scientific Proof of God's Necessity, which ends up being Scientific Proof of God's Existence, simply because evolution of any kind cannot do complexity, science, design, planning, manufacturing, field-testing, fine-tuning, programming, engineering, creation, intelligence, and psyche.

I do know that God created our neurotransmitter systems and our physical bodies in order to slow everything down for us, so that we can live it, experience it, learn from it, and remember having done so. But, thanks to evolution, genetic drift, genetic entropy, and random mutations, there's so many different things that can go wrong with our neurotransmitter systems that it's a miracle that they even work in the first place. I guess that I would have liked to have received from God something a bit more reliable and dependable than neurotransmitter receptors. I inherited a genetic defect in my serotonin system from my mother's side. People on that side of the family suffer from anxiety, OCD, depression, and addictions; and, I haven't been immune. I still have tons of unanswered questions and nowhere near enough time to figure it all out.

One thing that I do KNOW for sure is that there's no way in the universe that evolution (genetic drift), random mutations, and natural selection could have figured any of this out billions of years ago, because they don't have brains enough to do so. We human beings don't have brains enough and experience enough to figure out how all of this really works, so why on earth would anyone believe that evolution planned, designed, programmed, and created our genomes, our physical bodies, our physical brains, our eyes, and our neurotransmitter systems billions of years ago? It's physically impossible, because evolution, random mutations, and natural selection can't do design and creation. They can only do disease, chaos, entropy, death, cancer, and extinction.

Evolution, random mutations, and natural selection can't do science, reasoning, logic, engineering, fine-tuning, field-testing, prescience, teleology, planning, foresight, design, programming, manufacturing, complexity, and creation. There's no way that evolution can be our Designer and our Creator. It's physically impossible. There's no intelligence there in evolution, which is why the different types of evolution could never be our Designer and our Creator.

At least with further research and prayer, I finally started getting answers to some of the questions that I had. I still have a lot to learn, though. The more we learn, the more we realize how much we don't know. It also makes me wonder what questions I'm still failing to ask, because I get the sense that there are still many things that I'm missing and don't understand. I don't even know the questions that I need to ask; and, no asking, then no receiving. No seeking, then no finding. At least I finally got rid of My Materialism, My Nihilism, and My Atheism – which finally opened me up to Science 2.0 and Quantum Neuroscience, which were off-limits and out-of-bounds for me before I overcame my brainwashing and conditioning in Scientism and Scientific Naturalism.

I have been a computer scientist most of my life. I like to think that I know how computers work and how memory is stored at the physical level within a computer. As a result, I found the physical brain severely disturbing. For months, if not years, I couldn't put my finger on it; but, the physical brain has caused me a lot of cognitive dissonance. Something wasn't quite right. In fact, as the months passed, I found many things that aren't quite right about a physical brain.

Memory storage within the physical brain ended up being what disturbed me the most – that was the thing that I couldn't put my finger on at first. How are memories being stored within the brain? I figured that neurotransmitters should have something to do with it; but, I was wrong.

Every time I would take up this subject, the Neuroscientists would say, "We don't know how memories are being stored in the brain; but clearly, our memories are being stored within our brain because there's no place else that they can be stored." It was the same answer every time, "We don't know how your brain is storing your memories, but clearly it is. There's no place else to store your memories besides your brain. One of these days, we will figure out how it is being done." I had some studying and thinking to do.

We each have to start where we are, and then go on from there. I wish you well.

Action at a Distance is Proof of Concept

Action at a Distance is Proof of Concept where Quantum Mechanics or Transdimensional Physics is concerned, because Materialism, Naturalism, and Classical Physics by their very definition, in principle, cannot do Action at a Distance. According to Materialism and Naturalism, Action at a Distance does not exist and cannot exist, because it's physically impossible.

Anytime we witness or observe Action at a Distance, we are in fact witnessing some kind of Transdimensional, Non-Local, Spiritual, Quantum phenomenon or event. The proven and observed existence of Action at a Distance falsifies the fundamental tenets of Materialism and Naturalism.

The action potential is a dramatic redistribution of electrical charge across the membrane. Depolarization of the cell during the action potential is caused by the influx of sodium ions across the membrane, and repolarization is caused by the efflux of potassium ions. Let's apply some of the concepts introduced in Chapter 3 to help us understand how ions are driven across the membrane and how these ionic movements affect the membrane potential.

A neuron has sodium-potassium pumps, potassium ion channels, and sodium ion channels. The pumps establish ionic concentration gradients (differences in concentration) **so that K^+ is concentrated inside the cell and Na^+ is concentrated outside the cell.** (*Neuroscience*, pp. 80-84.)

BCP4

https://canvas.brown.edu/courses/851434/files/40426324/

http://mypsyche.us/wp-content/uploads/2017/12/BCP4.pdf

These ions are driven by some kind of invisible force; and, they move in a dramatic way! The action potential is actually driven by some kind of invisible Action at a Distance – some kind of intangible and invisible force or field. Who established these forces and these Rules of Engagement in the first place? Who designed a neuron to ACT this way? It can't be evolution (random mutations and natural selection) because evolution by definition in principle cannot do forces, fields, design, manufacturing, programming, teleology, nor Action at a Distance.

Since they each have different concentration gradients inside and outside the neuron, these ions interact with each other at a distance across the cell membrane of the neuron. There's some kind of invisible force driving them and causing them to interact with each other at a distance. These Na^+ and K^+ ions literally line up against the cell membrane of the neuron on opposite sides of the membrane fighting to get at each other. There's some kind of invisible force drawing them towards each other; and, this telekinetic force is working at a distance across the cell membrane in both directions!

Whenever we are dealing with invisible forces and invisible fields, we are in fact dealing with some kind of Transdimensional Physics or Quantum Mechanics, because Materialism, Naturalism, and Classical Physics by definition in principle cannot do Action at a Distance.

Physical matter was designed by God to REACT according to God's Laws or God's Commands – what most scientists call the Physical Laws and the Cosmological Constants.

That's what brings order out of chaos in our particular physical universe. Even though physical particles and physical molecules react, they do ACT. There is always Action at a Distance taking place between atoms and molecules. It's some kind of telekinetic force either drawing them towards each other or repelling them away from each other.

Nature's Psyche or the Intelligence within each atom of physical matter makes that particular atom ACT according to God's Commands, or the Physical Laws of our universe; and, Nature's Psyche within each atom has covenanted with God to obey God's Commands. Nature's Psyche does precisely what God Commands it to do, because it has promised to do so. God's Psyche or God's Intelligence tells Nature's Psyche within each atom and each molecule how to act and how to react to the Human Psyche's requests, demands, and commands. God's Psyche through Nature's Psyche is telling each atom on the periodic table and each molecule how to act, what to do, where to go, and what to become. There's no other way to get order and organization out of these things – thanks to random diffusion, entropy, and the second law of thermodynamics. God has to tell the atoms and molecules and proteins and living cells where to go, what to do, and what to become after they have been made or brought into existence by God; otherwise, these particles of matter are going to go nowhere and they are going to do nothing, simply defaulting to Classical Physics instead.

God has also commanded Nature's Psyche to REACT as best as possible to the Human Psyche's commands. That's how the Human Psyche is able to interact with and control its physical brain – through telepathic WiFi connections with Nature's Psyche, which resides within each physical atom that God has created or brought into existence.

All of this Action at a Distance takes place telekinetically and telepathically at the Transdimensional Level, Non-Local Level, or Quantum Level. Remember, telepathy is simply WiFi at the Quantum Level. WiFi is Action at a Distance! The wireless access point is interacting with the router wirelessly at a distance. Likewise, telekinesis is simply forces and fields interacting with each other at a distance at the Quantum Level. Magnetism is a type of telekinesis!

We understand precisely how WiFi and Radio and Magnetism – Action at a Distance – work at the macro physical scale; yet, most of us pretend that it doesn't exist and can't exist at the micro scale or the Quantum Transdimensional Non-Local Scale. We are wrong whenever we assume that there's no WiFi or Radio or Magnetism at the Quantum Level, because Action at a Distance has been caught in the ACT quadrillions of times in millions of different ways.

Electromagnetism, through magnets, IS Action at a Distance.

It's called the earth's magnetic field, and it exerts some kind of detectable invisible force between physical objects, at a distance. It's very real, very powerful, and extremely useful. The very existence of magnets and magnetism FALSIFIES the fundamental teachings of Materialism, Naturalism, and even Classical Physics because magnetism of any kind IS invisible intangible Action at a Distance.

The invisible, intangible, non-local, and quantum forces and fields are detected by what they do to the physical; therefore, WE KNOW that they are real even though we can't see them directly with our physical eyes. The Action at a Distance of Na^+ and $K+$ ions lining up telekinetically against opposite sides of the cell wall fighting to get at each other is very real, even though we can't see these invisible forces and fields with our physical eyes. Just because we can't see something with our physical eyes nor hear it with our physical ears doesn't mean that it doesn't exist. Lots of things exist that we can't see with our physical eyes nor hear with our ears. Our physical senses are limited to a very small part of the overall spectrum; and, we now know that to be true, even though we haven't always known that to be true.

Materialism and Naturalism and Atheism resist and fight against the discovery of the invisible, the intangible, the non-local, the non-physical, the transdimensional, the immaterial, the quantum, or the supernatural. It's all the same thing, and Scientific Naturalism fights against it all and tries to ban and censor it all, because it's supernatural or non-physical and its very existence falsifies Materialism and Naturalism and Darwinism.

Materialism and Naturalism are the greatest bane to scientific discovery that mankind has ever created. It's BAD SCIENCE, because it's based upon a denial of Reality. Materialism, Naturalism, and Atheism are based upon a refusal to look at evidence. That's the very definition of BAD SCIENCE! These people actively and deliberately censor, ban, reject, suppress, and even destroy evidence – any kind of evidence that they personally don't like or that successfully falsifies their theories, beliefs, and chosen ideas. I KNOW, because I used to be a Materialist, Nihilist, and Atheist. Thank God I finally found a better way to do science.

In the summer of 2017, I upgraded my science to Science 2.0. Science 2.0 allows all of the evidence into evidence, and it is based upon a preponderance of the evidence. Science 2.0 is based upon Phenomenology, the scientific study of all phenomena, events, experiences, and observations. Phenomenology is the scientific study of the Lived Experiences of the human race, including our science experiments, local observations, non-local observations, near-death experiences, and out-of-body experiences.

Science 2.0 is also based upon Action at a Distance or Transdimensional Physics, what we typically call Quantum Mechanics. Telepathy or the proven and observed Quantum Zeno Effect, telekinesis or invisible intangible forces and fields, Action at a Distance, Quantum Entanglement or Quantum Non-Locality, spirit matter or dark matter or tachyons, teleportation of physical matter or Quantum Tunneling, invisible intangible forces and fields such as photons magnets gluons bosons gravitons and dark energy, and Psyche or Quantum Non-Local Consciousness are a foundational fundamental part of Science 2.0, because these things have been observed, detected, lived, and experienced phenomenologically.

Unlike Materialism, Naturalism, Darwinism, Nihilism, and Atheism, Science 2.0 allows ALL of the evidence into evidence and is actually based upon a preponderance of the evidence. Science 2.0 is the way that science should have been done, but wasn't. Science 2.0 presents us with a different way and a better way of doing science. You will have to decide for yourself if it is so. You can lead a horse to water, but you can't make him drink. Likewise, you can lead an atheist to evidence, but you can't make him think.

For me personally, Quantum Neuroscience was a natural and fundamental out-growth of Science 2.0 – simply following the evidence where it leads me. I find Quantum Neuroscience and Science 2.0 extremely useful and interesting to study and ponder.

Denial gets us nowhere; and, nowhere is where many people want to go.

Einstein's rejection of Action at a Distance blocked him from being able to take his physics to the next level. Ask yourself a question, "Is your rejection and denial of Action at a Distance, Quantum Non-Locality, Quantum Tunneling (teleportation), the Quantum Zeno Effect (telepathy), Psyche, and Non-Local Consciousness preventing you from taking your Physics to the next level?" If so, then you have identified an opportunity for change and growth.

Exclusive reliance on Classical Physics, Materialism, and Naturalism is a Bad Thing – BAD SCIENCE – because it is based upon denial and a refusal to look at evidence. It prevents us from taking our science to the next level. I KNOW, because I used to be a Materialist, Naturalist, Nihilist, and Atheist. Been there, done that, and I KNOW where that road leads us. It's a dead-end – it was purposefully and deliberately designed to be that way.

Psyche or Non-Local Consciousness, Action at a Distance, Quantum Non-Locality, Quantum Tunneling, the Quantum Zeno Effect are proven and verified science – they have been observed, lived, experienced, and caught in the ACT. WE KNOW these things to be real and true, because collectively as a race we have lived them, experienced them, and observed them phenomenologically; so, the next question is, "How do we explain these things scientifically in a way that actually makes reasonable and logical sense?" Quantum Mechanics or Transdimensional Physics is the answer to the question. Quantum Mechanics is the answer to life, the universe, and everything.

It call comes down to information exchange or Intelligence.

There's not enough information storage capacity within our 3.2 billion base-pair genome to store the differentiation information, subtyping information, migration information, eventual destination, and interaction instructions of the 37.2 trillion to 100 trillion cells in our physical body. There's not enough information storage capacity in our 3.2 billion base-pair genome to command and control each and every ATP molecule and protein in our physical body. There's not enough information storage capacity in our 3.2 billion base-pair genome to store the subtyping information, location information, command information, and control information for the one quadrillion synapses in our physical brain – telling them what to do, what to become, how to grow, and how to draw in the axons and their growth cones to their specific location. That information has to be stored someplace else besides our genome, because there isn't enough information storage capacity in our genome for such a thing. It's not physically possible to store quadrillions of bytes of information within our 3.2 billion base-pair genome. So, where is all of that information store, and where is it coming from? It's stored in God's Database, and it's coming from God's Psyche or God's Intelligence.

WE KNOW from observation that lots of genetics goes into making the proteins that comprise and constitute a living physical cell; but, after that living cell has been made or come into existence through replication, something other than genetics takes over and tells that cell where to go, how to get there, and what to do and what to become when it gets there. A newly made cell could go anywhere; but, someone tells it where to go, how to get there, and how to act when it gets there. It's a miracle by Divine Design, not just a coincidence. Newly made cells KNOW precisely where they are going and what they are supposed to do when they get there.

Likewise, each protein is requisitioned by someone, and then it is told where to go and what to do and what to become after it has been made. Some people will interrupt at this point and say that there are molecule transporters and protein transporters within a living cell; but if so, that just pushes the problem one step further back. Somehow someone is telling those transporters where to go, what to pick up, and then where to take their cargo. Molecules floating free in the cytosol or cytoplasm can go anywhere in the cell; and according to Classical Physics, they will just sit there and go nowhere, unless someone tells them where to go and telekinetically makes them go there. Something or someone is enlivening or quickening the molecules in a living cell, commanding them telepathically and moving them telekinetically to where they need to go; and then, this same supernatural intelligence or psyche is actually telling them what to do when they get there.

Nature's Psyche provides instructions, command, control, purpose, and motivation for physical atoms and physical molecules. Nature's Psyche has covenanted to follow God's

Commands, or what we typically call the Physical Laws of the universe, including the intangible Quantum Laws, Transdimensional Laws, Non-Local Laws, or Spiritual Laws.

Quantum Mechanics and Quantum Neuroscience ARE a function of Psyche or Quantum Non-Local Consciousness – Action at a Distance. God's Psyche or God's Laws or God's Word of Command acting through Nature's Psyche brought order, organization, and light to chaos thereby forming or creating physical matter and this physical universe from pre-existing spirit matter. There is no order or organization without Psyche or Intelligence. Psyche is the Ultimate Causal Agent.

WE KNOW that Psyche, or The Spark, or Quantum Non-Local Consciousness exists, because it has been lived and experienced. Millions of different people have had Out-of-Body Experiences and Near-Death Experiences. WE KNOW phenomenologically from these Non-Local Experiences or Lived Experiences that ALL of our memories and experiences and feelings are store Non-Locally in the Transdimensional Quantum Realm; and, we get to experience them again during our Life-Review when we die. Our physical brain doesn't have anywhere near enough information storage capacity to store, remember, recall, and re-experience all of these memories and experiences and feelings. These Life-Reviews have to be stored someplace else besides our physical brain, because the physical brain is dead or offline when people are having and experiencing their Life-Reviews. This is logical common sense.

They call it Superior Autobiographical Memory; but, it is in fact Supernatural Autobiographical Memory.

https://www.youtube.com/watch?v=2zTkBgHNsWM

https://www.youtube.com/watch?v=en23bCvp-Fw

The scientists are surprised that these people are stable and sane.

Why?

It's because if all of these memories were actually stored within their physical brain, then it should all be simultaneously present and available at once, and the simultaneous presence of ALL knowledge and ALL information would drive any mortal physical person insane with information overload. Instead, all of their memories are compartmentalized and stored separately someplace else besides their brain, and their physical brains simply access this information one memory and one day at a time in sequential fashion. Our physical brain can handle one set of memories at a time, because it does so every day. Our physical brain couldn't handle nor store all of our memories, feelings, and experiences simultaneously without driving us insane.

Studying Memory Consolidation as well as Supernatural Autobiographical Memory convergently makes it clear that individual neurons, when they are designed properly and functioning properly, have the ability to address and to access the Human Psyche's Non-Local Database while at the same time adding information to it. Things like electroconvulsive therapy, neural degeneration, and blows to the head seem to disrupt this indexing and categorizing ability within the individual neurons. You can't download information from the cloud or from non-local storage if the physical device to which you are trying to download has been destroyed or does not exist anymore.

The current view of memory consolidation is that it continues for a very long time if not indefinitely. In other words, the evidence indicates that lasting memories become more and more resistant to disruption throughout a person's life. Each time a memory is activated [or accessed]**, it**

is updated and linked to additional memories. These additional links increase the memory's resistance to disruption by cerebral trauma such as concussion and ECS. (*Biopsychology*, p. 268.) [There's indexing and linking going on here in the background; and, not all of that is stored within the physical brain. If the neurons are destroyed, then they can no longer access the index and therefore can no longer access the memories; but, all of those memories are there when a person experiences a Life-Review after his physical brain is dead or offline. My memory systems were destroyed when I experienced Substance-Induced Psychosis Disorder; but, after I got sober, God slowly restored my memories to me, a process which wouldn't have been possible if our memories are stored only within our physical brain.]

Of course, the Materialists and Naturalists and Darwinists choose to believe that this indexing, linking, and consolidation are purely a physical process; but, these things can't be a purely physical process if people are successfully accessing and reliving their thoughts, feelings, experiences, actions, decisions, and choices after their physical brain is dead and offline during their non-local spiritual Life-Reviews. Science 2.0 and Quantum Neuroscience start with what we have experienced and observed and KNOW to be true; and, then they try to find a scientific explanation for what we already KNOW to be real and true through lived experiences or phenomenologically. Life-Reviews, Near-Death Experiences, and Out-of-Body Experiences are real and have truly happened to millions of different people; so now, let's try to find a scientific explanation for it all. That's the purpose of Quantum Neuroscience. Quantum Neuroscience is the scientific study of how the Human Psyche interacts with and controls its physical brain.

The scientists are looking for Superior Autobiographical Memory within the physical brain, but these memories and feelings and experiences do not exist within the physical brain. Instead, they are accessed and re-lived one at a time by the physical brain. Our physical brain is a Transceiver. Nature's Psyche within our physical brain transmits all of our experiences and feelings and memories telepathically into the long-term storage of our Non-Local Consciousness or Psyche. That's the way that God designed it to be. This is Action at a Distance. And, some people have the ability to retrieve and relive these memories and experiences and feelings telepathically from their long-term quantum non-local storage or database, while here in the flesh. It's a spiritual gift.

These people are able to experience their Life-Reviews at will, while still here in their physical body. These people just recall it, see it, feel it, and experience it telepathically as if they were there. It's a function of Psyche, not just the physical brain. WE KNOW this is so, because millions of people have experienced Life-Reviews when their physical brain was dead or offline. By definition, in principle or practice, our after-death Life-Reviews and their associated memories and feelings are not stored within our physical brain, because our physical brain is dead and offline when we are having our Life-Reviews. This is logical common sense.

Obviously, there are many different types of Memory Systems that get "wired" or mapped into our physical brain during neurodevelopment by Nature's Psyche. WE KNOW that they are there and exist because they have been observed, experienced, and documented.

Memory Systems

http://mypsyche.us/wp-content/uploads/2017/12/BCP24.pdf

One of the main purposes of Quantum Neuroscience is to try to figure out how the memories from all of our different physical Memory Systems get transferred into Non-Local

Storage and become Eternal Memories and Life-Reviews as a result. WE KNOW that Life-Reviews and Eternal Memories are there and exist, because they have been lived and experienced and observed by millions of different people. We start with what we KNOW, and then we try to find a scientific explanation for it. That's the way Quantum Neuroscience and Science 2.0 work.

Quantum Neuroscience and Science 2.0 are an attempt to explain scientifically what we already KNOW to be real and truly there through the lived experiences of the human race as a whole. Science 2.0 allows ALL of the evidence into evidence and then goes with a preponderance of the evidence. The preponderance of the evidence tells that we have Eternal Memories, Knowledge, Intelligence, Awareness, Consciousness, Sentience, and Life-Reviews after our physical brain is dead and/or offline. Quantum Neuroscience is an attempt to explain scientifically how ALL our physical experiences, memories, and feelings get transferred and stored into the Non-Local Non-Physical Quantum Storage that is associated with our Psyche or Quantum Non-Local Consciousness.

When it comes to Superior Autobiographical Memory or Supernatural Autobiographical Memory, these people may have been gifted with a superior Transceiver within their physical brain; or, their abilities may be purely a function of Psyche. It's hard to tell when it comes to the non-local or the spiritual, because the physical was designed by God to interface successfully with Nature's Psyche and the Human Psyche. It's hard to tell where one stops and the other begins. Psyche always seems to be there; whereas, the physical seems to come and go; but most of the time, they seem to overlap and interface with each other. The physical is necessary in order to have a physical experience; but, the spiritual or the non-local is necessary in order for us to be able to remember these physical events after our physical brain is dead, dissolved, and gone.

The thing we do KNOW for sure is that our physical brain doesn't have enough information storage capacity to store all of these memories, feelings, and experiences – that's why they have to be re-accessed and re-experienced one at a time; and, we KNOW for a surety that our after-death Life-Reviews are NOT stored within our physical brain because the physical brain is either dead or offline when people are having their out-of-body Life-Reviews. According to these people, it's all very orderly and organized and structured on both sides of the veil; and, this wouldn't be possible if Materialism and Naturalism were true.

Materialism, Naturalism, and Classical Physics cannot explain any of this; but, Transdimensional Physics, Quantum Mechanics, Quantum Telepathy, Action at a Distance, Non-Local Information Storage, Psyche, and Non-Local Consciousness certainly can. Psyche is an infinite singularity – it's theoretically capable of storing an infinite amount of data and information in something that has no physical size and takes up no physical space whatsoever.

Action at a Distance is Proof of Concept where Quantum Mechanics or Transdimensional Physics is concerned, because Materialism, Naturalism, and Classical Physics by their very definition, in principle, cannot do Action at a Distance. According to Materialism and Naturalism, Action at a Distance does not exist and cannot exist.

Anytime we witness or observe Action at a Distance, we are in fact witnessing some kind of Transdimensional, Non-Local, Spiritual, Quantum phenomenon or event.

Quantum Neuroscience is the scientific study or evidential study of how the Human Psyche interacts with and controls its physical brain.

Quantum Neuroscience and the study of Eternal Memories start with what WE KNOW to be real and true phenomenologically through direct observation and the lived experiences of the human race; and then, these new scientific disciplines try to find a logical, rational, scientific explanation for what we already know to be real and true. Rather than suppressing and destroying the evidence as Materialism and Naturalism do, Quantum Neuroscience and Science 2.0 start with a preponderance of the evidence and then they try to explain scientifically what we are observing, experiencing, living, re-living, and already know to be true as a race. Science 2.0 is a new and better way to do science – one that is based upon a preponderance of the evidence rather than the suppression of evidence. Science 2.0 allows ALL of the evidence into evidence. Science 2.0 is the way that science should have been done, but wasn't.

Quantum Control

Bear, M. F., Connors, B. W., & Paradiso, M. A. (2007). *Neuroscience: Exploring the Brain* (3rd ed.). New York: Lippincott Williams and Wilkins.

We have to take the physical as far down the rabbit-hole as we can possibly take it, before we are forced to switch over to Psyche and Quantum Mechanics for an explanation of what we are witnessing or observing or experiencing.

I learned to love *Neuroscience: Exploring the Brain*, because they take the physical as far as they can possibly take it; yet at times, they drop hints that there's something more to be found once we have run out of physical explanations for Neurodevelopment. As Darwinists, Materialists, and Naturalists, they often slip-up and take us into levels of science that are physically impossible and/or physically unknowable. Let me provide a few examples.

As neurons differentiate, they extend axons that must find their appropriate targets. Think of this development of long-range connections, or pathway formation, in the CNS as occurring in three phases: pathway selection, target selection, and address selection.

The three phases of pathway formation.

The growing retinal axon must make several "decisions" to find its correct target in the LGN [lateral geniculate nucleus]. **During pathway selection, the axon must choose the correct path. During target selection, the axon must choose the correct structure to innervate** [connect with]. **During address selection, the axon must choose the correct cells to synapse with in the target structure.** (*Neuroscience*, p. 697.)

Wiring the Brain:

http://mypsyche.us/wp-content/uploads/2017/12/BCP23.pdf

This is scary and forbidden stuff that they are talking about here. They are talking about concepts here that I have been talking about in my books and in my essays. Materialists, Naturalists, and Darwinists aren't supposed to talk about these things nor use this kind of language because evolution (genetic drift), random mutations, and natural selection by definition in principle cannot do and do not do decisions and choices. Evolution of any kind is supposed to be dumb and blind, not intelligent and purposeful.

Remember, any time you are compelled to use "decisions" and "choices" to explain what you are observing and experiencing and witnessing, you are in fact using Psyche or Intelligence to explain your scientific observations, because something like evolution, random mutations, or natural selection by definition in principle can't make choices and decisions, because evolution of any kind is always defined as being dumb and blind.

Anytime we observe or catch CHOICE in action, we have in fact isolated or identified Psyche, because Psyche or Intelligence is the Ultimate Causal Agent.

"The Ultimate Model of Reality: Psyche Is the Ultimate Cause"

https://www.amazon.com/dp/B071NC9JK6

"Putting Psyche Back into Psychology: Restoring Science to Consciousness"

https://www.amazon.com/dp/B071NC987S

"BioPsychoSocial: Including Psyche or Light into our Theoretical Models"

https://www.amazon.com/dp/B0713NDHVW

Let's observe carefully some of the other times that these people choose to use the words "decisions" and "choices" to describe what's happening during neurodevelopment.

Imagine for a moment that you must lead a growing retinal ganglion cell axon to the correct location in the LGN.

First, you travel down the optic stalk toward the brain. [Someone or something had to put the optic stalk there in the first place. Someone has to tell the growing axon to travel down the optic stalk rather than going someplace else.] **But soon you reach the optic chiasm at the base of the brain and must decide which fork in the road to take.** [Someone had to put the optic chiasm and the roads there is the first place.] **You have three choices: You can enter the optic tract on the same side, you can enter the optic tract on the opposite side, or you can dive into the other optic nerve.**

The correct path depends on the location in the retina of your ganglion cell, and on the cell type. [It also depends on KNOWING where each tract leads, which they don't mention. Choices and decisions are based upon knowledge and intelligence.] **If you came from the nasal retina, you would cross over at the chiasm into the contralateral optic tract; but if you came from the temporal retina, you would stay in the tract on the same side. And in no case would you enter the other optic nerve. These are examples of the "decisions" that must be made by the growing axon during** *pathway selection***.** [Making correct choices and correct decisions and choosing the correct path depends upon information, knowledge, and intelligence which are things that natural selection and random mutations cannot do. Remember, someone has to layout the paths or tracts in the first place, or they wouldn't exist.]

Having forged your way into the dorsal thalamus, you are now confronted with the choice of which thalamic nucleus to innervate. The correct choice, of course, is the lateral geniculate nucleus. This decision is called *target selection***.** [Somehow the growing axons have to KNOW in advance, or be TOLD when they get there, which thalamic nucleus is the correct nucleus for them to connect with. They have many to choose from, and they have to choose the right one. There's nothing random about any of this. It all involves decisions and choices. Each axon is targeting specific neurons! It's like trying to find a needle in a haystack!]

But finding the correct target still isn't enough. You must now find the correct layer of the LGN. You also must make sure that you sort yourself out with respect to other invading retinal axons so that retinotopy [a topographical map of the retina] **in the LGN is established. These are examples of the decisions that must be made by the growing axon during** *address selection***.** [Out of the one quadrillion synapses in our physical brain, each axon and each target neuron must communicate, decide, and choose which synapses to build, where to build them, and what types they should be. Topographical mapping, target selection, address selection, and synapse selection are where we quickly and easily overload the information storage capacity of our physical genome. 3D topographical maps, structural blueprints, target selection information, address selection information, and synapse differentiation have to be stored someplace else besides our physical genome, because our genome can't hold quadrillions of bytes of information.]

We will see that each of the three phases of pathway formation depends critically on communication between cells. This communication occurs in several ways: direct cell-cell contact, contact between cells and the extracellular secretions of other cells, and communication between cells over a distance via diffusible chemicals. As the pathways develop, the neurons also begin to communicate via action potentials and synaptic transmission. [These people take the physical as far as they can take it; but they stop whenever they run out of physical explanations, and talk about decisions and choices instead. Of course, being Materialists, Naturalists, and Darwinists, these people completely overlook and exclude Quantum Communication, Quantum Telepathy or "WiFi at the Quantum Level", Nature's Psyche, and the Human Psyche because such things aren't supposed to exist according to Classical Physics and Naturalism. The original cells blazing the original pathways or the original tracts, and dropping the breadcrumbs or the guidance cells along the way, had to rely upon Quantum Communication during neuron migration in order to know where they are supposed to go and what they are supposed to do while getting there. Of course, these people try to provide a physical explanation for these things whenever they can, and they should because our physical brain is a physical object after all; however, there are times when these people run out of physical explanations and are forced to resort to "decisions" and "choices" instead.] (*Neuroscience*, pp. 697-698.)

These people put "decisions" in quotes, because growth cones and growing axons aren't supposed to be able to make actual decisions and choices, according to the rules of Materialism, Naturalism, and Classical Physics.

Axons and dendrites aren't supposed to be able to make decisions and choices, but they do so all the time. They are alive, sentient, intelligent, aware, and KNOW precisely what they are doing and why they are doing it. We have run out of physical explanations, whenever we are forced to switch over to decisions and choices for an explanation. Whenever these scientists resort to "decisions" and "choices", they have indeed run out of physical explanations for what they are observing and experiencing, and have switched over to Quantum Non-Local Consciousness or Psyche instead – whether they realize it or not.

Of course, these people want you to believe that chemical evolution, genetic drift, random mutations, and natural selection designed and created and now command and control our molecular biology and neurodevelopment; but, that's physically impossible because any type of evolution by definition in principle cannot do Quantum Control, Decisions, and Choices. Instead, Materialism, Naturalism, and Darwinism make the claim that Quantum Phenomena or Supernatural Phenomena do not exist. There's no explanation to be had from Materialism, Naturalism, and Darwinism for Quantum Command, Quantum Control, Intelligence, Choices, Psyche, and Decisions. The Materialists, Naturalists, and Atheists paint themselves into a corner that they can't get out of. It just is what it is. I had to accept that, because I used to be a Materialist, Naturalist, Nihilist, and Atheist.

When it comes to Quantum Mechanics, Psyche, Transdimensional Physics, Quantum Telepathy, Quantum Telekinesis, Quantum Teleportation, Action at a Distance, Non-Physical Communication at a Distance, Non-Locality, the Immaterial, the Non-Physical, the Spiritual, Design, Teleology, Creation, Genetic Origination, Quantum Non-Local Consciousness, and Quantum Control, Materialism, Naturalism, Darwinism, and Classical Physics by design completely lack explanatory power. Materialism and Naturalism have been deliberately and knowingly castrated where Quantum Control, Transdimensional Physics, Intelligence, and Psyche are concerned. By definition, in principle, as a foundational axiom, evolution of any kind is supposed to be dumb and blind, which means that according to the Materialists, Naturalists, and Darwinists any type of evolution can't do Quantum Command and Quantum Control even if they were independent quantum objects to begin with which they are not. In other words, there's no Psyche or Intelligence there in evolution, so evolution can't do anything but entropy, chaos, and random destruction. Evolution of any kind can't do Quantum Command and Quantum Control.

Remember, anytime you employ the words "decisions" and "choices" to explain what you are observing and experiencing, you are in fact using Psyche and Intelligence to explain what you are experiencing and observing, because raw physical matter by definition in principle according to the Materialists and Naturalists cannot do decisions and choices.

If you have the time, read their chapter and notice how many times these people resort to "decisions" and "choices" while trying to describe neurodevelopment, and/or deliberately avoid Psyche and Intelligence while trying to describe neurodevelopment. Either way, it ends up pointing us directly to Psyche or Quantum Non-Local Consciousness every time, once we have run out of physical explanations. I love to study the science precisely where and when they run out of physical explanations for what they are seeing, observing, and experiencing. That's where the science starts to get really interesting. The non-local sciences are the final frontier in science. It doesn't get any better than that.

Wiring the Brain:

http://mypsyche.us/wp-content/uploads/2017/12/BCP23.pdf

There are physical explanations for these things, and then suddenly there are not. The physical is based upon and built upon the Non-Physical, or the Transdimensional, or the Quantum. That's just the way it is. That's where the evidence always leads us. I had to get used to it at first and resisted it at first, because I used to be a Materialist, Naturalist, Nihilist, and Atheist. I eventually chose to follow the evidence, which resulted in my abandoning my blind-faith in Materialism, Naturalism, Nihilism, and Atheism. I, my psyche, chose to be teachable. Now I see Signs of Psyche everywhere I go and within everything I choose to study and read, because now I'm looking for it and expecting to find it. Seek and ye shall find. Knock, and it shall be opened unto you. No seeking, then no finding. That's just the way it is. If you never sow, then you will never reap.

Remember, anytime you employ the words "decisions" and "choices" to explain what you are observing and experiencing, you are in fact using Psyche and Intelligence to explain what you are experiencing and observing, because raw physical matter by definition in principle according to the Materialists and Naturalists cannot do decisions and choices.

This is powerful science, once a person stops and really thinks about it.

Oh, it's happening, alright. Our brains really are processing data, running computer code, crunching numbers, transferring data, building cities, weighing decisions, and storing memories. It's just not happening at the physical level. It's all happening at the quantum level. The physical input of sensory data, and the physical output of motor commands, is the only thing happening at the physical level within the brain. It all happens at the end of an axon, where the switch is either turned ON or OFF. A synapse is a switch. Our brain is comprised of a 100 trillion switches or synapses. Our brain is also comprised of a 100 billion stand-alone switches or neurons, which are either ON or OFF. Furthermore, our brains are comprised of quadrillions of ion channels that are either OPEN or CLOSED. Everything within our brains reduces to a single BIT. There's nothing else there within our brain. Everything else is happening at the quantum level. All the important and interesting stuff is happening at the quantum level, outside our conscious awareness.

Memory Systems vs. Learning Systems

In their college textbooks, the Materialists, Naturalists, and Atheists who control our public education system teach that memory is the brain's ability to store and access the learned effects of experiences. These same people define learning as the brain's ability to change in response to experiences. They define memory consolidation as the transfer of short-term memories into long-term brain storage. They teach that our brain is the only place where our memories and learning are stored.

It's extremely important to these people that you choose to believe that our memories and learning are processed and stored exclusively in our brain and that all our memories and learning cease to exist when we die. Their chosen religion demands that the world believe that memories and learning exist exclusively within the brain and no place else. The Militant Atheists will kill you if you refuse to believe as they do. It's important to these people to believe that they will never be punished for their crimes, so that they can do whatever they want to do.

To these people, memory and learning are the same thing, because they define both as a product or epiphenom of the physical brain. These people conflate memory systems and learning systems, and they use these terms interchangeably as if they were the same thing. They teach that our brain does all of our psychological functions. They teach that all our memories and learning are processed and stored exclusively within our physical brain and nowhere else.

The multiple-trace theory teaches that memories are encoded in a distributed fashion throughout the hippocampus and other brain structures for as long as the memories exist. This thing is a black box; and technically, it's physically impossible. Our physical brain is made of stand-alone switches or bits. You can't store petabytes of software and RAM in a single bit. You can't transfer gigabytes of information through a single bit. It's physically impossible. A neurotransmitter receptor is a single bit. A synapse functions a single bit. A neuron is a single bit. Logic tells us that the claims of Materialism and Naturalism are physically impossible.

However, we have to turn to Near-Death Experiences (NDEs), Out-of-Body Experiences (OBEs), Shared-Death Experiences (SDEs), and Life Reviews for the observational evidence or scientific evidence needed to falsify the claims of Materialism, Naturalism, Darwinism, Nihilism, and Atheism. When we do, we soon discover that the Human Psyche still has its memories and that the Human Psyche continues to learn, long after its physical brain is dead and gone. The best way to defeat and debunk Materialism, Naturalism, and their derivatives is with observational evidence, experiential evidence, and empirical evidence; and, it's really easy to do, once you start accepting ALL the evidence into evidence. There is NO evidence supporting the primary assumptions or major premises of Materialism and Naturalism. Instead, ALL of the evidence that we have on hand as a race falsifies Materialism and Naturalism.

Where precisely within our neurons is our after-death Life Review being stored?

It is not!

Our after-death Life Reviews are being stored someplace else besides our physical brain, and so are our memories and learning and personality, which survive the death of our physical brain.

Furthermore, out-of-body travelers have observed that our physical body doesn't have experiences and doesn't form memories while our psyche is away from it. This implies that our memories are being formed by or experienced by our psyche, and then being stored someplace else besides our physical brain. This is especially true when it comes to the memories and learning which survive the death of our physical brain.

These types of observations and experiences are censored, blocked, ridiculed, mocked, banned, censured, and deleted by the Materialists, Naturalists, Nihilists, Darwinists, and Atheists. That's the way these people do science. Their only goal in life is to prevent you from discovering these types of non-local, non-physical experiences and observations.

During my research, I found it extremely useful to make a distinction between Memory Storage Systems and Learning Systems. The neuroscientists typically conflate or equate the two, greatly decreasing the explanatory power of their science in the process.

Learning Systems are like computer software and computer hardware. These things exist within our physical brains, and our chosen actions make or wire our physical brains in order to improve and enhance our physical functionality as a whole. Working memory, or short-term memory, seems to be a function or product of our physical brain as well, although we can't prove that it is so.

In contrast, there are plenty of indications and observational evidence suggesting that our declarative memories, particularly our episodic memories (memories for events), are not stored exclusively within our physical brain and that our physical brain doesn't have enough memory storage capacity to store all of our declarative memories or episodic memories simultaneously.

The scientists tried to discover where our memories are being stored within our physical brain. They were looking for memory engrams.

Engrams are theorized to be means by which memories are stored as biophysical or biochemical changes in the brain (and other neural tissue) in response to external stimuli. The existence of engrams is posited by some scientific theories to explain the persistence of memory and how memories are stored in the brain. The existence of neurologically defined engrams is not significantly disputed, though their exact mechanism and location has been a focus of persistent research for many decades. (Wikipedia.)

The existence of physical engrams or RAM in the brain is taken as a given and treated as an axiomatic law, even though their exact mechanism and physical location has never been observed nor discovered.

In fact, Karl Lashley and many others proved that physical engrams or memory engrams DO NOT EXIST. Based upon this discovery, the Materialists, Naturalists, and Neuroscientists concluded that our declarative and episodic memories are stored diffusely throughout our physical brain, which is physically impossible. It would take the intervention of Nature's Psyche using some type of telepathy in order to successfully store across the different neurons, synapses, or receptors within our brain. Each of these represents a single bit of information – OPEN or CLOSED – and it would take Nature's Psyche to turn these different receptors or synapses (bits) into bytes of computer programming, information, processing power, and memory storage.

Karl Lashley was a Behaviorist and Materialist; therefore, he was forced to start with their pre-chosen conclusion that ALL of our memories are stored exclusively within our physical brain, and then interpret every subsequent discovery accordingly.

Lashley discovered that our brain doesn't have memory engrams or RAM; but, because of his pre-chosen behavioristic bias, he was forced to concoct a physical explanation for memory storage anyway.

Lashley was originally in search of a single biological locus of memory or "engram". However, he ended up disproving his own theory, suggesting that memories were not localized in one part of the brain; rather, they were spread out through the cortex. As a result of this discovery, Lashley developed two separate theories, Mass action and Equipotentiality. Mass action refers to the idea that the rate, efficacy and accuracy of learning depend on the amount of cortex available. To be specific if cortical tissue is destroyed following the learning of a complex task, deterioration of performance on the task is determined more by the amount of tissue destroyed than by its location. Equipotentiality refers to the idea that one part of the cortex can take over the function of another part; within a functional area of the brain, any tissue within that area can perform its associated function. Therefore, to destroy a function, all the tissue within a functional area must be destroyed. If the area is not destroyed then the cortex can take over another part. These two principles grew out of Lashley's research on the cortical basis of learning and discrimination.

https://en.wikipedia.org/wiki/Karl_Lashley

Since there is no RAM or no memory engrams within the physical brain, you can't store any memories within them. So, what do you do? Since you can't shove the memories into a few neurons, then you telepathically and telekinetically spread the memories out throughout all the neurons in the brain, which is physically impossible.

It has been observed that you can cut out any part of the brain you want, and the consolidated retrograde memories remain. These types of memories, declarative consolidated memories, are not being stored within our brain. With something like Alzheimer's – one of my best friends has Alzheimer's – their short-term memory functionality is the first thing to go. Then other brain functions drop offline. However, ALL of his consolidated remote memories from before the Alzheimer's remain. You can damage as much of his brain as you like, anywhere you like, and his consolidated memories remain, suggesting that his consolidated memories aren't being stored within his brain.

Lashley and others observed that you can cut out any part of a person's brain that you want, and that person's long-term retrograde consolidated memories remain. From the naturalistic perspective, this observation suggests that ALL of our declarative memories and consolidated memories are being stored within each and every neuron; but, it's physically impossible to store petabytes of video and data within a single neuron.

Here we have finally run into something that is physically impossible.

Whenever we encounter science and observations that are physically impossible, then we are forced to look to Quantum Mechanics, Quantum Non-Local Consciousness, and Quantum Non-Locality for the scientific explanations that we need.

Observe carefully and notice that whenever the physical brain is damaged, functionality is lost and not necessarily the memories. The Materialists and Naturalists deliberately conflate or equate brain functionality with memory storage, in order to trick us and deceive us into believing that our memories and personalities cease to exist when brain functionality ceases to exist. This deception works on most people; otherwise, the Materialists and Naturalists wouldn't use it.

Global aphasia is a severe disruption of all language abilities. These people still remember the language. They just have a hard time using it because the functionality has been lost. If you can find a way to restore the functionality, then the memories and language abilities come back. You can use sodium amytal to temporarily eliminate their ability to speak; but, once the drug wears off, their ability to speak returns. The Materialists and Naturalists will say that these people lost their memories; but in fact, it was their functionality that was lost. Their memories remained intact during the whole experience, while they were unable to speak. Invariably, it's functionality that is lost during brain damage, and not the memories. The memories remain. Why is that? It's because memory and psyche are synonymous. If it's truly a memory, then it will survive the death of your physical brain. Cool, huh?

With Wernicke's aphasia, language comprehension is lost. They can speak, but it all comes out weird and non-sensical. These various language comprehension deficits, reading deficits, speech deficits, and writing deficits are unique to human beings. There are NO evolutionary precursors to reading and writing and record-keeping in the animal kingdom. Parrots can vocalize, but they don't always seem to know what they are talking about. Dogs can comprehend our language, but to a very limited extent. You can't have a heart-to-heart conversation with your dog and be absolutely sure that he or she understood you.

Williams Syndrome is a neurodevelopmental disorder that results in severe mental retardation accompanied by preserved language and social skills. In any type of neurodevelopmental disorder, we are looking at malfunctions and lost functionality; but, their memories seem to be just fine! In other words, they may have problems learning, but they remember if you've treated them badly or not. This same reality applies to Down syndrome.

During the year or two of substance-induced psychosis that I experienced from the dozen different prescription drugs they had me on, the functionality was lost, but I clearly remember who was treating me badly and whom I wanted to get away from. I didn't know how to turn on a television or play a DVD, anymore. But, I remember the people who helped to do so, and their motivation for doing so. Functionality is lost, but not necessarily the memories. However, there are huge chunks of that experience that I don't want to remember and don't want to think about; but, the memories are still there nonetheless. Think about it. My brain was totally fried. I was frizbot. Everything was malfunctioning and imbalanced. I could no longer add, subtract, multiply, or divide. I had to relearn all of that. Yet, the consolidated memories remain. The implication is that our memories aren't being stored within our brain. Learning functionality can be lost, yet the ability to retain memories remains. Learning systems seem to be something completely different than memory storage systems, where the physical brain is concerned.

Broca's area, in the inferior prefrontal cortex of the left hemisphere, is hypothesized to be the center of speech production. Notice carefully that these people still remember the language and understand the language, they just can't speak it anymore. The functionality was lost, not the memories. Broca's aphasia is broken functionality, NOT broken memories!

This observed reality seems to apply to ALL of the Cerebral Processing Units (CPUs) in our brain. Whenever one of the CPUs gets destroyed, the functionality or learning system is lost, but not the memories.

Karl Lashley observed that our memories seem to be distributed evenly throughout our whole brain. Near-Death Experiencers have noticed that our Life Review memories aren't stored anywhere within our physical brain. Whenever parts of the brain get destroyed, we are looking at lost functionality and lost learning systems, but not necessarily lost memories. It has been observed in coma patients and others that when the functionality comes back, then so does the memories. The memories aren't really lost. They are just inaccessible when the physical brain gets damaged or loses functionality.

For me personally, this just might be my greatest scientific discovery when it comes to Quantum Neuroscience. It sure changed the way I look at things. It had ripple effects. If our memories aren't being stored within our brain, then what else isn't being stored within our brain?

Visual completion is the filling in of a scotoma or blind spot by the brain. It's mathematical interpolation. But, is this functionality really taking place within the brain as the Materialists and Naturalists claim?

Why would I even ask such a thing? Of course, it's taking place within the brain! Where else could it take place?

I don't trust the Materialists, Naturalists, Darwinists, and Atheists anymore, because they seem to get everything wrong in the end. These people have a talent for picking the FALSIFIED and the FALSE and then choosing to believe in it. I used to be one of them, but I'm not there anymore.

When I finally realized and learned that NO memory, message, or programming code survives passage through a synaptic cleft, I was forced to ask questions such as, "Is visual completion functionality really taking place within the physical brain, or is that functionality being off-loaded non-locally to the cloud and being performed by Nature's Psyche instead at the quantum level?"

Clearly, the processing is being done and taking place; but, the million-dollar question is, "Where is it being done?" It's NOT being done within a synaptic cleft, because that's physically impossible!

So, where is it being done?

WE KNOW that there are NO memory engrams or RAM within our physical brain. Our memories seem to be getting stored non-locally in the cloud holographically as quantum waves by Nature's Psyche. But, what about our CPU functionality or processing-capabilities? Where's that taking place?

I wasn't ready to go there yet, when I first wrote parts of this book; but, I'm there now. It's physically impossible to do CPU functionality and computer processing through a synaptic cleft. A synaptic cleft terminates, randomizes, and scrambles everything that comes its way, including computer code or assembly language.

Clearly, our brains are doing CPU processing and memory storage – just NOT at the physical level. Most of this is taking place at the quantum level under the control of Nature's Psyche instead. That's where ALL the evidence has been pointing me – to the quantum level or the psyche level – because it's physically impossible to do CPU processing and RAM memory storage through synaptic clefts or within synaptic clefts. Furthermore, a neuron is nothing more than a hardware BIT – it's either ON or it's OFF. That's it! There's nothing more there at the physical level than ON or OFF.

You can't do terabytes of CPU processing and petabytes of memory storage within a single BIT or through a single BIT. It's physically impossible.

That's the way I have come to see it; but now, I'm going to try to explain why I came to see it that way in the first place. I put a ton of study and research into memory storage within the human brain. My research had repercussions.

Identifying and Locating Learning Systems

Introduction

This is a review of "Inactivation of Hippocampus or Caudate Nucleus with Lidocaine Differentially Affects Expression of Place and Response Learning" by Mark G. Packard and James L. McGaugh.

http://mypsyche.us/Inactivation-of-Hippocampus-with-Lidocaine.pdf

They are trying to verify that there are different types of learning systems within our physical brain; and with their experiment, they tried to confirm the location of two different learning systems – location-place learning, and stimulus-response habit-learning. They said that they were trying to resolve the place learning vs. response learning debate, even though they cited many other experiments which had already resolved this issue. Their experiment was important and unique because they used temporary lesions instead of permanent lesions, which successfully revealed many important details, which they themselves didn't even realize that they had discovered.

The Methods

They used standard stereotaxic surgery to implant cannulae (injection tubes) into either the hippocampus or caudate nucleus of 50 male rats. For two consecutive seven-day periods, they conditioned and trained the rats on a + maze preset as a T maze, in order to give the rats the opportunity to experience place learning and response learning. On the 8th and 16th days, they used temporary lesions (lidocaine injections) to selectively eliminate either the hippocampus or the caudate nucleus, and then observed the results of a single probe trial when they started the rats from the opposite side of the + maze, in order to test whether place learning or response learning had taken place.

The Result

The results confirmed that the hippocampus mediates spatial learning or place learning, and the caudate nucleus mediates habit learning or stimulus-response learning. Functional integrity of the hippocampus and caudate nucleus is necessary for the acquisition and the expression of these two types of learning. They also discovered that the hippocampus is involved in rapid acquisition of new information, whereas the caudate nucleus or the habit-forming system is slower and takes longer to kick in. If the caudate nucleus is disabled, then the rats reverted to place learning or spatial learning because they no longer had their habits to fall back on. However, once their habits were formed, the rats always went with their habits, unless their habit-module (the caudate nucleus) was temporarily offline.

Why is this important?

Our textbook, *Biopsychology*, by John Pinel repeatedly points out that neuro-surgeons can cut out whole lobes of the brain, yet the remote memories of childhood or highly-consolidated retrograde memories remain.

Remote memory is memory for events in the distant past. If they survive the death of your physical brain, then they are remote memories and long-term memories in every sense of the word.

My friend who has Alzheimer's Disease has pretty much lost his short-term memory or working memory, and his anterograde memory is diminishing except for highly significant events; but, no matter the amount of brain damage he receives, his retrograde memories of the decades before the onset of Alzheimer's Disease remain intact.

This combination of symptoms should not be possible if all of our memories are being stored exclusively within our physical brain. A careful study of those people and animals – who receive lobectomies, massive system-wide brain damage from neurodegenerative diseases, or temporary lesions to key memory facilitators in the brain – introduces the intriguing observation that our highly-consolidated, long-term, retrograde memories are not stored exclusively in our brain. Scientists repeatedly observe that our retrograde memories remain in place no matter how severely damaged our brain might become, which suggests the possibility that these consolidated memories are being stored someplace else besides the physical brain and that the structures within our brains are in fact transceivers and not just memory storage devices as typically claimed. This is important to me, because I used to be a vegetable – didn't know how to turn on a computer or a television. I lost all of my memories, except for my ability to read; but, over time my procedural, declarative, and retrograde memories were slowly restored to me.

I chose to pursue the details of this experiment because their vocabulary and descriptions come closest to matching with reality, of any science experiment concerning memory that I have encountered so far. They correctly identify the hippocampus and caudate nucleus as "learning systems" or "neural systems", and not as "memory storage devices". Most memory researchers that I encounter conflate or equate learning systems with memory storage devices; and, the empirical evidence doesn't justify such a leap of faith. Packard and McGaugh do conflate memory systems with learning systems, but careful reading reveals that they miraculously seem to manage to separate learning systems from memory storage devices as they should.

This is important because separating physical learning systems from "memories in non-local cloud storage" seems to most closely match with the observational evidence, experiential evidence, and memories obtained from Near-Death Experiences, Out-of-Body Experiences, and the Life-Reviews which people have when their physical brain is dead or offline.

The memory systems, hippocampus, and caudate nucleus are like WiFi adapters or computer modules – learning systems. Remove the WiFi adapter, and you no longer have access to Cloud Storage or Non-Local Memory Storage because you no longer have WiFi; but, the rest of the computer or brain continues to function just fine. Temporarily remove the hippocampus, and you no longer have spatial capabilities. Restore the hippocampus, and all the memories and learning come back automatically. Temporarily remove the caudate nucleus, and all of the habit-learning capabilities are gone; but, restore the caudate nucleus's functionality and all of the memories and associated prior-learning from habits comes back automatically.

Restoration of memories after temporary lesions implies that those memories weren't stored exclusively in the lesioned device, because lesioning is always assumed to delete the memories that are stored there. A careful reading of this experiment proves that the hippocampus and caudate nucleus are in fact computer chips, WiFi adapters, facilitators, transceivers, learning modules, or learning systems and NOT exclusively memory storage devices. The implication from all of this is that consolidated retrograde memories are stored someplace else besides the hippocampus and the caudate nucleus, because access to non-local memories – stored in the cloud – comes back whenever the lesioned devices are restored to functionality. Similar experiments with temporary lesions of other learning systems reveals the same thing, as does system-wide neurodegeneration, lobectomies, and severe brain damage.

At one point, Packard and McGaugh actually call the hippocampus and caudate nucleus "facilitators", and never do they call them memory storage devices. This distinction is extremely important. Packard and McGaugh don't even realize it, but they successfully discovered evidence suggesting that our consolidated retrograde memories are not stored exclusively within our physical brain. A careful reading of their experiment reveals the intriguing possibility that our non-local memories, stored off-site in the cloud, are soon restored and become accessible again, once the WiFi adapter or learning module comes back online and is restored to functionality. That's the benefit of using temporary lesions instead of permanent lesions. Functionality and access to memories and learning can be restored, if lesions are temporary.

What questions remain unanswered?

Obviously, further experiments need to be done and have been done to identify the location and purpose of other learning systems within the human brain. Packard and McGaugh even stated that further studies are needed; but despite that fact, I believe they successfully answered the questions that they originally set out to answer.

Of course, the questions I have always wanted answered ever since I first realized that there's not enough memory storage capacity within our physical brain to store a lifetime's worth of declarative or explicit memories is, "Where are our memories really stored, and how are our spirits able to access a lifetime's worth of memories after our physical brain is dead and gone?"

Is there anything that would make it a better experiment?

They started their experiment with the pre-chosen axiomatic conclusion and knowledge that the hippocampus controls cognitive and spatial learning and the caudate nucleus mediates stimulus-response or habit-forming learning. Technically, there was no reason for them to run this experiment, because they already knew the answers beforehand. Their time might have been better spent searching for a learning system that hadn't already been discovered.

It would have been really exceptional, though, if they themselves had figured out that their experiment demonstrates that our memories are not stored only within our brain, and that our memories and learning aren't really lost when our physical devices are temporarily disabled; but, they weren't looking for that, so they didn't make that discovery. Therefore, it falls upon someone like me to make that discovery for them, because I am looking for such a thing.

References

Packard, M. G., & McGaugh, J. L. (1996). Inactivation of hippocampus or caudate nucleus with lidocaine differentially affects expression of place and response learning. *Neurobiology of Learning and Memory*, 65, 65-72.

Pinel, J. (2014). *Biopsychology* (9th ed.). New York: Pearson.

"Modern biopsychology began in the mid-20th century. At that time, there was a major push to identify the areas in the brain where memories are stored. The search was largely championed by Karl Lashley, who wrote a famous review paper, *In Search of the Engram*, in which he described his fruitless efforts. Lashley and many who subsequently took up the search used the lesion method. If a particular structure were the storage site for all memories of a particular type, then destruction of that structure should eliminate all memories of that type that were acquired prior to the lesion.

No brain structure has shown this result! Lesions of particular structures tend to produce either no retrograde amnesia at all or retrograde amnesia for only the experiences that occurred in the days or weeks just before the surgery.

These findings have led to two major conclusions: (1) Memories are stored diffusely in the brain and thus can survived destruction of any single structure; and (2) memories become more resistant to disruption over time." (*Biopsychology*, pp. 277-278.) These findings also suggest the logical possibility that (3) our consolidated, long-term, retrograde memories are not stored in our physical brain anymore, because our consolidated retrograde memories remain even when most of our physical brain has been eliminated or destroyed.

Follow-Up

The effect of lidocaine injections on place learning observed in the present study was likely due to neural inactivation of the hippocampus.

http://mypsyche.us/wp-content/uploads/2017/12/Inactivation-of-Hippocampus-with-Lidocaine.pdf

It was due to the inactivation of the device and loss of functionality, not the loss of memories! Amnesia is defined as a loss of memories; but invariably, it ends up being a loss of functionality. When the functionality returns, the memories return as well, which means that the memories were never lost!

The little rats no longer have access to the Spatial Map that Nature's Psyche has mapped onto their brain. It's not about lost memories, it's about lost functionality. Once functionality returns, access to the Spatial Map returns, and thus the memories seem to return as well. Nature's Psyche maps different Learning Systems onto our brain as Maps of Physical Functionality. The brain is all about functionality, and not memory storage.

Packard and McGaugh portray the hippocampus and caudate nucleus as computer chips or computer modules or WiFi adapters, rather than memory storage devices, which I think reflects a more accurate reality of the situation. Remove the WiFi adapter, and you no longer have WiFi; but, the rest of the computer runs fine, including the on-board memory systems. Your cloud storage or your non-local cloud memories would go down, though, if you remove the WiFi adapter. In their research study, Packard and McGaugh called them "learning systems" rather than "memory storage devices", which I found extremely refreshing for a change.

You definitely don't want to copy and hand in this paper and these ideas to a PhD expert on memory research, because it's not going to go well for you if you do. I got a B on this particular assignment, when I did. Considering whom I was dealing with, I'm lucky to have done so well.

His response:

> **While these results do leave open the logical possibility that memories are not stored in the brain, that is unlikely to be the case.**

Well, it's also highly unlikely that our after-death Life Reviews are being stored in our brain.

Stalemate.

His comment failed to convince me that he is right.

He didn't get it, and couldn't see it, which I expected would be the case. You can't teach these people anything new, because they already KNOW that memories are stored exclusively in our physical brain and nowhere else. He even jumped to the other extreme and accused me of claiming that our "memories are not stored in the brain", and then punished me accordingly.

There's no getting past their KNOWLEDGE with something new, because they simply know better than you how things really work. He won't even consider the possibility, because he has been programmed and brainwashed through years of schooling and research not to. But, the truth is that it's physically impossible to store all of our memories within our brain.

His response reflects *all-or-nothing thinking*, which is a logic fallacy. He also *jumped to the conclusion* that I'm making the claim that our memories are not stored in our brain, which I'm not. He didn't get it and couldn't see it.

Once you discover Psyche, you can indeed make the mistake of assuming that all our memories are stored exclusively in Non-Local Cloud Storage, as I did at one time. All-or-nothing thinking can get you into trouble and prevent you from making new discoveries.

I think it would be a mistake to claim that our brains have no memory storage capabilities, because during my studies I have seen plenty of evidentiary proof that they do. I'm not trying to make such a claim, even though it can seem at times that I am.

But, one thing I do KNOW for sure is that our after-death life reviews are NOT being stored in our physical brain. In Shared-Death Experiences (SDEs), some healthy living people have experienced the life review of the dying or dead person. That's not physically possible, which provides sure and solid proof that our life reviews are not being stored within our physical brain. There's no way on earth that someone else's life review and memories could be stored within your physical brain!

Retrograde amnesia is loss of memory for events or information learned before the amnesia inducing brain injury. This is the one that seems like REAL memory loss, and not just a loss of physical functionality. Are certain brain structures mapped by Nature's Psyche to function as retrograde memory receivers or retrievers? Again, is it really memory loss, or is it lost functionality? From what I have been able to tell, the memories return once functionality is restored. Only when the memories are never stored in the cloud by Nature's Psyche, for later retrieval by the Human Psyche, do they seem to be forever irretrievable.

This is very confusing.

It's clear to me that it's physically impossible to store memories, thoughts, ideas, programming code, data, and information in a synaptic cleft, because a synaptic cleft randomizes and scrambles everything; and, the only message that gets through a synaptic cleft is ON or OFF. Every synapse on a neuron integrates or reduces to ONE hardware BIT, a neuron or a switch that is either ON or OFF.

Now for the really confusing part.

It's physically impossible to store memories within a physical brain; yet, a functional physical brain is needed if you want to consolidate memories, upload memories to the cloud, and then retrieve those memories from the cloud. The Human Psyche needs the structures in the physical brain in order to form memories, consolidate memories, and then retrieve those memories later on. Both during the transmission of the memories and the reception of the memories, the memories seem as if they have to pass through the physical brain or to register on the physical brain, but they aren't stored within the physical brain. If there is NO physical structure, such as a missing or broken occipital lobe, then there's nothing there for the Human Psyche to experience and remember.

Unless it registers on the brain, the Human Psyche doesn't have access to it. Even when it does register on the brain, such as in the case of working memory or short-term memory, if the physical machinery necessary for the uploading of those memories to the cloud and the retrieval of those memories from the cloud goes missing, then the Human Psyche still doesn't have access to those memoires, even if they spent some time in working memory.

Now that's really confusing!

Does working memory exist somehow at the physical level within the physical brain, or is it simply the fact that the frequency of the quantum waves for working memories are different than the frequencies used to store long-term memories, consolidated memories, and retrograde memories in the cloud? In the case of anterograde amnesia, why does the Human Psyche seem to experience the events and the memories, only to forget them all if the brain structure that consolidates them and uploads them to the cloud is missing? How can something be in the Human Psyche and then seem to get removed from the Human Psyche? Maybe it was never in the Human Psyche to begin with?

Thoughts and memories are quantum waves. If the thoughts and memories are there in the Human Psyche during working memory, then what removes those thoughts and memories during the consolidation phase whenever the anterograde functionality is broken? Maybe our memories are NEVER stored within the Human Psyche, but only within Nature's Psyche by Nature's Psyche.

How do you reconcile all of these different observations in a way that actually makes logical sense?

It's as if ALL of the working memories, retrograde memories, physical memories, anterograde memories, and all other memories are experienced by Nature's Psyche, processed by Nature's Psyche, uploaded by Nature's Psyche, and retrieved by Nature's

Psyche; but, only if the associated mapped physical functionality created by Nature's Psyche exists in the first place.

The Human Psyche seems to get ONLY what Nature's Psyche chooses to give it, and ONLY if the associated brain structures or Maps of Physical Functionality actually exist at the physical level to begin with. Maybe the Human Psyche is capable of functioning autonomously with the brain structures that it does have access to; but, Nature's Psyche, the physical machinery, and the Maps of Physical Functionality have to exist in order for Nature's Psyche to consolidate and upload the Human Psyche's working memories to the cloud for later retrieval. Or maybe the Human Psyche only has access to the consolidated memories that Nature's Psyche chooses to give it.

The Human Psyche seems to only have access to the physical buffer or the neurons that register all of the current most-recent input – working memory or working senses. If the Maps of Physical Functionality, which Nature's Psyche uses to consolidate and upload those memories to the cloud, are missing, then the memoires are lost; and, the Human Psyche no longer has access to them. Once the buffer is purged, then the Human Psyche no longer has access to that information, unless Nature's Psyche has found a way to store that information in the cloud; but, Nature's Psyche can't transfer working memories or current sensory input to the cloud if the structures that Nature's Psyche has mapped towards that purpose get damaged or remove.

There's a complex symbiosis among the Human Psyche, Nature's Psyche and the physical brain. It's difficult if not impossible to figure out the rules governing this symbiosis and how everything really works.

Once you figure it all out, be sure to let me know.

Do you see what I'm doing here? The principle is the important thing to learn, not necessarily the specifics.

I'm trying to adjust my theories in order to make them match with and explain ALL the observational evidence and ALL the proven FACTS. Science is observation and experience, or it should be.

In contrast, the Materialists, Naturalists, and Darwinists deny the evidence, reject the evidence, ridicule and ban and destroy the evidence; and then, these people make up an ad hoc just-so story out of thin air to explain what they think is happening or want to be happening.

I learn best through comparison and contrast.

One of these days, I hope to find a theory that matches with ALL of the evidence and explains everything that has ever been observed or experienced. Maybe I already have.

Learning is a test of what's being remembered

These different scientists successfully demonstrated that neurons are functioning more as memory access devices rather than as memory storage devices. Access to the memories and learning is immediately restored once the functionality of the brain region is restored.

The incomplete pictures test is a "test of memory" measuring the improved ability to identify fragmented figures that have been previously observed. Is it really memory that's being tested, or is it improved functionality? What is being programmed or mapped into us as we study pictures and try to memorize them? They actually say that they are measuring the "improved ability" or the improved functionality.

Clearly, there is some overlap between memory and functionality; so, how do we tell them apart when called upon to do so? Well, I finally found a way. If it survives the death of your physical body and physical brain, then it's a memory. If it doesn't survive the death of your physical body and physical brain, then it was physical functionality that you were looking at. How's that for a litmus test?

Memories are Psyche. Thoughts are Psyche. Choices are Psyche. They survive the death of our physical brain.

Our brain and psyche are symbionts. They are not either-or, nor are they all-or-nothing. They are symbionts and work together, at least until our physical brain is dead and gone.

It's impossible to determine where the brain's memory storage ends and all that non-local cloud storage begins; but, the brain is a physical device which means that it is going to have some physical limitations. It's physically impossible to store decades of our memories and experiences and video within our physical brain. It can't be done. But, most scientists can't see that because they have never run the numbers.

When it comes to Quantum Neuroscience, you are never going to get any of this from your college professors, which means that you are going to have to find it on your own and develop it for yourself if you want it.

You can't use Materialism, Naturalism, and Classical Physics to explain the physically impossible; but, you can definitely use Quantum Mechanics and Non-Local Consciousness to explain these things!

Thanks to Nature's Psyche and Quantum Mechanics, each individual atom and molecule within the human brain is alive with purpose and activity. It's alive, which is physically impossible!

The more I studied memory storage, memory engrams, and RAM, the more obvious it became that our brain is comprised of Cerebral Processing Units (CPUs) or Maps of Physical Functionality, and NOT memory engrams or RAM. In fact, repeated scientific experiments from Karl Lashley and others proved that our brain does not have universal memory engrams or consolidated RAM. Our brains are comprised of CPUs or Maps of Physical Functionality, not RAM. Over and over again, the evidence repeatedly tells me that our brains are all about functionality, physical functionality, and not necessarily memory storage. That's what I always end up with whenever I choose to follow the evidence to its logical conclusion.

There's NO visible signs of RAM anywhere within our brain; however, it is clear and obvious to me that our brain was designed to function as different types of Cerebral Processing Units (CPUs) or mapped by Nature's Psyche to perform physical functions at the command of Nature's Psyche. A CPU is a Map of sorts. However, there are NO wires in a physical brain; consequently, the CPUs within our brain are being mapped at the quantum level by Nature's Psyche rather than the physical level by mechanical wires.

Since there's no visible signs of RAM within our neurons and within our brain, that kind of tells me that maybe we should be looking for invisible signs of RAM non-locally in the cloud; and, there are indeed evidentiary indications that such a thing does in fact exist, especially where our after-death Life Reviews are concerned.

Of course, if you are a computer scientist as I am, you know that CPUs have cache and calculators have a stack, which are a type of working memory or short-term memory. It's clear and obvious that our brain has some type of working memory or short-term memory. Whether that memory or cache exists locally in our brain or non-locally in the cloud, we don't know, because there are NO visible signs of CPUs, RAM, wires, or cache within our neurons and glial cells.

Magnetoencephalography (MEG) us a technique for recording changes on the surface of the scalp produced by magnetic fields in the underlying patterns of neural activity. In other words, neurons are producing magnetic fields and could theoretically be communicating with each other through these magnetic fields. But, neurons have something even better that they can fall back on for communicating with each other – quantum waves and Nature's Psyche.

Magneto Resonance Imaging (MRI) is a procedure in which high-resolution images of the structures of the living brain are constructed from the measurement of waves that hydrogen atoms emit when they are activated by radio-frequency waves in a magnetic field. Notice that they are talking about quantum waves, atomic waves, magnetic waves, invisible waves, and radio waves. MRI is a quantum mechanical device. Its very existence and functionality FALSIFIES Materialism, Darwinism, and Naturalism. Memories and thoughts are quantum waves. The fact that our memories and thoughts are NOT scrambled by MRI is evidentiary proof that our thoughts and memories exist at a wave frequency that MRI does not touch nor interfere with. Our Human Psyche doesn't get erased by MRI, so that means that these two objects are not functioning on the same frequency.

Functional MRI (fMRI) is a magnetic resonance imaging technique for inferring brain activity by measuring the increased oxygen flow into particular areas of the brain. An fMRI has pretty good spatial resolution pinpointing where in the brain an event takes place; whereas, an EEG has good temporal resolution pinpointing when an event occurs.

When it comes to working memory, the best we can do is to use fMRI to observe what parts of our brain heat up whenever the Human Psyche chooses to think about something specific.

The BOLD signal is the blood-oxygen-level-dependent signal; and, it's recorded by fMRI and is related to the level of neural firing. These neurons are heating up and firing unilaterally. Whenever the Human Psyche is asked to think about something, process something, or pay attention to something, the Human Psyche literally reaches out telepathically and turns on or fires up the neurons necessary to accomplish the assigned task. Even when a person has his or her eyes closed, the visual processing centers within the brain will turn on and heat up and start firing when the human is asked to visualize in his mind's eye a particular scene, place, or object. It has been observed through fMRI that the Human Psyche is reaching out telepathically and turning on specific regions of its brain as needed.

My neuroscience teacher, a Materialist and Naturalist, told us that at first, he was afraid of getting an fMRI, because he thought that they would be able to read his thoughts and read his mind. He would be right, if our thoughts were actually produced by our physical brain. But, it's physically impossible to read and record our thoughts and our dreams with our physical instruments, because our thoughts and dreams exist as quantum waves at the quantum level, and they are being produced and experienced by the Human Psyche at the quantum level.

I mean, think about this logically as a scientist, and not as a Materialist and Naturalist. If our thoughts were indeed being produced by our brain and stored within our brain, then all of this magnetic resonance from fMRI would directly scramble and degauss our thoughts and memories if they were indeed being created by our brain and stored within our brain. But, since our memories are being created and stored at the quantum level by Nature's Psyche, our physical instruments can't touch our memories.

However, the scientists do use Transcranial Magnetic Stimulation (TMS) to disrupt brain activity in an area of the cortex by creating a magnetic field under a coil positioned next to the skull. The effect of the disruption on cognition is assessed to clarify the function of the affected area of cortex. By turning off an area of the cortex, they are trying to determine causation – what part of the brain causes what. Here again, we are looking at temporarily disrupted functionality, and not lost memories. The memory capabilities return when the functionality returns. In some of the science fiction that I watch, they actually used TMS to permanently fry parts of the brain and eliminate the subject's ability to empathize. Again, it's functionality that goes down, and not thoughts and memories. The person still had all of his thoughts and memories.

TMS and the other types of "invasive scanning" or lesioning don't touch Nature's Psyche, and they can't touch the Human Psyche – the ultimate causal agent. Nevertheless, whenever brain functionality goes down, it can indeed give the faulty impression that the Human Psyche has gone down or that the Human Psyche does not exist. Therefore, from the beginning we use observations, experiences, phenomenology, and empirical evidence to establish the existence of the Human Psyche; and then, we adjust our interpretations of the brains scans and the temporary lesions accordingly.

Access to certain types of memories can be temporarily lost by a loss in functionality, but the memories return or access to the map returns once the functionality returns. Our brain structures seem to be organized by Nature's Psyche as different Maps of Physical Functionality. Remove the physical structure, and access to the map is lost. Restore functionality, and access to the map comes back.

This seems to be the way our brains really work. In other words, it matches with the observational evidence, experimental evidence, and experiential evidence. Isn't it amazing how the truth is always consistent with every other truth? It all fits together perfectly, doesn't it?

A lot of the functionality of our brain seems to be taking place outside of our brain – or, invisibly within our brain within non-local space or transdimensional space, within all of that "empty" space between the nucleus of an atom and its electrons, which is something that I have started to call, "Nature's Psyche". An atom is 99.999% empty space; and, that's where Nature's Psyche resides, as well as some of our non-local cloud storage and non-local CPU processing. That's my opinion and best guess as a scientist; but, there's no way to prove me right and there's no way to prove me wrong – at least not with our physical instruments. The best we can do is to observe the neurons heat up or fire up, whenever the Human Psyche starts concentrating on something it finds interesting.

We have NO way of determining where our local cache and local memory storage ends, and where our non-local cloud storage begins, because Nature's Psyche handles all of these memory storage issues for us automatically completely outside the conscious awareness of our Human Psyche. Nature's Psyche knows when we are close to exceeding our physical capacity and off-loads things to the cloud for us automatically in order to compensate.

Since there are NO wires in our brain, a case can be made that there are no CPUs and RAM within our physical brain. Instead, it seems as if our brain structures are mapped by Nature's Psyche at the quantum level into different Maps of Physical Functionality.

Nature's Psyche and the Human Psyche use the Maps of Physical Input to assess the physical environment; and then, when the Human Psyche makes a choice or a decision, Nature's Psyche triggers specific neurons within its Map of Physical Output in order to instigate specific physical actions.

It's theoretically possible that NONE of our memories are being stored within our physical brain, at least not at the physical level. Who knows what's going on at the quantum level? When it comes to where and how our memories are being stored, indexed, and recalled, the best we can do is to let Nature's Psyche know what we consider to be most important by paying attention to it a lot.

Memory storage may just be the most interesting part of Neuroscience and Quantum Neuroscience, so I'm going to spend some time with it and see what I can learn.

Where Are Our Memories Stored?

Disable the functionality of the hippocampus, and spatial memories are lost. Restore functionality, and the memories come back. Disable the functionality of the caudate nucleus, and habits are lost. Restore the functionality, and they come back.

One possible interpretation for this evidence is that these brain structures are WiFi devices with access to non-local cloud storage. In contrast, if memories are really stored only in the physical brain, then whenever those memories are lost, they should never come back.

Whether our memories are stored in our physical brain or not, one thing is clear and obvious – the memories that show up in our Life Reviews after our brain is dead and gone cannot possibly be stored in our physical brain. They have to be stored on some kind of non-local cloud storage device.

It's theoretically possible that our brain has a hard drive of some sort. It at least has some kind of stack like a calculator – working memory – whether at the physical level or quantum level, we're not quite sure. If our brain does have some kind of local physical hardware memory storage, then it's obvious from our Life Reviews that the brain's hard drive is being backed up onto the cloud. Lose access to the hard drive; the memories come back when access is restored. Lose the hard drive, then the non-local cloud storage becomes the only way to get those memories back.

The question I have is, "If the brain structures are permanently damaged or removed thereby destroying the ability to create and remember and experience long-term memories, do the person's ongoing experiences and short-term memories and feelings make it into their Life Review? Or, are all of those memories, feelings, and experiences lost forever once a person's WiFi adapters have been destroyed?"

Is a functional brain needed to upload memories to non-local cloud storage?

We'll never know, because people with anterograde amnesia won't remember their Life Review and NDE should they have one, unless God were to actually download those memories directly into that person's brain and then create a permanent connection to those memories in their consolidated retrograde memory storage.

Actually, this isn't totally true, because I can think of an exception. People who are born blind can see when they die; and, their after-death Life Review contains video. A functional brain wasn't needed to produce the video for their after-death Life Review, which means that somebody else besides their brain and their Psyche was recording the video for their Life Review.

It's a mistake to conclude that all of our memories are being stored only within our brain. However, it may be a mistake to go too far the other direction, as well. I truly believe that the Maps of Physical Functionality, which Nature's Psyche creates within our physical brain, do indeed exist within our brain and are indeed stored within our brain. The Materialists and Naturalists erroneously call Maps of Physical Functionality "procedural memories" and "implicit memories"; but, they can't be memories because the Human Psyche isn't consciously aware of them. How can they be memories if you can't remember them?

According to William Buhlman and other out-of-body travelers, our physical body and spirit body aren't having experiences, asking questions, making memories, nor learning lessons. Our Psyche is. That reality and observation led me to erroneously believe that all of our memories are being stored in our Psyche. There's evidence that such is not the case, though. Our after-death Life Review is such a case. A Life Review wouldn't have to be shown to us, if it already existed within our Psyche. If it were stored within our own Psyche or our physical brain, then we would be able to call it up on-demand.

Do you see what's possible by choosing to allow ALL of the evidence into evidence? We actually learn something new and interesting whenever we do. It's certainly a lot more interesting than saying that all our memories are only stored within our brain, and that all our memories cease to exist when we die.

The Brain, Information Storage Capacity, and Memory

It's time to search for and present a new perspective on memory storage. The following quote from *Consciousness Beyond Life* by Pim van Lommel provided me with some help, making me feel as if I'm not alone in this venture.

> **The hypothesis that consciousness and memory are produced and stored exclusively in the brain remains unproven. For decades, scientists have tried unsuccessfully to localize memories and consciousness in the brain. It is doubtful whether they will ever succeed. At present science cannot explain how certain neural networks produce the subjective essence of thoughts and feelings because so far no neurophysiological study has identified any exact correspondence between specific neural activities and the specific content of memories, experiences, feelings, or thoughts.**

> **During a cardiac arrest the cerebral cortex, thalamus, hippocampus, and brain stem as well as all connections between them stop functioning, as we have seen, which prevents information from being integrated and differentiated – a prerequisite for communication and thus for the experience of consciousness. The experience of consciousness should be impossible during a cardiac arrest. All measurable electrical activity in the brain has been extinguished and all bodily and brain-stem reflexes are gone. And yet, during this period of total dysfunction, some people experience a heightened and enhanced consciousness, known as an NDE.**

> **The Brain, Information Storage Capacity, and Memory**

According to current knowledge, consciousness cannot be reduced to activities and processes in the brain. It is highly unlikely that thoughts and emotions are produced by brain cells. Above, we looked at the influence of electromagnetic fields on consciousness as well as the fact that information exchange between brain stem and cerebral cortex is a prerequisite for the experience of consciousness. The next logical question is how all the memories from a person's life can be stored and then recalled again together with their associated emotions. How do we explain short-term and long-term memory? How and where in the brain is this virtually unlimited amount of information stored? And how can this information be accessible at all times?

A single cubic centimeter of the cerebral cortex contains no less than a hundred million neurons, and because each neuron has at least a thousand synapses connecting it with surrounding neurons, each cubic centimeter has approximately 100,000,000,000 (10^{11}) synapses of dendrites that originate largely in other parts of the cerebral cortex. This means that the brain contains a total of about 10^{14} synapses. If one synapse contained one bit of information, brain function would require more than 100,000,000,000,000 (10^{14}) bits of information processing, which is far more information than the human DNA, our genetic code, can handle according to current knowledge. For this reason consciousness cannot be stored in our DNA, rendering a cell in our body and brain a highly unlikely producer for consciousness.

Simon Berkovich, a computer expert, has calculated that despite the brain's huge numbers of synapses, its capacity for storing a lifetime's memories, along with associated thoughts and emotions, is completely insufficient. At any waking moment during the day, there are approximately 10^{24} actions per second in the brain. Add to this the required capacity for long-term memory storage, and the total data storage capacity would have to be 3.10^{17} bits/cm^3, which, based on our current understanding of neuronal processes in the brain, is inconceivable.

[It's physically impossible to store 3.10^{17} bits of information within a cubic centimeter of brain matter, which can store only 10^{11} bits of information at most. We are talking about petabytes here, possibly at times approaching an exabyte per cubic centimeter if were talking about something like video or motor skills in the cerebellum. It's physically impossible to shove an exabyte of data into a terabyte of synapses; and, that's what you are looking at if all of our memories are stored within our physical brain – potentially exabytes of data storage needed to get the job done, over the typical lifetime of a human being. An exabyte is a million terabytes. And technically, due to the fact that all of our consolidated retrograde memories remain no matter what part of our brain we choose to cut out, the fact is that each neuron in our brain has to be storing an exabyte or two of memory in order to produce those kinds of results. That's physically impossible. Even if an exabyte ends up being an exaggeration, a petabyte certainly is not. A petabyte is a thousand terabytes, and it's physically impossible to store terabytes of information within a single neuron which basically has only an ON or an OFF state to begin with. A neuron either fires, or it doesn't. It doesn't do anything else in terms of information storage and information transfer that we can see with our physical instruments. A neuron either fires or it doesn't. How do you transfer and store petabytes of information through that single bit of the neuron either firing or not firing? I can't see it. Can you? If you figure it out, the Materialists and Naturalists would consider you to be a god, in line with evolution which they already consider to be a god.]

Neurobiologist Herms Romijn, formerly of the Netherlands Institute for Neuroscience, also demonstrated that the storage of all memories in the brain is anatomically and functionally impossible.

On the basis of these findings, we are forced to conclude that the brain has insufficient capacity for storing all memories with associated thoughts and feelings or retrieving capacity for stored information. Neurosurgeon Karl Pribram was equally certain that memories cannot be stored in brain cells, but only in the coherent patterns of the electromagnetic fields of neural networks. In his view the brain functions like a hologram. This hologram is capable of storing the vast quantity of information of the human memory. According to Pribram's holographic hypothesis, memories are stored not in the brain itself but in the electromagnetic fields of the brain. Pribram's hypothesis was inspired by the extraordinary experiments of psychologist Karl Lashley, who proved as early as 1920 that memories are stored not in any single part of the brain but throughout the brain as a whole. His experiments on rats showed that it did not matter which parts or indeed how much of the rats' brains were removed. The animals were still capable of carrying out the complex tasks that they had learned to do before the brain operations.

Earlier in this chapter I mentioned that the composition and cohesion of all brain structures, from molecules to neurons, is in constant flux, which raised a question about long-term memory. The debate about information storage and memory is further complicated by an article in *Science* with the provocative title "Is Your Brain Really Necessary?" This article was written in response to English neurologist John Lorber's description of a healthy young man with a university degree in mathematics and an IQ of 126. A brain scan revealed a severe case of hydrocephalus: 95 percent of his skull was filled with cerebrospinal fluid, and his cerebral cortex measured only about 2 millimeters thick, leaving barely any brain tissue. The weight of his remaining brain was estimated at 100 grams (compared to a normal weight of 1,500 grams), and yet his brain function was unimpaired. It seems scarcely possible to reconcile this exceptional case with our current belief that memories and consciousness are produced and stored in the brain.

The question is not just how short-term and long-term memory can function properly given the constantly changing synaptic connections in neural networks, but also how memory loss arises. As we get older our brains can atrophy as a result of Alzheimer's Disease or arteriosclerosis. Brain volume decreases when brain cells die and are no longer replaced, giving rise to damaged and less effective neural networks and slowly worsening dementia. Whereas long-term memory can remain intact for some time, short-term memory deteriorates, cognitive functions gradually decline, relatives are no longer recognized, and speech becomes more difficult or altogether impossible. These functions can also be lost after brain damage brought on by a cerebral hemorrhage, serious head trauma with permanent brain damage, long-term alcohol abuse, or encephalitis.

The obvious and correct conclusion must be that the brain has a major impact on the way people show their everyday or waking consciousness to the outside world. The instrument, the brain, has been

damaged, whereas "real" consciousness remains intact. Consciousness and the brain are interdependent, which is not to say that mental and emotional processes are identical with or reducible to cerebral processes. How else can we explain the fact that people with a severe form of dementia, or patients with chronic schizophrenia, sometimes can experience brief lucid moments ("terminal lucidity") shortly before they die? (*Consciousness Beyond Life*, pp. 184, 193-195.)

[Notice the truths being taught here. A synapse is a single bit of information – it's either OPEN or it's CLOSED. Alas, this same exact truth applies to neurons. They, too, end up being a single bit of information. They either fire, or they don't. ON or OFF. That's it!

ALL of the intelligence, information, sentience, consciousness, awareness, data, choices, decisions, flash RAM, CPU processing, differentiation, molecule assignments, messenger assignments, ATP and ADP assignments, ion control, neurotransmitter control, protein synthesis requests and destinations, synaptogenesis addressing, neurogenesis addressing, navigational logistics, developmental data, and 3D blueprint data associated with a specific neuron are STORED and PROCESSED someplace else besides the neuron, because all the neuron can do in the end is either ON or OFF, a single bit of information.

It's physically impossible to store petabytes of data, programming, command, control, intelligence, learning, and knowledge within a single BIT of information! The only noticeable adjustments to that single bit of information exchange through the neuron are an increase or a decrease in the frequency of firing a specific neuron along with an increasing or decreasing likelihood that a specific neuron will fire. The functionality, processing, memory storage, purpose, command, control, and overall wiring diagram of a specific neuron is stored and controlled outside of the neuron in the non-local non-physical realm. The neuron itself fires or it doesn't. That's it.

This is not what they teach you in neuroscience, but this is the way it really works. All they tell you in neuroscience textbooks is that we don't know how any of this works, but that it obviously does work. Well, Pim van Lommel is telling you how it works; and, so am I.]

Online source:

https://sites.google.com/site/crabtreescompendiumofesoterica/nibiru-2012-onward-related/afterlife-investigation-life-after-death/consciousness-beyond-life-the-science-of-the-near-death-experience

I got the impression that Pim van Lommel is centuries ahead of my college neuroscience teacher, when it comes to the memory storage capabilities of the human brain, even though my college neuroscience teacher claims to be an expert on the subject.

Why would that be the case?

It's because Pim van Lommel is looking for this information, and my college neuroscience teacher is not. No seeking, then no finding. That's just the way it is. If you don't believe it's possible, then for you personally, it's not possible. That's the way it works.

The truth is that it's physically impossible to store all of our memories within our brain. Yes, our brain shows some memory storage capacity, but not enough. Whenever we encounter the physically impossible, then that's when things start to get really interesting.

There are a few, but not many, who have caught onto these truths and realities.

Forsdyke, D. (2016). Memory: What Is Arranged and Where?

Despite fascinating studies on the input (acquisition) and output (recall) of human mental information, we know not how or where that information is stored for prolonged periods (long term memory). Storage in digital or analog forms could be chemical (e.g. like DNA) or physical (e.g. vibrations), and either molecular or submolecular. In the absence of evidence for localized or delocalized storage in our brains, the brain 'cupboard' must be deemed bare. Alternative locations are elsewhere in our body (corporeal), or outside our body (extra-corporeal).

Evidence on location is quite fragmentary and usually of negative nature: 1. Since brain cells have the same amount of DNA as other tissues, the storage form is unlikely to be DNA, and no comparable macromolecule has yet emerged. 2. Surgical interventions that delete parts of the brain do not remove specific information, the storage of which appears delocalized. 3. Brain size is not necessarily greater in those with phenomenal memories (savant syndrome). 4. Some of those with greatly decreased brain tissue volume can display normal or even advanced intelligence (see next chapter). These observations are consistent with extra-corporeal location. Drawing parallels from the internet – with desktop computers storing their information in a remote location ("cloud computing") – researchers from various disciplines are looking to physics for support of this strange idea.

[This is actually coming from a Darwinist and Evolutionist who is still trying to promote the theory of evolution as if it has some kind of use and validity. In other words, his claim is that evolution did all of this extra-corporeal locating, cloud computing, advanced intelligence, and memory storage, even though that's physically impossible. Like most of the others, he's giving evolution credit for things that it can't possibly do.]

https://books.google.com/books?id=VEgWDAAAQBAJ&pg=PA377&lpg#v=one page&q&f=false

Because I'm a computer scientist, I had already created my Non-Local Cloud Storage idea, for the storage of long-term consolidated memories, before I read this blurb from Forsdyke. I realized in recent months that Telepathy is WiFi at the quantum level; and, I know how WiFi works because I'm a computer scientist and a certified computer support specialist. There's no evidence of flash RAM and CPUs within each and every neuron or synapse, which would have to be the case if all of our memories and programming were being stored and done within our neurons or synapses.

Most people can't see this, and aren't willing to accept it or understand it; but, I certainly can. I've been studying computer science all my life, so I KNOW how these things work; and, our physical brain isn't anything like a computer. If there are CPUs and RAM within each of our neurons, then these CPUs and RAM are invisible, non-local, and non-physical. We can't see them nor detect them with our physical instruments. Remember, it's physically impossible to store petabytes of information within a single synapse, or neuron, or ion channel.

Neurons and synapses are not computers. They are bits – ON or OFF. But, just like a computer, ALL of the intelligence, foresight, programming, engineering, architecture, logistics, software, command, and control has taken place OUTSIDE of the computer – or outside of and separate from the neuron or the synapse. Unlike a computer, however, when it comes to neurons, all of the RAM, CPUs, software, and specialized processing seem to be OUTSIDE or missing from the neuron or separate from the neuron. The neuron's only physical superiority is the fact that the neuron has its own in-built power supply, the mitochondria, and a computer doesn't. You have to plug a computer into the wall; otherwise, it is dead.

A neuron's superiority apparently comes from all of its non-physical aspects; otherwise, a physical computer is vastly superior to a physical neuron. A neuron or a synapse is a machine, just like a computer; but, ALL of the intelligence or psyche or consciousness ultimately resides someplace else besides the neuron, synapse, or computer. In the case of neurons and synapses, ALL of their CPUs and RAM and intelligence reside someplace else besides the neuron or the synapse.

The analogies between the brain and a computer only start to come into play and hold true at the macro-scale. The different parts of the brain are similar to computer add-on cards or extension cards, each with a different assigned functionality.

And like a computer, it is obvious to any rational thinking person that the brain also had to have had an Intelligent Designer in order to bring it into existence in the first place. Computers don't spontaneously generate out of thin air, and neither do brains and genomes. A brain is hardware, just like a computer; and, they both need and require Designers and Creators and Manufacturers in order to get them going in the first place.

I'm not going to apologize for figuring this stuff out, because that's what we are supposed to do if we are scientists, or consider ourselves to be scientists, as I do.

Psyche IS the Placebo Effect, Choice, and the Quantum Zeno Effect. Voluntary behavior is a function of Psyche. Decisions and choices are a product of Psyche.

As I see it, we need to find ways to test the human psyche's telepathic and telekinetic influence on neurotransmitters, ions, ion channels, synapses, ATP molecules, proteins, genes, and neurons, like we have done with the Quantum Zeno Effect's influence on individual atoms.

It's that FIRST NEURON that fires, in a chain of action potentials, which is the important thing to study and learn about – that first neuron which unilaterally fires all on its own, out-of-the-blue, making your finger rise. There's NO physical cause for this! The neuron simply fires, and your finger rises. That's the thing we need to study. It all starts there. It starts with the Psyche's decision or choice to raise its finger. My finger doesn't rise, unless I tell it to do so.

Why does that specific neuron decide to fire or switch on? It hasn't received any input from any other physical object telling it to do so. It just fires, and your finger rises, whenever your Psyche wants your finger to rise. There's where the tires hit the pavement, in my humble opinion. That's the thing we should be trying to figure out. But, people aren't interested in that, because it's physically impossible.

Yet, the physically impossible is where everything starts to get interesting. Remember, it's physically impossible to store all of your memories in your brain.

Combine the FACT that it's physically impossible to store petabytes of memories within our brain, along with the FACT that someone is keeping an exabyte of memories and experiences for us non-locally in preparation for our upcoming Life Review, and it's reasonable to conclude that not all of our memories are being stored exclusively within our physical brain.

All of this memory consolidation – uploading and downloading from the cloud, including the writing of the data to our physical hard drive – takes place outside of our conscious awareness. It's being handled by Nature's Psyche, or God's Psyche, or the Angels of Heaven.

Since it has been observed and demonstrated that there is non-local cloud storage dedicated to each one of us and our upcoming Life Review, it really doesn't matter where Nature's Psyche decides to store our memories. Nature's Psyche can store our memories either within our physical brain or non-locally in the cloud, and the ultimate result is the same. Nature's Psyche knows when the brain's physical capacity is being exceeded, and it's now time to upload some of that to the cloud.

It also seems obvious from the experiential evidence that the material that's being stored for us in the cloud is vastly superior to anything that's being stored in our brain; so, I think it's obvious and clear and that it has been successfully proven that our memories are not stored exclusively within our brain. Since Nature's Psyche or God's Psyche is handling all of our local and non-local memory storage for us, outside of our conscious awareness, Nature's Psyche can store our memories anywhere it wants in any format that it wants.

I think I've made my case, but it's never going to be good enough to convince the Materialists, Naturalists, and Darwinists. I KNOW, because I used to be a Materialist, Nihilist, and Atheist, until people like Pim van Lommel and Jeffrey Schwartz caught my attention.

All Our Memories Are Stored in Our Brain

All of our memories are stored exclusively within our physical brain, because there's no place else that they can be stored.

Every neuroscience college textbook I have read tells us that our memories are stored exclusively in our brain, implying that when our brain dies then our memories come to an end. Even when they find experimental evidence that proves them wrong, they immediately return to their chosen conclusion that our memories are stored exclusively within our physical brain.

This is the million dollar question.

Are our memories stored exclusively in our physical brain, or are our memories backed-up off-site in some kind of non-local cloud storage?

Ultimately, this line of scientific research is important because as physical beings we have to identify the things that have a physical explanation, before we can learn to identify the things that don't have a physical explanation. Scientists are supposed to make discoveries. We have to explore the physical explanation for everything that we possibly can, before we can discover and identify the things that don't have a physical explanation.

Almost every memory researcher that I have ever encountered starts with the axiomatic conclusion that our memories are stored in our physical brain; and, they insert this chosen conclusion axiomatically as one of their premises or hidden assumptions into all of their science experiments and scientific research so that every study and experiment they do ends up confirming their initial conclusion that our memories are stored only within our physical brain. That ingrained process or methodology is called begging the question or circular reasoning, and it successfully prevents new discoveries.

Every neuroscience textbook makes the observation that consolidated retrograde memories are never lost, no matter what part of the brain a medical doctor chooses to cut out. You can cut out anything you want, and the consolidated retrograde memories remain. Their conclusion? They conclude that memories are stored diffusely throughout the whole physical brain, basically in every cell of the brain, so that as long as you have one brain cell left, then you still have access to all of your consolidated retrograde memories. Nobody has ever suggested any other explanation for memory storage.

The Materialists and Naturalists keep telling us that all of our memories are stored only within our physical brain, and that our memories cease to exist when we die. These people even go as far as to imply that all of our consolidated memories are stored in each and every neuron; but, I don't see how that's physically possible. I'm a computer scientist, programmer, web designer, and certified computer support specialist; and, I know for a fact that it's physically impossible to store petabytes of information within a single neuron

Nobody has ever suggested that our consolidated retrograde memories are being uploaded to Non-Local Cloud Storage, where they can be accessed and retrieved as long as we still have a single brain cell left in our head. Everyone just axiomatically assumes that all of our memories are being stored only within our physical brain.

If memory researchers continue to preface all of their research with their chosen conclusion that our memories are stored within our physical brain, then memory researchers and the scientists who study memory will never be able to figure out where our consolidated memories are really stored. That's the danger of starting with pre-chosen conclusions – it greatly decreases the chance that these people will ever discover anything new or ground-breaking.

Why is it important to identify the physical location of different learning systems?

Scientists are supposed to make discoveries, and follow the evidence wherever it leads them. We have to explore the physical explanation for everything that we possibly can, before we can discover and identify the things that don't have a physical explanation.

I just don't see how it's physically possible to store a lifetime's worth of consolidated retrograde memories in a single neuron or brain cell; yet, the experimental evidence suggests that is exactly what is happening – each one of the neurons in our brain is storing all of our consolidated retrograde memories.

How's that possible?

I can't devise a physical explanation for this that makes logical sense to me. Destroy as many neurons in your brain as you want, anywhere you want, and your consolidated retrograde memories will still be there. This truth and reality doesn't make sense from a physical perspective.

How can each physical neuron be storing all of our consolidated memories?

I don't see how it can be done. For me personally, it's just easier to visualize that Nature's Psyche within each physical atom of our physical body is storing all of our consolidated retrograde memories non-locally in the cloud for us, than it is to visualize and believe that each neuron has petabytes of physical RAM or flash memory hidden away within it.

Identifying the objects, processes, science disciplines, empirical evidence, and systems which don't have a physical explanation is where science really starts to get interesting; and, it all starts by trying to find physical explanations for the things that we observe.

As physical beings, we have to understand the physical before we can start to theorize and try to understand the quantum or the non-physical.

While developing Quantum Neuroscience, I observed that I always had to start with a physical explanation for what is happening, before I could identify the different things that don't have a physical explanation and that require a non-local or transdimensional explanation instead. In other words, we have to pursue the naturalistic or physical explanation for observed and experienced phenomena, before we can learn to identify the phenomena and events that don't have a physical explanation and that require a supernatural, transdimensional, quantum explanation instead.

It's extremely important to identify where different learning systems or learning modules are physically located with the brain so that we can do statistical modeling and learn to realize that there is not enough information storage capacity within a physical brain to store a life-time's worth of memories in these small physical modules or physical systems that we do identify. For example, how do you store a new declarative memory with all of its images, feelings, thoughts, data, and ideas within a new ion receptor on a post-synaptic neuron? I can't find a physical explanation for that, so I have to look for something else that makes logical sense to me in order to explain the long-term storage of new declarative memories.

The past year or two, I have been trying to upgrade my science – been trying to up my game. Is there a way to make science better?

Thinking critically to eliminate the pre-chosen conclusions and confirmation biases would make for better experiments over all. Allowing all the evidence into evidence would make for better theorizing and better hypothesizing, as well as potentially better experiments.

The Materialists, Naturalists, and Darwinists erroneously equate or confound learning systems with memory storage. These people always jump to the conclusion that all our memories are stored within our physical brain, without realizing that that's physically and statistically impossible.

Cutting edge scientists are starting to do the math, and they are now starting to realize that our physical brain has enough memory storage capacity to store about a seconds worth of our sensory input if that data is stored digitally as bits and bytes; and, our physical brain has enough memory storage capacity to store about seven minutes worth of memories and sensory input if our memories are stored holographically. There's nowhere near enough memory storage capacity within our physical brain to store all of our thoughts, sensory input, video, feelings, experiences, and memories for a whole lifetime.

Realizing that our physical brain and memory systems are transceivers, and not just data storage, would take our science and science experiments into the next century – a century where everyone knows that Materialism and Naturalism and Darwinism have been falsified by a lack of evidence or a failure to meet their burden of proof.

The scientists estimate that there are petabytes of memories being stored within your brain, meaning that there are petabytes of physical RAM within your brain. There are a few problems with that estimate.

First, there is NO visible RAM and NO visible CPUs within a neuron. A neuron is a single hardware BIT – it is either ON or OFF. That's it.

Second, there are NO wires within your brain, which means that there isn't even one functional BYTE of RAM anywhere within your brain, thanks to the randomizing effects of your synaptic clefts and the integrating effects of postsynaptic potentials.

Third, even if ALL of your 100 trillion synapses were wired together to function as BYTES of programming code and memory storage, there's not enough of them to store petabytes of RAM. Since your synapses are NOT wired together into BYTES of RAM, there's not even a single BYTE of functional physical RAM within your physical brain.

Fourth, it is physically impossible to put petabytes of physical RAM into a physical brain. A petabyte is at least 1,000 terabyte-size hard drives. Where are the thousands of hard drives within your brain? They don't exist! That's a serious problem.

Fifth, if your memories really were being stored as physical RAM within our physical brain, then ALL of those memories should be instantly available at all times, just like a hard drive. Technically, all of your memories should be simultaneously present at all times, if they really existed within your physical brain as physical RAM.

When it comes to vision and sound, we don't have instant replay! During actual usage, our memories have to be recalled from non-local storage in the cloud; and, most of the time we can't seem to do so. Our actual recall capabilities and our actual need to recall memories from the cloud are convincing evidence that our memories are NOT being stored in a physical format within our physical brain. The fact that all of our memories are not simultaneously and instantaneously available at all times is evidentiary proof that our memories are not being stored in a physical format within our brain, because a physical consensus reality is the most reliable type of reality that God has ever created.

The Kinds of Memories Stored within Our Brain

Memory storage IS the most interesting subject in neuroscience to study and learn about; and, as you soon will see, I've put a lot of time and effort into it.

The Materialists, Naturalists, Nihilists, Behaviorists, Darwinists, and Atheists start with the conclusion that ALL of our memories are stored within our brain; and then, these people adjust their observations and their interpretations of evidence accordingly. They can't prove that they are right, and we can't prove that they are wrong – at least not with our physical instruments.

It's easy to prove the Materialists and Naturalists wrong, however, if you are willing to allow Life Reviews, Near-Death Experiences (NDEs), Out-of-Body Experiences (OBEs), Shared-Death Experiences (SDEs) and other types of non-local experiences or spiritual experiences into evidence, which they are not willing to do, because that type of empirical evidence or observational evidence falsifies Materialism and Naturalism. When those go down, Darwinism and Atheism go down with them as fruit from the poisoned tree. Materialism and Naturalism are based upon a refusal to look at evidence.

Declarative Memories, memories for events and memories for facts, are typically what we think of whenever we think about the term "memory". Episodic memories are memories for events. Semantic memories are memories for facts.

Our Life Reviews seem to be comprised of Declarative Memories, specifically Episodic Memories; whereas, our Semantic Memories seem to be available to us after we are dead, before and after the Life Review.

It is obvious to me that our Life Reviews are not stored and cannot be stored within our physical brain. So, it makes me wonder what other memories aren't stored in our physical brain, and what memories are.

My research into neuroscience has led me to believe that our physical brain has no RAM or consolidated memory storage. Our Long-Term, Consolidated, Episodic Memories – or Life Review Memories – can't be stored within our physical brain at all, especially after our brain is dead and gone. And, since our brains have no RAM, it's also possible that our Consolidated Semantic Memories aren't stored within our brain either.

As a computer scientist, I personally would have assigned ALL of our Consolidated Memories and ALL of our Declarative Memories to RAM; but, our brains have no RAM or no memory engrams, so I have to find some other place to store our Declarative Memories besides our brain. Non-Local Cloud Storage seems to be the most logical solution.

Of course, just because I say that I'm encountering tons of evidence suggesting that NONE of our Long-Term, Consolidated, Retrograde, Declarative Memories are being stored within our brain, that doesn't necessarily mean that it is so. Likewise, just because ALL of our PhD scientists are telling us that our memories are only stored within our physical brain and that all our memories cease to exist when we die, that doesn't necessarily mean that it is so. Truth isn't a democracy, and truth can't be determined by taking a vote. It's a logic fallacy to try to establish the truth by taking a vote – the *most people* logic fallacy, or the *credentialism* logic fallacy, or the *appeal to authority* logic fallacy.

When it comes to memory storage, I have to fall back upon what WE KNOW as a race. In other words, I have to fall back on the observational evidence, experiential evidence, and empirical evidence.

WE KNOW that our after-death Life Reviews are NOT being stored within our brain. That's obvious. WE KNOW that it's physically impossible to store petabytes of video data, episodic memories, declarative memories, or petabytes of any other type of memory within our physical brain. That, too, is obvious. WE KNOW that we seem to have access to petabytes of memory, so those memories have to be stored someplace else besides our brain. Logical, but not so obvious.

Given what we already KNOW, it is logical to conclude that some of our memories are NOT being stored within our brain. Which ones? We don't know! But, if I were to make an educated guess, I would choose our video memories first, meaning our episodic memories; and then, I would go with the other memories that seem to survive the death of our physical body and physical brain, namely our intelligence or our semantic memories – our thoughts, choices, audio, feelings, personality, cherished sensations, and conversations. Clearly, anything that makes it into our Life Review and anything that survives the death of our physical brain WAS NOT stored exclusively within our brain but was stored in the cloud instead.

So, what is being stored within our brain, and how is it being stored?

That is the million-dollar question, isn't it?

Implicit memories are memories for which we have no conscious recall or recognition. Implicit memories are memories that we can't consciously remember.

Implicit memories are a bit of a misnomer. How can they be memories if we can't remember them?

Procedural memories are a special type of implicit memory that are associated with autonomic processes, reflexes, and motor skills that a person has developed through practice, or just by living life.

Are they really memories, if you can't remember them and aren't consciously aware of them? NO, they are NOT memories. They are functionality, not memories. They are Maps of Physical Functionality being produced at the quantum level by Nature's Psyche, which is why we aren't consciously aware of them and can't remember them.

I'm seeing tons of evidence suggesting that our implicit memories, procedural memories, or unconscious memories are being mediated or controlled by Nature's Psyche completely outside the conscious awareness of our Human Psyche. We are not aware of our implicit memories or unconscious memories or procedural memories, whether they are stored non-locally in the cloud or locally in our brain instead. It doesn't matter where our procedural memories are stored, because by definition in principle we aren't aware of our unconscious memories and implicit memories even if they are being stored within our brain. Tons of things are going on within our body and our brain that we are completely unaware of and have no conscious knowledge of; and, some of them could be taking place non-locally in the cloud for all we know.

So, what memories are being stored within our physical brain? We don't know. It seemed logical to me that anything that's not associated with our Life Review might in fact be stored within our physical brain.

However, it's also theoretically possible that NONE of our memories are stored within our brain – that all our memory storage and mental processing is taking place non-locally in the cloud and being mediated physically by Nature's Psyche completely outside our conscious awareness.

When I suggested the possibility of non-local storage to my neuroscience teacher, he told me, **"That is unlikely to be the case"**.

So, who is right?

I think we both are.

ALL of our memories, explicit and implicit, are being controlled by, mediated by, and stored by Nature's Psyche completely outside the conscious awareness of the Human Psyche. We have no idea how or where Nature's Psyche is storing our memories; but, we do seem to have two obvious solutions – within our physical brain and non-locally in the cloud.

My current theory and belief is that Nature's Psyche knows when our physical brain is about to overflow or exceed its "memory storage capacity" or its memory mapping capabilities; and then, Nature's Psyche off-loads or uploads some of those unused "memories" or maps to the cloud thereby freeing up some space on our physical hard drive as needed. This theory seems to fit ALL the evidence, from what I can tell.

When it comes to "procedural memories", it's my theory or belief that our Cache, or Working Memory, or Software is being developed and controlled by Nature's Psyche, and then being mapped onto our physical brain as improved, enhanced, and increased physical functionality or software.

It is also my theory that these Maps of Physical Functionality do indeed exist as registers or neurons within our physical brain at the physical level; but, these Maps of Physical Functionality are being created, mediated, monitored, and controlled at the quantum level by Nature's Psyche as needed and requested by the Human Psyche. The Human Psyche chooses to move a part of its physical body, and Nature's Psyche uses these Maps of Physical Functionality that it has created at the quantum level to accomplish the task at the physical level. The purpose of Quantum Neuroscience is to determine how the Human Psyche interfaces with and controls its physical body and physical brain. I believe that I have done so. These Maps of Physical Functionality that are being created by Nature's Psyche ARE the interface between the quantum level and the physical level that we have been looking for. These are the droids you've been looking for.

Clearly, petabytes of memories and programming code are being stored within our brain and accessed by our brain; but, that's physically impossible, which means that most of these petabytes of RAM and computer code are being stored and processed non-locally by Nature's Psyche at the quantum level, because it's physically impossible for all of this to be taking place at the physical level through our synaptic clefts. Thanks to the scrambling effect of synaptic clefts, distributed memories are physically impossible at the physical level; but, distributed memories are totally possible at the quantum level under the control of Nature's Psyche. Remember, Psyche can do the physically impossible at will.

When we finally discover and observe the physically impossible in action, then that's when science really starts to get interesting. Remember, Psyche can use quantum mechanisms to do the physically impossible.

Cerebral Processing Units and the Programming Model:

I have observed that our brain is comprised of many different Cerebral Processing Units (CPUs). CPUs have cache memory or working memory – of course, that doesn't mean that our brain has cache, but it's a theoretical possibility based upon the Cerebral Processing Unit Model that I have developed.

Our brain can't be the storage site for our Life Reviews; but as I see it, our brain can be the storage site for our brain functions, Cerebral Processing Units, or what the scientists call Procedural Memories – which are the hardware and programming necessary for motor skills, brain functionality, and learned computational tasks.

I'm a computer scientist; and, I've spent time programming computers. I've been very fond of the Programming Model for brain functionality, because it made sense to me; and, I'm not the only one.

The reticular formation is defined as a complex network of nuclei in the core of the brain stem that contains motor programs that regulate complex species common movements such as walking, swimming, and REM sleep.

For months if not years, I visualized the human brain as a collection of physical hardware (Cerebral Processing Units or CPUs) and software (programming). Programming requires a programmer, engineering requires an engineer, and manufacturing requires a manufacturer. I went with this Programming Model, until I found something better – Maps of Physical Functionality.

Once I finally realized that there are NO wires in our brain and that a synaptic cleft scrambles everything that comes its way integrating it all into a single hardware BIT, then I realized that there are no CPUs and RAM in our brain and found myself looking for a better model of brain functionality at the physical level. If you successfully demonstrate that there is no wired RAM within our brains, then you have also successfully demonstrated that there are no wired CPUs within our brain. Eventually it dawned on me that Nature's Psyche maps physical functionality into our brains at the quantum level, and then uses that physical functionality go get things done for us at the physical level.

In the meantime, while I was waiting to make this discovery, I used CPUs and the Programming Model to explain what I thought was happening in the physical brain. And, it was very important to me at the time to separate CPUs from RAM or memory storage. I could still visualize the different brain structures functioning as programmed CPUs while at the same time knowing that there is NO physical RAM within our brains. It was a workable solution, until it finally dawned on me that mapping physical functionality into a brain is basically the same thing as programming physical functionality into a brain. Then my transition to Maps of Physical Functionality was complete.

Now, let's return to the Programming Model and try to show what I did with it at the time.

The occipital lobe as well as the dorsal stream and the ventral stream look very much like CPUs dedicated to visual processing and interpretation. They are CPUs, not RAM or memory storage. The dorsal stream mediates the perception of where things are, and the ventral stream mediates the perception of what things are. Again, it seems to be more about functionality or perception, than memory storage. The ventral stream is about the conscious visual perception of physical objects. The dorsal stream is more about spatial location of objects.

The dorsal stream is the group of visual pathways (axons) that flows from the primary visual cortex to the dorsal prestriate cortex to the posterior parietal cortex, and its function is visually guided behavior and the spatial location of objects.

The dorsal stream is about functionality, and NOT memory storage. There isn't any video being stored within our brain.

How do we know?

Well, NONE of us have instant replay.

The only thing in our brain in terms of video is the current image impinging on the registers or neurons of our occipital lobe; but, that current image is being processed through the dorsal stream and ventral stream making it feel as if it's video that we are looking at. ALL of our 360-degree surround video for our upcoming after-death Life Review is being produced and stored for us by Nature's Psyche and the Angels in Heaven holographically at the quantum level non-locally within God's Database. NONE of that is being produced by our brain, nor stored within our brain.

Our Cerebral Processing Units (CPUs) are all about functionality, and NOT memory storage. I chose to separate the two, functionality and memory storage, because that's how I made ALL of my conceptual breakthroughs, and because the Materialists and Naturalists always choose to combine the two into one physical whole.

I eventually discovered that the Materialists, Naturalists, Darwinists, Behaviorists, Nihilists, and Atheists have a talent for picking and backing all the FALSEHOODS and FALSIFIED THEORIES and deliberately making these deceptions part of their belief system. Therefore, I realized that if I want to find and know the truth, then I have to be willing to reject everything that the Materialists and Naturalists have chosen to believe in. I use the Materialists and Naturalists as my barometer to help me detect and find every falsehood, because they have a talent for finding and embracing everything that is FALSE.

The amygdala looks very much like an emotion chip. Again, it's more like a CPU or computer chip than it is like RAM. It's about functionality, rather than memory storage. When you lose your amygdala, you lose your sense of fear or lose the ability to fear things, even though you still remember what things are. Kluver-Bucy Syndrome is the result of amygdala damage. It results in a lack of fear and hypersexuality. These people try to identify things by putting them in their mouth. They recognize or notice that things exist; and, they remember things; but, their ability to feel fear is gone. It's lost functionality, rather than lost memories.

The nucleus accumbens is a nucleus of the ventral striatum and a major terminal of the mesocorticolimbic dopamine pathway. The nucleus accumbens seems to function as our reward chip. A lot of our functionality is programmed or mapped directly onto our brain by Nature's Psyche at the quantum level. Once Nature's Psyche maps a specific function to a specific brain structure, then removing that brain structure removes that particular function. It doesn't have anything to do with memories, and everything to do with physical functionality.

Technically, there are NO wires within our brain, so it's not computer programming that we are looking at here, but instead Maps of Physical Functionality that are being created by Nature's Psyche at the quantum level.

Over and over again, all of the evidence suggests that the functionality can be lost; but, the memories remain. Our procedural memories, physical functionality, habits, skills, and implicit unconscious memories seem to be mapped directly onto our physical brain by Nature's Psyche; but, our declarative, episodic, consolidated, retrograde, semantic, factual, visual, long-term memories seem to be stored for us non-locally in the cloud, and then accessed for us by Nature's Psyche as needed.

Our consolidated memories seem to be extremely resilient and seem to survive the destruction of huge chunks of our brain, which shouldn't be surprising if they are indeed being stored for us non-locally in the cloud by Nature's Psyche as quantum waves at the quantum level. However, if the parts of our brain – that Nature's Psyche maps or assigns during neurodevelopment and neural mapping for the transmission of our memories to the cloud and the retrieval of our memories from the cloud – get destroyed, then the physical functionality associated with our memory systems will go down, and our memories will indeed seem to be lost.

Again, it's all about a loss of functionality, and not necessarily a loss of memories. Reversible lesions or temporary lesions are a method for temporarily eliminating the activity of a particular area in the brain while tests are being conducted. A few scientists have noticed that when certain parts of the brain go down, that person no longer has access to specific types of memories. However, once the functionality returns to normal, the memories return as well.

When I went through substance-induced psychosis, I seemed to lose most of my memories. I didn't know how to do things anymore. I didn't know how to turn on a television or play a DVD. I couldn't add, subtract, multiply, or divide. I had to have my wife fill out the medical forms for me, because I didn't even know if I was a smoker or not. It was gone. However, after I got sober, the functionality slowly returned, and my long-term memories slowly returned as well. Access to our memories is more about functionality than actual memory storage. Nature's Psyche is storing all of our memories for us non-locally in the cloud as quantum waves at the quantum level. Those memories don't disappear when our brain gets damaged or fried, but our access to those memories at the physical level certainly can disappear and cease to exist, at least until proper physical functionality is restored.

The explanatory power of memory storage and brain functionality is greatly improved by bringing Nature's Psyche, Quantum Mechanics, and the Quantum Level into the equation.

The densely populated cerebellum looks like a complex CPU dedicated to motor control. It's a CPU containing software, programming, implicit memories, or procedural memories – and not some kind of RAM storing our declarative memories and video data. Parts of our motor functionality seem to be mapped directly to our cerebellum by Nature's Psyche.

These different Cerebral Processing Units (CPUs) control our physical functionality, and they seem to be comprised of adaptive programming capabilities, meaning that their functionality can be remapped by Nature's Psyche. The scientific term for all of this is Procedural Memory or Implicit Memory. This is the type of memory that exists and is stored within our physical brain, or at least this is the type of memory that Nature's Psyche maps directly to our physical brain! Nature's Psyche can map specific memories to specific neurons within our brain – neurons like grid cells, place cells, and concept cells. The memories aren't really stored within the neurons, because that's physically impossible; but, when our memories are re-lived by the Human Psyche, Nature's Psyche fires those specific neurons triggering the associated physical feelings that go along with those memories.

Our brain is all about physical functionality (CPUs), and NOT memory storage (RAM). As I see it, our CPUs – hardware, software, and cache (working memory) – are being mapped onto our brain structures by Nature's Psyche at the quantum level. Our brain structures are comprised of different Maps of Physical Functionality that are being created by Nature's Psyche at the quantum level; and, our long-term consolidated memories seem to be getting stored directly into the cloud as quantum waves by Nature's Psyche at the quantum level. All of the physical evidence that I have come across points to this as being real and true, because it's physically impossible to store memories and programming code within a synaptic cleft at the physical level.

Procedural memory is a type of implicit memory (unconscious memory) and long-term memory which aids the performance of particular types of tasks without conscious awareness of these previous experiences.

Procedural memory guides the processes we perform and most frequently resides below the level of conscious awareness. When needed, procedural memories are automatically retrieved and utilized for the execution of the integrated procedures involved in both cognitive and motor skills, from tying shoes to flying an airplane to reading. Procedural memories are accessed and used without the need for conscious control or attention.

Procedural memory is created through procedural learning or, repeating a complex activity over and over again until all of the relevant neural systems work together to automatically produce the activity. Implicit procedural learning is essential for the development of any motor skill or cognitive activity.

https://en.wikipedia.org/wiki/Procedural_memory

Procedural Memories aren't the same thing as Declarative Memories!

The Materialists and Naturalists deliberately conflate CPUs (procedural memories) with RAM (declarative memories) within the physical brain; but, they aren't the same thing at all. The malleable hardware and software, the programming within the brain, are not the same thing as memory storage devices, even though the Materialists and Naturalists say that they are. The majority of the brain is dedicated to what they call "procedural memory" or CPU processing – the development and running of programs or software. There's precious little room left within the brain for "declarative memories," or memory storage, memory engrams, and RAM. The Materialists and Naturalists refuse to have anything to do with the quantum level or the psyche level, so you will never learn anything about that from them.

My neuroscience teacher wrote:

Nearly 70% of our brain has some role in visual processing.

You have got to read and study the Materialists and Naturalists carefully, because they are trying to trick you and deceive you.

Notice here that he actually (and correctly) says that 70% of our brain is dedicated to visual processing – NOT the storage of visual images. There's a monumental difference between CPU processing and flash RAM data storage. 70% of our brain is a CPU dedicated to processing visual input or visual stimuli. In other words, 70% of our brain is mapped by Nature's Psyche or dedicated by Nature's Psyche to our visual input.

CPUs typically have a bit of cache memory, but CPUs do not do long-term memory storage. Collectively, your brain is a CPU, a machine, a very nicely done machine, but a machine nonetheless. Your brain is comprised of different types of CPUs, each with an assigned function. There's no room left within your brain for RAM, or long-term memory storage. It's physically impossible to store petabytes of data within our brain; so, those petabytes of data, that are being stored, are being stored someplace else besides our brain.

It is obvious and clear that our after-death Life Reviews are NOT stored within our physical brain. So, what other types of memories are not stored within our physical brain?

And, what types of memories or functionality are stored within our physical brain?

These are the questions that I kept coming back to, over and over again, during my research. One chapter after another, while studying neuroscience, I kept coming back to these same questions (and same conclusions), as this essay will now reflect.

For a long while, I erroneously concluded that our brains are wired together as a collection of different CPUs. The Materialists and Naturalists talk about "brain wiring" all the time, so you have to at least be aware of the concept, even though it's not technically true.

However, I eventually was forced to face the fact that there are NO wires in our physical brain, at least NOT at the physical level. But, at this point in my research, I got a lot of mileage out of the CPU metaphor or analogy, as you will now see.

Likewise, memories are clearly being stored within our brain, but not necessarily at the physical level.

There's a distinction between the physical level and the quantum level, which is why we need something like Quantum Neuroscience in the first place.

For months, I went back and forth between the physical level and the quantum level. These essays reflect this reality.

Whenever I read a neuroscience book, they always talked about CPUs and RAM whenever they were describing the functionality of the physical brain. Then, whenever I started to research and study Quantum Neuroscience, I found myself gravitating towards Maps of Physical Functionality that are being created by Nature's Psyche at the quantum level within the physical brain. Eventually, the Quantum Neuroscience won out; but, it took a while.

Different Memory Systems:

So, what are the different types of memory – or the different types of memory systems – that we have?

Procedural memories – CPU programming or software – are definitely stored within our brain. It seems logical and clear to me that motor skills or procedural memories that are targeted at physical skills, physical processes, and physical functionality would be stored or encoded somewhere within our physical brain. Those types of physical learning or physical command and control aren't really needed after our physical body is dead and gone.

After you die, should God choose to place you into another physical body sometime in the future and you come from your mother's womb as a new-born baby, you are going to have to learn how to control that physical body all over again. Consequently, it seems logical to me that our procedural memories or motor skills are developed and stored within our physical brain. Our brains are comprised of CPUs and programmed hardware.

Meanwhile, voluntary control, or choice, or chosen behaviors are a function of Psyche; and, they continue after the physical brain is dead and gone, according to the empirical evidence and observational evidence from NDEs and OBEs. Likewise, the memories within our Life Reviews continue to exist after our physical brain is dead and gone, suggesting that these types of declarative memories are being stored someplace else other than the physical brain. It's logical, and it's obvious, for anyone who is willing to look and see.

Declarative memories are long-term explicit memories or conscious memories – facts (semantic memories) and events (episodic memories). These are memories for which you are consciously aware and can verbally describe or declare. The functionality and expression of these types of memories through the physical brain depend upon the hippocampus and the medial temporal lobe. If these physical devices are lesioned or eliminated, then the physical functionality for forming new declarative memories is lost; even though these people continue to remember their long-term consolidated retrograde declarative memories, which were stored diffusely or non-locally before the brain was damaged and ceased to work properly for the creation and expression of new declarative memories.

Any time the brain is lesioned or damaged, functionality is lost, but not necessarily the memories. Any memories that are stored non-locally in the cloud cannot be lost due to brain damage or lobectomies; but, functionality, processing power, and the brain's transceiver capabilities can certainly be lost if the brain is damaged.

I KNOW, though, that our after-death Life Reviews continue to be made and recorded, even if our physical brain is damaged. People who are born blind can see after they die; and, their Life Reviews have video, even though their brains never made any video during their lifetime.

Materialists and Naturalists deliberately conflate the hippocampus and medial temporal lobe with memory storage; but clearly, our consolidated long-term declarative memories are being stored someplace else besides our hippocampus or medial temporal lobe, because our consolidated declarative memories remain even after the hippocampus and MTL are eliminated or taken offline with temporary lesions.

After this discover, the Materialists and Naturalists declared that our long-term memories are being stored in the neo-cortex; but, even with prefrontal cortex damage or any kind of neo-cortex damage, the long-term declarative consolidated memories remain. So apparently, these types of long-term memories are being stored diffusely within each and every neuron of the brain; but, that's physically impossible. It's physically impossible to store petabytes, or even terabytes, of information within a single neuron. It can't be done.

As I see it, the primary purpose of our brain is to control our physical functions. The majority of the memory within our brain is procedural memory or skills memory – implicit memories that we aren't even consciously aware of. Technically, procedural memory is programming or learning systems. Motor skills, or knowing how to do things, are programmed directly into our brain.

By whom?

Any type of programming requires a programmer. This type of programming or brain re-wiring is not being produced by the Human Psyche, because we aren't even aware that it is taking place. The only other logical answer is that Nature's Psyche or God's Psyche is programming these functional changes and procedural memories into our brain, because we humans wouldn't know how to do so even if we tried.

If given the task, how would you program the knowledge and skills for riding a bicycle into a physical brain of someone who has never seen nor touched a bike? Would you be able to do so? Why not? This is beyond us, because we really don't know how the brain works, nor do we really know how the brain stores long-term memories and procedural memories. It seems logical that our procedural memories or physical functionality would be stored someplace within our brain, but we don't know where or how this is done. Obviously, evolution knows how to do all of this – according to the Materialists, Naturalists, and Darwinists – but we don't, because we're not as smart nor as skilled as evolution.

If Evolution is doing any of this, then Evolution is just another name for God, because it's physically impossible for Evolution to be storing our Life Reviews non-locally in the cloud.

By definition, in principle or practice, our declarative memories (memories of events and facts) are ephemeral and fleeting. Most of our sensory input passes through us, as well, without paying attention to it. We, our psyche, has to pay attention to sensory input in order to consolidate it within our memories. Even then, the process is less than efficient for most of us.

If I think of something new, I have to write it down immediately; otherwise, it is lost and gone. All of my books, of course, are the result of writing down what comes to mind from day to day as I study different things.

As I see it, the memory centers within our physical brain (if they exist) are dedicated primarily to our working memory or short-term memory. There's not enough memory storage capacity within our brain for anything else. And technically, procedural memories or motor skill memories aren't really memories – they are programming, or brain functionality, or software – and when properly programmed, they function automatically without us being consciously aware of them.

I know I'm splitting hairs here, and most of the neuroscientists will disagree with me; but, technically, it really isn't a memory unless we consciously remember it. It isn't a true memory if it is an implicit memory, or a procedural memory. It has to be an explicit memory, or a declarative memory, in order for it to be a conscious memory and a true memory; and, all of the empirical evidence keeps telling us that our Consolidated Declarative Memories, particularly our Episodic Life Review Memories, are NOT being stored within our brain.

In Summary:

Lots of learning takes place within our brain and physical body, a lot of which we aren't consciously aware of. These types of memory are called procedural memories, and implicit unconscious memories. There's a big difference between learning and memory, but the Naturalists and Neuroscientists tend to conflate the two. Procedural memories are learning systems, and not memory storage devices. Unconscious processes have been documented and proven to exist; but, they are completely different than conscious processes and conscious ideas which we tend to remember.

All these different physical processes, procedural processes, and implicit unconscious processes take place outside of the conscious awareness of the Human Psyche and even the physical brain, which suggests that these processes are being mediated and controlled by Nature's Psyche, or God's Psyche outside of our conscious awareness. The unconscious is a product of Psyche; but, which Psyche – Nature's Psyche or the Human Psyche? The unconscious seems to be a product of Nature's Psyche, because the Human Psyche and the physical brain don't seem to be consciously aware of any of it.

It's obvious and clear to me that our brain is being used primarily to map, organize, program, and control our physical skills and physical functionality, which leaves little or NO space for our declarative memories, or RAM, or memory engrams. Our brain is dedicated to our physical functionality, not our long-term declarative memory storage; and, declarative memory is what we typically think of whenever the scientists start talking about memories. However, it is important to realize that it's physically impossible to store petabytes of declarative memories within our physical brain. It can't be done. It's being done, but it's physically impossible, which sort of tells me that our Consolidated Episodic Memories or Life

Review Memories are being stored for us non-locally in the cloud by God's Psyche or Nature's Psyche outside of our conscious awareness.

Different parts of our brain are dedicated to different functions, resulting in different types of memory systems, or CPUs, or programming and therefore different types of learning or different types of learning systems. Remember, there's a huge difference between learning systems (programming of CPUs) and memory systems (RAM). Our physical brain is aimed primarily at procedural learning or skills learning. Our physical brain is comprised of learning systems, NOT memory storage devices! Our physical brain is comprised of CPUs, not RAM. The Materialists and Naturalists deliberately conflate the two, but they are wrong to do so. I'm a computer scientist and a certified computer support specialist, so I KNOW how these things really work; and, the Materialists and Naturalists don't. Our brains are comprised of different CPUs, not RAM! There's no sign of RAM within our physical brains. Karl Lashley proved that memory engrams or RAM do not exist within the physical brain; and, the Materialists and Naturalists have been ignoring that evidence ever since.

Let's examine some of the CPUs (Cerebral Processing Units) within our physical brain – what the Materialists and Naturalists call "procedural memories" or "motor skill memories". Let's start first with video memories or visual memories, because most of our brain is dedicated to visual processing.

Visual Processing:

"The human eye, a product of 600 million years of evolution." (*Biopsychology*, p. 132.)

Well, that explains everything, doesn't it?

That means that our evolutionary ancestors were blind for 600 million years before evolution got around to creating the human eye. Clearly, their blindness greatly accelerated their evolution, because it only took 600 million years for them to get eyes. Obviously, they were motivated to evolve, because all those things in the fossil record that could see were eating them alive. They were the fittest blind people of the bunch, so that's why evolution decided to give them eyes. Thank God evolution knew how to design and create eyes, or we would still be blind. It's good to be blind, because that means that evolution can help us to see. Look at all the different things I'm able to see thanks to evolution. I'm glad that evolution gave you eyes, or you wouldn't be able to read this book. What would we do without evolution? I'm glad that thing knew how to create proteins and genes out of thin air, or you and I wouldn't even be here. Then who would there be to buy and read my books?

I took the effort to critique and satire this in the hope that you will be able to see that the theory of evolution is nothing but a *fictional ad hoc just-so story*, a logic fallacy. You can embellish and interpret the theory of evolution any way you want, because it's a *fictional story* or a *scientific inference* that never happened in the first place.

If I say that our evolutionary ancestors were blind for 600 million years, then it was so. It's a fictional story after all. It can be any way I want it to be, and you can't say otherwise. It's my fictional story, so it has to be the way I say it was. If I say that evolution gave us fins for a billion years before it gave us arms and legs, then it was so. It's a fictional story after all. If I say that our plant ancestors had photoreceptors for eyes and photosynthesis, and that reality explains why our eyes have photoreceptors, then that explains how it really was, whether you want to call it homology or convergent evolution or serendipity. It's a fictional story after all. It can be any way that I want it to be, because I'm the author of this fictional story. It is whatever I say it is, because it's fiction. I'm making it up out of thin air, so it can be anything I want it to be. That's the way fiction works.

Because the whole thing is science fiction and we are making it up as we go along, the theory of evolution can be anything we want it to be. If he says that it took evolution 600 million years to make your eyes, then it was so, because it's a fictional story after all. If he says that apes, or monkeys, or rats, or frogs, or turtles, or plants, or viruses, or dragons, or unicorns, or space aliens, or flying spaghetti monsters were your ancestors, then it was so, because he's making up a fictional story out of thin air as he goes along. With the theory of evolution, you can have anything you want simply by making up a story and saying that it was so. This is a very powerful lesson to learn about the theory of evolution. Because the theory of evolution is fiction, you can have anything you want in your story simply by making it up out of thin air and then declaring axiomatically by fiat that it was so. That's the way fiction works.

Scientific inference is the fundamental method of biopsychology and of most other sciences – it is what makes being a scientist fun. This section provides further insight into the nature of biopsychology by defining, illustrating, and discussing scientific inference.

The empirical method that biopsychologists and other scientists use to study the unobservable is called <u>scientific inference</u>. Scientists carefully measure key events they can observe and then use these measures as a basis for logically inferring the nature of events that they cannot observe. (*Biopsychology*, p. 13.)

He even admits that making up stories and empirical evidence out of thin air is what makes being a scientist so much fun to begin with. He even admits that he's making up fictional stories and manufacturing empirical evidence ex nihilo, because it's fun to do so. He even calls his *ad hoc story telling* an empirical method, a fundamental method, and a scientific method. No wonders he believes in the theory of evolution. It's a fictional story that he helped to create and continues to promote. He never once realizes that *scientific inference* is a logic fallacy, and so is *fictional ad hoc story-telling*.

I learned because I got burned. I learned not to take the Materialists, Naturalists, Darwinists, Nihilists, Behaviorists, Psychiatrists, and Atheists on blind-faith anymore. I learned that many of these people are deliberately lying to us, and they know it. I learned that others are simply the blind trying to lead the blind. I learned to question and critique everything that they are telling us, because 97% of the time it is wrong. Oh, there's always that rare 3% of the time when they are right, and I have learned to look for that as well – the times when they slip up and actually tell us the truth without realizing it.

All you want is the truth. Everything else is worthless. I had to learn that as well.

Even Pinel slips up once in a while and falsifies Materialism, Naturalism, and the theory of evolution in the process. In this quote, he starts with the "most people" logic fallacy, and then tries to convince you that evolution designed and created it all. However, without realizing it, he actually ends up falsifying the theory of evolution.

Most people think their visual system evolved to respond as accurately as possible to the patterns of light that enter their eyes.

But, despite the intuitive appeal of thinking about it in this way, this is not how the visual system works. The visual system does not produce an accurate internal copy of the external world. It does much more.

465

My goal in this chapter is to help you appreciate the inherent creativity of your own visual system. (*Biopsychology*, p. 130.)

In other words, your visual system does much more than what evolution could have ever given it.

Without knowing it, Pinel is testifying of God's existence here, because evolution of any kind by definition in principle cannot do creativity. Evolution can't even do inheritance, because chemical evolution can't produce genes and proteins out of thin air. It's physically impossible. The very existence of genes and proteins is Scientific Proof of God's Existence, because evolution of any kind cannot design and create genes and proteins. Simple. Parsimonious. True.

It is the pinnacle of superstition, the last superstition, to believe that evolution (genetic drift), random mutations, and natural selection designed and created the first genes and proteins out of thin air. It's the pinnacle of superstition to believe that evolution designed and now controls all of this, including our visual system.

If I were to give you the task of designing and creating an eye from scratch, creating the proteins from scratch using quantum waves or telepathy, connecting it all to a person's brain, and writing the genome to reproduce the whole visual system on demand, could you do it? Would you know how to design and create each of the different types of cells and where to put them? Would it work when you are done?

NO?

Then why on earth would you ever believe that evolution (genetic drift), random mutations, and natural selection designed and created your eyes, brain, "brain circuits", and genome from scratch millions or billions of years ago? Design and creation of proteins, genes, eyes, and brains by evolution is physically impossible. Chemical evolution can't produce a single functional protein, let alone the matching gene to go with it. It's physically impossible for evolution to create proteins and genes from scratch. It can't be done, which means that it wasn't done. Evolution can't do the physically impossible; but, God's Psyche certainly can thanks to Quantum Mechanics or the Priesthood Power of God.

Remember, the genes for eye production and eye function did not exist until after God designed, programmed, and created your genome in the first place. Eyes are worthless, without a genome to reproduce them; and, evolution of any kind can't produce functional genes and proteins from scratch.

The Materialists, Darwinists, Naturalists, and Atheists want you to believe that evolution designed and created your eyes; but, that's physically impossible! Functional evolutionary precursors did not exist until after God designed and created the first eyes in the first place. When it comes to our eyes, brain, and visual system, evolution had nothing to do with them. Think people, think!

According to my neuroscience teacher, "70% of our brain is dedicated to visual processing." Clearly, this is where we must look, if we want to know what our brain is really doing! Dedicated by whom? Remember, dedication requires teleology, intention, meaning, and purpose, which are functions of Someone Psyche.

Let's start with the CPU Model. Notice, that 70% of our brain is dedicated to visual processing, NOT video storage! There's nowhere near enough memory storage capacity within our brain to store all of the petabytes of video that our brain receives and processes during its lifetime. It's physically impossible to store all of our video input within our physical brain. It can't be done. So, if any of that is being stored, it is being stored someplace else besides our brain. Our brains are CPUs – processors – not RAM or memory storage. Our brain is about functionality, not memory storage. There's no memory storage (RAM) within any of this – just the necessary cache within each of the CPUs needed to do visual processing.

According to the CPU Model, whenever a person lacks memories for something like video, the functionality was lost, not necessarily the memories. In fact, the long-term consolidated memories tend to remain, whenever one of these CPUs gets damaged or removed. The ability to use their brain to recall or retrieve their memories was lost. They lost functionality but not necessarily the memories, because the memories come back whenever functionality is restored. You destroy the visual system making the person blind, and he or she loses the functionality but still remembers what they have seen in the past. According to the CPU Model of brain functionality, the physical brain is comprised of CPUs, and not RAM. The memories are being stored someplace else besides the physical machinery. The only way this makes sense is if we define thoughts and memories as quantum waves.

By using the CPU Model alone, we can't safely conclude that NO memories are being stored within our brain, because the evidence doesn't support that conclusion. CPUs have cache or working memory. We are forced to move beyond the CPU Model, if we want to know what's happening within our brains at the quantum level or the psyche level.

Every neuroscience book has a chapter dedicated to our visual system.

Cones are visual receptors in the retina that mediate high acuity color vision in good lighting. Photopic vision is cone-mediated vision, which does color vision and predominates when lighting is good. Rods are visual receptors in the retina that mediate achromatic, low-acuity vision under dim light. Rods are used for dark adaptation. Astronomers use averted vision to see better through a telescope eyepiece, because the rods can pick up dim light much better than our cones can. The cones are in the center of our visual field, and they aren't of any real use when trying to see faint objects in dim light. Scotopic vision is rod-mediated vision, which predominates in dim light. A scotoma is a blind spot produced by damage or a disruption of the visual system.

Rods are fascinating objects. They were designed to emit a constant glutamate signal in the dark. A rod is constantly idling away sending a signal when it's dark. However, when any type of light lands on its rhodopsin sensors, the rod turns OFF and stops transmitting. The presence of light or the transduction of light turns the signal OFF. Many signals are transmitted through the nervous system by inhibition or turning OFF the signal. It seems kind of wasteful to me, though, that a rod would be constantly producing and releasing glutamate the whole time that it's in the dark and not being used. That means that the rods in your eyes are constantly idling away while you are asleep. Crazy, huh?

The truly fascinating thing is all the different neurons idling away in our brain (constantly firing) in the complete absence of any stimulus or physical input. It's called default mode. How can they fire without any incoming EPSPs to prime the pump and keep the action potentials going? Clearly all the different default mode neurons in our brain idling away or firing away while we are asleep were carefully designed to function like rods and to emit a constant signal in the complete absence of any sensory input. God's hand is obvious in all of this. There's no way in the universe that evolution could have designed and created all of this, because chemical evolution can't even create a single functional protein, let alone the matching gene to go with it.

The neurons that fire or turn ON in the complete absence of physical input are the ones that are of greatest interest to Quantum Neuroscience. How do they do that? Who is making them do that? Is it some kind of inbuilt programming or structural feature placed there by God, or is Nature's Psyche constantly firing the neuron while the brain is in default mode or delay mode? It's the exception to the rule that points us to what's wrong with the rule. It's the neurons that fire "out-of-the-blue" – without any physical input making them do so – that end up falsifying Materialism, Naturalism, and their derivatives.

Our output neurons function as individual stand-alone switches which are controlled by Nature's Psyche at the quantum level in order to produce physical effects for us at the physical level. Our input neurons and input cells function as individual registers and receptors at the physical level, converting physical input or physical stimuli into electrical waves and quantum waves. Thoughts and memories are quantum waves, and quantum waves survive the death of our physical brain. The Human Psyche can learn to read the electromagnetic waves and quantum waves coming into its body through its physical senses.

Broadcasting as we humans typically use this function involves video and audio. It also involves the use of waves! We have all experienced the situation where the brain stops broadcasting to the Human Psyche, and we are no longer consciously aware of what our brain and physical body are doing. This happens to us when we are asleep. The brain also stops broadcasting to the Human Psyche and memory consolidation during most epileptic seizures. These people don't remember anything from the time during their seizure, even though they appear to be awake. If I were to assign you the task of building a physical machine, an android, that is capable of broadcasting quantum wave information to the Human Psyche, and then at certain times capable of stopping those quantum level broadcasts, would you be able to make such a machine? Why not? What are the obstacles?

The opponent-process theory is the idea that a visual receptor or a neuron signals one color when it responds in one way (by increasing its firing rate) and signals the complementary color when it responds in the opposite way (by decreasing the firing rate).

Off-center cells are visual neurons that respond to lights shone in the center of their receptive field with "off" firing, and to lights shone in the periphery of their receptive field with "on" firing. This physical functionality is programmed or designed into them.

On-center cells are visual neurons that respond to lights shone in the center of their receptive fields with "on" firing, and to lights shone in the periphery of their receptive field with "off" firing. This physical functionality is programmed or mapped into them.

Simple cells are neurons in the visual cortex that respond or fire maximally to straight-edge stimuli in a certain position and orientation. Complex cells are neurons in the visual cortex that respond optimally to straight-edge stimuli in a certain orientation in any part of their receptive field. Complex cells can be used to sense motion. Who taught these neurons how to do this, or who makes them do this? Somehow this functionality was programmed or mapped into them by Nature's Psyche at the quantum level. There are no CPUs in a neuron at the physical level.

God's fingerprints are all over this thing! Intelligence and design are evident in every aspect of our visual system! Evolution can't do intelligence and design. It's physically impossible! Evolution can't do anything at the quantum level or the psyche level. Evolution is limited to the physical level.

Physical functionality has been mapped or programmed into each one of these neurons by Nature's Psyche or God's Psyche! We are NOT talking about CPUs here! We are talking about individual neurons – individual switches which are either ON or OFF. It's all determined by whether a specific neuron is ON or OFF. That's how our brains really work!

We use the CPU Model for convenience because people can relate to it; but, the whole of brain functionality reduces or integrates down to whether a specific neuron is ON or OFF. There is no CPU and no RAM within a neuron or a switch! It's physically impossible. The neuron is either ON or OFF – that's it! That's how our brains really work.

Nature's Psyche has mapped or programmed physical functionality into each one of the neurons in our brain, at the quantum level. Psyche works at the quantum level through quantum mechanisms and quantum waves. Nature's Psyche collapses the wave function thereby making the Human Psyche's choices physical and real. In contrast, evolution is limited to the physical level. Evolution can't collapse wave functions at the quantum level. By definition, in principle, evolution doesn't exist at the quantum level or the psyche level.

Parallel Processing at the physical level is the use of multiple CPUs networked together or wired together to enhance the speed and efficiency of software processing. In the physical brain allegedly at the physical level, parallel processing is the simultaneous analysis of a signal in different ways by the multiple parallel pathways of a neural network. Of course, at the quantum level, Parallel Processing is done however God and Nature's Psyche chooses to do it. The only thing we know for sure is that God's quantum computers at the quantum level are infinitely superior to our quantum computers at the physical level, which are vastly superior to our ordinary physical CPUs at the physical level. Imagine a quantum computer made out of quantum waves at the quantum level, accessible directly by God's Psyche and Nature's Psyche.

Nature's Psyche can map petabytes of memories to one specific neuron or atom, at the quantum level. Nature's Psyche can map the output of gigabytes or even petabytes of CPU functionality, parallel processing, and image processing software to one specific neuron or atom, at the quantum level. Nature's Psyche can map an exabyte of video to one specific neuron or atom, at the quantum level. With God's help, Nature's Psyche can map the output of an infinite number of quantum computers at the quantum level to one specific atom or neuron. We are looking at only one neuron, or one switch, or one atom at the physical level; but, an exabyte of memories, an exabyte of video, and an exabyte of CPU functionality or parallel processing can be mapped to that single atom or switch by Nature's Psyche at the quantum level.

Mapping at the quantum level is also bi-directional! Through mapping at the quantum level, whatever God knows, that single atom can know; and, whatever that single atom knows, God knows.

Powerful, is it not?

I'm trying to help you see what I have seen.

Through mapping at the quantum level, it's possible for an atom to be omniscient; and, it's also possible for God to be all-knowing. Since there is NO space-time or entropy in the quantum realm or the psyche realm, it's even possible to be prescient at the quantum level. Some people describe that during their Near-Death Experiences while in the presence of God, they know everything and they can also sense the future. The KNOW that some things will happen in their future, because they have already been ordained or pre-determined, and they KNOW that some things may or may not happen in their future depending upon what they choose to do.

Quantum Neuroscience turned out to be really amazing, once I finally decided to let it do its thing.

In contrast, we have this from the Darwinists and Naturalists:

"The human brain appears to have evolved from the brains of our closest primate relatives." (*Biopsychology*, p. 33.) **"The human eye, a product of 600 million years of evolution."** (p. 132.)

This is physically impossible!

It's physically impossible for our evolutionary ancestors to be blind for 600 million years and end up magically becoming the fittest that actually survive the experience. In the land of the blind, the one-eyed man is king. These people leave their brains behind at the door whenever they start talking about evolution. They don't even think about what they are saying. They just parrot what they have been told in school. The purpose of our public schools is to brainwash us into Materialism, Naturalism, and Atheism; and, it works.

It's physically impossible for the members of one species to produce fertile offspring by mating with members of other species. In other words, it's physically impossible for chimp-like ancestors to mate and produce genetically compatible male and female human offspring. It can't be done, which means that it wasn't done. Remember, it's physically impossible for chimp-like ancestors to mate and produce genetically compatible Mr. and Mrs. Mutant each and every time over millions of years of time, so that chimp-like ancestors can magically evolve into human beings. Magic, like evolution, is physically impossible. Creation ex nihilo, abiogenesis, spontaneous generation, chemical evolution, and macro-evolution are physically impossible. If they happened, they happened at the quantum level, and not the physical level! And, the very existence of the quantum level or the psyche level FALSIFIES Materialism, Naturalism, and the theory of evolution.

Now, let's switch over to the physical side and the CPU Model, for a comparison. Although not as powerful as the Maps of Physical Functionality Model, the CPU Model and the Programming Model have their uses and do give a good approximation of what's being done to our physical brain at the quantum level by Nature's Psyche.

The visual input comes from the retina, is relayed through the lateral geniculate nuclei within the thalamus, and then sent to the primary visual cortex where it is mapped, channeled, and organized. Once your brain has an image, then it sends it through the secondary visual cortices in an attempt to interpret the meaning or purpose of the image, and then on to the association cortex where the visual input is coordinated with other types of sensory input. All throughout this whole process, we are dealing with CPUs or video processing, NOT memory storage. If any of this video is being stored someplace, it's not being stored within our brain, except for buffering or temporary memory storage which we call cache or working memory. Our brain is dedicated to processing this information, not storing it. Making this distinction is critical, if we want to know what's truly going on within our brain.

Technically, though, there are no CPUs and no RAM within our brain because there are no wires within our brain. All of this apparent CPU functionality and RAM memory storage is taking place non-locally in the cloud, if it's taking place at all. We continue to use the CPU Model to describe the brain, because it is apparent that Nature's Psyche has mapped or programmed the different structures of our brain to function as if they were CPUs, even though that functionality is taking place at the quantum level rather than the physical level.

The primary visual cortex or visual map is located along the medial surface of each occipital lobe along the longitudinal fissure in the back of our brain.

The primary visual cortex is located in the posterior region of the occipital lobes, much of it hidden from view in the longitudinal fissure. Most areas of secondary visual cortex are located in two general regions: in the prestriate cortex and the inferotemporal cortex. Areas of association cortex that receive visual input are located in several parts of the cerebral cortex, but the largest single are is in the posterior parietal cortex.

The dorsal stream specializes in visual spatial perception. The ventral stream specializes in visual pattern recognition. The dorsal stream specializes in visually guided behavior. The ventral stream specializes in conscious visual perception. (*Biopsychology*, p. 151, 155.)

Notice that these things are processing visual input like CPUs do, not storing visual input. They are mapped for physical functionality. The dorsal stream in the posterior parietal cortex does location or spatial processing, and tries to identify where things are located. The ventral stream running through the inferotemporal cortex does object recognition, and tries to identify what objects are. Destroy the CPU, or destroy the machine, or destroy the map, and the functionality is lost, but not necessarily the memories. The different structures and cortices within our brain really are functioning as if they were CPUs or computers of some kind, even though that's physically impossible at the physical level thanks to the scrambling and integration effects of our synapses.

The inferotemporal cortex contains a secondary visual cortex that is involved in object recognition. Object recognition has been mapped to the inferotemporal cortex by Nature's Psyche. Remember, recognition is a function of Intelligence or Psyche. Even our image recognition software and hardware required a ton of intelligence to produce. There are NO wires or CPUs or RAM within our brains; yet, our image recognition capabilities – which are the result of an interaction between the Human Psyche and Nature's Psyche and God's Psyche – are infinitely faster and noticeably superior to anything man-made at the physical level that we have ever produced. When it comes to object recognition, Psyche is infinitely superior to any man-made physical machine.

Observation tells us that our declarative memories seem to be stored diffusely throughout the whole brain, which is physically impossible. It's physically impossible to store petabytes of video or petabytes of declarative memories within our physical brain. If any of this is being stored, it's being stored someplace else besides our brain. Our brain was designed to do processing, not memory storage. Our brain consists of CPUs or Maps of Physical Functionality, not RAM.

Finally, those who describe their Life Reviews describe it as 360-degree surround vision and often describe it from a third-person perspective; and, our brain never creates 360-degree surround vision or spherical vision, front and back, up and down. Clearly, when it comes to our Life Reviews, someone else is taking the 360-degree video or spherical video for us and storing that video for us, because it can't be produced and isn't being produced by our brain.

Interesting, is it not?

Our brain does video processing, NOT video storage. The video capture and video storage for our Life Reviews is being done someplace else besides our brain; and, it's being done in a 360-degree spherical format that our physical brain and physical body can't do.

Remember, our brain doesn't do and can't do the 360-degree surround vision, which is being done and then being stored within our after-death Life Reviews somewhere in the cloud. This is important to know and understand. Our brain can't be storing the video that shows up in our Life Review after we are dead, let alone make and store those types of 360-degree video.

Change blindness is difficulty perceiving major changes to unattended parts of a visual image, when the changes are introduced after brief interruptions between the presentations of the two images.

This is scientific evidence that our visual system functions as a buffer, NOT as memory storage!

The Human Psyche has to pay attention to visual images if it wants to remember any part of them.

Change blindness is scientific proof that our video images are NOT being stored within our physical brain where they can be re-analyzed and cross-comparison done. Our visual system is a buffer – first-in, then first-out – unless of course the Human Psyche actually chooses to pay attention to something within the visual field; and only then, the memories might start to form. Attention from the Human Psyche is essential for memory formation. Memories and thoughts are quantum waves, which are a product of the Human Psyche – at least where our own memories and thoughts are concerned.

Our brains are made up of Cerebral Processing Units (CPUs), and not RAM. The video memories or the spherical video data in our after-death Life Reviews are being stored someplace else besides our brain. Our brains can't even make that kind of data or memories. Our brains are functioning like CPUs, not RAM. Our brains are dedicated primarily to functionality or information processing, NOT memory storage. Nature's Psyche has mapped physical functionality onto the different structures within our brain at the quantum level. That's something that evolution can't do, because evolution doesn't work at the quantum level.

Science 2.0 allows ALL of the evidence into evidence. What did we just learn by allowing all of the evidence into evidence? We learned that the video for our Life Reviews cannot be produced by our physical brain, is NOT being produced by our physical brain, and therefore is NOT being stored within our physical brain. All of that 360-degree video processing for our Life Review is being done and processed by someone else somewhere else besides our physical brain. This is a monumental and significant discovery in neuroscience. Why? It's because 70% of our physical brain is dedicated to visual processing. We just learned that those parts of our brain are Cerebral Processing Units (CPUs), and NOT memory storage devices or RAM; and, this discovery has been repeatedly verified through scientific observations and science experiments.

Since Karl Lashley and others have proven that our brain doesn't have memory engrams (or RAM), it's theoretically possible that NONE of our video images are actually being stored long-term within our physical brain. If that's the case, then instantly we have opened up the theoretical possibility for our brain to be able to store and process everything else that comes its way. Think about it. The vast majority of the data or memories that are allegedly being stored within our brain – the petabytes of data – are video data. Yet, we KNOW that it's physically impossible to store petabytes of video data within our physical brain. It can't be done.

Since we now KNOW that the petabytes of video data for our Life Reviews are being produced someplace else besides our brain and being stored someplace else besides our brain, suddenly the brain starts to gain the capacity to at least theoretically begin to be able to store everything else that comes its way. It's still not physically possible to store terabytes of data within a single bit – within a single synapse or a single neuron; but, it does become possible to store and distribute CPU programming, or CPU processing, or CPU functionality across millions of neurons, or a 100 billion neurons, as long as we are not trying to store the data, or the memories, or the cache and RAM that come along with it.

Was the payoff, from allowing all of the evidence into evidence, worth it? It was for me. Of course, your mileage may vary, especially if you are one of those who wants to cease to exist when you die.

Sorry, but ALL of the observational evidence and experiential evidence tells us that you, your Psyche or Non-Local Consciousness, will go on – long after your physical brain is dead and gone. And, so will your Life Review. There's a Judgement Day or Life Review waiting for each and every one of us, whether we like the idea or not. It's unavoidable, because we don't have anything to do with it. Our Life Review is taking place and being built completely outside of our awareness and control, including the petabytes of video that go along with it. NONE of the video for our Life Review is being stored within our physical brain! It can't be.

What a relief!

It's nice to KNOW that our video data isn't being stored exclusively within our brain, since it's physically impossible to store petabytes of video within our brain to begin with!

This is a significant discovery in neuroscience, and MOST of the neuroscientists will reject it completely, because it's not physically possible for this to be true. And so it goes.

My neuroscience teacher rejected it, and slammed me hard for even suggesting it, because it's physically impossible for our memories to be stored anywhere else besides our physical brain. Every fool knows that!

It doesn't matter in the least to these people that it's physically impossible to store petabytes of video within a physical brain, because they know that they are right and that everyone else is wrong. Clearly, petabytes of data are being stored within your physical brain, so obviously, it must be physically possible to do so.

Whenever I have presented these types of analyses and models online, the response I have gotten from the Materialists, Naturalists, Atheists, and Darwinists is that I don't understand science and that I don't understand how these things work. Clearly, since petabytes are being stored within our physical brain, there must be some sort of physical process or physical mechanism for doing so. Our brains must be loaded full of physical RAM; and, all our memories are being stored within our synapses. When we die, all of our memories cease to exist.

I no longer have blind faith enough to be a Materialist and Naturalist.

I KNOW what it would take to shove a petabyte of RAM into your brain. I'm a computer scientist, so I KNOW something about these things.

Buying a petabyte of storage for YOURSELF?

First, you'll need a fridge,

Plus a few hundred thousand quid, a reinforced floor...

http://mypsyche.us/wp-content/uploads/2018/01/Putting-a-Petabyte-of-RAM-into-Your-Brain.pdf

https://www.theregister.co.uk/2013/01/09/storagebod_7_jan/

Would you be willing to pay me a few hundred thousand quid to put a petabyte of RAM or a petabyte of physical memory storage into your brain? I would have to put a singularity in there with it, in order to get it to fit within your brain.

Psyche is an infinite singularity – that should do the trick, I think. Let me put one of them in there, and you should theoretically have the memory storage capacity you need to get the job done for a lifetime.

Motor Skill Learning: Large portions of our physical brain are dedicated to controlling our physical body. This is logical common sense. The Materialists and Naturalists call this "procedural memory"; but technically, it's CPU programming or software being written onto hardware, and NOT memory storage. Once again, when it comes to our physical brain, we are looking at Cerebral Processing Units (CPUs), not RAM! Our brains are all about functionality, NOT memory storage. Memory storage requires a memory storage device, and there's little evidence of RAM, or dedicated memory storage devices, or engrams within the physical brain. In fact, Karl Lashley and others have proven scientifically that memory engrams do not exist. We don't have memory engrams or RAM within our physical brain. So, where are our declarative memories or Life Review memories being stored? Definitely not within our physical brain.

So, how much of our memory is stored within our brain?

We don't know. The thing we do KNOW is that our physical brain has nowhere near enough memory storage capacity to store the petabytes of data that our brains develop and process throughout our lives.

Remember, programming of any kind requires a programmer. So, who is doing all of the programming associated with all of our growing motor skills? Clearly, the Human Psyche isn't doing this programming, because the brain programming associated with our developing physical skills takes place completely outside our conscious awareness. Either Nature's Psyche or God's Psyche has to be doing the programming of our physical brain, because the Human Psyche wouldn't even know where to begin. The Human Psyche issues the commands or the requests; and, Nature's Psyche within each atom of our physical body does its best to reply or respond. Then, Nature's Psyche remaps our neurons and synapses accordingly.

According to the cognitive map theory, the main function of the hippocampus is to store memories of spatial locations. So, who or what is turning the stand-alone neurons of the hippocampus into a map? The neurons are completely separated from each other by synapses. It's physically impossible to turn them into a map. It would require Action at a Distance to turn neurons into a functional map. Action at a Distance is physically impossible, according to Classical Physics. It can't be done, unless of course you are willing to let psyche and quantum mechanics into the mix. Psyche and Quantum Mechanics were designed to do the physically impossible.

Likewise, it is physically impossible to store terabytes or petabytes of memories within our brain. Neurons are completely separated from each other by synapses. Neurons are single bits, and so are synapses and receptors – they are either on or off. ONLY Psyche could map-out specific neurons, synapses, and receptors to be used as RAM or memory storage. If any memories are being stored within our brain, it's Nature's Psyche who is allocating the quantum mechanisms, assigning the neurons, and putting the memories into them. It would be non-local cloud storage, and NOT physical RAM.

This is the most interesting science that you could possibly study, and most people don't give it a moment's thought, because they have chosen to believe that it doesn't exist.

The nitty-gritty of neurogenesis, synaptogenesis, and neural pruning takes place completely outside of our conscious awareness, so there's nothing here for us to remember. If memories of any of this are being stored anyplace, they are definitely being stored outside of our brain and outside of our conscious awareness. As far as I know, the details of our neurogenesis and synaptogenesis and physical development don't show up anyplace within our Life Reviews, so I conclude that these types procedural memories or learned skills aren't being stored anyplace besides within the structure or programming of our physical brain. It's functionality that's being generated by synaptogenesis and neurogenesis, NOT memory storage. Our brain doesn't have any memory engrams, which I take to mean that our brain doesn't have any RAM.

Over and over again, the Materialists and Naturalists deliberately conflate Learning Systems with Memory Systems. They do this so that they can say that ALL of our memories are stored within our physical brain. But Learning Systems (CPUs, hardware, programming, and software) are something completely different than Memory Systems (RAM, data storage, or memory storage).

Consequently, the Materialists and Naturalists repeatedly call these things "procedural memory" rather than "learning systems" so that they can say that ALL of our memories are stored within our brain. Their goal is to trick you and deceive you, making you believe that your psyche doesn't exist and making you believe that you and your memories cease to exist when you die. That's their goal, and exposing their deception is the only way to deal with it.

In my humble opinion, it's wrong to portray learning systems or CPU processors as memory storage devices, just so that you can say that psyche doesn't exist and that all our memories are being stored within our brain; but, that's what these people choose to do. I KNOW, because I used to be a Materialist, Naturalist, Nihilist, and Atheist. The goal is obfuscation, and not finding the truth. It's their religion, and they are fanatical about it. These people will trick you and deceive you by any means necessary, in order to get their way.

Over and over again, the Materialists and Naturalists tell us that the brain programs itself. In this, they are partly right and partly wrong. Nature's Psyche within each and every atom of the physical brain programs the brain and controls the brain. So, in a sense, the brain is indeed programming itself; but, programming requires a Programmer, and our physical brain is NOT a programmer. Our physical brain is the thing that gets programmed by Nature's Psyche, under the direction and help of God's Psyche. There's a difference here, an important difference; and, an understanding of all of this becomes available to us by allowing ALL of the evidence into evidence.

Procedural Memories: Procedural memories are physical skills, and they are being programmed or mapped into our physical brain by Nature's Psyche or God's Psyche – completely outside of our conscious awareness. Procedural memories are programming and/or mapping of our Cerebral Processing Units (CPUs).

Procedural memories are NOT memory engrams or RAM. The physical brain doesn't have memory engrams or RAM. Our brains have NO wires connecting our neurons. Procedural memories are software and physical functionality that's being encoded or mapped onto our physical brain by Nature's Psyche or God's Psyche at the quantum level, completely outside of our awareness and control. It either works, or it doesn't.

Are you starting to see how much we rely upon God, in order to get things done? Programming requires a Programmer, and where our physical brain and genome are concerned, that Programmer is God. Maps require Map Makers. This reality and truth effectively becomes Scientific Proof of God's Existence. Your genome is God's Signature.

It would seem logical to me to store our dynamic programming or acquired physical skills in perishable physical RAM, because we aren't going to need that kind of information after our physical brain is dead and gone.

All throughout the various neuroscience college textbooks, they talk about programs that are being stored within and run within our physical brain. Even though that's physically impossible, thanks to synaptic clefts, it is obvious and clear that Nature's Psyche is using the different parts of our brains as Cerebral Processing Units (CPUs) to process the programs that it has designed and customized for our brains.

There's no end to the number of programs and CPUs that Nature's Psyche seems to be running within our physical brains.

Reference memories are defined as general principles and skills that are required to perform a task. These are a type of enhanced programming or enhanced mapping. Technically, Nature's Psyche maps our brains for physical functionality, rather than programming our brains. One-to-on mapping more accurately describes what's really going on within our brain, because there are NO wires in our brain. Programming is more appropriately associated with computer chips that have their transistors wired together into a functional unit. Mapping is a better model for brain functionality; however, there are noticeable parts of our brains and nervous system – our reflexes – that seem to be hard-wired programming rather than being modifiable maps of physical functionality.

Nature's Psyche maps our sensory input to specific neurons in our brain; and, Nature's Psyche maps our physical functionality, our output, to specific neurons in our brains. Nature's Psyche has to do this mapping at the quantum level, because there are NO wires in our brains that can be used to program our brains at the physical level. The Cerebral Processing Units within our brain are maps of physical functionality, and not wired-together CPU chips and RAM chips.

Still, there are certain modules within our brain and nervous system, our reflexes, that seem to be programmed into us directly at the physical level; and obviously, there is going to be some noticeable overlap between our maps of physical functionality and our pre-programmed reflexes.

Free-running rhythms are circadian rhythms that do not depend on environmental cues to keep them on a regular schedule. That functionality is programmed directly into our brains by Nature's Psyche or God's Psyche.

Our visual system is chock full of programming! Our visual system is also mapped for specific physical functionality. Who knows where the physical ends and the non-local begins?

Completion is the visual system's automatic use of information obtained from receptors around the blind spot, or scotoma, to create a perception of the missing portion of the retinal image. That's the very definition of computer programming! It's mathematical interpolation that's taking place here! It's functionality, NOT memory storage! There's nothing there to remember coming from the blind spot! Over and over again, we observe that the brain does physical functionality, and NOT memory storage! Furthermore, we observe that physical functionality is being mapped onto the different structures within our brain by Nature's Psyche.

Complex cells are neurons in the visual cortex that respond optimally to straight-edge stimuli in a certain orientation in any part of their receptive field. Again, this is programming or functionality, or what the Materialists and Naturalists call procedural memories. Nature's Psyche has programmed these specific neurons to accomplish this specific task; and, they do. Remember, evolution can't do programming, but Nature's Psyche certainly can.

Contrast enhancement is the intensification of the perception of edges. It's programming! This functionality is programmed directly into the cell by Nature's Psyche.

Cryogenic blockade is the temporary elimination of neural activity in an area of the brain by cooling the area with a cryoprobe. The Materialists and Naturalists say that the memories have been lost. They are wrong! The memories and the programmed functionality come back after the brain thaws out and functionality is restored. While that are of the brain is frozen and non-functional, it no longer responds to the Human Psyche and no longer responds to Nature's Psyche either. If you cut out part of your brain, then it's no longer there to respond to Nature's Psyche or to your Human Psyche. Your memories remain, especially your Life Review memories; but, the functionality is completely lost.

These different Cerebral Processing Units (CPUs) in our brains are the result of programmed functionality and/or mapped functionality that's being done by Nature's Psyche or God's Psyche.

Nature's Psyche puts the synapses, the maps, and the invisible programming into place, but the synapses have to actually be used, or they disintegrate. This is called the critical period in neurodevelopment. Nature's Psyche puts the visual pathways into place; but, if they are never used, they eventually disintegrate. A critical period is a period during development in which a particular experience must occur in order for subsequent development in that area to continue. Priming must take place, or the pump isn't going to work. The programs and maps have to be used, or they get deleted.

The default mode is the pattern of brain activity that is associated with relaxed wakefulness, when an individual is not focused on the external world. The brain is simply idling. The brain is running on auto-pilot, or on built-in programming. Remember, any type of programming requires a Programmer! A map requires a Map Maker. So, who was the One who programmed your brain to do these functions? You didn't do it, because you aren't consciously aware of it. Evolution didn't do it, because evolution can't do programming and map making. Through a process of elimination, we find ourselves looking at Nature's Psyche or God's Psyche in order to answer this question. It's the only answer that makes logical sense.

Nature's Psyche has actually programmed specific neurons to fire or light up whenever the Human Psyche recognizes a specific person or a specific object. These are concept cells. These concept cells are located in the medial temporal lobe near the amygdala, and they seem to trigger an emotional response or visceral response whenever the Human Psyche recognizes someone or something that has become important to him or her.

Who assigns or maps these concepts or ideas to a specific concept cell?

The only logical answer is Nature's Psyche or God's Psyche.

The Human Psyche isn't doing it, because we aren't consciously aware of it. Evolution isn't doing it, because evolution can't do programming and processing and mapping.

The concepts and ideas, that the Human Psyche chooses to pay attention to and chooses to like, get programmed into or assigned to or mapped to these concept cells by Nature's Psyche; and, those cells light up or fire whenever the Human Psyche recognizes that specific object. The Materialists and Naturalists tell us that it is memory storage and computer processing that's taking place within this specific neuron; but, that's physically impossible. There's no physical RAM and no physical CPU visible within any of our neurons, even our concept cells. That single neuron can't be receiving messages or information from other neurons, because a synaptic cleft will scramble any message that you try to send through it. The assigned concept cell just unilaterally fires on its own, whenever the Human Psyche recognizes that specific object which has been assigned or mapped to the concept cell by Nature's Psyche.

A concept cell can theoretically be triggered by another cell; but then, who triggered that first cell when the Human Psyche recognized a familiar object? It's that FIRST cell – that triggers and causes a string of action potentials – which is physically impossible and that falls into the domain of Quantum Neuroscience. It is always that FIRST specific neuron, which fires unilaterally with NO input from any nearby cells, which ends up being the thing that's physically impossible and therefore requires some kind of Non-Local or Quantum explanation. Whenever Nature's Psyche reaches out telepathically and turns on a specific neuron, we have witnessed Action at a Distance, which is physically impossible and therefore a quantum process of some kind.

Procedural memories are implicit memories – they exist outside our conscious awareness.

Perseveration is the tendency to continue to make a formerly conditioned response or a habitual response, even though that response has subsequently proven to be incorrect. Perseveration is also the tendency to continue making a previously correct response that is currently incorrect.

The Wisconsin Card Sorting test is a test for perseveration that's the result of frontal-lobe damage. Perseveration is typically portrayed by the Materialists and Naturalists as lost memories, when in fact it's lost functionality. The Naturalists always conflate memories with functionality, because they want you to believe that all of your memories are stored within your brain so that you believe that all of your memories cease to exist when your brain dies. It is obvious that some sort of physical functionality has been mapped directly onto our physical brains, so these people conflate memories and functionality so that you are forced to believe that our memories are also mapped directly to our physical brain and nowhere else.

The thing they don't realize is that mapping of any kind requires a Map Maker, or an intelligent psyche. Whether we are talking about the Mapping of Physical Functionality or the Mapping of our Memories into non-local cloud storage, it's Nature's Psyche who is doing all of this mapping at the quantum level, completely outside the conscious awareness of the Human Psyche and our physical brain.

Conditioned taste aversion is an avoidance response developed by animals to the taste of food that has made them sick. The Materialists and Naturalists declare conditioned taste aversion to be memories of food that have made you sick. However, conditioned taste aversion could also be the result of a change in programming or functionality.

A neuron is a switch. It is either ON or OFF. The human brain is comprised of 85 billion to 100 billion neurons or switches, which have NO visible wires connecting them.

If I were to give you a map, how would you scan and store that map in a 100 billion stand-alone switches? You won't. You can't. It's physically impossible. There are no wires connecting those switches. Every switch is separated from every other switch by synaptic clefts, which will scramble any message that you might try to send through the cleft.

Yet, WE KNOW for a fact that Nature's Psyche is mapping whole areas of brain cortices to serve as maps for specific functions! That's physically impossible, but totally possible when the brain and the cortices and the maps are under the command and control of Nature's Psyche and quantum mechanisms. Notice how Nature's Psyche and Quantum Mechanics provide us with infinitely more explanatory power than we could ever achieve from Materialism and Naturalism alone. Quantum Mechanics is vastly superior to Classical Physics or Materialism.

When it comes to Quantum Neuroscience and trying to figure out how our brain really works, it requires converging operations wherein we use several different research approaches and several different types of science and evidence in order to solve a single problem. The subject of memory storage in our brain requires a lot of converging operations.

When it comes to Quantum Neuroscience, I have tried to go through the whole of neuroscience and explain how everything might be working at the quantum level, especially whenever it's physically impossible for it to work at the physical level. It's physically impossible to transmit messages through synapses, because a synaptic cleft will scramble any message that you try to send through it; therefore, we have to look for some other way that messages and programming are being transmitted and processed throughout our brains.

The first thing that comes to mind is WiFi, or Quantum Telepathy, or waves. Telepathy is WiFi at the quantum level; and, it works, even though it's physically impossible. Hook an EEG to your scalp, and what's the first thing that you observe? Waves! Waves are Quantum Telepathy, or WiFi at the quantum level. They exist, even though they are physically impossible. The neurons are generating waves, or someone is. The very existence of Quantum Waves, Quantum WiFi, or Quantum Telepathy falsifies Materialism and Naturalism.

Implicit Memories: Implicit memories are unconscious memories. It has been proven scientifically that the unconscious does indeed exist. So, what is it? If we choose to follow Science 2.0 and allow ALL of the evidence into evidence, it quickly becomes apparent that our unconscious memories are being handled by Psyche, particularly by Nature's Psyche and/or God's Psyche. Only when things become conscious, can our Human Psyche begin to deal with them and control them.

The unconscious is code for Psyche, particularly Nature's Psyche. If the Human Psyche is consciously aware of something, then obviously we and our brains are consciously aware of it too.

The innate immune system is our first line of defense. It acts near entry points to the body and attacks generic classes of molecules produced by a variety of pathogens.

Who is controlling this thing?

Remember, molecular control requires some kind of quantum mechanism at the quantum level; otherwise, atoms and molecules just sit there and vibrate under the effects of Brownian motion, random diffusion, and entropy.

Most of our autonomic nervous system and immune system is being run and controlled by Nature's Psyche or God's Psyche completely outside of our conscious awareness; however, we (the Human Psyche) can train ourselves to be aware of and to control some of these things. It is called mind-over-matter; and, when it takes place unconsciously or subconsciously, it's called the placebo effect or faith. See how much more becomes available to us by allowing ALL of the evidence into evidence, instead of dogmatically claiming that all of our memories are stored within our physical brain as the Materialists and Naturalists do! I like to think that there are possibilities.

Implicit memories and unconscious memories are being controlled by Psyche – particularly Nature's Psyche or God's Psyche – completely outside of our awareness and consciousness. Lots of times we can sense that it's there; but, we don't know what to do with it. Just give God permission to activate it, optimize it, improve it, and fix it. That's pretty much all we can do, because it's completely outside of our conscious control. We either live or die, depending upon what God's Psyche and/or Nature's Psyche chooses to do with us.

Behaviorism is all about prediction and control.

God's message to us is that none of us have complete control over our physical bodies, which means that none of us can predict what's going to happen to our physical bodies and when it's going to happen to us. The only thing we have control over is what we choose to do with the physical bodies that God has given to us. These realities and truths make the Behaviorists supremely uncomfortable – let me tell you – because I have lived it and experienced it first-hand; and, the complete loss of control and the complete loss of predictability was the ultimate worst thing that could have ever happened to me. It was an eye-opener, let me tell you!

When God or Nature's Psyche decides to take you out of the game, there's nothing you can do about it; so, you just as well get used to it, because that's the way it is.

The psychologists have developed tests for implicit memories. Implicit memories are memories that you can't remember.

The Materialists and Naturalists call these things "implicit memories" or "procedural memories" in an attempt to trick us and deceive us into believing that all of our memories are produced by our brain, stored within our brain, and cease to exist when we die. However, all of the empirical evidence or experimental evidence suggests that implicit memories and procedural memories are improved functionality and enhanced programming, and not actual memories. How can they be memories if you can't remember them? In contrast, we have tons of programming in this world that we aren't consciously aware of.

According to Materialism, Naturalism, and Classical Physics, your implicit memories and procedural memories are NOT being stored within your brain, or your brain would be able to remember them and would be consciously aware of them.

Of course, I'm splitting hairs here – same difference, and all of that. However, it is important to do so, because whenever we come across something that is physically unexplainable, such as memories that can't be remembered by our physical brain, then we have run into something that falsifies Materialism and Naturalism. You have to eliminate everything that is false, before you can find and know the truth.

The proven existence of implicit memories and procedural memories, which the physical brain and the human psyche aren't consciously aware of, provides evidentiary proof that Materialism and Naturalism are false, because someone somewhere psyche IS consciously aware of them, or they wouldn't exist.

Fascinating, is it not?

Repetition priming is used to develop implicit memories or hidden unconscious memories. A repetition priming test is a test for implicit memory capabilities.

Skeletal conditioning is the classical conditioning of a complex skeletal response.

Emotional conditioning is a specific learned behavior, or a procedure, commonly used in classical or Pavlovian conditioning research. It may also be called "conditioned suppression" or "conditioned fear response (CFR)".

The Behaviorists and Naturalists define conditioning as learning and memory; and, they define learning and memory as the same thing as well – as some type of physical process that happens within our physical brain and ceases to exist when we die.

My research into Quantum Neuroscience, brain functionality, and the after-death Life Review have led me to see memory and learning and conditioning differently than the way the Materialists and Naturalists have chosen to visualize and interpret these things. Conditioning could just as easily be Re-Mapped Physical Functionality that's being produced by Nature's Psyche at the quantum level. In other words, conditioning might not have anything to do with memory and learning, especially at the physical level, but might instead be a remapping of our physical functionality by Nature's Psyche at the quantum level.

Of course, the Materialists and Naturalists won't see it this way, because they have trained themselves not to. These people have undergone conditioning and brainwashing into Materialism and Naturalism, so at least they know how the process works and its results; but, it does have its drawbacks.

Nevertheless, I have observed that the Human Psyche can break its conditioning at will. Remember, the Human Psyche can trump its Nature and its Nurture at will. This is the primary lesson of the BioPsychoSocial Model of model of mental illness. Genes and Nature's Psyche provide us with our physical platform and physical functionality. The environment and those other psyches provides us with opportunities. And, the Human Psyche chooses what it wants to do with its genetic inheritance and the environment that it finds itself living within.

In conclusion, it's actually very difficult to make a fool-proof case demonstrating that our "procedural memories" and "implicit memories" are stored within our brain, because by definition in principle we are NOT consciously aware of any of these "memories" which actually suggests that they are NOT being stored within our brain. The Human Psyche is ONLY aware of what's happening within its brain. Since the Human Psyche isn't aware of its procedural memories and implicit memories; then clearly, they aren't happening within the brain.

So, what is happening within our brain?

The only thing that makes sense to me now is that Nature's Psyche is mapping physical functionality into our brain at the quantum level, and then using those Maps of Physical Functionality to do things for the Human Psyche at the physical level. Procedural memories and implicit memories are Maps of Physical Functionality built by Nature's Psyche at the quantum level, completely outside the conscious awareness of the Human Psyche and its brain.

Spatial Memory:

If I were to assign physical memory storage to the brain, the device that I would choose is the hippocampus, and the type of memory would be spatial memory.

Why?

It's because the hippocampus is larger in birds that hide their food for later consumption. And, the hippocampus is larger in London cabbies who have memorized the streets of London.

Of course, the logical question becomes, "Is that larger hippocampus the result of additional memories being stored within the hippocampus, or is it due to increased functionality?" The Materialists and Naturalists and Darwinists are united in saying that the larger hippocampus is the result of increased memory storage capacity and an increased number of memories being stored within it; and, all of those memories will cease to exist when the physical brain dies.

Some of the rest of us aren't so eager to jump to that conclusion, because we have observed that the physical brain has no RAM or memory engrams and that the different Cerebral Processing Units (CPUs) in the brain like the hippocampus represent enhanced functionality and NOT increased memory storage.

Still, if there is in fact RAM in your brain, the hippocampus is where it resides. All of our spatial memories or location memories are being processed or mapped through our hippocampus. The best case for memories being stored within the physical brain can be made through spatial memories and the hippocampus. Spatial memory is the best example of memories for functionality being stored within the brain.

Even then, there's a non-physical or quantum explanation as well. The hippocampus and spatial memory are also the best example of Nature's Psyche and its ability to Map Physical Functionality onto our brain at the quantum level. Nature's Psyche is actually mapping the spatial maps that we use most onto our hippocampus so that we actually have a visceral feel for where we are currently located and not just an intellectual belief.

The hippocampus plays some type of role in memory for spatial locations. Is the role some kind of functionality, or actual memory storage? We don't know, because there are NO visible wires connecting the hippocampus together into a functional whole. The functionality and memory storage associated with the hippocampus are taking place wirelessly and non-locally in the cloud.

How do we know?

WE KNOW because it is physically impossible to transfer messages and information through synaptic clefts. A synaptic cleft will scramble any message that you try to send through it. Therefore, if the hippocampus is indeed storing maps and locations for your favorite foods, that information isn't being stored within the synapses, as the Materialists and Naturalists claim.

Grid cells and place cells are my next choice for memory storage within the brain.

According to my neuroscience textbooks, "Place cells are neurons that develop place fields. They respond only when the subject is in a particular place in a familiar test environment." Specific place cells heat up and fire, whenever the Human Psyche and physical body are standing in a familiar place. How do they know? It's physically impossible for a lone neuron to turn on and fire up without any input from nearby neurons; yet, it does. How? There's only one way that this can be accomplished, and that's through the telepathic interaction of the Human Psyche, Nature's Psyche, and Quantum Mechanisms. Forces and fields are physically impossible. They are products of Quantum Mechanics and Psyche. Telepathy is WiFi at the quantum level, and the Materialists repeatedly tell us that telepathy and telekinesis are physically impossible. Subjectivity and recognition are a function of Psyche.

"Grid cells are entorhinal neurons that have multiple evenly spaced place fields." Grid cells seem to store place memories within these invisible place fields; and, that's physically impossible. It can't be done, unless we introduce Nature's Psyche and Quantum Mechanics into the equation. Then it can be done in spades for all eternity. Remember, there is no entropy in the non-local realm or spirit world. Entropy only exists within physical matter. Psyche, Non-Local Consciousness, Intelligence, spirit matter, forces, fields, and quantum mechanisms are PURE SYNTROPY. They have no expiration date built into them. They are eternal and everlasting. Entropy is a product of space-time, locality; and therefore, entropy is only associated with physical matter.

Mirror neurons are neurons that fire or turn ON both when a person makes a particular movement and when a person observes somebody else making the same movement. There's Intelligence or Psyche within these neurons. There's image processing CPU functionality and motion sensing functionality built into this single neuron; and, that's physically impossible. Who is doing all of the object recognition? The Human Psyche. Who is firing this specific neuron? Nature's Psyche. These terabytes of image processing, parallel processing, motion processing, sensorimotor processing, and object recognition processing have to be taking place at the quantum level, because it's physically impossible for all of this to be taking place within a single neuron or switch at the physical level. The neuron is either ON or OFF; so, where's all of this in-built functionality being processed and stored since it can't be done within a neuron at the physical level? Thank God we have the quantum level to fall back on, or we would be nothing but rocks and vegetables.

A mirror-like system is defined as areas of the cortex that are active both when a person performs a particular action and when the person observes somebody else performing the same action. The "person" is the Human Psyche. Nature's Psyche is the one who mapped this mirror-like functionality into these areas of the cortex. All of this functionality is taking place at the quantum level, because it's physically impossible do this kind of functionality through synaptic clefts. Synaptic clefts scramble and randomize everything that comes their way. The only thing that makes it through a synaptic cleft is OPEN or CLOSED. A neuron is either ON or OFF. You are never going to get image processing and parallel processing functionality through a synaptic cleft or within a single neuron or switch – at least not at the physical level.

It would seem logical to me to store spatial memories, place memories, and location memories within physically perishable RAM, because we aren't going to need any of those types of memories after our physical body is dead and gone. However, there is no RAM within a physical brain, because functional RAM cannot be constructed out of 100 billion stand-alone switches or neurons with no wires to connect them.

The Human Psyche pays attention to what it values most; and then, Nature's Psyche maps the brain accordingly. After all of my research, I have settled upon the conclusion that Nature's Psyche takes the 100 billion neurons or switches within our brain and organizes them into Maps of Physical Functionality at the quantum level. Neurons are switches with NO wires connecting them. If they are being organized or mapped into any sort of functionality or purpose, it's being done at the quantum level, because it's physically impossible to do it at the physical level.

Working Memory:

Since our physical brain is comprised of different Cerebral Processing Units (CPUs), we know from observing man-made CPUs that each CPU has cache, which is a type of memory that is used to aid the CPU in its processing functions.

It has been observed that each Cerebral Processing Unit (CPU) within our physical brain has its own cache, which it uses for processing the data that comes its way. Cache is NOT long-term memory storage. Cache is fleeting and reusable. There's NO consolidated declarative memories being stored within our cache.

Whenever you destroy a CPU within the brain, that functionality is lost, and then it becomes impossible for the Human Psyche to experience or process anything through that lost piece of machinery. Take down the video system, and then the Human Psyche can never get any video input through its physical brain. If the CPUs that were designed for processing events are destroyed, then the physical brain has NO experiences or events for the Human Psyche to experience and remember through its physical brain. It's the functionality that's always lost with brain damage, not necessarily the memories. Typically the consolidated declarative memories remain, no matter what part of the brain is eliminated or destroyed.

ALL of this new understanding and science became possible by choosing to go where everyone else refuses to go. It became possible by choosing to allow all of the evidence into evidence, and then by pursuing a preponderance of the evidence. Quantum Neuroscience is a new and different way of doing neuroscience. Clearly, it has its advantages. Its explanatory power is infinitely superior to anything that can be produced by the Materialists and Naturalists.

Declarative Memory:

Whenever anyone talks about memory, typically what they are actually talking about is declarative memories, which are memories for facts and events.

Declarative memory is a completely different animal. Parts of our physical brain are dedicated to the functions of recognition and recall, which are associated with declarative memories; but, this leave littles or NO room for the storage of episodic memories or declarative memories, because our brain is dedicated primarily to physical functionality and not memory storage.

I, personally, can't visualize how declarative memories – video, numbers, data, information, facts, smells, tastes, feelings, sensations, and events – can be stored within a single neurotransmitter, or a single ion receptor, or a single synapse, or even a single neuron. Each one of these things represents a single bit of information – ON or OFF, OPEN or CLOSED. It's physically impossible to store petabytes of data or declarative memories within a single bit, or even within a single cortex of a few million neurons or a few billion synapses. If any of this is being stored, it has to be stored someplace else besides our brain. Likewise, our after-death Life Reviews, which seem to consist of declarative memories, are definitely stored someplace else besides our physical brain.

Furthermore, by definition in principle, the synaptic cleft is supposed to be the pinnacle of chaos, random diffusion, and random chance. It shouldn't work, but it does. Someone or something has to be forcing order and organization onto our neurotransmitter systems and our physical brain. According to classical physics, our brain shouldn't work; but, clearly it does, so that means that we have to find some other explanation for its functionality besides Materialism, Naturalism, and Classical Physics.

I get the feeling that we still have no idea what's really going on within the synaptic cleft between synapses. The apparent results and functionality suggest that it's infinitely more organized and controlled than the random diffusion of neurotransmitters would ever allow or suggest. Something has to be forcing the right neurotransmitters to engage the correct receptors on the post-synaptic membrane at the right time, on schedule, all in a timely, orderly, and controlled fashion; otherwise, chaos would ensue, the synaptic cleft would plug up, and the neurons would burn out and die. Furthermore, it's physically impossible to store petabytes of data and programming within a synapse or a neuron. That data storage has to be taking place somewhere else, if it exists at all.

Random diffusion does not produce order. It produces entropy. Yet, where the physical brain and our neurotransmitter systems are concerned, God makes sure that some kind or order and organization is brought out of all that chaos, disorder, diffusion, and entropy. When presented with entropy, God can turn it into syntropy. The whole ends up being infinitely greater than the sum of the individual parts. Syntropy results in a reversal of entropy; yet, it's physically impossible to detect that syntropy taking place. We can only detect the entropy with our physical instruments, because entropy is one of the limitations of physical matter that God has placed into physical matter. There is no such thing as entropy in the spiritual realm or the non-local realm, because everything there is immortal and eternal. The non-local realm is ALL syntropy; but, none of it can be detected directly with our physical instruments. The best we can do is to infer that the non-local exists by observing the effects that it has on the physical atoms around us.

When it comes to this controversial "memory storage" issue, we are witnessing All-Or-Nothing Thinking on both sides of the aisle. That's simply not how it works!

Going to the one extreme, we need a new definition for memories. **If it survives the death of our physical body and physical brain, then it's a memory; otherwise, it's a physical reflex or a physical function or programming.**

Going to the other extreme, memories are typically defined as changes in the synaptic wiring of our brain.

Which definition for memory is correct?

They both are! This is not an either-or, all-or-nothing, situation. Remember that! It all depends upon which type of memory we are talking about.

There are many different memory systems which are represented exclusively by changed synapses and brain-rewiring; and obviously, these types of memories are indeed lost when the physical brain dies and rots away. Strokes, lobectomies, brain trauma, neurodevelopmental disorders, and neurodegenerative diseases represent this type of lost functionality, and therefore this type of memory loss. The lost functionality can indeed be described as a memory loss of a sort that's caused by brain damage. The functionality is lost, but the long-term consolidated declarative memories remain.

It is indeed logical to claim that when your brain dies and your body ceases to exist, these types of procedural memories or types of functionality and programming stored within the wiring of our brain and nervous system will indeed be lost. If God were to give you a new body, you would have to learn how to control it all over again.

In contrast, when it comes to long-term episodic declarative memories – memories for facts, intelligence, learned information, feelings, video impressions, smells, sounds, perceptions, decisions, choices, events, and experiences – which seem to be mediated or facilitated by different brain structures and which seem to show up in our Life Reviews after our body and brain are dead, it is obvious and clear that some kind of copy of these kinds of declarative memories is being made for us non-locally in the cloud, off-site, and outside of our brain.

The ultimate goal of the Materialists and Naturalists is to conflate and to confuse these different types of memories, by trying to make the assertion that our consolidated declarative memories are stored only within our physical brain, which simply is NOT the case! Their goal is to deceive you; and, it works.

The physical functionality, or the ability to use declarative memories, is mediated by the brain; and, parts of the brain may act as a buffer for processing and re-experiencing these declarative memories; but, there's plenty of evidence indicating that our consolidated declarative memories are being backed-up and stored outside of our brain because of our Life Reviews, because of retrograde retention of declarative memories after lobectomies, and because it's physically impossible to store petabytes of information within a quadrillion synapses or a hundred billion neurons, which have no obvious sign of RAM within them.

Remember, a synaptic cleft will scramble any message that you try to send through it; so, if any memories are being stored within your brain, they are being stored non-locally in the cloud, and they are being transmitted from one neuron to another though waves, WiFi, or Quantum Telepathy. Telepathy is WiFi at the quantum level.

Let's Run the Numbers:

One billion = 10^9 = 10 x 10 x 10 x 10 x 10 x 10 x 10 x 10 x 10.

One Hundred Billion = 10^{11}. The estimated number of neurons in the adult brain. A neuron represents one BIT of information. It is either ON or it is OFF. It either fires, or it doesn't. There's no identifiable CPU or RAM within a neuron. There's no room for petabytes of RAM within a neuron. It's physically impossible. Neurons also function as switches, gates, and registers. According to gate-control theory, signals descending from the brain can activate neural gating circuits in the spinal cord to block incoming pain signals. Gates are OPEN or CLOSED. Switches are ON or OFF. Registers either register an input signal, or they don't.

One Hundred Trillion = 10^{14}. This is the typical number given for the number of synapses within the adult human brain. A synapse represents one bit of information. It is either OPEN or CLOSED. Synapses aren't transistors, wires, RAM, nor CPUs. Synapses are gates. They are either OPEN or CLOSED. That's it. There's no room for petabytes of RAM within a synapse or an ion channel. It can't be done. It's physically impossible. It's also physically impossible to send a message through a synaptic cleft, because a synaptic cleft will scramble and randomize any message that you might try to send through it. The synaptic cleft is a randomizer or jamming device, not a transmitter. The Materialists and Naturalists claim that all of our programming, data, thoughts, and memories are distributed throughout our brain and stored in the synapses; but, that's physically impossible. You can't store anything in a synaptic cleft except random chaos, because it's physically impossible to do so. Remember, ALL the synapses on a neuron reduce or integrate to a single stand-alone BIT – a postsynaptic potential. A postsynaptic potential either reaches threshold and triggers an action potential, or it doesn't. ON or OFF. All the synapses on a neuron reduce to ON or OFF. That's it! ALL meaning and purpose is stripped away when it passes through a synaptic cleft and merges into a post-synaptic potential.

One Quadrillion = 10^{15}. The estimated number of synapses in a new-born child's brain, before consolidation and specialization and pruning. A synapse represents one bit of information. It is either OPEN or CLOSED.

One Petabyte = 10^{15} bytes. One petabyte is almost 10^{16} bits. It is estimated that petabytes of data are being stored within our brain; but, you can't slam 10^{16} bits of information into a single BIT, such as a neuron or a synapse or a neurotransmitter receptor. It's physically impossible. And, you can't slam 10^{16} bits into 10^{14} synapses or 10^{11} neurons. Once again, it's physically impossible. Since petabytes (plural) of data are allegedly being stored within our brain, those petabytes have to be getting stored someplace else besides our brain, because it's physically impossible to store petabytes of data within our brain. You can't shove petabytes of RAM into a physical brain, because it's physically impossible.

One Exabyte = 10^{18} bytes. The amount of memory that our physical brain would need to be able to store, so that the Materialists and Naturalists could safely say that ALL of our memories are being stored within our physical brain with plenty of room to spare. But, it's physically impossible to slam 10^{18} bytes of data (or 10^{19} bits of data) into 10^{14} synapses or 10^{11} neurons. It can't be done. It's physically impossible.

Anytime we run into something that is physically impossible, it points us directly to Quantum Mechanics, Quantum Non-Locality, and Quantum Non-Local Consciousness (Psyche), if we are seeking a scientific explanation for what we are witnessing or observing.

The book, *Consciousness Beyond Life*, explains what we are dealing with here. I quote:

> **A single cubic centimeter of the cerebral cortex has approximately 100,000,000,000 (10^{11}) synapses.** [Remember, a synapse represents a single bit of information – OPEN or CLOSED. It's physically impossible to store petabytes of declarative memories within a single bit or a synapse.]
>
> **If one synapse contained one bit of information, brain function would require more than 100,000,000,000,000 (10^{14}) bits of information processing.** [This leaves NO room for memory storage within the brain, because all of the neurons within the physical brain are dedicated to functionality or information processing, and not memory storage.]
>
> **The required capacity for long-term memory storage, and the total data storage capacity would have to be 3.10^{17} bits/cm³, which, based on our current understanding of neuronal processes in the brain, is inconceivable.** [It's physically impossible to store 3.10^{17} bits of information within a cubic centimeter of brain matter, which can store only 10^{11} bits of information at most.] (*Consciousness Beyond Life*, p. 194.)

Remember, it's physically impossible to store 3.10^{17} bits of information within a cubic centimeter of brain matter, which can store only 10^{11} bits of information at most. There's evidence that 3.10^{17} bits/cm³ of memories or information are indeed being stored someplace; but, since that amount of information can't fit within our physical brain, then clearly those memories are being stored someplace else besides our brain.

It has been repeatedly estimated by different scientists that petabytes of information are being stored within our physical brain, which is physically impossible.

Where are you going to put a petabyte of RAM, within a synapse or a neuron or even a brain? There's no sign of such a thing! And, how are you going to transmit that petabyte of RAM from one neuron to the next, one action potential or one bit at a time, in order to keep the whole thing updated and to retain all of our retrograde memories no matter which part of our brain gets cut out or dies? Remember, a synapse represents ONE bit of information – OPEN or CLOSED. It's physically impossible to store petabytes of information within that single bit of physical machinery, which means that we are looking for a non-physical explanation for the petabytes of data that our brains seem to be storing.

Likewise, a neuron effectively represents one bit of information, too, either ON or OFF. It's physically impossible to put petabytes of RAM within a single neuron. According to Uncle Google, a neuron only has 3×10^{14} atoms available per neuron; and, they are being used for many other things besides memory storage. It's physically impossible to shove 10^{16} bits of information into 10^{14} atoms! Since it's clearly being done, somehow, that reality means that it's being done by something other than physical processes, and it is being done someplace else other than the physical brain. The math doesn't lie!

We are only talking about RAM, or memory storage, here. Where are all the CPUs within each neuron that would be needed for processing that amount of data efficiently? Where are the billions of transistors? A neuron represents a single bit of data – ON or OFF. It's physically impossible to shove billions of transistors and petabytes of RAM into a single neuron, so that it can be the computer that the Materialists and Naturalists claim that it is. If all of that RAM and CPU processing are there within a single neuron, then it's definitely not there in a physical format because it's physically impossible to have petabytes of RAM and billions of transistors within a neuron.

Clearly the petabytes of data are being stored someplace, even though it's physically impossible to do so. Clearly the CPU processing is taking place and the equivalent of billions of transistors are needed, even though there are NO identifiable CPUs and transistors within our neurons.

ALL of the command, control, CPU processing, and RAM storage are taking place outside of the neurons, because neurons represent only ONE bit of information – they are either ON or OFF. That's it! Neurons either fire, or they don't.

It's physically impossible to store petabytes of declarative memories and data within a single bit of information. It's physically impossible to store billions of transistors within a single neuron and not be able to see them. It's physically impossible to store petabytes of information within an ion receptor or a single synapse – these things are either OPEN or CLOSED. Since it is physically impossible to do these things, we have to find some other explanation for how it's being done, and for how our Life Reviews are being preserved and then re-experienced after our physical body and physical brain are dead and gone.

Anytime you find something that's physically impossible, well that's where science really starts to get interesting! Is it not?

Do you want to find out how things truly work, or are you satisfied to remain in the dark for the rest of your life?

The only answer we ever get from the Materialists, Naturalists, Darwinists, and Atheists is that "evolution did it". These people are satisfied to know that evolution did it. That's the pinnacle of their science. Are you satisfied with that explanation? When given that explanation, the next question that arises is for me is, "How does evolution do it?" Well, evolution doesn't do it. We (our psyches) do it, or it doesn't get done.

There's a story that Abraham went into his father's shop and busted up most of the idols. His father made and sold idols to the public for them to worship. That's how he made his living. When confronted by all of the damage and the lost profits from his labors, his father grabbed Abraham and demanded to know who had done all of this damage and destruction. Abraham pointed towards the wall and replied, "The Big One." Obviously, it was the big idol who had destroyed all of the other smaller idols. Evolution is the Big One. Evolution is the idol that destroyed all the others.

And we all know that evolution designed and created your brain, that evolution is now controlling and programming your brain, and that evolution is now storing all the memories within your brain. It's physically impossible to store all of your memories within your brain, so evolution must be doing the job, since nothing else can.

Do you buy it?

Millions of people do.

Long-Term Potentiation:

The Materialists, Naturalists, Darwinists, and Atheists claim that ALL of our memories are stored exclusively within our brain, and that ALL our memories cease to exist when we die. Consequently, their obligatory task is to demonstrate how this is done, since they have made the claim. Typically, these people try to shift their burden of proof over to you and make you prove them wrong; but, don't allow them to do so. Demand an explanation.

Whole chapters within our neuroscience textbooks are dedicated to Long-Term Potentiation (LTP). The Materialists and Naturalists point to LTP as the mechanism by which new memories are added to and stored within our physical brain?

So, what is LTP?

LTP is an observed physical mechanism whereby a new ion receptor can be added to a specific post-synaptic neuron thereby increasing the efficiency of neurotransmission. LTP is a physical method that God has given to Nature's Psyche to add more ion receptors to a synapse.

ALL of the Materialists and Naturalists point to LTP as evidentiary proof that all of our memories are stored within our physical brain. They say, "Look, a new ion receptor was just added to that synapse, which means that a new memory was added to the brain."

This is one of the places where we see the conflation of physical functionality (CPUs) with memory storage (RAM) most prominently displayed among the Materialists and Naturalists. These people call LTP "memory storage". These people are making the claim that an ion receptor is RAM, or memory storage; but, they are wrong. An ion receptor is nothing more than increased functionality and improved efficiency. In ion receptor is a gate opened by a key.

I just can't see how the Materialists and Naturalists are able to shove megabytes of data, information, and memories into a single newly added ion receptor; but somehow, these people do. Hope springs eternal, I guess. But, it's physically impossible to put megabytes of data and memory into a couple of newly added ion receptors. It just can't be done. An ion receptor is not RAM! An ion receptor is a gate – OPEN or CLOSED; and at best, an ion receptor represents one bit of information – ON or OFF.

Alas, these people don't understand computer science and data storage. It is physically impossible to store megabytes of new memories in a single newly added ion receptor. An ion receptor is at best a single bit of information – it is either OPEN or CLOSED. It's physically impossible to store megabytes of new memories within a single bit, or an ion receptor. It can't be done, unless it's being done non-locally or non-physically in the spirit realm.

I guarantee you, if you can figure out how nature or evolution is shoving megabytes or gigabytes of data into an ion receptor, they will give you the Nobel Prize and celebrate your name until the end of time – as long as it's a physical mechanism. If it's a non-local or spiritual explanation, they will run you out of town on a rail, or crucify you in the square for blasphemy.

So, what is LTP, if it isn't memory creation and memory storage?

Well, what do we already KNOW?

We KNOW and have demonstrated repeatedly that our physical brain is comprised of different Cerebral Processing Units (CPUs) and that our physical brain has no RAM or dedicated universal memory storage. It has been proven scientifically that our brain has no memory engrams or RAM.

Furthermore, if LTP is the cause of memory consolidation and the location of memory storage in the brain, then artificially stimulated LTP or artificially created LTP should stimulate a flood of memories every time those neurons are stimulated by an electrode; but, it doesn't. The evidence repeatedly suggests that LTP is NOT memory storage.

So, what is LTP?

Long-Term Potentiation is enhanced functionality, or improved programming. At best, LTP represents computer programming, NOT memory storage. There's a big difference there between a CPU and RAM, which is why the Materialists and Naturalists deliberately try to obfuscate, conflated, confuse, and equate the two.

Their ultimate goal is to convince you that your Psyche does not exist and that none of your memories will survive the death of your physical brain, so that you don't ever have to worry about taking responsibility for your chosen actions. Kill somebody if you like, or rape somebody if you like, it doesn't matter, because you cease to exist when you die and none of it will be remembered anyway. This is what they have chosen to believe, and this is what they want you to believe.

But, the observational evidence and experiential evidence from Near-Death Experiences (NDEs), Shared-Death Experiences (SDEs), Out-of-Body Experiences (OBEs), Theophanies, and Non-Local Life Reviews tell us clearly and conclusively that the Materialists, Naturalists, Darwinists, Nihilists, Behaviorists, Determinists, Physical Reductionists, and Atheists are WRONG.

Our choice is either to go with the convenient fiction or the observational evidence; but, that's a choice you are going to have to make for yourself. Nobody else can do it for you.

Too Many Black Boxes

I'm a theoretician. I notice trends and consolidate them. It's just what I do.

I upgraded my science to Science 2.0

Science 2.0 allows all of the evidence into evidence, and then pursues a preponderance of the evidence. Nothing is considered out-of-bounds or off-limits when it comes to Science 2.0. It's a new and better way of doing science. It's the way that science should have always been done, but wasn't.

Science 2.0:

https://www.amazon.com/dp/B0771K6WTX

In this essay, my primary focus is Quantum Neuroscience and Quantum Mechanics. I want to discuss Black Boxes.

Black Boxes are any complex piece of equipment with hidden internal mechanisms, unknown and unknowable composition, or contents that are mysterious to the user.

The next article I wish to discuss delves into theoretical neuroscience. That's right down my alley. I'm a theoretician. That's just the way my mind works. I'm theorizing and thinking about everything all the time. Nothing is off limits for me anymore. That's why I had to abandon Materialism, Naturalism, Darwinism, and Atheism – because they are absolutely worthless for explaining how things are being done. My interest is in trying to figure out how everything works; and, you can't do that with Materialism, Naturalism, and Classical Physics.

During my research into Quantum Neuroscience, I came across a different method besides Long-Term Potentiation (LTP), which the neuroscientists are using to explain how memories are stored within the physical brain. They call this model, "Distributed Memory". Sounds good, until we start to break it down and examine what's really going on. Before long, we start to run into way too many black boxes. The residual Materialist and Naturalist within me doesn't like that. I experienced a ton of cognitive dissonance when I first read this article. Things just didn't add up for me. I didn't know it nor recognize it at first, but they were using way too many black boxes to get the magic done.

A Model of a Distributed Memory

Historically, most of the progress in neuroscience research has come from experimental studies. But today, theoretical neuroscience is playing an increasing role, and the use of computational models of neural systems is widespread. In some cases, a model can provide insights about the workings of a system that are otherwise hard to gain. One area is the study of systems for learning and memory.

We'll examine a simplified nervous system consisting of three sensory neurons (the inputs) and three postsynaptic neurons (the outputs). The inputs represent patterns of activity in visual afferents in response to the faces of three different people — Eric, Kyle, and Kenny. Ignoring the complexities of visual processing, let's assume each input excites all three output neurons — A, B, and C.

Before the system learns who Eric, Kyle, and Kenny are, each of the three output cells responds at roughly the same level to each of the inputs.

There is no way to tell, from the outputs of cells A, B, and C, which face was the stimulus at a given time.

Imagine that after being exposed to the three inputs repeatedly, the synaptic effectiveness or "strength" changes.

One possibility is that after learning, Eric's face activates only cell A, Kyle's face activates only cell B, and Kenny's face activates only cell C. We can tell which face is present at any time by looking to see which output neuron is active. All the relevant information that allows the system to recognize the face — the memory — is stored in the synaptic weights.

Part c represents a nondistributed memory because all the information about Eric, for example, is stored in the single synapse with output cell A. The system can recognize that Eric is present without even bothering to look at the outputs of cells B and C.

In an alternative system, learning the three faces would again alter the synaptic weights, but none of them would be zero. The synaptic changes that store the memories can make the inputs more or less effective; memory formation does not involve only increases in synaptic strength. This is a distributed memory system because the memory of each face is stored in three synapses.

In a real nervous system, many thousands of synapses might be involved. Note that recognition of one of the input faces requires comparing the strength of activity across all the output neurons — the memory is "distributed."

One reason the distributed system seems more realistic is that recordings from cortical neurons do not suggest individual neurons are specifically responsive to every image we recognize. Presumably, human recognition is based on the relative activation of many neurons.

An attractive feature of the distributed memory system is its relative immunity to catastrophic memory loss if output neurons die. In the nondistributed scheme, recognition of Kenny totally depends on the response of cell C. If this cell dies, say goodbye to Kenny. Contrast this with the distributed system. If cell B is lost, we can still recognize Kyle by comparing the responses of cells A and C.

The more neurons and synapses that are involved in the distributed memory, the lower the consequence of losing any single cell. This relative immunity to the effects of cell loss is a great advantage. Neurons in the brain die every day, and it is probably because of the distributed nature of memory that we don't suddenly lose memories for people and events. (*Neuroscience*, p. 736.)

http://mypsyche.us/wp-content/uploads/2018/01/Distributed-Memory.pdf

http://mypsyche.us/wp-content/uploads/2017/12/BCP24.pdf

Let's tear this thing apart and see what we end up with.

I'm all for computational models in theoretical neuroscience. If you have been paying attention to any of my other books and essays, you will have noticed that I have been using computational models all the way along to make my case. In fact, while studying neuroscience, I eventually realized that our physical brain is comprised of many different Cerebral Processing Units (CPUs). Among all of my many other skills, I'm a computer scientist and a certified computer support specialist.

So, what's this "system" that they repeatedly keep pointing us to in this article?

It's a black box of some kind.

When I say that the thalamus or hippocampus is a Cerebral Processing Unit (CPU), I'm pointing you to something that is observable and actually exists. They just talk about a hypothetical "system" – a black box – leaving you to use your imagination to fill in the details. I don't like that, personally, because *scientific inferences* are logic fallacies. The Materialist or Naturalist, who remains within me, doesn't like black boxes. He likes knowing how things really work.

Whenever they talk about their hypothetical "system", they are talking about some kind of distributed network that isn't physically visible and doesn't really exist in the physical realm. It may exist in the spiritual non-local realm, but clearly it doesn't exist in the physical realm. Hence, their hypothetical "system" is some kind of black box. They try to define it in physical terms in order to trick you into believing that it's some kind of physical process which actually exists; but in reality, it doesn't exist at all on the physical level.

Where are the wires linking the individual synapses on different neurons into functional bytes of RAM for memory storage? They don't exist! Where are the wires linking the different synapses on different neurons into bytes to be used as functional CPUs? They don't exist! If these wires do exist, they aren't found anywhere in the physical brain, because they are physically impossible.

Therefore, these "wires" must be some type of Non-Local WiFi or Telepathy at the quantum level, transmitting messages, data, functionality, and programming between the different molecules and synapses within the brain. This invisible wiring between the molecules, synapses, ionotropic receptors, and neurons is some sort of Black Box that these people just assume into existence for sake of argument; but, notice how they avoid talking about it directly. Its existence is implied, rather than explicitly stated.

Now, let's "ignore the complexities of visual processing" and put it into some kind of black box, and just assume that these three black boxes produce the kind of output necessary to trigger the three different neurons that we have in mind, in our example. We don't have to explain the complexities of visual processing. We just shove it into a black box and consider it explained.

Now, we are going to make our "system", our black box, LEARN how to tell the difference between Eric, Kyle, and Kenny. How? Who is doing the learning? How does this learning take place? We don't have to explain how, because it's a black box. It simply does what we want it to do, and we don't ever have to worry about how it does it. That's the beauty of a black box. It does whatever you want it to do. It LEARNS if you want it to learn.

All of these black boxes are taken as givens or premises.

Now, we are finally going to start to use our imagination and try to "visualize" how being exposed to the three black box inputs, repeatedly, ends up increasing synaptic effectiveness or ends up making strength changes in synapses. Again, this "strength change" or "increased synaptic effectiveness" is another black box which only exists in our imagination. We don't have to explain how this was done, because we used a black box;

and, a black box can do anything we want it to do. It's all in our imagination anyway, so relax and go with it.

Now that we have set up all these different black boxes, let's imagine the possibilities.

After our black boxes have LEARNED how to tell the difference between Eric, Kyle, and Kenny; then, Eric's face activates only cell A, Kyle's face activates only cell B, and Kenny's face activates only cell C.

Who chooses which face activates which cell? How does the neuron know which face is currently being viewed by the brain?

The black box, of course. I want black boxes, lots and lots of black boxes.

We can tell which face is present at any time by looking to see which output neuron is active. All the relevant information that allows the system to recognize the face — the memory — is stored in the synaptic weights.

Presto!

And there you have it.

Gigabytes of RAM stored in a synapse!

How?

Who is constructing all these different synaptic weights?

It doesn't matter. It's a black box, and black boxes can do anything.

ALL of our memories are stored in the synaptic weights – another black box. They don't have to explain how this is done or how the synapses are communicating or interacting with each other in order to establish these "weights", because they have used another black box in order to get the job done. They also don't have to explain how memories are being storied within the synapses or the neurons, because they have used a black box to explain it.

They also don't have to explain how each neuron now recognizes a different face. That explanation is supplied by three more black boxes.

Who teaches and then remembers the meaning of a synaptic weight? Who determines and assigns the meaning of a synaptic weight? Who does all the synaptic reorganization to establish these synaptic weights?

Another black box, of course.

So, where are all the CPUs, image recognition software, data, cache, RAM, hardware, intelligence, and functionality being stored in order to make these different synaptic weights possible?

In black boxes, obviously.

All the information about Eric, for example, is stored in the single synapse.

That's physically impossible, because a synapse at best represents one bit of information – either OPEN or CLOSED. A synapse is not RAM. It's a gate. But, that doesn't matter, because this "single synapse" is another black box. It can do anything we want it to do. So let it be written, so let it be done. We don't have to explain any of it, if we are using black boxes.

The "system" or black box also has recognition capabilities! Why? How? Because we said so! It's a black box, dummy. It does whatever we tell it to do. This black box KNOWS the answer automatically, and doesn't even bother to look at the outputs of cells B and C. It just knows these things.

Now that we have all of these black boxes in place doing their job, we are going to go for pay dirt!

In an alternative system, learning the three faces would again alter the synaptic weights, but none of them would be zero. The synaptic changes that store the memories can make the inputs more or less effective; memory formation does not involve only increases in synaptic strength. This is a distributed memory system because the memory of each face is stored in three synapses.

We start first with our black boxes who have LEARNED the three faces. Now we are going to alter the synaptic weights, or those other black boxes.

Why would it change? What's its motivation to change? Who is deciding that the synaptic weights are no longer going to be zero? Who is going to change the synaptic weights? How is this rebalancing accomplished? Who is going to implement this alternative system?

Why, it's another black box of course!

Meanwhile, the synaptic changes that store all these memories are also black boxes, because it's physically impossible to store megabytes of memories in a new neurotransmitter receptor and/or a new ion channel! But, you can store anything you want in a black box!

Now the memories of each face are stored in three different synapses (black boxes) by three different synaptic weights (black boxes). That's physically impossible; but, there you have it. It's totally possible because we used black boxes to get the job done. You can do anything you want with black boxes.

Who is doing this distribution of memories? Who chooses where to distribute them? Who determines the outcome and the respective weights? Who is doing all of the synaptic reorganization? How does this thing know which three synapses are storing that specific distributed memory? How does it know that it's three synapses, and not thirty or three thousand? We don't have to ask these questions, because we used black boxes to answer these questions. It's just done the way we want it done, because we said so, because we used black boxes.

There's too many black boxes in this Distributed Memory Black Box.

In the nondistributed scheme, recognition of Kenny totally depends on the response of cell C.

Who assigned Kenny to cell C? Who recognized Kenny in the first place so that it could assign Kenny to cell C?

We don't need to ask nor answer these questions because it was a black box who did all of this for us. Black boxes can do anything.

Memories should be lost when neurons die, but our memories aren't lost when our neurons die.

Why?

It's because our memories are stored in black boxes. You can't destroy memories that are stored a black box. It's physically impossible! These memories are stored non-locally in the cloud in black boxes, where they can never be destroyed nor eliminated.

In a distributed memory system, recognition of one of the input faces requires comparing the strength of activity across all the output neurons.

Who is comparing the strength of activity across all of the output neurons? It's required, after all, because they said so. How does it know which output neurons to compare, and which neurons to leave out of the comparison? Who is doing all of this recognition and comparison? How is this distributed (telepathic) recognition and comparison being done?

A BLACK BOX, of course. It's magic!

There's way too many black boxes in this thing! It's NOT a physical explanation at all.

The response that I have gotten online from the Materialists, Naturalists, Darwinists, and Atheists whenever I have done similar analyses is that I don't understand science and that I don't understand how these things work.

Ya think?

Of course I don't understand how these things work! They are black boxes. It's physically impossible to understand how a black box works!

These people are satisfied with this explanation for how our memories are being stored within our physical brain. It makes sense to them. Why are they satisfied? It's because they started with the pre-chosen conclusion that ALL of our memories are stored exclusively within our physical brain; consequently, it must be so.

Their pre-chosen conclusion is yet another black box!

Every time these people introduce a black box into their science, they are talking about something that ONLY exists at the quantum level, or spiritual level, or non-local level. It doesn't exist at the physical level; otherwise, they wouldn't need to use a black box to describe it. The Materialists and Naturalists refuse to think about this kind of quantum level stuff, because it falsifies Materialism and Naturalism – preferring instead to replace anything at the quantum level with a magical black box that they unilaterally define as being physical in nature and origin. In other words, they cheat and deceive themselves, in order to get what they want!

Let's summarize Distributed Memory, because it's extremely important to understand what this fictional story really is at the physical level. Let's start with the facts.

PRINCIPLES OF SYNAPTIC INTEGRATION

Most CNS neurons receive thousands of synaptic inputs that activate different combinations of transmitter-gated ion channels and G-protein-coupled receptors. The postsynaptic neuron integrates all these complex ionic and chemical signals and gives rise to a simple form of output: action potentials.

The transformation of many synaptic inputs to a single neuronal output constitutes a neural computation. The brain performs billions of neural computations every second we are alive.

As a first step toward understanding how neural computations are performed, let's explore some basic principles of synaptic integration. Synaptic integration is the process by which multiple synaptic potentials combine within one postsynaptic neuron.

The Integration of EPSPs

The most elementary postsynaptic response is the opening of a single transmitter-gated channel. Inward current through these channels depolarizes the postsynaptic membrane, causing the EPSP. The postsynaptic membrane of one synapse may have from a few tens to several thousands of transmitter-gated channels; how many of these are activated during synaptic transmission depends mainly on how much neurotransmitter is released.

EPSP Summation.

The difference between excitatory transmission at neuromuscular junctions and CNS synapses is not surprising. The neuromuscular junction has evolved to be fail-safe; it needs to work every time, and the best way to ensure this is to generate an EPSP of a huge size. On the other hand, if every CNS synapse were, by itself, capable of triggering an action potential in its postsynaptic cell (as the neuromuscular junction can), then a neuron would be little more than a simple relay station.

Instead, most neurons perform more sophisticated computations, requiring that many EPSPs add together to produce a significant postsynaptic depolarization.

This is what is meant by integration of EPSPs.

EPSP summation represents the simplest form of synaptic integration in the CNS. There are two types of summation: spatial and temporal. Spatial summation is the adding together of EPSPs generated simultaneously at many different synapses on a dendrite. Temporal summation is the adding together of EPSPs generated at the same synapse if they occur in rapid succession, within about 1–15 msec of one another. (*Neuroscience*, pp. 122-124.)

https://canvas.brown.edu/courses/851434/files/40426574/

http://mypsyche.us/wp-content/uploads/2017/12/BCP5.pdf

This is how your brain really works at the physical level!

ALL of the synapses on a single neuron are producing either Excitatory Postsynaptic Potentials (EPSPs) or Inhibitory Postsynaptic Potentials (IPSPs). ALL of the input coming into a single neuron, ALL of the EPSPs and IPSPs, integrate into a single OUTPUT that's called an Action Potential.

Do you understand what that truly means?

It means that ALL of the input, from the thousands of synapses on a neuron, gets compressed or reduced to a single BIT of output! The brain is a consolidator, not a diffusor. The brain isn't a CPU or RAM. A brain is simply a switch that turns physical functions ON or OFF. That's it! This means that someone outside of the brain has to be throwing the switches.

Can you understand now why I experienced a ton of cognitive dissonance and confusion when I first started to understand how this thing really works at the physical level? It became obvious and clear to me that the brain doesn't operate and can't operate as advertised – as functional RAM and CPUs at the physical level.

A neuron's default state is OFF.

Collectively as a single unit, all of the incoming EPSPs integrate or sum together to effectively depolarize a neuron, moving that neuron towards a threshold. Once the threshold is reached, the neuron turns ON and FIRES. A neuron is a switch. A neuron is either ON or OFF. That's it! That's the ONLY information, data, or memories that is getting sent through a neuron at the physical level – a single BIT of information that is either ON or OFF.

Are you already thinking what I'm thinking?

"That ain't gonna work!" You can't transmit a message or a memory like "Hello World" through a single hardware BIT – that's physically impossible. You also can't distribute gigabytes of memories through a single hardware BIT, even if you have a hundred billion hardware BITS, or switches, or neurons to work with. It's physically impossible.

It's physically impossible to transmit petabytes of data, or even a single byte of data, through a single BIT or switch that we call a neuron. And, it is physically impossible to store petabytes of memories within a single BIT or switch that we call a neuron. A hardware BIT is not a black box – it is physical, and it is real, and it is extremely limited. A BIT is limited to being either ON or OFF. That's all there is to it.

If the neurons in our brain are in fact connected together in some sort of Distributed Memory System, NONE of that functionality is taking place at the physical level because it's physically impossible. The ONLY way that Distributed Memory can be done within our brains is if it is being done wirelessly at the quantum level by Nature's Psyche through quantum mechanisms, such as WiFi or Telepathy between neurons and synapses.

Likewise, NO message, and NO memory, survives passage through a synaptic cleft. It's physically impossible. A synaptic cleft scrambles any message or memory that comes its way, at the physical level.

Because these people don't understand computer science and don't understand how these things really work, the people who created the Distributed Memory Model turned synapses into magical black boxes that store memories. But, that's physically impossible at the physical level, because ALL of the input from the synapses on a neuron is being integrated into a single Postsynaptic Potential! The neuron either fires, or it doesn't. At the physical level, ALL of those BITS of incoming information are being reduced to a single BIT of output called an Action Potential. It is summation and integration that we are looking at, when it comes to the physical level – and not distribution. At the physical level, we have Integration to a single BIT, and NOT Distributed Memories! Distributed memories are impossible at the physical level.

Do you understand the significance of this?

It's proof of concept for Quantum Neuroscience.

Clearly, all of our memories are indeed being distributed and stored throughout our brain. It's just NOT taking place at the physical level. It's taking place at the quantum level instead.

Obviously, our brains are processing terabytes of information, data, and programming code; and, our brains are storing petabytes of memories – just not at the physical level. It's all taking place at the quantum level, outside our conscious awareness.

At the physical level, ALL of that quantum processing and quantum memory storage gets reduced or integrated into a single BIT of physical information called an action potential, which triggers a single physical action. Once the Human Psyche makes a decision or a choice, Nature's Psyche triggers or fires THE specific neuron(s) that will make that action take place. According to the Orthodox Interpretation of Quantum Mechanics, it is Nature's Psyche who collapses the wave function thereby making the Human Psyche's requests and choices REAL and ACTUAL at the physical level.

Likewise, ALL of that vision or "video input" that our physical brain receives is being combined into a single image within our occipital lobe. The ONLY image being stored within our brain at the physical level is the current image. We don't have instant replay, because nothing else besides the current image is being stored and processed within our physical brain.

Video is not being stored within our physical brain – just the current image. If there is any video being stored for subsequent retrieval in an after-death Life Review, it is being created by Nature's Psyche or the Angels in Heaven, and it is being stored non-locally in God's Database for later use. Not storing anything within our physical brain in terms of video frees up a ton of space, allowing Nature's Psyche to map that single visual image onto the occipital lobes and to map other sensory input onto specific neurons, and allowing Nature's Psyche to map out our motor functions for future physical functionality. Most of our brain is dedicated to physical functionality as it should be. Everything else happens non-locally in the cloud.

Distributed Memory FAILS as a physical model at the physical level, because it is comprised of way too many Black Boxes. Each one of those Black Boxes represents something that is happening invisibly at the quantum level. However, Distributed Memory works perfectly at the quantum level, because it is a quantum process or a quantum mechanism.

Remember, Distributed Memory is physically impossible. It is better classified as black ops, or Quantum Mechanics instead.

Let's say for the sake of argument that there is a physical mechanism for storing our memories within synapses. It's still NOT going to work without Nature's Psyche to make it work! Distributed memories would have NO meaning, unless Nature's Psyche gives them meaning. Nature's Psyche would have to decide which synapses to distribute those memories among, remember the starting synapse, the ending synapse, and the sequence in between. Even if Distributed Memory did exist at the physical level, it would still have to be controlled, implemented, and remembered at the quantum level by Nature's Psyche. Information, memories, and even Distributed Memory are completely worthless unless they mean something to Someone Psyche. Distributed memories are completely worthless unless they have been organized and stored by Someone Psyche for later retrieval by Someone Psyche. You have to KNOW where the memory starts, where it ends, and the whole sequence in between, or Distributed Memory is worthless. Someone Psyche is needed to provide purpose, organization, structure, memory, and meaning!

The Atheists ridicule and mock faith-heads, never once realizing the monumental amount of blind-faith and ignorance that they themselves are operating under. I KNOW, because I used to be a Materialist, Nihilist, and Atheist. It dumbed me down, until I learned to let it go.

These people are serious about all of this. They really truly believe that their Model of Distributed Memory explains how our memories are being stored within our physical brain at the physical level.

But, I didn't see it that way, obviously. There's way too many black boxes.

Instead, this Distributed Memory Model explained why we desperately need Nature's Psyche, Quantum Mechanics, Non-Local Processing, and Non-Local Storage to play the role of all the different black boxes that they used in their model. In fact, their model points me directly to the fact that our memories and functionality aren't being stored exclusively within our physical brain but mostly in the cloud instead. Nature's Psyche and Quantum Mechanics are needed to perform the role of each one of the black boxes that they introduced in their model. That's just the way it is, because there's NO adequate physical explanation for Distributed Memory despite what the Materialists and Naturalists might claim. Black Boxes are Quantum Mechanisms, NOT physical realities.

It has to be Nature's Psyche or God's Psyche who is doing all the black ops, because the Human Psyche isn't consciously aware of all these different black box processes, nor is the Human Psyche experiencing them directly.

What do we observe and learn from all these different black boxes?

We observe that Nature's Psyche knows how to make a physical brain and knows how to operate a physical brain. Nature's Psyche has access to information that we don't have access to. It all comes down to information, knowledge, information processing, intelligence, information exchange, and information storage. Nature's Psyche has access to infinitely more information than the Human Psyche has access to.

According to the Orthodox Interpretation of Quantum Mechanics, the functionality of each one of these Black Boxes can only be fulfilled or implemented by Nature's Psyche. It's Nature's Psyche who collapses the wave function. All of this takes place outside of the conscious awareness of the Human Psyche. Each time you encounter a Black Box, you have in fact caught Nature's Psyche in the act – or God's Psyche in the act. There's no other logical explanation for a black box, because it's physically impossible to have a physical explanation for a black box, because then it would no longer be a black box if we did.

This is elementary, and essential.

This science is solid and convincing, because I used the scientific observations and scientific discoveries from the Materialists and Naturalists in order to make my case and prove my point. I didn't use any hocus pocus, smoke and mirrors, or black boxes to make my case.

It really is physically impossible to transmit petabytes of data through a single hardware BIT. It really is physically impossible to store petabytes of data within synaptic clefts, because synaptic clefts were designed to scramble or randomize every memory and message that comes their way. It really is physically impossible to store and process terabytes of hardware and software within the physical brain. It really is physically impossible to store petabytes of memories within synaptic clefts within the physical brain. Thanks to the randomizing effect of the synaptic cleft, it really is physically impossible for the 100 billion stand-alone neurons or switches in our brain to function unitedly as CPUs and RAM at the physical level.

Oh, I don't have any doubt that our memories are being distributed throughout our brain among the atoms of our brain. It's just not happening at the physical level, because it is physically impossible. How would you go about putting your memories into a single atom, if I were to give you the assignment to do so? Well, you wouldn't be doing it at the physical level, because it's physically impossible. You would have to do so at the quantum level, by converting your memories into Quantum Waves, and then storing those waves within that physical atom.

I made my point, and I used Materialism and Naturalism to do so.

This is a powerful discovery – a paradigm shift; and, it was made possible by taking Materialism and Naturalism seriously. It ends up being proof of concept for Quantum Neuroscience.

Black boxes abound!

A careful examination of Long-Term Potentiation (LTP) will reveal that it has a few black boxes of its own. Explaining memory storage within the human brain requires many different black boxes in order to get the job done. Watch them in your college textbooks, how they gush about LTP, thinking that they have found the smoking gun where ALL of our memories are being stored. But, it's a Black Box.

Why?

Well, LTP is a physical process that adds a new neurotransmitter receptor to a postsynaptic dendrite. It's physically impossible to shove megabytes of new memories into a brand new NMDA Glutamate Receptor or Calcium Ion Channel. It can't be done. It's physically impossible. An ion channel is a gate – OPEN or CLOSED – and therefore represents one bit of information. It's physically impossible to shove megabytes of information into a single bit. A newly added ion channel is not RAM, therefore it can't be new memory storage. The only way a new ion channel could be RAM is if it were linked into non-local RAM or telepathic RAM by Nature's Psyche. Anything that is done by Nature's Psyche is by definition, in principle or practice, a Black Box; and, it's physically impossible.

Since the Materialists and Naturalists say that it's being done – that new memories are beings stored in a new neurotransmitter receptor – they are employing a black box to get the job done, because it's physically impossible. Anytime the Materialists and Naturalists try to do something that is physically impossible – like shoving megabytes of newly formed memories into a neurotransmitter receptor that's going to be added sometime in the future – they are using a black box to work their magic.

Where are these newly acquired memories stored while the new ion channel is being made and added to the post-synaptic neuron? In a black box of course! And then later, after the new neurotransmitter receptor is added to the postsynaptic dendrite, who transfers those megabytes or gigabytes of new memories to that newly added ion channel? Another black box, obviously. And, where are all those megabytes of new memories being stored within that new receptor? In a black box, dummy! You can store anything you want and as much as you want in a black box.

LTP is comprised of a lot of different black boxes, who allegedly conspire together to create new physical RAM or new memory storage within the physical brain. We can see from the results that our memories are being stored someplace; but, we have to employ black boxes in order to explain how that is being done, because there are NO physically observable memory engrams or RAM within our physical brain.

Karl Lashley and others discovered that our physical brain doesn't have any memory engrams or RAM. Our brains don't have any dedicated memory storage. It doesn't exist. Well, that's a problem, isn't it? How are you going to resolve that one?

Karl Lashley was a Behaviorist and Materialist; therefore, he was forced to start with their pre-chosen conclusion that ALL of our memories are stored exclusively within our physical brain, and then interpret every subsequent discovery accordingly.

Lashley discovered that our brain doesn't have memory engrams, or RAM, or consolidated memory storage; but, because of his pre-chosen behavioristic bias, he was forced to concoct a physical explanation for memory storage anyway.

His theories of Mass Action and Equipotentiality were the result.

Mass Action: memory cannot be localized to a single cortical area, but is instead distributed throughout the cortex.

We have already established that Distributed Memories are the product of many different black boxes.

https://en.wikipedia.org/wiki/Mass_Action_Principle_(neuroscience)

Equipotentiality: The apparent capacity of any intact part of a functional brain to carry out the memory functions which are lost by the destruction of other parts of the brain. In other words, the brain can co-opt other areas to take over the role of the damaged part.

But, where does the brain store the "damaged memories" while it is co-opting other parts of the brain to store those memories that have been damage? In a black box of course!

https://en.wikipedia.org/wiki/Equipotentiality

Mass Action and Equipotentiality are basically the theoretical foundation for Distributed Memory within the physical brain. The previous Model of Distributed Memory was based upon Lashley's Mass Action and Equipotentiality; and, you already have some idea how many black boxes were used to make Distributed Memory seem realistic.

Mass Action and Equipotentiality are Black Boxes! They are the means by which all of our memories are stored or distributed throughout all of our neurons within our physical brain. So long as you have a single neuron within your brain, you still have access to the petabytes of data that are being stored within your brain. You can store as much as you need in a black box.

We don't have a logical physical explanation for how these things work; but somehow, all of our memories are telepathically distributed throughout all the neurons of our brain, so that you can cut out huge chunks of the brain and all of your memories will remain. Of course, that's physically impossible, which means that we have to use some kind of black box to explain how it is being done.

Clearly, all of our memories are being stored within our brain.

How is this done in actuality?

We don't have to explain how it is done, because we are using a black box to explain it. Clearly, it's being done, because it's clear that our memories are being stored someplace; and, since Materialism and Naturalism are the only acceptable answer, it's clear that that "someplace" has to be within our physical brain. But, that "someplace" is in fact a black box, because distributing memories telepathically is physically impossible.

How do we know that we are dealing with black boxes or the physically impossible?

Usually, common sense will tell us. Even the Materialists and Naturalists recognize it and realize it, whenever we are using some kind of black box or psyche to explain things. The thing they can't recognize is when they are doing it themselves.

We KNOW that it's physically impossible to store petabytes of data within our physical brain; and, we KNOW that it's also physically impossible to store terabytes of RAM within each of our neurons so that all of our memories can be distributed equally throughout our brain. We KNOW that something non-physical or "black box" has to be storing our memories, and then subsequently distributing them equally throughout our brain.

Furthermore, where are the memories stored, from the sections of the brain you cut out, before being re-distributed throughout the whole brain after you have cut out chunks of the brain? Within a black box of course! How are petabytes of RAM being stored within each and every neuron of our brain in this Distributed Model of memory storage? It's being done, obviously, so there must be a way that it's being done. Enter the Black Box; and, find out how it is done.

Since there is no RAM or no memory engrams within the physical brain, you can't store any memories within them. So, what do you do? Since you can't shove all the memories into a few neurons of RAM, then you have no choice but to telepathically and telekinetically spread the memories out, throughout all the neurons in the brain, which is physically impossible. But, it's a black box, which makes it possible. You can store anything you want in a black box, even the physically impossible.

Interesting, is it not?

Round, and around, and around we go, where it stops, nobody knows.

Memory storage is possibly the most interesting, and mysterious, subject that we can study in Neuroscience, because it's physically impossible to put petabytes of RAM into a physical brain.

Engram: a general term that refers to the unknown change in the brain that is responsible for storing memory.

That's the definition for a memory engram that was given in my neuroscience class. An engram is invisible memory storage in the brain. Does any of this sound familiar?

You guessed it! It's a black box.

We don't know how it's done, but clearly it's being done; so, we are going to call it an "engram" and make it responsible for storing all the memories in our brain. Even though it's physically impossible to store petabytes of RAM within a physical brain, clearly it's being done; so, we are going to call this thing an "engram" in order to explain how it's being done. Q.E.D. Asked, and answered.

Do you need to store petabytes of data?

Use an engram!

It can hold as much as you want to put into it!

Do you need to store googolplexes?

Use an engram!

It can hold as much as you want to put into it!

How's that possible?

It's a black box.

You can put anything you want into a black box, and as much of it as you want. You name it. You got it! It can store as much as you want it to store, and even more. But, don't call it a black box, call it an engram. People will believe you if you call it an engram.

Memory consolidation is a process by which memories are transferred from a mode of short-term storage to a mode of long-term storage.

What are these "modes"?

You guessed it.

They are black boxes.

Scientists don't even know where short-term storage or working memory is taking place in the brain. They guess that it's being done within the hippocampus and/or the pre-frontal cortex in some kind of distributed manner.

Karl Lashley and others have proven that our brains have NO long-term memory storage, or no RAM. Our brains don't have consolidated memory storage. If any memories are being stored in our brain, they are being stored in a distributed fashion, through mechanisms that the scientists can't see nor understand.

Memory consolidation is another black box!

First of all, the memories are distributed telepathically throughout the short-term "working storage" parts of our brain. It has to be done telepathically, because it's physically impossible to transfer gigabytes of coherent data through the ON-OFF bits of synapses. How do you tell the synapse when the message has started and when it has ended? A black box, obviously.

A synapse is a bit, a gate – it's either OPEN or CLOSED. Synapses are part of the overall CPU, Cerebral Processing Unit. They're not RAM or memory storage. You can't push gigabytes of data through a single bit without some kind of external Command and Control. Someone psychic has to be coordinating the distribution of our memories throughout the parts of our brain that are doing work on short-term memory; and, someone psychic or non-local has to be storing those memories non-locally so that it can distribute them throughout the various neurons in the brain that need them. It's some kind of black box doing all of this Command and Control, and Non-Local Storage.

Then, when Nature's Psyche decides what parts of your memories that you want to keep, it telepathically transfers those memories to long-term storage. So, where is this long-term storage located? There's no signs of it within our physical brain. Our brain doesn't have RAM or consolidated universal memory storage. Thank goodness! It's physically impossible to shove petabytes of RAM into a physical brain. Imagine what would happen if we tried to do so.

Buying a petabyte of storage for YOURSELF?

First, you'll need a fridge,

Plus a few hundred thousand quid, a reinforced floor...

http://mypsyche.us/wp-content/uploads/2018/01/Putting-a-Petabyte-of-RAM-into-Your-Brain.pdf

https://www.theregister.co.uk/2013/01/09/storagebod_7_jan/

Obviously, there are petabytes of data being stored within your physical brain in long-term storage. It's physically impossible, but it's being done. So, how is this being done? It's some kind of black box, obviously.

You can store as much as you need in a black box.

This is black ops that we are talking about here. It's all taking place completely outside our conscious awareness. The Human Psyche isn't even aware that these things are taking place. Petabytes of memory consolidation within our brain just is. It's a black box, because it's physically impossible.

Some people have suggested that memory consolidation is the transfer of conscious memories into unconscious memories. By definition, the unconscious exists outside of our conscious awareness. The Human Psyche isn't aware of the unconscious. The "unconscious" is Nature's Psyche – another black box.

Memory consolidation is an extremely fascinating subject to study, because it's physically impossible. Memory storage and memory consolidation is one of the most fascinating aspects of Quantum Neuroscience.

Transcription involves copying a gene from DNA onto a messenger RNA molecule, for future use in protein synthesis or gene expression. Transcription is a huge black box. The transcription factors, or transcription enzymes, receive telepathic commands telling them to go make a new protein. The Materialists and Naturalists never explain how this is done, because it's a black box. They don't have to explain it, if they use a black box. A black box can do anything.

The transcription factors are told what part of the DNA to unzip and which gene to read, in order to make a messenger RNA (mRNA) molecule. There's information exchange taking place here that's physically impossible. Nature's Psyche is telling these transcription factors where to go and what to do when they get there.

Nature's Psyche knows that the cell needs a new protein and where that new protein is needed. Nature's Psyche also knows what gene codes for that protein and where that gene is located. Nature's Psyche also knows how to issue the commands to the specific transcription factors who will be assigned to get the job done. At times, Nature's Psyche is prescient and proactive. It's way too efficient to be otherwise. All of this command and control is taking place invisibly or telepathically under the watchful eye of Nature's Psyche, who knows what needs to be done and when it needs to be done. It's all being done in a black box of some kind. This Black Box is Nature's Psyche, and Quantum Mechanics comprise the mechanisms that this Black Box uses to get the work done.

How does the mRNA get out of the cell nucleus, past the nuclear membrane, into the cytoplasm, and then on to the ribosome? Well, it can't be done through random diffusion, Brownian motion, entropy, and Classical Physics! It requires some kind of black ops to get the job done!

Even when the scientists can see something physical, like mRNA, that messenger molecule is receiving telepathic instructions telling it which ribosome to go to, where that ribosome is located, and how to get there. Even when the scientists can clearly see transcription factors in action, those transcription factors are receiving telepathic commands from Nature's Psyche telling them which part of the genome to unzip and copy. Eventually, the whole thing reduces to a black box; and, anything that reduces to a black box has reduced to something that is physically impossible; and, Nature's Psyche is the only thing we know of that can do the physically impossible.

Nature's Psyche ends up being the various different black boxes that the Materialists and Naturalists use in order to explain how things are getting done. Anytime they admit that they don't know how "this process" is being done, they are admitting that it's some kind of black box. They can see the physical results, but they can't see how it's done, because it is a Black Box or Nature's Psyche who is doing the job behind the scene.

Activation is the process of making a transfer RNA molecule. The whole activation process of transfer RNA (tRNA) is a black box. Someone is telepathically reading the mRNA molecule (which will arrive sometime in the future at a ribosome) telling the enzymes or activation factors who are making the tRNA molecule what specific amino acid and what specific codon to add to the tRNA molecule, so that this specific tRNA messenger will arrive in the correct sequence at the correct ribosome sometime in the future when that mRNA molecule finally gets there to its assigned ribosome for processing or translation. Prescience like this is some kind of quantum black box. Prescience, planning, command, and control are happening all the time invisibly within our cells, even though it is physically impossible.

By the time the tRNA molecule finally arrives at the designated ribosome for protein synthesis, gene expression, and the translation process, all of the heavy telepathic black box work has already been done. Action at a distance, telepathy, and the Quantum Zeno Effect are physically impossible. They are black boxes. We can see from the results that the work is being done; but, we can't see how that work is being done, because the work is physically impossible and because the work is being done by Nature's Psyche in a black box that we call Quantum Mechanics.

Quantum Mechanics exists to do the physically impossible. Quantum Mechanics is black ops – stealth command; and, Psyche is the Commander. It all takes place off the radar. The Human Psyche is completely unaware of it.

The mean number of atoms per amino acid is 19.20. Uncle Google.

By definition, in principle or practice, it is physically impossible to communicate with individual atoms and make them do what you want them to do. So, if you want to communicate with a specific atom and make it do something for you, how are you going to accomplish that task?

Imagine that the right number of 20 atoms just entered your cell at different locations, and you need them to combine together into an amino acid. How do you communicate with these atoms and let them know that you want them to get together and form an amino acid? Someone has to do this in order to form the amino acids in the first place. Someone has to know what type of amino acid needs to be created. Somebody has to put the amino acids together, or they wouldn't exist. So, I assign you to do it. How do you tell these atoms to come together and form an amino acid?

A black box of course. You can communicate anything you need to communicate through a black box.

It's physically impossible for atoms to communicate with each other, so we have to find a non-physical means of communicating with them and getting them to communicate with each other.

Telepathy is WiFi at the quantum level or atomic level. Psyche is Command and Control at the quantum level or atomic level.

Atoms communicate with each other and interact with each other through forces, fields, waves, and telepathy. Forces, fields, waves, telepathy, the Quantum Zeno Effect, and Psyche – at the quantum level or the atomic level – are black boxes. We know they work, but we can't use our physical instruments to see how they work. They just do.

Electrons are more like waves or clouds than they are like physical particles. Electrons are more quantum than physical, meaning that electrons are more transdimensional than physical.

In contrast, the nucleus within an atom – comprised of quarks and gluons – seems to be more localized and physical than wave-like and ethereal. Since gluons are forces and fields, that means that it is the quarks who are solidly located and fixed here in this physical reality, thus dominating the atom and making the atom into more of a physical object than a transdimensional object.

The various electron clouds, forces, fields, waves, and gluons are black boxes. We can't detect them with our physical instruments; but, we KNOW that they are there and working, by the effect that they have on the quarks and the nucleus of the atom.

The only way to communicate with quarks and make them do things for you is through forces, fields, and waves, which means Telekinesis, WiFi, Telepathy, or Radio Waves at the quantum level, because it's physically impossible to communicate with these quarks using our physical instruments to do so.

So you, Nature's Psyche, have told these 20 atoms to come together and form an amino acid. You need the job done now, so how do you get it done now? Do you wait for Brownian motion to bring these 20 atoms together from opposite sides of the cell? NO! That could take an eternity, and still never happen.

No, you use another black box – you reach out telekinetically and force those atoms together.

Or, you let the 20 atoms communicate with each other telepathically, select an agreed upon destination, and then quantum tunnel to that destination, whereupon they can combine into the desired amino acid.

Or, you can reach out with your mind and teleport them together.

The only way to Command and Control atoms at the quantum level is through Quantum Mechanics. It's called Quantum Mechanics, because it's comprised of quantum mechanisms for getting things done at the quantum level. Psyche is the ultimate causal agent. Psyche is Command and Control at the quantum level. The only way that Psyche can get anything done at the quantum level is through Quantum Mechanics.

Using Quantum Mechanics, you have successfully constructed the desired and needed amino acid. Congratulations. It worked. Quantum Mechanics is proven and observed and replicated science. It works! It's dependable.

Once you have the desired amino acid, then you can communicate with it and tell it where it is needed and where you want it to go.

How does it get there?

Another black box!

It doesn't have a tail, so it can't swim there.

Instead, the amino acid has to reach out with some kind of tractor beam and pull itself to the desired destination; or it has to propel itself with some kind of force or field to its destination; or it simply has to quantum tunnel to the desired target location. In other words, it has to do the physically impossible in order to get to where you want it to go in a timely fashion.

You can't sit around waiting for Brownian motion to do the job, because you might be waiting for all eternity. Random diffusion never gets anything done. It only maximizes entropy; and in this case, you need syntropy, not entropy. Physical matter is entropy. Chemical evolution is entropy.

Psyche is syntropy. Psyche is eternal. Psyche can never die or succumb to entropy, because it is pure syntropy. Since we KNOW for a fact that entropy exists because we have observed its effects, we also KNOW from the Quantum Law of Complementarity and from the fact that we exist that syntropy of some kind must exist. Syntropy has to exist, or we wouldn't exist. Psyche, or non-local consciousness, is syntropy. Get used to it. It exists.

Amino acids store energy, which implies that some kind of syntropy took place during their creation. Amino acids represent an increase in order and organization – syntropy. The manufacture of amino acids, however it is done, represents the exact opposite of entropy. Entropy tells us that the manufacture or construction of amino acids through Brownian motion or random diffusion is physically impossible. So, how are the correct amino acids assembled on demand? You need a black box to do this for you. You need syntropy.

Transmutation

What about transmutation? Transmutation is the conversion of one chemical element into a different chemical element – what they call Alchemy. Alchemy definitely qualifies as "black ops". With transmutation, the scientist is converting one type of atom into a different type of atom. Transmutation is technically a quantum process and is considered to be physically impossible.

Biological Transmutation

It's theoretically possible that living organisms could do alchemy or transmutation, through low-energy nuclear reactions (LENR). With the right enzyme or catalyst, a living organism should be able to do some kind of transmutation. It's theoretically possible.

http://e-catworld.com/what-is-lenr/

A scientist friend of mine told me that chickens, who are fed NO calcium, can still produce the calcium necessary to make their eggs. This is physically impossible, so I'm naturally skeptical of this claim. Calcium is an atom, an element. I can't find verification for this claim, that chickens can manufacture calcium; but, I know that the chickens will first pull the calcium from their bones, and then they might start eating rocks in the hopes of getting the minerals that they need to do their job.

I do know where this biological transmutation claim is coming from, though:

https://en.wikipedia.org/wiki/Corentin_Louis_Kervran

How would you convert one atom into a different atom? How would you convert a potassium atom into a calcium atom? Or, how would you convert H_2O into a calcium atom? You need some kind of black box, obviously. Maybe you could design and construct an enzyme or a catalyst that would do this for you. But, how would you make such an enzyme or catalyst in the first place? Enter the black box, and find out.

From what I can tell, biological transmutation has yet to be demonstrated or proven. It's definitely considered to be Fringe Science; and of course, the Materialists and Naturalists will call it a pseudo-science and label it as being physically impossible.

However, with the right enzyme or catalyst, biological transmutation is theoretically possible.

https://www.slideshare.net/lewisglarsen/lattice-energy-llc-lenr-transmutation-networks-can-produce-goldmay-19-2012

Larsen contends that bacteria in hydrothermal vents at the bottom of the ocean could be transmuting tungsten into gold or mercury. He also observes that this process could be duplicated in laboratories, possibly leading to biological LENR.

LENR Transmutation

LENR (Low Energy Nuclear Reactions) is an observed and proven form of transmutation. LENR is based upon the weak force.

It has been observed that catalytic converters in cars are transmuting Platinum into Gold through LENR.

https://www.slideshare.net/lewisglarsen/lattice-energy-llc-production-of-gold-via-lenr-transmutation-of-platinum-in-vehicular-catalytic-converters-sept-28-2016

The same group has also developed LENR processes that will transmute tungsten into gold and other precious metals.

http://www.kitco.com/ind/Albrecht/2014-02-25-Alchemy-2-0-Low-Energy-Nuclear-Reactor-Creates-Gold-and-Platinum.html

LENR transmutation remains controversial, of course; but, I know a person who has gotten it to work – someone besides Lewis Larsen. So, LENR transmutation is a verified technology, as far as I'm concerned. You just need to make the right lattice or provide the correct environment. I don't know how much the nano-lattice is costing them to produce and buy; and, I don't know how much energy they have to put into the system. Consequently, I don't know if commercial LENR transmutation is viable or not. I just know that there are a few people out there in the world who are having a lot of fun converting tungsten into platinum or gold.

Nuclear or Quantum Transmutation

In contrast to the controversial, nuclear transmutation is a proven reality, although it was controversial at one point in time as well.

A transmutation can be achieved either by nuclear reactions (in which an outside particle reacts with a nucleus) or by radioactive decay, where no outside cause is needed.

https://en.wikipedia.org/wiki/Nuclear_transmutation

Nuclear fusion and nuclear fission are based upon the strong force.

Carbon-14 or radiocarbon is constantly being created in the atmosphere by the interaction of cosmic rays with atmospheric nitrogen. Cosmic rays are doing transmutation! It's another black box, if you ask me; but, unlike most black boxes that I have encountered, this one actually makes logical sense and I can visualize it happening naturally.

Technically, these transmutations aren't physical reactions (chemical reactions), though – they are nuclear reactions or quantum reactions. There's a difference! Physical reactions or chemical reactions take place between atoms and molecules; and, they operate under the laws of Classical Physics. Nuclear reactions require some kind of Quantum Mechanics. Quantum Mechanics is a Big Black Box; and, it can do transmutation.

Quantum Mechanics is Black Ops

Technically, the strong nuclear force and the weak nuclear force are quantum mechanisms, rather than physical mechanisms or chemical reactions. The same can be said about the other forces and fields that we are aware of.

Quantum Mechanics can do syntropy, telepathy, telekinesis, teleportation, quantum tunneling, action at a distance, transmutation, and black ops at the quantum level or the atomic level. Nature's Psyche can do WiFi, Purpose, Teleology, Intelligence, Logistics, Causation, Command, Control, Syntropy, Action at a Distance, Communication, Telepathy, Teleportation, Telekinesis, and Quantum Mechanics at the quantum level or the atomic level. All of these things happen for us automatically outside of our conscious awareness, under the Command and Control of Nature's Psyche or God's Psyche. Quantum Mechanics is supernatural.

All of these things are black boxes, because they are physically impossible. The only thing I mentioned here that's physically possible is Brownian motion, random diffusion, chaos, or entropy. The rest of these processes or mechanisms are black boxes, and we can't see with our physical instruments how they are being done; yet, we know that they are being done, or we wouldn't exist.

Fascinating, is it not?

What's also fascinating is that the Materialists and Naturalists will NEVER be able to figure out nor understand any of this, because they deliberately prevent themselves from doing so. I KNOW, because I used to be a Materialist, Nihilist, and Atheist; and, all these things were beyond me at the time.

Quantum Neuroscience is the scientific study of how the Human Psyche interacts with, controls, and survives its physical brain. In order to be successful in this endeavor, we have to figure out what Nature's Psyche is doing and how Quantum Mechanics works. The Orthodox Interpretation of Quantum Mechanics as developed by Henry P. Stapp involves the interaction of the Human Psyche and Nature's Psyche in the collapse of the wave function. It's the collapse of the wave function that makes things physically real. Before that, they are just theoretical possibilities floating in the clouds.

Black Boxes are signs of Psyche and Quantum Mechanics!

The physical brain doesn't do anything but vegetate and idle, unless commanded by the Human Psyche to do something. Aside from idling in default mode, the brain doesn't do anything except provide a means for the Human Psyche to control its physical body. Humans aren't dancing to their DNA. The human brain is dancing to the Human Psyche, or it doesn't dance at all.

Neurodevelopment, or the construction of a physical brain from a zygote, has trillions of black boxes that we encounter and have to deal with all along the way. We can see from the results that the work is being done, but we have NO idea how the work is being done, because it is a Black Box or Nature's Psyche who is doing all that work for us behind the scenes, or behind the seen.

Neurons and glial cells migrate long distances to pre-chosen destinations after their birth. They then differentiate and specialize and undergo synaptogenesis after they reach their designated destination. These cells know precisely where each synapse is supposed to go, and what type of synapse it's supposed to be, so that functional Cerebral Processing Units (CPUs) are the eventual result of their migration, organization, and orchestration. How do a bunch of neurons and glial cells get together and form a thalamus, or hippocampus, or cerebellum? Why do they form a hippocampus in that specific location, and not a cerebellum? How do they do that? They're just cells after all! What's making them cooperate with each other and coordinate with each other in some kind of master plan? It's got to be some kind of black box doing the job! There's no other explanation.

Even when these things have guidance cells to follow or chemicals to sniff out, someone put those guidance cells or breadcrumbs there in the first place, or told that specific target neuron to start emitting a specific chemical and told that specific migrating neuron to sniff out that specific chemical. All of these things are taking place within black boxes behind the scenes. The only thing we can observe with our physical instruments are the physical results being produced by all of these different black boxes, who are intelligently coordinating their actions and their work invisibly behind the scenery.

Scientists have repeatedly observed that atoms and molecules that just sit there and do nothing in a dish literally come alive with purpose and activity simply by virtue of being placed within a living cell. This is observed science!

How's that possible?

It's possible by virtue of a couple of Black Boxes that we call Quantum Mechanics and Nature's Psyche. The physically impossible suddenly becomes possible once these atoms and molecules are placed under the command and control of Nature's Psyche residing within a living cell.

Now, we can hear the Naturalists, Materialists, and Darwinists complaining, "But, any fool knows that it's the DNA in the cell that's making all of these atoms and molecules behave!"

How? Why? What's so special about DNA?

It's just computer programming, software in hardware, a computer chip. DNA is just another molecule like all the other molecules.

DNA doesn't have telepathic powers, prescient powers, telekinetic powers, and psychic powers, does it?

DNA is just the Signature of God – it is computer programming or software – it isn't alive is it?

Aren't the little molecules and critters reading the DNA and transcribing the DNA? The DNA isn't doing anything to them – they are doing it to the DNA! DNA is the puppet on the string, while the other molecules are dancing around it doing their thing.

Actually, the whole cell is a symbiosis orchestrated by Nature's Psyche and enacted through Quantum Mechanics. It all takes place outside our conscious awareness, so the

Human Psyche isn't sentient of the details; but, Nature's Psyche or God's Psyche certainly is. Nature's Psyche is the boots on the ground, where a living cell is concerned. Quantum Mechanics is the weaponry. And, the molecules are conquered, deployed, or destroyed.

It has been repeatedly claimed that Quantum Mechanics is the best proven, most verified, and most used science on the planet. Ironically, Quantum Mechanics has never been observed directly; but, its effects on physical matter is always observed! Quantum Mechanics is a black box. Its effects are physically impossible, but they happen anyway. If they didn't, we wouldn't exist.

The theory of evolution is a Black Box as well, whether people realize it or not. Chemical evolution is physically impossible. So are abiogenesis, spontaneous generation, and macro-evolution. They are physically impossible, which means that they are black boxes. Black boxes can do anything, even evolution.

However, scientists have repeatedly demonstrated and proven that chemical evolution of proteins and genes is physically impossible. Even with a God-given ocean full of the correct amino acids, you could wait an eternity for a protein to come together naturally through random diffusion and Brownian motion; and, it still might not happen. If you can't make a protein with chemical evolution, you certainly can't make a matching gene to go along with it!

Random diffusion, Brownian motion, and chance are highly unreliable mechanisms for getting things done. God designed it this way to make it clear and obvious to all of us that we need Him and His influence, in order to get things done in this physical realm. Our very existence is Scientific Proof of God's Existence, because we shouldn't be here if random diffusion, Brownian motion, chance, and Classical Physics were the only thing in existence.

"Constituent cognitive processes are simple cognitive processes that combine to produce complex cognitive processes and that are assumed to be mediated by neural activity in particular parts of the brain." This is another Black Box that the Materialists and Naturalists are using in order to explain how the brain works. It's physically impossible to send messages through synaptic clefts, because synaptic clefts are randomizers and they scramble any type of message that you might try to send through them. These people just assume the functionality into existence by placing it into a black box, and then they put all of those black boxes into your brain, so that they can say that all of our cognitive processes are being mediated by our brains. All of these cognitive processes are physically impossible; but, that doesn't matter, because these people are using black boxes, which magically make all of these cognitive processes possible – simple, complex, and constituent. You can do anything with a black box!

Now, we are ready to discuss the heart of Neuroscience – our neurotransmitter systems. Neurotransmitter systems are black boxes. They shouldn't work, but they do.

From what we can see with our eyes and physical instruments, our neurotransmitter systems are based exclusively upon random diffusion, Brownian motion, entropy, and chance. Our neurotransmitter systems shouldn't work; but somehow, miraculously, they do. I have written essays describing how and why our neurotransmitter systems shouldn't work based upon Classical Physics, random diffusion, entropy, Brownian motion, and chance. These are highly unreliable mechanisms at the quantum level. Our neurotransmitter systems shouldn't work at all, based upon the mechanisms of Classical Physics. We should be constantly and randomly seizing all the time; but, we are not. We should be hallucinating and delusional all the time, but we are not.

Why?

The ONLY way to explain it is through Quantum Mechanics and Nature's Psyche.

Nature's Psyche is using Quantum Mechanics to force our neurotransmitter systems to work right and work efficiently. In other words, we have to employ Black Boxes in order to successfully and adequately explain how our neurotransmitter systems work; but boy, does this thing have explanatory power, once you see the light!

Even with Nature's Psyche in control of our neurotransmitter systems, there are still two observed and experienced ways through which our neurotransmitter systems can get messed up and cease to function properly.

The first way is to have too little neurotransmitter in the system.

If you don't have enough dopamine in the system, you are going to experience Parkinson's Disease. If you don't have enough serotonin in the system, you are going to experience depression, anxiety, and OCD.

Alas, if Nature's Psyche doesn't have the machinery (neurotransmitters) in the first place, then it can't do anything with that machinery to enable the functioning of our physical brain. Without the machinery, nothing happens, even when Nature's Psyche is in command. If you don't have any dopamine, then you are NOT going to have the associated functionality. If you are missing lobes of your brain, you are not going to have the associated functionality; and, there's nothing that Nature's Psyche can do about it. That's just the way it is!

Dopamine and serotonin are regulatory molecules. If you have no serotonin or dopamine, then you have no regulation of your brain functions. Simple!

The second way that things can go seriously wrong in our neurotransmitter systems is if we have too much neurotransmitter in our system.

Why?

Well, think about it logically based upon what we have learned so far about black boxes, Classical Physics, Quantum Mechanics, and Quantum Neuroscience.

What are the results of having too much glutamate in the system?

Strokes!

What are the results of having too much acetylcholine in the system?

Lung paralysis, asphyxia, and death!

What are the results of having too much dopamine in the system?

Schizophrenia!

What are the results of having too much serotonin in the system?

Heart disease!

In each one of these situations, there's too much of a good thing; and, it can kill you.

If you have too much dopamine in the system, then things start to happen randomly and automatically all on their own; and, hallucinations, delusional thoughts, psychosis, and schizophrenia are the result.

These two pages explain the different neurotransmitters involved in OCD.

For OCD, we are looking at Glutamate, GABA, serotonin, and dopamine as the culprits. Either the serotonin and dopamine aren't regulating things properly due to their non-existence, or the excitatory glutamate is overactive or the GABA isn't inhibiting things enough. There's too much random activity going on – obsessions and compulsions out of nowhere!

Schwartz explained that people afflicted with OCD engage in a wide variety of problematic behaviors — compulsive hand washing, door opening, repetitive checking of ovens and doors, even repeating the same word, phrase or sentence. The cause, at a neurological level, is hyperconnectivity between two brain regions, the orbitofrontal cortex and the caudate nucleus, creating a tidal wave of unfounded mortal fear and triggering habitual response as the only way to attain calm. But the worst part is that, despite recognition that all these thoughts and behaviors are irrational, the OCD sufferer feels driven to obey them, nonetheless.

http://discovermagazine.com/2013/nov/14-defense-free-will

OCD is caused by hyperconnectivity – too much random activity and too much neurotransmission going on. Under the control of Classical Physics, the brain is generating obsessions and compulsions against our will, out of thin air, because there are too many neurotransmitters or too much activity in the system. The only solution to OCD is to rewire the brain and eliminate the hyperconnectivity.

In other words, it's the randomness, the Classical Physics, which is causing our mental illnesses, disease, and death through neurotransmitter overload. If there's too much neurotransmitter in the system painting the targets on the postsynaptic neurons, Nature's Psyche can't keep up with it and regulate it properly. The Quantum Mechanics is actually swamped by the Classical Physics in this case. The neurotransmitters are being born faster than Nature's Psyche can shoot them down.

Fascinating, is it not?

This same truth and reality applies to Lewy bodies and amyloid plaques. Lewy bodies are clumps of proteins observed in the surviving dopaminergic neurons of Parkinson's patients. Amyloid plaques are proteins that overload the brains of Alzheimer's patients. These are literally proteins without a purpose. The intracellular communication and functionality has broken down. The production of proteins has run amok. Nature's psyche didn't request them and can't use them, or they are being produced by the molecular machinery to solve a problem that they can't fix. As the neurons die, they can't clean up after themselves. Once again, the Classical Physics and bad genetics are overrunning and overloading Nature's Psyche.

Autoreceptors are fascinating black boxes to study and learn about. Autoreceptors are a type of metabotropic receptor located on the presynaptic membrane and are triggered by a neuron's own neurotransmitter. Autoreceptors are used to tell the presynaptic neuron and terminal button not to release anymore neurotransmitters. Just like postsynaptic receptors, these presynaptic receptors have to be able to draw in their own neurotransmitters, defying the laws of entropy and random diffusion in order to do so. What's truly fascinating about autoreceptors is that they somehow magically know how long they are supposed to maintain contact with that neurotransmitter and therefore how long they should block the transmitter release process. How do they do that? It is as if the autoreceptors reach out telepathically sensing the postsynaptic membrane waiting for the postsynaptic membrane to receive the transmission, and then waiting for the synaptic cleft to be cleared of neurotransmitters, before disengaging the attached neurotransmitter thereby preparing for the release of the next batch of neurotransmitters. Autoreceptors were designed to slow-down the neurotransmission process. There's lots of magic or black-boxing going on where autoreceptors are concerned. Information exchange between neurons across the synaptic cleft is physically impossible, which means that we are looking at some kind of black box, telepathy, and quantum mechanics.

An annoying little black box for me has been the interneuron. Interneurons are neurons that have short axons or NO axons at all. Their function is to integrate neural activity within a brain structure. It's the interneurons with no axons or NO output that I find interesting and disturbing – the anaxonic neurons. An anaxonic neuron is a neuron where the axon cannot be differentiated from the dendrites. Some sources mention that such neurons have no axons and only dendrites.

You can check in any time you like, but you can never leave!

The anaxonic neurons seem to be there as dead-ends to stop action potentials. The Materialists and Naturalists would tell us that the anaxonic neurons are good places to store memories; but, I don't see how, since these things have NO output or axon. These things seem to be functioning as resistors or terminators. Only Nature's Psyche knows what it's really doing with these things.

The Materialists and Naturalists call axons "wiring"; and, I have been doing the same, even though it's technically a misnomer. The axons aren't functioning as wiring in the traditional physical sense that we visualize wiring, because each axon dead-ends at a terminal button, and then switches over to chemical transmission or neurotransmitters. Technically, with the randomizing and scrambling effect of the synaptic cleft, the physical brain shouldn't work reliably at all, but it does. Clearly, when it comes to brain functionality, there's more than meets the eye. It's as if these little buggers are alive and know precisely what they are doing and when they should do it. Each one of those neurotransmitters seems to be a little black box consisting of purpose, sentience, deliberation, choice, motility, and functionality.

Voltage-activated ion channels or voltage-gated ion channels are one of the most fascinating aspects of Neuroscience.

https://en.wikipedia.org/wiki/Voltage-gated_ion_channel

https://en.wikipedia.org/wiki/Gap_junction

https://en.wikipedia.org/wiki/Ephaptic_coupling

Voltage-activated ion channels open and close in response to changes in the level of the membrane potential.

There's a HUGE Black Box associated with ion channels and gap junctions that has NEVER been answered nor solved to my satisfaction. Can Nature's Psyche open ion channels and gap junctions directly at the quantum level completely ignoring the current membrane potential? It's theoretically possible, because Quantum Mechanics is Action at a Distance or Telekinesis.

It's that FIRST action potential, that FIRST neuron that is triggered or fired by Nature's Psyche directly, which is of greatest interest to Quantum Mechanics and Quantum

Neuroscience; and, I can't figure out how Nature's Psyche does that. One of the problems is that there are many different options or possibilities for how this is done. There are five main types of voltage-gated ion channels, and Nature's Psyche could theoretically open any one of them to get the job done. You could open the ionotropic receptors in the post-synaptic membrane on the dendrites flooding the neuron with EPSPs, thereby triggering an action potential. You could also open the calcium ion channels in the pre-synaptic terminal button thereby dumping a bunch of neurotransmitter into the synaptic cleft.

You could also open the right gap junctions or electrical synapses between an astrocyte and a neuron thereby triggering an action potential directly.

Another option is for Nature's Psyche to reach out telepathically and quantum tunnel the necessary neurotransmitters needed to open ionotropic receptors on the post-synaptic membrane. Open ion channel receptors typically produce EPSPs, which eventually lead to an action potential. This would be another way for Nature's Psyche to fire or turn on the specific neuron of its choice, in response to the Human Psyche's commands. The monoamine neurotransmitters in particular have to be maintained at a perfect balance in the synaptic cleft for proper brain function (serotonin, dopamine, epinephrine, and norepinephrine); and, these are the ones that I visualize as being quantum tunneled by Nature's Psyche to the correct receptors as needed, in order to start Action Potentials from scratch.

Ephaptic coupling is a form of communication within the nervous system and is distinct from direct communication systems like electrical synapses and chemical synapses. It may refer to the coupling of adjacent (touching) nerve fibers caused by the exchange of ions between the cells, or it may refer to coupling of nerve fibers as a result of local electric fields. In either case ephaptic coupling can influence the synchronization and timing of action potential firing in neurons. Myelination is thought to inhibit ephaptic interactions. – Wikipedia

Ephaptic coupling is the idea that neurons interact with each other and coordinate with each other through various different forces and fields and thoughts. This fits in well with Quantum Mechanics and Quantum Telepathy and the Quantum Zeno Effect, because Telepathy of any kind is WiFi or wireless communication at the Quantum Level. These invisible forces and fields can influence the "synchronization and timing of action potential firing in neurons"; and, Nature's Psyche works through invisible forces and fields at the quantum level.

Transpersonal communication is psyche-to-psyche communication, telepathic communication, and spirit-to-spirit communication at the quantum level – Quantum Telepathy, Quantum Waves, and the Quantum Zeno Effect. Nature's Psyche could be using transpersonal communication or ephaptic coupling in order to trigger action potentials in the target neurons that it wants to turn on or fire.

Transpersonal Psychology is the scientific study of Spirit-to-Spirit interaction, or Psyche-to-Psyche interaction, or "trans person" interaction. Any scientific discoveries in Transpersonal Psychology would help to provide some of the evidential and observational proof needed for something like Quantum Neuroscience. However, as you can tell, I was primarily interested in the Psyche-to-Physical interaction and the Physical-to-Psyche interaction, and that means Quantum Mechanics and interaction at the quantum level. Hence, Quantum Neuroscience was born. I wanted to understand the mechanisms behind life, the universe, and everything. I wanted to understand the physically impossible, and how it is being done.

According to the Orthodox Interpretation of Quantum Mechanics, from Henry P. Stapp, the Human Psyche makes a decision or a choice; and then, Nature's Psyche collapses the necessary wave functions in order to make that choice ACTUAL and physically REAL. Ion receptors and ion channels definitely function and reside on the boundary between the quantum level and the physical level. Maybe Nature's Psyche is just collapsing the necessary wave functions in order to get specific action potentials started.

We are definitely looking at black ops and black boxes here, because most of this is physically impossible.

So, maybe instead of finding THE ONE WAY that Nature's Psyche is triggering action potentials directly, it's possible that there are dozens of different ways that Nature's Psyche can use to get this job done.

Interesting, is it not?

I find the explanatory power of Quantum Mechanics and Nature's Psyche vastly superior to anything that the Materialists, Naturalists, Atheists, Darwinists, and Classical Physics can produce. The Materialists and Naturalists have no explanation for all of these different black boxes; instead, these people simply deny the existence these things – starting with Nature's Psyche and Quantum Mechanics, of course, and then moving on to God Himself. The Materialists and Naturalists put these things into black boxes, where they can't see them and don't have to deal with them.

This is powerful Philosophy of Science here. It gets right to the heart of the issue. This is where the tires hit the pavement in theoretical physics; and, all these black boxes point us directly to Quantum Mechanics and Nature's Psyche, because all of these different black boxes are physically impossible.

Whenever you encounter the physically impossible, you have in fact encountered yet another black box. We can see from the physical results that the work is being done and that the information is being transferred and used; but, we can't see how that work is being done or how that information is being exchanged, because it's all taking place within a black box of some kind. Anytime you encounter a black box, you have caught Nature's Psyche and Quantum Mechanics in the act. These black boxes reside within the Mind of God, because they are by definition, in principle, physically impossible.

Can I do theory, or what!

Some of you are going to love this one, and some of you are going to hate it.

C'est la vie!

Nevertheless, you've gotta love all these little boxes!

https://www.youtube.com/watch?v=LM8JhvfoqdA

It didn't come naturally for me, but I find black boxes an interesting subject to study and learn about. It was an acquired taste, with a massive payday at the end.

Mark My Words

Conclusions Regarding Memory Storage

It took me a few months (or a few decades depending upon how you choose to look at it) to stop trying to salvage "physical memory storage within the physical brain" or "RAM within the brain".

I then switched over to the CPU Model and concluded that our brain is made up of independent and semi-independent CPUs (Cerebral Processing Units); and for the most part, they function completely outside of our conscious awareness. In other words, the Human Psyche isn't even aware of most of the brain's functionality and processes. Whenever the brain is damaged, functionality is lost, not necessarily the memories. In fact, the consolidated memories or conscious memories typically remain, no matter how much of the brain gets damaged or removed. It's always functionality that is lost, not memories.

The brain is comprised of CPUs, and NOT memory storage or RAM. It has been proven that our brain doesn't have memory engrams or RAM. It's possible that most of what happens within our brain isn't stored anywhere for later retrieval, because it happens implicitly or unconsciously. And, there's convincing evidence that the vast majority of our memories, especially our video memories and consolidated declarative memories, aren't stored within our brain, because it's physically impossible to store petabytes of data within our brain.

It's obvious and clear that our after-death Life Review isn't being produced by and stored within our physical brain. Even people who are born blind have video within their after-death Life Reviews; and, that video certainly was not produced by their physical brain.

Repeatedly choosing to pay attention to something – a function of the Human Psyche – seems to be the KEY to getting Nature's Psyche to file that information away for us so that we can recall it again and again as needed. Nature's Psyche stores what's most important to us; whereas, we don't seem to have access to the things that we aren't paying attention to. Attention is the KEY. Attention must be paid, if we want to remember something and make it ours. Attention is a product of the Human Psyche.

Nature's Psyche mediates and controls our memory storage, and Nature's Psyche has two places where it can store our memories – locally within our physical brain and non-locally in the cloud. Nature's Psyche periodically off-loads our physical memories onto the cloud, in order to free up space on our physical hard drive as needed. All of this memory storage, memory consolidation, memory optimization, and freeing of space on our hard drive takes place outside the conscious awareness of the Human Psyche. We don't ever have to worry about our physical brain running out of memory storage, because Nature's Psyche makes sure that it doesn't, by periodically dumping our unused memories into the cloud.

Then when I had my major epiphany, I abandoned the CPU Model and switched over to Maps of Physical Functionality. Maps of Physical Functionality are infinitely more powerful and versatile, in my humble opinion. Maps of Physical Functionality work at both the quantum level and the physical level; and, they explain what's happening at every level of existence. Maps of Physical Functionality ARE the bridge or the interface between the quantum level and the physical level.

There's a significant reason why I switched to Maps of Physical Functionality. It's physically impossible to store memories within a synapse, because a synaptic cleft was designed to terminate, randomize, and scramble any memory, message, or programming code that comes its way, integrating ALL of that potential information into a single BIT that feeds directly into a single post-synaptic potential. Neurons are switches, a single BIT of hardware. Neurons are either ON or OFF. You can't make a CPU or RAM out of a single BIT.

If our memories are indeed being stored within our brain, they are being stored or mapped at the quantum level as quantum waves by Nature's Psyche, because it's physically impossible to store memories within a synapse at the physical level.

Nootropics are smart drugs that purportedly improve memory.

Even the college textbooks written by the Materialists and Naturalists declare that nootropics are a fraudulent hoax, and that smart drugs don't actually work. Now we know why. NONE of our long-term consolidated memories are being stored within our physical brain at the physical level. If there is indeed RAM within our brain, it ONLY exists at the quantum level, because Karl Lashley and others have proven that there is no physical RAM or no memory engrams within our physical brain.

Technically, programming code or software cannot survive a synaptic cleft either, which means that if our brains have different CPUs, the functionality for those Cerebral Processing Units is being mapped and controlled by Nature's Psyche at the quantum level as well.

Maps of Physical Functionality created by Nature's Psyche at the quantum level through quantum mechanisms seems to be the way that our physical brains are actually designed, organized, and built at the physical level so that they can be controlled at the quantum level by Nature's Psyche. So far, Maps of Physical Functionality is the best way to describe the different structures of the brain that I have found so far. It's the model that has the most explanatory power and the model that takes the whole of Quantum Mechanics, Quantum Non-Local Consciousness (Psyche), and Classical Physics into consideration.

Each Cerebral Processing Unit (CPU) within our brain is a Map of Physical Functionality that was mapped at the quantum level by Nature's Psyche. Our physical input registers onto neural maps within our physical brain, and both Nature's Psyche and the Human Psyche have direct access to the Maps of Physical Functionality that register our sensory input.

Our physical output consists of switches or neurons that are mapped onto and converge onto our primary motor cortex. Our primary motor cortex is yet another Map of Physical Functionality, the map that Nature's Psyche uses to trigger our physical output. Once the Human Psyche makes a choice and decides what it wants to do with its physical body, then Nature's Psyche collapses the appropriate wave functions firing the specific neurons that switch on or trigger our physical actions.

This is how our brains really work, both at the quantum level and the physical level.

That's Physically Impossible

Have you ever wondered why Classical Physicists, such as Albert Einstein, vigorously opposed Quantum Mechanics all of their lives? It's because Quantum Mechanics is supernatural in nature and origin. Quantum Mechanisms involve faster-than-light Action at a Distance, which is physically impossible. Quantum Mechanics is supernatural; and therefore, Quantum Mechanics falsifies Materialism, Naturalism, and their derivatives.

I used to be a Materialist, Nihilist, and Atheist. All of that was brainwashed or conditioned into me; and, I returned to it naturally and easily.

For months, if not years, I HATED not having a physical explanation for how the brain works. Eventually, though, I was forced to accept the fact that that is just the way it is. NO physical memory or physical message can make its way through a synaptic cleft and survive. It's physically impossible. Everything that is sent through a synaptic cleft is reduced to a single bit of information – ON or OFF – OPEN or CLOSED. The postsynaptic neuron either fires or it doesn't. That's it! It's physically impossible to transmit a message through a synapse. It can't be done, because any synaptic cleft scrambles everything that comes its way at the physical level.

You can see my dilemma, can't you?

It's physically impossible to provide a materialistic, naturalistic, and classical physics explanation for how our brains work. I was forced to turn to Quantum Neuroscience and Quantum Mechanics in order to explain how our brains work.

The Materialists, Naturalists, Darwinists, Nihilists, and Atheists will complain and say that all of this Quantum Neuroscience junk is physically impossible.

They are right, and that's the point!

The ironic thing is that these people don't realize that it's physically impossible to prove the primary assumptions or major premises of Materialism, Naturalism, Darwinism, Nihilism, and Atheism to be true. It's physically impossible to prove that the non-physical or the spiritual does not exist. It's physically impossible to prove that the supernatural, quantum mechanics, and psyche do not exist. Chemical evolution of proteins and matching genes from scratch is physically impossible. It's physically impossible to prove that the Non-Local Realm, Quantum Realm, Transdimensional Realm, or Spirit Realm does not exist. And, it's physically impossible to prove that God does not exist.

Even the Materialists and Naturalists are pushed up against the wall and have to take a leap of faith in order to believe what they have chosen to believe. The problem is that there is absolutely NO evidence to support their chosen beliefs. At least with Quantum Mechanics, there's tons of experimental evidence and observational evidence proving that quantum mechanisms are real and truly exist. Quantum Mechanics is one of our best-proven and most-used science disciplines on this planet. The very existence of Quantum Mechanics falsifies Materialism and Naturalism, because Quantum Mechanics is supernatural. Psyche, while employing quantum mechanisms, can do the physically impossible with ease.

Occasionally, the Materialists, Naturalists, Behaviorists, and Atheists will slip up and actually tell us the truth. Here's a rare but particularly important example.

The members of a species can produce fertile offspring only by mating with members of the same species. (*Biopsychology*, p. 27.)

In other words, it is physically impossible for a species to produce and give birth to a genetically compatible Mr. and Mrs. Mutant at the same moment in time. God designed it so that a species can only produce fertile offspring by mating with members of the same species. This is the ultimate truth; and, this quote actually came to us from a Materialist, Naturalist, and Darwinist who believes religiously in the theory of evolution. But, without realizing it, this scientific fact and scientific truth FALSIFIES Darwinism and the Theory of Evolution. In other words, macro-evolution of any kind – whether it be abiogenesis, spontaneous generation, or chemical evolution – IS physically impossible. Remember, macro-evolution is physically impossible. Two dogs cannot conceive and give birth to genetically compatible male and female cats. It's physically impossible. Likewise, two chimp-like ancestors cannot conceive and give birth to genetically compatible male and female human beings. It's physically impossible. Macro-evolution NEVER happened on this planet, because it's physically impossible; and, that truth FALSIFIES Materialism, Naturalism, and Darwinism.

Design and creation by evolution has NEVER been observed, because it is physically impossible for random chance or random mutations to design and create proteins, genomes, and life forms. In fact, evolution (genetic change), random mutations, and natural selection didn't even exist until after God designed and created the proteins, genomes, and life forms in the first place.

I finally discovered that by learning to identify the physically impossible, we open ourselves up to whole new realms of science, discovery, ideas, and information that were completely invisible to us before.

I used to be a Materialist, Nihilist, and Atheist. Clearly a change was in order, if I wanted to find and know the truth about life, the universe, and everything.

So as to improve the explanatory power of my science, I upgraded my science to Science 2.0. Science 2.0 allows ALL of the evidence into evidence, and then pursues a preponderance of the evidence. Science 2.0 is a new and better way of doing science that doesn't exclude anything from consideration. It's the way that science should have been done, but wasn't. With Science 2.0, I pull out all the stops. Nothing is off-limits or out-of-bounds. ALL of the evidence is allowed into evidence.

I think my greatest discovery from Science 2.0 and Quantum Neuroscience is something that I call "The Physically Impossible". I have taken to calling these things Black Boxes, because they are physically impossible.

Quantum Neuroscience is an attempt to explain the physically impossible.

My discovery and observation is that things like Evolution, Random Mutations, Natural Selection, Classical Physics, Naturalism, and Materialism CANNOT DO quantum mechanisms like telepathy (the Quantum Zeno Effect), telekinesis (Action at a Distance), teleportation (Quantum Tunneling), teleology (intelligence, purpose, attention, intention, volition, will, design, creation, choice, decisions, and psyche), and transpersonal psychology (Quantum Sociology, an After-Life, Life Reviews, Salvation of the Soul, and Psyche-to-Psyche Communication after we are dead). Materialism, Naturalism, and their derivatives can't survive the death of our physical body and physical brain. This is what I have discovered.

This was a major discovery for me personally, because I used to be a Materialist, Naturalist, Nihilist, and Atheist. For me, this discovery of Quantum Mechanics, the Human Psyche, and Nature's Psyche was an epiphany, and a major paradigm shift in my way of thinking about science and doing science. All of this newly discovered Quantum Mechanics is based upon Pim van Lommel's "non-local consciousness" concept, and Henry P. Stapp's

Orthodox Interpretation of Quantum Mechanics. For the first time in my life, I understand Quantum Mechanics, and that increased understanding came by getting rid of My Materialism, My Atheism, and My Nihilism. For me, this was a monumental scientific breakthrough – a paradigm shift in my way of thinking.

This essay pursues the physically impossible in relationship to Neuroscience and Quantum Neuroscience.

Ever since I discovered the physically impossible, I have started to collect examples of the phenomenon. The physically impossible is comprised of Quantum Mechanics, or what I like to call Black Boxes, and mediated by Nature's Psyche under the command and control of the Human Psyche. The Human Psyche makes the requests or the commands or the choices, and Nature's Psyche collapses the wave function making it REAL, if it is Lawful and God permits it.

It's fascinating to study the physically impossible. Here's some of what I have come up with so far.

God made it physically impossible for our brains to function as CPUs and RAM. A neuron is a switch – it is either ON or OFF. Our brains are comprised of 100 billion stand-alone switches with absolutely NO wires connecting them. With no wiring, it's physically impossible to construct a CPU or RAM out of a bag-full of switches. It can't be done. It's physically impossible. Transistors and switches are worthless if they don't have any wires connecting them. Our physical brains are worthless, because they have NO wires connecting their neurons together. It's physically impossible to send a message or a memory through a synaptic cleft, because a synapse scrambles everything that comes its way.

God made it physically impossible to send a message like "Hello World" through a synaptic cleft.

How?

There are over 100 different neurotransmitters. You should be able to spell anything you want with that, and then send it from one neuron to the next!

Well, there's a major problem that they always overlook.

Each synapse produces and transmits only two neurotransmitters at most, and sometimes only one. There's one of the regular ionotropic neurotransmitters, and sometimes there is a metabotropic neuropeptide, being passed through the synaptic cleft from one neuron to the next.

The most you can spell and send through a synaptic cleft is something like ON and OFF, which is precisely what we observe happening! This isn't theory that I'm talking about here. It has already been proven and observed. The most you can send through a synaptic cleft is ON and OFF. That's it!

We observe ionotropic neurotransmitter receptors producing Excitatory Post Synaptic Potentials (EPSPs), and on other synapses on possibly the same neuron we observe ionotropic neurotransmitter receptors producing Inhibitory Post Synaptic Potentials (IPSPs). EPSPs move the neuron switch towards ON, or an action potential. IPSPs move the neuron switch towards OFF. The ONLY thing being transmitted through a synapse by ionotropic receptors is ON or OFF, telling the postsynaptic neuron to either fire or not fire. That's it!

Occasionally, we observe neurotransmitters and neuropeptides in a synaptic cleft turning ON a metabotropic receptor on the post-synaptic membrane, which releases a G-protein into the postsynaptic neuron. The G-protein alters the functionality of the postsynaptic neuron, typically by adding more ionotropic receptors to the postsynaptic membrane thereby adjusting the synaptic weight that they are always talking about. But, the most that can be transferred through a synaptic weight is either a stronger ON signal or a stronger OFF signal, depending upon whether the synapse was designed to produce EPSPs or IPSPs.

The neuropeptide, the metabotropic receptor, the G-Protein, and the additional functionality made within the post-synaptic neuron is the closest that neurons get to communicating with each other at the physical level; and, the message is simply ON, meaning that a neuropeptide docks with a metabotropic receptor, which turns a G-Protein ON so that it can alter the functionality of the postsynaptic neuron.

The neurons regulate and measure ion flow, and they have internal processes that they use to adjust the ion flow. Sometimes, a post-synaptic neuron will send an endocannabinoid neurotransmitter across the synaptic cleft to its associated pre-synaptic neuron, getting the pre-synaptic neuron to alter the amount of neurotransmitter that it releases. This is a reverse transmission; and once again, the message is ON, telling the pre-synaptic neuron to make the pre-programmed adjustments.

It is Nature's Psyche who decides which neurotransmitter and neuropeptide a specific synapse is going to produce and transmit; and, it's Nature's Psyche based upon information from God's Database who decides where to place the different synapses connecting the axon of one neuron with the dendrites of a different neuron. There can be up to a quadrillion different synapses within a human brain; and, it's physically impossible to store the 3D mapping or 3D addressing locations, and synapse typing, for a quadrillion synapses within our genome which holds only 750 megabytes of information at most. This quadrillions-of-bytes of information is being stored non-locally in the cloud within God's Database, if it's being stored anywhere. It's physically impossible to store quadrillions of bytes of information within a 750-megabyte genome.

Remember, it's also physically impossible for 750 megabytes of genome programming to spontaneously generate out of thin air. Your genome is God's Signature; and, it's written all throughout your physical body. It's a long signature, but it gets the job done.

It gets infinitely more complex than this. I had to dumb it down so that I can understand it, which is what the Materialists and Naturalists also do in the college textbooks that they write for us. No matter its source, this whole thing was quite an epiphany for me; and, it was months in the making.

Remember, the most that makes it through a synaptic cleft is ON or OFF. A synaptic cleft is a gate. A synapse is a gate. OPEN or CLOSED. That's it! That's one single BIT of information. Remember, God made it physically impossible to send a message or a memory like "Hello World" through a synaptic cleft. It can't be done, which means that it isn't being done, no matter what the Materialists and Naturalists might claim.

If you have a memory of something like "Hello World", it's not getting through a synaptic cleft, and it's not being stored within a synapse, because that's physically impossible. A synapse can only transmit ON or OFF from one neuron to the next! A synapse is a gate or a switch, a single BIT of hardware, and NOT a memory storage device, CPU, or RAM. The Materialists and Naturalists will try to convince you otherwise, but they are wrong. I've noticed that they are always wrong.

God made it obvious that we need Him; otherwise, our brains wouldn't work. It's physically impossible for our brains to work. Brain function is a supernatural process or a quantum process. Thoughts, cognitions, and memories are supernatural, not physical. Remember that! It's impossible to detect and record our thoughts, cognitions, memories, and dreams with our physical instruments. You could say that it's physically impossible, because these things are spiritual objects and not physical objects.

It's cool, huh, to finally KNOW the truth!

I have been looking for this all of my life and didn't know where to find it. I had to get God to reveal it to me, because nobody else would. Thank God he was willing to reply to my questions and concerns.

Neurotransmitter Receptors:

https://en.wikipedia.org/wiki/Ligand-gated_ion_channel

http://mypsyche.us/wp-content/uploads/2018/01/Ligand-gated-ion-channel.pdf

"Up-regulation is an increase in the number of receptors for a neurotransmitter in response to a decreased release of that neurotransmitter." How does the post-synaptic neuron KNOW that the pre-synaptic neuron isn't releasing enough neurotransmitter, so that it can add a few more receptors to get the job done? There seems to be prescience and telepathy going on here; and, that's physically impossible. Why does a neuron choose to up-regulate rather than choosing to shut-down the synapse? Shutting down a synapse is also an option that the neurons can take – it's called pruning. How does the neuron know and decide that it's supposed to save the synapse rather than letting it go? There's intelligence and decision-making that's required here, and that's physically impossible.

Even the automatic "thermostats" that were programmed into the post-synaptic neurons were, well, programmed into the post-synaptic neurons by some intelligent being or psyche. It's physically impossible for neurons to have a psyche and to be intelligent, sentient, self-aware, conscious, and knowledgeable; but, they are. How do they do that? It's physically impossible. Well, there must be something going on here besides the physical.

Our after-death Life Review and our after-life are physically impossible.

Quantum Non-Locality, what the scientists typically call "Action at a Distance", is physically impossible. Therefore, Quantum Non-Local Consciousness, or Psyche, is physically impossible. I already mentioned that telepathy, telekinesis, and teleportation are physically impossible. Phase-shifting, levitation, magnetism, gravity, and dark energy are physically impossible. Technically, light and quantum waves are physically impossible. These things shouldn't exist, according to Materialism and Naturalism; but they do exist, and their existence falsifies Materialism and Naturalism.

Our neurotransmitter systems are physically impossible. They shouldn't work, but they do. That's physically impossible. They should gum up, cause a ton of signal noise, and produce an endless number of false starts; but, they don't. When it comes to neurotransmitter systems, there's more than meets the eye. According to Materialism, Naturalism, and Classical Physics, our neurotransmitter systems should be nothing more than entropy, random diffusion, and Brownian motion. Yet, thanks to Nature's Psyche and Quantum Mechanics, our neurotransmitter systems actually work, and work well. The only time they get messed up is when evolution has given us faulty mutated genes.

Targeting, quantum mechanics, and fine-tuning are physically impossible. Physical particles like atoms and molecules, when left on their own without any influence from Nature's Psyche, can ONLY do random diffusion, Brownian motion, entropy, chaos, and Classical Physics. Take these very same atoms and molecules, place them into a living cell or a living brain under the control of Nature's Psyche and syntropy, and these atoms literally come alive with purpose, motion, activity, and energy. That's physically impossible! But, it's completely possible through quantum mechanisms and psyche.

Ions and Neurotransmitters should function according to the Laws of Classical Physics with maximum diffusion, maximum entropy, and maximum chaos and disorder because they are physical particles. But, they don't act this way while inside a living brain and a living cell. Ions and Neurotransmitters function according to the Laws of Quantum Mechanics or Transdimensional Physics while inside a living brain and a living cell. That's physically impossible! Quantum Mechanics is one Huge Black Box that's physically impossible; yet, Quantum Mechanics is the best-prove and most-used science that we have, even though it's supernatural and physically impossible.

Within a living brain, under the control of Nature's Psyche completely outside our conscious awareness, there seems to be telepathy, telekinesis, and teleportation drawing our neurotransmitters to the different receptors on the presynaptic and postsynaptic neurons. It's always described as the very model of efficiency, which is only possible if each neurotransmitter and ion has a psyche, is conscious of where it is and where it is supposed to go, and uses quantum mechanisms to get there. It should be the very model of chaos and confusion, if classical physics, entropy, random diffusion, and Brownian motion were in control of our neurotransmitters.

Classical Physics can't explain what we are observing and witnessing within a living brain, and a living cell; but, telepathy, telekinesis, and teleportation (Quantum Mechanics) certainly can.

The functionality of ions, or charged atoms, is explainable by electrostatic pressure (magnetism) and their inherent drive to move down their concentration gradient (similar to osmosis). There's actually a physical explanation for these things that makes sense, assuming of course that you choose to define electromagnetic waves as a physical process.

However, with neurotransmitter molecules, we are looking at something completely different. It's as if neurotransmitters each have an invisible "magnet" that draws them to their specific receptors and nowhere else. They even rotate and align themselves to the target receptor so that they can dock with the post-synaptic receptor! Something is going on here that we can't see nor detect with our physical instruments. It's too efficient and reliable to be otherwise. The neuroscientists actually use the words "draw" and "transport" to explain how neurotransmitters get to the correct receptors in a timely fashion. But, who or what is driving this process or drawing in the neurotransmitters? All of this is physically impossible. Magnetic propulsion or telekinesis is physically impossible, and so is quantum tunneling or teleportation. Yet, quantum mechanics under the control of Nature's Psyche makes ALL of this possible; and, it all happens completely outside the conscious awareness of the Human Psyche and its physical brain.

Each neuroscience book has chapters dedicated to each of the five physical senses – vision, hearing, taste, smell, and touch.

So, how did evolution design, engineer, fine-tune, and perfect our physical senses?

It didn't! That's physically impossible.

How do we know?

WE KNOW because chemical evolution, random mutations, genetic recombination, and natural selection don't have any senses of their own to fine-tune, use, and perfect. You can't fine-tune, master, and perfect something that you don't have. It's physically impossible. Random mutations aren't using any physical senses to do their thing, and neither is genetic recombination. They are running blind, or they are using telepathy to get the job done. Telepathy is physically impossible.

Randomness by definition in principle or practice produces entropy, disorder, chaos, and death – NOT syntropy, order, fine-tuning, and life. Chemical evolution, spontaneous generation, or abiogenesis allegedly took place before there were any physical senses. Chemical evolution is never going to fine-tune and perfect your visual system, when it doesn't have one of its own to experiment with. And, natural selection is a lame duck and never did anything significant for us, except get you killed or selected against before you had time to copulate. There are NO creative powers within natural selection. Natural selection does extinction of species, NOT origin of species. Design, creation, and fine-tuning by evolution is physically impossible. It didn't happen, because it couldn't happen. It was physically impossible.

Almost every neuroscience and psychology textbook that I own tells us that evolution designed, created, and produced your genome, your eyes, and your brain. But, that's physically impossible!

How do we know?

WE KNOW because evolution (genetic change), random mutations, and natural selection didn't even exist, until after God designed, created, and produced the genomes and the living cells in the first place. Remember, evolution (genetic change) didn't exist, until after God designed and created the genomes, eyes, and brains. It's physically impossible, for something that doesn't exist yet, to design and create something that will someday exist in the future. Someone had to design and assign all of this functionality, because it's physically impossible for it to spontaneously generate out of thin air. Design and creation by chemical evolution, random mutations, and natural selection is physically impossible. This is Logic 101.

The physically impossible is the most interesting science to study and learn about; but, most people have chosen to believe that it doesn't exist, so they have nothing to study and learn about.

What would be your list for the greatest scientific discoveries of all time?

At the top of my list are the physically impossible!

The greatest discovery in the 19th century came in 1859, when Louis Pasteur discovered that spontaneous generation, abiogenesis, and macro-evolution are physically impossible. Louis Pasteur falsified Materialism, Naturalism, and the theory of evolution the very same year that Charles Darwin published "On the Origin of Species"; and, the Naturalists and Darwinists have been ignoring Pasteur's scientific discoveries ever since.

The greatest discovery of the 20th century came when they discovered the physically impossible – Quantum Mechanics and the causal role of Non-Local Consciousness or Psyche within Quantum Mechanics. Quantum Mechanisms and Psyche are physically impossible. According to the Materialists and Naturalists, these things don't exist because they are physically impossible. However, Quantum Mechanics is our most-observed, best-proven, and most-used science on the planet right now, because it works. Quantum Tunneling (teleportation), the Quantum Zeno Effect (telepathy and telekinesis), and Quantum Non-Locality or Action at a Distance (telekinesis and telepathy) have been observed, caught in the act, and proven to exist. The very existence of Quantum Mechanics and mind-over-matter falsifies Materialism and Naturalism.

The greatest discovery of the 21st century came when they discovered and proved that the chemical evolution of proteins and matching genes is physically impossible. In the same era, others discovered that design and creation by random mutations and natural selection is physically impossible. These were monumental scientific breakthroughs! These were paradigm shifts in our way of thinking about and doing science.

The odds of a functional protein developing by natural processes is 1 chance in 10^{164}.

https://www.youtube.com/watch?v=W1_KEVaCyaA

https://www.youtube.com/watch?v=cQoQgTqj3pU

https://www.youtube.com/watch?v=X7VqomfivC0

If you understand the math and the science, as well as the cheats or the God-given aspects of these models, then you simply KNOW that the chemical evolution of proteins and the matching genes to go along with them is physically impossible!

Given these odds, it's obvious and clear that the theory of evolution is the Ultimate Superstition. It takes an infinite amount of ignorance and blind-faith to believe in the theory of evolution, because chemical evolution is physically impossible.

The writing was on the wall; and consequently, I'm no longer a Materialist, Naturalist, Nihilist, and Atheist. The preponderance of the evidence convinced me that I was wrong.

The odds of a functional cell developing by natural processes are 1 chance in $10^{340,000,000}$. That's physically impossible!

https://www.youtube.com/watch?v=nuMvRExazAw

Why are these the greatest scientific discoveries of all time?

It's because they dispelled the myths, science fiction, superstitions, and falsehoods which have been spread-about throughout the science community for a few centuries now, by the Materialists, Darwinists, Naturalists, and Atheists. The discovery of the physically impossible replaced all of those naturalistic lies with truths, for anyone who is looking and willing to see. You have to eliminate all of the falsehoods, before you can find and know the truth. For me personally, that started by identifying the physically impossible, and then trying to find a scientific explanation for the physically impossible.

If you want to figure out how the brain really works, then you have to learn how to identify and explain the physically impossible. Careful study of neuroscience will reveal that a lot of the functionality within our brain is physically impossible. According to Classical Physics, our brains shouldn't work, but they do. When it comes to our brains, there's more than meets the eye.

Biopsychology is defined as the scientific study of the biology of behavior. Biopsychology is defined as Behaviorism. However, it is important to know that chosen behaviors are a function of Psyche or Non-Local Consciousness. It is our reflexes and instincts, which have been programmed into us, not our chosen behaviors! If you are doing science, it's extremely important to make a distinction between the physically possible (reflexes and heritable dispositions or traits) and the physically impossible (the Human Psyche choosing behaviors among many different options). Chosen behaviors are a function of Psyche, not biology. According to the scientific evidence from Near-Death Experiences (NDEs) and Out-of-Body Experiences (OBEs), we continue to make choices long after our physical body and physical brain are dead and gone, even though that's physically impossible.

Our brain stem regulates our reflex activities – heart rate and respiration. The brain stem regulates our inbuilt programming. A lot of this takes place automatically, completely outside the conscious awareness of the Human Psyche and the human brain.

Psyche or Mind is physically impossible. Cartesian Dualism is the philosophical position developed by Rene Descartes, who argued that the universe is composed of two elements, physical matter and mind. Every psychology book mentions Cartesian Dualism, usually to say that it is wrong or has been proven false.

If you believe in an immortal soul and a mortal body, then you probably subscribe to some version of Rene Descartes' mind/body solution: the existence of both body and mind (dualism), with a material body and an immaterial mind/soul. (*Biological Psychology*, p. 3.)

Cartesian Dualism actually does an excellent job of matching with reality, although I have adapted it quite a bit in order to make it match with ALL of the observational evidence and experiential evidence that I have on hand. Something is said to be scientifically accurate if it matches with reality. I have observed that spirit matter and physical matter are basically the same thing – they just exist in a different phase or different dimension from each other. Physical matter exists in space-time and has been infused with entropy. Physical matter has a beginning and it can theoretically have an end. Spirit matter is transdimensional and is pure syntropy, meaning that it doesn't age and is eternal.

I have observed that Psyche, Intelligence, or Quantum Non-Local Consciousness is something completely different. Psyche, thoughts, and memories are quantum waves or the vibrations within a string. Whenever the Psyche is separated from its spirit body and/or its physical body, its spirit body and its physical body don't have any experiences and don't make any memories. Psyche seems to be immaterial or without matter, which would make sense if Psyche is thoughts, memories, quantum waves, and the vibrations within a string.

The Placebo Effect is scientific evidence that mind-over-matter is real and truly happens from time to time. The Placebo Effect and the proven existence of Psyche or Mind falsifies Materialism and Naturalism. The one has lots of evidentiary support, and the other one doesn't have any evidentiary support. Of course, the Materialists and Naturalists will tell you that Psyche or Mind is physically impossible, which is precisely the point that Descartes was trying to make.

"Targeting" is code for Psyche!

The Materialists and Naturalists use the word "targeting" all throughout neuroscience. They never once realize that targeting or action at a distance is a product of Psyche and Quantum Mechanics. For example, "enhancers are stretches of DNA that control the rate of expression for target genes." This has Nature's Psyche and God's Psyche written all over it! Targeting requires intelligence of some kind!

When transcription factors or transcription enzymes target specific genes for transcription into mRNA molecules, that targeting process and transmission of intelligence takes place under the control of Nature's Psyche at the quantum level. The whole thing is physically impossible. There is NO physical mechanism that would allow individual atoms and individual molecules to communicate with each other and thereby interact and coordinate with each other. Communication and coordination between DNA molecules, protein molecules, amino acids, and individual atoms is physically impossible; yet, there's plenty of evidence that it's happening for real at the quantum level.

The adaptive immune system is the division of the immune system that mounts targeted attacks on foreign pathogens by binding to antigens on their cell membranes. Targeting and motivation requires intelligence or psyche; and, so does the recognition of antigens on cell membranes. There's signs of intelligence and psyche within these different cells, and that's physically impossible.

T-reg cells are regulatory T cells that protect the body from autoimmune disease by identifying and destroying T cells that engage in autoimmune activity. How do they know? Identification and targeting require Intelligence or Psyche. There's a Psyche or Intelligence within these cells. But, let's suppose for a moment that there is no Psyche in cells, then we know that it required some kind of Intelligence or Psyche to program the identifying, tracking, and targeting capabilities into these T-reg cells. There's no escaping Psyche or non-local consciousness. Without the presence of Psyche, molecules and atoms and dead cells just sit there doing nothing but vibrating under the effects of Brownian motion. Put these same molecules and atoms into a living cell, and they come alive with purpose and activity. Psyche makes all the difference.

Amacrine cells are retinal neurons whose specialized function is lateral communication. Horizontal cells are a type of retinal neuron whose specialized function is lateral communication. Specialized functionality does NOT spontaneously generate out of thin air, because that's physically impossible. Evolution of any kind cannot do complex specificity. Communication between neurons is physically impossible, because there are NO wires connecting neurons. Neurons are separated from each other by gaps called synapses. You can't transmit a complex message or memory between neurons, because it's physically impossible. All that a presynaptic neuron can do remotely is to open an ionotropic or metabotropic receptor on the postsynaptic neuron. An OPEN or CLOSED neurotransmitter receptor represents one bit of information, NOT a message, or a memory, or video. Neurons are batteries or switches, not CPUs and brains. Neurons are either charging or discharging – ON or OFF. It's physically impossible to transfer a message or memories through neurons and synapses! You can't shove gigabytes of data through a single hardware bit like a neurotransmitter receptor. It's physically impossible. Communication implies Psyche or Intelligence; whereas, the only thing being transmitted through a synaptic cleft is either ON or OFF. It's physically impossible to transmit bytes of information through a neuron, synapse, or receptor. A neuron is a switch. It is either ON or OFF.

This is where science starts to get interesting – whenever we are dealing with the physically impossible!

Even the Evolutionist relies upon Black Boxes or Nature's Psyche in order to make their fictional stories seem possible and real. By definition, "exaptation is a characteristic that evolved because it performed one function, but was later co-opted to perform a different function." Co-opted by whom? Random mutations and natural selection can't do co-opting! They don't work that way. Co-opting implies Intelligence and Psyche. The theory of evolution is science fiction – whole chunks of it are physically impossible!

"To evolve is defined as undergoing gradual orderly change." Random mutations can't do order and organization. It's physically impossible! Entropy, such as random diffusion and random mutations, can't do order, organization, syntropy, orderly change, and life. It's physically impossible. It's fascinating how we chose to build all of our science on something that didn't happen and is physically impossible – the evolution of genes and proteins from scratch.

The thing that gets the FIRST action potential started is the thing that can't be explained by physical matter, physical processes, and Classical Physics, because it has NO physical precursors. That's the thing that we should be studying and trying to understand –

the thing that triggers the FIRST action potential in a chain of action potentials, because that thing is physically impossible. Imagine it, reaching in with your mind and opening up all the ion channels on a specific neuron, in order to get a physical action started in the first place. That's physically impossible!

The quantum mechanism that gets the FIRST action potential started and fires off the first neuron in an action potential cascade is a phenomenon called attention, conscious observation, intention, choice, will, desire, intelligence, or psyche. Consciousness, observation, intention, choice, and attention are products of Psyche or functions of Psyche. Psyche is non-local consciousness – an infinite singularity. Psyche is physically impossible. Yet, Psyche or non-local consciousness is absolutely essential in order to make Quantum Mechanics work. Without Psyche, there would be NO quantum mechanisms.

See what you can do when you finally decide and choose to allow ALL of the evidence into evidence. Science 2.0 and Quantum Mechanics have infinitely more explanatory power than Materialism, Naturalism, and Classical Physics. Using Quantum Mechanics and Psyche, you can actually explain the physically impossible in a way that makes logical sense.

I have observed that chemical evolution, abiogenesis, spontaneous generation, and macro-evolution are physically impossible. It's physically impossible for atoms and amino acids to spontaneously generate into proteins and matching genes! It can't be done, ever, through the Laws of Classical Physics, entropy, Brownian motion, and random diffusion. The creation of proteins and matching genes is ONLY possible through the Laws of Quantum Mechanics under the direction of God's Psyche. Many people believe that Nature's Psyche is in fact God's Psyche. However, I prefer to visualize the processes or mechanisms of Quantum Mechanics as Nature's Psyche getting its power, influence, and knowledge from God's Psyche. Nature's Psyche covenanted with God to obey His commands or His laws. By obeying God's commands, God is able to give Nature's Psyche purpose, order, structure, life, light, and love.

The Human Psyche really doesn't have anything to do with the quantum processes that Nature's Psyche is using to control its physical body and physical brain, because these quantum processes all take place completely outside the conscious awareness of the Human Psyche and its physical brain. The thing the Human Psyche does do is to pay attention and then choose what to do with its physical body and physical brain, to which the Human Psyche has been assigned by God.

Technically, programming is physically impossible. Specifically, the spontaneous generation of programming out of thin air is physically impossible. It can't be done! Abiogenesis, spontaneous generation, and macro-evolution are physically impossible.

The spontaneous generation of hardware – computer chips and CPUs – is physically impossible. Likewise, the spontaneous programming of that hardware's programming is physically impossible. Programming is the product of Psyche or Intelligence. Stand-alone atoms and molecules can't do programming – it's physically impossible. Programming of any kind is a definitive and conclusive sign of Psyche or Intelligence. Programming is proof of concept where Quantum Neuroscience is concerned. The very existence of computer programming, such as our genome, falsifies Materialism and Naturalism, because it's physically impossible for genomic programming to spontaneously generate out of thin air.

A genome is a radically advanced piece of hardware and software programming. Functional genomes cannot spontaneously generate out of thin air. It's physically impossible. Your genome is God's Signature, and it is written all throughout your body. Your genome is the smoking gun that falsifies Materialism, Naturalism, Darwinism, and

Atheism. Your genome is convincing Scientific Proof of God's Existence. I used to be a Materialist, Nihilist, and Atheist; but once I realized what genomes really are, the very existence of genomes convinced me that I was wrong. Science itself ended up proving to me that God exists; and, I wasn't going to believe in God until I had scientific proof of God's existence, which I now do.

Let's talk about hardware and software.

A codon is a group of three consecutive nucleotide bases on a DNA or messenger RNA strand. Each codon specifies the particular amino acid that is to be added to an amino acid chain during protein synthesis. We are looking at programming here – CODE! It's called a codon, after all!

The spontaneous generation of programming code out of thin air is physically impossible. Not only did someone have to design the codons and decide what they mean; but, that same person also had to decide which of the 20 amino acids to attach to each of these 64 codons in order to build the proteins that he wanted. Then that person had to create the associated genes that would produce the desired proteins when those genes are decoded or translated. This someone cannot be evolution! It has been proven conclusively that chemical evolution cannot design and create proteins, nor the matching genes to go along with them; therefore, evolution didn't design the codons nor amino acids either. Evolution didn't do any of this, because it's physically impossible.

The genetic code is degenerate. Some amino acids are encoded by more than one codon, inasmuch as there are 64 possible base triplets and only 20 amino acids. In fact, 61 of the 64 possible triplets specify particular amino acids and 3 triplets (called stop codons) designate the termination of translation. Uncle Google.

Some intelligent psyche had to design and program this thing, or it wouldn't work; and, that's physically impossible according to the Materialists and Naturalists.

Color constancy is the tendency of an object to appear to be the same color, even when the wavelengths of light that it reflects change. Color constancy requires complex programming and processing. It's physically impossible for random mutations and natural selection to do complex programming, let alone physical processing of information.

Action Potentials – the firing of a specific neuron – is an All-or-None Response. The neuron either reaches threshold and fires, or it doesn't. The neuron is either ON or OFF, thus representing one bit of information! It's physically impossible to shove gigabytes of data through ONE bit of information storage or one bit of hardware. How do you let the "bit" know when the data stream starts and when the data stream ends? It's physically impossible. It can't be done! Transmitting gigabytes of data throughout a physical brain is physically impossible. But, it can be done telepathically by Nature's Psyche!

I'm a computer scientist and spent a decade as a computer programmer. I KNOW for a fact that a data stream or program needs some way to signal the start of the program and the end of the program. Distributed Memory is the hypothesis that our declarative memories are evenly distributed throughout our whole brain; but, that's physically impossible. Synapses and the associated neurotransmitter receptors represent a single bit of information at most – they are either OPEN or CLOSED. Likewise, neurons represent a single bit of information, because neurons are either ON or OFF. It's physically impossible to transmit a message, or a memory, or a program through a single bit. The ON would tell the synapse that the message has started, and the OFF would tell the synapse that the message has ended; but then, there would be NO message whatsoever between the ON and the OFF. It's physically impossible to distribute gigabytes of video and memories through synaptic clefts to selectively targeted neurons.

God has gone out of His way to make it obvious that it's physically impossible to transfer data through our brain, to store programs within our brain, and to store memories or RAM in our brain. How did God do this? He did this by constructing our brains exclusively with switches or bits, which are either ON or OFF. If there is something in our brain that is processing bytes of information at a time – start codes, end codes, and assembler codes – these processes and programs are invisible and exist only in the non-local realm or spirit realm.

It's completely and totally possible that Nature's Psyche is assigning specific bits on different neurons to function together as a byte of information, as human beings do with the computers that they manufacture and program. If Nature's Psyche is assigning specific bits on different neurons to function as bytes of data and software programming, then our brains could indeed function as a collection of Cerebral Processing Units (CPUs) or RAM, even though that's physically impossible with a collection of stand-alone bits or switches or neurons.

Our brain is a collection of stand-alone bits and switches. It shouldn't work, but it does. The fact, that our brains actually work and process information, is telling us that there's something going on besides the physically obvious. Through Quantum Mechanics and telepathy, Nature's Psyche can indeed do the physically impossible, including turning the stand-alone bits or switches in our brain into a functional computer, with CPUs and RAM.

Over and over again, ask yourself, "Who is spreading our memories and functionality throughout our brain, and how are they doing it?" It's impossible to find a physical explanation for how this is happening. You can't store computer programs and memories in a collection of stand-alone bits. It's physically impossible. The bits have to be given meaning by turning them into bytes, and this requires some kind of Psyche or Intelligence, an Engineer and a Programmer. Computers don't manufacture, program, and run themselves. Computers require some kind of Psyche or Intelligence to get all of this done. The same truth applies to a physical brain.

Knowledge of computer science is extremely important for understanding Quantum Neuroscience, just like knowledge of Quantum Mechanics is essential for making a functional computer chip. A neuron functions as a single bit, and so does a synapse and a neurotransmitter receptor. They are either ON or OFF. Remember, it is physically impossible to produce a functional computer with a collection of stand-alone bits. It can't be done. However, if someone external and separate from the computer – an engineer and programmer – were to assign specific bits, or neurons, or synapses, or receptors to represent bytes of information, then instantly, we find ourselves looking at a functional computer.

Our physical brain is comprised of stand-alone bits. It shouldn't work; but, it does. That means that there is an invisible programmer and engineer somewhere in the system, who is turning those individual bits into functional bytes of hardware, software, cache, and memory storage. There's no other way to get a collection of bits to function as a computer or a CPU. Someone, like Nature's Psyche or God's Psyche, has to be defining and assigning those individual bits to function collectively as bytes of hardware and programming; otherwise, our brains would do nothing and we would effectively be vegetables.

The brain is a CPU or computer. The brain needs a designer, manufacturer, programmer, and operator. God's Psyche was the designer and manufacturer. Nature's Psyche is the programmer. The Human Psyche is the operator or user. Where the brain is concerned, all of this is physically impossible; but, that doesn't matter, because that's how it is done. We have our own man-made computers to serve as proof of concept. I had to get used to this, but that's the way it is.

Cool, huh?

Anytime the neuroscientists mention programs and programming in the physical brain, they are talking about Programmers, Intelligence, and Psyche whether they know it or not. The spontaneous generation of hardware and programming is physically impossible; and, this truth will resurface repeatedly while studying different aspects of neuroscience, because it is clear to most people that someone programmed our Cerebral Processing Units (CPUs) to perform specific functions. This programming is ongoing and is happening right now within your brain, completely outside your conscious awareness. Just because you aren't aware of it, that doesn't mean that it doesn't exist.

The Materialists, Naturalists, Darwinists, Nihilists, Behaviorists, and Atheists refuse to think about these kinds of things, preferring instead to claim that they don't exist so that they don't have to think about them; but, Materialism and Naturalism are worthless, because they can't be used to explain the physically impossible.

Cognitions are higher intellectual processes such as thought, memory, attention, and complex perceptual processes. Cognitions are a function of the Human Psyche.

How do we know?

WE KNOW because of the scientific evidence provided by Near-Death Experiences (NDEs), Out-of-Body Experiences (OBEs), Share-Death Experiences (SDEs), and after-death Life Reviews. Science is observation, or at least it should be. Our cognitions, thoughts, memories, and perceptual processes continue to exist long after our physical brain is dead and gone. That is what we have observed. We claim to be scientists. We're supposed to figure out how these things work.

Remember, cognition is a product of the Human Psyche. Cognition survives the death of our physical brain. The Materialists and Naturalists put cognition into a black box and simply say that it's an epiphenom of the physical brain; but, that black box is in fact the Human Psyche, which has been observed to exist by out-of-body travelers. It's physically impossible to use our physical instruments to record our thoughts and our dreams, because ultimate our thoughts and dreams are being produced within the Human Psyche and NOT the physical brain. Psyche is the ultimate causal agent.

When trying to explain the physical brain and how it works, the neuroscientists typically describe the brain as a collection of programs. "Program" and "programming" are actually the words that these Materialists and Naturalists use to describe the purpose and functionality of the human brain. However, these people don't seem to realize that programs, hardware, and software require some kind of intelligent programmer; or, they wouldn't exist in the first place. Our genome is a programmed piece of hardware – software on a chip. Likewise, our brain is a programmable piece of hardware, and Nature's Psyche is the programmer. How do we know? WE KNOW because the Orthodox Interpretation of Quantum Mechanics, from Henry P. Stapp, teaches us that it is Nature's Psyche who collapses the wave function thereby making the infinite possibilities REAL and ACTUAL.

Psyche is the Ultimate Causal Agent. Psyche existed before it brought the first particles of physical matter into existence! Which came first, physical matter or psyche? Psyche came first, because Psyche created or organized physical matter out of pre-existing spirit matter. Of course, that's physically impossible; but, that's the way it was done nonetheless.

"Event-related potentials (ERPs) are represented by EEG waves that regularly accompany certain psychological events." ERPs are mind-over-matter, which is physically impossible! When the Human Psyche is having a psychological issue or a psychological event, Nature's Psyche literally reaches out telepathically and turns on the parts of the brain needed to process that psychological event. It has to be Nature's Psyche who is triggering these ERPs, because they happen completely outside the conscious awareness of the Human Psyche. Nature's Psyche collapses the wave function, according to the Orthodox Interpretation of Quantum Mechanics. The Human Psyche makes the request, or experiences the psychological issue or psychological event. This is Quantum Neuroscience 101. It all starts with Psyche and is mediated or controlled by Quantum Mechanisms, which is physically impossible.

I have had a very hard time visualizing how the firing of a neuron, after the accumulation of hundreds of EPSPs (Excitatory Post-Synaptic Potentials) finally force the neuron to threshold and push the neuron over into an action potential, could actually mean anything important. How can the accumulations of hundreds of EPSPs have any significant meaning? It seems more like the function of luck or random chance, to me.

However, when the Human Psyche reaches out with attention, desire, interest, or will and heats up a whole Cerebral Processing Unit (CPU) by paying attention to something that interests it, well now, that's highly significant, isn't it! That has meaning, doesn't it! It's the Human Psyche who gives its physical brain purpose and meaning; and, that's physically impossible.

Alternative splicing is a mechanism by which actions of individual genes can be controlled or edited so that one gene can produce two or more proteins. How could evolution or random chance produce such a thing? It couldn't! Not by random shuffling. Evolution or random mutations could definitely mess up alternative splicing, but evolution could NEVER produce it. It's physically impossible!

Aggregation is the alignment of cells within different areas of the embryo during development in order to form different structures. How would you make cells align three-dimensionally into different structures that have purpose and function, if I were to give you that task? It's physically impossible for the Human Psyche to reach out telepathically and make its thalamus form into a hippocampus instead. If I were to give you some neuron stem cells, how would you make them form into a physical brain, or form into a hippocampus to replace the hippocampus that you already have? Given our current level of knowledge, that's physically impossible. How are aggregation and differentiation being done? It looks like some kind of Black Box to me – something that is under the command and control of Nature's Psyche and that takes place through quantum mechanisms such as telekinesis, telepathy, and action at a distance, which are physically impossible.

The Darwinists, Materialists, and Naturalists teach that evolution designed and created your brain; therefore, these people should be able to turn a stem cell into a brain, or a heart, or a kidney, or a hippocampus at will. If evolution can do it, then they should be able to do it too. These people should be growing replacement organs for us in artificial wombs, so why aren't they? Evolution can do it, so why can't they? Something doesn't add up here, and my bet is on evolution. Evolution can't perform as advertised. For some reason, it's physically impossible.

Chosen behaviors are a function of the Human Psyche. Chosen behaviors are physically impossible! The atoms and molecules within our body can't choose our behaviors nor make our decisions for us. Instead, the atoms and molecules function automatically, doing their thing completely outside our conscious awareness, as if they aren't even there.

We aren't consciously aware of what Nature's Psyche is doing, or how it's doing it. Keeping track of Nature's Psyche is physically impossible!

http://www.simplifyinginterfaces.com/2008/08/01/95-percent-of-brain-activity-is-beyond-our-conscious-awareness/

If our brain were truly under the control of physical processes, then we should be consciously aware of all of them; but, we aren't. That observation suggests that our brain is under the control of Nature's Psyche, completely outside our conscious awareness.

Consciousness is the ultimate black box. Consciousness and cognition are a product of the Human Psyche, because they survive the death of our physical body and physical brain, according to the empirical evidence from Near-Death Experiences (NDEs) and Out-of-Body Experiences (OBEs). It's physically impossible for our thoughts, dreams, cognitions, and feelings to survive the death of our physical body and physical brain; but, Quantum Mechanics and Psyche make it possible.

Conscious awareness is the ability to perceive one's experiences, and it is typically inferred from the ability to verbally describe them. Conscious awareness happens while our spirit body is separated from its physical body and while our psyche is separated from its spirit body, which means that conscious awareness is ultimately a function of the Human Psyche.

Although seeming to be physically possible, our autonomic nervous system and immune system are Black Boxes, in that they happen or take place outside our conscious awareness under the command and control of Nature's Psyche. The Human Psyche and its physical brain are completely unaware of these physical processes and how they work. It all takes place sub-consciously or unconsciously; and, that's physically impossible! All throughout the history of psychology, the Materialists and Naturalists have resisted the "unconscious", because it's physically impossible. The unconscious and sub-conscious are a product of Psyche, and that's physically impossible.

Psychoactive drugs are drugs that influence subjective experience and behavior by acting on the nervous system. The P300 wave is the positive EEG wave that usually occurs about 300 milliseconds after a momentary stimulus that has meaning for the subject. Evolution can't do subjective, nor objective. Evolution can't do anything of value for us. It's physically impossible. Subjectivity and meaning are a product of the Human Psyche. We actually remember these things after our physical brain is dead and gone. I'm not going to apologize for noticing some of these things, because they are science and they are interesting. Identifying the physically impossible has become one of my most favorite things to do while studying science.

Nature's Psyche and the Human Psyche are physically impossible. Non-Local Consciousness is physically impossible. Quantum Mechanics is physically impossible. Choice, intention, and free will are physically impossible. Psyche and Quantum Mechanics can do the physically impossible, because they aren't physical mechanisms. Within my own mind, and on the written page, I kept trying to drill these concepts home, because they are physically impossible to understand and therefore take some getting used to. All of these things are black boxes, because they are physically impossible. The purpose of this essay is to put my thought processes into writing. As you can see, I have tried to come at the subject from every angle possible. I wanted Quantum Neuroscience to be a real science. I think I have succeeded, because I have observed that Quantum Neuroscience has infinitely more explanatory power than Materialism, Naturalism, Darwinism, and Classical Physics.

The neuroscientists frequently talk about the physical processes that have been programmed into us. For example, "sensorimotor programs are patterns of activity that are programmed into our sensorimotor system". These people actually used the word, "programmed" to describe this physical process. Programmed by whom? Evolution (random mutations and natural selection) cannot do programming. I KNOW, because I used to be a computer programmer. Programming is physically impossible for evolution, because evolution doesn't have hands and a mind. Programs don't write themselves. Even the programs that write programs – compilers – were also programmed by someone with a psyche or a mind. Evolution of any kind doesn't have a psyche or mind. These programs were written and compiled into our brain by Nature's Psyche, without our being consciously aware of them and the process. Programming is a black box. Programming is physically impossible without the presence of some kind of psyche, or mind, or intelligence.

Motor equivalence is the ability of the sensorimotor system to carry out the same basic movement in different ways that involve different muscles. They are talking about programming here; and, our programming scales up and transfers. If you practice using your hand to learn how to write, Nature's Psyche can take that learned functionality and use it to help you write with your other hand or to write with your foot or your teeth.

Instinctive behaviors are behaviors that occur in all like-members of a species, even when there seems to have been no opportunity for those behaviors to have been learned by experience. Instincts are in-built programming. Programming requires an intelligent Programmer or Psyche. Our autoimmune system, reflexes, and brain stem activities have been programmed into us by Nature's Psyche and God's Psyche. Remember, Psyche can use quantum mechanisms to do the physically impossible.

The sensitive period is the period during the development of a particular trait, usually early in life, when a particular experience is likely to change the course of that developmental process. The critical period is the period during development of a particular trait in which a particular experience must take place in order for subsequent developmental to occur. Physical functionality has to be primed and used; otherwise, it will deteriorate and cease to exist. Use it or lose it.

Instructive experiences are particular experiences or events that influence the direction of a genetic program of development. Permissive experiences are particular experiences or events that are necessary for a particular genetic program to be manifested. Genetic programs require some kind of intelligent Programmer. Genes and proteins don't spontaneously generate out of thin air. That's physically impossible.

"Apoptosis is programmed cell death. Apoptosis is cell death that's actively induced by genetic programs." Spontaneous generation of programs and genomes is physically impossible. Whenever the scientists talk about programs, they are talking about God's Psyche, because God was the only one around who could have done the programming necessary to create our genomes. Your genome is God's Signature. Your genome is Scientific Proof of God's Existence.

Where there's a wheel, there's a way. It's physically impossible for wheels to spontaneously generate. The same truth applies to watches, computers, cars, airplanes, skyscrapers, nano-machinery, genomes, proteins, and living cells. Only the presence of some kind of intelligence or psyche makes these things possible. Psyche can do mechanics at the quantum level. It's called Quantum Mechanics. God's Psyche or God's Intelligence is the only possible source for our genomes and proteins, because the spontaneous generation of these things is physically impossible. Remember, chemical evolution of proteins and their matching genes is physically impossible.

According to my neuroscience textbooks, "Place cells are neurons that develop place fields. They respond only when the subject is in a particular place in a familiar test environment." Specific place cells heat up and fire, whenever the Human Psyche and physical body are standing in a familiar place. How do they know? It's physically impossible for a lone neuron to turn on and fire up without any input from nearby neurons; yet, it does. How? There's only one way that this can be accomplished, and that's through the telepathic interaction of the Human Psyche, Nature's Psyche, and Quantum Mechanisms. Forces and fields are physically impossible. They are products of Quantum Mechanics and Psyche. Telepathy is WiFi at the quantum level, and the Materialists repeatedly tell us that telepathy and telekinesis are physically impossible. Subjectivity and recognition are a function of Psyche.

"Grid cells are entorhinal neurons that have multiple evenly spaced place fields." Grid cells seem to store place memories within these invisible place fields; and, that's physically impossible. It can't be done, unless we introduce Nature's Psyche and Quantum Mechanics into the equation. Then it can be done in spades for all eternity. Remember, there is no entropy in the non-local realm or spirit world. Entropy only exists within physical matter. Psyche, Non-Local Consciousness, Intelligence, spirit matter, forces, fields, and quantum mechanisms are PURE SYNTROPY. They have no expiration date built into them. They are eternal and everlasting. Entropy is a product of space-time, locality; and therefore, entropy is only associated with physical matter.

According to my neuroscience textbooks, "The Posterior Parietal Association Cortex is an area of the brain that receives input from the visual, auditory, and somatosensory systems; and, it is involved in the perception of spatial locations and the guidance of voluntary behaviors." The neurons, synapses, and neurotransmitter receptors are single bits of functionality. It is physically impossible to transfer visual, auditory, and somatosensory feelings through a single bit of hardware. Someone, external to the hardware which is represented by our brain and nervous system, has designated and assigned those single bits to function unitedly as bytes of programming, data, hardware, cache, and memory. The Orthodox Interpretation of Quantum Mechanics, from Henry P. Stapp, is telling me that this someone is Nature's Psyche, because it is Nature's Psyche who is collapsing the wave function and making everything real and actual. The Human Psyche is just the operator or the user of the physical brain and physical body. It is Nature's Psyche who is doing all of the actual work at the quantum level, completely separate from the conscious awareness of the Human Psyche.

Voluntary behaviors or chosen behaviors are a function of the Human Psyche. The Human Psyche is the operator or the user of these physical systems.

"Working memory is defined as temporary memory needed for the successful performance of a task that one is currently working on or studying." ONLY Nature's Psyche could turn the individual hardware bits in our brain into functional cache, RAM, and bytes of information. Nature's Psyche collapses the wave function making these things physically real. The ONE in this example who is working on the task, paying attention to the data, and choosing the outcome is the Human Psyche. The explanatory power of our science is greatly enhanced by including Nature's Psyche, the Human Psyche, and Quantum Mechanics into the mix. Psyche and Quantum Mechanics have infinitely more explanatory power than Materialism, Naturalism, Darwinism, and Classical Physics.

The angular gyrus in the left hemisphere plays a role in reading. The anterior cingulate cortex plays a role in the emotional reaction to painful stimuli. An association cortex receives input from more than one sensory system, and therefore plays a role in making associations between different sensory stimuli. The amygdala plays a role in emotion. Basal ganglia, striatum and globus pallidus, are subcortical nuclei (computer chips) that play a role in motor functions. The limbic system is a collection of interrelated nuclei and tracts that border the thalamus and play a role in emotion. Astrocytes play a role in the passage of chemicals from the blood into the neurons of the central nervous system.

Who mapped or assigned these roles to these specific brain structures? Who assigns purpose, meaning, and motivation to these brain structures?

Role-playing requires Actors! Anytime you encounter psyche functioning as an actor, you have found something that is physically impossible. It's not like there's a brain or CPU inside an astrocyte deciding what chemicals to let into a neuron and what chemicals to leave out. These Cerebral Processing Units (CPU) required an intelligent designer and creator for their existence in the first place; and now, they require some kind of actor or psyche for their continued functionality. They play a role, which means that they are intelligent and aware of their surroundings.

Nature's Psyche produces Maps of Physical Functionality with our brain, which determine the role or the function that each structure in the brain ends up playing.

Lateralization of function is the unequal representation or mapping of various psychological functions in the two hemispheres of the brain. Language and speech are typically lateralized or mapped to the left side of the brain by Nature's Psyche.

Nature's Psyche typically maps a completely different set of physical functionalities to the right hemisphere of the brain. The ability to recognize loss of function, visuospatially-oriented perception and behavior, and musicality all appear to be predominantly functions of the right cerebral hemisphere. Also, the ability to generate verbal inflections and to detect tone of voice appears to be localized or mapped to the right hemisphere. The specialized characteristics of the right hemisphere make it the seat of curiosity, synergy, experimentation, metaphoric thinking, playfulness, solution finding, artistry, flexibility, synthesizing, and risk taking. In addition, it is likely to be opportunistic, future oriented, welcoming of change, and to function as the center of our visualization capability. – Uncle Google.

The brain has been mapped by Nature's Psyche at the quantum level to do all of our psychological functions at the physical level.

Nature's Psyche decides what role each neuron in the brain will play. Nature's Psyche maps physical functionality and the necessary associated quantum level memories or quantum waves onto each neuron in the brain. These neurons and brain structures play a role, a role that was given to them by Nature's Psyche at the quantum level and made physically real by Nature's Psyche at the physical level.

The neuroscientists actually use the word "role" while describing what these CPUs or brain structures do. Role-playing requires an Actor. According to the Materialists and Naturalists, it's physically impossible for Psyche or Non-Local Consciousness to play the role of an Actor; but, that's precisely what's happening within our physical brain.

"The cerebrum plays a role in complex adaptive processes, such as learning, perception, and motivation." Roles require Actors! It's physically impossible for the Human Psyche and Nature's Psyche to play a role in the physical brain. So, who is the Actor playing the lead role in your cerebrum, and how is he doing so?

"The cerebellum is the metencephalic structure that has been shown to mediate the retention of Pavlovian eye-blink conditioning." Mediation requires a Mediator! So, who is the Mediator controlling all of these different brain functions?

Other types of cells, not just neurons, are designed and specialized to play a specific role.

Lymphocytes are specialized white blood cells that are produced in bone marrow and play important roles in the body's immune reactions. Phagocytes are cells such as macrophages and microglia that destroy in ingest pathogens. A macrophage is a large phagocyte that plays a role in cell-mediated immunity. Somebody Psyche designed and programmed physical functionality and motivational imperatives into these cells. We know it wasn't evolution because evolution has no psyche, and evolution can't do design and specialization. Evolution can't even produce a functional protein, let alone the matching gene to go with it.

The paraventricular nuclei are hypothalamic nuclei that play a role in eating. According to the neuroscientists, the lateral nucleus of the amygdala plays a major role in the acquisition, storage, and expression of conditioned fear. Melanopsin is a photopigment found in retinal cells that respond to changes in background illumination and play a role in synchronizing circadian rhythms. The medial preoptic area includes sexually dimorphic nuclei that play a key role in the control of male sexual behavior. The MT area of the cortex is located at the junction of the temporal, parietal, and occipital lobes; and, its function or role seems to be the perception of motion. The motor theory of speech perception is the theory that the perception of speech involves the activation of the same areas of the brain that are involved in the production of speech. The nucleus magnocellularis is the nucleus of the caudal reticular formation that plays a role in the relaxation of the core muscles during REM sleep and during cataleptic attacks. The reticular activation system plays a role in our arousal system. The striatum seems to play a role in memory for consistent relationships between stimuli and responses – habit formation. The suprachiasmatic nuclei of the medial hypothalamus control the circadian cycles of various body functions. The superior olives are medullary nuclei that play a role in sound localization.

We are talking about Maps of Physical Functionality here!

These are neurons that have been given a specific role or a specific function. Why there and not someplace else? Why are these locations consistent from one person to the next? Why do these neurons have that specific function and not some other function? They all have the same genes, so who gave them their specific role or function? Who mapped this physical function or this particular role onto these nuclei and these neurons?

Directed synapses are synapses in which the site of neurotransmitter release and the site of neurotransmitter reception are in close proximity. Non-directed synapses are synapses at which the site of neurotransmitter release and the site of neurotransmitter reception are far apart. These two are polar opposites. Who decides and chooses whether a synapse will be a directed or non-directed synapse? Well, it isn't evolution. Evolution of any kind can't do decisions and choices. According to the Materialists, Naturalists, Darwinists, and Behaviorists, choice is physically impossible.

Neurophysiology is the study of the functions and activities of the nervous system. The brain and nervous system are all about physical functionality. Who mapped all of this functionality onto the brain? It's physically impossible for our brain to function as CPUs, RAM, and Memory Systems, because there are no wires in our brain and the synaptic clefts scramble and randomize everything that come their way at the physical level. Yet, Nature's Psyche has mapped the human brain at the quantum level to function as if the brain were comprised of CPUs, petabytes of RAM, parallel processors, image processors, quantum processors, and eternal memory systems.

Neurotrophins are chemicals that are supplied to developing neurons by their targets to support their survival, growth, and migration. Nerve growth factor (NGF) is a neurotrophin that attracts growing axons of the sympathetic nervous system and promotes their survival. "Troph" means growth. Lots of growing axons are sniffing out NGF. Many different target neurons are emitting NGF. Who provides the specificity and decides which specific target neuron a specific axon will grow towards? Who designed these things and taught these developing neurons and target neurons what these chemicals mean? It wasn't evolution. Random mutations can't teach us anything – certainly not at the quantum level. Evolution can't do purpose, teleology, and meaning. Thoughts and memories are quantum waves.

NMDA receptors are glutamate receptors that play key roles in the development of stroke-induced brain damage and long-term potentiation at glutaminergic synapses. Who assigned these specific roles? Strokes are the result of runaway glutamate release binding to NMDA receptors. Why aren't there spasms, tremors, and false starts associated with a stroke? If neurons cause thoughts and store memories, then why isn't there a flood of random spontaneous thoughts and a random release of memories while building up to a stroke and during a stroke? If our memories are stored within our neurons, then why aren't our memories lost when our neurons die? Since there is NO visible RAM within our neurons and our synapses, where are our memories really being stored?

Neuropeptides are short amino acid chains comprising between 3 and 36 amino acids. Neuropeptide transmitters are peptides that function as neurotransmitters. About 100 have been identified. Neuropeptide Y is a peptide that is released both in the gut and by neurons of the arcuate nucleus, and it plays a role in hunger.

Who makes those specific cells release those specific neuropeptides? Who chooses when to do so? Who assigned the functionality? Who mapped, programmed, and designed all of this functionality into these neurotransmitters, neuropeptides, synapses, and neurons? Well, it wasn't evolution. Evolution of any kind can't do quantum mapping and quantum programming. Design is a product of Psyche; and, Psyche exists at the quantum level. Evolution by definition in principle can't do anything at the quantum level or psyche level. Evolution can't do the physically impossible.

Nature's Psyche (or God's Psyche) is the ONLY one that we know of who has the capability of mapping roles or mapping physical functionality onto neurons, synapses, and neurotransmitters at the quantum level. It can't be the Human Psyche, because the Human Psyche isn't consciously aware of these things. According to the Orthodox Interpretation of Quantum Mechanics, ONLY Nature's Psyche can activate specific areas of the brain by collapsing the necessary wave functions and thereby turning on the necessary neurons which it has mapped to that specific task or role or function. This is how our brain really works, both at the quantum level and the physical level.

WE KNOW that there is no RAM or memory storage within the physical brain, so that means that all memory storage is taking place non-locally in the cloud under the control of Nature's Psyche, because the Human Psyche isn't consciously aware of any of this functionality and memory storage taking place.

The dorsolateral prefrontal association cortex plays a role in the evaluation of external stimuli and the initiation of complex voluntary motor responses. This thing is comprised of different Maps of Physical Functionality. Volition is a function and product of the Human Psyche. Nature's Psyche collapses the wave function, thereby making things physically real. When the Human Psyche voluntarily chooses to move, Nature's Psyche directly triggers the correct neurons needed to get that "complex motor response" going in the first place. We know that it is Nature's Psyche who is doing all the triggering and the collapsing of the wave function, because the Human Psyche isn't consciously aware of any of these things.

The hypothalamus plays a role in motivated behaviors and controlling the pituitary gland. Nature's Psyche maps specific roles or specific functions onto these different brain structures at the quantum level. Our brain structures end up being different Maps of Physical Functionality, when Nature's Psyche gets done mapping functions or roles onto our brain. Our brain is all about physical functionality; and, that functionality or role-playing is mapped onto the brain at the quantum level by Nature's Psyche, completely outside the conscious awareness of the Human Psyche. Remember, the spontaneous generation of Maps out of thin air is physically impossible.

The Interpreter is a hypothetical mechanism that is assumed to reside in the left hemisphere and that continuously assesses patterns of events and tries to make sense of them. It's an intelligent machine with a psyche, purpose, and mission with in-built recognition software; but, that's physically impossible according to the Materialists and Naturalists. ONLY psyche can do interpretation, on both sides of the veil, both at the quantum level and the physical level. This hypothetical mechanism is some sort of Psyche or Intelligence.

Remember, if your Interpretation of Quantum Mechanics isn't telling you what the Human Psyche and Nature's Psyche are doing at the quantum level, and what quantum processes Psyche is using to do these things at the quantum level, then your Interpretation of Quantum Mechanics is worthless.

"The hippocampus is a structure of the medial temporal lobes that plays a role in memory for spatial locations." Who or what made all of these individual neurons and glial cells form together into a united whole that we call a hippocampus? There's not enough memory storage capacity within our genome to code for the entire blueprint of the human body. Some of that information and code has to be stored someplace else besides our genome. I have chosen to call this external storage, God's Database; and, Nature's Psyche has direct access to God's Database. This definition talks about some kind of "role" that the hippocampus is playing. Who assigned this role? Who makes the hippocampus perform this role? "Roles" require Actors! So, who is the Actor behind this role? My bet is on Nature's Psyche. Is this "role" some kind of CPU functionality, or is it memory storage? My conclusion is that this "role" is whatever Nature's Psyche needs it to be. Some of that role might be memory storage, and some of that role might be the execution of Psyche's software.

Nature's Psyche literally takes bits of individual hardware, and it turns those individual bits into a functional whole that we call a hippocampus or a Cerebral Processing Unit (CPU). There's NO way to get individual bits of hardware to function as a united whole, without some kind of intervention from external agents that we call designers, architects, engineers, programmers, manufacturers, and operators. This is Computer Science 101. When it comes to the physical brain, Nature's Psyche is the programmer and the manufacturer. Nature's Psyche collapses the wave function. God's Psyche was the designer, architect, and engineer. The Human Psyche is the operator or the user. Evolution – random mutations and natural selection – has nothing to do with it. Evolution can't touch these things. Evolution doesn't work at the quantum level.

Since it's physically impossible for evolution of any kind to do design, creation, and programming, what can evolution do for us?

Androgenic insensitivity syndrome is the developmental disorder in genetic males which is caused by a mutation to their androgen receptor gene which makes their androgen receptors defective, causing the development of a female body. Thank you evolution!

Alzheimer's Disease and schizophrenia have a strong genetic component. Huntington's disease is produced by a defective dominant gene. Phenylketonuria (PKU) is caused by a pair of mutated recessive genes. Thank you evolution! Mutations are the result of faulty DNA or faulty genes, which are a function of evolution. Mutations are the blessing or gift of genetic change or evolution.

Look at all the wonderful and marvelous things that evolution has done for us. Evolution can and does mess up the genome that God has given us.

Whenever the kids say that Materialism and Naturalism suck, or Darwinism stinks, that's physically impossible. Materialism, Naturalism, Darwinism, Behaviorism, Scientism, Determinism, Nihilism, and Atheism are philosophical concepts – religions – they really don't exist in any physical sense. Therefore, it's physically impossible for them to suck or stink. They aren't physical entities, after all.

However, whenever the Atheists and Naturalists say that God stinks or that God sucks, that is indeed physically possible, especially when it comes to the Biblical God Jesus Christ, because He really does exist and really does have a resurrected physical body. It is physically possible for the Biblical God to suck and stink, because He exists as a physical entity. Do you see how that works? The one exists, the other doesn't. Therefore, the one is physically possible, and the other one isn't.

Homology due to a common ancestor, analogous structures due to convergent evolution, and chemical evolution or macro-evolution are physically impossible, because evolution (genetic drift), natural selection, and random mutations cannot do design and creation and construction. Therefore, evolution can't do convergence and homology. It's physically impossible for any type of evolution to do so, because evolution of any kind doesn't really exist in any physical sense. Convergent evolution is a fictional ad-hoc just-so story – a black box – it doesn't really exist. It's a convenient fiction, and nothing more. There's no person or psyche out there called "evolution" who is doing all of this design, creation, functionality, homology, and convergence for us.

It takes a monumental amount of blind faith to believe in the creative powers of evolution. I never fully understood how much blind faith it really takes to believe in evolution, until I realized and accepted the fact that design and creation by random mutations and natural selection is physically impossible. Orderly, purposeful, and constructive change are physically impossible through random mutations. It can't be done; and, it takes an infinite amount of blind faith to believe that it can. The abiogenesis or self-assembly of proteins and matching genes is also physically impossible. Chemical evolution or spontaneous generation is physically impossible. Two chimp-like ancestors giving birth to two genetically compatible Mr. and Mrs. Mutant Human Beings is also physically impossible. It never happened, because it can't be done. It takes an infinite amount of blind faith to believe that it did. Understanding that evolution of any kind is physically impossible was the end of the theory of evolution, where I was concerned. I could no longer believe in it, because I knew better than to do so.

According to the Materialists and Naturalists, there's no sense believing in something that's physically impossible. This is ironic. Materialism, Naturalism, and Darwinism are self-defeating, because it's physically impossible to provide any evidentiary proof for their veracity and functionality. In order for evolution to work as advertised, it would require the intervention of God's Psyche and Quantum Mechanics to get the job done. Psyche and Quantum Mechanics can do the physically impossible. Alas, God has infinitely better ways for designing and creating life than physical evolution through natural processes; so, evolution is a dead-end no matter how we choose to look at it.

In all the college textbooks that I have read, they emphasize critical thinking. Critical thinking involves carefully assessing the strength of the evidence that's being presented to support an idea. In all the college textbooks, evolution is automatically exempt from criticism and critical thinking. You are not to question it. They cheat when it comes to the theory of evolution or Creation by Evolution, because it's physically impossible to find any evidence to support it. It's physically impossible to design and create anything new and functional through random mutations. It can't be done. Random mutations produce entropy, chaos, disease, birth defects, cancer, death, and extinction. Random mutations do not produce order, genomes, proteins, syntropy, programming, engineering, and life. It's physically impossible for them to do so.

It's fascinating how we chose to build all of our science on something that didn't happen and that's physically impossible – the evolution of genes and proteins from atoms. The evolutionary perspective is the approach to science that focuses on the environmental pressures that led to the evolution of the behaviors, brain, genome, and characteristics of current species. Given the FACT that evolution (genetic change), random mutations, and natural selection didn't even exist until after God designed and created the genes and the proteins in the first place, it's physically impossible for evolution to have designed and created our behaviors, brains, genomes, bodies, traits, and characteristics from scratch.

Chemical evolution can't synthesize a single functional protein, so it definitely can't synthesize or construct a functional gene to match that protein. One day I had to face reality. The chemical evolution of functional genes and proteins is physically impossible; therefore, we must find some other explanation for their origin and their existence.

Evolution doesn't exist as a living breathing entity. At best, the theory of evolution is a convenient fiction, science fiction, and a just-so ad hoc story concocted out of thin air. At worst, the theory of evolution is a deliberate deception and fraud designed to draw us away from the truth.

"A growth cone is an amoebalike structure at the tip of each growing axon or dendrite that guides growth to the appropriate target." Targeting requires intelligence, purpose, psyche, and life. Therefore, WE KNOW that evolution cannot be the source nor the cause of a growth cone's targeting mechanisms. There's no intelligence there in evolution. Evolution is random and blind; and, the growth of an axon is purposeful, targeted, intelligent, and deliberate. Evolution isn't controlling these targeting mechanisms; but, some type of Psyche or Intelligence is, even though that's physically impossible. Remember, we have evidentiary proof that Psyche can use quantum mechanisms to do the physically impossible. Quantum Mechanics is our most used and best proven science. Non-Local Consciousness and Quantum Mechanics are the pinnacle of science.

"The chemo-affinity hypothesis states that growing axons are attracted to the correct targets by different chemicals released by the target sites." How do they choose a specific target? There's more than one target site releasing chemicals during neurodevelopment. Precision targeting within the growing and developing brain is a product of Nature's Psyche. Targeting requires Intelligence or Psyche. Random mutations, Brownian motion, and random diffusion cannot do targeting! Classical Physics cannot do targeting, because targeting requires Action at a Distance. Even the sniffing out of chemicals by growth cones requires some kind of action at a distance or telepathic command and control, because there are millions of different chemicals within the brain. Which one are you going to sniff out? Why? What's your motivation?

The chemo-attractants and chemo-repellents have to be designed and created in the first place, which is something that evolution or random mutations cannot do. Then the axonal growth cones have to be programmed to grow towards the chemo-attractants and away from the chemo-repellents! Design and creation require Intelligence or Psyche. Programming also requires Intelligence or Psyche. Motivation, intention, or desire requires Intelligence or Psyche. Design, creation, fine-tuning, targeting, and programming are physically impossible through chemical evolution, random mutations, and natural selection; therefore, we must find some other explanation for the origin or source of these things. Clearly the targeting and the programming exists, or the neuroscientists wouldn't be using these words to describe what they are observing in the physical brain. But, programming and targeting are a function of Psyche, because they are physically impossible through abiogenesis, or spontaneous generation, or macro-evolution. When it comes to the physical brain, programming, fine-tuning, and precision targeting has to be done by Nature's Psyche through quantum mechanisms.

"Collateral sprouting is the growth of axon branches from mature neurons, usually to postsynaptic sites abandoned by adjacent axons that have degenerated." These mature neurons are taking initiative as if they are alive, conscious, and aware of their surroundings. It's as if these mature neurons have some kind of brain; but, that's physically impossible. Instead, we know that these mature neurons have a psyche that's telling them what to do and how to do it.

The reliable functionality and efficient functionality of neurotransmitter systems is physically impossible. They shouldn't work, but they do. Dopamine transporters are molecules or pumps in the presynaptic membrane of dopaminergic neurons that attract dopamine molecules in the synaptic cleft and deposit them back into the presynaptic neuron. There are many different types of transporters, for the neurotransmitters that have transporters. How do they work? I haven't been able to figure that one out. It's physically impossible to attract dopamine molecules without a mechanism for doing so. It seems to be some sort of telepathy or telekinesis at work, or a force field of some kind!

Neurotransmitter transporters bugged me to no end, because they are physically impossible. There's no way physically possible that entropy, Brownian motion, and random diffusion are going to attract dopamine molecules to them. Classical Physics doesn't work that way. Classical Physics can't do Action at a Distance! Clearly, transporters work, or they wouldn't exist; therefore, there must be some sort of quantum mechanism that's making them work, because their reliable functionality is physically impossible. Neurotransmitter systems are black boxes. According to Classical Physics, they shouldn't work, but they do.

https://en.wikipedia.org/wiki/Neurotransmitter_transporter

http://mypsyche.us/wp-content/uploads/2018/01/Neurotransmitter-transporter-Wikipedia.pdf

There are identifiable physical processes for getting a neurotransmitter through a membrane because both the neurotransmitter and membrane are physical in nature; but, there's no identifiable physical process for how the neurotransmitter transporters attract or draw in the randomly scattered neurotransmitters in the first place. The scientists leave that aspect of transportation unexplained as some kind of black box. Clearly there is a mechanism for getting the job done, but it's equally clear that it isn't a physical mechanism. We are looking at Psyche and Quantum Mechanics, instead of a physical mechanism and Classical Physics.

It's physically impossible for the Human Psyche to reach out telepathically and turn on specific Cerebral Processing Units (CPUs) whenever the Human Psyche is asked to pay attention to something specific, to run some mathematical calculations, to visualize an event, to think about something specific, or to do something specific. Yet, that's exactly what we observe happening in PET scans and fMRI scans whenever the Human Psyche is asked to visualize or think about something or to pay attention to something. The associated CPU turns on, fires up, and heats up in order to meet the demands of the request which the Human Psyche is making upon it. These areas of the brain heat up or fire up, even in the complete absence of physical stimuli, or physical input. They heat up simply because the Human Psyche was asked to think about something specific that required that part of the brain to accomplish that task. Attention must be paid, or the physical brain does nothing but idle. This reality and truth is proof of concept where Quantum Neuroscience is concerned.

Paired-image subtraction technique is the use of PET or fMRI to locate constituent cognitive processes within the brain by producing an image of the difference in brain activity associated with two cognitive tasks that differ in terms of a single constituent cognitive process. Here we actually catch the Human Psyche in the act of lighting up the brain by choosing to concentrate and pay attention to something specific.

Thoughts, memories, and cognition are quantum waves; and, where we are concerned, they are produced by the Human Psyche and we are consciously aware of them. Cognition is a product of the Human Psyche. It survives the death of our physical brain.

Even in the complete absence of visual input with a person's eyes closed, the visual processing centers of the brain will fire up or heat up whenever the Human Psyche is asked to picture something in its mind's eye or to visualize an event. The Cerebral Processing Unit is under the direct command and control of the Human Psyche, and that's physically impossible!

The Materialists and Naturalists hate this sort of stuff and deny its existence. There are branches of science which the Materialists, Naturalists, Darwinists, and Atheists completely reject, ban, block, censor, ridicule, mock, and actively fight against. Quantum Neuroscience will be one of them, because it's physically impossible. These people stand against anything and fight against anything that's physically impossible, including Quantum Mechanics and Psyche. Quantum Neuroscience isn't going to go over well with them, either, because it falsifies Materialism and Naturalism, by identifying and then explaining the physically impossible.

Remember, Quantum Mechanics, Non-Local Consciousness, and Psyche are physically impossible; yet, they have been proven to exist, through Near-Death Experiences (NDEs), Out-of-Body Experiences (OBEs), Shared-Death Experiences (SDEs), and after-death Life Reviews.

Ironically, spirit matter is physically possible. Spirit matter is just a different phase or different dimension of physical matter. They are the same thing – quantum objects! The concept of phase-shifting teaches us that two or more physical particles can exist in the exact same space at the same exact time, as long as they are out of phase with each other. Spirit matter and physical matter can exist in the same space, because they are out of phase with each other and exist in different dimensions. Your psyche and spirit body can exist in the same exact space at the same exact time as your physical body, because they are out of phase with each other. Phase-shifting is physically impossible according to the Laws of Classical Physics, but phase-shifting is an integral and inherent part of Quantum Mechanics.

Remember, spirit matter is the same thing as physical matter, but more primal or refined and phase-shifted. Dualism doesn't really apply to spirit matter and physical matter, because they are the same thing – quantum objects or matter. Holism is the correct philosophical concept that applies to spirit matter and physical matter. Dualism is more appropriately applied to Psyche and Matter, because these do indeed seem to be two completely different things with two completely different functions.

It would in fact be a *category error* logic fallacy to place Psyche and Spirit Matter into the same category, or to place Psyche and Physical Matter into the same category, which a lot of people try to do. Psyche or Non-Local Consciousness is a whole other animal. Psyche or Intelligence commands and controls spirit matter and physical matter through quantum mechanisms, which means that Psyche isn't spirit matter and Psyche isn't physical matter either.

Whenever we are talking about Cartesian Dualism and the Placebo Effect, we are talking about mind-over-matter, or psyche and matter; and, psyche is not the same thing as spirit matter. They are two different things. Psyche or Intelligence chooses and acts. Both spirit matter and physical matter react to Psyche's commands. It's physically impossible for matter to choose and act. Matter simply reacts. It is Psyche – Nature's Psyche, God's Psyche, and the Human Psyche – who is making all the choices and starting all the action.

Brains: I don't know how they do what they do, but they do.

A neuron is a battery, not a CPU. There's no detectable CPU in a neuron. A neuron is either charging or discharging, just like a battery. That's it! A neuron has NO detectable RAM for storing memories. A neuron is a switch – ON or OFF. It's as if all of our long-term consolidated memories are being stored someplace else besides our brain. That's what the observational evidence and experimental evidence repeatedly suggests.

If I were to give you 100 billion switches but NO wires, how would you turn them into functional CPUs or functional RAM? You can't without wires! It's physically impossible. But, let's suppose that you have WiFi or Quantum Telepathy built into each one of those switches, then how would you proceed to turn those switches into functional CPUs or RAM?

Well, for every neuron in the brain, you would have to go through the brain and connect eight switches (neurons) together wirelessly to form a functional unit called a byte, assign some sort of arbitrary but standardized meaning to each of the 256 different types of bytes that are possible, set the eight switches in each of the bytes to the correct meaning that you want that byte to represent (and keep doing so for over eighty years of time), and then force all of these different switches and bytes to function together as a united whole – either as a CPU or as RAM.

If brains are made up of 100 billion stand-alone switches, with no wires or connections between them, then why do brains work? There's NO physical reason why they should work, so why do brains work? The only answer I can give is that Nature's Psyche is assigning functionality and purpose to each of the neurons, or structures, or CPUs in the brain; and, Nature's Psyche is connecting those stand-alone switches together wirelessly into a functional whole. The mechanism that Nature's Psyche uses to get the job done is Quantum Telepathy or WiFi at the quantum level.

If you hook up an EEG to your scalp, what's the only detectable thing coming from your brain as a whole? Waves! Waves are WiFi at the quantum level! If the neurons in your brain are communicating with each other and sending complex messages to each other, they are doing so wirelessly through different types of waves. They are using WiFi at the quantum level or Quantum Telepathy. They definitely are NOT using synapses to send messages to each other, because synapses are randomizers. A synaptic cleft will scramble any message that you try to send through it! It's physically impossible to send messages of any kind through 100 billion stand-alone switches that have NO wires connecting them. It can't be done, so that means that it isn't being done.

When it comes to the brain, God gave us 100 billion switches to look at. 100 billion switches with NO wires connecting them. God made it extremely clear to every Computer Scientist, Materialist, Naturalist, Darwinist, and Atheist that our brains shouldn't work, because it's physically impossible to send messages of any kind from one neuron to the next through a synaptic cleft. It can't be done. A synaptic cleft is a randomizer, scrambler, and jamming device. A neurotransmitter receptor is also a bit – OPEN or CLOSED. The receptor receives the ON signal letting it know that the message has started, and then it receives an OFF signal letting it know that the message has ended; but, there's NO message! You can't send a message through a single stand-alone bit or switch. It's physically impossible.

So, how is the magic being done? Well, the only thing we KNOW for sure is that it isn't being done through the synaptic clefts, because it's physically impossible to send messages through synaptic clefts!

Have you ever wondered why Classical Physicists and Materialists, such as myself, vigorously opposed Quantum Mechanics all of their lives? It's because Quantum Mechanics is supernatural in nature and origin. Quantum Mechanisms involve faster-than-light Action at a Distance, which is physically impossible. Quantum Mechanics is supernatural; and therefore, Quantum Mechanics falsifies Materialism, Naturalism, and their derivatives. Quantum Mechanics is foreign to us because it's supernatural. Even when we do experience something supernatural, we think that we are hallucinating or losing our mind. We try to explain it away.

The Materialists and Naturalists refuse to think about the physically impossible, because these types of things falsify Materialism and Naturalism; but, the physically impossible is precisely where science starts to get interesting.

You are going to have to decide for yourself if any of this is useful to you, or not. We can't do that for you and we can't do that to you, because it's physically impossible. As the Oracle in the Matrix says, "You are going to have to make up your own damned mind what you want to believe."

Oh, I don't have any doubt that our memories are being distributed throughout our brain among the atoms of our brain. It's just not happening at the physical level, because it is physically impossible. How would you go about putting your memories into a single atom, if I were to give you the assignment to do so? Well, you wouldn't be doing it at the physical level, because it's physically impossible. You would have to do so at the quantum level, by converting your memories into Quantum Waves, and then storing those waves within that physical atom.

We have observed that it's Nature's Psyche who reaches out and fires the appropriate neuron whenever the Human Psyche finds something interesting. According to the Orthodox Interpretation of Quantum Mechanics, from Henry P. Stapp, it is Nature's Psyche who collapses the wave function. The Human Psyche simply chooses what it wants its brain to pay attention to. Nature's Psyche accomplishes the request or the task by collapsing the appropriate wave functions, thereby bringing the Human Psyche's desires into physical reality. That's how the Human Psyche interacts with and controls its physical brain – through Nature's Psyche and Quantum Mechanics. This is Quantum Neuroscience. It's really simple to understand, once a person chooses to do so. It's a lot simpler than trying to find a physical explanation for everything. That's physically impossible!

The Physical Brain Is a Transceiver

God has purposefully and deliberately gone out of His way to make it obvious to every computer scientist on this planet that it is physically impossible for neurons in the brain to communicate with each other.

Can you already hear the Materialists, Naturalists, and Atheists complaining, "But, there's the synapses, you idiot! The neurons are communicating with each other through their synapses."

Do you have any idea what a chemical synapse really is?

It's a randomizer!

Do you know what a randomizer is?

Well, let me try to explain it.

Let's say for example that you want to send the message, "Hello world", from one neuron to the next. How would you do so using synapses? Well, first of all, you would break the letters down into individual bytes, and then assign each letter of the phrase to a specific byte. You would also have a start code and an end code which would also be bytes of information, letting the receiving neuron know when it has received the start of the message and when it has received the end of the message.

Do you already see your first problem?

You do if you are a computer scientist.

There are NO bytes within neurons! A neuron is either ON or OFF – it's either charging or discharging. A neuron is a battery! A neuron represents a single bit of information! It's physically impossible to shove bytes of information into ON and OFF. It can't be done.

But now, for the sake of argument, let's pretend that you have been able to break down your message into individual bytes of information composed of different bits of neurotransmitter. Let's say that you have got it all lined up, with ON neurotransmitter and OFF neurotransmitter so that the neurotransmitter bits are lined up perfectly into bytes that spell the words "Hello World". Of course, there are no ON and OFF neurotransmitters. In fact, most neurons only produce one type of neurotransmitter, which would automatically represent the ON condition, thus leaving you with nothing to represent the off condition in your string of bits and bytes composed of neurotransmitters that line up to spell the words, "Hello World".

But, let's say that you have done it. You have lined up two different types of neurotransmitters, and you have lined them up in such a way as to spell the words, "Hello World", using the neurotransmitters as ON and OFF bits to spell the message. Now what do you do? Well, what does the neuron do with its neurotransmitters? The neuron uses pumps called transporters to pump its neurotransmitters into vesicles or bags filled with neurotransmitters. Only one type of neurotransmitter goes into each type of vesicle or bag. During the transportation process, while your message is being sucked into vesicles, it will be decomposed and disassembled into a bag of ON neurotransmitters and a bag OFF neurotransmitters. Well, there's the end of your message.

Okay, but let's say that you have magically created a special neurotransmitter vesicle that keeps your ON and OFF neurotransmitters lined up perfectly spelling the words,

"Hello World", each neurotransmitter representing a single bit of information, as they naturally do. So, what happens next? Well, what does the neuron do next with its neurotransmitter vesicles, when an action potential comes down the pipeline? A neuron dumps the contents of the vesicle into the synaptic cleft. Your lined-up message spelling "Hello World" will now be dumped randomly into the synaptic cleft. There is NOTHING in synaptic cleft to keep messages lined up in a specific order. A synaptic cleft is a randomizer! If there was a message spelled out in neurotransmitters within the presynaptic terminal button, it ceases to exist in the synaptic cleft!

NO message on earth can survive a synaptic cleft! It's physically impossible!

Each neurotransmitter will arrive randomly at the post-synaptic neuron, and some of the neurotransmitters will NEVER arrive. If there ever was a message composed of neurotransmitters in the presynaptic neuron, that message has been completely randomized by the time parts of it start to reach the post-synaptic neuron.

It's physically impossible to transfer a message through a synaptic cleft. It can't be done!

God has gone out of His way to make it totally and completely obvious that it's physically impossible to transfer strings of data or messages between neurons. The only thing that can be transferred between neurons, and between synapses, is ON and OFF – OPEN or CLOSED. That's it! The brain is as dumbed down as it can possibly get! Actually, for me personally, the brain was a monumental disappointment once I finally figured out how it really works. The brain doesn't work at all. It's physically impossible!

So, where is all of this putative processing power, and memory storage, and program execution, and intelligence taking place? Well, it's definitely NOT taking place in the ON and OFF bits of the neurons, synapses, and neurotransmitter receptors! That's physically impossible! You can't shove gigabytes of information through a single bit! You can't do gigabytes of computer processing within a single bit! The ON bit would tell the neuron that the message has started, and the OFF bit would tell the neuron that the message has ended; but, then there would be NO message. And, that's precisely how our physical brain works! The ON or firing neuron tells the brain that the message has started, and the OFF or the CLOSED neurotransmitter receptor tells the brain that the message has ended; but, there's NO message in between the ON and the OFF!

The ONLY way that our brains can work at all is if the messages between neurons are being sent telepathically, through WiFi, at the quantum level.

So, the Human Psyche chooses to think about something or to concentrate on something; and then, Nature's Psyche gets the message and reaches out telepathically and turns on the appropriate neurons and has them do the physical calculations or physical processing of the sensory input that they are receiving. But, is that really what's happening? No. It's physically impossible to do calculations and processing within a single bit.

Therefore, ALL of the functionality, message transfers, computer processing, and data storage within our brains has to be taking place non-locally in the cloud, with the ultimate result – YES or NO – ON or OFF – being downloaded directly to a specific neuron in the brain. Neurons can only do ON or OFF. They can't do complex computations and complex message transmissions. It's physically impossible, because ALL of the neurons are separated from each other by synaptic gaps. Synapses are randomizers, not computer processors!

Our brain is nothing but a transceiver.

The sensory data in our brain is transmitted to Nature's Psyche, and the cream of that reaches the conscious awareness of the Human Psyche. ALL of the processing, calculations, messages, and translation are done non-locally in the cloud by Nature's Psyche. The only thing the Human Psyche does on the input side of the equation is decide what all that telepathic information means to it personally. The Human Psyche does meaning. It tries to decide what all of it means.

When it comes to the output side of the equation, the Human Psyche decides what it wants to do with that information, and then transmits the request to Nature's Psyche. If the request is physically possible, Nature's Psyche reaches out telepathically to the specific neuron that is wired to handle the task, collapses the wave function or opens the correct ion channels, and your finger rises off the table.

This is how your brain really works. The neurons are either ON or OFF. They are either firing, or they are quiet. That's it! All of that organization into functional CPUs (Cerebral Processing Units), all of that transfer of petabytes of data, all of those petabytes of memory storage, and all of that massive parallel processing power is taking place non-locally in the cloud completely outside the physical brain, and completely outside the conscious awareness of the Human Psyche. The neurons either turn on, or they stay off. Nature's Psyche knows which neurons to turn on in order to accomplish complex tasks, and in what order they should be turned on; and, that's it. The neurons either turn on, or they stay off.

With all of that raw data, bits of data, coming into the different parts of the brain from one's physical senses, it's vetted, processed, and condensed by Nature's Psyche into something meaningful or useful, and then presented to the Human Psyche. The Human Psyche decides what it wants to do with that information. The Human Psyche makes a choice to lift its finger or to leave its finger in place – ON or OFF.

Once the Human Psyche has made a decision or a choice, Nature's Psyche, who wired the nervous system in the first place, reaches out telepathically and fires the specific neuron that lifts the specific finger which the Human Psyche wanted lifted.

That's it!

That's how your brain works.

ON or OFF!

Remember, it's physically impossible to store bytes of information within neurons, and it's physically impossible to transfer bytes of information across neurons or through neurons, because neurons are single bits – ON or OFF. Any information that one neuron might try to transfer to another neuron would get randomized within the synaptic cleft! God went out of His way to make it obvious that it's physically impossible to transfer bytes of information from one neuron to the next. The brain has NO physical mechanism for doing so! Even electrical synapses, or gap junctions, are either OPEN or CLOSED. That's it! Everything that's happening within the brain is happening in the Non-Local Realm or the Quantum Realm, outside our conscious awareness. Nature's Psyche feeds the Human Psyche consolidated sensory input, and the Human Psyche decides what it wants to do with that information. Once the Human Psyche makes a decision, then Nature's Psyche turns on the neurons that are needed to get the task done. ON or OFF!

That's it!

That's how your brain works.

Can you understand now why I was monumentally disappointed when I finally figured out how our brain really works? Our brain is nothing but stand-alone switches, which are either ON or OFF. There's no CPU there, and there's no RAM! There's nothing there but switches and relays. The neurons are either ON or OFF. From what I can tell, all of our memories are being stored non-locally in the cloud. The meaning, purpose, assignment, and task of each neuron is taking place in the cloud under the direction of Nature's Psyche. Nature's Psyche reaches out telepathically and turns on the neurons that it needs. It knows which neurons are needed to accomplish a specific physical task. The Human Psyche simply makes the request. Nature's Psyche collapses the wave function making it real.

It's physically impossible to store bytes of memories across different neurons, because most of the neurons don't have direct physical contact with other neurons. Even the ones that do, through gap junctions or electrical synapses, can only send one bit of information ON or OFF through that gap junction or electrical synapse. You can't shove gigabytes of data through a single bit – it's physically impossible.

This might in fact be my greatest discovery in Quantum Neuroscience. Our brains shouldn't work. It's physically impossible for them to work as they do. But, our brains work, thanks to Quantum Mechanics and Non-Local Consciousness.

We have to identify our physical memory systems and understand the limited storage capacity of our physical memory systems, before we find the incentive and desire to identify and understand the unlimited storage capacity of our non-local memory systems. Physical matter greatly decreases information storage capacity and density; and, our Psyche or Non-Local Consciousness has NO physical limitations.

By using convergent operations, and by converging our statistical modeling for the memory storage capacity of different physical memory systems with the observational and empirical evidence obtained from Near-Death Experiences (NDEs) and Out-of-Body Experiences (OBEs), we quickly learn that our Life-Reviews or Long-Term Memories are not stored within our physical brain but are stored non-locally or transdimensionally within our individual Psyche, in the Quantum Realm or Spirit World.

These observations mean that our physical brain is in actuality a Transceiver, and the different memory systems within our physical brain are different types of transceivers – transmitting and receiving different types of eternal memories to and from non-local storage. The proven existence (observed and experienced existence) of after-death Life-Reviews suggests that our spirit body is some kind of eternal memory storage device, and it also suggests that our Psyche or Intelligence is some type of infinite singularity capable of theoretically storing an infinite amount of information in something that has no physical size and takes up no physical space whatsoever.

(Editor's note: When I first wrote this, I hadn't discovered yet that our Life-Reviews are NOT stored within the Human Psyche, otherwise, we would have constant direct access to them after we are dead. Instead, I discovered that our Life Reviews are being stored in God's Database, and our Life Reviews are being produced by the Angels in Heaven. Our physical brains can't produce and don't produce the 360-degree surround vision that's experienced in our Life Reviews; and obviously, NONE of our Life Review memories are being stored in our physical brain. What was surprising to me was the discovery that NONE of our Life Review memories are being stored within the Human Psyche, either. Our memories are not our own. They are being kept for us by God! I deliberately left this error in place, because I used to believe that our Life Reviews were being stored in our own Psyche, which was a belief that got me into some trouble and created a lot of confusion later on. We can make some serious mistakes whenever we start jumping to conclusions.

When it comes to the non-local sciences, it's extremely important to pursue a preponderance of the evidence. That's the only way to get things right!)

All of these realizations and new discoveries become possible by allowing all of the evidence into evidence (particularly the empirical observational experiential evidence from NDEs and OBEs and Life-Reviews) and by realizing that our physical brain as a whole has nowhere near enough memory storage capacity to store all of the memories, thoughts, feelings, data, ideas, experiences, phenomena, images, and events that take place or are encountered during our mortal physical life.

With this realization, we are then able to see and understand that our physical memory systems act as processors and transceivers for different types of physical events or physical phenomena; but, they do not act as storage devices for our long-term memories or eternal memories, which continue to exist and are accessed during our Life-Reviews after our physical brain is dead or offline.

Realizing that most of our memories are stored in The Cloud – in some kind of organized and indexed non-local storage in the spirit world – helps us to understand why people with Superior Autobiographical Memory always access and relive their memories one-at-a-time, and also helps us to understand why these people are not driven insane by having ALL of their memories constantly bombarding them all at once. If ALL of their memories were stored within their physical brain, then ALL of their memories would be present simultaneously. It would be chaos. These people are sane because ALL of their memories are not stored in their physical brain but are stored and indexed non-locally, and are then accessed one-at-a-time in a manageable and realistic fashion.

Understanding that the physical brain is a transceiver rather than a memory storage device helps us to understand and explain how those with Superior Autobiographical Memory do what they do without going insane, helps us to understand why our physical brain is so efficient and doesn't get bogged down by the details, and helps us to understand how NDErs and OBErs are able to have and experience Life-Reviews after their physical brain is dead, offline, and gone. It also explains how we are able to forget things.

We wouldn't be able to forget anything if all of our memories were actually stored within our physical brain, because a physical reality is the ultimate and most dependable and most reliable consensus reality. We forget things because our long-term memories are stored someplace else besides our physical brain, and because most of us don't have the ability to re-access all those memories with our physical brain.

This is why it is important to understand where the different physical memory systems are located, what they do, and what they really are – namely, they are memory processors and memory transceivers, and not physical storage devices. With this knowledge, we can quickly see and understand that our physical brain with its various physical sub-systems is a transceiver that connects our non-local non-physical quantum psyche with our local physical environment. According to the empirical evidence from NDEs and OBEs, our physical brain is an interface and a transceiver – not a physical storage device. Anyone who studies memories should find this reality interesting, due to its vastly superior explanatory power.

Our physical brain is a transceiver, not a memory storage device. Our long-term memories or eternal memories are stored elsewhere besides our physical brain, according to our NDEs and Life-Reviews which take place when our physical brain is dead and gone.

Learning Systems Are NOT Memory Storage Devices

Over and over again, neuroscientists in our college textbooks keep conflating learning systems with memory storage devices.

It is obvious from the observational evidence that our physical brain has some sort of stack, or working memory, or short-term memory; and, if those memories aren't consolidated and stored someplace else, they are lost to us.

What isn't the least bit clear is where those consolidated long-term memories are being stored, once our working memory buffer or stack has been purged.

In all the different books about Neuroscience, they talk a lot about learning, memory, and amnesia. Those are some of the most interesting chapters on record, in my humble opinion, because there's a mystery behind them waiting to be solved.

You see, neurosurgeons keep cutting out huge chunks of people's brains in order to cure epilepsy or to remove an invasive tumor. Yet, no matter what the surgeons choose to cut out, remote memories of childhood or consolidated retrograde memories remain intact, which suggests that these consolidated retrograde memories are no longer being stored exclusively within the physical brain.

The repeated implication we get from these lobectomies is that ALL of our consolidated memories are being store within each and every neuron of our brain; but, that's physically impossible. There's no place within a neuron to store petabytes of data in a physical format; and, it's NOT the least bit clear how a single neuron can physically store a memory, learning, image, video, concept, idea, feeling, desire, or choice – let alone a life-time's worth of these things. WE KNOW that it's being done, because you can cut out any part of the brain that you want, and the person's consolidated memories remain intact. A lifetime's worth of memories are being stored in each and every neuron; but, that's physically impossible. I cannot visualize or imagine a way to store petabytes of data in a physical format within a single neuron or brain cell. Can you?

Amnesia is defined as a pathological loss of memory; but, over and over again, the evidence demonstrates that the memories are never lost. It's functionality that is lost, not the memories. The ability to access and use those memories are lost; but, the memories themselves are not lost. Anytime I encounter amnesia, I find myself asking, "Is it really a loss of memories, or just a loss of functionality?" ALL of the evidence that I have collected over the years suggests that amnesia is a loss of functionality, and not a loss of memories. Whenever the functionality is restored, the memories come back.

Infantile amnesia is the normal inability to recall events from early childhood. The Materialists and Naturalists will tell us that the memories were never formed and therefore don't exist. Based upon all the evidence, I have come to the conclusion that amnesia is a misnomer or misleading term. Amnesia doesn't seem to have anything to do with memories, and everything to do with a lack of functionality. In the case of infantile amnesia, Nature's Psyche hasn't yet mapped out the parts of the brain that consolidate and then store memories in the cloud. It's a lack of a specific type of functionality. Children remember the language they have learned and what their parents look like. Infantile amnesia really isn't a lack of memories. It's a lack of functionality.

Korsakoff's syndrome is defined as a neurological disorder that's common in alcoholics whose primary symptom is severe memory loss. But, is it really memory loss, or are we looking at lost functionality and a damaged brain instead?

A physical brain is a transceiver. The different neurons, molecules, and atoms within your physical brain are communicating with each other and Nature's Psyche at the quantum level through quantum waves. Nature's Psyche has mapped physical functions onto each structure within your brain. The parts of the brain that have been assigned and mapped by Nature's Psyche to transmit information to the cloud and to retrieve information from the cloud have to be there, or that particular functionality is lost. If the Map of Physical Functionality gets destroyed, then Nature's Psyche and the Human Psyche can't use that map to perform the physical functions that Nature's Psyche has assigned to that map and that group of neurons.

There's evidence suggesting that our Life Review is NOT being produced by our physical senses and our physical brain. The people who were born blind are able see during their Near-Death Experiences; and, they experience video of their past-life events during their Life Review. Clearly, the video in their Life Review was NOT created by their physical brain, because their brain lacked that particular Map of Physical Functionality. Their brain lacked the functionality, yet all of their memories remained. If it survives the death of your physical brain, then it really is a memory, isn't it? Your memories can and do survive, even though the physical functionality for storing and retrieving them is lost. This is what we have experienced and observed as a race.

Something like agnosia, which is the inability to recognize sensory stimuli, represents a loss of functionality, and not a loss of memories. Whenever the functionality is restored, the memories come back.

Simultanagnosia is the disorder characterized by the inability to attend to more than one thing at a time.

Visual agnosia is a failure to recognize visual stimuli. The recognition functionality is lost, but not necessarily the memories. Functionality can be lost, but the memories remain. There seems to be a distinct difference between our learning systems and our memory systems; yet, the Materialists and Naturalists deliberately equate the two.

People who go blind remember having had sight and what it was like, even though the functionality is lost.

Akinetopsia is an inability to perceive motion which results from damage to the dorsal visual pathway. These people remember being able to perceive motion in the past, and they remember the snapshots of people and objects that they currently perceive, even though they can no longer perceive motion now. Functionality is lost, not memories! The memories remain, no matter what type of functionality is lost, although the ability to access, form, and process those memories might be lost temporarily, due to temporary lesions and concussions.

I have concluded based on the empirical evidence that there really is no such thing as amnesia, just a loss of functionality, which only seems to produce memory loss. People who are born blind can see during their Near-Death Experiences, and their Life Reviews have video. That video and sight was NOT produced by their physical brain nor stored within their physical brain. These people experienced NO memory loss, just a loss of functionality, during their mortal lives. The memories are still there and are still being made and stored, even though the person might not have physical access to those memories due to some kind of brain damage or loss of functionality. The physical brain is a transceiver. If the brain gets damaged, its transmission and reception functionality are lost; but, the Angels in Heaven continue to keep a video and audio record for the person's upcoming Life Review, no matter how badly the brain gets damaged.

Anterograde amnesia is a loss of memory for events occurring after the amnesia-inducing brain injury. Anterograde amnesia is a loss of functionality. Other types of memories are still formed and stored, just not episodic memories. The person no longer has access to the events that he or she experiences, because the functionality for that process has been lost. However, the anterograde amnesia does nothing to stop or prevent that person's upcoming Life Review, because our Life Review isn't produced by our physical brain nor stored within our physical brain. It's also fully possible that NONE of the episodic memories are lost during anterograde amnesia, just access to them. It's possible that only functionality is lost, not the memories themselves. Restore functionality, and maybe all the memories will come back.

It has been observed using temporary lesions, that when the lesions wear off and the functionality is restored, then the memories come back. After a blow to the head, the memories tend to return as the brain heals. It's possible that there really is no such thing as amnesia or memory loss. According to the empirical evidence from Life Reviews, our memories are being stored for us, even when the brain is incapable of producing, experiencing, processing, and remembering those memories. It's possible that amnesia is nothing more than a temporary loss of brain functionality. All of the evidence that I have encountered continues to suggest that such is in fact the case. Memories aren't lost, just the access to them. The brain is a transceiver, and when the brain gets damaged, transception is lost. Functionality is lost, not the memories.

Transient global amnesia is sudden-onset severe anterograde amnesia and moderate retrograde amnesia for explicit episodic memories that is transient, typically lasting between 4 to 6 hours. Functionality returns once the "amnesia" has passed. The ability to transmit and receive memories to and from the cloud is temporarily lost; but, everything returns to normal once functionality is restored.

Aphasia is a brain-damage-produced deficit in the ability to use or comprehend language. Again, these people can remember the language; but, they can no longer speak it or comprehend it. Aphasia is a loss of functionality, not a loss of memories.

These observations led me to produce a new definition for memory.

If it survives the death of your physical body and physical brain, then it's truly a memory in every sense of the word.

The Human Psyche is slaved to the physical brain. The Human Psyche can only perceive what the brain is able to form. If the functionality of a system is lost, then the Human Psyche no longer has access to that function or system.

Over and over again, I have observed that the physical brain is comprised of Cerebral Processing Units (CPUs) and not memory storage or RAM. Whenever the brain is damaged, functionality or learning systems are lost, but not the memories. Learning systems (CPUs) are something completely different than memory storage (RAM).

Yet, the Materialists and Naturalists assure us that the physical brain is capable of storing petabytes of RAM, that memory storage and learning systems are the same thing, and that all of our memories or RAM cease to exist when our physical brain dies and ceases to exist. They are wrong. It's physically impossible to store petabytes of RAM within our brain; and, according to the empirical evidence from our after-death Life Reviews, our memories do not cease to exist when our brain dies. We also continue to live and learn after we die. I have observed that the Materialists and Naturalists are always wrong.

Clearly, the adult brain has access to petabytes of memories; but, how would you go about storing petabytes of data, video, memories, learning, and intelligence in a newly created and newly installed neurotransmitter receptor or ion channel as the Materialists and Naturalists claim is being done? It's physically impossible.

There isn't a physical explanation for how each neuron is able to store petabytes of video and data, so we have to look for a non-physical, supernatural, quantum explanation instead. We have no other choice. We are forced up against the wall on this one, and are forced to think out of the box if we want to solve it in any reasonable and logical fashion.

Thank God the thousands of after-death Life Reviews on record have given us a clue as to how and where our consolidated memories are being stored, because there isn't a physical explanation for how this might be done. Data storage in a physical format is highly inefficient; and, it's physically impossible to store a lifetime's worth of memories within a neuron, if that video, information, and data is being stored in a physical format of some kind.

Since WE KNOW that a lifetime's worth of consolidated memories are being stored in each and every neuron of our physical brain, and since WE KNOW that that is physically impossible, then the only logical conclusion is that our consolidated memories are being uploaded into and stored within a Non-Local Non-Physical Cloud Storage of some kind, so that our memories, learning, experiences, and choices are there waiting for us when it comes time for our Life Review, after we are dead and our physical brain is long gone. I see no other logical explanation for the evidence that we have on hand as a race.

You see, disabling a neuron doesn't necessarily eliminate, delete, and destroy the memories stored with it; and, neither does cutting it out and throwing it away! Those consolidated memories remain, no matter how much of your brain the neurosurgeons choose to cut out and destroy.

It's like holography!

You can take a hologram and tear it apart into little pieces, and each little piece contains the whole picture or the complete hologram. Each individual part contains the whole picture.

Based upon this observation, there is a theory that each physical atom within this universe contains within it ALL of the knowledge, data, memories, history, intelligence, experiences, and information in this physical universe as a whole. Each physical atom contains within it the sum-total of this physical universe's knowledge, information, data, memories, and experiences. Of course, all of this information is being stored holographically and non-locally as quantum waves of light within all that "empty space" within each atom, because it's physically impossible to have googolplexes of physical RAM within each and every atom on this planet. That amount of information has to be stored Non-Locally and Non-Physically because it's physically impossible to store that amount of information in physical RAM. This is just logical common sense.

Physical RAM and physical DNA are highly inefficient and wasteful ways of storing information and knowledge, because they take up a ton of space. In contrast, something like a Psyche or Non-Local Consciousness is theoretically capable of storing an infinite amount of information within something that has NO physical size and takes up NO physical space whatsoever, because each Psyche or Intelligence is an infinite singularity. Imagine it – an infinite amount of "areal density" or storage density within something that has no physical limitations whatsoever. How would you organize such a thing if given the task to do so? How would you touch it, access it, manipulate it, organize it, and use it? Telepathically, through the Quantum Zeno Effect, is the only thing that comes to mind. Remember, telepathy is WiFi at the quantum level.

Areal density is a measure of the quantity of information bits that can be stored on a given length of track, area of surface, or in a given volume of a computer storage medium. Generally, higher density is more desirable, for it allows greater volumes of data to be stored in the same physical space. Density therefore has a direct relationship to storage capacity of a given medium. Density also generally has a fairly direct effect on the performance within a particular medium, as well as price.

https://en.wikipedia.org/wiki/Areal_density_(computer_storage)

Over the past year, I have gone back and forth – memories are stored in our brain – memories are not stored in our brain – memories are stored within our brain. I've never been able to make up my mind.

Yes, a case can be made for the claim that memories are stored in the brain; but, a claim can also be made that it's physically impossible to store all of our memories in our brain.

Obviously, if our brain is comprised of Cerebral Processing Units (CPUs), then it's theoretically possible that memories are stored in the physical brain – like a stack in a calculator, or cache in a CPU, or RAM in a computer. But, it's impossible to know where local storage ends, and non-local cloud storage begins. Based upon this particular model, I settled upon the idea that Nature's Psyche is handling all of our memory storage for us, and whenever our local physical capacity is on the verge of being overloaded, then Nature's Psyche offloads the memories we aren't using to non-local storage in the cloud.

But, there are other models of memory storage that make just as much sense to me as this one.

Through quantum mechanisms, Nature's Psyche can turn neurotransmitter receptors or neurons into RAM just as easily as it can turn neurons and receptors into functional CPUs. Therefore, it really doesn't matter whether our brain is a CPU or RAM, because according to this particular theory for memory storage, it's all the same to Nature's Psyche. It's just receptors and neurons to Nature's Psyche, and Nature's Psyche can use them however it sees fit, as RAM, CPUs, or something else that we have yet to imagine.

The memory storage vs. learning systems debate – or the RAM vs. CPU debate – ceases to be an issue once we bring Nature's Psyche and Quantum Mechanics into the debate. This ends up being just another "unsolvable problem" created by Materialism and Naturalism which is automatically solved by getting rid of Materialism and Naturalism.

Of course, once I finally realized that it's physically impossible to send a message through a synaptic cleft, then I found myself leaning towards the belief that NO memories are stored within our brain. Based upon this model, the only thing in our brain is maps of functionality. Everything else is handled by Nature's Psyche outside our conscious awareness.

Based upon this Map Model, Nature's Psyche uses our physical brain to make different maps of physical functionality.

In our occipital lobe or striate cortex or primary visual cortex, our brain has a Retinotopic Map which is a one-to-one map of our retinas. This Retinotopic Map functions as working memory, or cache. The only thing stored in these registers or neurons is the current image that's coming in through our eyes, and then it's gone being replaced by the next image coming in through our eyes. This really does seem to be the way our brain functions on the physical level, through topography or maps.

If any of these visual images are being kept, they aren't being stored within our brain, because our brain doesn't have sufficient storage capacity to store our video memories. Our 360-degree video memories for our Life Review are being created and stored for us in God's Database by the Angels in Heaven, because our brain can't and doesn't produce those kinds of images. However, Nature's Psyche is also capable of off-loading the visual memories that the Human Psyche finds valuable or useful, into non-local cloud storage where they can later be retrieved as necessary.

One thing is clear. The memories that show up in our Life Reviews are not stored in our physical brain. We have to take that observational KNOWLEDGE and run with it, if we want to find and know the truth.

I do find the Map Model extremely useful and valuable, because it seems to match with reality, where physical brain functionality is concerned. Something is said to be scientifically accurate if it matches with reality.

There are NO physically observable CPUs or RAM in our brains, nor in our neurons, nor in our receptors. The idea that our brains are RAM, or memory storage, or memory engrams doesn't seem to match with reality. Our brains do indeed seem to be partitioned and function as separate Cerebral Processing Units (CPUs); but, the main problem with this model is that there are NO visible wires connecting the different switches in our brain like there are with our man-made CPUs. It's physically impossible to send a message through a synaptic cleft.

Consequently, the BEST model for brain functionality ends up being the Map Model, in my humble opinion. Different types of physical functionality map one-to-one to neurons and structures in the physical brain. This seems to be an observable fact.

Nature's Psyche maps physical functionality onto our brain both at the quantum level and the physical level, which is why it is called mapping or a map. The map is created at both levels by Nature's Psyche, which is why it works.

The Tonotopic Map refers to our auditory system. The auditory cortex is mapped by Nature's Psyche according to sound frequency. Nature's Psyche and the Human Psyche access this Tonotopic Map telepathically through quantum waves at the quantum level.

The Rhinotopic Map refers to our sense of smell. I've also seen it called the Chemotopic Map.

The Chemotopic Map is a topographical map of the various odors onto the olfactory bulbs.

New olfactory receptor cells are created throughout each individual's life to replace the ones that have deteriorated. Once created, the new receptor cells develop axons, which grow until the reach the appropriate sites in the olfactory bulb. Each new olfactory receptor cell survives only a few weeks before being replaced. How the axons from newly formed receptors scattered about the nasal mucosa find their target glomeruli in the olfactory bulb remains a mystery. (*Biopsychology*, p. 181.)

They KNOW where they are growing, 'er going.

How do the axons know where to go?

Targeting of any kind requires Intelligence or Psyche, because it requires intention, attention, selectivity, and choice. Random processes cannot do targeting!

Typically, whenever a Darwinist or Materialist or Naturalist talks about something remaining a mystery, they are talking about something that Nature's Psyche is doing at the quantum level in order to achieve physical functionality at the physical level.

The primary somatosensory cortex is mapped according to the surface of our body or our skin. The neuroscientists call this map the somatosensory homunculus. It's described as a somatotopic map. The gustatory system or our sense of taste is a part of the input side of the somatotopic map.

The motor homunculus is a somatotopic map of the primary motor cortex – the output side of the equation. Nature's Psyche KNOWS precisely which neuron it needs to trigger in order to get your finger to rise off the table.

It's logical to assume that this is the way our brain really works – as Maps of Physical Functionality. Each Cerebral Processing Unit (CPU) seems to be a Map of Physical Functionality. Each neuron seems to be assigned a physical function, by Nature's Psyche. Our brain seems to be organized into Maps of Physical Functionality by Nature's Psyche, both at the quantum level and the physical level. Maps of Physical Functionality end up being the bridge or the interface between the quantum level and the physical level, thanks to Nature's Psyche and its ability to collapse the wave function.

The Human Psyche reaches in to the sensory maps and experiences or perceives the different sensations or senses. Based upon that sensory input, the Human Psyche makes decisions and choices.

Then, whenever the Human Psyche makes a request, a decision, or a choice, Nature's Psyche through quantum waves or WiFi or telepathy reaches into the brain and triggers the neurons in the motor homunculus, which have been mapped to raise your finger off the table, and your finger rises. Nature's Psyche is the thing that mapped-out your brain in the first place. Therefore, it seems logical that Nature's Psyche is the thing that collapses the wave function, thereby making the Human Psyche's requests actual and real.

The Human Psyche isn't consciously aware of any of these Maps of Physical Functionality, but it slowly learns how to use them. Still, according to the Orthodox Interpretation of Quantum Mechanics from Henry P. Stapp, only Nature's Psyche can collapse the wave function both at the quantum level and the physical level. The Human Psyche simply makes choices. Nature's Psyche implements the plan, both at the quantum level and the physical level.

Interesting, is it not?

The "memories" that are actually remembered and stored within our brain are the Maps of Functionality – the software or the functionality – what they typically call implicit memories or procedural memories. If you knock out the Maps, then you destroy the functionality; however, the long-term consolidated memories remain, because they aren't stored within the brain. Our long-term consolidated declarative memories or Life Review memories remain no matter how much they cut out of our brain.

Fascinating, is it not, what becomes possible when we finally decide to allow all of the evidence into evidence and decide to take it seriously? We come a lot closer to explaining what's really happening within our brains, and we come a lot closer to actually explaining the observational evidence that we have on hand.

When I chose to pursue Quantum Neuroscience, I chose to try to adapt Neuroscience to the Orthodox Interpretation of Quantum Mechanics, and this Maps of Functionality Model ends up being the most useful result, in my humble opinion.

https://en.wikipedia.org/wiki/Topographic_map_(neuroanatomy)

http://mypsyche.us/wp-content/uploads/2018/01/Topographic-Map.pdf

As I see it, it is these different Topographic Maps, which are being created and stored within our physical brain by Nature's Psyche, that end up being the BEST explanation of brain functionality, both at the quantum level and the physical level. Whenever we, the Human Psyche, want our physical body to do something, Nature's Psyche reaches in to the appropriate neurons and fires them up, making what we want to happen an actuality or reality.

Of course, the Materialists and Naturalists will continue to tell you that all your memories are being stored in your synapses; but, they don't know what they are talking about, because it's physically impossible to store memories, video, and messages in a synaptic cleft or a neurotransmitter receptor. Synaptic clefts randomize and scatter everything that comes their way, integrating it all into a single Postsynaptic Potential (PSP).

Materialism, Naturalism, and Darwinism are the Ultimate Superstition, because these people have chosen to believe in the physically impossible. They have chosen to believe in magic or creation ex nihilo. Chemical evolution of proteins, and the matching genes to go along with them, is physically impossible. And, it's physically impossible for genes, random mutations, and natural selection to have designed and created the first genes, because genes didn't exist yet until after God created them in the first place. Likewise, it's physically impossible to store messages and memories within a synaptic cleft, neuron, or neurotransmitter receptor. No physical message can survive passage through a synaptic cleft. It's physically impossible. No messages or memories are being stored within our synapses. It can't be done. It's physically impossible.

But, it's obvious and clear that Nature's Psyche is mapping our physical functionality all throughout our brain, and then calling upon that physical functionality whenever we, the Human Psyche, want to get something done. Furthermore, Nature's Psyche is storing our memories and lived experiences wherever it wants to store them. We, the Human Psyche, have no say in the matter, because all of this mapping and memory storage takes place completely outside our conscious awareness. It's all being taken care of by Nature's Psyche behind the scenes, or behind the seen.

Molecular Mechanisms of Learning and Memory

Oh, it's happening, alright. Our brains really are processing data, running computer code, crunching numbers, transferring data, building cities, weighing decisions, and storing memories. It's just not happening at the physical level. It's all happening at the quantum level. The physical input of sensory data, and the physical output of motor commands, is the only thing happening at the physical level within the brain. It all happens at the end of an axon, where the switch is either turned ON or OFF. A synapse is a switch. Our brain is comprised of a 100 trillion switches or synapses. Our brain is also comprised of a 100 billion stand-alone switches or neurons, which are either ON or OFF. Furthermore, our brains are comprised of quadrillions of ion channels that are either OPEN or CLOSED. Everything within our brains reduces to a single BIT. There's nothing else there within our brain. Everything else is happening at the quantum level. All the important and interesting stuff is happening at the quantum level, outside our conscious awareness.

This chapter in *Neuroscience* helped me to decide, once and for all, what it is that I truly believe and where it is that I want to stand. I'm there now. Let me explain.

Molecular Mechanisms of Learning and Memory

http://mypsyche.us/wp-content/uploads/2017/12/BCP25.pdf

This chapter helped me to realize that the Neuroscientists are conflating or equating brain plasticity and synaptic plasticity with memory and learning; and, they aren't the same thing at all.

VERTEBRATE MODELS OF LEARNING

Let's summarize what we've learned from the invertebrate studies about the possible neural basis of memory:

1. **Learning and memory can result from modifications of synaptic transmission.**

2. **Synaptic modifications can be triggered by the conversion of neural activity into intracellular second messengers.**

3. **Memories can result from alterations in existing synaptic proteins.**

Keep these points in mind as we explore different types of activity-dependent synaptic plasticity in the mammalian brain. To illustrate, we focus mainly on synapses in the cerebellum and hippocampus, where current understanding is most complete. Keep in mind, however, that plasticity of synaptic transmission is widespread in the central nervous system (CNS), and relies on slightly different mechanisms at different types of synapse.

Synaptic Plasticity in the Cerebellar Cortex

In Chapter 14, we discussed the cerebellum in the context of motor control. We concluded that the cerebellum was an important site for motor learning, a place where corrections are made when the outcome of movements fails to meet expectations. Because these corrections are believed to be made by modifications of synaptic connections, the cerebellum has become a model system for the study of the synaptic basis of learning in the mammalian brain. (*Neuroscience*, p. 772.)

They are talking about synaptic plasticity or synaptic re-programming here; but, are they really memories? Do these synaptic changes survive the death of our physical body and physical brain? I see synaptic modification more as re-programming the physical brain or re-wiring the physical brain, in order to improve the functionality and reliability of the physical brain. Synaptic plasticity has nothing to do with our memories!

When I read this, suddenly I just KNEW that Neuroscience needs a better and more accurate definition for memories and learning; so, I created one.

A NEW DEFINITION FOR MEMORIES

If it survives the death of your physical body and physical brain, then it's truly a memory in every sense of the word, and learning has taken place.

Simple. Problem solved. No more conflating memories and learning with brain plasticity.

PLASTICITY is by definition in principle a reprogramming of our physical brain, improving the performance and functionality of our physical brain; and, we are not even aware that it's taking place, nor do we remember it, because it's automatic and autonomic.

I think it's time that scientists stop conflating brain plasticity with memories and learning, because they aren't the same thing at all. I think we have progressed far enough as a race during the past decade, that we are now ready and capable of making this change or adjustment.

By definition, in principle, brain plasticity doesn't survive the death of our physical body and physical brain; but, our memories and our learning most certainly do.

WE KNOW this to be true, thanks to the abundant and overwhelming empirical evidence from Near-Death Experiences (NDEs) and Out-of-Body Experiences (OBEs) that's become available to us during the past decade or so on YouTube, on the Internet, and in books. Millions of people have had NDEs and OBEs. By this point in history, 2018, NDEs and OBEs are observed science, experienced science, empirical science, and therefore proven science. We need to adjust our definitions and our other science disciplines accordingly so as to make them match with what we already KNOW to be real and true. The time has come. We can do it now. We have a preponderance of the evidence telling us to make this change.

If it survives the death of your physical body and physical brain, then it's truly a memory in every sense of the word, and learning has taken place. If it doesn't survive the death of your physical body and physical brain, then it's synaptic plasticity and NOT learning and memory. Memory, learning, and intelligence ARE a function of psyche, and not brain plasticity. Plasticity is a re-programming or re-wiring of the physical brain, and has nothing to do with memory or learning. We aren't even aware when our brain plasticity takes place, so there's nothing for us to remember or learn from the event.

This refinement in neuroscience is at the very heart and core of Quantum Neuroscience. Quantum Neuroscience is the scientific study of how the Human Psyche interacts with, controls, and survives the death of its physical brain.

Simple and clear.

It all fits now.

Good enough.

I'm pleased with the results, because it makes sense to me now; whereas, it didn't make sense to me before.

I just couldn't see how adding a new ion receptor to a synapse would magically manage to store a new declarative memory with all of its ideas, feelings, images, synchronicity, time-stamp, impressions, and thoughts. Now I don't have to try to figure it out, because now I KNOW that adding a new ion receptor improves the functionality and efficiency of the physical brain and does absolutely nothing to store our declarative memories, facts, feelings, thoughts, impressions, analysis, and remembered events into our psyche or non-local consciousness for future use in our Life-Reviews after our physical brain is dead and gone.

What the neuroscientists call "memory systems" or "learning systems" are in fact transceivers, or WiFi adapters of different types. If you pull the WiFi adapter out of your computer, then you no longer have access to Non-Local Storage or Cloud Storage. If you pull out the hippocampus, then you no longer have access to spatial memories and declarative memories; but, that doesn't mean that they no longer exist. You just don't have access to them anymore, or you can't process them and use them anymore. Pull out your caudate nucleus, and you no longer have access to your stimulus-response habits. It doesn't necessarily mean that they don't exist, you just don't have access to them anymore. In fact, it has been observed that if temporary lesions or temporary removal is used, then the spatial memories or place-location memories come back when hippocampal functionality is restored. Likewise, it has been observed that one's habits or stimulus-response learning comes back once the functionality of the caudate nucleus is restored. These different brain structures are transceivers or WiFi Adapters or computer processors, and NOT memory storage devices. The memories and learning are stored someplace else, because they come back after the functionality of these brain devices has been restored. That is what we have experienced and observed!

I experienced this myself, personally.

The medical doctors and psychiatrists had me cycling through a dozen different drugs on any given day. I was suffering from substance-induced psychotic disorder. It was chaos. It got so bad that I no longer knew how to turn on the television, turn on the computer, or play a DVD in the machine. I had to have someone else do that for me because I didn't know how anymore. I could no longer add or subtract. My motivation and desire to live was gone. My sense of right and wrong was gone. Most of my memories were gone. I was a vegetable. I was a zombie – one of the living dead.

I chose to go cold turkey, and I went through six months of withdrawal, psychosis, insanity, detachment from reality, and loss of procedural memories and declarative memories, before I finally got sober and started thinking rationally once again. If Materialism and Naturalism were 100% true, then NONE of my memories should have ever come back, because they were gone! However, after I got sober, God slowly restored my memories to me or gave me access to them once again. They weren't stored in my physical brain. They were stored someplace else, and God gave me access to them again over time. Nowadays, nobody would even realize that I used to be a zombie and a vegetable.

Like I said, it all makes sense to me now.

Now I KNOW how God was able to restore my memories to me, even though they had been gone for years from my physical brain. Our memories aren't stored in our physical brain. They are stored someplace else, because our memories survive the death and the damage of our physical brain. This is what I KNOW; and now, it's time to try to find a scientific explanation and a logical explanation for it all.

Awakenings

The movie, *Awakenings*, also confirms or verifies this reality – namely that our memories are stored Non-Locally in the Cloud, and not in our physical brain.

https://www.amazon.com/dp/B001DFUUPG/

https://en.wikipedia.org/wiki/Awakenings_(book)

https://en.wikipedia.org/wiki/Awakenings

Notice how their memories are restored to them once their brain functionality is restored. That technically wouldn't be possible if their memories were stored within their physical brain. Their memories of prior experiences and learning should be unusable if their memories were actually stored within their damaged and non-functioning physical brain.

Individual Case Studies

Let's run through a few quotes from *Neuroscience* and observe how differentiating between Brain Plasticity and Non-Local Cloud Storage helps us to explain what we already KNOW to be true from the Lived Experiences of the human race, including our collective Near-Death Experiences (NDEs), Out-of-Body Experiences (OBEs), Shared-Death Experiences (SDEs), and other types of Non-Local Experiences or Spiritual Experiences.

Let's do a review first.

Remember, enhanced explanatory power and a preponderance of the evidence are what we are going after with Quantum Neuroscience and Science 2.0.

Science 2.0 and Quantum Neuroscience are Empirical Sciences or Phenomenological Sciences, because they are based primarily on the Lived Experiences of the human race – all of our observations, experiences, phenomena, events, science experiments, repeatedly verified science like Quantum Mechanics, eye-witness evidence, and empirical evidence including our NDEs and OBEs and SDEs.

Because we are primarily physical beings in this mortal life, most of us can't achieve replicability nor reproducibility when it comes to our Non-Local Experiences or Spiritual Experience, but we can indeed OBSERVE the replication of Non-Local Experiences or the similarities between different NDEs and OBEs and SDEs. Science 2.0 is a better way of doing science, because it allows ALL of the evidence into evidence, and it then goes with a preponderance of the evidence. Science 2.0 allows the science experiments from Quantum Mechanics into evidence, rather than trying to sweep them under the carpet.

Remember, the BEST way to find and know the truth is to live it and experience it for yourself, or to choose to trust someone who has.

The second best way to find and know the truth is to eliminate ALL of the falsehoods, including everything that has been successfully falsified such as Materialism, Naturalism,

Darwinism, Nihilism, and Atheism. If you successfully eliminate everything that is false, then only the truth will remain.

Notice how I tend to start my different books by falsifying and eliminating Materialism, Naturalism, Darwinism, and Atheism. There's method to my madness – the Scientific Method. The Scientific Method was designed to eliminate everything that is false, and it's very good at doing that, if we allow it to do so. Remember, the second best way to find and know the truth is to eliminate everything that is false and everything that has been falsified, including Materialism, Naturalism, Darwinism, Nihilism, and Atheism. This is logic 101. It's elementary my dear reader. It works.

How often have I said to you that when you have eliminated the impossible [and the false], whatever remains, *however improbable*, must be the truth? — Sherlock Holmes.

If you choose to apply the Orthodox Interpretation of Quantum Mechanics to Neuroscience, then you end up with Quantum Neuroscience. The whole process starts by eliminating everything that is false and everything that has been falsified, including Materialism, Naturalism, and their derivatives.

Quantum Neuroscience

Quantum Neuroscience starts with what WE KNOW to be true phenomenologically – through science experiments, direct observations, eye-witness evidence, and the lived experiences of the human race; and then, we try to find a logical and rational scientific explanation for what we already KNOW to be true or have already experienced.

I have observed that when it comes to science experiments, astral projection, near-death experiences, out-of-body experiences, non-local experiences, spiritual experiences, non-local consciousness, action at a distance, non-locality, teleportation, phase-shifting, telekinesis, forces, fields, gluons, photons, gravitons, bosons, magnetism, electrons, quarks, space-time, dark matter, tachyons, spirit matter, dark energy, psyche, light, and other types of observed events or experienced non-physical phenomena, Quantum Mechanics or Transdimensional Physics does an infinitely better job of explaining what's going on than Materialism and Naturalism do. Materialism and Naturalism simply deny the existence of these things, and leave it at that. That's denial, not science! We never learn anything new through denial and self-deception. I KNOW, because I used to be a Materialist, Naturalist, Nihilist, and Atheist. These things are based upon a refusal to look at evidence.

Quantum Neuroscience is the scientific study of how the Human Psyche interacts with and controls its physical brain. For all practical purposes, Quantum Neuroscience is almost exclusively a human science.

Why?

It's because only human beings, human psyches, have the innate capability to record and share their Observational Experiences, Science Experiments, Non-Local Experiences, Transdimensional Experiences, Quantum Experiences, Out-of-Body Experiences (OBEs), Near-Death Experiences (OBEs), Shared-Death Experiences (SDEs), Psyche Experiences, Psychic Experiences, revelations from God, revelations of God or Theophanies, and other types of Spiritual Experiences.

Quantum Neuroscience is the study of how the Human Psyche interacts with and controls its physical brain, its spirit body, and spirit matter in the Non-Local Transdimensional Realms.

Quantum Neuroscience is primarily an Observed Science, an Experienced Science, an Experimental Science, an Experiential Science, an Empirical Science, an Eye-Witness Science, and a Recorded Science. Quantum Neuroscience is based totally and exclusively upon Observational Evidence, Experiential Evidence, and Empirical Evidence. In a very real sense, Quantum Neuroscience is the ultimate science. This means that Quantum Neuroscience is based upon Phenomenology. Phenomenology is the study of the Lived Experiences of the human race, including our Science Experiments, Quantum Experiences, Non-Local Experiences, OBEs, NDEs, SDEs, Theophanies, revelations from God, and other types of Spiritual Experiences. Phenomenology is also almost exclusively a human science.

The first book to send me down this path into the Experiential Sciences or the Phenomenological Sciences in 2015 was:

Buhlman, W. L. (1996). *Adventures Beyond the Body: How to Experience Out-of-Body Travel*. New York: HarperCollins Publishing.

There have been many others since then. His book opened me up to infinitely more. We each have a handful of books that really get through to us in the end, and this was one of them for me.

Remember, the BEST way to find and know the truth is to live it and experience it for yourself, or to choose to trust someone who has. William Buhlman was one of the first persons whom I encountered who has lived it and experienced it. I learned to trust him, because his out-of-body experiences are matched and exceeded by many others whom I have encountered and gotten to know since then, either in person or online. William Buhlman learned how to leave his physical body and explore the Astral Plane or the Non-Local Realms. While doing so, he tried to turn his out-of-body experiences into science experiments. He experimented with it and tried things out. He has also tried to teach others how to leave their physical body and explore the Non-Local Realms with their spirit body and their psyche or non-local consciousness. Quantum Neuroscience is based upon the observational evidence and experiential evidence (phenomenology) that's been obtained from various different Out-of-Body Experiences and exploration of the Non-Local Realms, or the Astral Plane, or the Spirit World. Phenomenology is the scientific study of events, phenomena, and experiences, including all of our Science Experiments, OBEs, NDEs, SDEs, and other types of Non-Local Quantum Experiences.

It's interesting that it took me 20 years to find William Buhlman's book. Why did it take me so long? It's because I wasn't looking for it and didn't believe that it existed. I wasn't ready for it. I spent time as a Materialist, Nihilist, Skeptic, and Atheist instead. I didn't believe that there could ever be any Scientific Proof of God's Existence; so for me personally, there wasn't any because I wasn't looking for it and because I didn't believe that it was possible. It was a self-fulfilling prophecy. No seeking, then no finding. That's just the way it works.

Quantum Neuroscience is way beyond fringe science – it's in a whole other dimension. Quantum Mechanics is Transdimensional Physics – the scientific study of how spirit matter and psyche function and work in the non-local realm, or transdimensional realm, or non-physical realm, or spirit world.

Quantum Neuroscience is the scientific study of how the Human Psyche interacts with and controls its physical brain.

There are hundreds of different interpretations of Quantum Mechanics. The BEST and most useful interpretation of Quantum Mechanics for explaining how the Human Psyche interacts with and controls its physical brain is the Orthodox Interpretation of Quantum Mechanics developed by John von Neumann, Henry P. Stapp, Mario Beauregard, and Jeffery M. Schwartz.

According to Orthodox Quantum Mechanics, the **Human Psyche** negotiates with or interacts with **Nature's Psyche** in order to trigger and control our physical responses. The Human Psyche queries or commands, and Nature's Psyche responds as well as it can. The Human Psyche was designed to act, to decide, and to choose. That's how the Human Psyche is identified and differentiated from Nature's Psyche. Nature's Psyche, which controls each atom and molecule of physical matter, was designed to respond or to react to the commands and requests of the Human Psyche.

"Quantum Physics in Neuroscience and Psychology: A Neurophysical Model of Mind-Brain Interaction"

https://www.researchgate.net/publication/7613549_Quantum_physics_in_neuroscience_and_psychology_A_neurophysical_model_of_mind-brain_interaction

http://www-physics.lbl.gov/~stapp/PTRS.pdf

http://mypsyche.us/wp-content/uploads/2017/10/PTRS.pdf

Quantum Neuroscience is a function of Psyche, whether we are talking about the Human Psyche or Nature's Psyche.

The physicists call Nature's Psyche the physical laws and the physical constants of the universe. The more creative among the bunch call it the Zero-Point Field of Light. The Astrophysicists seem to call this thing Dark Energy.

The New Agers call this thing the Quantum Sea of Light.

The theologians call Nature's Psyche the Mind of God, or the Light of Christ, or the Holy Spirit, or God's Psyche.

WE KNOW it exists because Out-of-Body Travelers have experienced it; and, so have the prophets of God.

According to Orthodox Quantum Mechanics, the **Human Psyche** negotiates with or interacts with **Nature's Psyche** in order to trigger and control our physical responses. The Human Psyche queries or commands, and Nature's Psyche responds as well as it can.

Some people equate Nature's Psyche with God's Psyche, but they are two different things. God covenanted with the psyche or intelligence within unorganized spirit matter offering to give it order, purpose, meaning, light, and life in exchange for its obedience to His commands. Matter of any kind has to obey God's Commands or God's Laws, or it will remain forever unorganized and purposeless. Physical matter is a highly ordered and organized type of spirit matter, which now follows God's Physical Laws. Only God knows how to convert spirit matter into physical matter. Each atom of physical matter had a beginning, which means that it had a Beginner or an Originator.

Each particle of physical matter has covenanted to obey God's Commands, which explains why Jesus Christ was able to do the things that He did while He was here on earth. God commanded Nature's Psyche within each atom and molecule of your physical body and physical brain to obey the commands of your Human Psyche to the best of their ability. Only God knows how to marry or bind a Human Psyche with a particular physical body and physical brain. We humans wouldn't know where to begin. The silver cord, which has been observed, seems to be the tie between our spirit body and our physical body. We have no idea what ties our Psyche or Non-Local Consciousness to our spirit body.

Quantum Neuroscience is an Experiential Science, Observational Science, Empirical Science, Human Science, and Phenomenological Science which means that it is based upon the Lived Experiences of the human race.

WE KNOW from the Lived Experiences of the human race that it is impossible to propel physical matter faster than the speed of light. God has placed a speed limit on the acceleration of physical matter, and there is no way for us to get around it. God deliberately slowed physical matter down, so that we can live it, experience it, learn from it, and remember having done so.

We also KNOW that there is no such thing as Parallel Worlds or Multi-Me's, because such a thing has never been lived nor experienced by anyone. We KNOW that the Parallel Worlds Interpretation of Quantum Mechanics is false, because it has been falsified by a complete lack of experience and observation.

However, we KNOW that Quantum Tunneling is real and truly happens. It has been observed and experienced. I, personally, have lived it and experienced it, so I KNOW that it is real and true. God can teleport physical matter – quantum tunnel physical matter – at will anywhere in this universe instantly, by stopping the passage of time for a physical object or temporarily removing all the entropy from a physical object. Quantum Tunneling doesn't involve acceleration. It involves teleportation instead. This is something that we can't do, because God prevents us from doing it; but, it's definitely something that God can do for us.

I was going 40mph. A car pulled out in front of me, and there was nowhere for me to go. I was right on top of that car. I remember hitting the brakes. Then I felt the whole of eternity fold into me, and time came to a complete standstill. I could sense everything around me 360 degrees, and I could semi-sense where I wanted to be. It was like I was a fly in amber, or I was in a bubble of protection. I had an eternity of time to sense, and feel, and analyze everything. Then snap. Time resumed, and I and my car were twenty or thirty feet away in the middle of a lawn. My car was completely stopped, and the engine had been turned off. I have NO memory of traveling the distance and NO memory of how I got there. I was just simply there. I had gone from 40 mph to 0 mph instantaneously. My car was running one second, and instantly stopped and off the next second. My foot was still on the brakes, and I still hadn't had time enough to actually turn the steering wheel. I was supernaturally calm. There hadn't been time enough for my adrenaline to start pumping. It's the strangest and most amazing thing that has ever happened to me and that I have ever experienced. It wasn't anything that I did. It was something that God did for me, in order to save me.

When it comes to Quantum Tunneling, it has been observed and experienced in our science experiments. Furthermore, some of us have lived it and experienced it in the flesh. It was nothing that I did. It was something that God did for me, in order to save me. God holds all the KEYS to Quantum Tunneling or Teleportation.

WE KNOW from experience that it is impossible to propel or drive physical matter faster than the speed of light. God designed it to be that way as part of the Physical Laws of our physical universe. But, there is a work-around that has been observed, lived, and experiences – Quantum Tunneling or teleportation of physical matter. Quantum Tunneling or teleportation is observed and proven science.

The reason that the atoms in your physical body and physical brain don't quantum tunnel away from you one at a time to the other side of the universe is because God commands them not to. When it comes to your physical body and physical brain and physical world, God is holding all of it together for you; otherwise, it would all quantum tunnel away on you until nothing remained.

By creating physical matter, physical genomes, physical bodies, physical brains, and physical laws, God created the Ultimate Consensus Reality. The beauty of a physical reality is that we can depend upon our physical body, our homes, our dog, our spouse, our children, our car, our computer, and our papers to be here tomorrow when we go looking for them. A physical reality is the Ultimate Consensus Reality. It's highly reliable and dependable. It's not going to dissolve and quantum tunnel away on you because God is preventing it from doing so. This is additional Scientific Proof of God's Necessity, which ends up being Scientific Proof of God's Existence. The fact that your computer, your car, your house, and your physical body doesn't dissolve and quantum tunnel away on you is Scientific Proof of God's Influence and therefore Scientific Proof of God's Existence.

This is really cool stuff, but only when you are ready for it and willing to accept it. It was impossible to accept and impossible to understand, when I was a Materialist, Nihilist, and Atheist. I had to upgrade my science before I was willing and able to go there.

Science 2.0 allows all of the evidence into evidence, and Science 2.0 gives preference and precedence to lived experiences, observed experiences, and eye-witness evidence over scientific inferences and philosophical speculation. Things like Materialism, Naturalism, Nihilism, Atheism, and the Parallel Worlds Interpretations of Quantum Mechanics are successfully falsified by a complete lack of observational and experiential evidence. In contrast, THE TRUTH is repeatedly and endlessly verified by observational evidence and experiential evidence over and over again, because THE TRUTH IS KNOWN by living it and experiencing it or by choosing to trust someone who has.

I experienced Quantum Tunneling first-hand. Others have experienced Phase-Shifting. My best friend has experienced Phase-Shifting, as an animal passed through his truck unscathed. According to phase-shifting, multiple particles of physical matter and spirit matter can occupy the same space if they are each in a different phase. If you phase-shift, you can walk through walls. Phase-Shifting has been lived and experienced by many different people. The Spirit World is right here where we are, just in a different phase. Likewise, multiple physical objects can occupy the same space, if each one of them is slightly out-of-phase from the others.

Some people have experienced revelations from the Biblical God Jesus Christ and/or have seen Him and touched Him during their NDEs and OBEs. My best friend is one of these. Many out-of-body travelers have experienced Quantum Telepathy – mind-to-mind or psyche-to-psyche communication. NDErs and OBErs routinely experience Quantum Telepathy, because it seems to be the standard and normal way of communicating in the Quantum Realms or the Non-Local Realms. These people can sense, perceive, and feel the love and peace radiating from the Being of Light or Jesus Christ while in His presence in the Non-Local Realm. My best friend is one of these people.

Quantum Neuroscience is based upon Quantum Mechanics or Transdimensional Physics, the study of how spirit matter and psyche function and work in the Transdimensional Realms or Non-Local Realms. The very existence of Quantum Mechanics, Transdimensional Physics, and Quantum Neuroscience FALSIFIES Materialism, Naturalism, Darwinism, Nihilism, and Atheism; so of course, the Materialists and Naturalists are going to fight, resist, and reject Quantum Neuroscience, or they are no longer going to be Materialists and Naturalists. That's just the way it is. Quantum Neuroscience is billions, if not trillions, of years ahead of the Materialists and Naturalists and Atheists, because these people are not ready to go there yet, because it involves the Mind of God and the Will of God. These people don't want to have anything to do with these things, and they get their wish.

You can lead a horse to water, but you can't make him drink. You can lead an Atheist to evidence, but you can't make him think.

Commentary Regarding Quantum Neuroscience

I used to be a Materialist, Nihilist, and Atheist because I considered myself to be a Scientist – and for other personal reasons as well. I understand the mind-set, because I have lived it and experienced it for myself.

It should be nothing but random chaos and entropy between the synapses according to the Laws of Classical Physics. Yet, everything seems to function perfectly and reliably.

How is that possible?

It's NOT possible according to Materialism, Naturalism, Random Diffusion, Random Chance, Entropy, and the Laws of Classical Physics.

It ONLY becomes possible through the Laws of Quantum Mechanics and the influence of Nature's Psyche. There is no other logical explanation. Psyche produces Order out of Chaos.

It has been estimated that the brain of a three-year-old child has about 10^{15} synapses (1 quadrillion), where the exchange of neurotransmitters takes place.

That means that there are at least some quadrillion different possible ways for the Human Psyche to interact with and control its physical brain, through all those different neurotransmitters, neuropeptides, and ions. That's a lot of fire-power at Psyche's command! Psyche or Non-Local Consciousness IS Command and Control.

Take note that transistors need barriers – neurotransmitters do not. Neurotransmitters are simply dumped into the synaptic cleft, where Psyche or the Quantum Zeno Effect can do with them as it pleases.

The synaptic cleft, neurotransmitters, ions, neuropeptides, and the neurons themselves are where the Human Psyche interacts with and controls its physical brain and physical body. It all takes place at the Quantum Level or the Spiritual Level, through Telepathy or the Quantum Zeno Effect. Quantum Mechanics is Spiritual Mechanics, or the way that Spirit Matter, Physical Matter, and Psyche really work, and work together.

Yet, the Materialists and Naturalists and Darwinists want you to believe that evolution (mutation/selection) designed and created all of this all by itself billions of years ago. What do you think? Is it possible? The thing I want to know is: How did this mythical creature, evolution, interact with and control everything at the quantum level telepathically through the Quantum Zeno Effect, when evolution doesn't even have a psyche or a soul? Evolution would have to be a God in order to do such a thing, wouldn't it?

As you can tell, there's no going back to My Materialism, My Naturalism, My Scientism, My Nihilism, and My Atheism. I know better now.

I wanted to understand Quantum Neuroscience – the way that the human brain functions at the quantum level.

The Human Psyche needs a mechanism through which to interact with and control its physical brain – the brain is a mechanical object after all. The neurotransmitter system is that mechanism by which the Human Psyche controls its physical brain. The synaptic cleft and the ion exchange within neurons gives the Psyche plenty of time to intervene, which wouldn't be possible if the brain were one continuous wired circuit like a computer or a car.

Psyche, or Non-Local Consciousness, or The Chooser is a spiritual process or a non-local process – Process 1. The brain is a physical process, based upon Process 3's control of

Process 2. In other words, the brain is a physical process based upon Nature's control of the quantum collapse of the wave function. Process 1 or the Human Psyche interacts with and negotiates with Process 3 or Nature's Psyche, in order to determine the physical reality (or neurotransmitter releases and neurotransmitter uses) which will manifest or come into existence as a causal result of this psychical, non-local, telepathic negotiation or exchange.

The action potential fires in the neuron, the sparks fly down the axon; but, someone still has to decide which packet (vesicle) of neurotransmitters and/or neuropeptides to release from among the hundreds of different neuropeptides and neurotransmitters that are available. That someone is Psyche or Non-Local Consciousness, acting through the Quantum Zeno Effect. The process is so efficient that we don't even have to think about it. It just happens automatically, at Psyche's command and control. The neurotransmitters, neuropeptides, and ions simply do what your Psyche wants them to do. If you want to be sick, you are going to be sick. If you want to be well and filled with light, you are going to be well and filled with light. Only competing independent entities can change that reality – viruses, bacterial infections, parasites, and physical injury or physical damage.

Telepathy is like radio waves and radio signals, but at a much smaller scale and a much higher frequency. There's nothing magical about it. The Quantum Zeno Effect – the telepathic interaction between the Human Psyche and Nature's Psyche – takes place at the level of vibrating strings, quarks, electrons, gluons, and bosons. That's why Psyche has no problem interacting with and controlling atoms, molecules, amino acids, DNA, proteins, ribosomes, ions, neurotransmitters, and neuropeptides. Remember, the smaller can reside within and control the larger. Psyche or Non-Local Consciousness is the smallest elementary particle of them all.

Process 1, or the Human Psyche, is the Actor, Agent, Prober, Questioner, Chooser, and Decider. Process 3, or Nature's Psyche, simply responds with a Yes or a No. The psyche within Nature or physical matter doesn't have an agentic capacity. Nature's Psyche simply responds by collapsing the wave function, based upon what's physically possible or allowed by God. The Psyche or Non-Local Consciousness in Nature or physical matter is alive, sentient, and aware; but, it was designed to respond and react – not act. Nature's Psyche or the physical matter has covenanted with God to obey God's laws – what we typically call the physical laws. That's how God brought order out of chaos, where our physical universe is concerned.

There were things that were designed to act, and things that were designed to be acted upon.

I find it fascinating that the Biblical God Jesus Christ revealed these truths to some of His chosen prophets over a thousand years ago.

2 Nephi 2: 13-14, 26:

11 For it must needs be, that there is an opposition in all things. If not so, my firstborn in the wilderness, righteousness could not be brought to pass, neither wickedness, neither holiness nor misery, neither good nor bad. Wherefore, all things must needs be **a compound in one**; wherefore, if it should be one body it must needs remain as dead, having no life neither death, nor corruption nor incorruption, happiness nor misery, neither sense nor insensibility.

12 Wherefore, it must needs have been created for a thing of naught; wherefore there would have been **no purpose** in the end of its creation. Wherefore, this thing must needs destroy the wisdom of God and his eternal purposes, and also the power, and the mercy, and the justice of God.

13 And if ye shall say there is no law, ye shall also say there is no sin. If ye shall say there is no sin, ye shall also say there is no righteousness. And if there be no righteousness there be no happiness. And if there be no righteousness nor happiness there be no punishment nor misery. And if these things are not there is no God. **And if there is no God we are not, neither the earth; for there could have been no creation of things, neither to act nor to be acted upon; wherefore, all things must have vanished away.**

14 And now, my sons, I speak unto you these things for your profit and learning; for there is a God, and he hath created all things, both the heavens and the earth, and all things that in them are, **both things to act and things to be acted upon.**

15 And to bring about his eternal purposes in the end of man, after he had created our first parents, and the beasts of the field and the fowls of the air, and in fine, all things which are created, it must needs be that there was an opposition; even the forbidden fruit in opposition to the tree of life; the one being sweet and the other bitter.

16 Wherefore, the Lord God gave unto man that he should act for himself. Wherefore, man could not act for himself save it should be that he was enticed by the one or the other.

22 And now, behold, if Adam had not transgressed he would not have fallen, but he would have remained in the garden of Eden. And all things which were created must have remained in the same state in which they were after they were created; and they must have remained forever, and had no end.

23 And they would have had no children; wherefore they would have remained in a state of innocence, having no joy, for they knew no misery; doing no good, for they knew no sin.

24 But behold, all things have been done in the wisdom of him who knoweth all things.

25 Adam fell that men might be; and men are, that they might have joy.

26 And the Messiah cometh in the fulness of time, that he may redeem the children of men from the fall. And because that they are redeemed from the fall they have become **free forever**, knowing good from evil; **to act for themselves and not to be acted upon**, save it be by the punishment of the law at the great and last day, according to the commandments which God hath given.

27 Wherefore, men are free according to the flesh; and all things are given them which are expedient unto man. And they are free to choose liberty and eternal life, through the great Mediator of all men, or to choose captivity and death, according to the captivity and power of the devil; for he seeketh that all men might be miserable like unto himself.

28 And now, my sons, I would that ye should look to the great Mediator, and hearken unto his great commandments; and be faithful unto his words, and choose eternal life, according to the will of his Holy Spirit;

29 And not choose eternal death, according to the will of the flesh and the evil which is therein, which giveth the spirit of the devil power to captivate, to bring you down to hell, that he may reign over you in his own kingdom.

30 I have spoken these few words unto you all, my sons, in the last days of my probation; and I have chosen the good part, according to the words of the prophet. And I have none other object save it be the everlasting welfare of your souls. Amen.

Your computer system or the wiring in your car or home is a "compound in one". There's NO Process 1 actor or chooser there. It simply reacts to your commands or your chosen actions, as you pound away on the keyboard or turn the steering wheel. Psyche provides purpose; whereas, physical matter and physical systems like computers simply react. Computers have no feelings about your choices, one way or the other. There's no intelligent actor there, and there never will be. A computer is a Responder or a Reactor – purely process 2 – the collapse of the wave function into a physical reality.

Process 1, or the Actor, or the Human Psyche puts free will and choice into the system. In contrast, Nature's Psyche (Process 3) and physical matter (Process 2) have no choice but to react, one way or the other. From what I can tell, if the Human Psyche's demands are physically possible, Nature reacts with a Yes command; otherwise, Nature reacts with a No command – and then the physical matter (Process 2) simply responds to Nature's command accordingly by collapsing the wave function into a single chosen reality. Technically, there's no chooser or actor in Processes 2 and 3. All of the choosing and decision making takes place within the Human Psyche and Process 1. This whole thing is Mind-Over-Matter.

When it comes to neurotransmitters and the physical brain, the Human Psyche is God. God-speed those neurotransmitters

God also designed it this way in order to increase redundancy. If the brain misfires, or doesn't fire, or stops functioning temporarily, the Human Psyche is still there to take up the slack and hopefully get it back on track.

Only when the physical brain is so extensively damaged that it's no longer functioning properly does Psyche lose complete control of the situation – physical death is often the result, because the Human Psyche no longer has control of its physical brain.

Why did God design it that way?

In order to give the Human Psyche lots of different places where it can directly control its physical brain.

Each Quantum Object – ion, electrolyte, neurotransmitter, neuropeptide, molecule, atom, protein, enzyme, ribosome, and cell – is a location where the Human Psyche can interact with and control its physical brain and physical body, through the Quantum Zeno Effect. In any new action sequence, someone has to trigger that first action potential, in order to get the whole chain of action potentials going. Imagine what you would have if the action potentials in your physical brain were firing randomly by chance as the Materialists and Naturalists claim. It would be chaos.

The Quantum Zeno Effect is where all of the magic takes place, except it isn't magic. The Quantum Zeno Effect is proven and verified science. The Quantum Zeno Effect is the telepathic connection between Process 1 and Process 3. The Quantum Zeno Effect is the telepathic link or psychic link between the Human Psyche and Nature's Psyche.

That's why our brain and our physical body isn't one continuous electrical circuit, because there would be nothing through which the Human Psyche could interact, if our brain were one continuous electrical circuit like a computer or a robot. The ion channel system, action potential system, and neurotransmitter system are needed in order to give the Quantum Non-Local Human Psyche locations to interact at the quantum level with its physical brain and to control that brain directly by starting action potentials whenever it needs to do so.

That's why God designed it that way!

We human beings can design and create electrical circuits, computer chips, and computers; but, it would take a God to be able to design and create a machine that can interact with and take instructions from the Human Psyche at the Quantum Level telepathically through the Quantum Zeno Effect. We humans wouldn't even know where to begin.

In fact, the naturalistic and materialistic scientists don't believe that Psyche and Telepathy exist, so why would these people even try to make a physical machine capable of interfacing with and interacting with Psyche or Non-Local Consciousness telepathically through the Quantum Zeno Effect? These people wouldn't even think to try.

So, why would God design a physical brain with a neurotransmitter system?

This is a case where the answer lies within the question, but only if the question is answered truthfully and accurately.

The Materialists and Naturalists and Darwinists ask similar questions, or at least they should.

"How did evolution design, create, field-test, and fine-tune our brains, our genomes, our physical bodies, and our neurotransmitter system all from scratch?"

Answer: Evolution didn't. It's impossible. Evolution, random mutations, and natural selection can't design and create anything at all. They are simply messengers of entropy, disease, death, and extinction. Mutation and selection can't design and create. It's impossible.

The Materialists, Naturalists, and Darwinists want you to believe that evolution designed and created your brain, your neurotransmitter system, your proteins, your genome, and your physical body. It's impossible. We humans can't even design and create such a complex system, so why on earth would anyone believe that evolution (random mutations and natural selection) designed and created our genomes and our brains? It's impossible. Evolution, random mutations, and natural selection didn't even exist until AFTER God designed and created and manufactured the genomes, brains, neurotransmitter systems, and life forms in the first place.

Since I wasn't getting what I needed from the Materialists and Naturalists, I had to discover and create Quantum Neuroscience for myself, so that I would have something that actually matches with the Observational Evidence, Experimental Evidence, Experiential Evidence, Eye-Witness Evidence, and Empirical Evidence found within the Lived Experiences of the human race.

Remember, Science 2.0 defines science as Observation, Experience, Knowledge, Fact, and Truth. Quantum Neuroscience is by definition in principle an Observed Science, Experimental Science, and Applied Science – no scientific inferences and philosophical speculations allowed.

IN SUMMARY – QUANTUM NEUROSCIENCE

I used to be a Materialist, Nihilist, and Atheist. I'm also a Scientist. I was trained to look at everything in terms of Classical Newtonian Physics. All of this has been ingrained into me and trained into me. You could say that I have been conditioned and brainwashed into Classical Physics, Materialism, and Naturalism, as have most of us.

According to the Laws of Classical Newtonian Physics, the physical brain should be nothing but Random Diffusion, Random Chance, Brownian motion, Maximum Entropy, Unorganized Chaos, and Clogged Full with Garbage within the Synaptic Clefts between the Synapses. Yet, the physical brain is observed to be and experienced to be dynamic and alive – the very Model of Efficiency.

How's that possible?

It's not possible, according to the Laws of Classical Physics, Materialism, Naturalism, Nihilism, and Atheism.

According to Classical Newtonian Physics and Naturalism, our physical brain and the cells within it should be just like any other jar of chemicals. Take ALL the chemicals and fluid between the synapses in the human brain and place them into a bottle or a jar. What have you got? I've got a jar of dirt! Since you are classier than I am, you've got a bottle of unflavored Gatorade! Take ALL the molecules and fluid floating within our neurons and glial cells, place that into a bottle or a jar, and you've got yourself some primordial soup. According to the Laws of Classical Physics, Naturalism, and Materialism, our physical brains should be nothing more than canisters of Gatorade – a random diffusion of chemicals.

I really had to up my game if I wanted to understand this thing.

Can you see now why I have had serious problems with synapses and our neurotransmitter system? I used to be a Materialist, Nihilist, and Atheist trained, or conditioned, or brainwashed into Classical Newtonian Physics just like any so-called "Good Scientist" should be.

The neurotransmitter system and synapses seemed highly inefficient to me, with too much room for error and quantum indeterminacy. I was thinking about it materialistically and naturalistically as a scientist and engineer. I equate efficiency with beauty. I was really frustrated with the neurotransmitter system and all of its apparent inefficiencies and needless complexity – all these little buggers just floating around between the synapses or terminal buttons willy-nilly – where chance and probability seemed to rule supreme. The frustration was real. I wrote in the margins of my college Neuroscience textbook, "Why did God design it this way?" I didn't get it, until I finally saw it.

At the time, I actually said a prayer and asked God why He did it that way; and suddenly, it came to me and I KNEW. That was quite an epiphany. I have been trying to put that single flash of insight into words ever since; so, here goes – once again.

The neurotransmitters are one of the ways that the Human Psyche interacts with and controls its physical brain through the Quantum Zeno Effect.

Simple, actually. Don't you just love parsimony?

All throughout my life, the neurotransmitter system seemed hopelessly inefficient to me, because I was thinking of it materialistically and naturalistically, which meant that I was seeing neurotransmitters as nothing but random chance, entropy, and the Second Law of Thermodynamics.

For over fifty-five years, I had never once looked at neurotransmitters from the perspective of the Human Psyche and the Quantum Zeno Effect. When I finally did, it dawned on me that the neurotransmitters are where the Human Psyche commands and controls its physical brain through the Quantum Zeno Effect, thereby effectively converting random chance into CHOICE.

The neurotransmitters are not a function of random chance as I had once assumed; but instead, the neurotransmitters are where the Human Psyche exercises its choices and its control over its physical brain. That was quite an epiphany, and it's already starting to bear fruit, tying together lots of other things that I didn't understand or know before. I once was blind, but now I see.

The Human Psyche can decide whether to release the neurotransmitters in the first place by opening the calcium ion channels in the synapse or terminal button, thereby forcing the release of the neurotransmitters and a subsequent chain of action potentials as a result. The Quantum Zeno Effect (telepathy from the Human Psyche) combines with the opening of calcium ion channels in the presynaptic terminal button and the subsequent release of neurotransmitters into the synaptic cleft, thereby triggering an action potential cascade sequence as a result. This is how the Human Psyche controls its physical brain, at the quantum level telepathically through the Quantum Zeno Effect.

The Quantum Zeno Effect is the KEY to understanding how the Human Psyche triggers action potentials and controls its neurotransmitter system. This is Quantum Neuroscience in action! This is the stuff that I wanted to learn in college; but, you will NEVER learn anything about this from the Materialists, Naturalists, Nihilists, and Atheists who write our college textbooks and run our public schools.

Instead, you have to turn to others, if you truly want to find and KNOW how the Human Psyche controls its physical brain and physical body. You have to turn to the Laws of Quantum Mechanics such as Quantum Non-Local Consciousness (the human psyche and nature's psyche), Quantum Non-Locality (the transdimensional realm or spirit world), the Law of Quantum Complementarity (physical matter and spirit matter), Quantum Entanglement (action at a distance or telepathy), the Quantum Zeno Effect (again, action at a distance or telepathy or remote control), and Quantum Tunneling (telekinesis and teleportation).

There is NO other way to explain how the Human Psyche controls its physical brain. Classical Physics can't explain it. In fact, Classical Physics, Materialism, Naturalism, Nihilism, and Atheism tell us that ALL these things are impossible and do not exist. BUT WE KNOW, from observation and experience and experimentation, that Classical Physics, Materialism, and Naturalism are wrong. These things are FALSE. Our physical brain really is under our command and our control. That is what we have observed, and that is what we have experienced.

Figure 8: Quantum Effects of Attention

The rules of quantum mechanics allow attention to influence brain function. The release of neurotransmitters requires calcium ions to pass through ion channels in a neuron. Because these channels are extremely narrow, quantum rules and the Uncertainty Principle apply. Since calcium ions trigger vesicles to release neurotransmitters, the release of neurotransmitter is only probabilistic, not certain. In quantum language, the wave function that represents "release neurotransmitter" exists in a superposition with the wave function that represents "don't release neurotransmitter"; each has a probability between 0% and 100% of

becoming real. Neurotransmitter release is required to keep a thought going; as a result, whether the "wash hands" or "garden" thought prevails is also a matter of probability. Attention can change the odds on which wave function, and hence which thought, wins.

Schwartz, Jeffrey M. *The Mind and the Brain: Neuroplasticity and the Power of Mental Force* (p. 364). HarperCollins. Kindle Edition.

There's NO need for there to be any neurotransmitters and synapses in a computer chip, computer, house, phone, or car because there's no need for telepathic control, telekinetic control, and psychical control where computers are concerned. We wouldn't know how to design and build such a thing anyway, if there were such a need. How do you marry a specific psyche to a specific brain? How do you tap into the telepathic connection between the Human Psyche and Nature's Psyche? You don't with computers. Instead, you simply build barriers between the transistors in order to prevent quantum tunneling from happening within your computer chips. Quantum tunneling is bad for computers and computer chips, so we put up barriers to prevent it from happening.

There are NO barriers between synapses in the physical brain, thereby giving the Human Psyche complete control and full rein to do as it pleases with all of those different neurotransmitters floating around between the synapses. Technically, the Human Psyche can control neurotransmitters at the quantum level telekinetically through the Quantum Zeno Effect and make those neurotransmitters go wherever it needs them to go, if enough attention or mental force or "word of command" is applied.

All of the physical particles in this universe have covenanted to obey God's Laws and God's commands; and, God has commanded the neurotransmitters within our physical brain to obey the commands of the Human Psyche to which they have been assigned. Simple.

I found it fascinating, when I finally realized that a Human Psyche theoretically has complete control over all of the different atoms, molecules, proteins, ribosomes, ions, neuropeptides, and neurotransmitters within its physical brain and physical body – at least when the physical body is functioning properly – the physical body and physical brain are machines after all with all the limitations of a physical object.

With a properly functioning physical body and physical brain, the Human Psyche can learn to have complete control over it all, through the Quantum Zeno Effect. Furthermore, Jeffrey Schwartz and team have documented and proven through Brain Scans that, with the right kind of Psyche-Therapy, the Human Psyche can actually rewire and normalize its physical brain.

Interesting, is it not?

Here, once again, we find God hiding in plain sight where the Materialists, Naturalists, and Atheists will never see Him and never find Him; and, we don't ever have to worry about the Materialists and Naturalists finding and learning how to use this kind of information, because they would never buy and read a book such as this anyway.

Now I finally had the answer to the question that had been nagging on me all my life – why was it done that way? Why all of those neurotransmitters and synapses? It all seemed so inefficient and pointless to me. I didn't like it. I wanted our physical brain to be like a computer – fast and completely reliable. I didn't understand at the time that God simply had other ideas and goals in mind, while He was designing and creating our neurotransmitter system, our genome, and our physical brain. God was successfully hiding from me within Quantum Science and Quantum Neuroscience, where I would never see Him and never find Him until after I finally started looking for Him.

Remember, the Quantum Zeno Effect is the telepathic connection between Process 1 and Process 3. The Quantum Zeno Effect is the telepathic or psychic link between the Human Psyche and Nature's Psyche. Psyche IS the Quantum Zeno Effect!

Computers are purely Process 2 – the wave function collapse of the physical matter. There's NO NEED for and no place within a computer chip where the Human Psyche can interact directly with the ions, neurotransmitters, neuropeptides, molecules, and atoms; consequently, the electrical circuits in a computer chip can be hard-wired and unchanging – the more unchangeable and reliable the better – with lots of reliable barriers between all of those different transistors to prevent quantum tunneling from taking place.

In contrast, the human brain has to be infinitely more dynamic and flexible and plastic because it's interacting directly with the Human Psyche and needs to be able to adapt on the fly.

The Human Psyche needs a mechanism through which to interact with and control its physical brain – the brain is a mechanical object after all. The neurotransmitter system is that mechanism by which the Human Psyche controls its physical brain, telepathically or remote-control through the Quantum Zeno Effect. The synaptic cleft and the ion exchange in the neurons gives the Psyche plenty of time to intervene, which wouldn't be possible if the brain were one continuous wired circuit like a computer or a car.

This is the information that I wanted to know and understand, because this is where God shows Himself to us and reveals Himself to us – at the quantum level, through our genome and neurotransmitter system. After learning about these things, it was clear and obvious to me that evolution (random mutations and natural selection) flying blind could have NEVER designed, created, manufactured, and fine-tuned our neurotransmitter system at the quantum level or spiritual level. I already KNEW that evolution could never have designed and created our genome; but now, I also KNOW that evolution could never have designed, created, and fine-tuned our physical brain and our neurotransmitter system. The writing was on the wall.

Thankfully, we human beings will never build a machine whose personality, or intelligence, or psyche is capable of surviving the death of its physical body. Pull the plug and smash the machine to bits, and that is literally the end of it. It's dead Jim. It's not going to be showing up in the afterlife as a unique person or personality with a Life Review. Depending upon who does the poll and the types of questions asked, it has been estimated that most of our scientists don't believe in the existence of a psyche, soul, non-local consciousness, or afterlife; so, these people are never going to build a machine that has a psyche and an afterlife. They wouldn't know how to do so, even if they tried.

I have watched *AI*, *I Robot*, the *Terminator Movies* and series, the *Matrix* movies, and *Ex Machina*. Artificial intelligence – it's impossible. Sentient computers are impossible. We wouldn't know how to design and create such a thing, even if they were possible; and, raw physical matter doesn't spontaneously come alive.

There's no way that we human beings could ever design and create a machine that's capable of interacting with a Human Psyche telepathically through the Quantum Zeno Effect – we don't know how God marries a specific psyche with a specific brain. Most of us don't even believe that telepathy and the human psyche exist and are real, so how could we ever create a physical machine that interacts with a Human Psyche telepathically through the Quantum Zeno Effect? We can't. Only God could do such a thing! We wouldn't even know where to begin; and, we are infinitely more intelligent and prescient than evolution, random mutations, and natural selection could ever be. We actually have a brain.

A physical machine can either survive the death of its physical body, or it can't. We human beings are such a machine. Our computers and robots are not. We have Life Reviews after we die, and continue to function and live after we die – our computers and robots do not.

Artificial Intelligence or an Artificial Psyche is impossible for humans to construct. We don't know how to make such a thing and marry it with non-local consciousness.

Penrose, R. (1994). *Shadows of the Mind: A Search for the Missing Science of Consciousness*. New York: Oxford University Press.

Rychlak, J. F. (1991). *Artificial Intelligence and Human Reason: A Teleological Critique*. New York: Colombia University Press.

Notice how each human being is given a small piece of the puzzle. If we combine together everything we KNOW and have EXPERIENCED as a race, we start to approach THE TRUTH – as long as we are willing to eliminate everything that is false and has been falsified such as Materialism, Naturalism, Darwinism, Nihilism, Behaviorism, Determinism, Scientism, and Atheism.

There are NO synapses and neurotransmitters within a computer, because a computer doesn't have a Psyche that it needs to interact with and take instructions from at the quantum level. Instead, we actually have to make our computer circuits, computer chips, and transistors large enough and wide enough so that our computers don't accidentally develop a mind of their own and crash, through quantum interference and quantum tunneling. We need lots of barriers in computer chips, not open synaptic clefts where quantum mechanics reigns supreme.

Remember, the smaller can dwell within, interact with, and control the larger. The Psyche is smaller than strings, quarks, gluons, bosons, and electrons; therefore, the Human Psyche can dwell within, interact with, and control these things telepathically through the Quantum Zeno Effect. For all intents and purposes, Psyche IS the Quantum Zeno Effect.

Neurotransmitters, Synapses, Ions, and Action Potentials!

Why did God design it and build it this way?

So that there would be quadrillions of different locations where the Human Psyche can interact with and control its physical body and physical brain at the quantum level. Remember, Psyche functions at the quantum level or the spiritual level, and not the physical level. There has to be a quantum interface between Psyche and the Physical Brain; otherwise, we would be vegetables and nothing but mindless computers and robots, in desperate need of programming. The Ions, Electrolytes, and Neurotransmitters ARE that quantum interface between the Human Psyche and its physical brain.

That's why God did it that way!

This reality and truth becomes obvious and clear once a person understands and accepts Process 1 and Process 3 of the Orthodox Interpretation of Quantum Mechanics, as well as understanding and accepting the reality of the Quantum Zeno Effect or the Telepathic Effect taking place between Process 1 and Process 3, and what it really truly means where quantum control at the quantum level is concerned.

God designed and created our neurotransmitter system and the synapses the way He did so that the Human Psyche would have quadrillions of different locations where it can interact with and control its physical brain at the Quantum Level through the Quantum Zeno Effect.

Cool, huh?

Well, at least I think it is.

I learn best through comparison and contrast.

The author of my college Neuroscience textbook is a Materialist, Naturalist, Darwinist, and Atheist. He never mentioned Quantum Neuroscience nor any of the things that I mentioned here in this section – Quantum Tunneling, Quantum Entanglement, Quantum Non-Locality, and the Quantum Zeno Effect. He seemed to be totally and blissfully unaware of Quantum Mechanics.

Instead, he tries to explain everything from the perspective of Materialism, Naturalism, Classical Physics, and Evolution – what he calls an Evolutionary Perspective. He has a whole chapter dedicated to the Theory of Evolution and the Evolutionary Perspective, where he repeatedly states that all of that non-physical, immaterial, and spiritual stuff isn't real and doesn't exist. He periodically and authoritatively asserts that the intelligent and cutting-edge neuroscientists and biopsychologists are all Darwinists and believe in the Theory of Evolution. He really truly wants you to believe that evolution (random mutations and natural selection) designed, created, and now oversees and controls your genome, your physical brain, your complete physical makeup, your behavior, your future, and your neurotransmitter system through the observed and proven laws of Classical Physics, Materialism, Naturalism, and Darwinism.

He's completely clueless when it comes to Quantum Mechanics and seems to have completely rejected the stuff, if he ever knew about it in the first place. Instead, he explains everything within the physical brain in terms of Classical Physics, Materialism, Naturalism, and the Theory of Evolution. Whenever he gets to the truly interesting bits – the quantum mechanical bits – he just skips over them or effectively brushes them off by trying to explain them in terms of Classical Physics, if he even bothers to give an explanation. Occasionally, he will say that we don't know how this or that particular process works, but he repeatedly assures us that evolution does. He simply describes everything in terms of Classical Physics and the Theory of Evolution, and he gives evolution all the glory. I found that I had to turn to other people and supplement my studies in order to learn about and understand Quantum Neuroscience and how everything functions at the quantum level. My eyes were opened, and now I see.

We human beings could never design and create a machine that takes commands from and interacts with a Human Psyche at the quantum level, so why on earth would anyone believe that evolution did it? They believe, simply because they want to believe, because there's no other logical reason to do so.

It's their religion or their creed.

Kalat, J. W. (2008). *Introduction to Psychology* (9th ed.). Belmont, CA: Wadsworth, Cengage Learning. **"The brain is the product of Evolution."** (p. 12).

Pinel, J. (2014). *Biopsychology* (9th ed.). New York: Pearson. **"Genetic endowment is a product of its evolution."** (p. 24).

These people teach and preach that your brain and your genome are the product of Evolution; but, that's impossible.

Why?

Think about it logically.

It's impossible because evolution, random mutations, and natural selection didn't even exist, until AFTER God designed and created and produced the brains and the genomes in the first place. That's a no-brainer!

The Darwinists, Materialists, Naturalists, and Classical Physicists want you to believe that evolution (random mutations and natural selection) designed and controls all of this. It can't, because evolution cannot function as a mental force or a psychic force like Psyche can.

Our genome and our neurotransmitter system only make sense from a perspective that's based upon Quantum Mechanics, Non-Local Consciousness, God's Psyche, and the Quantum Zeno Effect.

But, mention any of this in a public college, and watch what happens. You will be put down.

The Materialists, Naturalists, Darwinists, Nihilists, Behaviorists, and Atheists respond by trying to ban, censor, and block this type of information in an attempt to prevent you from discovering it and knowing that it exists. That's the way these people do science; and, it works.

I have been a scientist all of my life, but it took me over fifty-five years to discover this stuff, because nobody in our public classrooms in our public colleges is teaching this kind of science – Quantum Neuroscience.

Instead, these people provide a materialistic interpretation, a Classical Physics interpretation, and an evolutionary interpretation of "quantum mechanics" and "quantum mechanical processes"; and then, these people move on. They stack the deck, tip the scales, and load the dice so that it always comes up Darwinism, Naturalism, Atheism, and Classical Physics every time; and then, they preach it and teach it axiomatically, as if it's the God-given absolute truth. Self-deception works; and, it works every time – even on the materialistic and naturalistic PhDs. I KNOW, because I used to be a Materialist, Nihilist, Skeptic, and Atheist.

You can have any conclusion you want simply by *affirming the consequent* or *begging the question*, as the Materialists and Naturalists do. But, things get very complicated when one chooses to bring phenomenology – experiences with and observations of the Non-Local Realm or Quantum Realm – into the equation. Then creative thinking is finally called for and even demanded.

The Materialists, Naturalists, Darwinists, Nihilists, Behaviorists, and Atheists have a reason for being afraid of psyche, non-local consciousness, the quantum zeno effect, telepathy, telekinesis, psi, near-death experiences, out-of-body experiences, and other types of non-local experiences or spiritual experiences, because the observation and the experience of these things FALSIFIES Materialism, Naturalism, Darwinism, Nihilism, Behaviorism, Scientism, and Atheism.

Quantum Neuroscience falsifies these false philosophies.

Cool, huh?

Well, at least I thought it was when I finally discovered these things.

I wanted to find and know the truth. Now I have.

Of course, your mileage may vary, and you might in fact have a completely different agenda. To each their own! We reap what we sow. We get what we pursue. Such is life.

Even after all of this, I get the feeling that we humans still have NO clue as to how things really work between the synapses of our physical brain at the quantum level.

Why?

True knowledge and true understanding always results in practical applications.

If we humans really understood how things work at the quantum level or the psyche level, then we would be designing and creating machines that could interface with and interact with the Human Psyche at the quantum level.

In contrast, most of our scientists refuse to believe that psyche, nonlocal consciousness, and quantum telepathy or quantum control even exist; so, how on earth could they ever design and create a physical machine capable of interacting with a Human Psyche at the quantum level through the Quantum Zeno Effect or something similar? They are blocked from doing so because of their unbelief, ignorance, and lack of understanding.

These people are going to have to figure out first how the Human Psyche controls its physical brain, before they can start developing practical applications for that knowledge. We still have a long ways to go as a race, unless you know something that I don't.

God holds all the KEYS to this one.

I believe that God has checks and balances and barriers in place preventing us from designing and creating sentient machines whose personality or intelligence is capable of surviving the death of its physical body. Our own ignorance and unbelief will prevent us from creating such a sentient machine, with a psyche of its own. The same reality applies to Teleportation or Faster-than-Light Travel. These things exist – Psyche and Quantum Tunneling – but God is preventing us from binding them or marrying them to our physical machines that we make.

New Age – Spirituality for Atheists

Buddhism is spirituality for the Atheists, and so is New Age.

The New Age material provides proof of concept where non-locality, non-physicality, spirituality, and the after-life are concerned.

The New Age material is identified whenever you encounter a Near-Death Experience (NDE) or Out-of-Body Experience (OBE) wherein the person makes no contact whatsoever with God, or the Being of Light who is Jesus Christ. In the New Age material, you are God. You are a law unto yourself. You're just out there in the cosmos doing your own thing.

New Age denies the existence of God. New Age is spirituality for Atheists.

Many people have observed that the New Age material is mediated, controlled, and presented by Satan, demons, or evil spirits while pretending to be Angels of Light or pretending to be Gods – assuming that these people even encounter Angels or Gods during their Astral Projections. The goal is to use spiritual things to distract you and turn you away from God.

The New Age material is a combination of the philosophies of men, mingled with truths and actual experiences. It can sometimes be difficult or impossible to tell with the New Age stuff where the actual experiences end, and the fictional story begins.

New Age is big on the Law of Attraction and the claim that YOU ARE GOD. They sometimes talk about the God Factor, and you soon realize that you are the God Factor. It's all about you. You are the focus of attention. New Age is spirituality for the Atheists. New Age is pantheism – the claim that everything is God. It's all about you, and what you can get out of the universe.

The following video is probably the pinnacle of the New Age NDEs. You can kind of sense that it's fictional or fabricated – that it's an amalgamation of many different NDE-type experiences rolled into one. This video is pedantic and preachy. It kind of reminded me of Carl Sagan's *Cosmos* series on television – a wonderful piece of fictional story-telling. Warning: you have to turn off the video on this one, and just listen to it, because the video was designed to hypnotize you, or to brainwash you, or to give you an epileptic seizure. The video is horrific. If you pay close attention, you will also notice that parts of this story contradict the truths that I have discovered during my research. That's the nature of the New Age material.

https://www.youtube.com/watch?v=Wjch_Qi_nio

Another NDE from this New Age crowd.

https://www.youtube.com/watch?v=6JUOOxdDezw

It's a cute little story.

As I said, the New Age crowd are big on the Law of Attraction, and MONEY is the primary thing that they are trying to attract to themselves. The purpose for their productions is to make money. They are trying to make a living off of spirituality and astral projection. It's what they do. If you send them a dollar or two, then you have helped them to accomplish their primary purpose for making these things.

The main flaw in the New Age material is that you have no idea where the actual first-hand experiences end and the fictional story-telling begins. Of course, it's all presented as being real and true, just like any good fiction story is.

My best friend is a Seer. Seers are prophets of God, who have seen and conversed with Jesus Christ. My friend had an interesting lesson or two to teach me. Here is one thing that my friend taught me.

Satan means adversary.

Satan is the master of the Star Trek holodeck experience. While you are out-of-body and during your Near-Death Experiences (NDEs), Satan can and will create for you the most interesting holodeck experiences that you can possibly imagine – galactic tours, meeting other people on other planets, faster than light travel, talking with stars and planets, beautiful vistas, alien beings, fictional past lives, relatives and descendants whom you have never met before, a fictional soulmate who is out there in the cosmos waiting for you to come and find her and merge with her, levels for you to progress through, and a whole host of other fascinating distractions. It's perfect, it's beautiful, and as long as you never find God and never encounter God during your OBE or NDE, then Satan has succeeded.

Satan's whole purpose and goal is to keep you away from God, and if he can do so during your OBEs and NDEs, then Satan has succeeded beautifully. As long as you never find God, Satan has won.

The New Age people preach and teach that God does not exist, and that you are God; and consequently, during their travels of the Astral Plane, these people never find God and never encounter God. For them personally, God does not exist, because they never go looking for Him and consequently they never find Him. These people become a law unto themselves – traveling through Satan's holodecks enjoying all of the wonders and beauties that Satan has prepared for them. This is how Satan entraps the good people and leads them astray. As long as he can keep them away from God, he wins.

If it leads you away from God, His Kingdom, His Church, His Prophets, and His Saving Ordinances, then it's from Satan. It has been designed to distract you, help you to enjoy the trip, and to keep you away from God; and, it works, especially for the people who aren't looking for God or don't want to have anything to do with God.

The evil and wicked people end up burning and suffering in hell, because they chose to go there. The good and lukewarm people end up in Satan's holodeck experiencing lots of wonderful distractions. The seekers and the desirous end up in the presence of God. The Kingdom of God will be filled with the willing – those who have gone looking for God, wanted to find God, and wanted to be with God.

For those who don't want to have anything to do with God, there's plenty of other places for them to go and lots of different things for them to experience after they die.

Whatever you want, you will get. Those who go to hell chose to be there; and, the same truth applies to those who chose to go to the Kingdom of God and God's presence.

The New Age Travelers actually experience these different things while out-of-body, during their NDEs, and during Remote Viewings. They are experiencing and enjoying Satan's holodeck. Anything they can imagine is possible. If you want to experience aliens, you can experience aliens. If you want to visit people living in the moon, or people living on Mars, they will be there waiting for you. If you want to have a ride on highly advanced spaceships, alien space ships, and UFOs, that's available to you for the taking. If you want to experience being a tree, or someone of the opposite sex, or what it's like to be a monkey, you can experience a life-time of that. You can experience what it is like to be a God and a creator of worlds. If you want to experience sex with lots of beautiful willing soulmates or those seventy virgins you have been promised, then that's available also. Anything you want, you can have it and experience it to your heart's content. It's real, just like anything on the Star Trek holodeck was real.

All of this is out there waiting for you in Satan's holodeck, and as long as it keeps you away from God, Satan has succeeded.

By far, my most favorite piece of New Age is "Adventures Beyond the Body: How to Experience Out-of-Body Travel" by William Buhlman. It is spirituality for Atheists; and, I used to be an Atheist, Materialist, Nihilist, and Naturalist, so this book had a powerful impact on me when I first read it. William Buhlman has no interest in organized religion or God. He's just out there exploring the Astral Plane every day – a lone warrior doing his own thing, treating the whole thing like a science experiment. It was right up my alley, and a good way to get my feet wet when it comes to all this out-of-body stuff.

The Rosicrucians are handing out all of their books for free on Kindle. For these people, the Astral Plane is like a computer game, and the goal is to level up or achieve higher levels of spirituality – become the Grand Poohbah of the tenth level or whatever. It's about bragging rights. I'm better than you are, because I'm a higher level than you are. These people also explore places like Atlantis and Lemuria – virtual reality worlds or fictional worlds within Satan's holodeck. You can spend a lifetime in these places if you want to, have a family, and everything else. As long as it keeps you away from God, it has accomplished its purpose. The whole goal of all of this is to keep you distracted, so that you don't waste your time looking for and finding God.

Martin, B. Y., & Moraitis, D. (2014). *Communing with the Divine: A Clairvoyant's Guide to Angels, Archangels, and the Spiritual Hierarchy*.

Martin, B. Y., & Moraitis, D. (2010). *Karma and Reincarnation: Unlocking Your 800 Lives to Enlightenment*.

I enjoyed Barbara Martin's books. Again, the goal is to level up, power up, learn to see auras and arch-angels, develop mad skills, and achieve Enlightenment, which will do a good job of distracting you and keeping you away from God and His Church.

Buddhism is New Age also. You spend your life sitting there, meditating, and contemplating your breath or your navel. Again, the goal is to level up or power up and reach the next level of Enlightenment. It's a distraction designed to take your mind off of God and place it somewhere else instead. Buddhism is different, though, in that they teach that your spirit, psyche, or soul ceases to exist when you die. With Buddhism, you spend your life trying to achieve enlightenment, only to have it all lost when you die and cease to exist.

Buddhism and New Age are spirituality for Atheists, and this stuff will do an excellent job of keeping you away from God, if that's your primary goal in life. They were designed to do so, and they work. As a person who spent most of his life avoiding God, I have learned to recognize a distraction when I see one.

"Adventures Beyond the Body: How to Experience Out-of-Body Travel" by William Buhlman.

This is one of my many Reviews PROMOTING the book "Adventures Beyond the Body" by William Buhlman. This book single-handedly puts an end to Materialism!

This book is one of my top-ten most favorite books because of all the different things that it taught me about spirit, spirituality, and the non-physical realms. This is one cool book! I have purchased multiple paperback copies for me and my friends, and I also own a Kindle copy. My intent here is to point EVERYONE to this book and tell them to read it, especially if they have any lingering feeling that Materialism or Naturalism might be true.

Spirit or the spiritual has to be experienced first-hand in order to know that it is real and true. There's really no other way, unless you are willing to trust other people's experiences of the spiritual, which I am willing to do. I have learned to trust this person's spiritual experiences, because they fit in extremely well with everything else that I have learned about spirit and spirituality so far.

The Copyright for this book permits "quotations embodied in critical articles and reviews". So, that's what I am going to do here. I'm embodying quotations in this CRITICAL ARTICLE and this REVIEW of his book, "Adventures Beyond the Body" — except it should be noted that I'm NOT being critical of Buhlman and his book, but instead trying to promote it. I USE his books to CRITICIZE the Materialists and the Atheists who mock and ridicule spiritual things in their "critical reviews" online.

Because Buhlman doesn't seem to be promoting any particular Organized Religion, he comes across as an unbiased authority on spirit and out-of-body travel, which some Atheists and Materialists might actually appreciate, since most Atheistic Materialists are deathly afraid of Organized Religion. Buhlman seems to take a secular approach to spirituality, which is really cool if you happen to be someone who is on the fence about all of this.

For PROMOTIONAL purposes and preemptive defense, here are a couple of my most favorite quotes from Buhlman's book which are germane to the topic at hand. Whenever I'm taking ridicule and heat from the Materialists and Atheists online, I love to drop these quotes and this book on them:

> Twenty years ago I firmly believed that the physical world we see and experience was the only reality. I believed what my eyes told me — life possessed no hidden mysteries, only countless forms of matter living and dying. The facts were clear; there was no evidence or proof of nonphysical worlds or our continued existence after death. I questioned the intelligence of anyone softheaded enough to accept the illogical concepts of heaven, God, and immortality. In my mind these were fairy tales created to comfort the weak and manipulate the masses. For me, life was simple to understand: the world consisted of solid matter and form, and the concepts of life after death and heaven were feeble human attempts to create hope where none existed.

> I possessed the arrogant knowledge of a man who judges the world with his physical senses alone. I supported my conclusions with the overwhelming observations provided by science and technology. After all, if something mysterious was there, science would certainly be aware of it.

> My firm convictions of reality and life continued until June of 1972. During a conversation with a neighbor, our discussion turned to the possibilities of life after death and the existence of heaven. I proceeded to present my agnostic viewpoints

with vigor. To my surprise my neighbor didn't contest my conclusions; instead, he related an experience that he had had several weeks before. One evening just after drifting to sleep, he was shocked to discover himself floating above his body. Completely awake and aware, he became frightened and instantly fell back into his physical body. Excited, he told me it wasn't a dream or his imagination, but a fully conscious experience.

Intrigued by his experience, I decided to investigate this strange phenomenon for myself. After several days of research, I discovered numerous references to out-of-body experiences throughout history. With some searching I found a book on the topic that actually described how out-of-body experiences are induced. The entire subject seemed extremely weird, and I considered the book the result of an overly active imagination.

Out of curiosity, I decided to try one of the out-of-body techniques before sleep. After repeated daily attempts, I began to feel a little ridiculous. In three weeks the only thing I experienced out of the norm was an increase in my dream recall. I became more and more convinced that this entire subject was nothing more than an intense or vivid dream stimulated by the so-called out-of-body techniques.

Then, one night about eleven o'clock I drifted to sleep during my out-of-body technique and began dreaming that I was sitting at a round table with several people. They all seemed to be asking me questions related to my self-development and state of consciousness. At that moment in the dream I began to feel extremely dizzy, and a strange numbness, like from Novocain, began to spread throughout my body. Unable to keep my head up, I passed out, hitting my head on the table. Instantly I was awake, fully conscious, lying in bed facing the wall. I could hear an unusual buzzing sound and felt somehow different. Extending my arm, I reached for the wall in front of me. I stared in amazement as my hand actually entered the wall; I could feel the vibrational energy of it as if I was touching its very molecular structure. Only then did the overwhelming reality hit me, *My God, I'm not in my body.*

Excited, my only thought was, *It's real. My God, it's real!* Lying in bed, I stared at my hand in disbelief. When I tried clenching my fist, I could feel the pressure of my grip; my hand felt completely solid, but the physical wall in front of me looked and felt like a dense, vaporous material with form.

Determined to stand, I began to move effortlessly to the foot of my bed, my mind racing with the reality of it all. Standing, I quickly touched my arms and legs, checking to see if I was solid, and to my surprise I was completely solid, completely real. But around me, the familiar physical objects in my room no longer appeared completely real or solid; instead, they now looked like three-dimensional mirages. Glancing down, I noticed a large lump in my bed. Amazed, I could see that it was the sleeping form of my physical body silently facing the wall.

As I focused my vision on the opposite side of the room, the wall seemed to fade slowly from view. In front of me I could see a wide, green field extending far beyond my room. Looking around, I noticed a figure silently watching me from about ten yards away. It was a tall man with dark hair, a beard, and a purple robe. Startled by his presence, I became frightened and instantly "snapped back" into my physical body. With a jolt I was in my body, and a strange feeling of numbness and tingling faded as I opened my eyes. Excited, I sat up, my mind exploding with the realization of what had just occurred. I knew it was absolutely real, not a dream or my imagination. My entire ego awareness had been present.

Suddenly, everything I had ever learned about my existence and the world around me had to be reappraised. I had always seriously doubted that anything beyond the physical world existed. Now my entire viewpoint changed. Now I absolutely knew that other worlds do exist and that people like myself must live there. Most important, I now knew that my physical body was just a temporary vehicle for the real me inside, and that with practice I could separate from it at will.

Excited about my discovery, I grabbed a pen and paper and wrote down exactly what had occurred. A flood of questions filled my mind. Why is the vast majority of the human race unaware of this phenomenon? Why aren't the various sciences and religions investigating it? Is it possible that this unseen world is the "heaven" referred to in religious texts? Why isn't our government exploring this apparent parallel energy world? Is it possible that our overwhelming dependence on physical perceptions has led us to overlook an incredible avenue of exploration and discovery?

As the initial shock of my first experience sank in, I realized that my life could never be the same again. The more I pondered the significance of my experience, the more profound I realized it to be. All my agnostic beliefs had been swept away in a single night. I knew that I had to reappraise everything that I had learned since childhood, everything that I had assumed to be true. My comfortable conclusions about science, psychology, religion, and my existence had obviously been based on incomplete information. I felt excited, but also uneasy — my familiar concepts of reality no longer seemed relevant. Increasingly, I felt in a void. On several occasions when I talked to friends about my experience, they found it too bizarre to take seriously. In 1972 the term *out-of-body experience* had not even been coined; back then, the most common description was astral projection. No one that I knew at the time had even heard of astral projection, and if you told people you had left your body, they immediately thought that you were on drugs or losing your mind. I quickly discovered that I had to keep my experiences to myself or face some degree of disbelief and even ridicule.

After my first out-of-body experience, my mind was overflowing with endless possibilities and questions. Desperate for information and guidance, I spent several weeks in libraries and bookstores searching for whatever knowledge was available on the topic. I quickly found that little was available; only a handful of books had been written on the subject, and some of these were decades old and out of print. By the end of July 1972, I realized that I was on my own.

I decided to focus on the one technique that had worked for me before. This technique involved visualizing a physical location that I knew well as I drifted off to sleep. As before, I pictured my mother's living room with as much detail as possible. At first it seemed difficult, but after a few weeks I could picture the room's details with increasing clarity; the furniture, patterns in fabrics, textures, even small imperfections in wood and paint began to be clear in my mind. I realized that the more I pictured myself within the room interacting with the physical objects, the more detailed my visualizations would become. With practice I learned to physically walk around the room and memorize specific items that it contained. I also learned the importance of "feeling" the environment with my mind: the feel of carpet on my feet; the sensation of sitting in a chair, walking, turning on a lamp, or even opening the door. The more detailed and involved I was within my visualization, the more effective were my results. Although it was challenging at first, after a while it became fun to make my visualizations come alive in my mind. At this point I decided to keep a journal to record my out-of-body experiences.

Buhlman, William L., "Adventures Beyond the Body: How to Experience Out-of-Body Travel" (pp. 3-8). HarperCollins. Kindle Edition.

The Materialists, Darwinists, and Atheists can be extremely stupid, ignorant, inexperienced, uneducated, and dense when it comes to the parts of SCIENCE that they have chosen not to believe in. I know, because I used to be one of them.

—

Journal Entry, October 2, 1982

> I hear the buzzing, engine-like sounds and will myself out-of-body. I step to the bedroom door and automatically request "Clarity now!" My vision improves and I step through the door, into the living room. Still feeling a little out of sync, I verbally repeat my request with more emphasis, "Clarity now!" I feel my awareness and vision snap into place.
>
> My thoughts are clear and I make a verbal demand, "I need to see the form I'm in now!" Instantly I feel an intense sensation of being drawn within myself. I'm suddenly different, weightless as though I'm floating in space. As I look forward I see a sparkling, bluish white form. For some reason, I seem to know that I'm looking at my nonphysical body from a different perspective. I stare in amazement at this form before me that shines and flows with energy and light. It looks like an energy mold created from a million tiny points of light; it radiates a bluish glow but appears to have a defined outer structure. The body of light before me is naked and is identical to my physical form. Even though my body looks firm, there is a noticeable energy motion and radiation present. I can see what appears to be an ocean of blue stars throughout my body. It's difficult to describe because the stars are stable, yet moving at the same time; the light and energy of my body appear to change and flow almost like the waves of an ocean.
>
> As I stare at the body of light, it hits me that I must be in another body. Yet I can't perceive any form or substance; I'm like a viewpoint in space without shape or form of any kind. As I reflect upon my new state of being, I feel a sensation of rapid motion and I'm instantly back within my physical body.
>
> Lying still and reviewing my experience, I'm struck by an inescapable conclusion: I must possess multiple energy-bodies. The form I just experienced was noticeably lighter (less dense) than even my second nonphysical body. I realize that the traditional view of our possessing two bodies — a physical body and a spiritual body — is far too simplistic; we are much more complex than this. Just as there are multiple nonphysical energy dimensions within the universe, each of us must consist of multiple energy-bodies or vehicles of expression.
>
> Now I seriously wonder just how many nonphysical bodies or forms this involves. I suspect that there must be one within each dimension of the universe and that all of these are interrelated and connected, just as the physical body is connected to its first nonphysical (spiritual) body.

Buhlman, William L., "Adventures Beyond the Body: How to Experience Out-of-Body Travel" (pp. 34-35). HarperCollins. Kindle Edition.

END OF QUOTES.

—

If you want to know more, then go out and buy a copy of William Buhlman's book for yourself.

In my book here, the most I can do is to tell you where to go and look; and, I do make it a point to PROMOTE the books and the people that I found most helpful and useful during my research. Clearly, I have given this whole thing some study and thought. This book from William Buhlman gave me solid irrefutable evidence that Materialism is FALSE. Whenever an Atheist or Materialist is being caustic, mean, and rude online, I love to drop these two quotes on them. They don't know what to do with them except to mock them and dismiss them out-of-hand; or even better, they don't respond at all.

These two quotes from William Buhlman have become an essential part of my defense package whenever I find myself engaging the Borg (Materialists) online.

The Darwinists and Materialists have accused me of carpet bombing the threads that I started online. These two quotes are a couple of bombs that I love to drop on them from time to time!

Take note that William Buhlman has other books for sale, but this one was my favorite one so far.

William Buhlman discovered first-hand that Materialism and Atheism are false.

The only way to KNOW that the spiritual is real is to experience it first-hand for yourself, or to trust someone who has experienced it first-hand. I have had trust issues for most of my life, so it would have been nice to have had some first-hand spiritual experiences of my own like these that Buhlman has had. Unfortunately, most Atheistic Materialists like me block ourselves from pursuing and having spiritual experiences like Buhlman's because we don't believe that such a thing is possible, so we never even try it. Materialism is a curse that ends up being self-confirming and self-reinforcing — believing Materialism to be true makes it true in one's own life.

For most Materialists and Atheists, they need some kind of jump-start or extreme experience to get them into spirituality and pursuing God. But, once you have had some convincing spiritual experiences of your own, then typically you don't want to go back to the limitations and darkness and blindness of Materialism.

It can be a very interesting (and sometimes scary) to experience that internal paradigm shift from Materialism to Spirituality or Spiritualism. It's like a light going on, and then suddenly you just KNOW that Materialism is false and why it is false. When you see the Light, you just know that your whole life before as a Materialist and an Atheist was completely false and a self-deceptive lie. It can be a bit of a shock to the system at first; but then, you find yourself getting very excited about all of the new and interesting things that have suddenly opened up for you.

Once you know that Materialism is false, then you KNOW that Darwinism is false, because Darwinism is a Materialistic philosophy. Once you know that Darwinism is false, then you KNOW that goo-to-you Darwinian evolution is FALSE also because the Theory of Evolution is fruit from the poisoned tree.

When it comes to Science of any kind, Materialism IS the ultimate bane of that Science. Materialism IS the poisoned tree from which all falsehoods are produced. Materialism ALWAYS produces the WORST explanation or interpretation of scientific evidence.

I know that some of this makes the Materialists and Atheists and Darwinists uncomfortable, but who cares? What have the Materialists and Atheists done for us? They take without giving anything of value in return. What good have their lies and self-deceptions done for any of us? What thing of value have they actually contributed to our intellectual and spiritual growth? What thing of value have they done for society? During the 20th century, they exterminated at least 200 million of their opponents in the name of Darwinism (Fascism) and/or some kind of Militant Enforced Atheism (Communism). Materialism has not been good for society — it's too selfish to be of any value to the rest of us. And, Materialism IS BAD SCIENCE!

Creation by Purely Materialistic Processes is scientifically impossible, because Physical Processes require some kind of intelligent being to get them going, coax them or force them in the right direction, and then deliberately weed out the failures while keeping the successes. Scientists are required to DO SCIENCE or to DO MANUFACTURING in order to make Materialistic Processes organize into new and useful things.

Spirit beings or living beings design, program, plan, engineer, create, manufacture, and deploy new things — NOT MATERIALISTIC PROCESSES! There's no brain, no mind, no intelligence, and no hands for manipulation and manufacturing when it comes to purely physical processes or random luck. Spirit or sentience or life designs and creates new things, not random material processes. This IS a Scientific Reality, thus making Materialism scientifically FALSE.

If Materialism is FALSE, then God must of necessity exist in order to do all of the different things and all of the science that needed to be done which "Materialism" or random material processes could never have done. This is logical common sense. This is reasoning, deduction by a process of eliminating False Premises, and even science!

This Scientific Reality can be hard to put into words at times, but random material processes and random chemical reactions cannot design and create anything from scratch. It requires some kind of intelligent mind in order to design and create something new from scratch, and actually make it work. Anytime living breathing Scientists DO SCIENCE in a science lab, then it is NO longer Creation by Evolution or Creation by Natural Processes or Creation by Chance, but is instead Creation by Intelligent Beings. Intelligent Beings can design and create and manufacture new and unique things — Evolution, Chance, Chemical Reactions, and Mindless Directionless Physical Processes cannot.

Darwinian molecules-to-man evolution IS scientifically impossible. Darwinian ape-to-man evolution IS scientifically impossible. Creation by Chance IS scientifically impossible. Abiogenesis IS scientifically impossible. Creation of new functional genomes through step-by-step Mutation/Selection IS scientifically impossible. Macro-evolution of any kind IS scientifically impossible. Creation of new functional genomes by Micro-evolution or Macro-evolution IS scientifically impossible. Creation by Evolution IS scientifically impossible. Technically, Design and Creation by any kind of Mindless Materialism or any kind of Mindless and Directionless Materialistic Process IS scientifically impossible. My goal here is to eliminate the impossible.

How often have I said to you that when you have eliminated the impossible, whatever remains, *however improbable*, must be the truth? — Sherlock Holmes

When you have eliminated Evolution of every kind from the design and creation process, then what remains? It's elementary my dear Watson.

Follow the evidence!

Keep the best and get rid of all the rest!

The Readiness Potential and Libet's Experiments Prove that Psyche and Free Will Do Not Exist

Benjamin Libet proved that Psyche and Free Will do not exist.

Back in the 1980s, the American scientist Benjamin Libet made a surprising discovery that appeared to rock the foundations of what it means to be human. He recorded people's brain waves as they made spontaneous finger movements while looking at a clock, with the participants telling researchers the time at which they decided to waggle their fingers. Libet's revolutionary finding was that the timing of these conscious decisions was consistently preceded by several hundred milliseconds of background preparatory brain activity (known technically as "the readiness potential").

The implication was that the decision to move was made nonconsciously, and that the subjective feeling of having made this decision is tagged on afterward. In other words, the results implied that free will as we know it is an illusion — after all, how can our conscious decisions be truly free if they come after the brain has already started preparing for them?

For years, various research teams have tried to pick holes in Libet's original research. It's been pointed out, for example, that it's pretty tricky for people to accurately report the time that they made their conscious decision. But, until recently, the broad implications of the finding have weathered these criticisms, at least in the eyes of many hard-nosed neuroscientists, and over the last decade or so his basic result has been replicated and built upon with ever more advanced methods such as fMRI and the direct recording of neuronal activity using implanted electrodes.

These studies all point in the same, troubling direction: We don't really have free will. In fact, until recently, many neuroscientists would have said any decision you made was not truly free but actually determined by neural processes outside of your conscious control.

Dr. Christian Jarrett. (February 2016).

Neuroscience and Free Will Are Rethinking Their Divorce:

http://nymag.com/scienceofus/2016/02/a-neuroscience-finding-on-free-will.html

http://mypsyche.us/wp-content/uploads/2017/11/Neuroscience-and-Free-Will-Are-Rethinking-Their-Divorce.pdf

My bachelor degree was in Psychology.

Many of the Materialists, Naturalists, Behaviorists, and Atheists – who wrote the college textbooks that I was forced to buy, read, and memorize – used Benjamin Libet's experiment to prove that Psyche and Free Will do not exist. It is November 2017 as I write this, and these were books that were published and/or updated during the past five years. The Materialists, Naturalists, Behaviorists, and Determinists have gotten a lot of mileage out of Libet's experiment for a few decades now. These people use Libet's experiment to prove that Free Will and Psyche do not exist, as if such a thing were actually possible.

As a lifelong Scientist, Theorist, Logician, and Skeptic, I could immediately sense that there was something wrong with their chosen interpretation of the experimental evidence. It took a bit of study, but I was finally able to identify many different ways to falsify the Materialists' and Naturalists' chosen conclusions, as well as identify many of the logic fallacies they were using to make their case. Now let me mention some of the steps I went through in order to falsify their chosen interpretations of the evidence that was produced by Libet's experiment.

WE KNOW from the Lived Experiences of the human race, including our Near-Death Experiences and Out-of-Body Experiences, that the Human Psyche or Non-Local Consciousness does in fact exist.

But, the Materialists and Naturalists use Benjamin Libet's experiments with the readiness potential to prove that Free Will and the Psyche do not exist.

How are they able to do that, since WE KNOW from observational evidence and experiential evidence that Psyche or Non-Local Consciousness does in fact exist and does in fact survive the death of our physical brain?

How do they do that – use Libet's experiment to prove that Psyche and Free Will do not exist?

Quite easily actually.

The Materialists and Naturalists start with the assumption that the first step of the process, the Readiness Potential, is caused by the physical brain. In other words, they assume right from the beginning that Psyche or Non-Local Consciousness does not exist. And then, these people assume that all subsequent effects are also the product of our physical brain. In other words, these people *beg the question* and *affirm the consequent*. That's how these people prove all of their theories and ideas true, including the Theory of Evolution – by *affirming the consequent* and *begging the question*. They literally insert their chosen conclusion into all of their arguments as one of the premises and as empirical proof that their chosen conclusion is true. That's begging the question.

Begging the question is when the conclusion is assumed to be true within the premise of an argument. In other words, these people are using their chosen conclusions as evidence and proof – as premises – in all of their arguments and interpretation of evidence. The Materialists and Naturalists and Atheists always complain that other people are using logic fallacies in order to make their case, never once realizing that the Materialists and Naturalists have been using logic fallacies for centuries in order to prove that their theories and ideas are true.

The Materialists and Naturalists in general have never studied the Philosophy of Science; and therefore, they don't know that the Scientific Method is actually based upon a logic fallacy that's called *affirming the consequent, jumping to conclusions*, or *begging the question*.

What these people do is start with the assumption that the Theory of Evolution is true or the assumption that Psyche and Free Will do not exist; and then, these people insert their chosen conclusion into all of their arguments as one of their *hidden premises*. Thereby, ALL of their arguments and interpretations of evidence end up proving their chosen conclusions, because they are actually using their chosen conclusions as evidence and proof that their chosen conclusions are true, within the premises of all their arguments. This process and logic fallacy is known as *begging the question* or *affirming the consequent*; and, this is the logic fallacy that the Materialists and Naturalists use to prove all of their ideas and theories true. Most of these people don't even realize that they are *begging the question* and *jumping to conclusions*, because these people have been taught to do so in our public universities and have been taught that it is Good Science. Only the study of the Philosophy of Science can falsify their claims and prove them wrong.

Rychlak, J. F. (1981). *A Philosophy of Science for Personality Theory* (2nd ed.). Malabar, FL: Robert E. Krieger Publishing Company.

For a good and honest introduction to the Philosophy of Science, see also:

Slife, B. D. & Williams, R. N. (1995). Science and Human Behavior. In *What's Behind the Research? Discovering Hidden Assumptions in the Behavioral Sciences*, (pp. 167–204). Thousand Oaks, CA: SAGE Publications.

http://mypsyche.us/wp-content/uploads/2017/04/Science.pdf

Remember, the ultimate goal in all of this is to find and know the truth. If it's a lie, then ultimately it has no value to us.

It can be extremely fascinating to study the psychology of the Materialists, Naturalists, Darwinists, Nihilists, and Atheists. These people are in denial; and, denial is one of the most common defense mechanisms. When I was a Materialist, Nihilist, and Atheist, I had some severely messed-up psychology, which can be fascinating to look back on and study.

Eventually I discovered that many, if not most, of the Materialists, Naturalists, Darwinists, Nihilists, and Atheists are actually trying to trick us and deceive us. The truth cannot be found nor established through deceptions and lies. I learned to distrust the Materialists, Naturalists, and Darwinists because I had caught them in dozens of different lies.

Through *affirming the consequent* and *begging the question*, you can prove anything you want to be true. You simply start with your chosen conclusion, and then insert your chosen conclusion into all of your arguments as one of the premises and as evidentiary proof that your chosen conclusion is true.

This is the *scientific inference* logic fallacy in action. These people literally use their chosen conclusion as empirical evidence proving that their chosen conclusion is true; and, most of them don't even realize that they are doing so. These people also use their scientific inferences or personal interpretations of the evidence as empirical evidence and proof that their theories and ideas are true. Using your chosen conclusion as evidence and as one of your premises is a logic fallacy; and, so are *hidden assumptions*.

If you notice carefully, the Materialists and Naturalists and Darwinists demand that the Creationists and Faith-Heads meet and exceed their burden of proof through a preponderance of the evidence and follow the rules of science, but these same Materialists and Darwinists make a *special exception* or a *special pleading* where Materialism, Naturalism, and Darwinism are concerned. Instead, the Materialists and Naturalists shift their burden of proof to you and demand that you prove them wrong – which is really easy to do once you KNOW all the different logic fallacies that they use to make their case and prove their point. *Special pleading* is also a logic fallacy, which the Materialists and Naturalists use to demand that Materialism, Naturalism, Darwinism, and Atheism be exempt from the rules of science and the rules of evidence. These people have to make this *special pleading* and shift their burden of proof to you, because there is NO evidence to support their claims.

Remember, you can have anything you want and prove it to be true as the Materialists and Naturalists do, by *begging the question* and *affirming the consequent*. This reality and truth is one of the reasons why Science 2.0 defines science as Knowledge, Observational Evidence, Experiential Evidence, Eye-Witness Evidence, and Empirical Evidence. Observational Evidence and Lived Experiences should always be given precedence and priority over scientific inferences, personal interpretations of evidence, and affirmed consequents, especially if a person wants to avoid *begging the question*.

Some of the Materialists, Naturalists, Darwinists, Nihilists, Behaviorists, and Atheists have no problem with *begging the question* and *affirming the consequent*. They do so deliberately and knowingly, because they are trying to trick you and deceive you. I have encountered a few of these people online, who are deceivers and liars. I have caught them in their lies. However, I believe that the greater majority of the Materialists and Naturalists have no idea that they are *begging the question* and have no idea that they have been tricked and deceived. I had no idea when I was a Materialist, Nihilist, and Atheist that most of my ideas were based upon logic fallacies, because the materialistic and naturalistic PhDs in our public colleges and public schools never teach us that the Scientific Method is based upon a logic fallacy called *affirming the consequent*. In fact, I don't think that most of them even know that it is so.

Besides the fact that the Materialists, Naturalists, Darwinists, Behaviorists, Nihilists, and Atheists are always *affirming the consequent* and *begging the question* whenever they use the Scientific Method and their *scientific inferences* to prove that their theories and ideas are true, what other evidence do we have to falsify the claims of the Materialists and Naturalists?

Well, we have subsequent experiments to fall back on.

I, myself, have run such an experiment and observed the result. I have chosen to raise my finger, and the finger raises – readiness potential followed by an action potential – just like in Libet's experiments. However, I took the whole thing one step further. At times, when my finger begins to raise after the readiness potential has been primed and the action potential triggered, I have actually intervened, stopped the process, and forced my finger down to the table instead. In other words, the Human Psyche can actually intervene in the process and reverse the process even after the finger has started to lift off the ground. Try it, it works. Your finger starts to rise, and you (your psyche) can block it or stop it, at will.

Luckily, for those who find this state of affairs philosophically (or existentially) perplexing, things are starting to look up. Thanks to some new breakthrough studies, including one published last month in Proceedings of the National Academy of Sciences by researchers in Germany, there's now some evidence pointing in the other direction: The neuroscientists are backtracking on past bold claims and painting a rather more appealing account of human autonomy. We may have more control over certain processes than those initial experiments indicated.

The German neuroscientists took a different approach from past work, using a form of brain-computer integration to see whether participants could cancel a movement after the onset of the nonconscious preparatory brain activity identified by Libet. If they could, it would be a sign that humans can consciously intervene and "veto" processes that neuroscience has previously considered automatic and beyond willful control.

The participants' task started off simply enough: They had to press a foot pedal as quickly as possible whenever they saw a green light and

cancel this movement whenever they saw a red light. **Things got trickier when the researchers put the red light under the control of a computer that was monitoring the participants' own brain waves. Whenever the computer detected signs of nonconscious preparatory brain activity, it switched on the red light. If this preparatory activity is truly a signal of actions that are beyond conscious control, the participants should have been incapable of responding to these sudden red lights. In fact, in many cases the participants were able to cancel the nonconscious preparatory brain activity and stop their foot movement before it even began.**

Now, there was a point of no return — red lights that appeared too close (less than about one-quarter of a second) to the beginning of a foot movement could not be completely inhibited — there simply wasn't time for the new cancellation signal to overtake the earlier command to move. But still, the principle stands — these results suggest at least some of the activity identified by Libet can, in fact, be vetoed by conscious will.

Dr. Christian Jarrett. (February 2016).

Neuroscience and Free Will Are Rethinking Their Divorce:

http://nymag.com/scienceofus/2016/02/a-neuroscience-finding-on-free-will.html

http://mypsyche.us/wp-content/uploads/2017/11/Neuroscience-and-Free-Will-Are-Rethinking-Their-Divorce.pdf

In other words, Benjamin Libet's early experiments in 1980 weren't sensitive enough nor fast enough to detect what was really going on; and then based upon incomplete evidence, the Materialists and Naturalists made a *scientific inference* that Psyche and Free Will do not exist. Remember, *scientific inferences* are also logic fallacies, because they are a form of *begging the question* or *affirming the consequent*, which are a form of *circular reasoning*.

Benjamin Libet's experiment and initial chosen conclusions were flawed.

Likewise, the Materialists' and Naturalists' chosen conclusions don't follow from the evidence. These people fail to meet their burden of proof through a preponderance of the evidence, because the preponderance of the evidence falsifies their claims and proves that their chosen conclusions are false.

The following article explains some of what's wrong with Libet's experiment as well as some of what's wrong with Libet's initial conclusions.

"Benjamin Libet and The Denial of Free Will: How Did a Flawed Experiment Become So Influential?" Posted Sep 05, 2017 by Steve Taylor PhD.

An even more serious issue with Libet's experiment is that it is by no means clear that the electrical activity of the "readiness potential" is related to the decision to move, and the actual movement. Some researchers have suggested that the readiness potential could just relate to the act of paying attention to the wrist or a button, rather than the decision to move. Others have suggested that it only reflects the expectation of some kind of movement, rather being related to a specific moment. In a modified version of Libet's experiment in which participants were asked to press one of two buttons in response to images on a computer screen, the participants showed "readiness potential" even before the images came up on the screen,

suggesting that the readiness potential was not related to the decision of which button to press.

Notice: I corrected the grammar and punctuation in this quote so that it made more sense to me.

Benjamin Libet and the Denial of Free Will:

https://www.psychologytoday.com/blog/out-the-darkness/201709/benjamin-libet-and-the-denial-free-will

http://mypsyche.us/wp-content/uploads/2017/11/Benjamin-Libet-and-The-Denial-of-Free-Will.pdf

How ironic!

The Readiness Potential is simply the result of the Human Psyche's decision or choice to pay attention to its finger and the clock. The Human Psyche is getting its physical brain ready to do the task which the experimenter has assigned. There's no way to tell when the decision to raise the finger was made, because there's no change in the readiness potential when the decision is finally made and because the decision to look at the clock and report the time always takes place after the finger has been raised.

The Libet experiment and the chosen interpretations of that experiment from the Materialists and Naturalists are fundamentally flawed; yet, the Materialists and Naturalists continue to preach and teach dogmatically and religiously within all their college textbooks that Benjamin Libet's experiment proves that Psyche and Free Will do not exist.

The chosen conclusions of the Materialists, Naturalists, Darwinists, Behaviorists, Nihilists, and Atheists have been falsified many different ways and by many different science disciplines. Materialism, Naturalism, Darwinism, Behaviorism, Physicalism, Reductionistic Atomism, Determinism, Scientism, Nihilism, and Atheism have been FALSIFIED by Observational Evidence, Experiential Evidence, Empirical Evidence, Experimental Evidence, Eye-Witness Evidence, Non-Local Evidence, and the Lived Experiences of the human race.

If you eliminate everything that is false and everything that has been falsified, eventually only the truth will remain.

The materialistic and naturalistic conclusion that Free Will and Psyche do not exist is based upon a faulty interpretation of Libet's experiment. These people literally equate the Readiness Potential with the Physical Brain and the Brain's Decision to move, never once realizing that the Human Psyche can veto that Readiness Potential at will. In other words, the Materialists and Naturalists are *jumping to conclusions* and *begging the question*, once again. It's what they do! The science itself has actually falsified the Materialists' chosen conclusions, although it has taken "science" three or four decades to do so, where the Libet experiment is concerned. The Materialists and Naturalists were not motivated to falsify their chosen conclusions, after all.

Further evidence eventually comes along that always ends up falsifying the claims of the Materialists, Naturalists, and Atheists, which is why Materialism, Naturalism, Darwinism, and Atheism are based upon a refusal to look at evidence – because ALL of the observational evidence and experiential evidence proves the Materialists and Naturalists wrong. If done correctly and interpreted correctly, our science experiments also end up falsifying Materialism, Naturalism, and their derivatives such as Darwinism. The Materialists and Naturalists and Darwinists who control the peer-review process, always censor, ban, and block any evidence that successfully falsifies Materialism and Naturalism and

Darwinism, so that none of that evidence ever makes it into our science journals. These people load the dice and stack the deck right from the start, so that everything comes up proving or ends up proving that Materialism and the Theory of Evolution are true. That's how these people do science – by *begging the question* and *affirming the consequent*.

Furthermore, Jeffrey Schwartz and Benjamin Libet himself eventually concluded that the Readiness Potential is actually controlled by the Unconscious. Since I KNOW for a fact from the Eye-Witness Evidence and the Lived Experiences of the human race that Psyche or Non-Local Consciousness does exist, I now always define the Unconscious or Nonconscious as Psyche, based upon the Observational Evidence and Experiential Evidence. Furthermore, the Unconscious has always been defined as Psyche, right from the very beginning of the Science of Psychology, which was originally the Study of the Human Psyche. Both Freud and Jung defined the Unconscious as an integral part of Psyche.

Free Will and Free Won't

Libet thus produced the first experimental support for the version of free will that Richard Gregory famously called "free won't." At first glance, the detection of a readiness potential before consciousness of the wish to act appears to bury free will: after all, cortical activity leading to a movement is well under way before the subject makes what he thinks is a conscious decision to act. The neuronal train has indeed left the station. If free will exists, it seems to be like a late passenger running beside the tracks and ineffectually calling, "Wait! Wait!" Yet Libet does not interpret his work as proving free will a convenient fiction. For one thing, the 150 or so milliseconds between the conscious appearance of will and the muscle movement "allow[s] enough time in which the conscious function might affect the final outcome of the volitional process," he notes. Although his results have been widely and vigorously debated, one interpretation with significant experimental support is this: there exists conscious cerebral activity whose role may be "blocking or vetoing the volitional process so that no actual motor action occurs," as Libet wrote in 1998. "Veto of an urge to act is a common experience for individuals generally." It is also, of course, the essence of mindfulness-based OCD treatment and reaffirms Sherrington's insight that "to refrain from an act is no less an act than to commit one": thus, "free won't."

Experiments published in 1983 clearly showed that subjects could choose not to perform a movement that was on the cusp of occurring (that is, that their brain was preparing to make) and that was preceded by a large readiness potential. In this view, although the physical sensation of an urge to move is initiated unconsciously, will can still control the outcome by vetoing the action. Later researchers, in fact, reported readiness potentials that precede a planned foot movement not by mere milliseconds but by almost two full seconds, leaving free won't an even larger window of opportunity. "Conscious will could thus affect the outcome of the volitional process even though the latter was initiated by unconscious cerebral processes," Libet says. "Conscious will might block or veto the process, so that no act occurs." Everyone, Libet continues, has had the experience of "vetoing a spontaneous urge to perform some act. This often occurs when the urge to act involves some socially unacceptable consequence, like an urge to shout some obscenity at the professor." Volunteers report something quite consistent with this view of the will as wielding veto power. Sometimes, they told Libet, a conscious urge to move seemed to bubble up from somewhere, but they suppressed it. Although the possibility of moving gets under way some 350 milliseconds before the subject experiences the will to move, that sense of will nevertheless kicks in 150 to 200 milliseconds before the muscle moves — and with it the power to call a halt to the proceedings. Libet's findings suggest that free will operates not to initiate a voluntary act but to allow or suppress it. "We may view the unconscious initiatives for voluntary actions as 'bubbling up' in the brain," he explains. "The conscious will then selects which of these initiatives may go forward to an action or which ones to veto and abort.... This kind of role for free will is actually in accord with religious and ethical strictures. These commonly advocate that you 'control yourself.' Most of the Ten Commandments are 'do not' orders." And all five of the basic moral precepts of Buddhism are restraints: refraining from killing, from lying, from stealing, from sexual misconduct, from intoxicants. In the Buddha's famous dictum, "Restraint everywhere is excellent."

The evolution of Libet's thoughts about his own experiments mirrors that of neuroscience as a whole about the reality of volition. Libet had long shied from associating his findings with free will. For years he refused even to include the words in his papers and resisted drawing any deeper conclusions from his results. At the 1994 "Toward a Scientific Basis of Consciousness" conference (Tucson I), Libet was asked whether his results could be used to support the existence of free will. "I've always been able to avoid that question," he demurred. But in later years he embraced the notion that free will serves as the gatekeeper for thoughts bubbling up from the brain and did not duck the moral implications of that. "Our experimental work in voluntary action led to inferences about responsibility and free will," he explained in late 2000. "Since the volitional process is initiated in the brain unconsciously, one cannot be held to feel guilty or sinful for simply having an urge or wish to do something asocial. But conscious control over the possible act is available, making people responsible for their actions. The unconscious initiation of a voluntary act provides direct evidence for the brain's role in unconscious mental processes. I, as an experimental scientist, am led to suggest that true free will is a [more accurate scientific description] than determinism."

This may seem an enfeebled sort of free will, if it does not initiate actions but only censors them. And yet the common notion of free will assumes the possibility of acting otherwise in the same circumstances, of choosing not to perform actions that tempt us each and every day.

Schwartz, Jeffrey M. *The Mind and the Brain: Neuroplasticity and the Power of Mental Force* (pp. 306-308). HarperCollins. Kindle Edition.

(HarperCollins generously permits quoting from their books for review and promotional purposes. For me personally, this book from Jeffrey Schwartz is the textbook for Quantum Neuroscience. It doesn't get better than this.)

Benjamin Libet is typically portrayed – in the naturalistic and atheistic college textbooks that I own and have been forced to read – as having successfully proven that Psyche and Free Will do not exist. He did no such thing; and, Libet seemed to have mellowed with age and experience, according to Schwartz, and was no longer dogmatically Deterministic, if he ever was.

This whole Libet Controversy has been a prolonged Lived Experience for me, as I follow it from one deterministic and behavioristic college textbook to the next. It was refreshing to finally get some truth on the matter, for once. I now KNOW that the Materialists and Naturalists as a whole are trying to trick us and deceive us, so I find it totally refreshing and enjoyable whenever I actually encounter the truth in any scientific subject that I come across. The truth is out there; but, you won't get it from the Materialists, Nihilists, Darwinists, Naturalists, Behaviorists, Determinists, and Atheists.

I now give preference or priority to Observational Evidence, Lived Experiences, and Experiential Evidence over scientific inferences, personal interpretations, sophistry, and philosophical speculation. Many of the Materialists, Naturalists, Darwinists, Behaviorists, Nihilists, and Atheists are deliberately and knowingly trying to trick us and deceive us; and, the others have been innocently and unknowingly tricked and deceived.

Remember, Materialism and Atheism of any kind are based upon a refusal to look at evidence – any evidence that successfully falsifies Materialism, Naturalism, and Darwinism. *Refusing to admit evidence into evidence* is also a logic fallacy, and one that the Materialists and Naturalists rely upon to make their case and meet their burden of proof. These people literally use their ignorance as evidence that their chosen conclusions are real and true; and, they try to shift their burden of proof to you and demand that you prove them wrong. That's dogma and religion and blind faith, not science! These people are *jumping to conclusions*; and, many of them don't even know that they are doing so.

The Smoking Gun – Proof that Our Memories Aren't Stored in Our Psyche

Remember, I was indoctrinated, conditioned, brainwashed, and programmed into Materialism, Naturalism, Scientism, Skepticism, Agnosticism, and Darwinism for the first fifty years of my life. I think that way. I am that way. I have always been that way. It's my default position. I naturally fall back on it whenever anything confusing or controversial comes up. I've got blinders on, and it takes a bit of work to see through it.

Oh, the Materialists and Naturalists are going to love this one.

After writing most of this book, I was reading the *Biopsychology*, *Neuroscience*, and *Biological Psychology* books, and I came across split-brain patients. These people have two separate brains, which at times work at cross-purposes from each other. The very existence of these people – the way that they function and work – was telling me that everything I wrote in this book is wrong. My Materialism and Naturalism prevailed.

Here's a sampling from these books as well as what I wrote in the margins of those books.

In some of the human split-brain studies, the two hemispheres initiated conflicting behaviors, apparently because they were thinking differently. In one task, a patient was asked to arrange a group of blocks to match a pattern on a small card. He was told to use only his right hand (left hemisphere), which is not generally good at this type of task. As the right hand struggled to arrange the blocks, the left hand (right hemisphere), which knew how to do it, reached in to take over. Only the restraint of the experimenter kept the left hand from pushing the right one out of the way to solve the puzzle. Another patient studied by Gazzaniga would sometimes find himself pulling his pants down with one hand and pulling them up with the other. These bizarre behaviors make a strong case that there are two independent brains controlling the two sides of the body. (*Neuroscience*, p. 629.)

It also makes a strong case for the claim that our memories and learning are stored only within our brain, and not some kind of universal ethereal psyche.

Left Hemisphere Language Dominance.

Although split-brain humans are normal in most every way, there is a striking asymmetry in their ability to verbalize answers to questions posed separately to the two hemispheres. For instance, numbers, words, and pictures visually presented only in the right visual field are repeated or described with no difficulty because the left hemisphere is usually dominant for language. Likewise, objects that can be manipulated only by the right hand (out of view of both eyes) can be described. These findings would be entirely unremarkable except for the fact that such simple verbal descriptions of sensory input are impossible for the right hemisphere.

If an image is shown only in the left visual field or an object is felt only by the left hand, the person will be unable to describe it and will usually say that nothing is there. An object could be covertly placed in a patient's left hand, and there would be no verbal indication of even noticing. This absence of response is a consequence (and demonstration) of the fact

that the left hemisphere controls speech in most people. If you think about the implications for split-brain people, you'll realize that they have an unusual existence. Following their surgery, they are unable to describe anything to the left of their visual fixation point: the left side of a person's face, the left side of the room, and so on. What is startling is that this doesn't seem to disturb them. (*Neuroscience*, p. 630.)

There's NO communication between the hemispheres through the Psyche, which suggests that psyche doesn't exist and that our memories are stored only within our physical brain. The evidence here repeatedly suggests that memories are stored in our physical brain and accessed from our physical brain, and not from some united universal Psyche.

Human split-brain patients seem to have in some respects two independent brains, each with its own stream of consciousness, abilities, memories, and emotions.

Before I recount some of the key results of the tests on split-brain humans, let me give you some advice. Some students become confused by the results of these tests because their tendency to think of the human brain as a single unitary organ is deeply engrained. If you become confused, think of each split-brain patient as two separate individuals.

In the neurophysiological laboratory, major discrepancies in what the two hemispheres learn can be created. As you are about to find out, this situation has interesting consequences. (*Biopsychology*, p. 400.)

There's NO psyche or consciousness here – no non-local consolidation of memories, learning, and knowledge – just two separate machines which no longer have a hardline connecting each other. The brain is a physical machine, and nothing more. If all of our memories are stored in our Psyche, then both hemispheres of split-brain patients should have access to the same information, but they do not. Sharing of information should take place through the Psyche, but it doesn't. So, what's really going on?

Cats with both their optic chiasms and corpus callosums transected learned the discrimination at a normal rate with one eye blindfolded, but they showed no retention whatsoever when the blindfold was switched to the other eye. (*Biopsychology*, p. 398.)

In other words, the Psyche hadn't learned anything, which suggests that the Psyche doesn't exist! No retention implies no memories and no learning. The memories for this task were stored only in the physical brain, and NOT the Psyche. Well, that just falsified all my theories and ideas on the subject!

Ouch!

Split-brain cats showed no evidence that information stored in one hemisphere was available to the other hemisphere.

Further, behaviors learned by one hemisphere could not be performed by the other hemisphere. In fact, split-brain animals could learn different behaviors in each hemisphere, suggesting the cats were functioning as if they had two separate "brains".

After the initial training, the patch was switched to the other eye. The transfer did not affect performance in the control animals, but it profoundly affected the experimental subjects. Following the switch, the performance

of the experimental animals dropped to a chance level (i.e., results governed by chance alone), and the cats relearned the task at the same rate as naïve (untrained) animals. There was absolutely no evidence for any previous learning in the split-brain cats. (*Biological Psychology*, p. 470.)

There was absolutely no evidence for any previous learning in the split-brain cats. This shouldn't be the case if our memories are stored in the Psyche and then downloaded from the Psyche. Something is seriously wrong with my theories and my ideas. Our memories and our learning aren't being stored in our Psyche after all; so, that means that they must be getting stored within our brain(s). These animals have two brains, and two completely different sets of memories, which means that their memories are only being stored within their brains and not in some ethereal psyche.

Is the human psyche compartmentalized? I don't see how. Maybe the spirit body is? But, I don't see any way to save this. This evidence suggests that the Human Psyche and that animal psyches do not exist.

In the chimeric figures test, and many other tests with split-brain patients, the Psyche doesn't know that each hemisphere is seeing and experiencing different things. This clearly demonstrates and proves that our memories are not being stored within our Psyche. Apparently, the Psyche doesn't exist after all; and, all of our memories are being stored within our brains and nowhere else.

Observe, the Materialists and Naturalists will pull this out of context and use it as proof that Psyche does not exist. It's good stuff, if that's your ultimate goal in life. But, I warn you, I employed a couple of serious logic fallacies in order to make my case, just like every Materialist and Naturalist does. Maybe you can see a few of them already if you have had any training in logic and the Philosophy of Science.

After reading this material, my initial natural gut reaction was to conclude that everything I had written in this book, *Quantum Neuroscience*, is wrong. In other words, I fell back on my ingrained Materialism and Naturalism out of habit, as if it had been conditioned into me.

Obviously, I had some thinking, praying, and studying to do, because something didn't add up here. Clearly, there is something important that I am missing.

Yes, a few of you will see through the flaws in my reasoning immediately, but I wasn't one of them. I had some work to do, and a few modifications of my theories and ideas to make.

As physical beings, we naturally prefer Materialism, Naturalism, Classical Physics, Physical Limitations, and Scientism, because they make logical sense to us, are observable by anyone not just the righteous, and are replicable on demand – because they are based upon the ultimate consensus reality, a physical reality. It's extremely easy to conflate brain modules with memory storage modules, because we want it to be so. It's very easy to conclude from the physical evidence alone that Psyche or Non-Local Consciousness does not exist.

In contrast, with Quantum Mechanics, Action at a Distance, and Quantum Non-Local Consciousness, we start to see the hand of God in action. Quantum Mechanics from beginning to end is Supernatural, and it simply doesn't make any sense to the Materialists and Naturalists because it's beyond them.

The physical brain is a machine, a network of different computers and calculators. Our man-made calculators have a stack or memory; and, our computers have RAM, hard drives, and now cloud storage for memory. It's obvious that the physical brain stores memories, and therefore has some level of memory storage capacity.

Even if we were to conclude that our brain is comprised of Cerebral Processing Units (CPUs) with no RAM (memory engrams) whatsoever within our brain, computer scientists like me KNOW that CPUs have cache or working memory. This is a model that I love a lot, because it has a lot of explanatory power. Consequently, based upon this model of the brain, it's obvious to me that each CPU within our brain has working memory or cache at its disposal.

The million dollar question is, "Are all of our memories stored in our physical brain, and do our memories cease to exist when our physical brain dies? Or, is there more to be found in terms of memory storage or cloud storage in the Quantum Realm or the Supernatural Realm?" Memory storage may just be the most interesting subject in neuroscience to study.

For me personally, the solution to this contradiction or problem came from starting with what we already KNOW to be real and true, and then diving in and studying in much greater detail the after-death Life Review. We have to go back to what we KNOW, before we can begin to solve contradictions like the one I encountered here with split-brain patients. It starts by allowing ALL of the evidence into evidence, and then going forward from there.

Life Review

From the evidence being provided by Split-Brain Patients, it seems as if NONE of our memories are being stored within the Human Psyche, or the Human Psyche doesn't exist.

So, how do we resolve this conundrum?

I can only tell you how I resolved it. I had to return to basics. I had to return to what I KNOW to be real and true. I had to return to and study the after-death Life Review.

WE KNOW from the collective Lived Experiences of the human race that our personality, spirit, psyche, or intelligence survives the death of our physical body and physical brain, according to the empirical evidence, experiential evidence, and observational evidence from Near-Death Experiences (NDEs), Out-of-Body Experiences (OBEs), and Shared-Death Experiences (SDEs).

This KNOWLEDGE implies that our memories or our intelligence survives the death of our physical brain. But, how many of our memories survive, and where are those memories stored?

For a year or two now, I have tended to visualize the Human Psyche as an infinite singularity – infinite memory storage capacity within something that has NO physical size and takes up NO physical space whatsoever. By that definition of Psyche, it just seemed natural to me that ALL of our memories, experiences, feelings, learning, and knowledge would be stored in our Psyche. However, observing split-brain patients repeatedly tells us that such is not the case.

So, where are our memories stored in the Non-Local Quantum Realm, or do all our memories cease to exist when our physical brain dies? Something has to give. Either none of our memories are stored non-locally in the cloud, or we don't have immediate access to the ones that are.

The Life Review during a Near-Death Experience

https://www.youtube.com/watch?v=6ONqq1d2QUs

Notice carefully how she had a Life Review, and in her Life Review, she saw and experienced the ripple effect of her good choices and the loving acts of kindness that she did for others. She experienced their feelings, and the way that her choices and their positive feelings from her choices spread into the lives of their children, friends, and family. She experienced their memories and feelings, as well as her own.

Then she said that she got to experience her father's Life Review. She experienced someone else's Life Review – saw and felt things that she herself had never lived nor experienced personally.

Then I saw my problem and found my answer.

I had erroneously chosen to believe that the Life Review is stored in the Human Psyche. Our Life Review is stored in God's Psyche, or Nature's Psyche, or God's Database – non-locally in the cloud – and then it is linked with or cross-referenced with other people's Life Reviews which are also stored in the cloud or the Supernatural Non-Local Quantum Realm.

The memories that survive our death are NOT stored within our Psyche. They are stored someplace else by someone else besides our Psyche. Consequently, the Psyche doesn't have access to these things at will. God owns our Life Review and our memories. We don't.

Most of us do not have live access or dynamic access to our Life Reviews, especially the ripple effect of our own personal Life Review. Our Life Review is out of our control and in the hands of God, as is so much of our existence and life. Our Life Review memories are NOT being stored within the Human Psyche. They are being produced by Nature's Psyche and stored within God's Database instead.

Nature's Psyche has access to infinitely more information and knowledge than the Human Psyche. Nature's Psyche actually has more access to our memories than the Human Psyche does. In fact, the Human Psyche is limited exclusively to what its physical brain can give it. If the physical brain isn't working right, the Human Psyche is going to get faulty input, because Nature's Psyche has been commanded by God not to share its superior knowledge and information with the Human Psyche. The only way the Human Psyche can get any information in addition to what its physical brain gives it is if we get that information through revelation from God and the Holy Spirit. The Human Psyche is slaved to the physical brain, and only gets its input from the physical brain. The Human Psyche isn't consciously aware of what Nature's Psyche is doing, nor is the Human Psyche aware of how Nature's Psyche is storing its memories and experiences.

Split-brain patients don't have dynamic and live access to their Life Reviews, Nature's Psyche, or their Non-Local Cloud Storage, so these people really do have two brains or two machines that their single Psyche is trying to coordinate and manage as best it can. The Human Psyche is ONLY getting information from its physical brain. The Human Psyche doesn't have access to the information Nature's Psyche is generating. These people are indeed going to have problems sharing information between these two computers and coordinating everything between these two computers, now that the hard line between these machines has been cut.

When it comes to the physical realm, the Human Psyche only experiences what registers upon the neurons of its physical brain. With NO axons feeding in compensating or modifying information from the other hemisphere, the Human Psyche can only sense and control the physical input that each separate hemisphere is giving it, not matter how confusing things might get to be.

Have you ever tried to synchronize a research paper or a computer program that is being written on many different machines by many different people? That's what we are dealing with in the case of split-brain patients – one psyche trying to coordinate the information and actions of two different machines with their add-on cards or extension cards.

All of our choices and the ripple effects of our choices are being recorded and cross-referenced by God, the Angels, and Nature's Psyche within the cloud as our own unique personal Life Review; but as a general rule, we won't have access to any of that information until after our physical brain is dead or offline and the angels present to us our Life Review including its ripple effects. The angels or God wouldn't have to present our Life Review to us, if all of that information was already stored within the Human Psyche and its physical brain.

Ring, K., & Cooper S. (1999). *Mindsight: Near-Death and Out-of-Body Experiences in the Blind* (2nd ed.). Kearney, NB: Morris Publishing.

Within this book, *Mindsight*, people who were born blind report experiencing video, as well as sounds, feelings, tastes, smells, and touch during their Life Review. People who were born blind can see during their Near-Death Experiences. People who are born blind experience video or movies of their life during their Life Review which means that this part of their Life Review was NOT created by their physical brain nor stored within their physical brain. This is evidentiary proof and observational proof that the claims of Materialism and Naturalism are wrong.

It's obvious from the evidence that our Life Reviews are NOT being produced by our brain nor stored within our brain. Furthermore, it's obvious and clear that our physical brain can't store petabytes of video and other sensory information. Our Life Reviews are being stored someplace else besides our brain, and someplace else besides the Human Psyche. The Human Psyche doesn't have access to the memories that are being created and stored by Nature's Psyche. The Human Psyche is slaved to the physical brain, and the Human Psyche only has access to the information that registers on the neurons of the physical brain.

Therefore, we can safely conclude based upon the evidence and common sense that the majority of our memories are not stored within our brain nor within our Psyche, but are being store non-locally in the cloud by Nature's Psyche instead. Consequently, the claim that our memories are stored exclusively within our brain and cease to exist when our brain dies has been FALSIFIED by the empirical evidence at our disposal.

Angels above us are silent notes taking

Of every action; then do what is right!

https://www.lds.org/music/library/hymns/do-what-is-right?lang=eng&_r=1

Someone else is keeping our Life Reviews for us; and, they aren't being stored within our own Psyche or Spirit Body. Our Life Reviews aren't being produced by our physical brain, nor stored within our physical brain either. Our Life Reviews are being produced by Nature's Psyche or the Angels in Heaven and stored within God's Database, instead. You can't touch this! See what becomes possible by allowing ALL of the evidence into evidence? You can finally figure out what's really going on.

Observational evidence and experiential evidence have marvelous, wonderful, and powerful explanatory power which Materialism and Naturalism and Atheism completely lack. Now you can see why I upgraded my science to Science 2.0, which allows all of the evidence into evidence and pursues a preponderance of the evidence. You can also see why I'm no longer a Materialist, Nihilist, and Atheist. For me personally, the evidence hit critical mass, and now there's no going back to my former ignorance and limitations.

John 14: 26:

But the Comforter, which is the Holy Ghost, whom the Father will send in my name, he shall teach you all things, and bring all things to your remembrance, whatsoever I have said unto you.

There you have it – the true reality of the situation. The Holy Ghost is a recording device. He's a Recorder. That's one of His functions. One of His main purposes is to bring things to our remembrance.

We were designed not to remember everything we have ever done or experienced. As physical beings, we had limitations placed upon us, especially memory storage limitations. Our Life Reviews are NOT being stored within our physical brain, nor are they being stored within our individual spirit body or psyche. Our physical brains don't have enough memory storage capacity to store our Life Reviews, which is why we forget things and have to have them brought to our remembrance again by the Holy Ghost.

A Life Review wouldn't be necessary if ALL of our memories were being stored in our Psyche or within our Spirit Body. It would ALL be present before us simultaneously if it were stored within our own Psyche or Spirit Body. The fact that it isn't instantly available to us is telling us that it isn't being stored within our own Psyche.

This is a good thing, because our physical brains can glitch, become damaged, and fail. Someone else is keeping the good stuff in store for us. When necessary, the Holy Ghost can bring it to our remembrance.

This also seems to mean that Satan and the evil spirits can bring to our remembrance all the times that we have been hurt, angered, hateful, and offended. Someone on that team is also keeping a record of all the bad that we do.

The Life Review during a Near-Death Experience

https://www.youtube.com/watch?v=6ONqq1d2QUs

Notice how she said that God or the Being of Light is able to filter out the garbage. The Holy Ghost deals in the good things, the acts of kindness and love, including their ripple effects.

Satan and the evil spirits deal in the garbage – the stuff you want to forget, repent of, and learn to forgive.

That's how you tell the difference between the two.

I can tell you from personal experience that inspiration and revelations from Satan feel exactly the same as revelations and inspiration from the Holy Ghost; but, the message is completely different. The Holy Ghost brings comfort, joy, compassion, love, and peace. The peace is simply amazing. For the first time in my life, I'm at peace. It took over fifty years to achieve, but I'm at peace now. Peace, forgiveness, compassion, friendship, charity, hope, trust, and love are gifts from the Holy Ghost through the Atonement of Christ.

In contrast, temptations or revelations from Satan feel as if the Holy Ghost is telling you to burn the house down, to kill yourself, to beat your children to death, to divorce your wife, that there's nothing wrong with adultery or pornography, to stop going to church, that you can handle your addictions, to lie in order to protect yourself, that there's nothing wrong with taking what you need, that the pain and anguish will never end unless you kill yourself, to get even and to get revenge, to hate that bastard because he wronged you, and to quit Christianity because it's pointless and worthless.

Revelations from Satan feel just like revelations from the Holy Ghost; but, the message is completely different. We need to learn to discern the difference between the two.

If you are doing wrong and being evil and addicted to drugs, it opens you up to the buffetings of Satan so that you can start getting your revelations and inspiration from Satan and from the evils spirits or demons who follow him and are promoting his cause, which is to make you as miserable as they are.

It just is what it is, and I had to learn to get used to it.

I had to be sober and had to be trying to do good for over a year, before the Holy Ghost came into my life and started giving me revelations and insights instead of getting all of my input from Satan.

If you open the door to Satan, he will come right in.

In contrast, you kind of have to prove yourself and invite Him in for a year or two before the Holy Ghost will start dealing with you on a regular basis. You also have to be willing to do whatever the Holy Ghost is encouraging you to do – repenting of your sins and praying for the ability to forgive and love – before it makes any sense for the Holy Ghost to even try to help you. There's no logical sense for God to give you a revelation if you have no intention whatsoever to do anything with it. You have to be willing to act on it, use it, and incorporate it, or it makes no sense for God or the Holy Ghost to give it to you.

It can be demonstrated and has been repeatedly demonstrated that memories and knowledge and programming are indeed being stored within the physical brain. The existence of working memory and physical functionality within the physical brain is a proven fact.

So, where are our after-death Life Reviews being stored? And, how is that functionality being mapped or programmed into our brains?

The Life Review can't be stored within our brain, because our brain is in principle dead and gone whenever we and others are experiencing our Life Review.

Furthermore, evidence from split-brain patients indicates that either our memories are NOT being stored within our Psyche, or our brain doesn't have access to the memories that are being stored within our Psyche. Eventually I concluded that the Human Psyche and the physical brain are synonymous. They function as a unit. And, both the Human Psyche and the physical brain do NOT have access to the memories that are being stored in Nature's Psyche or being stored by Nature's Psyche in the cloud.

Once I started to study this subject in earnest, the preponderance of the evidence pointed me to a startling conclusion. Our Life Reviews are NOT stored within our Psyche. I had just assumed that our long-term declarative memories or Life Reviews were being stored within the Human Psyche, our Psyche; but, I was wrong.

God and the Angels wouldn't have to show us our Life Review, if our Life Review and the associated long-term declarative memories were being stored within our Psyche. We would have immediate and constant access to that information all the time, if our declarative memories, consolidated memories, and Life Review were being stored within our Psyche. The observed FACT that we (our Psyches) don't have on-demand access to our Life Review suggests clearly and conclusively that our Life Review or declarative consolidated memories are being stored someplace else besides our Psyche and our brain.

The Human Psyche isn't aware of the things that Nature's Psyche and the physical brain are unable to provide it. The Human Psyche is slaved to the physical brain or limited by the physical brain. Unless we choose to invite the Spirit of God into our lives, the Human Psyche is limited to the physical level; whereas, Nature's Psyche seems to be functioning purely on the quantum level, which is why we aren't consciously aware of what Nature's Psyche is doing. Yet, through quantum waves or telepathy, Nature's Psyche is fully aware of the Human Psyche's desires, choices, and commands. Even though the Human Psyche isn't monitoring nor sensing Nature's Psyche, it is clear that Nature's Psyche is monitoring and listening to the Human Psyche. The Human Psyche makes all the decisions and choices; and, it's Nature's Psyche who collapses the wave functions and fires the neurons, which make those choices and decisions physically real and actual.

Fascinating, is it not?

Well, I think it is.

Remember, if your Interpretation of Quantum Mechanics can't explain to you how the Human Psyche interacts with and controls its physical brain through negotiations with Nature's Psyche at the quantum level, then your Interpretation of Quantum Mechanics is worthless.

Refusing to look at evidence and *destroying evidence* are logic fallacies. Materialism, Naturalism, Darwinism, Scientism, Nihilism, Behaviorism and Atheism are based upon a refusal to look at evidence. These pseudo-sciences are based upon deliberate and known logic fallacies. That's why I upgraded my science to Science 2.0. Science 2.0 allows ALL of the evidence into evidence, and then Science 2.0 pursues a preponderance of the evidence. Science 2.0 uses the tried and true Burden of Proof Methodology that's used in a court of law, allows ALL of the evidence into evidence, and then pursues a preponderance of the evidence.

This is something that the Materialists and Naturalists can't do and won't do, because the proven and verified existence of Quantum Mechanisms such as Telepathy (the Quantum Zeno Effect), Telekinesis (Action at a Distance), Teleportation (Quantum Tunneling), and Psyche (Thoughts, Memories, Quantum Waves, and the Mind-Over-Matter Placebo Effect) FALSIFIES Materialism, Naturalism, and their derivatives such as Darwinism, Behaviorism, and Atheism. You can't remain a Materialist and Naturalist once you KNOW the truth.

Signs of Psyche in Split-Brain Patients

Uploading our memories to non-local cloud storage is physically impossible according to Materialism, Naturalism, Darwinism, Atheism, and Classical Physics; and, that's where things start to get really interesting.

By returning to the basics of what we already KNOW to be real and true as a race, I was then able to go back into these same textbooks and see signs of psyche even within split-brain patients.

These observations of split-brain patients actually contradict what these different neuroscientists chose to lead with. The Materialists and Naturalists always start with the conclusion that ALL of our memories are stored exclusively within our brain, and that we and our memories cease to exist when we die. They start with this conclusion, insert it as evidence or as the hidden premise in all of their arguments, and then adjust everything else accordingly. That's the way these people do science, by *begging the question* and *affirming the consequent*; and, it works. Millions of people fall for it, as I did.

Feelings of emotion appear to be readily passed between the hemispheres of most split-brain patients. (*Biopsychology*, p. 403.)

This observation does imply that communication through the Psyche is taking place; however, there are at least two ways to interpret this observation – both equally valid.

Either the Human Psyche is passing the emotions and feelings between the hemispheres of split-brain patients, or the commissure that shares emotions and feelings between hemispheres wasn't severed when the corpus callosum was severed. Which is it? I don't know; but, it doesn't automatically rule out Psyche as a possible answer. I'm no longer jumping to conclusions like I was before. Instead, I'm now trying to allow all of the evidence into evidence, and then see what I can make of that.

It has been observed that our feelings and emotions survive the death of our physical body and physical brain; consequently, it's not unrealistic to hypothesize that our emotions are being controlled or mediated by the Human Psyche and not necessarily the physical brain. Emotion is not purely a physical phenomenon. In the non-local realm, emotion is a product of Psyche or Non-Local Consciousness.

Emotions are passed between hemispheres in split-brain patients, which means one of two things – either the emotion commissure wasn't cut in split-brain patients, or emotion is a function of Psyche. And in this case, it's actually possible that both answers are correct.

I have observed that anytime we observe CHOICE, we are in fact observing Psyche of some kind in action. Our choices aren't made by our genes or our environment. Our choices are NOT a function of Nature and Nurture. Our choices are a product of our Psyche. Pure and simple! Your Psyche can trump or override your Nature and your Nurture at will. That's just the way it works.

The Case of Peter, the Split-Brain Patient Tormented by Conflict

At the age of 8, Peter began to suffer from complex partial seizures. Antiepileptic medication was ineffective, and at 20, he received a commissurotomy, which greatly improved his condition but did not completely block his seizures. A sodium amytal test administered prior to surgery showed that he was left-hemisphere dominant for language.

Following surgery, the independent mischievous behavior of Peter's right hemisphere often caused him (his left hemisphere) considerable frustration. He (his left hemisphere) complained that his left hand would turn off television shows that he was enjoying, that his left leg would not always walk in the intended direction, and that his left arm would sometimes perform embarrassing, socially unacceptable acts (e.g. striking a relative).

In the laboratory, Peter (his left hemisphere) sometimes became angry with his left hand, swearing at it, striking it, and trying to force it with his right hand to do what he (his left hemisphere) wanted. (*Biopsychology*, pp. 402-403).

Wow, this is probably the most fascinating split-brain case study on record.

There's a "he" there, his Psyche, who actually KNOWS that his right hemisphere is misbehaving!

It's as if his right hemisphere is being controlled by a trickster, or a poltergeist, or an evil spirit. It's as if his right hemisphere has been possessed by some kind of demon.

There's a single person here who is getting upset at the malfunctions and glitches being produced by his right hemisphere. It's like he is suffering from split-personality, or multiple personality disorder, or dissociative identity disorder. It seems as if his right hemisphere has been possessed.

Fascinating!

Of course, the Materialists and Naturalists will assure us that this Dual Possession or Dual Psyche is an illusion. They will tell us that science has proven that psyche or mind is a product of our physical brain, and nothing more. Consequently, in this case, they will tell us and have told us that Peter simply has two brains, which explains all the weirdness that's happening to him. Peter's right hemisphere is just following its own unique autonomic programming.

The only way to know for sure that the Materialists and Naturalists are wrong is by allowing ALL of the evidence from Near-Death Experiences (NDEs), Out-of-Body Experiences (OBEs), and Shared-Death Experiences (SDEs) into evidence, and then by pursuing a preponderance of the evidence.

If you are willing to allow all of the evidence into evidence, then it starts to become clear that Peter has a single dominant psyche, or person, or intelligence who KNOWS when his other side is misbehaving or not doing what he wants it to do. Peter's experience is indeed strange. It's as if he, his Human Psyche, is controlling his left hemisphere, and Nature's Psyche or an evil spirit is controlling his right hemisphere; yet, he KNOWS it.

A Split-Brain Patient:

https://www.youtube.com/watch?v=ZMLzP1VCANo

There seems to be ONE person or one psyche here, who is getting a bit confused by the conflicting input and results.

Of all the books that I own, *Biopsychology* by John Pinel has the best and most comprehensive treatment of the split-brain phenomenon. Apparently, he was as fascinated by the subject as I was.

In the other case studies that Pinel presents, it's clear that there is ONE person or ONE psyche, who is trying to coordinate the input and the actions of two separate machines or two separate Cerebral Processing Units (CPUs). It's also becomes clear that there is NO common universally accessible memory storage device or RAM within the physical brain.

Some of these people get confused and frustrated by the conflicting information or input. The patient wouldn't get confused unless there is a single dominant Psyche in the patient to get confused by the malfunctioning and non-communicating machinery.

It has been observed that split-brain patients can successfully do two different things at once. But soon, I find myself asking, "Who is controlling and coordinating and choosing the actions of these two separate machines?" It also quickly becomes clear, from some of the case studies, that the Human Psyche can coordinate the input from two separate machines in order to accomplish a single goal. These observations suggest some kind of overriding command and control system. Will or choice is a function of Psyche – always has been and always will be. Where there's a will, there's a psyche.

In most split-brain patients, the right hemisphere does not seem to have a strong will of its own; the left hemisphere seems to control most everyday activities. (*Biopsychology*, p. 402.)

In a couple of the case studies presented by Pinel, the patient KNOWS that he or she has a split-brain, and that single person develops ways to beat the system! This is called Cross-Cuing and the Helping-Hand Phenomenon. Under the direction of the Human Psyche, the side of the brain that knows the answer reaches out and helps the other side of the brain to choose the correct answer. In one of the examples, the patient reached out with his left hand and wrote the answer on the back of his right hand so that the other side of his brain would know the answer too.

Studying split-brain patients seems to tell us that the left hand doesn't always know what the right hand is doing – or does it? Sometimes their hand seems to have a mind of its own. However, due to a single unified Human Psyche, through cross-cuing and the helping-hand phenomenon, the left hand reveals that it KNOWS that the right hand is failing, and then takes steps to correct the situation.

There's evidence here in these case studies that communication between severed hemispheres is taking place under the direction of or through the command and control of a single person or psyche. Apparently, I was too hasty in my decision to eliminate psyche from the equation. There are signs of psyche, even within split-brain patients.

Karl Lashley and others have proven scientifically that our brain does not have memory engrams or RAM. When it comes to the physical brain, we are looking at CPUs, and not RAM. The physical brain is about functionality, not memory storage. Each CPU has its own cache or working memory; but, there is NO universal RAM storage within our physical brain accessible by all the CPUs within our brain.

Whenever the physical brain is lesioned or damaged, functionality is lost, but the memories typically remain, although the memories sometimes become inaccessible. In the case of temporary lesions, though, the memories always come back whenever the functionality is restored. There's two ways to interpret these observations. Either our consolidated memories are being stored non-locally in the cloud and access to them is lost when functionality is lost; or, our consolidated memories are stored within our physical brain but access to them is lost when functionality is lost.

So, which is it? It could be a little bit of both, actually. We don't know where brain storage capacity ends, and non-local cloud storage begins. We just KNOW that it's physically impossible to store petabytes of data within our brain; so, some of that memory storage has to be taking place somewhere else besides our brain.

The only way to resolve this contradiction is to observe that the consolidated memories which end up in our after-death Life Reviews can't possibly be stored within our physical brain, because our brain is typically dead, offline, or gone when we are experiencing our Life Review. This suggests that at least some of our memories are being uploaded to the cloud and stored within the cloud, rather than all of them being stored within our physical brain.

Since all of this consolidation and uploading and downloading takes place outside of our conscious awareness, it's clear that the Human Psyche isn't in charge of this process. Our consolidated memories and Life Reviews are being handled by Nature's Psyche and/or God's Psyche completely outside of our control. Nature's Psyche and/or God's Psyche decides whether to store our memories within our physical brain or to store them non-locally in the cloud. It all happens automatically outside of our conscious awareness.

Fascinating, is it not?

I finally have an explanation for what's really going on.

There are payoffs that come from choosing to allow all of the evidence into evidence and from choosing to pursue a preponderance of the evidence.

Although Nature's Psyche is keeping a dynamic, indexed, and current record of all of our thoughts, experiences, feelings, choices, decisions, and actions, apparently our working memory or the Human Psyche doesn't have immediate access to these things at will.

Instead, one of the functions of the Holy Ghost is to help us remember things, to bring things to our remembrance, and to help us to access our long-term non-local cloud storage where these memories are located and being stored.

Apparently, storage of memories is taking place, not information exchange, where our non-local non-physical off-site cloud storage is concerned.

Remember, the Holy Ghost brings things to our remembrance. This wouldn't be necessary if all of our memories were stored in our spirit body and accessible to our Psyche at will; and, this wouldn't be necessary if all of our memories were stored in our physical brain and accessible at will.

According to the Orthodox Interpretation of Quantum Mechanics, it is Nature's Psyche who collapses the wave function, thereby making the Human Psyche's choices and decisions physical, actual, and real. This means that it is Nature's Psyche who is triggering our motor neurons making them fire and making our physical bodies move.

At some point in the process, the desires of Nature's Psyche for our physical body has to transition from the non-local to the physical; and, it does so through the physical media of DNA, genes, brains, and protein synthesis – under the guidance of God's Psyche and God's Database.

At some point there has to be a physical causation for our physical existence and our physical actions. At some point in the process, we have to transition from the spiritual to the physical. The Human Psyche makes the choices and the requests. Nature's Psyche works directly on the physical matter in order to accomplish these physical tasks in the physical realm thereby making them physically real.

All of this reminds me of the science fiction stories where all of your conversations and a video image of all of your physical actions are being stored as Quantum Waves within the glass of your windows.

And then, you discover that the rocks are recording all of your thoughts, feelings, choices, desires, and decisions.

If any of this is true, and it most likely is given what we know about Quantum Mechanics and Quantum Waves, then that means that even if our memories are not being stored within the Human Psyche and its physical brain, our memories are definitely being recorded and stored within the glass and the rocks by Nature's Psyche instead.

There's plenty of evidence suggesting that our long-term consolidated memories are not being stored within the Human Psyche nor within the human brain. If so, this reality would explain a lot of what we see in split-brain patients, but still make a Life Review possible after we are dead.

If Nature's Psyche is capturing, recording, and storing our memories in the atoms, rocks, water, and glass around us, then that would explain how our Life Reviews are being maintained; and, it would also explain why the Human Psyche and the physical brain aren't consciously aware of any of these things.

The "unconscious" is a function and a product of Nature's Psyche. Our physical brain and the Human Psyche aren't consciously aware of the unconscious.

Oh, I don't have any doubt that our memories are being distributed throughout our brain among the atoms of our brain. It's just not happening at the physical level, because it is physically impossible. How would you go about putting your memories into a single atom, if I were to give you the assignment to do so? Well, you wouldn't be doing it at the physical level, because it's physically impossible. You would have to do so at the quantum level, by converting your memories into Quantum Waves, and then storing those waves within that physical atom.

The brain is a physical machine that Nature's Psyche uses to get things done for the Human Psyche in the physical realm. If the genes and proteins are bad, or the physical brain is broken, then there's nothing that Nature's Psyche can do about it. Nature's Psyche can't make broken machinery work, no matter how much the Human Psyche might want it to.

Nature's Psyche works perfectly, but it is evolution or random mutations that breaks everything and messes everything up. Bad genes produce bad proteins, and bad proteins are nothing but broken machinery. Nature's Psyche can't do anything about the broken machinery. We actually have to turn to God (God's Psyche), if we want to have our broken machinery fixed for us.

Jumping to Conclusions

I had concluded that since our memories are not being stored within our Psyche and retrieved from our Psyche, then our Psyche does not exist.

So, what was my flaw that got me into this problem in the first place? Can you see it? Do you understand it?

I did what any Materialist and Naturalist does upon finding evidence that our memories are NOT being stored within our Psyche – I immediately jumped to the conclusion that Psyche does not exist. Jumping to Conclusions is a logic fallacy, as are Scientific Inferences. They fall under the umbrella of the logic fallacy that has been called Begging the Question, which means that you are using your pre-chosen conclusion as evidentiary proof that your chosen conclusion is true. You are Affirming the Consequent. That's what the Materialists, Naturalists, and Darwinists do, without ever realizing it. It's the way they do science, by Jumping to Conclusions, Affirming the Consequent, and Begging the Question. That's the way that they prove to themselves that the Theory of Evolution is true. This is very powerful Philosophy of Science here, and they are completely clueless about it all.

I made some *Conflation Errors* and *Category Errors*. On the spiritual side, I erroneously equated the Human Psyche with a memory storage device concluding that our memories are stored within our Psyche. On the physical side, I erroneously equated the physical brain with memory storage or RAM. I assumed that if it was making it into the physical brain, then it was automatically making it into the Human Psyche. It seemed like a logical conclusion to me; but, such has proven not to be the case.

Interesting, is it not?

We always run into problems whenever we employ *Category Errors*. I placed the physical brain into the category of long-term consolidated memory storage devices (hard-disk drive). On the other side of the equation, I placed the Human Psyche into the same category (a solid state drive). Why do *Category Errors* always mess us up? It's because they involve *Jumping to Conclusions*!

Functionality seems to be the primary purpose of both the Human Psyche and the physical brain – NOT memory storage; consequently, we run the risk of making serious judgment errors anytime we choose to categorize either the physical brain or the Human Psyche as memory storage devices.

I also employed *All-or-Nothing Thinking*. Since I had convincing evidence that our memories are not being stored within our Psyche, I jumped to the 'nothing' end of the equation and concluded that our long-term memories are not being stored anywhere but within the brain – just like any Materialist or Naturalist does. I jumped to the conclusion that there is NO non-local storage and NO psyche whatsoever – *all or nothing*!

Logic fallacies result in faulty reasoning and faulty conclusions.

On the other end of the spectrum, I had also *assumed* that since there's a copy of our memories, learning, and experiences being made in non-local storage in the cloud for our Life Review, that those memories were being stored within our own Personal Psyche and should therefore be immediately available to our own Human Psyche on demand. Such is not the case.

Since then, I have acquired and/or remembered evidence indicating that our Life Reviews are not stored in our own Personal Psyche, but in Nature's Psyche or God's Database instead. This discovery resolves the situation, because it fits ALL the evidence we have on hand. Our Life Reviews, or our Long-Term Consolidated Memories that survive the death of our brain, are NOT being stored within our Psyche – they are being stored non-locally somewhere else instead.

Can you see any other flaws in my logic, or anything else that I had gotten wrong?

Well, I learned to do so.

I *refused to allow all of the evidence into evidence* which is a logic fallacy.

Assumptions involve *jumping to conclusions* and *begging the question*. *Scientific inferences in the absence of evidence* is a logic fallacy. I had erroneously *assumed* that information can be and should be passed through the Human Psyche to both hemispheres of split-brain patients. It seemed like a reasonable *scientific inference* to me. However, the preponderance of the evidence tells us that this *assumption* is wrong. It doesn't work that way.

Why?

It's because whenever the Human Psyche is bound to and residing within a physical body and physical brain, all of the evidence suggests that that psyche is completely subservient to and dependent upon the physical body and the physical brain for its information. If the physical brain isn't consciously aware of something at the physical level, then the Human Psyche residing within it isn't consciously aware of it either. And, this process even seems to be bi-directional. If the Human Psyche finds something humorous in the right hemisphere of the brain, the left hemisphere in split-brain patients doesn't know what it is that its psyche finds so amusing. In other words, the physical brain seems to be dominant in this symbiosis; and, the physical brain is the limiting factor in this symbiosis. If the physical brain doesn't know something, then the Human Psyche doesn't know it either. The Human Psyche and physical brain are effectively synonymous.

Evidence also suggests that the right hemisphere understands complex pictures despite its inability to say so. In one experiment, a subject was shown a series of pictures in her left visual field, and at one point a nude photo appeared in the series. When the researcher asked what she saw, she said nothing, but then she began to laugh. She told the researcher that she didn't know what was funny, but that perhaps it was the machine used in the experiment.

The results of these split-brain studies demonstrate that the two hemispheres can function as independent brains and that they have different language abilities. (*Neuroscience*, p. 631.)

http://mypsyche.us/wp-content/uploads/2018/01/BCP20.pdf

If the brain(s) doesn't know what's going on, then the Human Psyche doesn't know either. It can only guess.

If either hemisphere is unaware of some fact or piece of information, then the Human Psyche is also unaware of it. In this relationship, the physical brain rules, not the psyche. The Human Psyche is simply along for the ride. The Human Psyche and physical brain are synonymous, unless God is invited to reveal things to us directly at the quantum level.

Disruption of consciousness takes place during sleep and during different types of epileptic seizures. Information flow to the Human Psyche goes offline whenever brain functionality goes offline. The Human Psyche seems to go offline with it. Clearly, the Human Psyche is slaved to the physical brain while residing within the physical brain. When it comes to the physical realm, the Human Psyche can only know what the physical brain is consciously aware of. God deliberately crippled us so that we could learn and grow from the experience.

Fascinating, is it not?

The Human Psyche is slaved to the physical body and the physical brain; and, this reality seems to be bi-directional. It's physically impossible to use our physical instruments to record our thoughts and our dreams, because our thoughts and dreams are being produced within the Human Psyche and NOT the physical brain. Thoughts and memories are quantum waves, which is why they survive the death of our physical body and physical brain.

Although the Human Psyche can incorporate physical sensations into our dreams, our night visions are not the result of physical stimuli and physical input. Our dreams are lived-out and experienced within the mind's eye of the Human Psyche.

How do we know?

First, WE KNOW by the fact that we dream despite the fact that our physical body is paralyzed during the dreaming process and there is little or NO physical sensory input coming into our brains while we are dreaming. We also KNOW because it's physically impossible to record our dreams with our physical instruments.

Our brains are over-rated and given way too much credit, especially when it comes to our memories, thoughts, and dreams.

It's amazing how combining ALL of the evidence eventually leads you to the truth, as long as you are willing to follow the evidence wherever it might lead you. Following the evidence got rid of many of my pre-conceived notions.

So, what else have I begun to learn, or what else have I already forgotten?

I have observed that wicked spirits – who go through Hell, Gehenna, or Purgatory and atone personally for their own sins – are purged or emptied by the experience. Their memories are gone, and they become innocent again, so that they can be reborn again as new and innocent people. They don't remember their former lives in mortality. They might know that they had lived previous lives in mortality, but they can't remember anything about their former lives. Seth is one of those, as explained in Jane Robert's "Seth books". She latched onto him through an Ouija board, which gives us a hint as to the type of being he is and the type of being he used to be. There are many indications that he had been purged.

https://www.amazon.com/Jane-Roberts/e/B000APH242/

I Am NOT a Creationist! So What Am I?

https://www.amazon.com/dp/B071XTM8XY

Our Life Reviews are not stored in our own personal psyche, but in Nature's Psyche or God's Database instead. If all of that information were stored in our own psyche, then after we are dead, all of that information should be constantly present and constantly available, which it isn't. Our Life Reviews have to be shown to us or presented to us, which clearly suggests that they are not stored in our own personal psyche. Our memories really aren't ours. They belong to God instead.

After we die, we don't automatically remember our pre-mortal existence and all of our previous lives if we have had them. God has to reveal them to us, or we have no memory of them. God is the holder of our memories, not us.

Interesting, is it not?

After making this discovery, I have gone through some of these chapters within this book and adjusted my views accordingly.

This reality also explains why the Psyche of split-brain patients doesn't have immediate access to the memories, learning, and experiences being created and uploaded to God's Database in non-local storage from both sides of their brain. Our memories aren't stored in our own Personal Psyche. They are stored in God's Database instead.

Well, that explains why we so easily forget things, and why God has to keep reminding us of them. We don't have immediate access nor direct access to God's Database where our memories are actually stored. The right hand really doesn't know what the left hand is doing, most of the time. Our physical brain has severely limited memory storage capacities; and, the buffer is being purged all the time. Our Psyche doesn't have direct access to our memories, on demand. We gain access to our memories through God's permission. Tomorrow really is a new day.

Yes, the physical brain does indeed have some memory storage capacity, but nowhere near enough to get the job done all by itself. It's physically impossible to store petabytes of memory and data within a physical brain. And yes it is true, most of our Long-Term Consolidated Memories are NOT being stored within the Human Psyche – not because the Human Psyche doesn't exist, but because our Non-Local Consolidated Memories are being stored someplace else instead within God's Database. It's Nature's Psyche who decides whether our Consolidated Memories are going to be stored within our physical brain or non-locally in the cloud, because ALL of this takes place completely outside our conscious awareness.

According to the Orthodox Interpretation of Quantum Mechanics, the Human Psyche makes requests, makes decisions, and chooses what to pay attention to; and, Nature's Psyche does its best to comply with the demands by collapsing the appropriate wave functions making it real, true, and actual. Whenever the Human Psyche chooses to pay attention to something, Nature's Psyche ramps up or turns on the parts of the brain that Nature's Psyche has dedicated to those specific purposes.

The dedication of the Cerebral Processing Units (CPUs) to specific tasks or functions takes place completely outside our conscious awareness, which means that the dedication of the physiology to specific functionality is done by Nature's Psyche or God's Psyche, and NOT the Human Psyche. According to the Orthodox Interpretation of Quantum Mechanics, it's Nature's Psyche who collapses the wave function producing the actuality or the reality, and NOT the Human Psyche. The Human Psyche only chooses what it wants to pay attention to, while trying to ignore everything else. It all comes down to choice; and, choice is a product or function of Psyche.

All you want is the truth, because it always explains everything. The truth can be had ONLY by allowing all of the evidence into evidence, and then by pursuing a preponderance of the evidence. With the internet, we have enough information and evidence at our fingertips, that this actually becomes possible for us now; whereas, it wasn't before. God and Psyche are now coming out of hiding, and the Naturalists can no longer stop it, like they did before.

You are going to have to decide for yourself if I know what I'm talking about, or not. That's not something that I can do for you.

Supernatural Autobiographical Memory

It seems to me that our physical brain has access to all of the CURRENT physical sensory information that remains functional, as long as the brain remains functional; and, the Human Psyche has access to that information too.

However, when our memories are consolidated and uploaded to Nature's Psyche or God's Database, our Psyche and our brain no longer have direct access to that information. We have access to parts of the index, but most of us don't have access to the actual feelings, sensations, and experiences. But, those with Superior Autobiographical Memory do have direct access to the actual experiences and not just the index.

There are some people who do seem to have direct access to their memories and experiences and feelings which are being stored within God's Database. It's called Superior

Autobiographical Memory, but I tend to call it Supernatural Autobiographical Memory. They actually re-experience and relive their memories and feelings and sensations whenever they recall them. The FACT that all of these recalled memories and feelings and sensations are NOT simultaneously present all at once implies that they are NOT stored within the physical brain and that they are being downloaded and re-experienced one-at-a-time from someplace else besides the brain. Examine the evidence and decide for yourself whether you agree with my observations, or not.

Superior Autobiographical Memory:

https://www.youtube.com/watch?v=2zTkBgHNsWM

https://www.youtube.com/watch?v=en23bCvp-Fw

The scientists are surprised that these people are stable and sane.

Why?

It's because if all of these memories were actually stored within their physical brain, then it should all be simultaneously present and available at once, and the simultaneous presence of ALL knowledge and ALL information would drive any mortal physical person insane with information overload. Instead, all of their memories are compartmentalized and stored separately someplace else besides their brain, and their physical brains simply access this information one memory and one day at a time in sequential fashion. Our physical brain can handle one set of memories at a time, because it does so every day. Our physical brain couldn't handle nor store all of our memories, feelings, and experiences simultaneously without driving us insane.

The Materialists, Naturalists, and Atheists will tell you and assure you that these people's memories are being stored exclusively within their physical brain; and therefore, their brains must be vastly superior to ours. But, what do the Naturalists and Atheists truly know? They are making things up out of thin air just like everyone else; and, they don't have a clue as to how Psyche and God's Database really work; so, why should their philosophies and opinions be given precedence over everything else?

Savants and people with eidetic memory have stated that they see the answer in their mind's eye or see the image in their mind's eye. They have to recall the information from the cloud where it is being stored, because they can't store all of that information simultaneously within their brains.

Savants who are able to do supernaturally impossible calculations within their mind aren't actually doing any calculations in the traditional sense. Most of them will state that they will see the answer in their mind. It's given to them without any calculation on their part. They just download the answer from the cloud. I find that reality fascinating, even if others do not.

The Turning Point

It was slow in coming, but the following quote was the turning point for me. There's no going back now.

Pinel, J. (2014). *Biopsychology* (9th ed.). New York: Pearson.

Modern biopsychology began in the mid-20th century. At that time, there was a major push to identify the areas in the brain where memories

are stored. The search was largely championed by Karl Lashley, who wrote a famous review paper, *In Search of the Engram*, in which he described his fruitless efforts. Lashley and many who subsequently took up the search used the lesion method. If a particular structure were the storage site for all memories of a particular type, then destruction of that structure should eliminate all memories of that type that were acquired prior to the lesion.

No brain structure has shown this result! Lesions of particular structures tend to produce either no retrograde amnesia at all or retrograde amnesia for only the experiences that occurred in the days or weeks just before the surgery.

These findings have led to two major conclusions: (1) Memories are stored diffusely in the brain and thus can survived destruction of any single structure; and (2) memories become more resistant to disruption over time. (*Biopsychology*, pp. 277-278.)

These findings also suggest the logical possibility that (3) our consolidated, long-term, retrograde memories are not stored in our physical brain anymore, because our consolidated retrograde memories remain even when most of our physical brain has been eliminated or destroyed.

Obviously, I didn't understand the significance of this during my first or second readings; but, eventually after yet another re-reading, it dawned on me that this is the KEY to understanding what's really going on.

Our long-term consolidated memories are not being stored in our physical brain. If they were, then ALL of them would have to be stored in each and every neuron; and, that's physically impossible. Furthermore, if all of our long-term consolidated memories were stored within our brain, then they would ALL be instantaneously and simultaneously present at all times; and, such a thing would drive us crazy. Can you imagine the confusion if it were all in there and all present all at once?

There's no sign of petabytes of RAM, within each and every neuron; yet, that RAM has to be there if all of our consolidated memories are being stored diffusely within our physical brain, so that we can cut out any part of our brain and our consolidated memories remain. If our consolidated memories are being stored in our brain, then according to the observational evidence, they have to be stored in each and every brain cell in our brain; and, that's just physically impossible.

After reading, understanding, and accepting this quote as being real and true, I just simply KNEW that our consolidated memories are being uploaded to non-local storage in the cloud. There's no going back, now that I KNOW.

However, there's plenty of evidence suggesting and demonstrating that our long-term consolidated memories are NOT being stored within our Psyche or Non-Local Consciousness, either. So, where are they being stored? That leaves Nature's Psyche and God's Database. Our memories are not our own. They belong to God.

Concept Cells

For me personally, Concept Cells or Jennifer Aniston Neurons put it over the top into the realm of the physically unexplainable or physically impossible.

In fact, the first couple of times that I was introduced to Concept Cells, the concept was over my head and beyond me, and my brain skipped right past it, because it's not physically possible; and, I'm always looking for the physical explanation for things. I've trained myself to do so. Only after the third reading or so – days later – did it dawn on me what's really happening here.

In one of the first neurons to be studied in this way, the neuron fired in response to images of the actress Jennifer Aniston, but not to the 80 other images.

Jennifer Aniston neurons are highly selective. Each neuron responds to only a small number of test objects, or individuals – often only one could be found.

Without question, the most remarkable feature of Jennifer Aniston neurons is that they respond to ideas or concepts rather than to particulars, which is why they are also known as concept cells.

For example, a Halle Berry neuron responded to all photos of the actress (even when she was dressed in her Cat Woman costume), to her printed name, and to the sound of her name. (*Biopsychology*, p. 277.)

That's not possible!

A neuron represents a single bit of information – it is either ON or OFF, it either fires or it doesn't. A neuron is NOT a CPU. There's no CPU visible within any neuron. If such a thing exists within a neuron, then it exists non-locally in the spirit realm. It's physically impossible for the object recognition to be taking place within that single concept cell. So, who or what is doing the object recognition? Who or what decides that you are in fact looking at a picture of Jennifer Aniston, and not something else? Who or what chose and selected that specific neuron to fire whenever you are thinking about Jennifer Aniston? Choice and selection are a function of Psyche, not raw physical matter. So, how does the information get specifically to the Jennifer Aniston concept cell and no other, making that specific cell fire? They are kind of vague about all of this. There's no mention of any nearby neurons firing so as to make the Jennifer Aniston cell fire. I guess I'm going to have to do a bit more research and see if I can figure out how the magic is being done.

How does a single specific neuron KNOW that you, your psyche, is looking at a specific person or object? There's intelligence, sentience, consciousness, telepathy, and information exchange going on here, not just the random dumping of neurotransmitters. It fired unilaterally – the first and only cell to fire. Where did it get its information from? Other cells nearby weren't dumping neurotransmitters onto it to get it to fire, or were they? If they were, then that means that there are Jennifer Aniston neurotransmitters out there. No, the information is being sent telepathically to that specific neuron causing that specific neuron to fire. I don't see any other logical explanation.

In one case, a neuron that invariably responded to the Sidney Opera House responded to photos of the Bahai Temple in India. When questioned about it later, the patient said she thought the Bahai Temple photos were photos of the Sidney Opera House. (*Biopsychology*, p. 277.)

This is physically impossible. Her psyche KNEW! Her psyche KNEW why that specific neuron was firing to photos of both the Bahai Temple and the Sidney Opera House. Her psyche KNEW and that specific neuron KNEW, which implies some kind of telepathic communication taking place between the two, because raw physical matter by definition in principle cannot do telepathy or Action at a Distance. Her psyche was actually reaching into her physical brain and making that specific neuron fire, whenever she looked at pictures of the Bahai Temple or Sidney Opera House. There's no other logical explanation that I can think of to explain this. Whether that specific neuron was triggered directly, or other specific neurons were triggered in order to trigger that specific neuron, the result is the same. We are witnessing mind-over-matter in some form or another.

Of course, the common sense answer and solution to these problems that we get from the Materialists, Naturalists, Atheists, and Darwinists is that evolution is doing all of this; but, I don't see how. Evolution, by definition in principle, is a physical phenomenon; and, physical phenomena cannot do telepathy or Action at a Distance. Psyche, by definition in principle, certainly can, because it has been caught in the act of doing so.

Over and over again, I ask myself how this information is being processed, stored, and then transferred to that specific neuron or concept cell through neurotransmitters. I don't see how. It's physically impossible. There's no way to store megabytes of information within a neurotransmitter. A neurotransmitter isn't a memory storage device. It wasn't designed to be such. A neuron either fires, or it doesn't. On or off! There's no way to transmit megabytes of data through a single on or off switch, let alone target addressing for a specific neuron within the physical brain. There's too much specificity and complexity, too much information being exchanged, and way too much processing power and efficiency going on here, which cannot be explained by the random dumping of neurotransmitters into synapses.

Jennifer Aniston Neurons: What's in a Name?

March 21, 2014

By John Pinel

It is not often that I encounter a scientific article that excites me, but the article published by Rodrigo Quiroga in *Nature Reviews Neuroscience* (August 2012, Volume 13, 587-597) did just that. In his article, Quiroga describes his discovery of Jennifer Aniston neurons in the medial temporal lobes of human patients. These neurons are interesting to biopsychologists because they likely play a major role in certain kinds of human memory, but I think that just about anybody would find them interesting: Few things are more fascinating to humans than the human brain, and Jennifer Aniston neurons are particularly cool.

Quiroga got the opportunity to record neural activity from neurons in the medial temporal lobes (hippocampus, amygdala, medial temporal cortex) of patients who were suffering from severe epilepsy. Prior to the surgical removal of their epileptic foci, electrodes were implanted in their medial temporal lobes to precisely locate the foci. This provided Quiroga and his colleagues with the opportunity to study the response patterns of these neurons.

Remarkably, these neurons responded to concepts rather than to the particulars of stimuli. For example, one of the first neurons to be investigated responded to 7 different photos of Jennifer Aniston, but did not respond to photos of 80 other people or objects. Many more neurons of this type have now been identified: for example, neurons have been identified that selectively respond the Halle Berry, the Sydney Opera House, Diego Maradona, and mother Theresa. Remarkably, these neurons also responded to the printed and spoken names of the particular concepts that they encoded, not just photographs of them. These various human temporal lobe concept neurons have been termed "Jennifer Aniston neurons" after the first such neuron to be discovered.

Quiroga emphasized two points about the selectivity of various Jennifer Aniston neurons. First, it is clear that there is more than one neuron in each brain encoding a particular concept. It has been estimated that humans have concepts for 10,000 to 30,000 things, and if only one neuron responded to each, it is unlikely that the particular neuron that responded to a concept could be identified during the time allowed for testing. Second, it has been discovered that although each Jennifer Aniston neuron is not totally selective. For example, it was subsequently discovered that the neuron that responded to Jennifer Aniston also responded to another person: Lisa Kudrow, Jennifer Aniston's co-star in the well-known television series, "Friends."

The discovery of Jennifer Aniston neurons clearly ranks as an important neuroscientific discovery: it is a striking example of how experience influences brain function, and it provides important clues about how the human brain retains concepts. Also, the idea that a single neuron can respond reliably to the image of a particular person or to the sound or sight of her name is thought provoking – a good topic of conversation among friends.

Be that as it may, I must admit that the name itself played an important role in attracting my interest in Jennifer Aniston neurons: Not many neuroscientific phenomena are named after television or movie personalities. Using Jennifer Aniston's name for human medial temporal lobe concept neurons is good fun—and I have never found fun and good science to be mutually exclusive. More importantly, this name is easy to remember and immediately reminds every one of the observations that led to the discovery. Thus, generations of students and scientists will benefit from the name.

I wonder whether Jennifer Aniston knows that an important class of human neurons is named after her. If she does, does she fully appreciate their significance?

http://biopsyc.com/aniston/

This is insane! Radically advanced technology!

Once again, in my humble opinion, we have moved beyond what is physically possible into the realm of Psyche and Quantum Telepathy instead. I see no other way to make logical sense of this phenomenon. I certainly can't see how evolution is doing all of this; and, "evolution" is always the explanation that they give whenever they come across anything that doesn't have a physical explanation. Observe for yourself, and see if it is not so.

Of course, they will try to find physical explanations for this thing, and they should. But, I'm having a hard time seeing it. So far, I haven't found any of the physical explanations satisfying, because there are black boxes all along the way, which state what happens but not how it happens. Synapses are added or pruned away in order to form a memory. Or an additional ion receptor is added to a synapse in order to consolidate a memory. Fine. But how does adding a synapse or an ion receptor upload and store gigabytes of video, data, and information?

Check out some of the original articles for yourself, and see if you can figure out how the magic is done. They never say how it's done. They just assume that it's some sort of physical process. What else could it be? It can't be Psyche or God or Quantum Mechanics, because those aren't allowed; so, it has to be some sort of physical process. Sometimes they will admit that they don't know how these things work, but they assure us that there's a physical explanation to be had somewhere along the line, because everyone knows that psyche, spirit, and God do not exist.

Neuroscientists ardently debate two alternative theories of how memories are encoded in the brain. One theory contends that the representation of a single memory — the image of Luke Skywalker, for instance — is stored as bits and pieces distributed across millions or perhaps billions of neurons. The alternative view, which has gained more scientific credibility in recent years, holds that a relatively few neurons, numbering in the thousands or perhaps even less, constitute a "sparse" representation of an image. Each of those neurons will switch on to the image of Luke, whether from a distance or close-up. Some but not all of the same group of neurons will also fire to the related image of Yoda. Similarly, a separate set of specific neurons activates when perceiving Jennifer Aniston.

https://www2.le.ac.uk/centres/csn/publications-1/Publications/scientificamerican0213-30.pdf

Great, the correct neurons switch on and fire whenever you are looking at Luke or Jennifer. But, how do they KNOW that you are looking at Luke or Jennifer, and not at something else? How does that information get into them? There should be other neurons firing around them to get that information into them; but, it's not happening from what I can tell. The correct neurons just "switch on" or fire out-of-the-blue, selectively and specifically, depending upon what you are looking at. It's magic. There's a huge black box here, a magical black box, whether they can see it or not.

Furthermore, who is picking and assigning that specific neuron to Jennifer Aniston? Whoever assigned that specific neuron to Jennifer Aniston had to be able to recognize Jennifer Aniston for itself, and then turn around and pick a specific neuron to assign to Jennifer Aniston. There's intelligence, sentience, and awareness going on here within these black boxes – not just random mutations.

And, where are the RAM chips within each neuron, which are allegedly storing these bits and pieces of information? I didn't know that neurons had RAM, image processors, and image processing software within them. They often call neurons "processors", but they are cells just like any other cell. Where's the CPU, since they are processors? There is no RAM, and there is no CPU within a neuron. It's not there! There's just genes, proteins, ions, amino acids, and ATP molecules who somehow magically KNOW precisely what they are supposed to be doing and when. If there are CPUs and petabytes of RAM in neurons, then they are invisible and transdimensional, which is physically impossible so that can't possibly be the answer.

Notice how they always limit themselves to two possibilities – macro-physical or micro-physical. They never consider the THIRD POSSIBILITY – non-physical or non-local.

I just can't see how concepts, ideas, images, feelings, and their associated memories can be stored in a single neuron or concept cell, let alone a single ion receptor or ion channel or neurotransmitter – not in a physical format at least. There's no way to put gigabytes of physical RAM into an ion, neurotransmitter, ion receptor, or ion channel. Likewise, there's no way to put petabytes of physical RAM into a single neuron, and then program that specific neuron to respond only to certain images or photographs or concepts.

Imagine it, that single concept cell has a complex image processor, a video card or physics card, image recognition software, along with all of the associated data and video stored within it; and, it KNOWS that it's supposed to fire whenever its owner, the Human Psyche, looks at or is told about Jennifer Aniston or Lisa Kudrow.

How long does it take our computer networks and image processing software to parse through millions of pictures looking for a specific person? Yet, our psyche (the Human Psyche) and the associated psyche within that specific neuron or concept cell (Nature's Psyche) does this instantly and accurately every time.

As our thoughts rely on constructions we make about the external world, both perception and memory are based on the meaning we attribute to what we sense or recall. This attribution of meaning is subjective: it involves abstraction or, in other words, extracting relevant features and leaving aside an immense number of details. In this Perspective I argue that the recently identified 'Jennifer Aniston neurons' — or 'concept cells' — in the MTL are the pinnacle of this abstraction process and provide the conceptual representation of stimuli that underlies declarative memory functions.

https://pdfs.semanticscholar.org/bbf8/306f969719ecaf1d4fd1971979cbac8cc237.pdf

Even more amazing is the observed and experienced FACT that your psyche or spirit will continue to recognize Jennifer Aniston and everyone else, long after your physical body and physical brain are dead and gone! How is all of this information getting from our physical brain into our Life Reviews? The only answer that makes sense to me is that it's being done wirelessly, through WiFi or telepathy. Telepathy is WiFi at the quantum level.

Our thoughts, perceptions, and memories are a function of Psyche, because we continue to have thoughts, continue to perceive things, and continue to remember things long after our physical brain is dead and gone, according to the empirical evidence from Near-Death Experiences, Out-of-Body Experiences, and Shared-Death Experiences. Attributions and subjectivity are a function of Psyche, because they involve discrimination or choice. Anytime we encounter choice, we have in fact encountered Psyche or Intelligence.

Obviously, our memories are being stored someplace and complex video processing is taking place somewhere. This is the given! We obviously have memories, and it's clear that the human brain seems to be a superior image processor in comparison to our manmade computers.

But, what isn't clear is how all of this information is being transmitted from neuron to neuron one neurotransmitter or action potential at a time, how all these terabytes of information and memories and software are being stored physically within each neuron, where the CPUs and video processors are located within each neuron, how each neuron is getting assigned a specific task or a specific concept to manage, how ALL of our consolidated memories are being stored in each and every neuron, and how all of this information is being uploaded into Non-Local Cloud Storage for our upcoming Life Review.

Do you ever get the sense that there's more that we don't understand than there is that we do understand? Do you ever get the sense that there are more things that don't have a physical explanation than there are that do have a physical explanation?

This really is totally insane!

We are looking at radically advanced technology here, far beyond anything that we human beings can possibly imagine, let alone design and create for ourselves. I'm a computer scientist, and I have some semblance of an idea what's involved when it comes to image recognition software and the processing power needed to run it and parse the gigabytes of data in a timely fashion. I've written sorting routines in the past, and I understand sequential processing. It takes a long time to sort through thousands of records in a database looking for a specific one, let alone trying to sort them into some kind of useful order.

The Human Psyche, and Nature's Psyche within the neuron, go directly to the correct answer every time; and, there's no delay taking place while the neuron sorts through gigabytes of physical RAM searching for the correct answer. We are witnessing processing power, parsing software, image recognition software, and data storage capacities far beyond anything that's physically possible; and, it's all taking place within a single neuron or concept cell.

As I see it, the memories of Jennifer Aniston – the data, the images, the image processing software, and the image processors – have to be stored within that specific neuron, but in some kind of non-physical format. Or, all of this information has to be stored someplace else in the cloud on the universal internet, and then that particular concept cell is linked wirelessly to that non-local storage and those non-local processors.

All of this has to be happening wirelessly or telepathically because we don't have Ethernet cables running throughout our body. We have individual, completely separate, neurons – with neurotransmitters floating between them. How would you go about transferring an image of Jennifer Aniston from the occipital cortex where it was constructed to that specific concept cell or neuron? I can visualize how an image of Jennifer Aniston is constructed from a retinal image within the occipital cortex, because there's a physical explanation for that; but then, how does that completed image get compressed and how is that completed image downloaded into a specific neuron one neurotransmission or one action potential at a time? How does the occipital cortex know where to send that image?

And, how would you go about converting neurotransmitters arriving randomly at synapses on the Jennifer Aniston neuron into a pictorial image of Jennifer Aniston? How would you know that you are supposed to? All neurotransmitters do is trigger the release of G-proteins, or trigger EPSPs and Action Potentials. Neurotransmitters fire the neuron, and increase the rate of firing or decrease the rate of firing through G-proteins and the number of ion receptors on the synapse. Neurotransmitters don't transfer digital video or digital information – bits and bytes; and, there's no indication that neurotransmitters do our thinking for us. Even "electrical synapses" are transferring ions between neurons and astrocytes, NOT bits and bytes.

There really is NO information exchange taking place between neurons in terms of bits and bytes. Either the ion channels open, or they don't. Either the action potential fires, or it doesn't. On or off is what the neuron does. One bit, on or off. That's it! The neuron doesn't do petabytes of information processing, petabytes of digital transfers to other neurons, and petabytes of digital data storage. All of this data processing, information exchange, and memory storage seems to be taking place somewhere else besides our neurons.

Even then, even if we were to find a physical explanation for the connectivity problems and transmission problems and storage problems, someone still has to recognize Jennifer Aniston and then assign that specific neuron to Jennifer Aniston, so that it can fire or "switch on" whenever your Psyche is looking at Jennifer Aniston.

Object recognition has been mapped to these concept cells by Nature's Psyche. Remember, recognition is a function of Intelligence or Psyche. Even our image recognition software and hardware required a ton of intelligence to produce. There are NO wires or CPUs or RAM within our brains; yet, our image recognition capabilities – which are the result of an interaction between the Human Psyche and Nature's Psyche and God's Psyche – are infinitely faster and noticeably superior to anything man-made at the physical level that we have ever produced. When it comes to object recognition, Psyche is infinitely superior to any man-made physical machine.

The functionality of concept cells is mapped into them by Nature's Psyche. It has to be Nature's Psyche, because the Human Psyche isn't consciously aware of any of this.

I'm seeing signs of psyche or signs of intelligence written all over this one, even if others are not.

It's kind of obvious and clear that this whole thing involves Psyche and Quantum Telepathy. Ain't got no strings on me!

The Brain – A Monumental Disappointment

I used to be a Materialist, Nihilist, and Atheist.

When I took up Neuroscience, I was looking for a physical explanation for how our brains work; and, I was expecting to find it. I was looking for a physical explanation for everything; and, my Neuroscience teachers told me that I was going to receive it.

What did I find instead?

I discovered that if our brains really are functioning as CPUs and RAM, as they clearly and obviously seem to be, our brains aren't doing so on the physical level, because it's physically impossible to transfer bytes of information and megabytes of memories through a synaptic cleft. The ONLY thing you can get through a synaptic cleft is ON or OFF – as well as MORE-ON or MORE-OFF. It's physically impossible to transmit "Hello World" through a synaptic cleft, because everything gets reduced to ON or OFF when it's sent through a synaptic cleft.

There's a physical explanation for how our brains do ON and OFF. But, there is NO physical explanation for how our brains are transmitting gigabytes of memories from one neuron to the next, nor is there any physical explanation for how our brains are doing terabytes of CPU processing and petabytes of RAM storage through all those 100 trillion synaptic clefts that our brains are comprised of. You can't process terabytes of software or computer code through a synaptic cleft, nor can you store petabytes of memories within a synapse as the Materialist and Naturalists claim is being done. It's physically impossible.

Complex stimuli are normally perceived as integrated wholes, not as combinations of independent attributes. How does the brain combine individual sensory attributes to produce integrated perceptions? This is called the *binding problem*. (*Biopsychology*, p. 164.)

Actually, the binding problem is caused by Materialism, Naturalism, Darwinism, and Atheism. The binding problem is solved by realizing that everything feeds into one single central Overlord or Psyche, the Human Psyche. The brain doesn't combine sensory input into integrated perceptions – the Human Psyche does. The brain handles the sensory input, or the physical input. The Human Psyche handles the perceptions. The binding problem, the free will vs. determinism problem, the nature vs. nurture problem, and the mind-brain problem are immediately solved by getting rid of Materialism, Naturalism, and their derivatives. All of the different unsolvable problems in science are immediately eliminated and solved by getting rid of Materialism, Naturalism, and Darwinism.

Instead of finding a physical explanation for everything while studying Neuroscience, I discovered the physically impossible. It was a monumental disappointment for me at first, until I finally accepted the fact that that's the way it is.

The residual Materialist and Naturalist within me found the brain extremely disturbing!

Neurons are switches – stand-alone switches – ON or OFF. They aren't even wired to each other. They just sit there emitting chemicals.

For the few of you who might have gone through what I went through, after studying Neuroscience, Abnormal Psychology, Psychology, and the Brain for a year or more, all I can say is that it has been a monumental disappointment.

There's NO supercomputer Command and Control center like there's supposed to be, as the binding problem so adequately attests.

There's no Seat of Consciousness, and nothing essential, within the brain.

You can cut out any part of your brain that you want, and your long-term consolidated retrograde memories remain. Nothing is essential, when it comes to memory storage in the brain; and, that's physically impossible!

There's NO digital transfer of data and video going on throughout our brain. There's only ON or OFF where each neuron is concerned. It either fires or it doesn't. There's really not much there to work with.

The whole thing is highly inefficient! It has to dump a million glutamate neurotransmitters into a synapse in order to guarantee that your finger rises, and then clean up the mess afterwards to prevent the synapse from plugging up and the brain from burning out.

The only part of our brain that is even remotely like a computer or CPU is our Cerebellum; and, it can be programmed by our Psyche to do gymnastics, juggling, basketball, and other finely-tuned physical motions with our physical body. Your retinas transfer an image into your occipital lobe that your Psyche can learn to recognize, read, and interpret. It's a video card or a monitor, not necessarily a CPU. Our occipital lobe functions as a monitor or a buffer, and NOT as video storage. The ONLY thing in there is the current image coming from your eyes. Our inferior temporal cortex and medial temporal lobe may or may not be in charge of consolidating our memories and uploading them into non-local cloud storage. ALL of this functionality and memory storage is taking place at the quantum level, if it's taking place at all.

Materialists, Naturalists, and Atheists claim that there's no such thing as Psyche or Non-Local Conscious, and that ALL of our memories are stored exclusively in our physical brain. We don't know for sure if this is true, even though many neuroscientists pretend that they do. ALL of the observational evidence suggests that the Materialists and Naturalists are wrong.

Oh, it's happening, alright. Our brains really are processing data, running computer code, crunching numbers, transferring data, building cities, weighing decisions, and storing memories. It's just not happening at the physical level. It's all happening at the quantum level. The physical input of sensory data, and the physical output of motor commands, is the only thing happening at the physical level within the brain. It all happens at the end of an axon, where the switch is either turned ON or OFF. A synapse is a switch. Our brain is comprised of a 100 trillion switches or synapses. Our brain is also comprised of a 100 billion stand-alone switches or neurons, which are either ON or OFF. Furthermore, our brains are comprised of quadrillions of ion channels that are either OPEN or CLOSED. Everything within our brains reduces to a single BIT. There's nothing else there within our brain. Everything else is happening at the quantum level. All the important and interesting stuff is happening at the quantum level, outside our conscious awareness.

Did I mention that there's no noticeable Command and Control within the physical brain?

If your Psyche decides that it wants to raise your finger, then your Psyche reaches out telepathically into your brain and literally triggers the neuron that raises your finger, and your finger rises. That's it! There's nothing more! No processing, no calculating, nothing! Your Psyche triggers the specific neuron that raises your finger, and your finger rises.

Sensory input is only slightly more interesting. You, your psyche, chose to watch every episode of *Friends*. You liked it and fell in love with Jennifer Aniston. It was important to you personally. Now, whenever a picture of Jennifer Aniston registers on your occipital cortex and temporal cortex, your Psyche who is monitoring the contents of its brain eventually realizes that it's looking at or sensing a picture of Jennifer Aniston; and then, your Psyche reaches out telepathically and fires off its Jennifer Aniston concept cell, which is near the amygdala, dumping a bunch of neurotransmitters into your brain producing either a physical sensation of recognition or a physical sensation of pleasure and joy. That's it!

There's no blazing fast super-computer in your brain grinding away at light speed doing image processing and mathematical calculations. There aren't petabytes of physical RAM in your brain storing all of your memories. When it comes to the neurons in your brain, they are either ON or OFF. They either fire, or they don't. There's nothing physical in your brain telling them to fire. Either your Psyche recognizes a picture of Jennifer Aniston and fires off a concept cell, or it doesn't.

Even your pain reflexes are mediated through your spinal cord. You accidentally step on a tack, and your spinal cord moves your foot out of the way.

We have a clock in our suprachiasmatic nucleus that regulates our circadian rhythms. There are machines or regulators in our brain stem that automate our breathing, heart-beat, and digestion. But, these are machines, thermostats, and not super-computers. And it's not the least bit clear how or if all these separate switches or neurons are functioning together as a united whole at the physical level. Thanks to our synaptic clefts, if any part of our brain is functioning like a CPU or RAM, that functionality is taking place at the quantum level, and not the physical level.

Our long-term consolidated memories aren't stored in our physical brain and can't be stored in our physical brain, because our brain has NO physical RAM. You can cut out any part of your brain that you want, and your consolidated memories remain.

However, if you cut out the part of your brain that registers vision, then you lose your vision. If you cut out the part of your brain that coordinates your movements, then you are either spastic or paralyzed. If you cut out the part of your brain that registers experiences or cognitive learning, then it becomes impossible for your Psyche to learn and register anything new through your brain. If you don't have the machinery, then your Psyche can't use it.

Trans-neuronal degeneration of a neuron is caused by damage to another neuron which is linked by a synapse. How does the pre-synaptic neuron know when the post-synaptic neuron dies, goes missing, or is offline? Sometimes the neurons before and after a missing neuron will sacrifice themselves or kill themselves. Other times, the neurons before and after a dead neuron will try to establish a connection with each other. Who decides whether the neurons will kill themselves or connect with themselves? There is an Intelligent Agent here in this process that's making all of these decisions or choices. I call it Nature's Psyche.

The neurotransmitter system is the most unreliable, inefficient, and glitchy thing that God has ever made. Logic tells us that it shouldn't work, but somehow miraculously it does, as if God is propping it up and making it work despite all the obstacles. Our neurotransmitter systems are always going wrong causing most of our mental illnesses; and, if glitchy neurotransmitters don't get you, then faulty genes will. ALL of our heritable illnesses are caused by random mutations – thank you evolution!

Like everyone else, I went into neuroscience hoping to find solutions to some of my ongoing problems. I found none. The monoamine neurotransmitters are notoriously impossible to balance and get right. Get one gene wrong, and they don't work at all. It's nasty. On my mother's side, our serotonergic systems are messed up; and, the problems are passed from generation to generation through our genes. On my mother's side, we also have the gene that makes us easily and quickly addicted to practically anything. There's no solution for that, either.

Our neuroscientists and neurosurgeons don't know anything. The only thing they can do is cut out the parts of the brain that aren't working right, which they do. As my abnormal psychology textbook so adequately stated, when it comes to neurodegeneration and neurodevelopmental abnormalities, the prospects for any kind of treatment or cure is abysmally bleak.

Neurodegeneration is permanent, and typically a death sentence. Until we master genetic engineering and replace all of our bad genes which evolution gave us with good genes that work right, we are not going to do anything to solve neurodevelopmental abnormalities. Even then, genetic engineering is not going to do anything to help us – only our children. There's no way to repair a brain structure that developed abnormally. It has to develop correctly from the beginning, which means having good genes from the beginning. All of our heritable illnesses come from evolution or random mutations. That's the only thing we get from evolution. Evolution is never going to turn us into X-men. It doesn't work that way. Evolution is the extinction of species by means of natural selection – all the way. Actually, natural selection doesn't do anything either, except get us killed or selected against. It's the random mutations that are promoting and hastening our eventual extinction.

But we have stem cells!

Stem cells seem to be hit and miss. Sometimes the results are indeed miraculous; and other times, the stem cell treatments actually do more harm than good. The problem is that we don't know anything. We don't know how these things really work.

Each individual neuron can form thousands of links with other neurons in this way, giving a typical brain well over 100 trillion synapses (up to 1,000 trillion, or a quadrillion, by some estimates). Functionally related neurons connect to each other to form neural networks (also known as neural nets or assemblies).

Current studies estimate that the average adult male human brain contains approximately 86 billion neurons. As a single neuron has hundreds to thousands of synapses, the estimated number of these functional contacts is much higher, in the trillions (estimated at 0.15 quadrillion).

Some estimates conclude that a child's brain has 100 billion neurons and a quadrillion synapses, before pruning and consolidation takes place. Uncle Google.

Brains: I don't know how they do what they do, but they do.

A neuron is a battery, not a CPU. There's no detectable CPU in a neuron. A neuron is either charging or discharging, just like a battery. That's it! A neuron has NO detectable RAM for storing memories. A neuron is a switch – ON or OFF. It's as if all of our long-term consolidated memories are being stored someplace else besides our brain. That's what the observational evidence and experimental evidence repeatedly suggests.

If I were to give you 100 billion switches but NO wires, how would you turn them into functional CPUs or functional RAM? You can't without wires! It's physically impossible.

But, let's suppose that you have WiFi or Quantum Telepathy built into each one of those switches, then how would you proceed to turn those switches into functional CPUs or RAM?

Well, for every neuron in the brain, you would have to go through the brain and connect eight switches (neurons) together wirelessly to form a functional unit called a byte, assign some sort of arbitrary but standardized meaning to each of the 256 different types of bytes that are possible, set the eight switches in each of the bytes to the correct meaning that you want that byte to represent (and keep doing so for over eighty years of time), and then force all of these different switches and bytes to function together as a united whole – either as a CPU or as RAM.

If brains are made up of 100 billion stand-alone switches, with no wires or connections between them, then why do brains work? There's NO physical reason why they should work, so why do brains work? The only answer I can give is that Nature's Psyche is assigning functionality and purpose to each of the neurons, or structures, or CPUs in the brain; and, Nature's Psyche is connecting those stand-alone switches together wirelessly into a functional whole. The mechanism that Nature's Psyche uses to get the job done is Quantum Telepathy or WiFi at the quantum level.

If you hook up an EEG to your scalp, what's the only detectable thing coming from your brain as a whole? Waves! Waves are WiFi at the quantum level! If the neurons in your brain are communicating with each other, sending complex messages to each other, and storing memories, they are doing so wirelessly through different types of waves. They are using WiFi at the quantum level or Quantum Telepathy. They definitely are NOT using synapses to send messages to each other, because synapses are randomizers. A synaptic cleft will scramble any message that you try to send through it! It's physically impossible to send messages of any kind through 100 billion stand-alone switches that have NO wires connecting them. It can't be done, so that means that it isn't being done. ON or OFF is the only thing that gets through a synaptic cleft. That's one BIT of information.

When it comes to the brain, God gave us 100 billion switches to look at. 100 billion switches with NO wires connecting them. God made it extremely clear to every Computer Scientist, Materialist, Naturalist, Darwinist, and Atheist that our brains shouldn't work, because it's physically impossible to send messages of any kind from one neuron to the next through a synaptic cleft. It can't be done. A synaptic cleft is a randomizer, scrambler, and jamming device. A neurotransmitter receptor is also a bit – OPEN or CLOSED. The receptor receives the ON signal letting it know that the message has started, and then it receives an OFF signal letting it know that the message has ended; but, there's NO message! You can't send a message through a single stand-alone bit or switch. It's physically impossible.

So, how is the magic being done? Well, the only thing we KNOW for sure is that it isn't being done through the synaptic clefts, because it's physically impossible to send messages and memories through synaptic clefts!

The only thing the physical brain does is give the Human Psyche a way to control its physical body. Everything else is being handled by Nature's Psyche. Each part of the brain is assigned a specific physical function by Nature's Psyche and God's Blueprint. If you cut-out or destroy part of the brain, then you lose that functionality. It's all about functionality! Cut out the areas that control short-term working memory, and your working memory is gone. Cut out the occipital lobes, and you are blind, because your Psyche no longer has the part of its brain that processes vision. Cut out the audio cortex, and your audio card is gone; and then, your Psyche can no longer sense sounds through your physical brain. Cut out the part of your brain that processes events and experiences, and then your Psyche receives NO information letting it know that your brain and body have been experiencing events. That's the way it works.

Over and over again, we observe that functionality, and not consolidated memories, is lost whenever a part of the brain is cut out or destroyed. The different parts of our brain are extension cards or add-on cards – video cards and audio cards – and not memory storage devices. There are no RAM chips in our physical brain. All of our consolidated memories seem to be stored someplace else besides our brain. Cut out the part of your brain that consolidates and experiences events, and suddenly your Psyche has no memory of your body experiencing any events, because technically your body hasn't experienced any events. It's not a matter of lost memories. It's all a matter of lost functionality.

This is the KEY!

The neuroscientists erroneously equate brain modules with memory storage devices, starting with the axiomatic conclusion that all of our memories are stored in our brain, and nowhere else. They are wrong. Yes, our brain has some memory storage capabilities, but nowhere near enough storage capacity to store a lifetime's worth of memories, video, and experiences.

Medial temporal lobe structures that are critical for long-term memory include the hippocampus, along with the surrounding hippocampal region consisting of the perirhinal, parahippocampal, and entorhinal neocortical regions.

The medial temporal lobes (near the sagittal plane) are thought to be involved in encoding declarative long term memory. The medial temporal lobes include the hippocampi, which are essential for memory storage, therefore damage to this area can result in impairment in new memory formation leading to permanent or temporary anterograde amnesia.

https://en.wikipedia.org/wiki/Temporal_lobe

Our inferior temporal cortex and medial temporal lobe are brain modules that were designed to do object recognition. They are not memory storage devices as the scientists repeatedly claim. They have no observable flash RAM within them. There are NO image processors, no CPUs, no video cards, and no flash RAM in a Jennifer Aniston cell or concept cell. If you destroy these things, then your Psyche can no longer use your brain to do object recognition; but, all of your consolidated retrograde memories remain. Destroy the device, and functionality is lost, not necessarily the memories.

Scientists have repeatedly observed that if you restore the functionality, as in the case of temporary lesions, then the ability to form new declarative memories comes back. When it comes to brain modules, it's all about functionality. These things have no flash RAM for storing memories. Brain modules are not solid state drives (SSDs); and, neurons are not computers. Neurons are batteries, and nothing more. Neurons are either charging or discharging. That's it!

All of the truly interesting bits – the petabytes of RAM, the banks of parallel image processors, the developmental blueprint for locating and type-casting synapses, and even the neuron-to-neuron communication system ARE ALL located within Nature's Psyche. There's none of that in our physical brain.

The physical brain is really disappointing once you realize what it truly is. It's a machine that is commanded and controlled by the Human Psyche, through Nature's Psyche and Quantum Mechanics. That's it.

The physical brain idles when the Psyche isn't present. Out-of-body travelers have observed that it's the Psyche – that immaterial viewpoint floating in space – who does all of our experiencing and remembering, and NOT our spirit body, and certainly not our physical body. Whenever your Psyche is away from your physical body, your physical body has no experiences and forms no new memories. Whenever your Psyche is separated from its spirit body, your spirit body has no experiences and forms no new memories. This is what has been experienced and observed.

The brain is not a memory storage device, and neither is our spirit body. It's all done within our Psyche, or Nature's Psyche; and, according to after-death Life Reviews, Nature's Psyche or the Angels in Heaven are backing up and making a separate copy of our Psyche's experiences, choices, and memories.

Ironically, world-renowned neurosurgeon Wilder Penfield discovered and revealed all of this, over fifty years ago; but, nobody was listening to him, because he wasn't telling them what they wanted to hear.

The Materialists and Naturalists are looking for the smoking gun, and they will until the end of time, and never find it because it doesn't exist. It would have huge neon signs all over it if it did exist. The power-drain would be monumental. It would be hotter than Hades, and soon our physical brain would melt-down into slag, if such a thing existed here in this physical realm. It would be the CPU to end all CPUs. You couldn't miss it if it existed. Given the amount of image processing and image recognition that our brain allegedly does and given the nearly instantaneous speed at which it does it, that part of our brain should be glowing hotter than the sun, but it isn't.

This thing doesn't exist, which tells me that when it comes to our physical brain, most of our image processing, memory storage, command, and control is taking place in the Quantum Realm, the Transdimensional Realm, or the Non-Local Realm where our Psyche originates and resides.

Of course, I don't expect you to agree with me. Nobody will. But, I decided to tell it as I see it. Our physical brain is very disappointing, I must confess, because it was supposed to be infinitely more than what it actually is. Our physical brain doesn't even survive the death of our physical body, but our Psyche certainly does. Everything about you resides in your Psyche, not your physical brain. That's just the way it is. That's the way it has been observed to be. The collective experiences or the lived experiences of the human race tell us that it is so.

Enter Quantum Neuroscience!

John C. Eccles and Wilder Penfield

Eccles, J. C. (1994). *How the SELF Controls Its BRAIN*. New York: Springer-Verlag.

By 1994 and even during the decades before, Sir John Eccles was using Quantum Mechanics to explain how the SELF or the Psyche controls its physical brain. These ideas aren't new. They are proven and verified science. I'm simply applying this knowledge directly to neuroscience itself, as should have been done decades ago.

--

Penfield, W. (1975). *The Mystery of the Mind*. Princeton. New Jersey: Princeton University Press.

https://muse.jhu.edu/book/39368

http://www.chabad.org/library/article_cdo/aid/113106/jewish/Appendix-5-Neurology-Medicine-and-the-Soul.htm

http://mypsyche.us/wp-content/uploads/2017/10/Neurology-Medicine-and-the-Soul.pdf

This book resolves the Mind-Brain Problem, using Scientific Experiments and Scientific Observations to do so. Wilder Penfield has the distinction of getting there first.

By using scientific experimentation, scientific methods, brain stimulation, and observation of the results, the Mind-Brain Problem was officially solved in 1975 by Wilder Penfield. Others have replicated and confirmed Wilder Penfield's findings; but, he got there first, because he was brave enough to report what he was finding.

The Materialists and Naturalists spend all of their time suppressing and ridiculing this type of observational evidence in an attempt to explain it away or convince us that it does not exist.

Remember, other neuroscientists like John Eccles, Mario Beauregard, and Eben Alexander have come to the very same conclusions, but Wilder Penfield has the distinction of getting there first. These men tend to quote Wilder Penfield and this book, *The Mystery of the Mind*, in many of their books and presentations.

The following is some of what they quote:

Is the mind merely a function of the brain? Or is it a separate but closely related element?

Throughout my own scientific career, I, like other scientists, have struggled to prove that the brain accounts for the mind.

(*The Mystery of the Mind*, Preface.)

This is the million-dollar question and the ultimate quest.

Is there any evidence of the existence of neuronal activity within the brain that would account for what the mind does?

Before venturing to answer, it may be of interest to refer again to action that the mind seems to carry out independently, and then to reconsider briefly our experience with stimulation of the cortex of conscious patients and our experience of what effects are produced by epileptic

discharge in various parts of the brain. This should give some clue if there is a mechanism that explains the mind.

"(a) What the Mind Does

It is what we have learned to call the mind that seems to focus attention. The mind is aware of what is going on. The mind reasons and makes new decisions. It understands. It acts as though endowed with an energy of its own. It can make decisions and put them into effect by calling upon various brain mechanisms. It does this by activating neuronal mechanisms. This, it seems, could only be brought about by expenditure of energy.

(b) What the Patient Thinks

When I have caused a conscious patient to move his hand by applying an electrode to the motor cortex of one hemisphere, I have often asked him about it. Invariably his response was: "I didn't do that. You did." When I caused him to vocalize, he said: "I didn't make that sound. You pulled it out of me." When I caused the record of the stream of consciousness to run again and so presented to him the record of his past experience, he marveled that he should be conscious of the past as well as of the present. He was astonished that it should come back to him so completely, with more detail than he could possibly recall voluntarily. He assumed at once that, somehow, the surgeon was responsible for the phenomenon, but he recognized the details as those of his own past experience. When one analyzes such a "flashback" it is evident, as I have said above, that only those things to which he paid attention were preserved in this permanently facilitated record.

(c) What the Electrode Can Do

I have been alert to the importance of studying the results of electrode stimulation of the brain of a conscious man, and have recorded the results as accurately and completely as I could. The electrode can present to the patient crude sensations. It can cause him to turn head and eyes, or to move the limbs, or to vocalize and swallow. It may recall vivid re-experience of the past, or present to him an illusion that present experience is familiar, or that the things he sees are growing large and coming near. But he remains aloof. He passes judgment on it all. He says "things seem familiar," not "I have been through this before." He says, "things are growing larger," but he does not move for fear of being run over. If the electrode moves his right hand, he does not say, "I wanted to move it." He may, however, reach over with the left hand and oppose his action.

There is no place in the cerebral cortex where electrical stimulation will cause a patient to believe or to decide...

I am forced to conclude that there is no valid evidence that either epileptic discharge or electrical stimulation can activate the mind...

The mind seems to act independently of the brain in the same sense that a programmer acts independently of his computer...

For my own part, after years of striving to explain the mind on the basis of brain-action alone, I have come to the conclusion that it is simpler (and far easier to be logical) if one adopts the hypothesis that our being does consist of two fundamental elements. If that is true, it could still be true that energy required comes to the mind during waking hours through the highest brain-mechanism.

Because it seems to me certain that it will always be quite impossible to explain the mind on the basis of neuronal action within the brain, and because it seems to me that the mind develops and matures independently throughout an individual's life as though it were a continuing element, and because a computer (which the brain is) must be programmed and operated by an agency capable of independent understanding, I am forced to choose the proposition that our being is to be explained on the basis of two fundamental elements. This, to my mind, offers the greatest likelihood of leading us to the final understanding toward which so many stalwart scientists strive...

The nature of the mind presents the fundamental problem, perhaps the most difficult and most important of all problems. For myself, after a professional lifetime spent in trying to discover how the brain accounts for the mind, it comes as a surprise now to discover, during this final examination of the evidence, that the dualist hypothesis seems the more reasonable of the two possible explanations.

Since every man must adopt for himself, without the help of science, his way of life and his personal religion, I have long had my own private beliefs. What a thrill it is, then, to discover that the scientist, too, can legitimately believe in the existence of the spirit!

(From Wilder Penfield, *The Mystery of the Mind*.)

Source:

http://www.chabad.org/library/article_cdo/aid/113106/jewish/Appendix-5-Neurology-Medicine-and-the-Soul.htm

Wilder Penfield set out to prove Materialism and Naturalism true; but, he ended up FALSIFYING Materialism and Naturalism instead.

This is the smoking gun – catching psyche in action completely separate from the physical brain. This effectively solves the Mind-Brain Problem.

However, if science continues to be defined as Materialism, Naturalism, Darwinism, Nihilism and Atheism, then science will NEVER be able to solve the Mind-Brain Problem, because Materialism and Naturalism cause the Mind-Brain Problem, not solve it.

But if instead, we choose to upgrade our science to Science 2.0 and choose to define science as Observation, Experience, Knowledge, Truth, Fact, and the Lived Experiences of the human race, then we quickly realize that the Mind-Brain Problem has already been solved by the collective Lived Experiences of the human race as a whole, including those presented to us by Wilder Penfield.

Good enough.

Since I have upgraded my science to Science 2.0, in all things I choose to go with the observational evidence rather than the philosophical guesswork of the Materialists and

Naturalists. I simply find and present the observational evidence, and then leave it up to you my reader to decide for yourself what to make of it. Wilder Penfield made me a believer in the non-local or the non-physical, but your mileage may vary because you are a completely different person than I am.

Of course, these observed and proven realities show us the desperate need that we have for something like Quantum Neuroscience, so that we can figure out what's really going on.

There are others out there who are smarter and much more experienced than I am; and, they could probably do a much better job of introducing and explaining Quantum Neuroscience than I can – if they actually wanted to do so, which apparently they don't. Therefore, it's left up to someone like me to introduce it to the world, since nobody else is going to.

Mark My Words

Neuroscience from a Quantum Neuroscience Perspective

All I was ever looking for was a physical explanation for why our brain and neurotransmitter systems work. All I ever found was physical explanations why they shouldn't work.

Now that you know where I have come from and where I am going with Quantum Neuroscience, let's run some Neuroscience through the process and see what we end up with.

I will use *Neuroscience: Exploring the Brain* as my text, because it's freely available online if you know where to look for it; and, it's relatively inexpensive to get a used hardcopy of the book if you want one.

https://www.amazon.com/dp/0781760038/

Bear, M. F., Connors, B. W., & Paradiso, M. A. (2007). *Neuroscience: Exploring the Brain* (3rd ed.). New York: Lippincott Williams and Wilkins.

This essay will focus mainly on chapters 2 through 6 of their book. Let's see what we can do with them.

Neuroscience: Exploring the Brain.

https://canvas.brown.edu/courses/851434/modules

BCP2

https://canvas.brown.edu/courses/851434/files/40331111/

http://mypsyche.us/wp-content/uploads/2017/12/BCP2.pdf

BCP3

https://canvas.brown.edu/courses/851434/files/40426262/

http://mypsyche.us/wp-content/uploads/2017/12/BCP3.pdf

BCP4

https://canvas.brown.edu/courses/851434/files/40426324/

http://mypsyche.us/wp-content/uploads/2017/12/BCP4.pdf

BCP5

https://canvas.brown.edu/courses/851434/files/40426574/

http://mypsyche.us/wp-content/uploads/2017/12/BCP5.pdf

BCP6

http://mypsyche.us/wp-content/uploads/2017/12/BCP6.pdf

BCP7

https://canvas.brown.edu/courses/851434/files/40676532/

http://mypsyche.us/wp-content/uploads/2017/12/BCP7.pdf

Remember, there are two different types of behavior: reflexes or pre-programmed instincts, and chosen behaviors. Reflexes, pre-programmed behaviors, and the autonomic nervous system are the result of Nature's Psyche working on and working through the physical matter in our brains and bodies, outside our conscious awareness. From our perspective, these things happen automatically. Chosen behaviors are a function and product of the Human Psyche. We are consciously aware of our chosen behaviors, because we have chosen to pay attention to them.

The "unconscious" or "sub-conscious" is a function of Nature's Psyche, because we aren't consciously aware of these processes. The "conscious" is a product and function of the Human Psyche. According to the Orthodox Interpretation of Quantum Mechanics, by Henry P. Stapp, only Nature's Psyche has the ability to collapse the wave function, thereby making the Human Psyche's wishes, desires, and choices PHYSICAL and REAL. This is Quantum Neuroscience. This explains how the Human Psyche is able to interact with, control, and survive its physical brain.

I'm very impressed by the explanatory power of Quantum Neuroscience. Let's make some comparisons, so that you can see what I mean.

Chapter 1:

These people are primarily trying to figure out how complex psychological phenomena – such as thought, cognition, intellectual processes, attention, learning, memory, motivation, perception, and emotion – are being produced by brain activity. In Quantum Neuroscience, we recognize and accept that these types of complex psychological phenomena are a function of the Human Psyche, because they survive the death of our physical brain. Quantum Neuroscience is an attempt to put Psyche back into psychology. We try to figure out what's happening at the quantum level or the psyche level, and how the brain is used by Nature's Psyche and the Human Psyche to get things done at the physical level.

Psychology was originally the study of the Human Psyche. In the early 1900's, the Behaviorists, Materialists, and Naturalists prevailed; and, they changed the meaning of psychology to the study of overt behavior. Fifty years later, the humanists tried to bring "internal processes" back into the picture, such as learning, memory, motivation, perception, and emotion. Most of the humanists remained Materialists, Naturalists, and Atheists; but, they did cease to be Behaviorists and Determinists, and did try to bring choice and self-actualization back into the picture. Transpersonal Psychology is the new branch of humanism that is trying to bring Psyche or Spirit back into psychology, as well as some type of Phenomenology.

The only thing useful in Chapter 1 of _Neuroscience_ is the introduction to the Scientific Method. An introduction to the Scientific Method is always useful. We have to know how things work at the physical level, before we can explore how things work at the quantum level.

Why?

It's because identifying the physically impossible points us directly to Quantum Mechanics and Quantum Non-Local Consciousness (Psyche), for a scientific explanation of what we are observing and experiencing. Where the Scientific Method and Classical Physics end, Quantum Mechanics and Quantum Neuroscience begin. As long as we are aware of this reality, we can derive a lot of benefit from the Scientific Methods.

The Scientific Process

Neuroscientists of all stripes endeavor to establish truths about the nervous system. Regardless of the level of analysis they choose, they work according to a scientific process, which consists of four essential steps: observation, replication, interpretation, and verification.

Observation.

Observations typically are made during experiments designed to test a particular hypothesis. For example, Bell hypothesized that the ventral roots contain the nerve fibers that control the muscles. To test this idea, he performed the experiment in which he cut these fibers and then observed whether or not muscular paralysis resulted. Other types of observation derive from carefully watching the world around us, or from introspection, or from human clinical cases. For example, Broca's careful observations led him to correlate left frontal lobe damage with the loss of the ability to speak.

Replication.

Whether the observation is experimental or clinical, it is essential that it be replicated before it can be accepted by the scientist as fact. Replication simply means repeating the experiment on different subjects or making similar observations in different patients, as many times as necessary to rule out the possibility that the observation occurred by chance.

Interpretation.

Once the scientist believes the observation is correct, he or she makes an interpretation. Interpretations depend on the state of knowledge (or ignorance) at the time the observation was made and on the preconceived notions (the "mind set") of the scientist who made it. As such, interpretations do not always withstand the test of time. For example, at the time he made his observations, Flourens was unaware that the cerebrum of a bird is fundamentally different from that of a mammal. Thus, he wrongly concluded from experimental ablations in birds that there was no localization of certain functions in the cerebrum of mammals. Moreover, as mentioned before, his profound distaste for Gall surely also colored his interpretation. The point is that the correct interpretation often remains unrecognized until long after the original observations were made. Indeed, major breakthroughs are sometimes made when old observations are reinterpreted in a new light.

Verification.

The final step of the scientific process is verification. This step is distinct from the replication performed by the original observer. Verification means that the observation is sufficiently robust that it will be reproduced by any competent scientist who precisely follows the protocols of the original observer. Successful verification generally means that the observation is accepted as fact. However, not all observations can be verified, sometimes because of inaccuracies in the original report or insufficient replication. But failure to verify usually stems from the fact that additional variables, such as temperature or time of day, contributed to the original result. Thus, the process of verification, if affirmative, establishes

new scientific fact, and, if negative, suggests new interpretations for the original observation.

Occasionally, one reads in the popular press about a case of "scientific fraud." Researchers face keen competition for limited research funds and feel considerable pressure to "publish or perish." In the interest of expediency, a few have actually published "observations" that were never made. Fortunately, such instances of fraud are rare, thanks to the scientific process. Before long, other scientists find they are unable to verify the fraudulent observations and begin to question how they could have been obtained in the first place. The material you will learn in this book stands as a strong testament to the success of the scientific process. (*Neuroscience*, pp. 15-16.)

I upgraded my science to Science 2.0, which allows ALL of the evidence into evidence, and then pursues a preponderance of the evidence. Science 2.0 pursues a quantum mechanical explanation for the physically impossible. The explanatory power of Quantum Neuroscience is vastly superior to anything that we can get from Materialism, Naturalism, Behaviorism, Atheism, and Classical Physics.

We study and learn the physical basis for how things work, until we run into things that have NO physical explanation. When dealing with the physically impossible that has NO physical explanation, we turn to the observational evidence and experiential evidence from Quantum Mechanics, Quantum Non-Local Consciousness (Psyche), Near-Death Experiences (NDEs), and Out-of-Body Experiences (OBEs) in order to explain what we are observing and experiencing. We can actually use Quantum Mechanics and Psyche to explain the physically impossible. This is how Science 2.0 works. Science 2.0 is an evidentiary science, observational science, and empirical science. Under Science 2.0, Lived Experiences and Phenomenology are given precedence over philosophical speculation and wishful thinking.

Science 2.0, Quantum Mechanics, and Quantum Neuroscience are observational sciences, and experiential sciences. Observations and experiences take precedence over philosophical speculation and guesswork. Science 2.0 allows all of the evidence into evidence, including introspection and spiritual experiences. Observation is the only way we make discoveries that are real and truly exist.

When dealing with the non-physical, spiritual, non-local, and the quantum, replication on demand isn't always possible. Instead, we are forced to turn to the tried and true Burden of Proof Methodology, and we are obligated to meet our burden of proof through a preponderance of the observational evidence and experiential evidence. Yes, Quantum Mechanics has been replicated on demand and proven to be true. However, Lived Experiences such as Near-Death Experiences and Out-of-Body Experiences and Life Reviews can't always be replicated on demand. Instead, when it comes to these types of first-hand quantum experiences, we turn to the Burden of Proof Methodology and pursue a preponderance of the evidence. In other words, we use Science 2.0, which allows ALL of the evidence into evidence, and then pursues a preponderance of the evidence. The demand for physical proof is the fundamental flaw of Materialism and Naturalism, because all you have to do is find something that is physically impossible but happening anyway, and you have successfully FALSIFIED Materialism and Naturalism.

They mention design.

Design and creation by evolution has NEVER been observed, because it is physically impossible for random chance or random mutations to design and create proteins, genomes, and life forms. In fact, evolution (genetic change), random mutations, and natural selection didn't even exist until after God designed and created the proteins, genomes, and life forms in the first place.

They mention replication.

For all the different observations and experiences that can't be replicated on demand, we must go with a preponderance of the evidence and a burden of proof methodology, just like in a court of law. We have to make our case through a preponderance of the evidence.

They mention interpretation and verification.

When it comes to the chemical evolution of proteins and genes, the spontaneous generation of life forms, the abiogenesis of cells and viruses, the design and creation by natural selection, and the creation of functional genomes by random chance and random mutations, there is NOTHING to interpret and NOTHING to verify, because NO observations of these things have ever been made.

In other words, there is NOTHING to verify because NO observations have been made and therefore there are NO protocols to follow, where the theory of evolution is concerned. NO observations and NO verification means that the Theory of Evolution, Materialism, Naturalism, Darwinism, Nihilism, Behaviorism, and Atheism have been successfully FALSIFIED.

A careful study of the Philosophy of Science will reveal that "interpretation" is the fundamental foundational flaw within the Scientific Method. For every observed fact or event, there is theoretically an infinite number of different interpretations that can be given to that single event or scientific observation. The *affirming the consequent* logic fallacy is the ultimate result. Faulty interpretations abound in Materialism, Naturalism, Darwinism, Nihilism, Behaviorism, and Atheism. Even though you are using a Scientific Method, by *affirming the consequent*, you can have any conclusion or interpretation you want and actually prove that it is true.

It's physically and logically impossible to verify scientific observations and scientific interpretations by *affirming the consequent*, or *verifying the consequent*. You can't do verification by using the Scientific Method and *affirming the consequent*. This is the second primary flaw in the Scientific Method. By *affirming the consequent*, you can verify anything you want and prove that it is true. That's how the Materialists, Naturalists, and Darwinists prove that the theory of evolution is true; and, they don't even know that they are *affirming the consequent* or *begging the question* in order to make their case.

Remember, you can't use Materialism and Naturalism to prove that Materialism and Naturalism are true. Using Materialism and Naturalism to prove that Materialism and Naturalism are true is the very definition of *circular reasoning* and *begging the question*, which are logic fallacies. Yet, that's precisely how the Materialists and Naturalists prove that Materialism, Naturalism, and the Theory of Evolution are true – by *begging the question* and using *circular reasoning*. Their whole science is based upon a wide variety of different logic fallacies. They literally trick themselves and deceive themselves into believing that their Materialism, Naturalism, Darwinism, Nihilism, Behaviorism, and Atheism are real and true.

This is Philosophy of Science 101.

Thanks to the *affirming the consequent* logic fallacy that's built directly into the Scientific Method, verification is absolutely worthless if you want to find and know the truth. You can't use the Scientific Method to verify the truth of anything.

Instead, you have to use *negating the consequent* to FALSIFY your theories and your ideas. That's the way the Scientific Method should be used if you want to find and know the truth. I have used *negating the consequent* and the Scientific Method to FALSIFY Materialism, Naturalism, Darwinism, and Atheism. It's really easy to do, once you know how it works. The Scientific Methods should be used to falsify theories, not verify them. It's impossible to verify theories using the Scientific Method; but, you can falsify anything that is false by using the Scientific Methods and *negating the consequent*. That's how I was able to use the Scientific Method to FALSIFY Materialism, Naturalism, Darwinism, and Atheism.

Remember, the BEST way to find and know the truth is to live it and experience it for yourself, or to choose to trust someone who has.

The second best way to find and know the truth is through a process of elimination, wherein you use the Scientific Method to falsify everything you can by *negating the consequent*. It's an arduous task; but, if you successfully eliminate everything that has been falsified and everything that is false, then ONLY the truth will remain. It's elementary my dear reader. You can start the process of elimination by falsifying Materialism, Naturalism, Darwinism, and Atheism. That will get you the most mileage and traction.

They mention scientific fraud.

This is precisely what happened in the case of the Theory of Evolution. For 150 years they have been publishing observations that were never made!

Fraudulent observations abound in Materialism, Naturalism, and especially the Theory of Evolution. The whole thing is a fraud, because there is NO observational evidence supporting the primary assumptions or major premises of Materialism, Naturalism, Darwinism, Behaviorism, Nihilism, and Atheism.

The Materialists, Naturalists, Darwinists, and Atheists always get the Scientific Method wrong, because they don't understand the Philosophy of Science and how *affirming the consequent* is working against them to trick them and deceive them.

I discuss these truths in much greater detail at the beginning of this book and within my other books, and give actual examples; so, I won't do so here.

When it comes to Quantum Mechanics, it's impossible to provide direct physical proof of concept – everything is inferred from the effects that quantum mechanisms have on physical matter. We can and do observe the effects of Quantum Mechanics. In fact, Quantum Mechanics is our best-proven and most-used science, because it works, even though it is spiritual or non-local in nature and origin. Remember, where the Scientific Method and Classical Physics end, Quantum Mechanics, Psyche, and Quantum Neuroscience begin. Ironically, Quantum Neuroscience was already a proven science, BEFORE I coined the term and developed the discipline. Quantum Neuroscience was just sitting there waiting for someone to find it, use it, and develop it. All the scientific experimentation and proof of concept has already been done.

Quantum Mechanics and Quantum Neuroscience are applied physics at their very best, in my humble opinion. The only thing you have to be careful with is to make sure that you get an Interpretation of Quantum Mechanics that isn't based upon and tainted by

Materialism and Naturalism, because Materialism and Naturalism cannot do the physically impossible; but, the correct Interpretation of Quantum Mechanics certainly can.

When it comes to the "Neuroscience" part of Quantum Neuroscience, I still choose the book *The Mind and the Brain: Neuroplasticity and the Power of Mental Force* by Jeffrey Schwartz as the best introduction to practical neuroscience, applied neuroscience, and clinical neuroscience that I have discovered so far. This book is Neuroscience the way it should be done! It's actually useful neuroscience, and it's completely compatible with everything that I have discovered and developed concerning Quantum Neuroscience.

When it comes to the "Quantum" part of Quantum Neuroscience, I still choose *Consciousness Beyond Life: The Science of the Near-Death Experience* by Pim van Lommel as the ultimate best introduction to Quantum Mechanics and Non-Local Consciousness of any book on the planet. This book is too good to be true; yet, there it is. It exists; and, it actually explains what's really going on when it comes to Quantum Non-Local Consciousness, Psyche, or Mind. I haven't found anything better.

With a complete understanding of those two books and a bit of intelligence, you should be able to design and develop your own version of Quantum Neuroscience that is consistent, logical, makes sense, and actually works.

When it comes to an official Interpretation of Quantum Mechanics, which matches perfectly with Quantum Neuroscience and explains everything that is going on within the mind-brain interface, the Orthodox Interpretation of Quantum Mechanics from Henry P. Stapp is the ONLY one that explains everything you need to know in a logical and useful fashion – the interplay between the Human Psyche, Nature's Psyche, and the collapse of the wave function.

Any interpretation of Quantum Mechanics that is based upon Materialism, Naturalism, and Classical Physics is absolutely worthless when it comes to Quantum Neuroscience and trying to explain how the Human Psyche interacts with, controls, and survives its physical brain; and, most of the official interpretations of Quantum Mechanics listed on the Wikipedia are based exclusively on Materialism, Naturalism, and Classical Physics; and they by definition in principle deny the existence of Psyche or Non-Local Consciousness a priori.

You can't use Materialism, Naturalism, and Classical Physics to explain how the Human Psyche works; but, by using the Orthodox Interpretation of Quantum Mechanics from Henry P. Stapp, you certainly can. All you want is something that works and makes logical sense! You will find what you are looking for in the Orthodox Interpretation of Quantum Mechanics and Quantum Neuroscience.

Chapter 2:

Chapter 2 of *Neuroscience* focuses on the physical aspects of Neurons and Glial Cells within the physical brain.

Quantum Neuroscience starts by identifying the physically impossible, and then it continues by trying to find a scientific explanation for the physically impossible.

You need an Interpretation of Quantum Mechanics that explains what the Human Psyche and Nature's Psyche are doing at the quantum level. If your Interpretation of Quantum Mechanics doesn't do that, then it's worthless. You also need an Interpretation of Quantum Mechanics that explains how God's Psyche does His science and His work through quantum mechanisms. If your Interpretation of Quantum Mechanics can't do that, then it's worthless.

In this chapter, they mention how the cells in our nervous system are assembled into circuits that mediate sensation, perception, movement, speech, and emotion. Assembly requires an Assembler.

Maps require a Map Maker. Differentiation requires a Differentiator. Intention requires and Intender. Targeting requires Intention. Laws require a Law Giver. Blueprints and Designs require an Architect. Engineering requires an Engineer. Migration requires a Migrator. Manufacturing requires a Manufacturer. Machinery and Proteins require an Engineer and Mechanic. Computers, CPUs, RAM, Genomes, and Software require Engineers, Computer Scientists, and Programmers. Logistics requires a Logician. Fine-Tuning requires a Fine-Tuner. Science requires a Scientist. Intelligent Results require Intelligent Designers and Intelligent Manufacturers. These things don't just spontaneously generate out of thin air at the physical level. Evolution can't write words, and evolution can't write programming code. This is Logic 101.

They state that your DNA contains the blueprint for your entire body.

This claim is false.

How do we know?

WE KNOW, because it's physically impossible to store petabytes of blueprints within the 750 megabytes that our genome can hold.

The most useful part of this chapter deals with Protein Synthesis or Gene Expression.

The "reading" of the DNA is known as gene expression. The final product of **gene expression** is the synthesis of molecules called **proteins**, which exist in a wide variety of shapes and sizes, perform many different functions, and bestow upon neurons virtually all of their unique characteristics. **Protein synthesis**, the assembly of protein molecules, occurs in the cytoplasm. Because the DNA never leaves the nucleus, there must be an intermediary that carries the genetic message to the sites of protein synthesis in the cytoplasm. This function is performed by another long molecule called **messenger ribonucleic acid**, or **mRNA**. Messenger RNA consists of four different nucleic acids strung together in various sequences to form a chain. The detailed sequence of the nucleic acids in the chain represents the information in the gene, just as the sequence of letters gives meaning to a written word.

The process of assembling a piece of mRNA that contains the information of a gene is called **transcription**, and the resulting mRNA is called the transcript. Protein-coding genes are flanked by stretches of DNA that are not used to encode proteins but are important for regulating transcription. At one end of the gene is the **promoter**, the region where the RNA-synthesizing enzyme, *RNA polymerase*, binds to initiate transcription.

The binding of the polymerase to the promoter is tightly regulated by other proteins called **transcription factors**. At the other end is a sequence of DNA called the **terminator** that the RNA polymerase recognizes as the end point for transcription.

In addition to the non-coding regions of DNA that flank the genes, there are often additional stretches of DNA within the gene itself that cannot be used to code for protein. These interspersed regions are called **introns**, and the coding sequences are called **exons**. Initial transcripts contain both introns and exons, but then, by a process called **RNA splicing**, the introns are removed and the remaining exons are fused together. In some cases, specific exons are also removed with the introns, leaving an "alternatively spliced" mRNA that actually encodes a different protein. Thus, transcription of a single gene can ultimately give rise to several different mRNAs and protein products.

Messenger RNA transcripts emerge from the nucleus via pores in the nuclear envelope and travel to the sites of protein synthesis elsewhere in the neuron. At these sites, a protein molecule is assembled much as the mRNA molecule was: by linking together many small molecules into a chain. In the case of protein, the building blocks are **amino acids**, of which there are 20 different kinds. This assembling of proteins from amino acids under the direction of the mRNA is called **translation**.

The scientific study of this process, which begins with the DNA of the nucleus and ends with the synthesis of protein molecules in the cell, is known as *molecular biology*. (*Neuroscience*, pp. 30-31.)

The scientific study of the gene expression process at the quantum level is also an integral part of Quantum Neuroscience.

WE KNOW that it is physically impossible for chemical evolution to design and create functional proteins along with the matching genes to go with them. So, evolution didn't create gene expression, and evolution isn't doing the gene expression either. We have to look someplace else for an answer as to how this is being done.

Who tells the transcription factors which protein is needed by the cell and which gene will produce that specific protein? More importantly, who tells the transcription factors where that specific gene is located in the genome?

If I gave you the task to perform, how would you talk to a molecule, tell it where to go, and what to do when it gets there? How would you teach a molecule what it needs to know? Well, you certainly aren't going to be talking with a molecule at the physical level, because that's physically impossible. Instead, you have to communicate with an atom or a molecule telepathically through quantum waves at the quantum level. There is no other way to get this done. Only the smaller can dwell within the larger, controlling and communicating with the larger. Psyche and quantum waves are smaller than physical atoms. It's elementary!

Who taught the transcription factors how to recognize a promoter and terminator? Who put the promoters and terminators there in the first place? It wasn't evolution because evolution can't design and create at the quantum level, where atoms and molecules reside.

How did these transcription factors learn that alternative splicing is available for certain genes? Who taught these transcription factors how to read a gene at the quantum level?

How do the transcription factors know the difference between the exons and the introns, and how do these transcription enzymes know the start and end points for the introns and exons?

The transcription factors seem to be alive, intelligent, conscious, sentient, and aware.

There are clear signs of Intelligence or Psyche behind these different transcription factors or transcription enzymes, and that's physically impossible.

We are talking about quantum mechanisms here, and NOT physical mechanisms.

A case can be made that the transcription factors are more intelligent than the whole human race combined. Have you memorized the location of each of the 20,000 genes in the human genome? The transcription factors have. Have you memorized which protein(s) each of the genes in the human genome can be used to make? The transcription factors have. Have you memorized the genomes of all the other species on this planet? Their transcription factors have.

Depending on how you choose to look at it, it took the human race as a whole anywhere from 5 to 50 years to map the human genome; and, we still don't have clue what most of the genes actually do.

Either the transcription factors have ALL of this information memorized, or the transcription factors have direct access to God's Database where ALL of this information is stored! Can you read God's Database and use it to determine where each of the genes in the human genome is located, and use it to learn what protein(s) each of the genes can be used to make? The transcription factors can. So why can't you?

Let's face the facts. The transcription factors are infinitely more intelligent and knowledgeable than we are. Those things could kill you if they wanted to, and there's nothing you could do to stop them. Scary huh?

What is the difference between proteins synthesized on the rough ER (endoplasmic reticulum) and those synthesized on the free ribosomes? The answer appears to lie in the intended fate of the protein molecule. If it is destined to reside within the cytosol of the neuron, then the protein's mRNA transcript shuns the ribosomes of the rough ER and gravitates toward the free ribosomes. (p. 33.)

They talk about mRNA "gravitating towards free ribosomes" or traveling towards ribosomes. Who is doing the intention, command, and control thereby making the mRNA molecule travel through the pores of the nucleus and then gravitate towards the free ribosomes? How does the mRNA gravitate? What is this "gravitation force" and how does it work? How does the mRNA know telepathically at a distance which ribosome is free and which ribosome it should gravitate towards? Who tells them where to go and how to get there?

The mRNA molecules and tRNA molecules and rRNA molecules are like little remote-controlled cars. Remember, telepathy or quantum waves are radio waves at the quantum level. Someone is telling these messenger molecules where to go, how to get there, and what to do when they get there.

We are talking about quantum mechanisms here, and NOT physical mechanisms.

Targeting is a function of Psyche. An mRNA molecule's shunning one type of ribosome while gravitating towards a different type of ribosome is proof of Intelligence or Psyche. The mRNA molecule selects the type of ribosome to bind with based upon the future fate of the protein that is going to be produced. Selectivity, and prescience or foreknowledge, are signs of Psyche and Intelligence. Somebody put this information or knowledge into these mRNA molecules.

It is not surprising that neurons are so well endowed with rough ER, because, as we shall see in later chapters, special membrane proteins are what give these cells their remarkable information-processing abilities. (*Neuroscience*, p. 34.)

Information-processing abilities are a function of Psyche or Intelligence. Information-processing requires Intelligence or Psyche. It's physically impossible for these mRNA molecules to be so intelligent, cunning, and smart.

Obviously, we are talking about quantum mechanisms here, and NOT physical mechanisms. Remember, while using Quantum Mechanics, Nature's Psyche can do the physically impossible.

All of this intention, command, control, intelligence, targeting, and gravitation is physically impossible, because we are dealing with a molecule here that seems to be moving towards a targeted ribosome and then docking with that ribosome all on its own. Ships like mRNA molecules require some kind of propulsion system; but, there isn't one available at the physical level, so it has to acquire its propulsion from the quantum level. This is Quantum Neuroscience in action.

Adenosine triphosphate (ATP) is the cell's energy source. When the mitochondrion "exhales," 17 ATP molecules are released for every molecule of pyruvic acid that had been taken in. ATP is the energy currency of the cell. The chemical energy stored in ATP is used to fuel most of the biochemical reactions of the neuron. (p. 35.)

After the ATP molecules are exhaled randomly into the cytoplasm, who or what tells the ATP molecules where to go, how to get there, and what to do when they get there? Again, these little ATP molecules seem to be functioning like remote-controlled cars. They KNOW precisely where they are supposed to go and what they are supposed to do when they get there.

The process of joining small proteins to form a long strand is called polymerization; the resulting strand is called a polymer. Polymerization and depolymerization of microtubules and, therefore, of neuronal shape can be regulated by various signals within the neuron.

One class of proteins that participate in the regulation of microtubule assembly and function are microtubule-associated proteins, or MAPs. Among other functions (many of which are unknown), MAPs anchor the microtubules to one another and to other parts of the neuron. Pathological changes in an axonal MAP, called tau, have been implicated in the dementia that accompanies Alzheimer's Disease. (p. 35.)

Who is telling the different proteins to polymerize or depolymerize? What's their motivation? Who tells them to depolymerize and when to do so? Who sends and regulates ALL these signals? These people actually said that various signals are being sent to regulate the shape and function of a neuron. Who taught the molecules within these cells how to communicate with each other, and who taught these molecules what these various signals mean?

Signal-processing requires intelligence or Psyche. There are even Master and PhD degrees available in signal-processing! But here, we are talking about signal-processing at the quantum level between molecules, through quantum waves and under the control of Nature's Psyche.

Remember, evolution, random mutations, and natural selection can't do quantum waves and signal-processing at the quantum level.

Anytime you catch the Materialists, Darwinists, and Naturalists using the word "unknown" to describe a physical process, most of the time that's code for something that's happening at the quantum level or psyche level. Anytime a Materialist or Naturalist admits that he doesn't know how a particular physical function works, WE KNOW that he is most likely talking about something that is being controlled at the quantum level through quantum waves by Nature's Psyche. The Materialists, Naturalists, and Atheists aren't always lying to us. There really are things that they don't know.

Because dendrites function as the antennae of the neuron, they are covered with thousands of synapses. The dendritic membrane under the synapse (the postsynaptic membrane) has many specialized protein molecules called receptors that detect the neurotransmitters in the synaptic cleft. (*Neuroscience*, p. 41.)

Detection of molecules by molecules implies action at a distance, a quantum mechanical function, not a physical function. Only Psyche or Intelligence can do functionality! Functionality at any level doesn't spontaneously generate out of thin air, despite what the Materialists and Naturalists might claim.

There's NO physical entity in the synaptic cleft called "evolution," who is moving the neurotransmitters and degradation enzymes around with tractor beams at the quantum level. Evolution doesn't work at the quantum level.

An essential role of astrocytes is regulating the chemical content of this extracellular space. For example, astrocytes envelop synaptic junctions in the brain, thereby restricting the spread of neurotransmitter molecules that have been released. Astrocytes also have special proteins in their membranes that actively remove many neurotransmitters from the synaptic cleft. A recent and unexpected discovery is that astrocytic membranes also possess neurotransmitter receptors that, like the receptors on neurons, can trigger electrical and biochemical events inside the glial cell. Besides regulating neurotransmitters, astrocytes also tightly control the extracellular concentration of several substances that have the potential to interfere with proper neuronal function. For example, astrocytes regulate the concentration of potassium ions in the extracellular fluid. (p. 47.)

Whoa!

Astrocytes regulate the chemical content of extracellular space and synaptic clefts telepathically through quantum waves and tractor beams. Telepathic oversight! Can you do telepathic oversight of the contents of your synaptic clefts? Why not? Astrocytes can! Astrocytes build walls around specifically chosen synapses in order to keep the neurotransmitters from escaping. Astrocytes have specialized tractor beams in their membranes that actively remove neurotransmitters from the synaptic cleft. Astrocytes have sensors or detectors. Astrocytes KNOW telepathically where all the different molecules in the synaptic cleft are located and what they are doing.

Remember, there's NO physical entity in the synaptic cleft called "evolution," who is moving the neurotransmitters and degradation enzymes around with tractor beams at the quantum level. Evolution doesn't work at the quantum level. But, Astrocytes seem to be working at the quantum level.

Whoa!

Astrocytes are regulating and tightly controlling ions or atoms within the extracellular fluid! That's physically impossible! Regulation, tractor beams, and sensors are physically impossible at the atomic level or physical level! But, at the quantum level while using quantum waves, Nature's Psyche can indeed do regulation, atomic manipulation, tractor beams, sensors, and quantum tunneling with ease. Through quantum waves, Nature's Psyche can even do electrical and biochemical events if it wants to.

Is that cool, or what!

Beam me up Scotty!

I got all worked up and excited when I first discovered electrical synapses and gap junctions!

I thought they were wires! I was wrong.

The Materialists and Naturalists call them electrical synapses in order to trick you into believing that they are wires – wires that are connecting the different neurons in our brain into functional CPUs, RAM, and Memory Systems.

These are not the wires that you are looking for.

Yes indeed, gap junctions are direct connections between neurons; but, they are not functioning as wires.

Gap junctions (and electrical synapses) are the same thing as ion channels (and chemical synapses). They simply let ions through to either excite or inhibit the Postsynaptic Potential (PSP) in the target neuron. That's it! They either succeed in turning the target neuron ON, or they don't and the target neuron remains OFF. A neuron is nothing more than a single hardware BIT, a switch – it is either ON or OFF. There's nothing more there to see at the physical level. ALL of that supernatural and miraculous functionality and memory storage within a neuron is taking place at the quantum level under the command and control of Nature's Psyche, because it's physically impossible at the physical level.

> **Cells connected by gap junctions are said to be electrically coupled. Transmission at electrical synapses is very fast and, if the synapse is large, fail-safe. An action potential in the presynaptic neuron can produce, almost instantaneously, an action potential in the postsynaptic neuron** [through gap junctions and electrical synapses]. **Electrical synapses are sometimes found between sensory and motor neurons in neural pathways mediating escape reflexes.** (*Neuroscience*, p. 104.)

In a different book, the author mentioned that some Astrocytes are connecting themselves directly to the neurons that they control with gap junctions or what have been dubbed electrical synapses. The Astrocytes can then push ions and small molecules through gap junctions into nearby neurons, in order to trigger action potentials in those target neurons. Gap junctions really aren't wires. They are gates. They are either OPEN or CLOSED. So, who is opening and closing those gates, and why? Gates require a Gatekeeper. That's the REAL question, "Who decides when these gap junctions will open, and when they will close?"

As mentioned, the opening of specific gap junctions can be triggered in some cases by an action potential in the presynaptic neuron; but, that's not the only purpose or use for gap junctions and electrical synapses. It's the gap junctions that open out of the blue under the command and control of Nature's Psyche that interest us the most, where Quantum Neuroscience is concerned.

When the Human Psyche decides that it wants to raise its finger, it's theoretically possible that Nature's Psyche can reach in telepathically to the Astrocyte that is connected to the neuron that raises your finger, and that Astrocyte unilaterally opens the gap junctions attached to the correct neuron and floods the target neuron with an Excitatory Post-

Synaptic Potential (EPSP) sufficient enough to trigger the neuron that raises your finger, and your finger rises. This is yet another way that Nature's Psyche can get that FIRST action potential in a chain of action potentials started, whenever it needs to do so. Of course, not every neuron has an Astrocyte connected to it with gap junctions; but, the ones that do have yet another method by which Nature's Psyche can get an action potential started within that specific neuron.

Reaching in telepathically and firing the neuron of your choice is physically impossible; but, it's really easy to do at the quantum level if you are Nature's Psyche.

I don't think they really know what Astrocytes are doing, yet.

I actually read someplace that Astrocytes trigger action potentials in target neurons, supposedly by depolarizing the target neuron with a huge EPSP. However, I have come across evidence that suggests the polar opposite – a braking mechanism.

The membrane potential (Vm) of astrocytes is more negative than that of neurons. For example, astrocytes have a Vm of about -85 mV, whereas neuronal membrane potential is about -65 mV. — Uncle Google.

So, what would naturally happen if we were to open up the gap junctions between an Astrocyte and a specific neuron?

Wouldn't the neuron get hyperpolarized or move towards -85 mV? It should, shouldn't it; all other things being equal? Of course, the equilibrium potential of the different ions would have a powerful effect, and I don't know the internal constitution and equilibrium potential of the ions in an Astrocyte. I don't remember any of these people spending a lot of time talking about and analyzing the innards of Astrocytes. Astrocytes and their functionality are still black boxes.

However, if an Astrocyte has a way of hyperpolarizing a target neuron either through gap junctions, equilibrium potentials, or some other mechanism, Nature's Psyche could use an Astrocyte to turn OFF a neuron that it doesn't want to fire; thereby stopping an action that the Human Psyche no longer wants to take place.

I have no idea what an Astrocyte is actually doing. For all I know, an Astrocyte could have one mechanism for triggering an action potential in a target neuron and a completely different mechanism for stopping an action potential in a target neuron.

There's a lot more than this to be discovered or found at the quantum level. I just hit the highlights in this chapter – the things that caught my attention the most.

It's at the molecular level and the atomic level where we see Quantum Mechanics and Quantum Neuroscience in action. The quantum level is where the tires really hit the pavement! Those various different forces, fields, and quantum waves can do the physically impossible.

Remember, evolution can't do the physically impossible. It's physically impossible for something that doesn't physically exist to trigger physical processes at will. That's why Materialism, Naturalism, Darwinism, and Atheism are self-defeating. It's physically impossible for them to do anything, because they don't have a physical body through which they can do it. Psyche or Non-Local Consciousness can do the physically impossible; but, evolution has NO intelligence or psyche.

The Materialists, Naturalists, Darwinists, and Atheists gloss over these various different quantum mechanisms and psychic mechanisms by using words like "travel" and "gravitate" and "regulate", trying to trick you into believing that these are physical processes when in fact they are invisible, non-physical, immaterial forces, fields, and quantum mechanical waves under the control of Nature's Psyche. In fact, the very existence of quantum waves, gravitation, magnetism, and action at a distance FALSIFIES Materialism and Naturalism, which is why these people don't want you to learn anything about quantum waves and quantum mechanisms. The Materialists, Naturalists, Darwinists, Nihilists, and Atheists skip over these quantum mechanical processes as if they don't exist; but, they exist alright, in spades!

Remember, if your Interpretation of Quantum Mechanics can't explain what Nature's Psyche is doing at the quantum level, and how Nature's Psyche is doing it, then your Interpretation of Quantum Mechanics is totally worthless.

In summary, I want to emphasize that you have to know how everything works at the physical level, before you can successfully identify the physically impossible, and then start to explore things at the quantum level or psyche level. Quantum Neuroscience starts with the physically impossible, and then tries to find a scientific explanation for it. You can't successfully use Materialism, Naturalism, and Classical Physics to explain the physically impossible; but, you can definitely use Quantum Mechanics and Quantum Non-Local Consciousness (Psyche) in order to explain the physically impossible in a scientific manner.

This is the lesson and the message of Quantum Neuroscience. Rinse and repeat!

Quantum Neuroscience has been there staring us in the face for the past century, and nobody noticed it, because they weren't looking for it and didn't want it.

Chapter 3:

Chapter 3 of *Neuroscience* focuses on Ions and the Neuron's Resting Potential.

As we move further and further away from Psyche, Quantum Waves, Quantum Non-Locality, the Quantum Level, Telepathy, the Quantum Zeno Effect, Quantum Tunneling, Action at a Distance, Molecules, and Atoms, we move away from Quantum Neuroscience and transition over to Physical Neuroscience and the things that actually have a physical explanation.

At the physical level, we tend to see signs of Intelligence or Psyche only within the designed functionality which has been programmed into our proteins and genes by God's Psyche. At the physical level, it's all about physical functionality and trying to figure out where ALL of that physical functionality came from and why it works. It has been proven scientifically that evolution (genetic changes), random mutations, natural selection, and chemical evolution can't do the design and creation of functional proteins, let alone making the matching genes to go with those proteins. It's physically impossible for chemical evolution to make proteins and genes. It can't be done, which means that it wasn't done, despite what the Materialists and Darwinists might try to tell you. It's physically impossible to do the physically impossible; and, evolution of any kind is a physical process. Physical processes can't do the physically impossible. This is Logic 101.

Quantum Neuroscience starts by identifying the physically impossible, and then it tries to find a scientific explanation for the physically impossible.

Information is encoded in the frequency of action potentials of individual neurons, as well as in the distribution and number of neurons firing action potentials in a given nerve. This type of code is partly analogous to Morse code sent down a telegraph wire; information is encoded in the pattern of electrical impulses. Cells capable of generating and conducting action potentials, which include both nerve and muscle cells,

are said to have excitable membrane. The "action" in action potentials occurs at the cell membrane.

Thus, the *firing frequency* of action potentials reflects the magnitude of the depolarizing current. This is one way that stimulation intensity is encoded in the nervous system. Although firing frequency increases with the amount of depolarizing current, there is a limit to the rate at which a neuron can generate action potentials. The maximum firing frequency is about 1000 Hz. (*Neuroscience*, pp. 53, 79.)

For once, they actually say something that's theoretically possible, both at the physical level and the quantum level. Information can indeed be encoded in frequencies or waves, and then transmitted as frequencies or waves, including the quantum waves that Nature's Psyche is using to communicate with the neurons and to control the neurons.

The electrical charge of an atom depends on the difference between the number of protons and electrons. When this difference is 1, the ion is said to be *monovalent*; when the difference is 2, the ion is *divalent*; and so on.

Ions with a net positive charge are called *cations*; ions with a negative charge are called *anions*. Remember that ions are the major charge carriers involved in the conduction of electricity in biological systems, including the neuron. The ions of particular importance for cellular neurophysiology are the monovalent cations Na^+ (sodium) and K^+ (potassium), the divalent cation Ca^{2+} (calcium), and the monovalent anion Cl^- (chloride). (*Neuroscience*, p. 54.)

Ions are atoms with the "wrong" number of electrons. Technically, ions and electrons use quantum mechanisms and function at the quantum level. It's physically impossible to make an atom do anything at the physical level. Any reaction taking place at the physical level among atoms is being facilitated or performed at the quantum level by quantum mechanisms of some sort. This is crucial information which the Materialists and Naturalists formally reject and choose to ignore, because its very existence falsifies Materialism and Naturalism.

In order to perform their many functions in the neuron, different proteins have widely different shapes, sizes, and chemical characteristics. (p. 56.)

What's their motivation? Who designed and built these proteins to fulfill their assigned function? Who assigned the function? Who is making all of these different proteins coordinate their actions into a functional unified whole? Who is making them perform?

Evolution of course!

There's no other explanation at the physical level!

But, what about the quantum level? Who is making these proteins work together at the quantum level? The Materialists, Naturalists, and Darwinists deliberately ignore this question, pretending that the quantum level or the psyche level does not exist.

Still, it's a legitimate question for a REAL Scientist, "What quantum mechanisms are being used during protein synthesis and during the deployment or the use of proteins within the physical body and physical brain?"

I'm not going to apologize for thinking that Quantum Mechanics, Action at a Distance, or Transdimensional Physics is cool.

> **The distribution of electrical charge across the membrane. The uneven charges inside and outside the neuron line up along the membrane because of electrostatic attraction across this very thin barrier. Notice that the bulk of the cytosol and extracellular fluid is electrically neutral.** (*Neuroscience*, p. 63.)

These atoms or ions are interacting with each other at a distance, at the quantum level, through invisible quantum waves of some kind. There are many different types or frequencies of quantum waves, including electrostatic waves. This is Proof of Concept where Quantum Neuroscience is concerned.

The very existence of invisible quantum waves falsifies Materialism and Naturalism. Thoughts and memories are quantum waves. So are bosons, gluons, magnetism, electrostatic attraction, gravity, and dark energy. ALL of these things involve action at a distance at the quantum level! All of these things are quantum mechanisms happening at the quantum level or the psyche level. All of these things are quantum waves. ALL of these things are also represented by some kind of vibrating thought, or vibrating string, according to string theory. Vibrations are quantum waves!

This is how everything fits together at the quantum level within the physical brain.

You can't find the truth by rejecting the evidence.

The Materialists and Naturalists and Darwinists deliberately reject this evidence, which is why these people end up with something less than the truth.

ALL of these different ions, electrons, pumps, and membranes are working together at the quantum level in order to get things done for us at the physical level.

Everything works flawlessly at the quantum level or psyche level under the command and control of Nature's Psyche. However, the whole thing can and does fall apart at the physical level thanks to evolution or random mutations.

> **In recent years, it has become increasingly clear that many inherited neurological disorders in humans, such as certain forms of epilepsy, are explained by mutations of specific potassium channels.**

> **There are several different sodium channel genes in the human genome. Differences in the expression of these genes among neurons can give rise to subtle but important variations in the properties of the action potential. Recently, single amino acid mutations in the extracellular regions of one sodium channel have been shown to cause a common inherited disorder in human infants known as *generalized epilepsy with febrile seizures*.**

> **Mutations slow the inactivation of the sodium channel, prolonging the action potential. Generalized epilepsy with febrile seizures is a *channelopathy*, a human genetic disease caused by alterations in the structure and function of ion channels.** (*Neuroscience*, pp. 71, 89.)

All of our physically-caused mental illnesses seem to be caused by malfunctioning genes, proteins, neurotransmitters, synapses, and neurons at the physical level. These are matter-over-mind malfunctions which have been given to us by evolution or random mutations.

Break the physical machinery, and there's nothing that Nature's Psyche can do about it. Currently, we have to turn to God or God's Psyche or God's Priesthood if we want our broken genes and broken proteins fixed for us at the quantum level or the psyche level. Fixing things at the quantum level also fixes them at the physical level. Cool, huh?

Chapter 4:

Chapter 4 of *Neuroscience* focuses on the Action Potential.

Remember, atoms, ions, and electrons can only be controlled, manipulated, modified, and coordinated at the quantum level through some kind of quantum wave. Typically all of this takes place under the command and control of Nature's Psyche (or God's Psyche) completely outside the conscious awareness of the Human Psyche. Occasionally, though, the Human Psyche learns how to detect or observe the effects of quantum mechanisms on physical atoms and physical molecules at the physical level.

It all works together, both at the quantum level and the physical level, into one unified whole, which means that the effects of Quantum Mechanics or Action at a Distance can be replicated and verified through scientific experimentation at the physical level at will.

As we move further and further away from atoms, molecules, proteins, and genes, we move further and further away from the quantum level towards the physical level. We can actually create physical explanations for physical events and physical processes happening at the physical level; but, that doesn't change the fact that quantum mechanisms (or Priesthood Mechanisms) were used at the quantum level by Natures Psyche (and God's Psyche) to give the atoms and the molecules their physical functionality in the first place.

This is Quantum Neuroscience that we are talking about here – an attempt to explain what Nature's Psyche is doing for us at the quantum level in order to give us our physical functionality at the physical level.

This chapter focuses primarily on the physical mechanisms that are being observed and recorded at the physical level by our physical instruments. Remember, ALL of these physical mechanisms are facilitated and made possible by quantum mechanisms at the quantum level under the command and control of Nature's Psyche (or God's Psyche). NO physical particle and NO physical mechanism comes into existence without the command, control, and intervention of Someone Psyche at the quantum level or psyche level. That's just the way it is.

You can mess up physical functionality by messing with the physical mechanisms, such as genes and proteins.

The Effects of Toxins on the Sodium Channel. Toshio Narahashi, working at Duke University in the early 1960s, made the seminal discovery that a toxin isolated from the ovaries of the puffer fish could selectively block the sodium channel. *Tetrodotoxin (TTX)* **clogs the Na^+-permeable pore by binding tightly to a specific site on the outside of the channel. TTX blocks all sodium-dependent action potentials, and therefore is usually fatal if ingested.** [Receptor blockers can indeed kill you, just as I suspected.] **Nonetheless, puffer fish are considered a delicacy in Japan. Sushi chefs train for years, and are licensed by the government, to prepare puffer fish in such a way that eating them causes numbness around the mouth. Talk about adventurous eating!**

TTX is one of a number of natural toxins that interfere with the function of the voltage-gated sodium channel. Another channel-blocking

toxin is *saxitoxin*, produced by dinoflagellates of the genus *Gonyaulax*. Saxitoxin is concentrated in clams, mussels, and other shellfish that feed on these marine protozoa. Occasionally, the dinoflagellates bloom, causing what is known as a "red tide." Eating shellfish at these times can be fatal, because of the unusually high concentration of the toxin.

In addition to the toxins that block sodium channels, certain compounds interfere with nervous system function by causing the channels to open inappropriately. In this category is *batrachotoxin*, isolated from the skin of a species of Colombian frog. Batrachotoxin causes the channels to open at more negative potentials and to stay open much longer than usual, thus scrambling the information encoded by the action potentials. Toxins produced by lilies (*veratridine*) and buttercups (*aconitine*) have a similar mechanism of action. Sodium channel inactivation is also disrupted by toxins from scorpions and sea anemones.

What can we learn from these toxins? First, the different toxins disrupt channel function by binding to different sites on the protein. Information about toxin binding and its consequences have helped researchers deduce the three-dimensional structure of the sodium channel. Second, the toxins can be used as experimental tools to study the consequences of blocking action potentials. For example, as we shall see in later chapters, TTX is commonly used for experiments that require the blocking of impulses in a nerve or muscle. The third and most important lesson from studying toxins? Be careful what you put in your mouth! (*Neuroscience*, p. 89.)

Notice that these scientists have concluded that information is transmitted through the frequencies or the waves associated with the action potentials. Quantum Neuroscience also theorizes that thoughts and memories are transmitted through quantum frequencies or quantum waves at the quantum level, and that thoughts and memories are stored as quantum waves at the quantum level.

Shut-off the action potentials, and the physical body eventually dies, because the physical organism stops breathing.

Overload the synapses and neurons with glutamate, and strokes are the result. Strokes can also kill the neurons, the physical body, and the physical brain.

Having no action potentials can kill you; and, having too much action, too much stimulation, or too many action potentials can also kill you.

According to these people, the sodium-potassium pumps continue working at cross-purposes during an action potential. The action potential has to be significant enough to actually counteract the sodium-potassium pumps.

They deliberately start with Materialism and Naturalism, which means that they end up with Materialism and Naturalism.

We now come to the step where this information is distributed and integrated by other neurons in the central nervous system. This transfer of information from one neuron to another is called synaptic transmission, the subject of the next two chapters.

It should come as no surprise that synaptic transmission, like the action potential, depends on specialized proteins in the neuronal membrane.

Thus, a picture begins to emerge of the brain as a complicated mesh of interacting neuronal membranes. Consider that a typical neuron with all its neurites has a membrane surface area of about 250,000 µm². The surface area of the 100 billion neurons that make up the human brain comes to 25,000 m² — roughly the size of four soccer fields. This expanse of membrane, with its myriad specialized protein molecules, constitutes the fabric of our minds. (*Neuroscience*, p. 98.)

Remember, spirit matter and physical matter are basically the same thing, except they exist in a different phase or different dimension. Thanks to quantum mechanisms, it's totally possible for two or more atoms or particles to occupy the same space at the same time, as long as they are out of phase with each other. Our psyche, our spirit body, and our physical body are out of phase with each other, which means that they can occupy the same space at the same time. A spirit body can walk through a physical body unphased. Spirit matter and psyche are pure syntropy. They don't age. They are eternal and everlasting because they are pure syntropy. God takes a particle of spirit matter and infuses it with space-time or entropy in order to convert it into a particle of physical matter. God slows down that spirit particle to sub-light speeds in order to convert it into physical matter. God changes its phase or its dimension.

Remember, thoughts and memories are quantum waves being produced, transmitted, recorded, and stored at the quantum level by Someone Psyche, which explains why all of our thoughts and memories are able to survive the death of our physical brain. Our mind is something other than our brain!

The Materialists and Naturalists deliberately conflate Psyche at the quantum level with our physical brain at the physical level. It's what they do.

The authors announce in advance that they are going to spend the next couple of chapters trying to explain how our quantum thoughts and quantum memories are being distributed and stored as physical matter, physical receptors, and physical synapses within our physical brain – which is physically impossible. Mapping and storing quantum thoughts and quantum memories as quantum waves within a physical atom is totally possible at the quantum level – it happens all the time; but, it's completely impossible at the physical level with physical atoms and physical synapses and physical neurons.

Remember, the smaller can dwell within and control the larger. Have you ever heard of vibrating strings and string theory? The vibrations or quantum waves are smaller than the strings, which means that the vibrations or quantum waves can dwell within and control the strings.

The vibrations within a string are quantum waves. Thoughts and memories are quantum waves, which means that thoughts and memories are the vibrations within a string. The string is the physical component, and the vibrations are the psyche component. Psyche IS thoughts and memories, which means that Psyche is quantum waves or the vibrations within a string. The smaller can dwell within and control the larger. Atoms can communicate with each other through quantum waves or telepathy. Nature's Psyche can communicate with and control atoms at the quantum level through quantum waves or telepathy. Telepathy is WiFi at the quantum level. Strings on the physical side such as quarks are fixed by God and remain constant. Only God can change them. The strings in the middle seem to have a bit more flexibility – gluons, bosons, photons, gravitons, electrons, leptons, and dark energy – half in and half out. The strings on the psyche side are alive and infinitely variable. They may not be strings at all. In fact, they may be the different vibrations within a string – infinitely flexible and variable. These are our thoughts, memories, and Psyche.

This is the KEY to understanding Quantum Neuroscience and how the Human Psyche interacts with Nature's Psyche in order to command and control its physical brain. Go with the truth and the preponderance of the evidence, and forget about everything else. Or in my case, use the truth and the preponderance of the evidence to debunk and falsify everything else.

I'm not going to apologize for figuring this stuff out, because it's really cool if you think about it.

I just unified physics at every level. Of course, the person who will get credit for all of this is the person who figures out all the mathematics behind it; but, I have given them a target to shoot towards.

Chapter 5:

Chapter 5 of *Neuroscience* focuses on Synaptic Transmission.

This is where ALL of their theories, hopes, and ideas fall apart at the physical level, because what they are proposing in terms of information transmission through a synaptic cleft and distributed memory storage within synapses is physically impossible. It ONLY becomes possible if you are willing to bring the quantum level or the psyche level into play.

Remember, Someone Psyche has to store the memories in the cloud while the synapses are being built and re-configured; and, then Someone Psyche has to distribute those memories across different synapses, if those memories are indeed distributed across different synapses. Of course, with the introduction of Someone Psyche into the equation, information and memories no longer need to be transmitted through synaptic clefts, and memories no longer need to be stored in a distributed fashion among our physical synapses.

This is the very heart of Quantum Neuroscience – the thing that makes it valuable to us in the first place. Quantum Neuroscience has infinitely more explanatory power than Materialism, Naturalism, Darwinism, Behaviorism, and Classical Physics.

They start their discussion by mentioning that there are two types of synapses – chemical synapses and electrical synapses.

Electrical synapses and gap junctions have been one of my most favorite subjects to study in Neuroscience. There's a lot of speculation that goes on; and, I have done my share in some of the other essays in this book, so I won't do so here.

The thing that they typically don't mention is that both gap junctions and ion channels are GATES, which means that they are either OPEN or CLOSED. The ion flow from every synapse on a neuron merges into a SINGLE Postsynaptic Potential (PSP). Any memory or message that evolution will try to send through a synaptic cleft will get reduced or integrated into a single PSP. It's physically impossible to distribute our memories among different synapses at the physical level. A neuron is effectively a switch – it is either ON or OFF. That's it.

At the physical level, a neuron is a switch. You can't pass messages and memories through a switch at the physical level. You can't store messages or memories within a synapse or synaptic cleft at the physical level. A neuron is a SINGLE hardware BIT; and, it is either ON or OFF. There are NO wires connecting our neurons. Our neurons are NOT organized into functional BYTES at the physical level. There are NO wires in our brains.

You can't construct functional CPUs, RAM, and Memory Systems out of a 100 billion stand-alone switches or neurons, if you have NO wires to connect them. It's physically impossible. If there are functional CPUs, RAM, and Memory Systems within our brain, they exist at the quantum level and NOT the physical level.

This is the thing that they NEVER mention in any of their college textbooks. Either they haven't made the connection, or they don't want you to make the connection; but either way, neurons are stand-alone switches at the physical level and nothing more. If neurons have any other purpose besides being a physical switch, that purpose is given to them at the quantum level by Nature's Psyche.

Even though their commentary is restricted to the physical level only, their commentary about electrical synapses on pages 103-105 and page 108 can be fascinating to read about, especially if you have never heard about electrical synapses and gap junctions.

BCP5

https://canvas.brown.edu/courses/851434/files/40426574/

http://mypsyche.us/wp-content/uploads/2017/12/BCP5.pdf

Now they talk about the chemical synapses – starting with the one we know the most about.

> **Neuromuscular synaptic transmission is fast and reliable. An action potential in the motor axon always causes an action potential in the muscle cell it innervates. This reliability is accounted for, in part, by structural specializations of the neuromuscular junction. Its most important specialization is its size — it is one of the largest synapses in the body. The presynaptic terminal also contains a large number of active zones. In addition, the post-synaptic membrane, also called the *motor end-plate*, contains a series of shallow folds. The presynaptic active zones are precisely aligned with these junctional folds, and the postsynaptic membrane of the folds is packed with neurotransmitter receptors. This structure ensures that many neurotransmitter molecules are focally released onto a large surface of chemically sensitive membrane.**

> **Because neuromuscular junctions are more accessible to researchers than CNS synapses, much of what we know about the mechanisms of synaptic transmission was first established here. Neuromuscular junctions are also of considerable clinical significance; diseases, drugs, and poisons that interfere with this chemical synapse have direct effects on vital bodily functions.** (*Neuroscience*, pp. 110-111.)

The neuromuscular synapses are a case where God stacks the deck and loads the dice so that it can't miss.

These things are custom-designed to maximize the efficiency of the random broadcasting of neurotransmitters into the synaptic cleft.

I did calculations based upon input from other books and other articles, and I estimated that at a neuromuscular synapse up to one million neurotransmitters get dumped into the synaptic cleft following an action potential in the pre-synaptic neuron. This is a case where the postsynaptic membrane is literally painted with neurotransmitter to make sure that the message gets through. In those books, they give evolution all the credit for designing and creating our neuromuscular junctions; but of course, I choose to look at it from a different angle because it has been proven scientifically that evolution of any kind cannot do design and creation.

To me, a neuromuscular synapse is a witness of how inefficient and unreliable chemical synapses and neurotransmission really is. Think about it. The thing has to dump a million neurotransmitters into the synaptic cleft just to guarantee that your finger rises off the table. That's overkill, but it's necessary because neurotransmission is notoriously inefficient and unreliable. Neurotransmission shouldn't even work; but, miraculously somehow it does.

Not only do synaptic clefts function as randomizers and scramblers of any memories or messages that come their way; but also, a chemical synapse functions as a brake or dampener slowing everything down. If there is not enough of a post-synaptic potential (PSP) to trigger an action potential, then ALL of the input coming into the neuron will be completely eliminated and neutralized by the sodium-potassium pumps. Thanks to synaptic clefts, all of the input on a neuron gets integrated into a SINGLE post-synaptic potential, thus making a neuron nothing more than a switch at the physical level. If the input on a neuron isn't sufficient enough to generate an action potential – the ON state of a neuron – then all of that input is completely eliminated by the sodium-potassium pumps as if it never happened and never existed. It's physically impossible to pass memories, messages, and information through a neuron, because it's ALL being integrated into a single PSP or it's ALL being erased and eliminated by sodium-potassium pumps.

I eventually came to the conclusion that God designed it this way so that we would KNOW that it is physically impossible to make functional CPUs, RAM, and Memory Systems out of a bunch of stand-alone hardware BITs that have NO wires connecting them. God is deliberately pointing us to the quantum level or the psyche level for the answers that we seek. It's like God put up a big sign saying, "Your brain doesn't work at the physical level and can't work at the physical level!"

If our memories are being distributed among the synapses within our brain as the Neuroscientists and Naturalists claim, then it's happening at the quantum level, because it's physically impossible at the physical level.

Interesting, is it not?

Next, they talk about the Neurotransmitters and the principles of chemical synaptic transmission at the physical level. Obviously, they never mention or talk about the quantum level – leaving it up to someone like me to do so.

Neurotransmitter Receptors and Effectors

Neurotransmitters released into the synaptic cleft affect the postsynaptic neuron by binding to specific receptor proteins that are embedded in the postsynaptic density. The binding of neurotransmitter to the receptor is like inserting a key in a lock; this causes conformational changes in the protein, and the protein can then function differently. Although there are well over 100 different neurotransmitter receptors, they can be classified into two types: transmitter-gated ion channels and G-protein-coupled receptors. (*Neuroscience*, p. 115.)

As I have so adequately stated a hundred times, ALL of the input through the ion channels gets merged or integrated into a single Postsynaptic Potential. There's nothing going on here except integration.

This means that the G-proteins end up being the most fascinating thing to study when it comes to neurotransmission. G-proteins are also switches that turn ON or turn OFF additional functionality within a neuron. God designed the many different G-Protein systems to give Nature's Psyche the ability to re-program, re-map, and/or alter the functionality of a specific neuron at the physical level. G-proteins trigger the release of Effectors, which have an observable physical effect on the physical functioning of a specific neuron.

G-protein is where the Naturalists and Darwinists got the idea that evolution designed and programmed the brain so that the brain can reprogram itself. The thing that these people don't realize is that it's physically impossible for chemical evolution to produce a functional protein, let alone the matching gene to go with it. Since chemical evolution is physically possible, that means that chemical evolution never took place – at the physical level at least. This means that if any chemical evolution took place, it ALL took place at the quantum level or the psyche level.

These people do an excellent job describing what's happening at the physical level, thus leaving it up to you and me to figure out what's happening at the quantum level or the psyche level. Command and control – quantum mechanisms – happen at the quantum level. Choice, selection, attention, intention, and decision-making – thoughts and memories – take place at the psyche level.

It all comes down to choice!

Anytime that you encounter "choice," you have in fact caught Psyche in the act at the quantum level.

How do we know?

WE KNOW because the Materialists, Naturalists, and Behaviorists are united in their observation that "choice" is physically impossible at the physical level. The atoms have no choice but to follow and obey the physical laws (God's Laws or God's Commands). If choice is taking place, then we KNOW that it's taking place at the quantum level, because it's physically impossible for choice to happen at the physical level.

Fascinating, is it not?

You should be aware that the same neurotransmitter can have different postsynaptic actions, depending on what receptors it binds to.

Autoreceptors.

Besides being a part of the postsynaptic density, neurotransmitter receptors are also commonly found in the membrane of the presynaptic axon terminal. Presynaptic receptors that are sensitive to the neurotransmitter released by the presynaptic terminal are called *autoreceptors*. Typically, autoreceptors are G-protein-coupled receptors that stimulate second messenger formation. The consequences of activating these receptors vary, but a common effect is inhibition of neurotransmitter release and, in some cases, neurotransmitter synthesis. This allows a presynaptic terminal to regulate itself. Autoreceptors appear to function as a sort of safety valve to reduce release when the concentration of neurotransmitter in the synaptic cleft gets too high. (pp. 118-119.)

Who chooses what sub-type of receptor to place into a post-synaptic membrane? A decision or choice has to be made here. How is that choice made? How is it implemented? Who mapped or programmed these "consequences" into the system? Who designed and manufactured all of this? Design and engineering require intelligent choices.

Regulation requires intelligence and timing.

Who decides that the concentration of neurotransmitter in the synaptic cleft is too high? Who is measuring the concentration of neurotransmitter in the synaptic cleft? HOW is it measuring the concentration of neurotransmitter in the synaptic cleft? How does it KNOW these things?

We are looking at Someone Psyche in order to get all of this done in an efficient and timely manner. Neurotransmitters are molecules. The ONLY way to count them, regulate them, and manage their consequences is at the quantum level, because there are NO wires connecting them at the physical level.

Furthermore, the concentration of neurotransmitters in the synaptic cleft should NEVER get too high if they are being completely cleaned out after each action potential. There's a dichotomy here that they NEVER discuss!

Most of the time they mention that the idea that all of the neurotransmitters are cleaned out of the synapse by transporter pumps in the pre-synaptic terminal button or degradation enzymes, after each action potential. If that's really true, then there should be NO need for autoreceptors.

However, the professional articles also talk about undirected synapses and the concentration of monoamine neurotransmitters (Serotonin, Dopamine, Norepinephrine, and Epinephrine) having to be maintained at a specific level for proper and balanced brain functioning. The concentrations for monoamines can indeed get to be too high or too low within a synaptic cleft thereby requiring additional regulation in terms of drugs or medication.

Too much serotonin causes heart disease. Too little causes OCD, anxiety, and depression. Too much dopamine causes schizophrenia and psychosis. Too little dopamine causes Parkinson's Disease. The monoamines have to be properly balanced for optimal brain function.

This also leads to another severe problem or conundrum.

If monoamines are being maintained at a constant level in the synaptic cleft, who or what prevents them from randomly firing action potentials willy-nilly thereby producing a ton of signal noise or constant muscle spasticity due to ALL the false positives that would be the natural result of maintaining monoamines at a specific level within the synaptic cleft.

Someone Psyche would actually have to monitor the synaptic clefts where monoamines are being stored – making sure that no false starts happen by using repulsor beams to push the neurotransmitters away from the receptors, and also using action at a distance or quantum tunneling to teleport the monoamine neurotransmitters to the post-synaptic receptors when Nature's Psyche actually wants to trigger an action potential within that specific post-synaptic neuron. All of this is physically impossible, which means that it's happening at the quantum level.

The monoamine systems work too well and too reliably for them to be under the control of Classical Physics, Brownian motion, entropy, and random diffusion. There should be signal noise and false starts to the max; but, there isn't. Someone Psyche is intervening and preventing it from happening.

This is Quantum Neuroscience, which means that I'm going to spend some time trying to figure out what's happening at the quantum level.

Neurotransmitter Recovery and Degradation

Once the released neurotransmitter has interacted with postsynaptic receptors, it must be cleared from the synaptic cleft to allow another round of synaptic transmission.

One way this happens is by simple diffusion of the transmitter molecules away from the synapse.

For most of the amino acid and amine neurotransmitters, however, diffusion is aided by their reuptake into the presynaptic axon terminal. Reuptake occurs by the action of specific neurotransmitter transporter proteins located in the presynaptic membrane. Once inside the cytosol of the terminal, the transmitters may be enzymatically destroyed, or they may be reloaded into synaptic vesicles.

Neurotransmitter transporters also exist in the membranes of glia surrounding the synapse, which assist in the removal of neurotransmitter from the cleft.

Another way neurotransmitter action can be terminated is by enzymatic destruction in the synaptic cleft itself. This is how ACh is removed at the neuromuscular junction, for example. The enzyme acetylcholinesterase (AChE) is deposited in the cleft by the muscle cells. AChE cleaves the Ach molecule, rendering it inactive at the ACh receptors.

The importance of transmitter removal from the cleft should not be underestimated. At the neuromuscular junction, for example, uninterrupted exposure to high concentrations of ACh after several seconds leads to a process called desensitization, in which, despite the continued presence of ACh, the transmitter-gated channels close. This desensitized state can persist for many seconds even after the neurotransmitter is removed. The rapid destruction of ACh by AChE normally prevents desensitization from occurring. However, if the AChE is inhibited, as it is by various nerve gases used as chemical weapons, the ACh receptors will become desensitized and neuromuscular transmission will fail . . . thereby paralyzing the respiratory muscles of its victims.

The immense chemical complexity of synaptic transmission makes it especially susceptible to the medical corollary of Murphy's Law, which states that if a physiological process can go wrong, it will go wrong. When chemical synaptic transmission goes wrong, the nervous system malfunctions. Defective neurotransmission is believed to be the root cause of a large number of neurological and psychiatric disorders. (*Neuroscience*, 119-122.)

These people can't sense it because they have trained themselves not to look for it; but, I sense a huge amount of Quantum Mechanics taking place during neurotransmission. It's as if the whole process has been oiled and finely-tuned at the quantum level. There is also a ton of design and fine-tuning that has been done at the physical level. Evolution can't do design and fine-tuning, which means that Someone Psyche had to design and create our neurotransmission systems for us, or they wouldn't exist.

The diffusion model stinks in my humble opinion. Why doesn't this diffusion trigger false starts and signal noise? Random diffusion is not reliable. It should produce a ton of signal noise. I guess if the receptors are few in number, then the sodium-potassium pumps can negate the signal noise that random diffusion would constantly produce.

Now, let's discuss the Active Processes for the termination of neurotransmission.

How does the "neurotransmitter recovery and degradation" system KNOW that the neurotransmitter has interacted with the postsynaptic receptors, and that it is now time to clear the synaptic cleft? It has to wait for interaction or neurotransmission, before clearing the synaptic cleft. Timing is critical!

They actually say here (and I've never seen it said anywhere else) that the post-synaptic muscle cells actually deposit the AChE degradation enzyme into the synaptic cleft after successful neurotransmission; but, how are the presynaptic Reuptake Transporters told that the post-synaptic neuron has received the transmission and that it's now time to turn on and remove the neurotransmitters from the synaptic cleft? Some kind of Quantum Telepathy or Quantum Sensors have to be transmitting this information to the pre-synaptic neuron at the quantum level.

There's also another massive problem that comes into play where the "degradation enzyme" route is concerned, that they never seem to talk about. How does the AChE degradation enzyme get removed from the synaptic cleft, so that it doesn't interfere with and stop the next action potential to come along?

We are talking about molecules here. The ONLY way to interact with, communicate with, and control molecules is at the quantum level through some kind of quantum wave, quantum force, or quantum field. There are NO wires connecting these molecules within the synaptic cleft; and, even if there were, there would still have to be Someone Psyche there in the synaptic cleft generating and interpreting the messages in the first place.

You simply cannot escape the quantum level, if you want to know what's really happening within your physical brain.

Whether we are talking about left-over neurotransmitters or degradation enzymes that have now done their job, these thing have to be removed from the synaptic cleft at the quantum level in preparation for the next action potential.

Furthermore, if the ACh neurotransmitters stay in the synaptic cleft too long, then desensitization occurs and you stop breathing. If there are too many glutamate neurotransmitters in the synaptic cleft, the post-synaptic neurons get destroyed and a stroke is the result. When it comes to neurotransmission, you don't want to mess with this thing, because it can kill you if you do.

My favorite section in this book is the section entitled "Principles of Synaptic Integration" on pages 122 through 130. Once I finally realized how EPSP and IPSP and the Post-Synaptic Potentials really work, suddenly I KNEW that it's physically impossible to store memories within a synapse or a neuron at the physical level. I have covered this extensively in other essays, so I won't do so here. Instead, I'll just type up some of the notes that I wrote in the margins of my book.

Thousands of synaptic inputs REDUCE or INTEGRATE to a single neural output called an Action Potential. This represents thousands of BITS of input reduced to one BIT of output.

At default, the neuron is OFF. All of the thousands of synapses on the neuron produce a single summation Excitatory Postsynaptic Potential (EPSP) that moves towards threshold and the triggering of a single Action Potential output. Once a neuron turns ON, then immediately it turns OFF again.

This reality and truth makes it physically impossible to distribute memories among synapses at the physical level, and makes it physically impossible to transmit memories through neurons at the physical level. This reality and truth makes it physically impossible to store our memories within our physical brain.

This reality and truth also makes it physically impossible to make or construct CPUs, RAM, and Memory Systems within our physical brain. You can't make CPUs and RAM out of a bunch of stand-alone switches with NO wires connecting them. It's physically impossible. If there are any CPUs, RAM, and Memory Systems within our brain, they exist at the quantum level because it's physically impossible for them to exist at the physical level.

Once an Action Potential chain has started, Nature's Psyche has a way to turn things OFF quickly, through inhibitory neurons and IPSPs.

The evidence within this chapter, when properly understood, completely negates their concluding remarks.

> **We will describe more examples of synaptic modulation and their mechanisms. However, you can already see that modulatory forms of synaptic transmission offer an almost limitless number of ways that information encoded by presynaptic impulse activity can be transformed and used by the postsynaptic neuron.**
>
> **You'll see that the chemistry of synaptic transmission warrants all this attention because defective neurotransmission is the basis for many neurological and psychiatric disorders. And virtually all psychoactive drugs, both therapeutic and illicit, exert their effects at chemical synapses.**
>
> **In addition to providing explanations for aspects of neural information processing and the effects of drugs, knowledge of chemical synaptic transmission is also the key to understanding the neural basis of learning and memory. Memories of past experiences are established by modification of the effectiveness of chemical synapses in the brain. Material discussed in this chapter suggests possible sites of modification, ranging from changes in presynaptic Ca^{2+} entry and neurotransmitter release to alterations in postsynaptic receptors or excitability. As we shall see in Chapter 25, all of these changes are likely to contribute to the storage of information by the nervous system.** (*Neuroscience*, pp. 130-131).

Notice that when it all goes wrong, it goes wrong at the physical level, and NOT the quantum level. God permits things to go wrong at the physical level, so that we can learn and grow from the experience. It's painful at times, but pain is our greatest motivator and our greatest teacher.

If you were actually paying attention in this chapter, we've already seen that it's physically impossible to store memories and information within a synapse and neuron. They talk about encoding information and memories within synaptic modulations; but, that's physically impossible at the physical level. Where are the new memories being stored, while the synapses are being modulated? They NEVER even ask that question.

Thoughts and memories are quantum waves.

How do we know?

WE KNOW because our thoughts, memories, and personality survive the death of our physical brain, according to the tons of empirical evidence from Near-Death Experiences, Out-of-Body Experiences, Shared-Death Experiences, and after-death Life Reviews.

Furthermore, modulated synapses have absolutely NO meaning unless Nature's Psyche is giving it meaning! Even modulated and encoded information has NO meaning unless Someone Psyche is giving it meaning. Distributing memories or information among synapses is physically impossible. If our memories are indeed distributed among our synapses, it's being done at the quantum level by Nature's Psyche, because ONLY Nature's Psyche can KNOW which synapses contain the memories, where the message or information starts and ends, and what that message actually means, because Nature's Psyche is the ONLY one who can read, transmit, and store quantum waves at will.

The authors talk about "modification". God designed and programmed our neurons to be able to adjust themselves or re-map themselves at the physical level, just like an artificial intelligence. Programming of any kind, even self-adjusting programming, required Someone Psyche in order to bring it into existence in the first place.

Chapter 6:

Chapter 6 of *Neuroscience* focuses on Neurotransmitter Systems.

What motivates the Neurotransmitters and Neuropeptides to move towards the post-synaptic receptors rather than diffusing evenly within the fluid of the synaptic cleft? In animations, the Neurotransmitters and Neuropeptides are always shown as going straight to their target. Not only do they have to move towards their target, but they also have to rotate into the correct orientation so that they can dock with the post-synaptic receptors. Study the logistics of these things. It's as if they are intelligent and alive.

The enzyme's activity is regulated by various signals in the cytosol of the axon terminal. (*Neuroscience*, p. 145.)

Everything within the cytosol, including the "signals", eventually reduces to atoms and molecules; yet, these atoms and molecules continue to find ways to communicate with each other. That's physically impossible, which means that these "signals" eventually reduce to quantum waves at the quantum level. Think about it logically. The ONLY way atoms can communicate with each other and interact with each other is through quantum waves at the quantum level.

There are forces, fields, or quantum waves coming off these various molecules; and, that's how they communicate with each other and interact with each other successfully. Who is making these forces, fields, quantum waves, and signals at the quantum level?

How do these enzymes KNOW what they are supposed to do and when they are supposed to do it? It says above that they are regulated. Regulated by whom? Regulation requires Intelligence and Someone Psyche. Through these "signals" or quantum waves, the enzymes are actually informed as to whether there is too much neurotransmitter in the system, or not enough. The enzymes actually shut down their activity if there is too much neurotransmitter; and, the enzymes actually increase their activity if there's not enough. It's as if these enzymes are intelligent and alive; but, that's physically impossible! Remember, Someone Psyche can do the physically impossible through quantum mechanisms.

Whether we realize it or not, the very existence of physical atoms, physical molecules, neurotransmitters, neuropeptides, proteins, genes, and genomes IS Scientific Proof of God's Existence, because ALL of these things would not exist if God did not exist. Your genome is God's Signature, and your physical body is God's handiwork. They wouldn't exist without God.

How do we know?

WE KNOW because scientists have repeatedly observed and repeatedly proven that it's physically impossible for chemical evolution, random mutations, and natural selection to design and create functional proteins and the matching genes to go along with them from scratch. It can't be done, which means that it wasn't done. In fact, evolution (genetic drift), random mutations, and natural selection didn't even exist until after God designed and created the physical atoms, the proteins, and the genes in the first place.

Neurotransmission – this is where ALL of their theories, hopes, and ideas fall apart at the physical level, because what they are proposing in terms of information transmission through a synaptic cleft and distributed memory storage within synapses is physically impossible. It ONLY becomes possible if you are willing to bring the quantum level or the psyche level into play.

Remember, Someone Psyche has to store the memories in the cloud while the synapses are being built and re-configured; and, then Someone Psyche has to distribute those memories across different synapses, if those memories are indeed distributed across different synapses. Of course, with the introduction of Someone Psyche into the equation, information and memories no longer need to be transmitted through synaptic clefts, and memories no longer need to be stored in a distributed fashion among our physical synapses.

This is the very heart of Quantum Neuroscience – the thing that makes it valuable to us in the first place. Quantum Neuroscience has infinitely more explanatory power than Materialism, Naturalism, Darwinism, Behaviorism, and Classical Physics.

It all comes down to choice!

Anytime that you encounter "choice," you have in fact caught Psyche in the act at the quantum level.

How do we know?

WE KNOW because the Materialists, Naturalists, and Behaviorists are united in their observation that "choice" is physically impossible at the physical level. The atoms have no choice but to follow and obey the physical laws (God's Laws or God's Commands). If choice is taking place, then we KNOW that it's taking place at the quantum level, because it's physically impossible for choice to happen at the physical level.

Materialism, Naturalism, Darwinism, Behaviorism, and even Atheism are self-defeating. Atheism is creation from nothing by nothing, which is impossible any way that we choose to look at it. Creation Ex Nihilo is Atheism, which is impossible. Evolution can't design and create physical atoms, molecules, neurotransmitters, neuropeptides, proteins, and genes, because evolution by definition in principle can't do anything at the quantum level or the psyche level. Choice is physically impossible according to the Materialists, Naturalists, Behaviorists, and Atheists, which means that ALL of our choices are taking place at the quantum level. Choice is a function of Psyche, and according to these people, Psyche is physically impossible. If Psyche exists, it only exists at the quantum level or the psyche level.

See what you can derive when you choose to take these people seriously and then choose to use ALL of the evidence that we have on hand as a race to falsify their ideas! The very existence of Quantum Mechanics and quantum mechanisms at the quantum level or the psyche level FALSIFIES Materialism, Naturalism, and their derivatives.

The thing you need to know most about neurotransmitters and neuropeptides is that it's physically impossible for random mutations and chemical evolution to design, engineer, create, manufacture, and then deploy most of them if not all of them.

I've discussed neurotransmitters extensively in other essays within this book, so I won't do so here. Instead, let's return to the basics of Quantum Neuroscience. The basics have teeth!

The *ORIGIN* video from Illustra Media does some mathematical modeling demonstrating that the odds of a functional protein developing by chance are 1 chance in 10^{164}; and, that's with God taking the time to fill an ocean completely full of the right 20 amino acids with the right chirality, and with God forcing 150 of those amino acid sequences to line up every second so that they can see if they will fold up into the right usable protein.

Technically, there's no logical reason for that ocean full of the right 20 amino acids to deliberately line up 150 amino acids in sequence every second in order to try out all of the 10^{164} possible combinations in a timely fashion – naturally, or through undirected processes. God is going to have to provide the ocean full of amino acids in the first place, sustain that ocean full of amino acids in existence for trillions of trillions of trillions of trillions of years, and force 150 of those amino acids to line up in sequence every second throughout those trillions of trillions of years JUST TO PRODUCE a single protein "naturally". The odds of that whole process happening by random chance, without God's intervention, is ZERO. There is NO CHANCE!

https://www.youtube.com/watch?v=W1_KEVaCyaA

https://www.youtube.com/watch?v=cQoQgTqj3pU

https://www.youtube.com/watch?v=X7VqomfivC0

The odds of a functional cell developing by natural processes are 1 chance in $10^{340,000,000}$.

https://www.youtube.com/watch?v=nuMvRExazAw

The current estimated age of this universe is 13.8^9, or 13.8 billion years. In comparison to 13.8 billion years, something like $10^{340,000,000}$ is synonymous with IMPOSSIBLE or INFINITY, because there is NO CHANCE that any living cell could ever develop by chance through natural processes in 13.8 billion years of time. In fact, there is technically NO CHANCE that a single functional protein could develop by chance in 13.8 billion years, even if we were to start with a God-given ocean full of the right 20 amino acids with the right chirality, no destructive oxygen, and no sunshine to destroy that particle of meat once it comes together.

Now let's say that, thanks to God's intervention and help, chemical evolution has managed to create a single functional protein "naturally" after trillions of trillions of years of time, NOW chemical evolution has to create the matching gene to go along with that protein; but, the mechanisms for doing so do NOT exist on Amino Acid World. That protein will eventually disintegrate, and we'll have to start all over again with NO gene in place to successfully replicate that protein. Chemical evolution of proteins along with their matching genes is physically impossible. It can't be done, which means it wasn't done. God has an infinitely better way of making functional proteins along with their matching genes – it's called Quantum Mechanics or the Priesthood Power of God.

There's NO WAY in the multi-verse that evolution of any kind can produce a functional protein, let alone the matching gene to go with it. It's physically impossible. It can't be done.

The Theory of Evolution, Materialism, Naturalism, Darwinism, Behaviorism, and Atheism ARE DEAD. Long live Quantum Mechanics, Quantum Neuroscience, and the Priesthood Power of God!

Chapter 7:

Chapter 7 of _Neuroscience_ focuses on the Structure of the Nervous System.

In this chapter, they finally start discussing our physical anatomy.

The thing to remember about physical anatomy is that every machine in the physical body was designed at the quantum level by God's Psyche, and is now implemented, mediated, or brought into existence by Nature's Psyche. Physical machinery doesn't just spontaneously generate out of thin air. That's physically impossible. Remember, physical machinery and physical particles are brought into existence at the quantum level by God's Psyche, and then these physical machines are commanded, controlled, and coordinated by Nature's Psyche completely outside the conscious awareness of the Human Psyche.

Nature's Psyche is infinitely more intelligent and has access to infinitely more knowledge and information than the Human Psyche does. We aren't consciously aware of what Nature's Psyche is doing. Nature's Psyche is operating on a different frequency than the Human Psyche. The only thing the Human Psyche can do is to choose and decide what it wants to do with the physical body that God has given it.

Remember, if your Interpretation of Quantum Mechanics can't explain what Nature's Psyche is doing at the quantum level and how Nature's Psyche does it, then your Interpretation of Quantum Mechanics is worthless.

Quantum Neuroscience starts by identifying the physically impossible, and then it continues by trying to find a scientific explanation or a quantum mechanical explanation for the physically impossible. By using quantum waves and quantum mechanisms, Nature's Psyche can do the physically impossible.

The interesting thing to remember about physical atoms is that their origin is physically impossible. It's physically impossible for something non-physical to design and create something physical. So, where did the first physical atoms come from? Every physical atom had a beginning, which means that they had a Beginner. The origin of the very first physical atoms actually FALSIFIES Materialism and Naturalism. Think about it!

God had to use His intelligence or psyche to design and originate the first physical atoms; and then, God's Psyche had to use quantum waves to bring those first physical atoms into existence. But, that's physically impossible!

Neuroplasticity – Wiring the Brain

The dendrites in our brain transmit post-synaptic potentials. The axons in our brain transmit action potentials. In this case, a "potential" is an electrical current comprised of ions, or electrically charged atoms. The Materialists and Naturalists call dendrites and axons "wires". They do so in order to give you the erroneous impression that your whole brain is wired together into a single functioning whole like a CPU or RAM, which technically isn't true at all. To counteract this deception, I will occasionally make the statement that there are NO wires in your brain. They, of course, will disagree with this statement and point you directly to dendrites and axons in order to counter that statement. However, thanks to the synaptic clefts and synapses that separate every neuron and astrocyte, each neuron is functionally and electrically separated from every other neuron and astrocyte by synapses, which effectively means that there are NO wires in your brain. In other words, your brain is NOT wired together into a single CPU or RAM like the Materialists and Naturalists claim that it is. Technically, your brain is NOT comprised of circuits like these people claim that it is. A wire is a single, usually cylindrical, flexible strand or rod of metal. Wires are used to bear mechanical loads or electricity and telecommunications signals. There's NOTHING like this within a physical brain. There are NO wires in your brain.

Now that you know where I have come from and where I am going with Quantum Neuroscience, let's run some Neuroscience through the process and see what we end up with.

I will use *Neuroscience: Exploring the Brain* as my text, because it's freely available online if you know where to look for it; and, it's relatively inexpensive to get a hardcopy of the book if you want one.

Bear, M. F., Connors, B. W., & Paradiso, M. A. (2007). *Neuroscience: Exploring the Brain* (3rd ed.). New York: Lippincott Williams and Wilkins.

This essay will focus exclusively on chapter 23 of their book. Let's see what we can do with it.

BCP23

http://mypsyche.us/wp-content/uploads/2017/12/BCP23.pdf

Classical Physics, Materialism, and Naturalism are at a serious disadvantage compared to the explanatory power of Quantum Mechanics. When it comes to explanatory power, Quantum Mechanics is vastly and noticeably superior to anything that the Materialists and Naturalists can provide.

There's nothing wrong with finding a physical explanation for brain functionality, because a physical explanation of long-term memory storage within your physical brain can't and doesn't eliminate your after-death Life Review, which will take place after your brain is dead and gone. But, I personally didn't find the physical explanations for long-term memory storage within the physical brain convincing, after examining all the evidence that I could get my hands on.

In contrast, finding a demonstrable, observed, and proven mechanism for transferring brain memories, brain data, and brain processes into non-local, spiritual, cloud storage – such as the Quantum Zeno Effect and any type of telepathic or telekinetic Action at a Distance – is in fact extremely dangerous, harmful, and wrong for the Materialists and Naturalists, because the observe and proven existence of Quantum Tunneling, Action at a

Distance or Quantum Nonlocality, and the Quantum Zeno Effect FALSIFIES Materialism and Naturalism. They won't stand for it!

You want to find as many physical explanations for brain functionality and memory storage within the brain, because when you finally run out of physical explanations, then that's when things really start to get interesting!

Chapter 23 of their *Neuroscience* book is about the physical wiring within our brain. Wiring in our brain is a *misnomer* and *category error* logic fallacy, because there are NO wires in our brains connecting our neurons together. These people are literally trying to trick you into believing that all of the neurons, or switches, or transistors within our brains are wired together into a functional network, CPUs, and RAM. The whole thing is based upon a *category error* logic fallacy and a whole lot of trickery and deception. It's magic, sleight of hand, bait and switch. It's cheating, but that's the way the Materialists, Naturalists, Darwinists, and Atheists do science, so it's okay. It's real and it's true because they say so – these people have PhDs after all, so they know how things really work.

Obviously, I didn't fall for their deception, because I too know better.

Now, it's up to you to decide which one of us knows BESTEST.

I'm very impressed by the explanatory power of Quantum Neuroscience. Let's make some comparisons, so that you can see what I mean.

Chapter 23:

Chapter 23 of *Neuroscience* focuses on "Wiring the Physical Brain".

The title of this chapter is a misnomer and oxymoron.

There are NO wires within the physical brain; so, it's physically impossible to get our brains wired together into functional CPUs and RAM.

If there is any wiring within our physical brain, it ONLY exists at the quantum level outside our conscious awareness.

These Materialists and Naturalists call it "wiring" in order to trick you and deceive you into believing that it's some kind of physical process, when it is not.

Still, the wiring metaphor is a useful analogy that I have often used in order to try to make sense of the physical brain and how it works. We have to run with what we have got and with what we have been given. We need some way to develop a common frame of reference so that we can discuss it. Consequently, I'm going to go with it and see what I can make of it.

We have seen that most of the operations of the brain depend on remarkably precise interconnections among its 100 billion neurons.
(*Neuroscience*, p. 690.)

They ask how such precise wiring can arise.

Evolution did it. There's no other explanation given to us.

But, it's physically impossible for evolution to reach into our brains and wire our brains, especially at the quantum level where our brains are wired, if they are wired.

Precision requires come kind of intelligence or psyche. And technically, the neurons in our brains are NOT interconnected, certainly NOT by wires. Everything ends at the synaptic cleft; and, a synaptic cleft scrambles every message and every memory that comes its way – at the physical level at least. Who knows what's happening at the quantum level? We only know what's happening at the physical level; and, at the physical level each one of our neurons is singing, "Ain't got no strings on me!"

They mention genetic programs that allow growing axons to detect correct pathways and select correct targets.

> **Here we take another look at brain development, this time to see how connections are formed and modified as the brain matures. We will discover that most of the wiring in the brain is specified by genetic programs that allow axons to detect the correct pathways and the correct targets. However, a small but important component of the final wiring depends on sensory information about the world around us during early childhood. In this way, nurture and nature both contribute to the final structure and function of the nervous system.** (p. 691.)

Detection implies action at a distance.

They are talking about genetic programs that didn't spontaneously generate out of thin air, because it's physically impossible for them to do so. Who wrote the genetic programs in the first place? Programs require Programmers. Who designed these detectors? Detectors require Designers. At the molecular and atomic level, detectors also require some kind of Telepathy, Intelligence, or Action at a Distance. Who decided and chose the correct pathway and the correct target? Targeting of any kind requires Psyche or Intelligence. So does choice! Choice is the ultimate sign of Psyche, because Psyche is the ultimate causal agent.

They talk about cellular ballet and the choreography of cell proliferation.

Ballet and choreography require a Choreographer.

There's signs of Psyche or Intelligence ALL over Cell Proliferation.

> **The fate of the newly formed daughter cells depends on a number of factors. Curiously, a ventricular zone precursor cell that is cleaved vertically during division has a different fate than one that is cleaved horizontally. After vertical cleavage, both daughter cells remain in the ventricular zone to divide again and again. This mode of cell division predominates early in development to expand the population of neuronal precursors.**

> **Later in development, horizontal cleavage is the rule. In this case, the daughter cell lying farthest away from the ventricular surface migrates away to take up its position in the cortex, where it will never divide again. The other daughter remains in the ventricular zone to undergo more divisions. Ventricular zone precursor cells repeat this pattern until all the neurons and glia of the cortex have been generated.** (p. 691.)

Why?

What's their motivation?

Who is overseeing and controlling this whole process?

Who decides when it is time to switch from vertical cleavage to horizontal cleavage?

Who decides whether the daughter cell is going to be a neuron or a glial cell?

Who decides where to send the daughter cells into the cortex? They can go anywhere they want, theoretically.

The information, foreknowledge, planning, teleology, logistics, mapping, signal-processing, and intelligence is off the scale. It really can't be measured accurately, because it can't be detected at the physical level. We are still very much at the quantum level or the psyche level here, when it comes to Cell Proliferation, Cell Migration, and Cell Differentiation!

It gets a lot worse, before it gets better.

In humans, the vast majority of neocortical neurons are born between the fifth week and the fifth month of gestation (pregnancy), peaking at the astonishing rate of 250,000 new neurons per minute. Recent findings suggest that although most of the action is over well before birth, the adult ventricular zone retains some capacity to generate new neurons.

However, it is important to realize that once a daughter cell commits to a neuronal fate, it will never divide again. How does the cleavage plane during cell division determine the cell's fate? Remember that all of our cells contain the same complement of DNA we inherited from our parents, so every daughter cell has the same genes.

The factor that makes one cell different from another is the specific genes that generate mRNA and, ultimately, protein. Thus, cell fate is regulated by differences in gene expression during development. Gene expression is regulated by cellular proteins called transcription factors. If transcription factors, or the "upstream" molecules that regulate them, are unevenly distributed within a cell, then the cleavage plane can determine which factors are passed on to the daughter cells. For example, proteins called notch-1 and numb migrate to different poles of ventricular zone precursor cells. When the neuron divides vertically, the notch-1 and numb proteins are partitioned symmetrically. However, when the cell divides horizontally, notch-1 goes with the daughter that will migrate away, while numb remains with the cell that will divide again. Research suggests that notch-1, "unopposed" by numb, activates the gene expression that causes the cell to cease dividing and migrate away from the ventricular zone.

Mature cortical cells can be classified as glia or neurons, and the neurons can be further classified according to the layer in which they reside, their dendritic morphology, and the neurotransmitter they use. Conceivably, this diversity could arise from different types of precursor cell in the ventricular zone. In other words, there could be one class of precursor cell that gives rise only to layer VI pyramidal cells, another that gives rise to layer V cells, and so on. However, this is not the case. Multiple cell types, including neurons and glia, can arise from the same precursor cell. Because of this potential to give rise to many different types of tissue, these precursor cells are also called neural stem cells.

The ultimate fate of the migrating daughter cell is determined by a combination of factors, including the age of the precursor cell, its position within the ventricular zone, and its environment at the time of division.

Cortical pyramidal neurons and astrocytes derive from the dorsal ventricular zone, whereas inhibitory interneurons and oligodendroglia derive from the ventral telencephalon. The first cells to migrate away from the dorsal ventricular zone are destined to reside in a layer called the subplate, which eventually disappears as development proceeds. The next cells to divide become layer VI neurons, followed by the neurons of layers V, IV, III, and II.

(*Neuroscience*, pp. 691-693.)

This is insane! Especially when you start to consider what they are not saying!

These daughter cells are making commitments.

Why?

What's their motivation?

My daughters weren't willing to make commitments, at least not with me.

Who is choosing the specific genes that generate the mRNA and ultimately the proteins that will be produced in a specific cell? In other words, who is choosing the cell fate which is regulated by differences in gene expression during development?

During each aspect of Cell Differentiation, someone is making this choice! Choice is a function and product of Psyche. So are commitments.

Proteins migrate, transcription factors migrate, and cells migrate.

Who's telling them where to go?

Oh, I can tell them where to go, but that doesn't mean that they will actually go there.

Who designed all of these different genes and proteins in the first place?

Functionality, purpose, design, creation, and teleology are a product of Psyche; and, Psyche does its work through quantum waves at the quantum level. Evolution, random mutations, and natural selection have NO access to the quantum level; consequently, the different types of evolution can't do design and creation – command and control.

Scientists have repeatedly proven that it's physically impossible for chemical evolution to produce functional proteins, let alone the matching genes to go along with them.

Again, who decides when it's time to switch from vertical cleavage to horizontal cleavage? I like cleavage as much as the next guy (it's supposedly genetically programmed into me); but, somebody has to decide how it's supposed to be done to make it look right and work right.

Who taught the upstream molecules how to regulate transcription factors? They're molecules! Where's their brain? They ain't got one!

Notice how these people put all these different things into a black box and leave them unexplained. These processes can't be explained in physical terms because their choices and actions are taking place at the quantum level – hence the black boxes. Anytime you see the Materialists, Naturalists, and Darwinists using black boxes to explain a physical process or the chosen actions of atoms and molecules, these people are in fact talking about quantum mechanisms or non-physical mechanisms that aren't supposed to exist according to Materialism and Naturalism.

"Multiple cell types, including neurons and glia, can arise from the same precursor cell. Because of this potential to give rise to many different types of tissue, these precursor cells are also called neural stem cells."

When stem cells divide, they divide into a daughter cell and another stem cell. Why? Who's telling them to do this or making them do this? How does a stem cell decide which type of daughter cell to make?

Then, within the brain, daughter cells migrate to pre-chosen destinations. Upon reaching their destination, they differentiate again into a specific type of neural cell or glial cell. Choice, targeting, and differentiation are products of Intelligence or Psyche.

Who is telling these daughter cells what type of daughters they should become? I couldn't tell my daughters what to do; so, who's getting away with it at the quantum level within our physical brain? Daughters of these stem cells can become anything they want; and, some of them are choosing to become neural stem cells. Then these neural stem cells can become any type of neuron or glial cell that they want to become; so, who is helping them to make the choice?

Well, it can't be evolution, because evolution can't do telepathy and choice, because telepathy and choice are physically impossible according to the Materialists, Naturalists, Behaviorists, and Determinists.

They talk about fate.

When it comes to daughter cells and migrating cells, who is choosing their eventual fate? Who is doing all of this cell differentiation?

Choice is a function of Psyche, because according to the Materialists and Naturalists and Behaviorists, choice is physically impossible! That's the whole message of Materialism, Naturalism, and Behaviorism – that choice of any kind is physically impossible. Choice is the product of Psyche, and Psyche can do the physically impossible at will.

At some point in this whole process, we find ourselves staring into the face of Nature's Psyche and God's Psyche! God's Signature is written all over this thing, and we've just begun.

They haven't even mentioned the wires yet!

You remember the wires, don't you, the wires that don't actually exist?

The sources of cortical cells. Proliferation of cortical pyramidal neurons and astrocytes occurs in the ventricular zone of the dorsal telencephalon. However, inhibitory interneurons and oligodendroglia are generated in the ventricular zone of the ventral telencephalon; consequently, these cells must migrate laterally over some distance to arrive at their final destination in the cortex.

Cell Migration

Many daughter cells migrate by slithering along thin fibers that radiate from the ventricular zone toward the pia. These fibers are derived from specialized radial glial cells, providing the scaffold on which the cortex is built. The immature neurons, called neuroblasts, follow this radial path from the ventricular zone toward the surface of the brain.

Recent studies indicate that some neurons actually derive from radial glia. In this case, migration occurs by the radial movement of the soma within the fiber that connects the ventricular zone and pia. When cortical assembly is complete, the radial glia withdraw their radial processes. Not all migrating cells follow the path provided by the radial glial cells, however. About one-third of the neuroblasts wander horizontally on their way to the cortex. (pp. 694-695.)

I sort of lost it when I first learned that glial cells and neurons migrate. It was simply too much for a former Materialist and Atheist to handle.

I'm a computer scientist. I understand instinctively the amount of information storage capacity needed to store terabytes of command and control computer code, petabytes of location addressing for individual neurons and glial cells, a petabyte of differentiation information, and then petabytes of memories or RAM.

The amount of information needed to get a 100 billion neurons and a 100 billion glial cells to migrate to the correct location in the brain (I've seen estimates as low as 85 billion neurons and 85 billion glial cells), and then differentiate into the correct type of cell when they get there, and then to correctly form the proper synapses with the correct neurotransmitter type(s) is on the order of petabytes – even with the help of radial glial cells. Remember, these migrating cells actually differentiate into structurally complete and fully functional Cerebral Processing Units (CPUs), such as the thalamus, hypothalamus, hippocampus, cerebellum, amygdala, striatum, and dozens of different cortices.

How do they do that?

It requires petabytes of information!

Where do they get the information and knowledge from?

A migrating daughter cell in the brain can become any type of neuron or glial cell that it wants to be, so who decides what type of cell it's going to be when it reaches its pre-chosen destination; and, how is that message delivered? And, who tells this cell when it has reached its destination; and, who tells this cell how many synapses to make and where to make them?

We are talking about petabytes of information here. Potentially up to an exabyte of RAM would be needed to store all the 3D blueprints, location addressing, differentiation typing, molecular command, molecular control, and all the ongoing maintenance information for the development and duration of a physical body. And, it is estimated that another exabyte would be needed to store all of our memories, especially if Nature's Psyche is keeping a video record of our memories non-locally in the cloud for us, as seems to be the case considering what we have learned from after-death Life Reviews.

Think of it! Potentially an exabyte of "RAM" non-locally in the cloud dedicated to your physical body, and another exabyte in the cloud dedicated to your memories.

It's physically impossible to store that amount of information within a physical genome! It's physically impossible to store that amount of information in a physical brain! In fact, it's physically impossible to store petabytes of RAM in evolution. Evolution doesn't exist as a physical entity, so you can't store any programming, data, or memories within evolution. It's physically impossible. In fact, you can't even store a byte of information, or a bit of information, in evolution.

Remember it's physically impossible to store petabytes of command and control and differentiation information within our 750 megabyte genome. So, where is all of this information being stored, and how is it being accessed?

The only thing I could ever come up with that made logical sense to me is that ALL of this exabytes of information is being stored as quantum waves holographically in God's Database in the non-local realm; and then, that information is being accessed as needed by Nature's Psyche, who collapses the wave function thereby deciding what type of cell a daughter cell will eventually become, and where it will become that type of neural cell or glial cell at the end of its migration.

There's no other logical explanation for what we are observing, because we are observing the effects and the results of the physically impossible, whenever it comes to Cell Proliferation, Cell Migration, Cell Differentiation, Neurodevelopment, and the Physical Development and Physical Maintenance of the Physical Body.

The attempt to explain the physically impossible in terms of quantum mechanisms is Quantum Neuroscience in action. In my humble opinion, science doesn't get any more interesting than this.

When you finally identify the physically impossible, then that's when science really starts to get interesting.

They talk about cell migration and the inside-out pattern of cortical development.

This is fascinating material, and was a total surprise when it was first presented to me. Neural cells and glial cells are born at the center around the neural tube, then each one of them migrates up through the current layers of the developing cortex to the top layer of the developing cortex. ALL of the neurons and glial cells travel or migrate to their chosen destination.

The inside-out pattern of cortical development involves orderly waves of tangential and radial migrations, progressing systematically from the deepest layer to ever more progressive higher layers of cortical development. The cells that are born at the end of the process have to travel the furthest to get to their pre-selected destinations.

Inside-out development of the cortex.

The first cells to migrate to the cortical plate are those that form the subplate. As these differentiate into neurons, the neuroblasts destined to become layer VI cells migrate past and collect in the cortical plate. This process repeats again and again until all layers of the cortex have differentiated. The subplate neurons then disappear.

Not all migrating cells follow the path provided by the radial glial cells, however. About one-third of the neuroblasts wander horizontally on their way to the cortex.

This orderly process can be disrupted by a number of gene mutations. (Neuroscience, p. 695.)

Who is choosing their destination and their eventual destiny? Where did the neurons of the subplate get their instructions and assignment telling them how to differentiate from the neurons of the cortical plate? Who controls the construction of a thalamus? The neurons and glial cells are migrating one at a time. How do they coordinate with each other when they reach their destination, because they don't have any synapses yet? Who teaches them or tells them where to build their dendrites, axons, and synapses after they have reached their destination so that they actually form into the types of neurons and glial cells that produce a functioning thalamus or a functioning hippocampus?

Someone had to plan and design all of this, because evolution and random mutations simply mess it up. Nature's Psyche performs flawlessly every time. The ONLY thing that can mess up neurodevelopment is the faulty genes and faulty proteins that evolution or random mutations have given us. This orderly process can be severely messed up by a number of different gene mutations!

Do you think that it requires a bit of intelligence and motivation in order to get this neurodevelopment process to work correctly every time? If I were to assign you the task of migrating newly born neurons and glial cells to destinations of your choosing and assign you the task of organizing them into a functional cerebellum, hypothalamus, amygdala, thalamus, striatum, hippocampus, and cortices when these cells get to the destination you have chosen for them, would you be able to fulfill this task?

Why not?

Evolution can do it, so why can't you?

God is trying to make it obvious and clear that evolution could never and would never design such a thing as migrating glial cells and migrating neurons! It's orderly and systematic, rather than random. Glial cells migrate tangentially. They launch out into the ocean and find their destination all on their own. Ain't got no strings on glial cells.

As we have seen, most cortical neurons are born in the ventricular zone and then migrate along radial glia to take up their final position in one of the cortical layers. Thus, it seems reasonable to conclude that cortical areas in the adult brain simply reflect an organization that is already present in the ventricular zone of the fetal telencephalon. According to this idea, the ventricular zone contains something like a microfilm record of the future cortex, which is projected onto the wall of the telencephalon as development proceeds.

The idea of such a cortical "protomap" was originally based on the assumption that migrating neuroblasts are precisely guided to the cortical plate by the network of radial glial fibers. If migration is strictly radial, we might expect that all the offspring of a precursor cell would migrate to exactly the same neighborhood of the cortex. However, as mentioned above, one-third of all neuroblasts stray considerable distances as they migrate toward the cortical plate. This finding initially seemed to be at odds with the protomap hypothesis. (*Neuroscience*, 696.)

Notice here that they are speculating that our brains are organized according to different Maps of Physical Functionality.

So, who laid down this map in the first place? Evolution of any kind can't do mapping, because mapping and targeting require Intelligence or Psyche; and, there's no intelligence behind evolution.

Who taught the newly born cells how to read this map? It requires Intelligence or Psyche in order to be able to read a map.

Who chooses the final destination of their migration? Remember, every newly born neuron and glial cell – neuroblast – migrates!

Who tells these newly differentiated neurons and glial cells how to connect together into a functional thalamus, hippocampus, amygdala, striatum, and cortices? Who programmed and designed and mapped this whole thing?

The only logical answer that I can come up with is Nature's Psyche or God's Psyche. WE KNOW that the Human Psyche didn't make these blueprints and maps, because the Human Psyche isn't consciously aware of any of these things.

In fact, they introduce this possibility – that Nature's Psyche is controlling all of this neurodevelopment – without even realizing that they have done so.

Now, they are going to talk about the wires that aren't really wires – the axons and the dendrites.

As neurons differentiate, they extend axons that must find their appropriate targets. Think of this development of long-range connections, or pathway formation, in the CNS as occurring in three phases: pathway selection, target selection, and address selection.

The three phases of pathway formation.

The growing retinal axon must make several "decisions" to find its correct target in the LGN.

1. During pathway selection, the axon must choose the correct path.

2. During target selection, the axon must choose the correct structure to innervate.

3. During address selection, the axon must choose the correct cells to synapse with in the target structure. (*Neuroscience*, p. 697.)

Choice is a function of Psyche or Intelligence!

Contrary to what the Materialists, Naturalists, and Darwinists claim, evolution of any kind can't do selection! Selection requires Intelligence, Psyche, Memory, and Choice. Intelligence is Psyche. Memory is Psyche. Choice is Psyche. All of these things survive the death of our physical brain. Memories, thoughts, and choices are comprised of quantum waves at the quantum level. Evolution can't do intelligence, memory, and choice, because evolution has NO psyche. Evolution, Materialism, Naturalism, and Classical Physics can't do anything on the quantum level. They are purely physical processes.

Evolution by definition in principle cannot DO choice! Choice signifies Intelligence or Psyche. According to the Materialists and Naturalists, psyche and choice are physically impossible.

All along the way, a growing axon is making choices and picking specific targets, especially the pioneer axons. Who taught the axons how to do all of this? How do the axons know where they are supposed to go? Who tells the axons where to make synapses?

The only answer that makes logical sense to me is Nature's Psyche, or God's Psyche. Even if you want to say that our genes are doing all of this, which is physically impossible, our genes were created by God's Psyche or God's Intelligence because it is physically impossible for chemical evolution to create functional proteins, let alone the matching genes to go along with them.

With this model of brain development, they introduced the possibility that Nature's Psyche is controlling all of our neurodevelopment; but, then they switch right back to Materialism and Naturalism by choosing to call axons and dendrites "wires".

Do you realize that all of the neuroscientists tell us that our brains are wired together into functional CPUs, RAM, and Distributed Memories?

It's a deceptive lie to claim that the neurons in our brain are wired together into CPUs, RAM, and Distributed Memory, because there are NO wires connecting the neurons in our brain. Our neurons are stand-alone switches, and nothing more. They are either ON or OFF.

We have to eliminate every falsehood and myth before we can find and know the truth. The second greatest myth that we have gotten from the Materialists and Naturalists through neuroscience is that the neurons in our brains are wired together into functional CPUs, RAM, and Distributed Memory. It isn't so. Their claim is science fiction.

The title for Chapter 23 in *Neuroscience* is "Wiring the Brain".

It's fascinating how they dedicate a whole chapter to something that is false and doesn't even exist – wiring within your brain! These people do the same thing with Design and Creation by Evolution. This is science fiction, and nothing more. Evolution can't do design and creation, and there are NO wires in your brain!

After studying everything that they had to offer me and going through their concentrated brainwashing program, the Materialists, Darwinists, Behaviorists, Naturalists, and Atheists did manage to convince me that Nature's Psyche has indeed wired our brain together into functional CPUs and RAM and Maps – just not at the physical level.

A chemical synapse is a terminator – an electrical dead-end. A chemical synapse is like unplugging your computer from the electrical outlet, removing your WiFi adapter, and/or unplugging your RJ-45 ethernet cable from your DLS Router or Cable Modem. A synaptic cleft terminates, randomizes, scrambles, and then integrates every memory and every message that comes its way – at the physical level at least. Who knows what's happening at the quantum level or psyche level? Only Nature's Psyche knows. Wouldn't it be fascinating to be able to listen in on and understand the Psyche-to-Psyche communication taking place at the quantum level between all the different atoms and molecules within your brain? Wouldn't it be fascinating to be able to read and decipher quantum waves? Then we would really KNOW how things work, wouldn't we? Alas, the Human Psyche has NO access to any of this information exchange, taking place at the quantum level between the atoms and molecules within our brain.

You are going to have to decide for yourself what to do with this information, because it's physically impossible for me to force you to accept it and force you to make sense of it.

Remember, according to the Materialists and Naturalists and Behaviorists, psyche and choice are physically impossible, which means that psyche and choice do not exist. The only thing left for these people is to claim that our brains are wired together into functional CPUs, RAM, and Distributed Memories. But, their claim is FALSE. There are NO wires or metal rods in our brain conducting electrical currents from one neuron to the next.

The Growing Axon

Once the neuroblast has migrated to take up its appropriate position in the nervous system, the neuron differentiates and extends the processes that will ultimately become the axon and dendrites. At this early stage,

however, the axonal and dendritic processes appear quite similar and collectively are still called neurites. The growing tip of a neurite is called a growth cone.

The growth cone is specialized to identify an appropriate path for neurite elongation. The leading edge of the growth cone consists of flat sheets of membrane called lamellipodia that undulate in rhythmic waves like the wings of a stingray swimming along the ocean bottom. Extending from the lamellipodia are thin spikes called filopodia, which constantly probe the environment, moving in and out of the lamellipodia. Growth of the neurite occurs when a filopodium, instead of retracting, takes hold of the substrate (the surface on which it is growing) and pulls the advancing growth cone forward.

Obviously, axonal growth cannot occur unless the growth cone is able to advance along the substrate. An important substrate consists of fibrous proteins that are deposited in the spaces between cells, the *extracellular matrix*. Growth occurs only if the extracellular matrix contains appropriate proteins. An example of a permissive substrate is the glycoprotein laminin. The growing axons express special surface molecules called integrins that bind laminin, and this interaction promotes axonal elongation. Permissive substrates, bordered by repulsive ones, can provide corridors that channel axon growth along specific pathways.

Travel down such molecular highways is also aided by fasciculation, a mechanism that causes axons growing together to stick together. Fasciculation is due to the expression of specific surface molecules called cell-adhesion molecules (CAMs). The CAMs in the membrane of neighboring axons bind tightly to one another, causing the axons to grow in unison.

The growth cone.

The filopodia probe the environment and direct the growth of the axon toward attractive cues.

Fasciculation.

The bottom axon grows along the molecular "highway" of the extracellular matrix. The other axons ride piggyback, sticking to one another by the interaction of cell-adhesion molecules (CAMs) on their surfaces. (pp. 698-699.)

This is how it really works, according to the neuroscientists.

Of course, the first question that comes to mind is, "Who laid out or mapped out this extracellular matrix?" It's the question that drives us. It's the question that brought you here. You know the question, just as I did. "What is the Matrix?"

Who put this matrix or map together in the first place? Matrices and maps don't just produce themselves out of thin air as the Materialists and Naturalists claim.

Who put the attractive cues there in the first place, and who tells each axon which attractive cue to pursue? Cues require some kind of Intelligence or Psyche in order to get them to work right.

Highways require planning and construction. Who laid out the highway in the first place? Highways don't just spontaneously generate out of thin air!

Integrins, laminin, CAMs, and growth cones require design, construction, and then deployment – which are a function or product of Intelligence or Psyche. It's as if these things are alive; but, that's physically impossible.

Design, migration, targeting, selection, specialization, choice, and differentiation require Intelligence or Psyche – command and control. According to the Materialists, Naturalists, and Darwinists, evolution of any kind can't do specified complexity or specialization or choice, because everything within evolution is supposed to be the result of random chance.

Imagine if we turned your brain development over to evolution or random chance. What would be the result? You would end up with a jar of Gatorade for a brain – random diffusion and Brownian motion.

Evolution isn't alive, so it's physically impossible for evolution to create life.

They say that growth cones are specialized.

Specialized by whom?

Evolution can't do specialization! Evolution can't do programming, design, or creation.

The growth cones are like blood-hounds, specially designed to sniff out a particular trail. The question is, "Who laid out the trail or deposited the breadcrumbs in the first place?"

The only answer we get from the Materialists and Naturalists is evolution; but, evolution can't do anything at the quantum level where all of this choice, decision-making, selection, targeting, differentiation, and thought is taking place.

The Materialists, Darwinists, and Naturalists refuse to think about these kinds of things; but, they are true nonetheless. I see signs of Psyche, Intelligence, Mapping, and Guidance all over axonal growth.

Of course, the Materialists and Naturalists always leave out Psyche and Quantum Waves, which get the first cells there, differentiated, and then broadcasting cues in the first place. These people also leave out the Psyche and Quantum Waves, which get the extracellular matrix built in the first place. The Materialists and Naturalists resort to Black Boxes in order to explain how this is done – typically by refusing to talk about it in the first place.

Then they talk about axon guidance.

Guidance requires an intelligent Guide!

They talk about intermediate targets that an axon grows towards.

Who put these intermediate targets into place?

They also talk about guidance cues.

Who chooses, makes, and positions these guidance cues?

At some point in the process, we always run out of physical explanations!

Chemoattraction and chemorepulsion.

(a) The protein netrin is secreted by cells in the ventral midline of the spinal cord. Axons with the appropriate netrin receptors are attracted to the region of highest netrin concentration.

(b) The protein slit is also secreted by midline cells. Axons that express the protein robo, the slit receptor, grow away from the region of highest slit concentration. Up-regulation of robo by axons that cross the midline ensures that they keep growing away from the midline.
(*Neuroscience*, p. 700.)

Who is telling the cells to secrete netrin or slit?

Obviously, it's evolution who is doing all of this, because there is no other explanation that can be given at the physical level.

Who designed, created, placed, and commanded the secretion of these different guidance factors? Who tells an axon to move towards one of them, and which one on which target cell to move towards? Targeting and selection require some kind of Intelligence or Psyche.

Even something, with a solid physical explanation like chemo-attraction and chemo-repulsion, eventually runs into something that requires decision-making, choice, and a selective change in functionality – up-regulation. Regulation of any kind requires Intelligence or Psyche – thought and planning.

At some point in the process, we always run out of physical explanations!

Thoughts and choices are quantum processes performed at the quantum level by some kind of Psyche. Psyche explains who is making all of these choices, because evolution can't do choice. Only Psyche can do thoughts and choices.

This is Quantum Neuroscience that we are talking about here. We're trying to figure out what Nature's Psyche is doing at the quantum level in order to get our brains built, mapped, and then functioning properly.

We're not done yet.

They are now going to talk about mapping, which is precisely what Nature's Psyche is doing at the quantum level with our physical brain! Of course, they are going to deliberately exclude psyche from the equation because they are talking about neuroscience; but, I'm not because I'm talking about Quantum Neuroscience.

Establishing Topographic Maps.

Let's return to the example of the growing retinogeniculate axon. These axons grow along the substrate provided by the extracellular matrix of the ventral wall of the optic stalk. An important "choice point" occurs at the optic chiasm. Axons from the nasal retina cross and ascend in the contralateral optic tract, while axons from the temporal retina remain in the ipsilateral optic tract. From our discussion so far, we can infer that nasal and temporal retinal axons must express different receptors to cues secreted at the midline.

Once the axons from the retinas are sorted out at the midline, they continue on to innervate targets such as the LGN and the superior colliculus. Sorting of the axons occurs again, this time to establish a retinotopic map in the target structure. If we accept the notion that axons differ depending on

their position in the retina (as they must, to account for the partial decussation at the optic chiasm), then we have a potential molecular basis for the establishment of retinotopy. This idea, that chemical markers on growing axons are matched with complementary chemical markers on their targets to establish precise connections, is called the chemoaffinity hypothesis. (pp. 700, 702.)

Do you know of any maps that just spontaneously generated out of thin air?

Of course not!

Maps require some kind of intelligent Mapmaker, like Nature's Psyche!

Imagine what your retinotopic map would be like if we turned it over to evolution and random mutations? You would be blind! Evolution can't do intelligence or mapping. Mutations cause ALL of our heritable diseases – NOT make us better than we were before. Adam and Eve had perfect God-given genomes, which is one of the reasons why they lived so long; and, evolution has been messing it up ever since.

Now they are going to talk about synapse formation or synaptogenesis.

Who chooses the specific target of an axon and a synapse? Supposedly it's the same person doing both tasks, isn't it?

Let's say that I give you an axon and a dendrite, and I assign you the task of creating a synapse between them. Could you do it? How would you tell these two neurites where to build the synapse that you want them to build? How would you tell them what type of synapse to build? How would you tell these two neurites what sub-type of neurotransmitter that you want them to transmit, and what type of neuropeptide you want them to transmit, if any? How would you tell the post-synaptic membrane how many neurotransmitter receptors to build and deploy? Remember, the synapse hasn't been built yet, so there is nothing there physically in place to transmit this information between these two neurites at that specific location.

The level of information exchange needed to properly differentiate and place one quadrillion synapses into a child's body is simply off the chart. Even if you were to only transmit two bytes of information with quantum waves between the axon and dendrite for each synapse that you want them to form, you are looking at petabytes of information exchange at the quantum level taking place, just for synaptogenesis alone! Remember, it's physically impossible to store petabytes of information within our 750 megabyte genome. It can't be done, which means that it isn't being done.

This mapping information is being stored someplace else besides our genome and our brain! It's physically impossible to store petabytes of information within a genome or a brain, especially a brain that hasn't been built yet.

Some of the scientists have estimated that there are one quadrillion synapses in a new-born child's body – a blank slate, if you will. Then, as the child uses its brain and body, those synapses get pruned away until we end up with something like 100 trillion synapses in an adult brain.

As the Human Psyche tries to use its brain and body, Nature's Psyche produces Maps of Physical Functionality by eliminating the synapses that aren't being used. This is what's really happening at the quantum level where our physical brain is concerned.

Imagine that you are a General, and I have given you the task of handling ALL of the logistics for one quadrillion soldiers. First, you have to produce them, recruit them, and deploy them. In the case of these synapses or soldiers, you have to train them and use them by conducting some kind of on-the-job training. You have to feed them and keep them alive, while you try to figure out which ones can do the job and which ones cannot. You have to make accommodations for them. And, as the productivity numbers come in, you have to fire the ones that aren't working or aren't working right. In the case of synapses, you have to disassemble them, and in the case of soldiers you would have to kill them because you have no place else that you can send them.

If I were to give you this task, could you do it? Would you succeed?

Why not?

According to the Materialists, Naturalists, and Darwinists, evolution is doing all of this all of the time, so why can't you? It's only a quadrillion solders or synapses that we are talking about here. That shouldn't be too hard to handle, should it?

What?

"It's physically impossible", you say.

Apparently evolution, random mutations, and natural selection can do the physically impossible, so why can't you?

What?

Are you saying that evolution can't do the physically impossible?

Well then, we have a serious problem, don't we? We better get Houston on the line.

Remember, random mutations are a function of chance and entropy, and NOT choice. Natural selection is also largely a function of luck or chance.

Natural selection was designed and created by the Darwinists to remove psyche and choice from the equation.

Evolution can't do choice. Evolution can't do psyche. Evolution does entropy, random diffusion, degradation, mutations, disease, cancer, death, and extinction.

Once all of the targeting and choosing and selection associated with synaptogenesis has been done by Nature's Psyche, then it's finally time for the molecular machinery that God designed and built to come into play. Remember, all of this takes place outside the conscious awareness of the Human Psyche, both at the quantum level and the physical level.

Before I get started with synaptogenesis, let me ask, "Who or what keeps the body alive and functioning before the synapses are formed?" The Materialists, Naturalists, Darwinists, and Neuroscientists tell us that all of our intelligence, memories, motivation, learning, and consciousness resides within our synapses.

The best description of the synaptogenesis process at the physical level comes from *Neuroscience: Exploring the Brain*.

Synapse Formation.

When the growth cone comes in contact with its target, a synapse is formed. Most of what is known about this process comes from studies of the neuromuscular junction. The first step appears to be the induction of a cluster of postsynaptic receptors under the site of nerve-muscle contact. This clustering is triggered by an interaction between proteins secreted by

the growth cone and the target membrane. At the neuromuscular junction, one of these proteins, called agrin, is deposited in the extracellular space at the site of contact. The layer of proteins in this space is called the basal lamina. Agrin in the basal lamina binds to a receptor in the muscle cell membrane, called muscle-specific kinase, or MuSK. MuSK communicates with another molecule, called rapsyn, which appears to act like a shepherd to gather the postsynaptic acetylcholine receptors (AChRs) at the synapse. The size of the "flock" of receptors is regulated by another molecule released by the axon, called neuregulin, which stimulates the receptor gene expression in the muscle cell.

The interaction between axon and target occurs in both directions, and the induction of a presynaptic terminal also appears to involve proteins in the basal lamina. Basal lamina factors provided by the target cell evidently can stimulate Ca^{2+} entry into the growth cone, which triggers neurotransmitter release. Thus, although the final maturation of synaptic structure may take a matter of weeks, rudimentary synaptic transmission appears very rapidly after contact is made. Besides mobilizing transmitter, Ca^{2+} entry into the axon also triggers changes in the cytoskeleton that cause it to assume the appearance of a presynaptic terminal and to adhere tightly to its postsynaptic partner.

Similar steps are involved in synapse formation in the CNS [Central Nervous System], but these may occur in a different order, and they definitely use distinct molecules. Microscopic imaging of neurons in tissue culture reveals that filopodia are continually being formed and retracted from neuronal dendrites seeking innervation. Synapse formation begins when such a dendritic protrusion reaches out and touches an axon that might be passing by. This interaction appears to cause a preassembled presynaptic active zone to be deposited at the site of contact followed by the recruitment of neurotransmitter receptors to the postsynaptic membrane. In addition, specific adhesion molecules are expressed by both presynaptic and postsynaptic membranes that serve to glue the partners together.

Steps in the formation of a neuromuscular synapse.

1. The growing motor neuron secretes the protein agrin into the basal lamina.

2. Agrin interacts with MuSK in the muscle cell membrane.

3. This interaction leads to the clustering of ACh receptors in the postsynaptic membrane via the actions of rapsyn. (*Neuroscience*, pp. 702-704.)

You probably can't even see it, if you have been brainwashed into Materialism and Naturalism like I was; but, these people have these atoms and molecules interacting with each other and communicating with each other.

If I were to give you the task of communicating with atoms and molecules and telling them what to do, how would you do so? How do atoms and molecules communicate with each other?

There is ONLY ONE logical answer, where atoms are concerned. They communicate with each other at the quantum level using quantum waves or telepathy in order to do so.

There is NO other explanation possible, because it's physically impossible for atoms to communicate with each other at the physical level.

Who initiates contact, and why?

They have these molecules inducing, contacting, triggering, interacting, depositing, binding, communicating, shepherding, gathering, regulating, stimulating, mobilizing, causing, adhering, seeking, recruiting, gluing, acting, and the using distinct molecules.

These are ACTION verbs. Who is doing all of these choices and performing all of these actions? Well, it isn't evolution, because evolution can't choose nor act. Evolution is comprised of physical reactions.

Action requires purpose, teleology, foresight, and intelligence. Physical matter was designed to react – Brownian motion, random diffusion, entropy, and classical physics. ONLY Psyche or Intelligence can ACT – choice, syntropy, intelligence, and quantum mechanisms.

If I gave you the task of using distinct molecules, how would you go about doing so? The migrating Ach Receptors are fascinating. How do they do that? Why do they do that? What's their motivation?

The basal lamina seems to be some sort of mesh, like the extracellular matrix. Who put this matrix together in the first place? Matrices, laminae, proteins, and receptors don't just spontaneously generate out of thin air. That's physically impossible!

Let's say I give you the task of creating and producing the first functional protein. How would you do so?

Well, first of all, you would have to design the thing atom-by-atom and be relatively confident that it's going to work right when you are done, or it isn't going to be worth your time. You are going to have to gather together the correct number and types of atoms.

Next, you will have to make the correct atoms come together into the right amino acids that you need for the task. Then you would have to make the amino acids line up in a specific order, connect or dock in a specific way, and then fold up into a 3D structure in a specific manner. How would you communicate all of this information to these atoms and molecules? The only way you could do so is at the quantum level with quantum waves or telepathy. How would you make these individual atoms and molecules come together and line up properly? Again, the only way would be to use some kind of Action at a Distance, or Telekinesis, or Quantum Tunneling at the quantum level. You would have to use quantum mechanisms to make these atoms and molecules come together into the proper sequence and structure. Remember, it's physically impossible to do any of this at the physical level, because you don't have any proteins or enzymes yet because you are making the first one.

These same realities and truths apply directly to the atoms and molecules that are being used to make synapses. There aren't any synapse-making machines out there, stamping out synapses as it goes along the axon! The atoms and molecules are doing all the "choosing" and "action" and "quantum mechanisms" and "syntropy", under the command and control of Nature's Psyche. Take the very same atoms and molecules and drop them into a jar outside the living brain, and you would have Gatorade, Brownian motion, random diffusion, and entropy.

Do you see how this works?

It doesn't work, unless Nature's Psyche chooses to get involved and God's Psyche lets it do so.

Someone Psyche was given the task of producing the first physical atom, the first synapse, the first physical protein, and the first physical gene to go along with that protein; and, that Someone Psyche had to use quantum mechanisms and intelligence in order to get the job done. Proteins and genes don't just spontaneously generate out of thin air. It's physically impossible. They require Someone Psyche working at the quantum level in order to bring them into existence.

These very same truths apply to the atoms and molecules that are being used to make synapses. Synapses don't just spontaneously generate out of thin air. That's physically impossible. Someone Psyche is putting them together at the quantum level.

During synaptogenesis, there's communication going on – a sort of Morse code.

Who taught these newly born cells and growing axons this language or means of communication? They are born knowing how and when to communicate using these various different chemical signals and quantum waves. These cells are alive, sentient, conscious, and aware. During migration, they KNOW where they are going and what to do when they get there. Who is teaching them and telling them how to do all of this? This interaction or communication is going in both directions! That means that these cells or neurons were born knowing how to communicate with each other.

If I were to give you the task of communicating with a neuron and making it do something specific, how would you accomplish that task?

Well, you could do it through quantum waves or chemicals – BUT ONLY if you KNOW its language or communication system. You have to KNOW what the quantum waves and chemicals mean to a neuron, or they aren't going to do you any good.

During synaptogenesis, who chooses the molecules to use, the synapse type, and the synapse location? Who chooses or decides what a synapse means? Who maps physical functionality onto a synapse? It has to be Someone Psyche at the quantum level, because evolution and classical physics can't do choice.

When the neurotransmitters are made and dumped into the cerebral spinal fluid, who tells them where to go and who helps them to get there? Remember, molecules under the control of Classical Physics just sit there and vibrate under the effects of Brownian motion, or they randomly diffuse evenly throughout the fluid if they are ionized and motivated to do so.

Ordering, mobilizing, triggering, seeking, interaction, coordination, pre-assembly, manufacturing, expression, guidance, recruitment, constructing matrices, and reaching out to the correct target are a function of Psyche or Intelligence. Someone Psyche is telling these target cells what chemical to secrete or what process to extend, and telling the growing axons and dendrites what chemical signals to grow towards or to avoid. It's as if these things are alive, which is physically impossible according to the Materialists and Naturalists.

They spend some time talking about the physical mechanisms associated with cell death and synaptic pruning, synaptic capacity, synaptic segregation, synaptic convergence, synaptic competition, synaptic plasticity, and synaptic transmission. These things are described at the physical level; but, someone has to be triggering or activating them at the quantum level. Synaptic Weights and Distributed Memory are quantum level functions, if they exist at all, because the ONLY message or memory that survives a synaptic cleft at the physical level is ON or OFF – OPEN or CLOSED.

Someone Psyche is deciding which synapses and neurons to eliminate and kill; and, Someone Psyche is deciding which synapses and neurons to strengthen and preserve. Someone Psyche KNOWS which synapses and neurons are being used, and which ones are not. This person or personality may or may not have physical mechanisms for accomplishing these tasks of synaptogenesis and synaptic pruning; but, if the Psyche does have physical mechanisms for accomplishing some of these processes, it still has to choose whether to use these physical mechanisms or not. Someone Psyche literally has to choose whether to preserve and strengthen a neuron, or to kill it dead through apoptosis.

This Someone Psyche has to be Nature's Psyche or God's Psyche, because the Human Psyche isn't consciously aware of these things. When it comes to boots-on-the-ground and quantum mechanisms, we are most likely looking at Nature's Psyche. I believe that God's Psyche doesn't get involved unless it has covenanted to do so and/or received permission to do so. Praying to God and asking for His help and making covenants with God are the only way to get God to help us, whether we are talking about the quantum level or the physical level. God has promised each one of us not to interfere in our lives unless we give Him permission to do so. If we don't want to have anything to do with God, He will grant our wish every time. God only goes where He is wanted.

They spend some time in this chapter talking about long-term potentiation (LTP). LTP is an enduring facilitation of synaptic transmission – enhanced neurotransmission. The Materialists, Naturalists, and Neuroscientists equate LTP with long-term memory storage; but, I never bought it, because I know for a fact that synapses integrate everything that comes their way at the physical level into a single BIT of information causing the neuron to be either ON or OFF. LTP looked more like reprogramming to me – reprogramming with additional or enhanced physical reliability and increased functionality. ALL of the mapping or programming that takes place in the brain under the command and control of Nature's Psyche was designed to either increase or decrease the physical functionality of the brain. Nature's Psyche uses that physical functionality to get things done for us, when we request them.

I have discussed this topic in other essays, so I won't do so here. All I will say is that Someone Psyche had to program or map this physical functionality or molecular functionality into the system at the quantum level, or it wouldn't exist.

The atoms and molecules don't do anything unless Someone Psyche makes them act, makes them combine, makes them coordinate, and makes them do something. Atoms and molecules have to be forced to do things, or they do nothing. If I gave you the task of forcing an atom to do something, how would you accomplish that assignment? Well, you would have to use some sort of quantum mechanism, especially if you are wanting that atom to interact with and engage other atoms and molecules in an effective and productive manner.

They start with science fiction and end with science fiction.

We have seen that the generation of circuitry during brain development occurs mostly before birth and is guided by cell-to-cell communication through physical contact and by diffusible chemical signals. Nonetheless, while most of the "wires" find their proper place before birth, the final refinement of synaptic connections, particularly in the cortex, occurs during infancy and is influenced by the sensory environment. Although we have focused on the visual system, other sensory and motor systems are also readily modified by the environment during critical periods of early childhood. In this way, our brain is a product not only of our genes, but also of the world in which we grow up.

The end of developmental critical periods does not signify an end to experience-dependent synaptic plasticity in the brain. Indeed, the environment must modify the brain throughout life, or there would be no basis for memory formation. In the next two chapters, we'll explore the neurobiology of learning and memory. We will see that the mechanisms of synaptic plasticity proposed to account for learning bear a close resemblance to those believed to play a role in synaptic rearrangement during development. (*Neuroscience*, p. 722.)

They start with science fiction and end with science fiction.

There are NO wires in our brains. There's NO circuitry in our brains. It's physically impossible to pass memories or information through a synaptic cleft. It's physically impossible to store memories within a synapse or synaptic cleft. They completely ignore the communication that's taking place at the quantum level through quantum waves. They completely ignore Quantum Mechanics – the best-proven and most-used science on this planet.

Our brain is a product not only of our genes and the world in which we grow up, but also Nature's Psyche and the choices of the Human Psyche. They completely and deliberately ignore the role of Nature's Psyche, the Human Psyche, and God's Psyche in all of this; but remember, the brain is also a product of the Human Psyche's choices.

It's time to up our game to Quantum Mechanics, Psyche, and quantum mechanisms at the quantum level.

Remember, if your Interpretation of Quantum Mechanics cannot explain what Nature's Psyche and the Human Psyche are doing at the quantum level and how they are doing it, then your Interpretation of Quantum Mechanics is worthless. This means that a materialistic and naturalistic interpretation of Quantum Mechanics is pointless and worthless, because it completely lacks the explanatory power that we need to explain what's happening at the quantum level under the command and control of Nature's Psyche.

The Human Psyche pays attention to what it values most; and then, Nature's Psyche maps the brain accordingly. After all of my research, I have settled upon the conclusion that Nature's Psyche takes the 100 billion neurons or switches within our brain and organizes them into Maps of Physical Functionality at the quantum level. Neurons are switches with NO wires connecting them. If they are being organized or mapped into any sort of functionality or purpose, it's being done at the quantum level, because it's physically impossible to do it at the physical level.

There's NO mechanism within a neuron at the physical level for storing and processing memories. If the neurons and neurotransmitter receptors really are connected together as BYTES of programming code, data, and memories, it's all being done at the quantum level by Nature's Psyche, because there are NO visible wires connecting neurons and receptors together at the physical level.

Neuroplasticity – Memory Systems

Memory Systems is the most interesting and most confusing subject to study in Neuroscience. It ends up being nothing but smoke and mirrors, but it can take months or years to figure out that it is so.

Now that you know where I have come from and where I am going with Quantum Neuroscience, let's run some Neuroscience through the process and see what we end up with.

I will use *Neuroscience: Exploring the Brain* as my text, because it's freely available online if you know where to look for it.

Bear, M. F., Connors, B. W., & Paradiso, M. A. (2007). *Neuroscience: Exploring the Brain* (3rd ed.). New York: Lippincott Williams and Wilkins.

This essay will focus exclusively on chapter 24 of their book. Let's see what we can do with it.

BCP24

http://mypsyche.us/wp-content/uploads/2017/12/BCP24.pdf

Classical Physics, Materialism, and Naturalism are at a serious disadvantage compared to the explanatory power of Quantum Mechanics. When it comes to explanatory power, Quantum Mechanics is vastly and noticeably superior to anything that the Materialists and Naturalists can provide. The Orthodox Interpretation of Quantum Mechanics, by Henry P. Stapp, is an infinitely better method for describing memory storage than anything you will ever get from Naturalism, Materialism, and Classical Physics. That's just the way it is.

I spent months studying the Philosophy of Science so that I could know and understand how the Materialists, Naturalists, Darwinists, and Atheists are working their magic and making their case. It was well worth it – a total eye-opener.

The Materialists, Naturalists, Darwinists, Nihilists, Behaviorists, and Atheists start with the pre-chosen conclusion that ALL of our memories and brain functionality are stored and processed within our physical brain and nowhere else. Then, these people insert this pre-chosen conclusion into all of their arguments and proofs as one of their hidden premises, using their pre-chosen conclusion as evidentiary and empirical proof that their pre-chosen conclusion is real and true. This is a logic fallacy known as *begging the question*. The whole of Materialism, Naturalism, Behaviorism, and Atheism are based upon this logic fallacy.

Furthermore, the Materialists, Naturalists, Darwinists, and Atheists unknowingly use the *affirming the consequent* version of the Scientific Method to make their case and prove their point. *Affirming the consequent* is a serious logic fallacy, but *affirming the consequent* is the very foundation of Scientific Naturalism and Reductionistic Materialism. Their whole science is based upon *affirming the consequent*.

Negating the consequent is the ONLY version of the Scientific Method that is philosophically and logically sound. In fact, I have successfully used *negating the consequent* to FALSIFY Materialism, Naturalism, Darwinism, and Atheism. It's powerful, it's convincing, and it works. It's extremely easy to do, once you know how *negating the consequent* works. In other words, I have actually used the Scientific Method to FALSIFY Materialism, Naturalism, Darwinism, and Atheism by *negating the consequent*.

Ironically, Quantum Neuroscience is already an observed and proven science. Quantum Mechanics is our best-proven and most-used science on the planet. Quantum Neuroscience was already a proven science before I ever coined the term. It was just sitting there waiting for someone to find it.

Chapter 24 of their *Neuroscience* book is about physical memory systems or physical RAM.

These people start with the conclusion that all our memories are stored exclusively within physical matter within our physical brain; and, even when they demonstrate otherwise, they go right back to their pre-chosen conclusion that all our memories are being stored within our brain and nowhere else. They turn physical memory storage into a *self-fulfilling prophecy*, which is a logic fallacy, because it is based upon *circular reasoning* which is also a logic fallacy.

I upgraded my science to Science 2.0, which allows ALL of the evidence into evidence, and then pursues a preponderance of the evidence.

We study and learn the physical basis for how things work, until we run into things that have NO physical explanation. When dealing with the physically impossible that has NO physical explanation, we turn to the observational evidence and experiential evidence from Quantum Mechanics, Quantum Non-Local Consciousness (Psyche), Near-Death Experiences (NDEs), and Out-of-Body Experiences (OBEs) in order to explain what we are observing and experiencing. We can actually use Quantum Mechanics and Psyche to explain the physically impossible. This is how Science 2.0 works. Science 2.0 is an evidentiary science, observational science, and empirical science. Under Science 2.0, Lived Experiences and Phenomenology are given precedence over philosophical speculation and wishful thinking.

Nowhere is the need for Quantum Mechanics, Quantum Non-Local Consciousness (Psyche), and Quantum Neuroscience more obvious and clear than memory storage?

Why?

It's because it's physically impossible to store petabytes of data and memories within a physical brain! It's physically impossible to do terabytes of physical hardware and terabytes of software execution within a physical brain. It's also physically impossible to transfer gigabytes of data or even a byte of data through an ionotropic receptor or synaptic cleft. It's physically impossible to store petabytes of memories or megabytes of data within a receptor or synaptic cleft. These people say that that is what is happening at the physical level within our brains – all of our memories are being stored physically within our brain and all of our CPU processing is being done physically within our brain; but I don't buy it, because it's physically impossible. It's physically impossible to do the physically impossible. If we want to do the physically impossible, we have to turn to Quantum Mechanics and Psyche in order to get the job done.

It has been observed with PET and fMRI that the Human Psyche decides what it is going to pay attention to, and the necessary neurons heat up or ramp up accordingly. The Human Psyche has its own memory stores, and it keeps reminding itself what it wants to pay attention to; and, then under the direction of Nature's Psyche, the brain fires and wires accordingly.

The "wiring" within our brains is being done wirelessly at the quantum level by Nature's Psyche, completely outside our conscious awareness. There are NO physical wires within our brain; but, there's plenty of evidence that a lot of WiFi or wireless processing is taking place within our brains at the quantum level.

By paying attention to things, the Human Psyche chooses what it wants to remember, and those memories stay with us even after our brain is dead and gone. It's the Ultimate Memory System. A physical memory system doesn't survive the death of our physical body and physical brain. Remember, the Human Psyche makes requests and pays attention; and, Nature's Psyche fires and wires the physical brain accordingly, at the quantum level.

This is cools stuff, but only if you are looking for it and willing to accept it when you find it.

On page 736, they present a Model for Distributed Memory in order to explain how our memories are being stored physically within our physical brain. Look at all of the Black Boxes that they employ, just so they can get a physical, distributed, memory engram made out of physical synapses. They never once realize that a synapse is a randomizer and scrambler. NO message or memory survives a synaptic cleft, besides ON or OFF. They never once realize that it's physically impossible to store petabytes of memories within a randomizer, scrambler, or synaptic cleft. They are completely blind to all of this, because they are using lots of different black boxes in order to make their case. That's wishful thinking, not science.

I analyze and dissect their Model for Distributed Memory in great detail in a different essay, so I won't do so here. It's sufficient for my purposes simply to make notice that it exists.

They talk about localization of declarative memories in the neocortex. That's their story, and they are sticking with it. Even when they find evidence that contradicts their pre-chosen conclusion, they go right back to their claim that our declarative memories are being stored physically within our brain. They all do that! You can't change what they choose to do. You can only debunk it and demonstrate why it is wrong.

What's wrong with localization of declarative memories in the physical brain?

It's physically impossible, and it doesn't match with the observational evidence and experimental evidence from Quantum Mechanics. Their pre-chosen conclusions are based exclusively on Materialism, Naturalism, and Classical Physics. The fact that these people have a pre-chosen conclusion is also a violation of science itself; but, that's the way these people do science.

I'm very impressed by the explanatory power of Quantum Neuroscience. Let's make some comparisons, so that you can see what I mean.

The Human Psyche pays attention to what it values most; and then, Nature's Psyche maps the brain accordingly. After all of my research, I have settled upon the conclusion that Nature's Psyche takes the 100 billion neurons or switches within our brain and organizes them into Maps of Physical Functionality at the quantum level. Neurons are switches with NO wires connecting them. If they are being organized or mapped into any sort of functionality or purpose, it's being done at the quantum level, because it's physically impossible to do it at the physical level.

Chapter 24:

Chapter 24 of *Neuroscience* focuses on Memory Systems.

This chapter is heavy-duty material from beginning to end.

We have Psyche, Intelligence, Quantum Waves, and Quantum Mechanisms written all over our Memory Systems from start to finish. It's unavoidable, even though it's physically impossible.

As a former Materialist, Naturalist, Nihilist, and Atheist, this chapter is so painful that it hurts, especially when you start looking at it from a quantum perspective.

BCP24

http://mypsyche.us/wp-content/uploads/2017/12/BCP24.pdf

—

They start with their conclusion that ALL of our memories are stored within our neural hardware; and, they never stray away from their pre-chosen conclusion. Once again, they start this chapter with science fiction, which means that they are going to end up with science fiction.

> **Brain lesions differentially affect different types of remembered information, suggesting that there is more than one memory system.**

> **Learning and memory are the lifelong adaptations of brain circuitry to the environment. They enable us to respond appropriately to situations we have experienced before.**

> **In this chapter, we discuss the anatomy of memory — the different parts of the brain involved in storing particular types of information.**

> **In Chapter 25, we will focus on the elementary synaptic mechanisms that can store information in the brain.**

> **Learning is the acquisition of new information or knowledge. Memory is the retention of learned information. We learn and remember lots of different things, and it is important to appreciate that these various things might not be processed and stored by the same neural hardware. No single brain structure or cellular mechanism accounts for all learning. Moreover, the way in which information of a particular type is stored may change over time.** (*Neuroscience*, 726.)

This chapter was painful for me – it caused a lot of cognitive dissonance. At times, I wanted to believe what they were telling me; but, I was plagued by nagging doubts. I vacillated back and forth for months, never able to make up my mind and never able to figure out what is really true.

They start the chapter by trying to distinguish between Declarative Memories (explicit memories) and Nondeclarative Memory (procedural memories or implicit memories).

I already KNEW about Karl Lashley's experiments with memory, and the conclusion that some people had drawn – that there are NO memories stored within our brain. There are no memory engrams in our brain, which means that there is no RAM in our brain, which means that our memories might not be stored within the neural hardware at all. Our brains might not be storing the information at all, but accessing the information from non-local cloud storage instead.

> **What was the effect of the size and location of the lesion? Interestingly, Lashley found that the severity of the deficits caused by the lesions (both learning and remembering) correlated with the size of the**

lesions but was apparently unrelated to the location of the lesion within the cortex. Based on these findings, he speculated that all cortical areas contribute equally to learning and memory. (*Neuroscience*, 732.)

They deliberately watered-down Lashley's discoveries, choosing instead to use Lashley's observations as the basis for their Distributed Memory Model and theory. They really had no other choice in the matter, because they had started with the conclusion that ALL of our memories are stored within our brain, and that ALL of our memories cease to exist when we die.

I chose to interpret Lashley's discoveries differently. That's the problem with the scientific method – for every scientific discovery, there are theoretically an infinite number of different plausible interpretations or explanations that can be given for that single scientific discovery.

I chose to interpret this discovery as demonstrating that we are looking at damaged machinery and not necessarily lost memories. It's the functionality that was lost by lesioning, not the memories.

They started this chapter with pre-chosen conclusions. I too had already started to draw some conclusions of my own before I read this chapter

I had chosen to believe that Lashley's experiment had successfully demonstrated that there are NO memory engrams and therefore no RAM within our physical brain. Although I hadn't put it into words yet, I was also starting to sense that it's physically impossible to pass memories and information through a synaptic cleft, which means that it's physically impossible to store memories within a synaptic cleft. In other words, synaptic mechanisms CANNOT be used to store information and memories in the brain.

I was already in trouble, because ALL of their models in this chapter and in the other neuroscience books were based on the idea that ALL of our memories are being stored within our synapses at the physical level. I was starting to sense that that's physically impossible.

Based upon our after-death Life Reviews as a race, I had come to the conclusion that our declarative memories (memories for facts and events) survive the death of our physical brain; but, our procedural memories (motor skills and maps of physical functionality) might not survive the death of our brain because our procedural memories aren't really needed when our brain and body are dead and gone.

When I started reading this chapter, I was visualizing "learning" as CPU programming or "procedural memories". I still visualized declarative memories as RAM; but, I KNEW that there are NO memory engrams or RAM within the physical brain that we can observe, which meant that the "RAM" storing our memories existed as some kind of non-local storage in the cloud.

I had come to the conclusion that each structure in the brain functions as a unique Cerebral Processing Unit or CPU. I stuck with that model for months; and, I really liked it. It worked for me. It took me months to figure out that it is wrong, because the Neuroscientists use similar models all throughout their books as well.

However, the evidence eventually got through to me.

I finally realized after months of researching and writing this book that if it's physically impossible to store a byte of memory or information in a synaptic cleft and it's physically impossible to pass a byte of information or memory through a synaptic cleft, then it's also physically impossible to process a byte of computer code through a synaptic cleft.

In other words, if it can be demonstrated that there is no RAM within our brains, then you have also successfully demonstrated that there are no CPUs in our brain. Suddenly, I found myself looking at a blank slate, not knowing what to do with it. The writing was on the wall. There are NO wires, RAM, CPUs, or Distributed Memories within our brain, because these types of things are physically impossible to implement through chemical synapses.

I was in need of a New Model for brain functionality and memory storage – one that actually matches with ALL the evidence that we have on hand as a race. But, I wasn't there yet! I was still mired down in the Cerebral Processing Units and Brain Programming Model that I liked so much.

I wrote at the time that it makes logical sense that procedural memories are stored in our physical brain, because they control physical processes. Procedural memories are programming. In contrast, declarative memories are backed-up in non-local consciousness in the cloud, because they survive the death of our physical brain.

I asked, "Do these brain structures encode for the physical functionality of non-local memory storage and retrieval?" I was convinced that our long-term memories and after-death Life Reviews are NOT being stored within our physical brain! I was starting to visualize the physical brain as a physical transceiver for our non-local memories, which are being stored in the cloud in God's Database.

Over and over again from one book to the next, I kept expecting them to tell me for real how our memories are being stored within our brain at the physical level by physical processes. They would get me convinced that some of our memories are indeed being stored within our physical brain at the physical level; but then, it would invariably dawn on me at some point during my research that their physical explanation for physical storage of memories is in fact physically impossible.

Observe the back-and-forth that I went through for months on end, by examining what I wrote in the margins of my books.

The preponderance of the evidence points to functionality. There doesn't seem to be any RAM within the brain. Nondeclarative memory or implicit memory, what they sometimes call procedural memories, look like programming to me, not memories.

Declarative memories are easy to form and are easily forgotten. There is no clear limit to the number of declarative memories the brain can store. Nondeclarative memories require repetition and practice over long periods of time. (*Neuroscience*, p. 727.)

Why does there seem to be no limit to our declarative memories? It's because they are being stored non-locally in the cloud and not in our physical brain. Whenever our hard-drive starts to get full, Nature's Psyche off-loads the declarative memories that we aren't using into the cloud and stores them off-site or non-locally for us instead. Declarative memories are stored within non-local cloud storage.

In contrast, our "procedural memories" look more like physical functionality to me than memories, which leads me to believe that our procedural memories are programming that is being mapped into our brain by repetition and practice over long periods of time. Programmed by whom? Mapped by whom? Well, it can't be the Human Psyche, because we aren't consciously aware of any these processes – both at the physical level and the quantum level. It has to be Nature's Psyche who is doing this programming and mapping for us. According to the Orthodox Interpretation of Quantum Mechanics from Henry P. Stapp, Nature's Psyche is the one who collapses the wave function thereby turning all of those quantum waves, thoughts, and memories into a Map of Physical Functionality and a Map that is physically real at the physical level.

As proof of concept that all of our memories are being stored only within our brain, they present the case of an individual with eidetic memory or photographic memory.

How did he do it?

S. described several factors that may have contributed to his great memory. One was his unusual sensory response to stimuli — he retained vivid images of things he saw. When shown a table of 50 numbers, he claimed it was easy to later read off numbers in one row or along the diagonal because he simply had to call up a visual image of the entire table. Interestingly, when he occasionally made errors in recalling tables of numbers written on a chalkboard, they appeared to be "reading" errors rather than memory errors. For instance, if the handwriting was sloppy, he would mistake a 3 for an 8 or a 4 for a 9. It was as if he were seeing the chalkboard and numbers all over again when he was recalling the information. (p. 728.)

Well, this is proof that our memories are being stored within our brain, isn't it?

ONLY if you start with the conclusion that our memories are stored in our physical brain and no place else.

There are other interpretations that make just as much logical sense as the physical explanation. In fact, some of them make more sense.

This image was stored someplace else besides his brain.

How do we know?

Well, the brain processed this image but didn't actually store it; otherwise, he wouldn't have to "call it up" – it would just always be there! He wouldn't have to recall any of these images is they were already there within his brain. Subtle, I know, but true nonetheless.

The people with Superior Autobiographical Memory experience the same "recall process". They have to deliberately recall the memories from cloud storage, because their memories are not being stored within their physical brain. If their memories were stored within their physical brain, then ALL of their memories would be simultaneously present all the time, and it would be information overload and possibly kill them or drive them insane.

The people with Eidetic Memory and Superior Autobiographical Memory are sane, which means that their memories are NOT being stored within their brain.

If you destroy the part of the brain that God has designed and Nature's Psyche has mapped to do the "call up" or "recall" of memories, it will make it seem as if the brain itself was storing those memories or images, even though it was not.

Working memory is a temporary form of information storage that is limited in capacity and requires rehearsal. It is often said that working memory is information held "in mind." These different digit spans in different modalities are consistent with the notion of multiple temporary storage areas in the brain. (*Neuroscience*, p. 729.)

Conclusion, working memory is information that's being held in the physical brain.

Notice how they slipped up and actually called "working memory" something that is "held in the mind", or held within the Psyche. You are not supposed to do that if you are a Materialist or Naturalist. Of course, they put "in mind" into quotes so that you would know that it's not actually real – just a metaphor or analogy.

But then, they go back to their pre-chosen conclusion that working memory is information that's being held in the physical brain.

I, too, was convinced for months that our working memories are being stored within our physical brain as some sort of cache set aside for each one of our Cerebral Processing Units (CPUs).

It was only when I finally realized that it's physically impossible to store memories within a synaptic cleft and physically impossible to pass memories through a synaptic cleft, did I finally realize that even our working memory is not being stored within our physical brain.

I wrote, "If there were petabytes of physical RAM within our brain, the scientists would have discovered it by now. However, they discovered that our brain has NO memory engrams or RAM."

So, what's happening if NONE of our memories are being stored within our physical brain? Well, the only other logical explanation is that Nature's Psyche is processing and storing our memories for us non-locally in the cloud, and then mapping the memories (to which we pay the most attention) to specific neurons within our brain. Whenever the Human Psyche tries to access or recall a certain set of memories, Nature's Psyche retrieves those memories from the cloud and fires the appropriate neurons that go along with them.

We are looking at Maps of Physical Functionality, and NOT physical RAM.

They talk about the Search for the Memory Engrams within our brain.

They subsequently use Karl Lashley's discovery that there are NO memory engrams or no RAM within our brain as "proof of concept" that our memories are being distributed equally throughout our brain and being stored within our synapses. The result was their Model of Distributed Memory on page 736. In other words, they *beg the question* and *jump straight to their pre-chosen conclusion* that ALL of our memories are stored within our brain and no place else. They even use their pre-chosen conclusion as evidence that their pre-chosen conclusion is right.

Despite their claims to the contrary, I really truly believe that Karl Lashley and others proved that there are NO memory engrams or RAM within a physical brain; and, I went forth accordingly.

I wrote a lot in the margins of my book debunking their ideas one-by-one as I went along. Their oft' repeated claim that ALL of our memories are being stored in our synapses and/or distributed among our synapses doesn't hold water, because it's physically impossible to store memories within a synaptic cleft and to pass memories through a synaptic cleft. You can't do the physically impossible at the physical level.

I wrote, "These people conclude from the beginning that all our memories are stored within our brain and no place else. However, they are wrong. Lashley's experiments are often interpreted as proof that we don't have memory engrams! Furthermore, destroying cells or deactivating cells does not destroy the memories – it eliminates the functionality of those cells instead."

They introduce Hebb's Cell Assembly Model of memory storage within the physical brain. The only way that model would work for storing memories is if the different neurons within the assembly were communicating with each other telepathically at a distance at the quantum level, because there are NO wires in our brain at the physical level.

Stated another way, inferotemporal cortex is both a visual area and an area involved in memory storage. Further evidence that inferotemporal cortex is involved in certain types of memory storage comes from physiological experiments in which the response properties of individual neurons are examined. For example, recordings made from IT neurons suggest that they may encode memories of faces. (Neuroscience, pp. 734.)

HOW?

How are the inferotemporal cortex and concept cells doing image recognition and object recognition? It's physically impossible, because there are NO wires, CPUs, and RAM within our brain; and, NOTHING survives passage through a synaptic cleft except ON or OFF.

The ONLY thing that makes any sense to me is that the Human Psyche is doing the recognition – which is a function of the Human Psyche – and then, Nature's Psyche is firing the appropriate neurons which it has mapped to those specific memories, images, and objects.

I can explain this thing at the quantum level or psyche level; but, I can't explain it at the physical level, because NOTHING makes it through a synaptic cleft except ON or OFF and because NOTHING makes it through a neuron except ON or OFF. Neurons are switches. It's physically impossible to pass a byte of memories or information through a neuron; and, it's physically impossible to store a byte of memories or information within a neuron. Neurons are switches, not CPUs or RAM.

These people repeatedly conflate the task or the physical function with memory storage! They are not the same thing at all. The true test would be to use temporary lesions for their experiments. If the memories come back, then that implies that the memory or skill isn't stored in the physical brain and that that part of the brain has been set aside by Nature's Psyche as a facilitator or mediator or a learning module, and NOT a memory storage device. In other words, Nature's Psyche has mapped that particular non-local functionality and those particular non-local memories to those specific neurons. Then Nature's Psyche triggers those particular neurons whenever that particular functionality or set of memories is needed by the Human Psyche.

I wrote in the margin of my book, "The neuron has to be getting input or data – processed facial data – before it can respond. How does gigabytes of data and programming code get transferred or transmitted through synapses that are either OPEN or CLOSED? Where are the CPUs, which are allegedly doing all of this complex image processing?"

There aren't any wires in our brain, which means that there aren't any CPUs within our brain, unless Nature's Psyche is actually mapping the different structures of our physical brain to function as CPUs at the quantum level.

The first time new faces are seen, the cell responds at about the same moderate level to all of them. However, with a couple of additional exposures, the response changes, such that some faces evoke a significantly greater response than others. The cell is becoming selective in its response to these new stimuli as the researcher watches it. With

continued presentation of the same group of faces, the response of the neuron to each pattern becomes more stable.

Changing responses of a cell to unfamiliar faces. When the four faces are presented for the first time, there is a moderate response to each. With subsequent presentations, the cell becomes more responsive to faces 1 and 2 and less responsive to faces 3 and 4. (*Neuroscience*, p. 735.)

OUCH!

There's intelligence, selectivity, choice, and psyche being done here within this neuron! That's physically impossible!

So, where are all the CPUs, RAM, and gigabytes of image processing software within this neuron? Why does this this specific neuron choose to become more responsive to certain faces and less responsive to other faces? Who is doing all this choosing and selectivity for this neuron? Why does it become more responsive to faces 1 and 2 and less responsive to faces 3 and 4? Who is making this choice? How is it being done? All of this is physically impossible within a neuron that is only capable of functioning as a single hardware BIT with either fires or doesn't!

The ONLY logical explanation is that all of the processing, functionality, memory storage, and mapping for this one specific neuron is being done non-locally in the cloud by Nature's Psyche, completely outside the conscious awareness of the Human Psyche.

When bird experts and car experts view pictures of both birds and cars, the brain responses are different. The bird experts have areas of extrastriate visual cortex that are significantly more activated by images of birds than by other objects, such as cars. (p. 735.)

So, is it the cortex that's doing the object recognition, or is the cortex simply responding to the Human Psyche's recognition of the object?

Obviously, these people have concluded a priori that it's the cortex that is doing all of this CPU processing of image recognition software and the memory storage to go along with it; but, that's physically impossible, because it's physically impossible to pass a gigabyte of programming code through a synaptic cleft and there are NO wires and no CPUs and no RAM within a physical brain.

They have concluded in advance that ALL of our memories are being stored within our physical brain and no place else, even though that's physically impossible. Now, it's time for them to present their Model of Distributed Memory, because they believe that they have made their case that all of our memories are being stored within our synapses as synaptic weights. I dissected this thing in great detail in other essays, so I won't do so here. Their Model of Distributed Memory is nothing but smoke and mirrors – one physically impossible black box after another.

Their ultimate goal was to explain Karl Lashley's observations that the memories still exist even when that particular part of the brain is cut-out or removed.

In this model they distribute our memories among the synapses within our physical brain and store our memories within these synapses. That's physically impossible. There are NO wires in our brain connecting our neurons together into some kind of distributed network! The ONLY way that Distributed Memory could ever work is if it is being done by Nature's Psyche at the quantum level.

WE KNOW form Near-Death Experiences (NDEs) and Shared-Death Experiences (SDEs) that our thoughts, memories, and personalities survive the death of our physical brain; so, the real question becomes, "How do our memories and experiences get uploaded to non-local space or the spirit realm?" The only logical answer that I can come up with is that thoughts and memories are quantum waves of some kind; and, our thoughts and memories are being stored as quantum waves in God's Database by Nature's Psyche or the Angels in Heaven completely outside the conscious awareness of the Human Psyche.

It took months of study, research, and thought; but for me personally, this became the smoking gun proving that it's physically impossible to store our memories in a distributed fashion among all the different neurons and synapses in our brain, because it's physically impossible to pass even a single byte of information through a synaptic cleft and have it survive the journey. Neurons are switches – they are either ON or OFF. It's physically impossible to store even a single byte of memories or information within a neuron. If you were to try to do so, all of that information is going to be reduced or integrated into a single BIT of information – either ON or OFF. That's what's really happening at the physical level within our physical brain.

If our memories are really being distributed evenly throughout our brain, it's all being done and handled at the quantum level by Nature's Psyche, completely outside the conscious awareness of the Human Psyche. Quantum waves are thoughts and memories after all; and, our memories and thoughts survive the death of our physical brain according to the empirical evidence from Near-Death Experiences, Out-of-Body Experiences, and our after-death Life Reviews.

They believe that they have made their case and proven that our memories are stored in a distributed fashion among our synapses; and, I KNOW that they have failed to make their case because I KNOW that it's physically impossible to pass memories through a synaptic cleft and neuron and have those memories survive intact during the journey. ALL synaptic input into a neuron gets reduced or integrated into a single Post-Synaptic Potential that either ends up firing the neuron or doesn't. The neuron ends up being a single hardware BIT. It's either ON or OFF.

They persist and I persist, and never once do we come to a consensus. In fact, there are occasionally moments in their book when they prove that their Distributed Memory Model is faulty, weak, and FALSE. For example:

Electrical Stimulation of the Human Temporal Lobes

One of the most intriguing and controversial studies implicating the neocortex of the temporal lobe in the storage of declarative memory traces involved electrical stimulation of the human brain [by neurosurgeon Wilder Penfield].

Are these people re-experiencing events from earlier in their life because memories are evoked by the electrical stimulation? Does this mean that memories are stored in the neocortex of the temporal lobe? Those are tough questions! One interpretation is that the sensations are recollections of past experiences. The fact that such elaborate sensations were obtained only when the temporal lobe was stimulated suggests that the temporal lobe may play a special role in memory storage.

However, other aspects of the findings do not clearly support the hypothesis that engrams are being electrically activated. For instance, in some of the cases in which the stimulated part of the temporal lobe was removed, the memories that had been evoked by stimulation in that area could be evoked by stimulation somewhere else. In other words, the memory had not been "cut out." Also, it is important to appreciate that complex sensations were reported only by a minority of the patients, and all of these patients had abnormal cortex associated with their epilepsy.

There is no way to prove whether the complex sensations evoked by temporal lobe stimulation are recalled memories. However, it is clear that the consequences of temporal lobe stimulation and temporal lobe seizures can be qualitatively different from stimulation of other areas of the neocortex. Now we'll take a closer look at the structure of the temporal lobes and the elements within them that are strongly implicated in learning and memory. (*Neuroscience*, pp. 737-738.)

It's always the exception that falsifies and debunks the rule!

In ALL of their different science experiments, the memories are NEVER "cut out", which means that the memories are NOT being stored within the physical brain itself.

Whenever they cut out the part of the brain that was triggering a specific set of memories, Nature's Psyche just remaps those specific memories to a different part of the brain. Those memories are being stored as quantum waves in the cloud, and Nature's Psyche can map those non-local memories to our physical brain in any way that it sees fit. That's how the neurosurgeons are able to cut out any part of our brain that they want, and our long-term consolidated memories remain.

Memories and thoughts are quantum waves. Quantum waves are produced by Psyche. It's physically impossible to produce a quantum wave at the physical level!

Nature's Psyche can map our long-term quantum memories stored in the cloud to any specific physical neuron that it sees fit to map those memories; thereby, thanks to the re-mapping capabilities of Nature's Psyche and the non-local cloud storage of our quantum memories by Nature's Psyche, our memories can seem to move from one physical neuron to the next however Nature's Psyche deems fit to migrate or re-map our quantum memories from one physical neuron to the next.

Memories and thoughts are quantum waves, and they survive the death of our physical brain if they are truly memories. Nature's Psyche can remap our quantum memories to any physical neuron that it sees fit. Whenever the Human Psyche chooses to remember something specific, Nature's Psyche triggers the associated neurons that it has mapped to those specific quantum memories and experiences. Remember, quantum memories or quantum waves survive the death of our physical brain, because they are NOT physical in nature and origin.

They mentioned "declarative memory traces" being stored within our physical brain. But, is it really memory storage at the physical level? Or, is it the fact that the quantum memories that Nature's Psyche has mapped to that specific neuron are being triggered and then downloaded from non-local cloud storage whenever that specific neuron is stimulated by electricity?

Which model has the most explanatory power?

If the memories are really being stored at the physical level, then why do they remain when that part of the brain is cut-out? The physical model can't explain how cut-out memories remain, but the quantum model definitely can!

The quantum model has infinitely more explanatory power than the physical model!

Who is choosing which image or experience to download into that person's brain whenever that specific neuron is given electrical stimulation? Why that memory and not some other memory? There's selectivity or choice going on here that's physically impossible. Atoms, molecules, and neurons can't do choice according to the Materialists, Naturalists, and Behaviorists. Choice is physically impossible according to these people.

If you want choice that actually works, you have to turn to Nature's Psyche, the Human Psyche, and God's Psyche at the quantum level, because choice is physically impossible at the physical level according to the Materialists, Naturalists, and Behaviorists.

Memories, thoughts, and choices are quantum waves being produced by the Human Psyche and stored in the cloud for us by Nature's Psyche.

By cutting out different parts of the brain and having the quantum memories remain, these scientists repeatedly prove that our quantum memories are NOT being stored within our physical brain. They repeatedly prove that there are no memory engrams and no physical RAM within our brains! They repeatedly prove that our quantum memories are being stored for us non-locally in the cloud by Nature's Psyche, and that Nature's Psyche can map those non-local memories to any physical neuron that it wants to. They repeatedly prove that it's mapped functionality and quantum memory downloads that are being activated by electrical stimulation, and NOT a memory engram or RAM.

They emphasize that there's no way to prove that those recalled memories are being stored within that part of the brain. In fact, they provided convincing evidence that those recalled memories are not being stored within that part of the brain that subsequently gets cut out by the surgeon.

There's no way to prove that they are right, and there's no way to prove that I am wrong – at least not at the physical level. You can't use Materialism, Naturalism, Darwinism, Nihilism, Behaviorism, and Atheism to prove anything of value – not even at the physical level.

However, out-of-body travelers (OBErs) and Near-Death Experiencers (NDErs) form new memories, have new thoughts, and experience new and interesting environments at the quantum level or the psyche level, even while their physical brain is dead or offline. Thoughts and memories are quantum waves, and these people actually remember these experiences should they live through their ordeal and their physical brain regains functionality.

It's impossible to prove that I am right or that I am wrong at the physical level; but, it's totally possible to prove that I am right at the quantum level or the psyche level, which is why the Materialists and Naturalists and Atheists deny the existence of the quantum level or the psyche level. These people have no other choice, because the proven existence of Quantum Mechanics, Quantum Non-Local Consciousness, the quantum level, and the psyche level FALSIFIES Materialism, Naturalism, Darwinism, Nihilism, Behaviorism, and Atheism.

The Materialists and Naturalists are determined to make sure that all your memories are stored within your brain and no place else, so that your memories and their responsibility towards you cease to exist when we die. They have an agenda. They are trying to avoid the judgments of God and the punishment for their sins. They want to keep sinning in the hope that they will never be called upon to make an accounting for their sins. They will indeed get their wish if they can somehow get their memories, our memories, and God's memories to cease to exist when our physical brains die. They never once realize that memories and thoughts are quantum waves, and that it's physically impossible to destroy a quantum wave.

Amazing, is it not?

Examine their motives in order to figure out what they are trying to do and why they are trying to do it.

Our explanatory power goes ballistic whenever we choose to bring the quantum level and the psyche level into play. Nature's Psyche can do the physically impossible at will, at the quantum level; and, Nature's Psyche can map any quantum wave or memory that it wants onto a specific neuron at the physical level. The smaller can reside within and control the larger. This reality and truth explains how concept cells work – both at the quantum level and the physical level.

In contrast, the Materialists and Naturalists cripple themselves and eliminate their explanatory power by limiting themselves exclusively to the physical level.

Thoughts, memories, and choices are quantum waves being produced by Psyche at the quantum level. They survive the death of our physical body and physical brain.

Memory storage and memory access at the quantum level by Nature's Psyche ends up becoming the very heart of Quantum Neuroscience. Memory storage was always the most interesting subject in Neuroscience to study and try to understand; and, it remains so even at the quantum level – especially at the quantum level.

They persist, so I persist. Persistence is not futile, especially when you are dealing with the quantum level. The physical level is transitory and ephemeral; but, the quantum level is eternal and everlasting.

Oh, this whole book is going to be extremely painful for the Materialists and Naturalists – both physically and spiritually. But, I can soothe my conscience with the knowledge that they will never buy and read this book.

The explanatory power of Quantum Neuroscience is through the roof, which is why I like it so much.

I have infinitely more weaponry in my arsenal than the Materialists and Naturalists do, because I can demonstrate that they are wrong both from the physical level and the quantum level. In fact, I can always prove them wrong from the quantum level; and sometimes, I can even prove them wrong from the physical level. In return, they can't do anything to touch me, because Materialism, Naturalism, Darwinism, Nihilism, Behaviorism, and Atheism don't work at the quantum level or the psyche level. These falsified philosophies ONLY exist at the physical level, and most of the time they don't even work at the physical level. Most of our physical processes at the atomic and molecular level actually require help from the quantum level and the psyche level, in order to make them work as advertised in the first place.

Materialism, Naturalism, and their derivatives can't survive the evidence. All of the evidence that we have on hand as a race proves that Materialism and Naturalism are false; and, both Materialism and Naturalism have NO evidence to support their major premises or primary assumptions which claim that the psyche level and the quantum level do not exist. The very existence, the proven existence, of quantum mechanisms FALSIFIES Materialism, Naturalism, and their derivatives.

They will persist in trying to convince us that all of our memories are being stored exclusively within our physical brain at the physical level; and, I will persist in demonstrating that that is physically impossible both from the physical level and the quantum level.

Let's get on with it.

The monkeys studied by Klüver and Bucy following temporal lobectomy had a peculiar way of interacting with their environment, in addition to a host of other abnormalities. The monkeys explored their room by placing objects in their mouth. If an object was edible, they would eat it; if it wasn't, they would drop it. However, their behavior suggested that they did not have basic perceptual deficits. In the words of Klüver and Bucy, they exhibited "psychic blindness" — even though they could see things, they did not appear to understand with their eyes what the objects were. They would repeatedly go back to the same inedible objects, put them in their mouth, then toss them aside. This problem with object recognition is probably related to memory function in the temporal lobe. (*Neuroscience*, p. 738.)

The temporal lobe is a machine that has been mapped by Nature's Psyche to do object recognition, not memory storage. They remembered that they were supposed to find something to eat, and they proceeded to do so. They also remembered whether things are edible or not, once these things were in the mouth.

Memories and thoughts are quantum waves. I don't think they need translation in order to be stored non-locally in the cloud. Memories survive the death of our physical body and physical brain, so they don't really need a brain in order to form or come into existence. Consequently, I don't think the brain is being used to record memories into the cloud. However, I do believe that whenever the Human Psyche wants to recall something that is important to it, Nature's Psyche has to download those memories from the cloud into the brain so that their physical components become accessible once again.

Due to the missing machinery, the Monkey's Psyche isn't getting the information that Nature's Psyche has mapped to the missing machinery; so, the monkey has to try a different method for identifying edible food. There really is a psychic blindness. Nature's Psyche has mapped object recognition to the temporal lobe. Just like they say, it's a "memory function" and NOT actual memory storage. The memory of what these objects are is being stored someplace else, because they KNOW whether the object is edible or not once they get the object inside their mouths.

The whole purpose of Quantum Neuroscience is to try to figure out what Nature's Psyche is doing for us at the quantum level. Its looks like to me that Nature's Psyche is mapping physical functionality into our brains for us at the quantum level. According to the Orthodox Interpretation of Quantum Mechanics, Nature's Psyche is the thing that collapses the wave function thereby making the Human Psyche's desires and commands physical and real.

The human brain is comprised of 100 billion stand-alone switches, with NO wires connecting them. How are you going to create CPUs and RAM out of a bag full of switches with no wires to connect them? You're not! It's physically impossible. If there are any CPUs or RAM within the human brain, they are being mapped into the human brain at the quantum level by Nature's Psyche. This is the way our brains really work – both at the quantum level and the physical level. The preponderance of the evidence tells us that it is so.

They discuss the case of H. M. and his medial temporal lobe amnesia. Medial temporal lobe amnesia is a loss of physical functionality due to damage or removal of the medial temporal lobes. Its major identifying feature is anterograde amnesia for explicit memories combined with preserved intellectual functioning. The physical machinery, which

Nature's Psyche has assigned to experience physical events and to consolidate those events as quantum waves or memories into long-term non-local storage or God's Database, has been destroyed. Without the physical machinery for storing and retrieving those types of memories, it becomes physically impossible to do so. However, Nature's Psyche and the Angels in Heaven will continue to make a record of the events; but, Nature's Psyche isn't obligated to share what it knows with the Human Psyche. In fact, it seems as if God has commanded Nature's Psyche not to do so.

These are fascinating realities that become obvious once we finally choose to allow all of the evidence into evidence.

> **Milner has said it appears that H.M. forgets events as quickly as they occur. If he is told to remember a number and is then distracted, he will not only forget the number, he will also forget that he had even been asked to remember one.**

> **To be clear about the nature of H.M.'s amnesia, we must contrast what was lost with what is retained. He remembers his childhood, so long-term memories formed before the surgery and his ability to recollect past events were not destroyed. His working memory is also normal. For instance, with constant rehearsal he can remember a list of six numbers, although any interruption will cause him to forget. He simply has an extreme inability to form new declarative memories.**

> **He has learned a very small number of declarative facts since his surgery. For example, he can recognize and name a few people who became famous after his surgery, such as President John Kennedy. He also learned the floor plan of a home he moved to after the surgery. These rare remembered facts probably result from extensive daily repetition.**

> **H.M. is also able to learn new tasks (i.e., form new procedural memories). For example, he was taught to draw by looking at his hand in a mirror, a task that takes a good deal of practice for anyone. The odd thing is that he has learned to perform new tasks, despite the fact that he has no recollection of the specific experiences in which he was taught to do them (the declarative component of the learning).**

> **The characteristics of H.M.'s amnesia reinforce the idea that the neuroanatomy and neural mechanisms underlying procedural and declarative memory, as well as short-term and long-term memory, are not identical. In our search to understand the role of the medial temporal lobe in learning and memory, we are led to focus on the processing and consolidation of new declarative memories.** (*Neuroscience*, 739.)

His childhood memories were stored someplace else besides his brain.

Memories are quantum waves, and it's physically impossible to store memories and information and data in a synaptic cleft at the physical level. If any memories are being stored in our synapses, they are being stored at the quantum level by Nature's Psyche.

Any consolidated memories from before the surgery are apparently mapped into the cloud by Nature's Psyche, because you can cut out any part of the brain you want, and those types of retrograde memories seem to remain intact.

The machinery was damaged, not the memories themselves. The memories or quantum waves that show up in our after-death Life Review cannot be touched nor changed by physical brain damage. Likewise, his retrograde childhood memories cannot be removed, because they are now being stored and accessed someplace else besides the machinery that was removed.

Nature's Psyche had mapped motor functions and skills (procedural memories and implicit memories) to parts of his brain that were not removed. So, it's logical that those Maps of Physical Functionality should remain intact.

Nature's Psyche had mapped declarative memory FUNCTIONALITY to the parts of the temporal lobe that were removed, meaning that going forward, H. M. wasn't able to use that map and that part of his brain for declarative memory formation and retention. All of his pre-surgery declarative memories remained, because they had already been stored non-locally in the cloud by Nature's Psyche, and can supposedly be accessed from any part of the brain.

Ironically, if the physical model were actually true, and all of our memories are being stored within our brain and no place else, then it should be physically impossible for H. M. to form any new declarative memories. Yet, unlike most accounts of his story, these people actually mention instances where H. M. formed and stored new declarative memories. That's physically impossible! So, how was it done? The only thing that makes logical sense is that Nature's Psyche found a way to map the declarative memories that were most important to H. M. into his brain somewhere else besides the temporal lobe, despite the fact that H. M. no longer had the physical machinery needed to form and store new declarative memories.

Whenever the map or machinery is damaged, the consolidated memories remain, suggesting that the memories or quantum waves are being stored someplace else besides the physical machinery.

In the medial temporal lobe, a group of interconnected structures appears to be of great importance for declarative memory consolidation.

The effects of temporal lobectomy, and particularly the amnesia of H.M., make a strong case that one or more structures in the medial temporal lobe are essential for the formation of declarative memories. If these structures are damaged, severe anterograde amnesia results. (pp. 740, 741.)

If you damage or remove the Map of Physical Functionality, then both the Human Psyche and Nature's Psyche can't use that particular Map for any type of physical functionality, including the experiencing and remembering of the associated physical experiences. Wipe out the Map that Nature's Psyche created for visual functionality, and suddenly it becomes impossible for the Human Psyche and Nature's Psyche to use that Map to see at the physical level. Wipe out the Map that Nature's Psyche created for the experiencing and remembering of events, and suddenly it becomes impossible for the Human Psyche and Nature's Psyche to use that Map to experience and remember events or declarative memories that are physical in nature and origin.

This is simple and parsimonious to understand, once a person chooses to do so.

Notice that these people are actually stating that the medial temporal lobe is being used or mapped for the "formation" of declarative memories and the "consolidation" of declarative memories, NOT the storage of declarative memories. If the declarative memories were actually stored within the medial temporal lobe as the Materialists and Naturalist claim, then ALL of his retrograde declarative memories should be gone; but, they are not.

Memories and thoughts are quantum waves, produced by Someone Psyche.

The declarative memories and episodic memories that show up in our after-death Life Review definitely are NOT stored within our physical brain at the physical level.

What other memories or quantum waves aren't stored within our brain at the physical level?

A physical brain isn't needed for storing, indexing, and consolidating quantum waves, which means that we can continue to have thoughts, memories, and experiences long after our physical brain is dead and gone.

It's clear to me from the evidence that Nature's Psyche has mapped the medial temporal lobe to the task of memory consolidation or quantum wave consolidation. If the Map of Physical Functionality is missing, then physical experiences can no longer be mapped or consolidated into the cloud.

The question I have is whether these consolidated memories or declarative memories FAIL to get stored into Non-Local Consciousness or Nature's Psyche or God's Database in the cloud, or not. I don't know. I can visualize it happening either way; but, I have changed my beliefs as additional evidence has come to my attention.

The Human Psyche ONLY knows what makes it into its physical brain, because the Human Psyche is slaved to the physical brain for the duration of the person's mortal life. The human brain and the Human Psyche are in sync, or in some kind of a symbiosis. The Human Psyche can only know what comes to it through its physical brain, unless of course God were to reveal something additional to the Human Psyche. Revelations from God are the reason why the prophets of God seem more intelligent and wise than the rest of us are.

Nature's Psyche KNOWS infinitely more and has access to infinitely more information than the Human Psyche does; and, Nature's Psyche is NOT obligated to share what it knows with the Human Psyche. In fact, Nature's Psyche has been commanded by God not to share what it knows with the Human Psyche. Nature's Psyche has direct access to God and to God's Database. It's theoretically possible for Nature's Psyche to know everything. The only thing Nature's Psyche can't do is that it is forbidden to repair the damaged genes and proteins that evolution or random mutations have given to us. Evolution is entropy or the second law of thermodynamics – random mutations; and, Nature's Psyche is forbidden from redeeming us from the fall of Adam and Eve.

I KNOW that someone else besides our brain is making our after-death Life Review for us. I assume that these declarative memories are being stored by Nature's Psyche or the Angels in Heaven in God's Database in the cloud, whether there is brain damage or not. NONE of our Life Review memories and events are being stored within the Human Psyche or the physical brain, because the Human Psyche doesn't have instant access to these things. It's obvious that our Life Review memories can't be stored in our physical brain and then expected to survive the death of our physical brain.

I get the impression that everything in the physical brain links with and uploads to non-local storage in the cloud. The problem comes when the Map of Physical Functionality isn't formed in the first place or gets destroyed somewhere along the way.

I believe that Someone Psyche is storing all of our quantum waves or memories for us in the cloud, whether we have the necessary physical machinery or not. People who were born blind can see after they are dead, and they have video in their Life Review. The physical machinery wasn't necessary for the production of their Life Review. However, the physical machinery was absolutely needed if the Human Psyche wanted to be able to see during its mortal life. Remember, the Human Psyche only has access to what the physical brain is able to produce and/or register. Nature's Psyche has no such limitations, and Nature's Psyche isn't obligated to share with the Human Psyche that it knows.

During one's mortal life, the Human Psyche ONLY has access to what's being registered within its physical brain – unless of course God intervenes and does something more for the Human Psyche than what it can get from its physical brain.

Attention, intention, deliberation, and choice from the Human Psyche is needed in order to get Nature's Psyche to map improved and enhanced physical functionality onto the physical brain for us. Nature's Psyche heats up or fires up the parts of our brain that are mapped to the physical objects and physical concepts that we choose to pay attention to. Nature's Psyche makes the Maps of Physical Functionality for us at the quantum level, and then Nature's Psyche uses those Maps of Physical Functionality to get things done for us at the physical level. ALL of this takes place outside the conscious awareness of the Human Psyche. The Human Psyche is only consciously aware of what registers upon its physical brain, unless the Human Psyche gets some revelation and help from God.

I hope this detailed analysis will prove helpful to someone.

They actually go through every part of the brain where they think physical memories are being stored physically within the brain. However, it is physically impossible to pass even a single byte of information or memories through a synaptic cleft; and, it's physically impossible to store memories in a synaptic cleft. If memory storage is taking place within the brain, it isn't taking place at the physical level, because it's physically impossible to store memories in a physical brain.

Since thoughts and memories are quantum waves, you can store quantum waves anywhere you want, including an atom or a physical brain. Quantum Mechanics has infinitely more explanatory power than Materialism, Naturalism, Darwinism, Nihilism, Behaviorism, Atheism, and Classical Physics.

Next, they discuss spatial memory, working memory, the hippocampus, and place cells.

When they destroy the hippocampus in rats that have been trained to run a maze, their long-term cloud storage remains intact, but their short-term working memory gets destroyed. The new spatial memories for which arms to avoid can't be stored in the hippocampus, because there is no hippocampus. Their "efficiency device" or Map of Spatial Functionality has been destroyed, and NOT the memory storage.

There's NOTHING for the rats to remember or experience at the physical level, because the spatial mapping machinery or the spatial mapping registers have been destroyed.

Once again, thoughts and memories are quantum waves; and, quantum waves can be stored anywhere that Nature's Psyche wants, even within a single atom or non-locally in the cloud.

Place Cells

In a fascinating series of experiments begun in the early 1970s, John O'Keefe and his colleagues at University College London showed that many neurons in the hippocampus selectively respond when a rat is in a particular location in its environment. (*Neuroscience*, pp. 746-747.)

Oh yes, place cells along with concept cells are some of the most interesting neurons to study. These things are alive, sentient, and aware.

Remember, neurons are stand-alone switches. There are NO wires connecting them into some kind of physical CPU or RAM.

How does the place cell KNOW that the rat is in a particular location? It's firing unilaterally with NO apparent input – no physical input at least.

Who assigns that specific place or location to that specific neuron?

There's choice or specificity or selectivity going on here. Choice and selectivity are functions of Psyche. They are quantum waves of some kind!

Who designed these place cells to perform this specific function? What makes them choose to respond? How do they KNOW that they are in a familiar environment or place? Who programmed them or mapped them for this specific functionality?

The neuron "selectively responds" which means that it unilaterally turns on whenever the rat is the specific location that Nature's Psyche has assigned to that specific neuron or place cell. According to the Orthodox Interpretation of Quantum Mechanics, it is Nature's Psyche who collapses the necessary wave functions and turns on specific neurons for us. It would also be Nature's Psyche who is mapping specific locations or places onto specific neurons or place cells.

Performance in the radial arm maze, discussed above, may utilize these place cells, which code for location. Of particular importance in this regard is the finding that the place fields are dynamic. For instance, let's say we first let a rat explore a small box, and we determine the place fields of several cells. Then we cut a hole in a side of the box so the animal can explore a larger area. Initially, there will not be any place fields outside the smaller box. But after the rat has explored its new expanded environment, some cells will develop place fields outside the smaller box.

These cells seem to learn in the sense that they alter their receptive fields to suit the current environment. It's easy to imagine how these sorts of cells could be involved in remembering arms already visited in the radial arm maze. And, if they are involved in running the maze, it certainly makes sense that performance is degraded by destroying the hippocampus.

Whether or not there are place cells in the human brain is not known. However, PET imaging studies show that the human hippocampus is activated in situations involving virtual or imagined navigation through the environment. (*Neuroscience*, p. 748.)

Code requires a Programmer; and, so does mapping. If you are going to map or encode physical functionality into a specific neuron or place cells, it's going to require a Map-Maker or a Programmer like Nature's Psyche in order to do so. The Human Psyche isn't consciously aware of any of these things, because they are taking place at the quantum level and not the physical level.

These are quantum level maps that Nature's Psyche is making. Memories and thoughts are quantum waves.

These specific stand-alone neurons or switches are learning and remembering; but, that's physically impossible. If I were to give you a light switch, how would you make it learn and make it remember things? You wouldn't – not at the physical level at least. A switch is either ON or OFF; and, a neuron is a switch.

There's no visible CPUs or RAM within a neuron.

A neuron is a battery – it's either charging or discharging. That's it!

To say that a neuron or place cell is learning and remembering is like saying that your car battery or flashlight battery is learning and remembering. It's physically impossible!

If these neurons are in fact learning and remembering, then it's happening at the quantum level under the command and control of Nature's Psyche, because it's definitely not happening at the physical level.

Place fields are quantum waves.

Place cells are dynamic, meaning that they can be reused and remapped by Nature's Psyche to refer to a different location than the one that they referred to before.

These neurons aren't altering their receptive fields, or place fields, or quantum waves, because that's physically impossible. Nature's Psyche is the one who is altering their fields or quantum waves at the quantum level.

Remember, neurons are switches. They are either ON or OFF.

If I were to give you a bag full of switches and NO wires, how would you hook them together into functional Memory Systems, CPUs, and RAM? You wouldn't. Not at the physical level at least! In order to get CPUs, RAM, and Memory Systems out of 100 billion stand-alone switches or neurons, you are going to have to get Nature's Psyche to "wire" them together for you at the quantum level. There's no other way it can be done.

Remember, thoughts and memories are quantum waves or vibrating strings. They are produced, stored, and accessed at the quantum level by some kind of Intelligence or Psyche. Thoughts, memories, quantum waves, vibrating strings, Intelligence, and Psyche survive the death of your physical body and physical brain because they exist at the quantum level and not the physical level. There is NO entropy at the quantum level or spiritual level, which means that nothing ages and dies in the quantum realm or the spirit realm or the psyche realm. The quantum level or psyche level is pure syntropy, which means that it has NO physical limitations whatsoever. In contrast, evolution is limited to the physical level. Evolution can't do anything at the quantum level, because evolution has no Intelligence or Psyche.

I learn best through comparison and contrast, which is why I use it so much in my writings.

They spend some time discussing the physical functionality of the brain, never once realizing that this physical functionality is being mapped and put into place by Nature's Psyche at the quantum level.

Our primary point is that the hippocampus is particularly active in this spatial navigation task with humans, just as it is in rats. The caudate activation is thought to reflect movement planning. (*Neuroscience*, p. 749.)

Through brain scans, they can see parts of the brain being turned on or activated.

Someone Psyche is reaching out telepathically and activating or turning on the hippocampus and caudate nucleus! It happens when the Human Psyche chooses to pay attention to something specific that has been mapped by Nature's Psyche into the physical brain for that specific purpose.

The brain simply idles in default mode, until the Human Psyche decides to pay attention to something specific.

The caudate nucleus is typically mapped by Nature's Psyche for use in stimulus-response habit formation. The hippocampus is typically mapped by Nature's Psyche for use in spatial navigation.

In Olton's original studies using the radial arm maze, he described the result of hippocampal lesions as a deficit in working memory. The rats were not able to retain recently acquired information concerning arms already explored. Thus, working memory might be one aspect of hippocampal function. This would explain why the rats with lesions could avoid going down arms that never contained food but still not remember which arms they had recently visited. Presumably after training, the information about no-food arms was saved in long-term memory, but working memory was still required to avoid the arms where food had already been retrieved. (*Neuroscience*, p. 749.)

The rats always remember their long-term consolidated memories despite the lesions. How's that possible, if all of our memories are being stored within our brain and no place else?

Nature's Psyche maps and then repeatedly re-maps the hippocampus for spatial navigation; and, the net effect is some kind of working memory that is being implemented and constantly remapped at the quantum level by Nature's Psyche. Remember, it's physically impossible to map memories onto the physical brain at the physical level, thanks to the randomizing and scrambling effects of the synaptic cleft. If spatial memories are being mapped onto our physical brain, it's taking place at the quantum level, because it's physically impossible at the physical level.

They mention long-term storage. Karl Lashley and others have proven that there is NO long-term storage, or NO memory engrams, or no RAM within a physical brain; consequently, we KNOW that our long-term memories are being stored in the cloud in God's Database, because our long-term memories, thoughts, and personality survive the death of our physical brain according to the empirical evidence from Near-Death Experiences (NDEs), Shared-Death Experiences (SDEs), and our after-death Life Reviews.

Cohen and Eichenbaum describe the function of the hippocampus, in conjunction with other structures of the medial temporal lobe, as involving relational memory. The basic idea of relational memory is that highly processed sensory information comes into the hippocampus and nearby cortex, and memories are formed in a manner that links all the things happening at the time.

Interconnectedness is a key feature of declarative memory storage. (*Neuroscience*, 749.)

It's being dynamically indexed and linked – just like our after-death Life Review!

How?

That's physically impossible!

ALL of this linking and interconnectedness is physically impossible at the physical level thanks to the randomizing, scrambling, and integrating effects of the synaptic clefts.

Oh, all of this interconnectedness, linking, mapping, and programming is taking place, all right! It's just taking place at the quantum level or the psyche level, because it's physically impossible at the physical level.

THIS!

Spatial navigation. In this experiment, a rat navigates its environment based on (a) a spatial map or (b) a series of relational memories.

To navigate its environment, a rat could use either a mental map of space or relational memories associated with environmental cues. The distinction between the spatial map and relational memory hypotheses is illustrated in Figure 24.19. If there is a spatial map, one would expect place fields to be ordered in the hippocampus as the locations are in space, much like the retinotopic receptive fields in visual cortex. Experts disagree about the extent to which hippocampal place cells provide such an organized map of the entire area around the animal. In a relational memory scheme, neurons encode information about place as a series of simple associations between nearby objects and concurrent sounds and smells. (*Neuroscience*, p. 750).

This is where my idea – for Maps of Physical Functionality being produced by Nature's Psyche at the quantum level – came from. The relational memory scheme is physically impossible, because it's impossible to encode information into neurons at the physical level thanks to the scrambling effects of our synaptic clefts. In contrast, it's totally possible for Nature's Psyche to map and then remap physical functionality onto parts of the physical brain at the quantum level.

Without realizing it, these people are talking about "fields" or quantum waves! They are talking about "spatial maps," an "organized map", and a "mental map of space". They are talking about Quantum Neuroscience. They are saying all of this without actually saying it, because all of this Quantum Neuroscience stuff is physically impossible. They refuse to go into the physically impossible; but, the physically impossible is where science really starts to get interesting.

The hippocampus is mapped for spatial location and spatial discrimination at the quantum level by Nature's Psyche. The hippocampus is NOT a physical memory storage device, because it's physically impossible to store BYTES of physical memory within a physical brain.

When I read their Spatial Map Hypothesis, suddenly I KNEW that Nature's Psyche is mapping Spatial Maps into specific place cells within our brain at the quantum level. My Maps of Physical Functionality Theory was born. Of course, the whole thing was primed by comments such as the following:

The retina-geniculate-striate system is retinotopic; each level of the system is organized like a map of the retina. This means two stimuli presented to adjacent areas of the retina excite adjacent neurons at all levels of the system. (*Biopsychology*, p. 141.)

Each level of the system is organized like a map, including the quantum level or the psyche level! How do the axons know how to produce and maintain the map? These maps scale up as the body grows and the brain develops. How do the axons do that? How do the axons know how to do that? Mapping of any kind requires a Mapmaker! Maps don't spontaneously generate out of thin air. That's physically impossible. Someone Psyche is overseeing all of this at the quantum level so that it turns out right at the physical level. This is the best explanation for what we are observing.

> **He discovered that the human primary somatosensory cortex is somatotopic – organized according to a map of the body surface. This somatotopic map is commonly referred to as the somatosensory homunculus.** (*Biopsychology*, p. 174.)

How does it do that?

This mapping is NOT being done by the physical matter at the physical level.

How do we know?

We know that Nature's Psyche is doing all of this mapping for us, because if a person with hydrocephalus doesn't have a primary somatosensory cortex at the physical level, Nature's Psyche can still map that physical functionality onto a different part of the brain if it chooses to do so.

Functionally normal people with only 3% of a brain end up being proof of concept for Quantum Neuroscience, as far as I'm concerned. They lack 97% of the physical functionality, but they are still functioning normally. That's physically impossible; but, completely possible when Nature's Psyche and Quantum Mechanics are in control of the situation.

The amazing thing about Maps of Physical Functionality is that Nature's Psyche can choose to map that functionality to different parts of the brain, especially when the part of the brain traditionally selected for the task is missing or severely damaged. Maps are versatile. They can be mapped to anything by the Mapmaker, Nature's Psyche.

Maps of Physical Functionality was also put into my mind by chapters such as this one, entitled "The Mapmakers", from *The Mind and the Brain: Neuroplasticity and the Power of Mental Force* by Jeffrey M. Schwartz:

> http://publicism.info/psychology/mind/6.html

May the Schwartz be with you!

> **Although the content of consciousness depends in large measure on neuronal activity, awareness itself does not.... To me, it seems more and more reasonable to suggest that the mind may be a distinct and different essence.** — Wilder Penfield, 1975.

I simply KNEW what was going on within our brain – both at the quantum level and the physical level – once I understood and accepted the Maps of Physical Functionality that Nature's Psyche is actively and dynamically mapping onto our brain both at the quantum level and the physical level.

Maps of Physical Functionality is the BEST way to describe brain functionality – both at the quantum level and the physical level. It covers everything.

Maps of Physical Functionality are infinitely scalable.

Nature's Psyche can put a detailed map of London into your hippocampus when you need it. Then Nature's Psyche can re-map your hippocampus with a 3D spatial map of your home and where everything is located within your home and yard. Then Nature's Psyche can re-map your hippocampus with a map of your favorite fishing holes when that is needed. Nature's Psyche can even help you to build a mental map of the innards of a living cell within your hippocampus if you want it to do so. Change locations or change your area of study, and Nature's Psyche simply changes the map. Powerful!

Through quantum mechanisms, Nature's Psyche can download a detailed map of London into your hippocampus and your place cells, assuming that you spent the effort trying to memorize it in the first place.

Through quantum mechanisms, Nature's Psyche can map petabytes of video or non-local memory storage onto a single atom or a single neuron or a single concept cell at the quantum level, if Nature's Psyche chooses to do so.

Through quantum mechanisms, Nature's Psyche can map the quantum output of exabytes of Parallel Quantum Computer Processing into a single atom or a single neuron or a single brain region if it chooses to do so.

This is Quantum Neuroscience. This is COOL SCIENCE! It's radically advanced compared to anything that the Materialists, Naturalists, Darwinists, Behaviorists, and Atheists can give us.

I hope you will forgive me for finding all of this so fascinating; but, somebody had to find it fascinating or it wouldn't have gotten done.

When we bring the quantum level and Nature's Psyche into play, BOTH **the spatial map and the relational memory hypotheses** become possible at the quantum level or psyche level. Nature's Psyche chooses how to map our memories – spatially or relationally. Everything becomes possible at the quantum level under the command and control of Nature's Psyche.

We know that Nature's Psyche is choosing what each specific neuron will respond to, because all of this mapping of physical functionality is taking place outside the conscious awareness of the Human Psyche.

> **Evidence also suggests that place cells may encode relational information of types other than spatial location. A simple example is that the responses of neurons with place fields are sometimes also affected by other factors, such as the speed or direction the rat is moving.**

> **The researchers found that some neurons in the hippocampus became selectively responsive for certain pairs of odors. Moreover, the neurons were particular about which odor was at which port — they would respond strongly with odor 1 at port A and odor 2 at port B, but not with the odors switched to the opposite ports. This indicates that the response of the hippocampal neurons relates the specific odors, their spatial locations, and the fact that they are presented separately or together. It was also shown that hippocampal lesions produce deficits on this discrimination task.**
> (*Neuroscience*, p. 750.)

How and where is all of this encoding or mapping taking place? Encoding, mapping, and programming are a function of Someone Psyche!

These neurons are being mapped and then remapped as needed. That's physically impossible, because a neuron is nothing more than a single hardware BIT. If there's any mapping, or programming, or encoding taking place within a neuron, it's taking place at the quantum level or the psyche level.

Who is doing the selection? Selection is a function of Someone Psyche.

Who mapped this physical functionality onto these specific neurons?

https://www.ncbi.nlm.nih.gov/pmc/articles/PMC2572713/

https://en.wikipedia.org/wiki/Place_cell

Like every other structure within the physical brain, the hippocampus is comprised of millions of stand-alone switches or hardware BITS that we call neurons. It's physically impossible to store petabytes of 3D spatial information within the 65 million neurons typically associated with the hippocampus, amygdala, and entorhinal cortex. It's physically impossible for a hippocampus to function as a CPU, RAM, or a Memory System because there are NO wires within our brain. But, what do we observe? We observe the hippocampus functioning as if it is an extremely complex CPU Memory System with tons of RAM and a massive hard-drive backup. That's physically impossible, but it's happening anyway!

How's that being done? It's physically impossible. It can't be done, yet it's being done.

The very existence of something like the hippocampus – its functionality, versatility, scalability, and sheer processing power – is proof of concept where Quantum Neuroscience is concerned. The whole thing is physically impossible. It's physically impossible because stand-alone switches can't function as CPUs, RAM, and Memory Systems; yet, there it is anyway.

Since the functionality of the hippocampus is clearly there and evidently real, and since the whole thing is physically impossible thanks to the randomizing and integrating effects of our synaptic clefts, it's clear to me that all of this functionality is being produced, deployed, implemented, and operated at the quantum level or the psyche level by Nature's Psyche. We KNOW that it has to be Nature's Psyche mapping all of this functionality into the hippocampus at the quantum level, because the Human Psyche isn't consciously aware of any of these things.

The Human Psyche makes the choice and pays the attention; and, Nature's Psyche produces the functionality both at the quantum level and the physical level. That's the way our brains work.

Remember, if your Interpretation of Quantum Mechanics cannot explain what the Human Psyche and Nature's Psyche are doing at the quantum level to produce all of the brain functionality that we seem to be witnessing at the physical level, then your Interpretation of Quantum Mechanics is absolutely worthless. The Orthodox Interpretation of Quantum Mechanics from Henry P. Stapp explains what the Human Psyche and Nature's Psyche are doing at the quantum level in order to produce everything that we are witnessing and experiencing at the physical level.

Remember, the physical neurons have to be there, or Nature's Psyche will have nothing to map. Obviously, if the Map of Physical Functionality is removed at the physical level, then it can't be used at the quantum level by Nature's Psyche to map different things into our physical brain. The quantum level is the basis or the foundation of the physical level; but, the physical level has to be there if we want to have physical functionality at the physical level.

Due to its superior explanatory power, versatility, scalability, and sheer functionality, Maps of Physical Functionality end up being the BEST way to describe brain function – BOTH at the quantum level and the physical level – because Maps of Physical Functionality bridge or interface ALL levels of existence into one united whole and work at every level of reality or existence.

If Nature's Psyche needs to map non-local RAM storage or a whole Memory System into the place cells in the hippocampus, it can at will. If Nature's Psyche needs to map zillions of quantum computers to the neural input from our visual system and then have the output of these zillions of quantum computers dump into specific neurons within our hippocampus or concept cells, Nature's Psyche can do so at will.

Powerful, is it not?

Procedural Memory

The basal ganglia are important for the control of voluntary movements. (*Neuroscience*, p. 751.)

What the Neuroscientists call "procedural memories" are associated with and involved with voluntary movements. These authors spend some time explaining the experimental evidence that describes how procedural memories are being mapped into our physical brain.

Voluntary movement involves choice, which is a function of Someone Psyche.

How do we know?

WE KNOW because choice is physically impossible according to the Materialists, Naturalists, Darwinists, Behaviorists, and Atheists. Whenever choice is observed, we automatically KNOW that it's taking place at the quantum level under the command and control of Someone Psyche. Attention, intention, and choice are the smoking guns where Psyche is concerned.

Some of these scientists noticed that certain types of working memory continue to function when the hippocampus is cut out, so suddenly they were looking for other places in the brain where working memory can be stored. They settled on the Neocortex, particularly the Prefrontal Cortex.

Of course, if you are distributing a person's memories all throughout the Neocortex of that person's brain through some kind of distributed memory system, as the Materialists and Naturalists are trying to do, then it's theoretically possible to cut out anything from the brain and still have all your long-term consolidated memories remain. If you put ALL of your memories into each and every neuron in your brain, then as long as you have one neuron in your brain, you still have all your memories. Of course, distributed memory is physically impossible; but, they don't know that. They don't understand how the brain really works. They even admit as much when pushed into a corner and forced to do so.

We have seen that working memory may be one of several memory functions of the hippocampus, but there appear to be many structures in the brain that temporarily store information for working memory.

Because it is so well developed in humans, prefrontal cortex is often assumed to be involved in those characteristics that distinguish us from other animals, such as self-awareness and the capacity for complex planning and problem solving. One reason for thinking that prefrontal cortex may be involved in learning and memory is that it is interconnected with the medial temporal lobe and diencephalic structures previously discussed. (*Neuroscience*, p. 754.)

Under the CPU Model of Brain Functionality that I and many others developed (before I knew better), the physical brain is comprised of multiple different Cerebral Processing Units (CPUs); and, as any computer scientist like me knows, CPUs have cache or working memory. Under this CPU Model, working memory is CPU cache. With many different CPUs within our brain, there would be many different types of working memory within our brain.

The CPU Model has a few weaknesses.

Self-awareness, perception, complex planning, thought, and problem solving – it has been observed that we maintain or enhance these abilities after our brain is dead and gone.

Furthermore, the CPU Model of Brain Functionality doesn't work at the physical level thanks to the scrambling and integrating effects of our synaptic clefts. If there are indeed Cerebral Processing Units or CPUs within our brain, they are being implemented and controlled at the quantum level by Nature's Psyche because they are physically impossible at the physical level. We literally have to bring Nature's Psyche and Quantum Mechanics online if we want to have CPU functionality and RAM storage within our physical brain.

The Prefrontal Cortex and Working Memory

This cell responds strongest during the delay period, when there is no visual stimulus.

The neurons in prefrontal cortex have a variety of response types, some of which may reflect a role in working memory. Figure 24.26 shows two response patterns obtained while a monkey performed a delayed-response task. The neuron in the top trace responded while the animal saw the location of the food, was unresponsive during the delay interval, and responded again when the animal made a choice (Figure 24.26a). The response of the neuron simply correlates with the presentation of the stimuli.

Perhaps more interesting is the response pattern of the other neuron, which fired only during the delay interval (Figure 24.26b). This cell was not directly activated by the stimuli in the first or second interval in which it [the monkey] saw the food wells. The increased activity during the delay period may be related to the retention of information needed to make the correct choice after the delay (i.e., working memory).

This cell responds strongest during the delay period, when there is no visual stimulus. (Neuroscience, p. 756.)

Who is motivating or forcing this specific cell to keep firing, to keep responding, or to stay ON during the delay period in the complete absence of input or visual stimulus? How does it fire with NO incoming EPSPs to trigger an action potential? Where's it getting the energy from, in order to constantly generate action potentials with no EPSPs or Postsynaptic Potentials to provide that energy?

Clearly, this neuron has been designed, programmed, and mapped to function this way. Designed, programmed, and mapped by whom?

Here's additional proof of concept for Quantum Neuroscience. What this cell is doing is physically impossible; but, it's doing it anyway.

Firing neurons with NO physical input from EPSPs is physically impossible, according to everything we know about neurons.

Someone Psyche has placed information and functionality and energy into that single neuron, and that Someone Psyche keeps the neuron firing during the delay period – in the complete absence of visual input.

Someone Psyche is reaching out telepathically through quantum waves and keeping this neuron firing.

Why?

What does it accomplish?

All of this happens outside our conscious awareness, which means that it's being handled for us by Nature's Psyche at the quantum level. The Human Psyche is only aware of what registers on its physical brain, unless God is invited to give us additional information above and beyond our physical senses.

This isn't the only place where this supernatural and miraculous brain functionality is happening.

It's called default mode.

The physical brain idles in default mode, when there is NO physical input – typically while we are asleep and dreaming. Who keeps all of these billions of neurons firing in default mode, with NO physical input, while we are sleeping or while we are away traveling the Astral Plane? It is those neurons – which fire or turn ON in the complete absences of EPSP input – that we should be paying attention to and studying the most. The whole thing is physically impossible according to what we know about neurons. There's NO physical cause for these neurons to turn ON and FIRE; yet, they do.

Can you see it and understand it?

Most people can't; but, over years of study and research, I have trained myself to do so.

The very existence of default mode verifies Quantum Neuroscience and falsifies Materialism, Naturalism, Darwinism, and Behaviorism, because default mode has to be implemented and sustained at the quantum level by Nature's Psyche, because there is technically NO physical input to keep it going. That miraculous and supernatural default mode functionality, which keeps us alive at the physical level while we are sleeping, was mapped or programmed into our brain at the quantum level by Nature's Psyche and is used by Nature's Psyche to keep us alive while we are asleep and have no physical input to keep our neurons going.

Furthermore, while we are awake, our physical brain just idles in default mode unless the Human Psyche chooses to pay attention to something specific or chooses to do something specific. ONLY after the Human Psyche chooses to pay attention to something specific or chooses to do something specific DOES Nature's Psyche ramp up or heat up the specific parts of our brain that Nature's Psyche has mapped for that specific physical functionality.

Maps of Physical Functionality explain everything, both at the quantum level and the physical level. It doesn't get any better than that, in my humble opinion. All I ever wanted was to know how everything works.

Notice as these people move through this "Memory Systems" chapter, things seem to become ever more progressively impossible at the physical level; yet, they assure us that evolution is doing all of this for us at the physical level. That too is physically impossible.

The ONLY time that evolution (genetic drift or genetic change) had anything to do with your life and your creation was during the Genetic Recombination that took place while your sperm and your egg were being made; and, we know that Genetic Recombination was programmed into us by God, because it didn't exist until after God designed and created it in the first place. Remember, chemical evolution can't design and create a functional protein, let alone the matching gene to go along with it. Chemical evolution of genes and proteins is physically impossible, which means that it didn't happen, because it couldn't happen.

They've got a good thing going here.

Imaging Working Memory in the Human Brain.

Human brain imaging experiments suggest that numerous areas in the prefrontal cortex are involved in working memory.

Both experiments looked for brain activity during the delay interval while the subject had to hold information in mind. In the first experiment, this was information about faces; in the second experiment, it was information about spatial location.

Brain areas showing significant working memory activity are summarized in Figure 24.27. Six areas in the frontal lobe showed significant sustained activity during the delay period, suggesting a role in working memory.

Three areas exhibited stronger sustained activity for facial identity than spatial location, one area was more responsive to spatial memory, and two areas were active equally in facial and spatial memory tasks. An interesting unanswered question is whether working memory for other types of information is held in the same or different brain areas. (*Neuroscience*, pp. 756-757.)

Notice carefully that it was the human subject, the Human Psyche, who was told that it had to hold this information or these Working Memories in mind, during the observed delay period. These people even said as much. However, they never once make the connection between the Human Psyche and Working Memory; but, it's there hidden in plain sight within the evidence, waiting for anyone to find it who wants to find it.

Remember, it was the Human Psyche who was tasked by the experimenters to keep all of that information "in mind" during the observed delay period. If any of that Working Memory is being stored within the brain, it's being mapped and stored by Nature's Psyche at the quantum level for later use at the physical level. It's always Psyche or Mind who is trying to hold our thoughts, memories, information, or quantum waves in store for later use by the physical brain. Remember, thoughts and memories are quantum waves that are being produced, transmitted, received, processed, and stored by Someone Psyche at the quantum level or the psyche level. How do we know? We KNOW because our thoughts and memories and personality survive the death of our physical brain according to the tons of empirical evidence from Near-Death Experiences (NDEs), Shared-Death Experiences (SDEs), and after-death Life Reviews.

The Psyche is controlling all of this, not the brain. The brain is simply responding – or in the case of delay neurons, the brain is simply idling in default mode.

These authors continued to use brain scans to document other areas of the Neocortex, where there are parts of the brain that heat up and fire during the delay period, when there is NO physical input to keep those neurons firing. They assign Working Memory to these areas of our physical brain and the delay periods within our physical brain, never once realizing that it's physically impossible to store megabytes of Working Memory within a single hardware BIT, whether that BIT is ON or OFF. They never once realize that it's physically impossible to pass even a single BYTE of memory through a synaptic cleft and have it come out the other side intact. They never once realize that Nature's Psyche is mapping this Working Memory onto these specific parts of the brain and turning on these specific parts of the brain whenever Nature's Psyche needs this type of physical functionality within our physical brain. They just see parts of the brain turning on during the delay period and *jump straight to the conclusion* that it is those specific parts of our brain that are storing our all our Working Memory at the physical level within our physical brain, even though technically that's physically impossible.

Area LIP and Working Memory

Cortical areas outside the frontal lobe have also been found to contain neurons that appear to retain working memory information.

The responses of many neurons suggest that they are involved in a type of working memory.

The neuron begins firing shortly after the peripheral target is presented; this seems like a normal stimulus-evoked response. But the cell keeps firing throughout the delay period in which there is no stimulus, until saccadic movements finally occur. Further experiments using this delayed-saccade task suggest that the response of the LIP neuron is temporarily holding information that will be used to produce the saccades.

Other areas in parietal and temporal cortex have been shown to have analogous working memory responses. These areas seem to be modality specific, just as the responses in area LIP are vision specific. This is consistent with the clinical observation that there are distinct auditory and visual working memory deficits in humans produced by cortical lesions. (*Neuroscience*, p. 757.)

We shouldn't be too surprised by this physical observation once we have discovered Maps of Physical Functionality and how they really work – both at the quantum level and the physical level. There's nothing preventing the Human Psyche from developing and having Maps of Physical Functionality of its own – what these people are calling Working Memory.

It's fascinating how we all conflate Memory with the turning-on of certain switches within our physical brain; but, we do so because we have never been told anything in school about Nature's Psyche and the quantum level. You can't use the information if you don't know anything about it; and, the Materialists, Naturalists, and Atheists who are our PhD teachers in our public schools go out of their way to make sure that we know nothing about Nature's Psyche, the Human Psyche, the quantum level, and the psyche level. It's what they do, and they are very good at it. The whole purpose of our public education system is to indoctrinate us into Materialism, Naturalism, Darwinism, Nihilism, Behaviorism, and Atheism. Our public schools are temples and shrines to Materialism, Naturalism, and Atheism. It works, because that's what they set out to do.

The authors began this chapter with science fiction, and they end with science fiction.

Multiple brain systems are used for memory storage. What remains unclear is how memory types defined by psychological testing are

correlated with anatomical structures. **Declarative memory depends on the hippocampus and related structures, procedural memory involves the striatum, and working memory traces are found in many brain locations.**

We've said that engrams may exist in temporal lobe neocortex, among other places. But what is the physiological basis for the memory storage?

It is thought that memories are ultimately stored in structural changes in neocortex. Our brain is constantly undergoing rewiring, to a certain degree, to adapt to life's experiences. The nature of these structural changes in the brain, which underlie learning and memory, is the subject of Chapter 25. (*Neuroscience*, p. 758.)

Their concluding remarks end the chapter with their continued fixation on their pre-chosen belief that ALL of our memories are being stored within out physical brain and no place else.

Everything they say in these concluding remarks is FALSE.

There are NO wires within our brain; and even if there were, who is rewiring our brain, how do they know how to do this rewiring, and how are they doing so? Who is controlling the growth of our dendrites and axons, as well as neurogenesis and synaptogenesis? Who is making all of these different structural changes within our brain? When it comes to all these different types of synapses, neurotransmitter sub-types, and neurotransmitter receptors, who is doing all the typing, differentiation, addition, and subtraction? Somebody has to know what they are doing at the quantum level or the molecular level; or, we would end up with a jar of dirt for a brain.

I've got a jar of dirt!

It's physically impossible to store ALL of our memories within our physical brain. It's physically impossible to pass even a BYTE of memory through a synaptic cleft and have it survive intact at the other end of the transmission. Thanks to the scrambling, randomizing, and integrating effects of every synapse in the brain, it's physically impossible to distribute our memories among our synapses at the physical level. There are NO memory engrams within our brain, meaning that there is no physical RAM within our brain. There are no CPUs and RAM and Memory Systems within a neuron. It's physically impossible for neurons to learn or remember anything, because neurons function and exist as single stand-alone hardware BITS or switches at the physical level. They are either ON or OFF. It's physically impossible to use our brain systems for storing memories. There is NO physiological basis for memory storage!

The authors suggest a science experiment that involves using a microelectrode to test if a specific neuron is storing long-term memories. They never once realize that it's physically impossible to store bytes of long-term memories within a single hardware BIT, which we call a neuron.

The authors suggest a science experiment where a specific neuron in the visual cortex is detected responding to faces, and then you are tasked to determine whether that neuron is involved in perception and the storage of memories for faces. They assign "perception" to all of the CPUs within that specific neuron; and, they assign "storage of memories" to all of that RAM within that specific neuron. They never once realize that it's physically impossible to do gigabytes of parallel CPU processing within a single hardware BIT that we call a neuron. Likewise, they never once realize that it's physically impossible to store even a single BYTE of video or memory within a single hardware BIT that we call a neuron. They never once realize that there are no CPUs and no RAM and no Memory Systems within a neuron.

A neuron is like a battery or a switch. It's physically impossible to store long-term memories as physical bytes within a battery, or a switch, or a single hardware BIT like a neuron.

However, if we are willing to allow Quantum Mechanics and Nature's Psyche into play, then instantly it becomes possible to do tons of CPU processing at the quantum level and exabytes of RAM memory storage at the quantum level, and then map the output from ALL of that to specific neurons within a physical brain at the physical level.

They end the chapter as they began; and, I too end their chapter as I began by asking, "Is there really a physiological basis or physical basis for memory storage in the brain, especially when we are talking about our after-death Life Review?"

We humans are trying to control Nature at the physical level, and we are trying to maximize Nature's output at the physical level; but, it all backfires on us, because we don't know what we are doing at the quantum level or the molecular level where all the physical functionality really begins and takes place. We are constantly focused on the physical level, never once realizing that there is a quantum level or a psyche level that has to be taken into consideration.

God made it physically impossible to store memories within our brain at the physical level so that we would KNOW that He is needed if we want to have memories of any kind, and so that we would KNOW that He exists. God is hiding in plain sight where nobody can see Him, unless they are actually looking for Him.

God permits homosexuality to exist, so that any rational and sane person would automatically KNOW that evolution, random mutations, and natural selection are FALSE. Evolution is NOT functioning properly as advertised. There should be no homosexuals in any species on this planet, if the theory of evolution were really true.

I find these realities and truths amazing. Of course, your mileage may vary, especially if you are at war with God and hate Him with a passion.

Conclusion

Because these people refuse to engage at the quantum level or the psyche level, these people begin their chapter with science fiction and end their chapter with science fiction. These people don't let the facts get in the way of their fiction.

According to the Materialists and Naturalists, it's physically impossible to store memories and information within an atom.

So, when does it become physically possible to store a memory?

It becomes physically possible to store a memory when Someone Psyche chooses to map certain BITS of physical matter into a single functional unit or BYTE of physical matter. Then it becomes physically possible to store a memory within physical matter.

Alas, all of the ion receptors on a neuron function as a single hardware BIT – they are either OPEN or CLOSED. ALL of the synapses on the neuron are merged or integrated into a single BIT of information called a Postsynaptic Potential (PSP). Consequently, a neuron is effectively a single hardware BIT – it is either ON or OFF. That's it!

There's NO mechanism within a neuron at the physical level for storing and processing memories. If the neurons and neurotransmitter receptors really are connected together as BYTES of programming code, data, and memories, it's all being done at the quantum level by Nature's Psyche, because there are NO visible wires connecting neurons and receptors together at the physical level.

There are no BYTES within a physical brain! They don't exist!

And, for those BYTES to actually exist "within" a physical brain, Someone Psyche would have to map the individual BITS within your brain to function together as BYTES; and, ALL of this mapped functionality would have to take place at the quantum level under the command and control of Someone Psyche, in order for it to exist in the first place because it's physically impossible to implement at the physical level.

As I studied Neuroscience, my primary emphasize was to try to figure out how the brain is storing our memories in a physical format. The only thing I got out of the PhD experts on the subject is, "We don't understand the mechanism for storing memories within the brain, but obviously your memories are being stored within your brain because there's no place else to store them." Whenever they said that they didn't understand the mechanism for memory storage in the brain, they meant that they didn't understand the physical mechanism for storing memories in a physical format within the brain.

I have been a computer scientist most of my life. I like to think that I know how memory storage works at the physical level. Can you imagine how severely disturbing it was for me not to be able to find any physical mechanism for storing memories within a physical brain? I couldn't accept it at first. It had to be wrong. But over and over again, ALL of the evidence and research kept pointing me to the same conclusion, "We don't know how memories are being stored in a physical format within the brain, because our memories are NOT being stored in a physical format within our physical brain, because it's physically impossible to do so."

Oh, I fought it and resisted it for months. It couldn't be true. Yet, I was never able to find a convincing explanation for how our memories are being stored in a physical format within our physical brain. I never found an answer. Eventually, I was forced to conclude my search and was forced to accept the fact that our memories are NOT being stored in a physical format within our brain, because it's physically impossible to store memories within a neuron, and because it's physically impossible to transfer memories through a synaptic cleft and have them arrive at the other end intact.

Clearly, our memories are being stored within our brain, even though that's physically impossible. I was left with one overriding conclusion. Yes indeed, ALL of our memories are being stored within our brain; but, they are being stored at the quantum level and not the physical level, because it's physically impossible to store physical memories in a physical format within a physical brain. You can't even store a single BYTE of programming code, data, or memory within a physical brain at the physical level, because there are NO wires in your brain connecting the physical BITS into functional BYTES. Eventually, I was forced to accept the evidence and what it was trying to teach me. There is NO physical RAM within a physical brain; and, our memories are NOT being stored in a physical format within our brain. Good enough!

If there are memories being stored within the trillions of hardware BITS within our physical brain – ionotropic receptors, metabotropic receptors, and neurons – those memories are being stored at the quantum level by Nature's Psyche, because it's physically impossible to store BYTES memories within a single stand-alone hardware BIT like a neuron.

I made my point. Now I leave it up to you to decide for yourself if you agree with me or not. It doesn't matter either way, because it doesn't change the facts.

Learning, Memory, and Amnesia: How Your Brain Stores Information

Thoughts and memories are quantum waves or the vibrations within strings. That's how our thoughts and memories are able to survive the death of our physical brain.

Learning experiences are selected or chosen into existence by the Human Psyche. If I (my psyche) don't choose to sit down and study something, then I will never learn anything new. That's just the way things work. Quantum Neuroscience was chosen into existence by my psyche and through God's inspiration, because I chose to pursue this line of research and I asked for God's help while doing so. That's the way it worked.

I couldn't find a convincing physical explanation for how our memories are being stored in a physical format within our physical brain.

There's NO mechanism within a neuron at the physical level for storing and processing memories. If the neurons and neurotransmitter receptors really are connected together as BYTES of programming code, data, and memories, it's all being done at the quantum level by Nature's Psyche, because there are NO visible wires connecting neurons and receptors together at the physical level.

John Pinel in his book, *Biopsychology*, has dedicated a chapter to all the memory storage that is being done within your brain, entitled "Learning, Memory, and Amnesia: How Your Brain Stores Information".

His book ends up being a lot weaker than the *Neuroscience* book, because Pinel persists in his chosen belief that evolution created your brain and is now running your brain, controlling your brain, and storing all your memories in your brain, which is physically impossible.

The *Neuroscience* authors did a better job handling and discussing Learning and Memory, because they didn't rely as much on evolution to get the job done. This is good, since evolution can't do anything at the quantum level anyway.

But no matter which college textbook you choose to study, they ALL teach that learning is new CPU programming within the brain and memory is new RAM within the brain. They define a priori (from the beginning before doing any actual science) learning and memory as products or functions of the physical brain. What they teach is demonstrably false; but, it isn't obviously false, which is why they are able to trick us and deceive us.

> **Learning and memory are two ways of thinking about the same thing: Both are neuroplastic processes; they deal with the ability of the brain to change its functioning in response to experience. Learning deals with how experience changes the brain, and memory deals with how these changes are stored [within the brain] and subsequently reactivated. Without the ability to learn and remember, we would experience every moment as if waking from a lifelong sleep.** (*Biopsychology*, p. 260.)

He starts with faulty conclusions, which results in faulty interpretations in the middle, and ends with faulty conclusions at the end as well. That's the way the Scientific Method works – garbage in, garbage out.

Pinel examined case studies from individuals who had parts of their brains damaged or removed. He used these case studies as documentary proof that our memories have a neural basis and that our memories are stored within our brain.

As a result, I ended up asking the same question over and over and over again, **"Is it really lost memories that we are observing in this case study, or is it lost functionality?"** Invariably, it ended up being lost functionality and NOT lost memories. The memories always returned once the functionality returned, which means that the memories really weren't lost after all.

The Materialists and Naturalists and Behaviorists deliberately equate or conflate learning and memories. Due to the evidence, especially due to the OBSERVED FACT that all of our memories or quantum waves survive the death of our physical brain, I learned to separate "memories" from "learning". When the brain is damaged, learning or functionality is lost; but, the memories remain. Whether we can access those memories or not, due to the lost functionality, is another question that has to be dealt with and studied. The empirical evidence led me to believe that the memories or quantum waves are always being stored someplace by Nature's Psyche; but, the memories can't always be recalled due to lost functionality within the physical brain. Rinse and repeat!

Nature's Psyche has been commanded by God not to share what it knows with the Human Psyche. Nature's Psyche knows a lot more and has access to infinitely more information than the Human Psyche does. Therefore, the Human Psyche only knows that comes to it through its physical brain, unless of course God can be persuaded to reveal something more to the Human Psyche who asks God for additional input.

Periodically, Pinel implies that evolution designed and created our Memory Systems.

John Pinel had a much harder time making me a believer, because he chose to use evolution to explain all the physical functionality within our brain, including memory storage. I KNEW he was wrong from the very beginning, because I KNEW why he was wrong. Evolution of any kind can't design and create functional proteins, genes, genomes, eyes, or brains. It's physically impossible. It's physically impossible for chemical evolution to design and create a functional protein, let alone the matching gene to go along with it. It can't be done, which means that it wasn't done!

Simple. Parsimonious. Checkmate!

I had FALSIFIED the theory of evolution before I ever started reading Pinel's book.

Remember, evolution can't do anything at the quantum level, which means that evolution is completely worthless when it comes to anything that involves Psyche at the quantum level or the psyche level.

Now, let's examine this chapter and see what else Pinel got wrong.

All of the Materialists, Naturalists, Darwinists, Nihilists, Behaviorists, and Atheists want you to believe that all your memories are stored only within your physical brain and that all of your memories and YOU cease to exist when you die. That's it. That's what they are trying to prove and that's what they want you to believe. These people have to get you to deny the existence of God, Psyche, and the Quantum Level or Non-Local Realm; but, that's really easy to do when it comes to most of us. Once you are in denial, they have got you where they want you.

Since Pinel rejects the existence of the quantum level or the psyche level, he can't use it to explain what's being observed at the physical level. He has to make up a story out of thin air instead, which he does.

Every chapter on "memory" uses the case study from H.M. – Henry Molaison – not Herman Munster. Due to epilepsy, H.M. had his medial temporal lobes removed. Ever since, the psychologists and neuroscientists have studied what memory systems he lost and what memory systems he retained. None of his retrograde consolidated declarative memories were lost; but, his ability to form, consolidate, and then recall new declarative memories was lost. Once again, careful examination of his case demonstrates that NONE of his consolidated memories were lost; but, the functionality or the ability to form new consolidated declarative memories was lost. They called it anterograde amnesia.

His case demonstrated that Nature's Psyche typically maps memory consolidation for our declarative memories to our medial temporal lobes. Consolidation involves moving our short-term memories into long-term storage in the cloud. When that physical functionality is lost, it makes it physically impossible for the Human Psyche to recall those long-term memories from the cloud or anywhere else that Nature's Psyche might have stored them.

The Human Psyche ONLY has access to what comes to it through its physical brain. If the parts of the brain that consolidate memories for us is destroyed, the Human Psyche no longer has access to that information. I believe that Nature's Psyche continues to store all of that information for us in the cloud; but, the Human Psyche no longer has access to it, because that part of the physical brain has been destroyed. Either the consolidation part of the brain is destroyed or the recall part of the brain is destroyed; but either way, the physical effect is the same and the memories or quantum waves seem to be gone.

From the physical side, medial temporal lobe removal seems to result in anterograde amnesia, the inability to form new declarative memories. Memory consolidation, recall of new memories and experiences, and conscious awareness anterograde declarative memories seem to be lost Again, the evidence from other case studies suggests that the physical functionality is lost, and NOT the memories or the quantum waves.

Amnesia is defined as lost memories; but, I learned that "amnesia" is a misnomer. The consolidated memories or quantum waves are NEVER LOST; however, brain functionality, including the parts of the brain that Nature's Psyche uses to recall our memories from the cloud, can indeed be lost due to brain damage. It's also possible for brain damage to remove the parts of our brain that Nature's Psyche has assigned to the task of consolidating our memories and uploading them to the cloud. It is physical functionality that is lost, and not the memories or the quantum waves.

I believe that Nature's Psyche continues to store ALL of our quantum waves or memories or thoughts in the cloud for us; but, Nature's Psyche isn't obligated to share what it knows with us. Therefore, if we lose functionality within a part of our brain, the physical functions that Nature's Psyche has mapped to that part of the brain will be lost, unless we can persuade Nature's Psyche to remap that functionality someplace else.

Thanks to Nature's Psyche and the Quantum Maps of Physical Functionality that Nature's Psyche maps onto our physical brains, there really is a neural basis to memories, learning, and brain functionality. Destroy the part of the brain to which Nature's Psyche has mapped a specific physical function; and, that specific physical function is lost. It may seem as if the memories or quantum waves are lost as well, but it's the physical functionality that is lost and NOT the quantum waves or memories. If the parts of the brain that Nature's Psyche has assigned to memory recall or memory storage are lost, it can seem as if the memories were lost, when in fact it was the physical functionality that was lost and not necessarily the memories. Anytime the physical functionality comes back, the memories come back as well. That's what has been observed.

Remember, brain functionality is needed in order for the Human Psyche to have a physical experience and remember it. Physical memory functions can be lost, but the quantum waves or memories cannot. However, the Human Psyche ONLY has access to what it can glean from its physical brain. If the physical brain can no longer consolidate declarative memories or can no longer recall declarative memories, then the Human Psyche no longer has access to that information, when that information is moved out of short-term storage into long-term storage. The emphasis on "memory functions" is the correct emphasis, instead of focusing exclusively on memory storage as the Materialists and Naturalists do.

Our after-death Life Review memories are not stored within our physical brain, so what other memories are not stored within our physical brain?

While examining H.M. and the other case studies, I repeatedly asked, "Are this patient's anterograde declarative memories really lost or will they show up in his after-death Life Review?" I wrote: It would be interesting to know how much of their "lost memories" show up in their Life Review. "All that God wants", was my initial conclusion.

There are two possibilities here: The memories were stored in the missing device, or the missing device accessed those types of episodic and explicit memories from non-local long-term cloud storage. It makes you wonder how much of it will make it into their after-death Life Review.

Based on the empirical evidence, I eventually concluded that the physical functionality can be lost; but, the thoughts, memories, or quantum waves are never really lost. People who are born blind can see during their Near-Death Experiences; and more importantly, people who were born blind experience video during their Life Review. In their Life Review, they can actually see the rooms and the people that they are interacting with. This is empirical proof that our Life Review is NOT being produced by our physical brain, and it is also evidentiary proof that our thoughts, memories, or quantum waves survive the death of our physical brain. We don't need a functional visual system nor a functional brain in order to have video and memories during our Life Review.

Look at what you can accomplish by choosing to allow ALL of the evidence into evidence. I've accomplished the physically impossible here, simply by choosing to allow all of the evidence into evidence, and then pursuing a preponderance of the evidence.

Whenever the brain is offline, it isn't making and consolidating memories for the Human Psyche to experience and remember. Such a thing happens to us when we are asleep, and when our spirit body and psyche are away exploring the Astral Plane. The physical brain limits what the Human Psyche can know and recall; but, our Life Reviews are handled separately by Nature's Psyche or the Angels in Heaven, and don't depend upon our physical brain for their existence.

> **Today there is a wide measure of agreement, which on the physical side of science approaches almost to unanimity, that the stream of knowledge is heading towards a non-mechanical reality; the universe begins to look more like a great thought than like a great machine. Mind no longer appears as an accidental intruder into the realm of matter; we are beginning to suspect that we ought rather to hail it as a creator and governor of the realm of matter.** – James Jeans

Memory begins to look more like a great thought than like great RAM storage. Remember, thoughts and memories are quantum waves or the vibrations within strings. They survive the death of our physical brain. It's truly a memory if it survives the death of your physical brain.

Explicit memories are declarative memories. Implicit memories are procedural memories and unconscious memories. The conscious stuff is run through the Human Psyche. The unconscious automatic stuff is handled by Nature's Psyche.

Pinel mentions a few case studies where the individual regains some of the memories that he or she lost. This suggests that the memories weren't lost, just the functionality. Restore the brain functionality, and the memories come back! This reality, truth, and observation is proof of concept where Quantum Neuroscience is concerned.

Pinel persists, so I persist.

The hippocampus and related structures play a role in memory consolidation. (*Biopsychology*, p. 269.)

Was the hippocampus designed and mapped by Nature's Psyche to consolidate memories and then store them someplace else in the brain, or was the hippocampus designed to upload and download memories from non-local cloud storage? Either way, destroy the hippocampus, and it will seem as if the memories were lost.

He talks about memory engrams and new memory engrams being produced in the hippocampus. He talks about memories being linked to memory engrams within our brain. How is this done? He doesn't say. These links have to be made at the quantum level by Nature's Psyche, because it's physically impossible to do so at the physical level, because there are NO wires in our brain. Lose the link or the index or the brain functionality, and obviously access to the memory is lost.

The physical brain doesn't have memory engrams or physical RAM. There are NO wires in our brain. Over months of time, they would repeatedly get me convinced that memories are really being stored within our physical brain; but, it was an illusion. It's physically impossible to store even a BYTE of information within a synaptic cleft, an ion receptor, or a neuron. A neuron is a hardware BIT – ON or OFF. That's it. You can't store a BYTE of information in a physical BIT. It's physically impossible. And, there are NO wires in our brain connecting the neurons or BITS within our brain together into functional BYTES. There are no BYTES within our physical brain. They don't exist.

I couldn't find a convincing physical explanation for how our memories are being stored in a physical format within our physical brain.

Despite evidence to the contrary, Pinel equates amnesia with memory loss and then points at the different types of brain damage and neurodegeneration as proof that amnesia and memory loss are being caused by brain damage, which he then concludes provides proof that our memories are only stored within our physical brain and no place else. He uses the *conflation* logic fallacy or the *category error* logic fallacy in order to make his case. He equates memory loss with brain damage, and he tries to brush over the fact that our memories aren't lost when our brain gets damaged.

The classical theory of memory consolidation is Hebb's theory. He argued that memories of experiences are stored in the short-term by neural activity reverberating (circulating) in closed circuits. These reverberating patterns are susceptible to disruption – for example, by a blow to the head – but eventually they induce structural changes in the involved synapses, which provide stable long-term storage. (*Biopsychology*, p. 268.)

How?

It's physically impossible to transfer even a single byte of information through a synaptic cleft or neuron, and have it survive intact during the journey.

These "reverberating memories" sound an awful lot like quantum waves or thoughts to me – at the quantum level at least.

Hebb's theory is physically impossible at the physical level; but, it's definitely possible at the quantum level. Memories and thoughts are quantum waves reverberating through the quantum realm or the spirit realm.

Pinel periodically ruins his evidence and his presentation with his evolutionary perspective, and I can't help making the occasional comment in the margins of my book.

Presumably, the implicit system was the first to evolve.

Presumably, the evolution of explicit memory systems provided for the flexible use of information. (*Biopsychology*, p. 264.)

The theory of evolution is nothing but faulty presumptions. With all of his different presumptions, it's clear that he's making up a fictional *ad hoc just-so story* out of thin air as he goes along.

It has been observed that our thoughts, memories, and personality survive the death of our physical brain. That's what has been experienced and observed. This reality is empirical knowledge and a phenomenological fact.

There is NO presumption or guesswork in my statement of FACT. I'm simply sharing what has already been observed and experienced first-hand by millions of different people.

Can you see the difference?

Pinel mentions electroconvulsive shock (ECS) under his evolutionary perspective. Evolution is NEVER going to be doing ECS on the brains of humans – not in a million years!

Pinel mentions evidence that ECS can disrupt memory consolidation producing retrograde amnesia for recently acquired memories. How long did the retrograde amnesia last? Was it permanent? He never says, probably because he never bothered to find out. The memories were missing from the physical brain, and that's all that he was interested in.

The most convincing evidence for memories being stored within the brain came from these ECS treatments, which selectively eliminated recent retrograde memories and apparently nothing else. However, he never mentions if the memories returned after the ECS wore off, nor does he ever successfully explain how memories are being stored within our brain.

He talks about islands of memory that survive our most traumatic events. I have islands of memory for my suicide attempt. Most of it is missing; but, bits of it remain. Memory loss is the primary side effect of the sleeping pills that I overdosed on. That part of my brain was shut down for a while; but, the ability to form new memories returned once the brain functionality returned.

Their discussions of memory consolidation at times made me wonder if repeated use, recall, and attention forces these consolidated memories to actually be stored within the Human Psyche. Does the Human Psyche have its own memory storage systems? That doesn't seem to be the case. Observation of split-brain patients seems to suggest that the Human Psyche doesn't have its own memory storage system, because with the split-brain patients, their right hemisphere doesn't know what the left hemisphere is doing, and vice versa. There's no information exchange between hemispheres taking place through the Human Psyche; and, Nature's Psyche has been commanded by God not to share what it knows with the Human Psyche, so there's no help there either.

Over and over again the Materialists, Naturalists, and Darwinists conflate physical functionality with memory storage. The rat remembers to check for food, but after medial temporal cortex lesions, the rat can no longer do object recognition. He erroneously calls it "object recognition memory"; but, the rat is remembering just fine. It's just that the rat can no longer do object recognition which is a physical function.

There's NO mechanism within a neuron at the physical level for storing and processing memories. If the neurons and neurotransmitter receptors really are connected together as BYTES of programming code, data, and memories, it's all being done at the quantum level by Nature's Psyche, because there are NO visible wires connecting neurons and receptors together at the physical level.

The Materialists, Naturalists, and Darwinists deliberately (but sometimes unknowingly) use the *begging the question* logic fallacy to prove that the theory of evolution is true.

They look at all the physical brains and then state, "Look what evolution created." That's *begging the question* or *jumping to conclusions*, which are logic fallacies. Likewise, before doing any actual science, they start with the conclusion that ALL of our memories are being stored within our physical brain and no place else. The thousands of after-death Life Review on record tell us clearly and conclusively that the Materialists and Naturalists are wrong; but, these people refuse to look at the evidence. *A refusal to look at evidence* is also a severe logic fallacy. Naturalism and Atheism of any kind are based upon a refusal to look at evidence. These people selectively ban, block, censor, ridicule, mock, and destroy any evidence that falsifies Materialism, Naturalism, Darwinism, Scientism, Nihilism, Behaviorism, and Atheism. Their actions are religious fanaticism or dogmatism, and NOT science. Science, true science, allows ALL of the evidence into evidence, and then pursues a preponderance of the evidence.

> **Consistent with the view that the hippocampus plays a role in spatial processing is the fact that many hippocampal neurons are place cells.** (*Biopsychology*, p. 275.)

Consistent with the CPU Model that I started out with, he used the word "processing" instead of memory storage. At the time, I had concluded the "processing" is the correct term for brain functionality, and NOT memory storage. Of course, he also mentions that the hippocampus does working memory and reference memory in the preceding section of his book; so, he's not as consistent separating brain functionality from memory storage as I started to be.

Every now and then, he will mention a postsynaptic neuron that fires without any input from a presynaptic neuron. He never once realizes how significant this really is – Nature's Psyche picking a specific postsynaptic neuron to fire and collapsing the necessary wave functions in order to make it so. This unique functionality is extremely important to emphasize and understand.

> **As the rat familiarizes itself with the environment, many hippocampal neurons acquire a place field in it – that is, each fires only when the rat is in a particular part of the test environment. Each place cell has a place field in a different part of the environment.** (*Biopsychology*, p. 276.)

These place cells are firing unilaterally without any apparent physical input from nearby neurons. Who or what gives these neurons a place field or a quantum field? Who assigns a specific place in the environment to each specific place cell? There's no RAM within a neuron, so how is this information being stored within the neuron? How does the neuron KNOW that the rat is now in the correct place and it's time for the neuron to start firing?

It can't be evolution, because evolution or random mutations can't do anything at the quantum level with forces and fields.

Someone Psyche is assigning a specific place or location to each grid cell and place cell, and then dynamically accessing that information or triggering that information when needed.

Researchers have shown that the firing of a rat's place cells indicates where the rat "thinks" it is in the test environment, not necessarily where it actually is. (p. 276.)

Someone Psyche is reading the rat's mind, and then triggering the appropriate place cells that match with where the rat thinks it is. It can't be evolution, because evolution or random mutations can't read a person's mind or psyche.

Grid cells are entorhinal neurons that each have an extensive array of evenly spaced place fields, producing a pattern reminiscent of graph paper. The distance between the evenly spaced place fields is flexible; in experimental animals kept in smaller environments, the fields are closer together. The even spacing of the place fields in the grid cells could enable spatial computations in hippocampal cells.

Other types of neurons in the entorhinal cortex are associated with spatial location. For example, *head direction cells* are tuned to the direction of head orientation, and *border cells* fire when the subject is near the borders of its immediate environment. (*Biopsychology*, p. 276.)

How do these cells KNOW the location of the individual, the head orientation of the individual, and the nearness of the individual to a border? Who is making these cells respond or fire? Who is mapping this physical functionality onto these specific neurons? Where is the information being stored, how is it being transmitted to the neuron, and how is the neuron processing that information when it receives it? The ONLY thing a neuron can do at the physical level is to FIRE or to stay OFF. So, all of this enhanced functionality has to be taking place at the quantum level under the command and control of Nature's Psyche. There's no other logical explanation that I can come up with, since there are NO wires in our brains connecting the BITS together into some sort functional whole at the physical level.

The hippocampus does give the illusion that spatial memories are being stored within the hippocampus at the physical level; but, that's physically impossible, thanks to the randomizing and integrating effects of our synaptic clefts which reduce ALL of the physical input coming into a neuron into one single Postsynaptic Potential or one single BIT of information.

Concept cells also give the impression that they are storing memories, even though that's physically impossible within a single hardware BIT such as a neuron.

When it comes to concept cells, place cells, grid cells, head direction cells, and border cells, where is all of the object recognition and location recognition being processed, since there are no CPUs and no RAM visible within a neuron at the physical level? Once again, the only answer that makes logical sense is that ALL of this CPU processing, data manipulation, memory storage, object recognition, and location recognition is taking place at the quantum level under the command and control of Someone Psyche, because it's physically impossible to do all of this at the physical level within a single hardware BIT that we call a neuron.

I couldn't find a convincing physical explanation for how our memories are being stored in a physical format within our physical brain.

Pinel dedicates a section entitled "Where Are Memories Stored?" to telling us where the different types of memory are being stored within our physical brain, even though it's physically impossible to store even a single united physical BYTE within our brain. From the beginning, he jumped to the conclusion that all of our memories are stored within our brain; and then *begging the question*, he continues forward using his pre-chosen conclusion as evidentiary proof that our memories are stored within our brain and no place else. He never budges from his pre-chosen conclusion, even though our after-death Life Reviews are evidentiary proof that his pre-chosen conclusion is wrong.

I discuss this section elsewhere, so I will just list a few of my repetitive notes, to demonstrate my thought processes.

It's not the least bit clear how a neuron can store a thought, memory, idea, image, feeling, concept, or desire. There is NO visible RAM within a neuron. A neuron is a single hardware BIT. It is either ON or OFF. That's it. You can't store a message or a memory within ON or OFF. It's physically impossible.

Pinel talks about storing memories of visual input within the inferior temporal cortex and perirhinal cortex; but, that's physically impossible. We don't have instant replay, which means that our visual memories are not being stored within our brain. We have video in our Life Review, which suggests that our visual memories are not being stored within our brain. If our memories really were stored within our physical brain on some sort of hard drive or solid-state drive, then ALL of our memories would be instantly available at any time just like a hard drive; but, since they are not, it's clear that our memories are not being stored at the physical level within our physical brain. Even if our neurons and synapses were wired together into functional bytes at the physical level, the brain still wouldn't have enough physical memory storage capacity to store the petabytes of video that show up in our after-death Life Review. It's more likely that these cortices are being mapped by Nature's Psyche to facilitate the uploading of physical visual experiences to the cloud and then used later on to facilitate the downloading of visual experiences from the cloud.

The observed fact that consolidated retrograde memories remain, after the lesioning of any part of the brain that we want to cut out, is convincing evidence that our long-term retrograde memories are no longer being stored within our brain. Our brain structures are memory facilitators, and NOT memory storage devices. Brain structures facilitate physical experiences; but, the thoughts and memories are stored as quantum waves rather than physical BYTES within the brain. The evidence suggests that our brain structures are memory retrieval devices, and NOT memory storage devices.

The prefrontal cortex seems to be functioning as working memory and not memory storage. The "memory regions" within our brain facilitate the creation of memories at the physical level, but not the storage of memories at the physical level.

The cerebellum is thought to participate in the storage of memories of learned sensorimotor skills through its various neuroplastic mechanisms. (*Biopsychology*, p. 279.)

It is programming or mapped physical functionality that he is talking about here, and NOT the storage of declarative memories. The Materialists and Naturalists deliberately conflate the two – brain functionality and memory storage – in an attempt to convince us that our memories are being stored within our brain and no place else. It's extremely important to these people that all your memories cease to exist when you die. They don't want you to remember that they have been lying to you and trying to trick you and deceive you.

Every now and then, he will mention a postsynaptic neuron that fires without any input from a presynaptic neuron. He never once realizes how significant this really is – Nature's Psyche picking a specific postsynaptic neuron to fire and collapsing the necessary wave functions in order to make it so. This unique functionality is extremely important to emphasize and understand.

He talks about the brain being dynamic. Dynamic means sentient, aware, conscious, adaptable, active, vibrant, forceful, self-motivated, and alive – which are ALL functions of the Human Psyche. Observe what the physical brain does when it's dead and there is no Human Psyche present telling it what to do.

He mentions transcription factors being activated by neural activity. Transcription factors are molecules. Transcription factors are free-roaming. Someone Psyche has to demand the production of specific transcription factors, and then tell them where to go and what to do when they get there.

Like everyone else, he spends a lot of time attributing Long-Term Potentiation (LTP) to memory storage within the physical brain. However, LTP is a change in brain functionality and NOT memory storage. It's always a change in brain functionality that we are observing and experiencing at the physical level, and NEVER a change in memories or memory storage capacity.

Synaptogenesis and Long-Term Potentiation really don't explain how our brain stores memories and information at the physical level, because it's physically impossible to store even a single BYTE of information within a synaptic cleft, because a synaptic cleft will scramble, randomize, and eventually integrate everything that comes its way into a single BIT of information called a Post-Synaptic Potential. LTP adds new ion receptors to the postsynaptic membrane. Ion receptors are gates – OPEN or CLOSED. It's physically impossible to store a BYTE of memory within a gate or a single hardware BIT. Synapses, synaptic clefts, neurons, and LTP fail to provide memory storage at the physical level; but, who knows how Nature's Psyche is using them at the quantum level.

In the margins of my book, I made many comments about his LTP Models that I don't mention anywhere else, so I will mention them here.

Artificially stimulated or artificially induced LTP is worthless as a demonstration of physical memory storage within the brain, unless it can be demonstrated that artificial LTP actually produces a flood of memories or learned behaviors to ensue.

The evidence in one experience actually demonstrates that artificially induced LTP shuts down learning and memory! LTP and ECS can actually give people epilepsy. They conflate LTP with memory storage; but clearly, LTP is re-programming or altered functionality, and not the storage site of new declarative memories as they claim. LTP is a change in functionality and has nothing to do with memories. Procedural memories are a change or an improvement in physical functionality. LTP seems to "store" procedural memories or implicit memories physical skills that have been practiced and learned. Technically, procedural memories and implicit memories are NOT memories, because the Human Psyche isn't aware of them and can't remember them. How can they be memories, if you can't remember them? Procedural memories and implicit memories are changes in physical functionality, and not memories per se.

LTP results in changes of the synapse, particularly an addition of more ionotropic receptors to the postsynaptic membrane. I still can't visualize how changes in a synapse can store new memories at the physical level. However, I can easily visualize how changing a synapse can increase (or decrease) the functional efficiency of a synapse, thereby resulting is skill changes, improved physical functionality, and the improved physical reliability of a particular skill or ability. LTP improves physical functionality, and has nothing to do with our memories.

After he finishes discussing LTP, I wrote, "I still can't see how adding a new ionotropic receptor and a few more ions to a neuron could ever store a memory." It has nothing to do with memories. It's improved physical functionality that we are looking at with LTP.

<u>Conclusion</u>

Over and over again, month after month, my Neuroscience teachers would get me convinced that memories are being stored within our physical brain; and, each time I would finally poke holes in their theories and successfully explain why each one of their theories is physically impossible. I'm a computer scientist, so I KNOW how memory storage works at the physical level; and, there's nothing like that within our physical brain.

I couldn't find a convincing physical explanation for how our memories are being stored in a physical format within our physical brain.

After all my research and wishful thinking, I reluctantly realized that there is NO physiological explanation or physical explanation for memory storage within the brain that I found convincing and believable. I eventually grasped the fact that, due to the way that God made our brains from bottom to top as stand-alone BITS, it's physically impossible to store even a single BYTE of memory in a physical format within our physical brain. For me personally – a former Materialist, Nihilist, and Atheist – that was a major scientific discovery and a significant epiphany. Of course, this will be the discovery that everyone will resist the most – I KNOW, because it's the one that I resisted the most.

It's physically impossible to store physical memories in a physical format at the physical level within the physical brain. It can't be done, which means that it isn't being done. However, when we choose to let Nature's Psyche and quantum mechanisms in to play, suddenly we notice that thoughts and memories are quantum waves, and Nature's Psyche can theoretically store ALL of our memories or quantum waves within a single physical atom within our physical brain, if Nature's Psyche wants to do so. Then we finally realize that Quantum Neuroscience has infinitely more explanatory power than Materialism, Naturalism, and Classical Physics.

Nature's Psyche maps physical functionality onto our brain at the quantum level. If the functionality goes down, then there's nothing for the Human Psyche to experience and remember from that part of the brain. The Human Psyche is subservient to the physical body and physical brain. Who knows what's happening at the quantum level where Nature's Psyche is concerned? The ONLY thing we do know for sure is that the memories return when the physical functionality returns, which means that the memories were never really lost in the first place. Only the functionality was lost.

If the physical functionality goes down, then the Human Psyche isn't going to be able to get anything from that part of its brain while the functionality remains down. That's just the way it is. If the damage is permanent, sometimes it can take a while for Nature's Psyche to remap that functionality to a different part of the brain; and sometimes, Nature's Psyche never remaps that functionality to a different part of the brain, especially when the brain is in physical decline all across the board. At some point Nature's Psyche decides that it's time to stop trying and just let the whole thing go.

There's NO visible RAM within ions, ion receptors, synaptic clefts, or neurons. If it is there, it exists non-locally or non-physically at the quantum level, since it clearly doesn't exist at the physical level.

The Human Psyche pays attention to what it values most; and then, Nature's Psyche maps the brain accordingly. After all of my research, I have settled upon the conclusion that Nature's Psyche takes the 100 billion neurons or switches within our brain and organizes them into Maps of Physical Functionality at the quantum level, which Nature's Psyche then uses to get things done for us at the physical level. Neurons are switches with NO wires connecting them. If our neurons are being organized or mapped into any sort of functionality or purpose, it's being done at the quantum level, because it's physically impossible to do so at the physical level.

Remember, it's physically impossible to store even a single BYTE of memory or programming code within a single hardware BIT; and, there are NO wires within your brain connecting the trillions of different hardware BITS into even one single hardware BYTE. There isn't even a single BYTE of physical RAM or physical memory storage within your physical brain at the physical level. Thanks to our synaptic clefts, it's physically impossible for a physical BYTE to exist at the physical level within the physical brain. That means that all the petabytes of memory within your brain exist at the quantum level, and not the physical level.

Thanks to Karl Lashley, we have known for 100 years that there are NO memory engrams (no RAM) within a physical brain, a fact that they have tried to ignore or gloss-over ever since it was discovered. Instead, they speculated that our memories are distributed throughout all of our brain. According to the Behaviorists, our memories have to be stored someplace within the brain, because there's nowhere else to store them. However, Distributed Memory is physically impossible at the physical level, thanks to the scrambling, randomizing, and integrating effects of our synaptic clefts. Oh, our memories are indeed being "distributed" all throughout our brain; but, it's all being done at the quantum level by Nature's Psyche, because it's physically impossible to distribute memories throughout the brain at the physical level.

There's NO mechanism within a neuron at the physical level for storing and processing memories. If the neurons and neurotransmitter receptors really are connected together as BYTES of programming code, data, and memories, it's all being done at the quantum level by Nature's Psyche, because there are NO visible wires connecting neurons and receptors together at the physical level.

Never once could I find a convincing physical explanation for how our memories are being stored in a physical format within our physical brain. It's physically impossible to store memories in a physical format within a physical brain. It can't be done, which means that it isn't being done.

Fascinating, is it not?

Quantum storage of memories as quantum waves at the quantum level by Nature's Psyche is one of the main discoveries of Quantum Neuroscience; and, it will probably be the discovery that they will resist the most. I KNOW, because it was the discovery that I resisted the most. I used to be a Materialist, Naturalist, Nihilist, and Atheist, so I know how it goes. We tend to resist these kinds of things, even when the preponderance of the evidence is telling us that we are wrong.

Neuroplasticity – Molecular Mechanisms of Learning and Memory

Now that you know where I have come from and where I am going with Quantum Neuroscience, let's run some Neuroscience through the process and see what we end up with.

I will use *Neuroscience: Exploring the Brain* as my text, because it's freely available online if you know where to look for it.

Bear, M. F., Connors, B. W., & Paradiso, M. A. (2007). *Neuroscience: Exploring the Brain* (3rd ed.). New York: Lippincott Williams and Wilkins.

This essay will focus exclusively on chapter 25 of their book. Let's see what we can do with it.

BCP25

http://mypsyche.us/wp-content/uploads/2017/12/BCP25.pdf

Classical Physics, Materialism, and Naturalism are at a serious disadvantage compared to the explanatory power of Quantum Mechanics. When it comes to explanatory power, Quantum Mechanics is vastly and noticeably superior to anything that the Materialists and Naturalists can provide. The Orthodox Interpretation of Quantum Mechanics, by Henry P. Stapp, is an infinitely better method for describing learning and memory storage than anything you will ever get from Naturalism, Materialism, and Classical Physics.

I spent months studying the Philosophy of Science so that I could know and understand how the Materialists, Naturalists, Darwinists, and Atheists are working their magic and making their case. It was well worth it – a total eye-opener.

The Materialists, Naturalists, Darwinists, Nihilists, Behaviorists, and Atheists start with the pre-chosen conclusion that ALL of our memories and brain functionality are stored and processed within our physical brain and nowhere else. Then, these people insert this pre-chosen conclusion into all of their arguments and proofs as one of their hidden premises, using their pre-chosen conclusion as evidentiary and empirical proof that their pre-chosen conclusion is real and true. This is a logic fallacy known as *begging the question*. The whole of Materialism, Naturalism, Behaviorism, and Atheism are based upon this logic fallacy.

Furthermore, the Materialists, Naturalists, Darwinists, and Atheists unknowingly use the *affirming the consequent* version of the Scientific Method to make their case and prove their point. *Affirming the consequent* is a serious logic fallacy, but *affirming the consequent* is the very foundation of Scientific Naturalism and Reductionistic Materialism. Their whole science is based upon *affirming the consequent*.

Negating the consequent is the ONLY version of the Scientific Method that is philosophically and logically sound. In fact, I have successfully used *negating the consequent* to FALSIFY Materialism, Naturalism, Darwinism, and Atheism. It's powerful, it's convincing, and it works. It's extremely easy to do, once you know how *negating the consequent* works. In other words, I have actually used the Scientific Method to FALSIFY Materialism, Naturalism, Darwinism, and Atheism by *negating the consequent*.

Ironically, Quantum Neuroscience is already an observed and proven science. Quantum Mechanics is our best-proven and most-used science on the planet. Quantum

Neuroscience was already a proven science before I ever coined the term. It was just sitting there waiting for someone to find it.

Chapter 25 of their *Neuroscience* book is about the molecular mechanisms of learning and memory.

I started my research by asking some serious questions.

How does adding a new synapse, or a new neurotransmitter receptor on a specific post-synaptic neuron, consolidate a memory or turn a memory into a long-term stored memory?

There doesn't seem to be an answer to this question in terms of Materialism, Naturalism, and Classical Physics. These people pretend that there is, but I didn't find any of their explanations or models convincing, because they contained way too many assumptions, fudges, and black boxes. These people are secretly and unknowingly using Quantum Mechanics or Black Boxes in order to make their case – a case and pre-chosen conclusion which claims that all of our memories are being stored within our physical brain and nowhere else. That's *begging the question* and *cheating*; but, I sincerely believe that most of these people don't even know that they are cheating, because they have never studied the Philosophy of Science and don't know anything about the *begging the question* and the *affirming the consequent* logic fallacies.

Wilder Penfield's direct stimulation of neurons triggered a flood of memories.

Based upon what I now know to be true, this scientific observation makes me think that specific neurons, specific synapses, and specific postsynaptic receptors are being mapped onto or indexed onto specific non-local experiences or memories by Nature's Psyche; and, when the Human Psyche requests a specific declarative memory from Nature's Psyche, the specific neuron or neuro-receptor is triggered by Nature's Psyche, which then causes Nature's Psyche to also download a specific experience from Non-Local Consciousness or God's Database and broadcast its feelings, thoughts, and memories throughout the physical brain.

This is all hypothetical of course, but it's definitely more of an answer that what we get from the Naturalists, who say that adding a synapse or adding a post-synaptic receptor creates and consolidates a memory through the synaptic cleft and synapse.

How does that newly created synapse or newly added receptor differ from the quadrillions that were created and added before it? It doesn't from a physical perspective! So, how can new and unique memories be added to one of these things? Where are all the new memories being stored BEFORE that new synapse is built or BEFORE that new ionotropic receptor is added to a synapse? In some kind of quantum black box, of course, because you can store anything you want in a black box until you have the physical hardware in place to contain it. These people have to use quantum mechanisms to store the memories temporarily while the new synapses are being built and the new receptors are being added. They don't seem to realize this fact; but, that's the way it is! The memories have to be stored someplace while the new synapses are being built and the new receptors are being added. Only Quantum Mechanics and Psyche provide a mechanism for how this is being done.

Where's all the cell differentiation information, memories, programming, CPU processing power, RAM, and brain functionality being stored BEFORE the first synapses are built, and the first ionotropic and metabotropic receptors are added to that synapse? Well, it's NOT being stored within our synapses as the Materialists and Naturalists claim, because there are NO synapses yet.

There's more going on here than Classical Physics and Naturalism. We are seeing signs of the Supernatural or Quantum Mechanics.

Declarative memories are non-local spiritual phenomena – a function and product of Psyche – because they survive the death of our physical brain according to the empirical evidence from Near-Death Experiences (NDEs) and Out-of-Body Experiences (OBEs).

Declarative memory creation and declarative memory consolidation within the physical brain needs and requires a non-local, quantum, spiritual, psychic explanation because it's physically impossible to pass gigabytes of memories through a synaptic cleft, especially a synapse that doesn't exist yet.

The Materialists, Naturalists, and Atheists NEVER think about any of this, because they have started with the pre-chosen conclusion that all of our memories are being stored physically within our physical brain; and then, these people use this pre-chosen conclusion as evidentiary proof that their pre-chosen conclusion is real and true. These people are never going to look at anything from the quantum level or psyche level, because they have convinced themselves that the quantum level or the transdimensional non-local level does not exist. Self-deception works, and it works every time, especially when it comes to PhD Materialists, Naturalists, and Darwinists.

Fascinating, is it not?

I'm very impressed by the explanatory power of Quantum Neuroscience. Let's make some comparisons, so that you can see what I mean.

Chapter 25:

Chapter 25 of *Neuroscience* focuses on the Molecular Mechanisms of Learning and Memory.

Molecular mechanisms of learning and memory? This is a bit of an oxymoron, because molecules don't learn, and molecules don't remember; or, do they? Well, molecules certainly don't learn, and they certainly don't remember at the physical level. But, what about the quantum level or the psyche level?

Research suggests that synaptic transmission can actually direct local protein synthesis in some neurons. In Chapter 25, we will see that synaptic regulation of protein synthesis is crucial for information storage in the brain. (*Neuroscience*, p. 45.)

They start with the physically impossible, and they end with the physically impossible. Nevertheless, this chapter is heavy-duty material from beginning to end.

We have Psyche, Intelligence, Quantum Waves, and Quantum Mechanisms written all over our Learning Systems from start to finish. It's unavoidable, even though it's physically impossible.

As a former Materialist, Naturalist, Nihilist, and Atheist, this chapter is so painful that it hurts, especially when you start looking at it from a quantum perspective.

In the interest of space, I'm just going to quote what I wrote in the margins of my book, rather than quoting the whole chapter and then making comments after each quote.

BCP25

http://mypsyche.us/wp-content/uploads/2017/12/BCP25.pdf

—

We may soon know the physical basis of memory. [And someday pigs will fly.] (*Neuroscience*, p. 762.)

Then we'll have to figure out the spiritual basis for memory, for the times when we no longer have a brain.

Declarative memories are memories for facts, events, and thoughts.

Distributed memories are memories that have been distributed in a weighted fashion among two or more synapses in the physical brain.

They tell us that ALL of our memories are being stored within our synapses.

This is another one that bugs me severely!

I can't visualize how altering synapses could ever possibly store a memory or a thought.

Where are the memories and thoughts being stored while the synapses are being altered or rebalanced?

The thought takes place BEFORE any synaptic alteration, and then supposedly the synapse is altered to store that thought or memory. On the other hand, there is NO reason to alter synapses BEFORE the memory or thought takes place, because doing so would feed faulty and fake memories into the system that never happened and never existed.

This is the chicken and egg paradox! It can't be resolved at the physical level through physical mechanisms.

Furthermore, I can't visualize how adding a neurotransmitter receptor could ever possibly store a thought or memory. You can't shove megabytes of memories or thoughts into a single newly added neurotransmitter receptor or ion channel, at least NOT at the physical level. And, where are those thoughts and memories being stored BEFORE that new ion channel is added to the synapse in order to store them? I guess those new memories are being store non-locally in the cloud at the quantum level, while an ion channel is being added to store them at the physical level.

This is the classical chicken and egg paradox! It can't be resolved at the physical level through physical mechanisms. It can only be solved at the quantum level, where our thoughts and memories are formed in the first place.

Since the thoughts take place BEFORE the synaptic alteration, since our thoughts can't be recorded nor detected by our physical instruments, and since our thoughts and memories continue after our physical brain is dead and gone, it's clear and obvious to me that our thoughts and memories really have nothing to do with our synapses and physical brain.

Recall that declarative memories have a certain ethereal quality. (p. 762.)

Whether they realize it or not, "ethereal" is code for non-physical, non-local, and spiritual. Our thoughts and memories are ethereal or spiritual or non-local in nature and origin. They are not being created by our physical brain, and they are not being stored within our physical brain. Memories and thoughts are quantum waves, which is why they survive the death of our physical brain.

These people don't think about any of this quantum level stuff, because they don't want to. Its very existence falsifies Materialism and Naturalism.

A lexical procedure is an experimental procedure that involves reading aloud and observing the physical functionality that is based upon specific stored information about written words. The Materialists and Naturalists just assume that this information is being stored within the physical brain; but, that's physically impossible, because you can't store memories and information within a synaptic cleft.

Procedural memories have characteristics that make them more amenable to investigation. Besides the fact that these memories are particularly robust, they can be formed along simple reflex pathways that link sensations to movements. Procedural learning involves learning a motor response (procedure) in reaction to a sensory input. (*Neuroscience*, p. 762.)

This is a *conflation error* or *category error* logic fallacy that bugs me a lot.

How can "procedural memories" and "implicit memories" be memories if the Human Psyche isn't aware of them and can't remember them?

They are talking about procedural skills or motor skills – physical functionality – which are in fact the result of programming, training, and conditioning. It's not memory in the traditional sense of the word. It's improved functionality. You could say that practice is improving the wiring or the programming, which would indeed improve the functioning of a motor skill. In actuality, though, there are NO wires within our physical brain, so there's NO way to improve the wiring or the programming either, at least not at the physical level.

Scary huh?

We've just entered the Twilight Zone!

The logic errors that they employed in this chapter helped me to make up my mind and consolidate what it is that I truly believe.

They start this chapter by discussing the brain mechanisms for procedural learning, or the learning of new physical skills; but soon, they *conflate* or *fuse* it all with declarative memories which are something else entirely. Whenever we think of memory, we always think of declarative memories, and NOT our physical learning systems. It's a *category error* logic fallacy to equate the two! And, all throughout this chapter they use this logic fallacy in order to make their case. They don't even realize that they are doing so, because they have been trained all of their lives to use these specific logic fallacies in order to do their science and make their case. They call it the scientific method, and it's all based upon the *affirming the consequent* logic fallacy.

Remember, procedural learning is programming or new Maps of Physical Functionality, and NOT memories. How can they be memories if we aren't consciously aware of them? Maps of Physical Functionality are built and then used by Nature's Psyche at the quantum level, completely outside the conscious awareness of the Human Psyche and its physical brain.

The physical machinery is necessary for learning and experiencing physical realities and physical behaviors. Our spinal cord and brain stem control our reflexes and autonomic programming. God or Nature's Psyche programmed Maps of Physical Functionality into our spinal cord and brain stem that keep our body running in default mode, while our spirit and psyche are away exploring the Astral Plane. While separated from its physical body, your spirit body is attached to your physical body by a silver cord or a silver thread. Sever the cord, and your physical body dies.

743

A reflex isn't memory. A reflex is designed and programmed into us by Nature's Psyche, completely outside the conscious awareness of the Human Psyche. Autonomic systems are the result of programming by Nature's Psyche.

In contrast, chosen behaviors are a function of the Human Psyche, because we continue to choose and make decisions long after our physical brain is dead and gone. The Human Psyche chooses what it wants to do with the physical body and physical brain that God has given to it.

This is Quantum Neuroscience, which is something completely different than what the Materialists and Naturalists try to feed us.

They talk about an L7 motor neuron, and an L29 interneuron; and then, they suggest that your memories and learning systems are being stored in the way that these neurons are wired together.

So, who is wiring these things together in such a way that they actually store memories and learning within the wiring itself?

Well, it isn't evolution, like they claim.

L7 has been given a name. It's a recurring character, which means that its existence, positioning, wiring diagram, and functionality is stored someplace in God's Database or God's Official 3D Blueprint, which isn't stored within our genome. This is programming and functionality that we are looking at, NOT memories! This particular function has been mapped directly to the L7 motor neuron by Nature's Psyche completely outside the conscious awareness of the Human Psyche. This functionality can't be a memory, because there's nothing there for the Human Psyche or the physical brain to consciously remember.

L29 has been given a name. It's a recurring character, which means that its existence, positioning, wiring diagram, and functionality is stored someplace in God's Database or God's Official 3D Blueprint.

Somewhere there is a 3D Map for all of these various neurons, muscle cells, axons, and the various types of synapses which feed into these neurons and muscle cells; and, this 3D Map of Physical Functionality can't be stored within the 750 megabytes of our genome, because our genome codes primarily for proteins and NOT the layout, positioning, and differentiation of cells, axons, and synapses. It's physically impossible to shove petabytes of 3D Mapping Functionality into 750 megabytes of genome!

All throughout this chapter, they try to make a case suggesting that ALL of your memories are being stored in your synapses, and all your learning is being stored in "the way your axons are wired together". So, where are your memories and learning being stored while the new synapses are being made and the new axons are being wired together? Non-locally in the cloud I suppose. Where else are you going to store them?

However, their theories are fundamentally flawed to begin with.

There are NO wires in our brains; and, it's physically impossible to store memories in a synaptic cleft.

A synaptic cleft randomizes, scrambles, and then integrates everything that comes its way. ALL of the synapses on a neuron get integrated into a SINGLE post-synaptic potential (PSP), which either reaches a threshold triggering an action potential, or it doesn't. Every synapse on a neuron reduces or integrates to a single BIT, a neuron, which is either ON or OFF. A neuron is a switch, a single hardware BIT. You can't store memories within a single hardware BIT. It's physically impossible. And, you can't run computer code or "learning" through a single hardware BIT, either. It's physically impossible.

Brain functionality, learning, and memories are physically impossible – at the physical level at least. But, that reality doesn't stop these people from trying to make up a story that seems to fit the facts.

They put in a lot of effort trying to show how the L29 interneuron and G-Proteins are used to do classical conditioning, or what they call new learning and new memories.

Again, learning and memories are a misnomer, because we are looking at new functionality, new associations, and new mapping – NOT memories – especially when it comes to what they call "procedural memories". There's nothing happening here that the Human Psyche and physical brain can remember directly or consciously, so it really isn't memory that we are looking at. It's a new Map of Physical Functionality being put together for us unconsciously by Nature's Psyche.

At the end of all their speculation, they finally provide a sobering thought.

It is important to realize, however, that none of the *Aplysia* studies have uncovered the cellular basis of learning, only some *cellular correlates* of learning. Although the studied synaptic changes can accompany behavioral learning, they may not be essential for learning to occur. In other words, there may be (and, in some instances, definitely are) additional mechanisms that contribute to habituation, sensitization, and classical conditioning. Thousands of neurons in the *Aplysia* nervous system are active during even the simplest reflexes, and synaptic changes related to learning are likely to be widely distributed among them. Nonetheless, the invertebrate studies have been invaluable for identifying candidate molecular mechanisms for learning and memory. (*Neuroscience*, p. 771.)

So, what are these additional mechanisms they are talking about?

They are quantum mechanisms or Quantum Mechanics.

They have run out of physical explanations!

There's NO physical entity in the synaptic cleft called "evolution," who is moving the neurotransmitters and degradation enzymes around with tractor beams at the quantum level. Evolution doesn't work at the quantum level.

The DNA in the nucleus of a cell isn't exactly a physical brain. Transcription factors, who chose the protein to produce and find the associated gene within the genome, are smarter and more intelligent than we are; but, they don't have a brain. They have something better – a Psyche or Non-Local Consciousness with direct access to God's Database.

Remember, Nature's Psyche can do the physically impossible at the quantum level, through quantum waves or quantum mechanisms. Nature's Psyche can push around and manipulate atoms and molecules at the quantum level through quantum waves and quantum mechanisms.

So, our memories and learning aren't really being stored at the cellular level, because that's physically impossible.

Consequently, these people switch to Synaptic Plasticity instead.

But, are these synaptic changes really memory?

Are we consciously aware of them? NO! Do they survive the death of our physical body and physical brain? NO! Well then, they can't be memories or learning, can they?

If it survives the death of your physical body and physical brain, then it's truly a memory in every sense of the word, and learning has truly taken place.

At this point they talk about the cerebellum and how it is "wired".

That thing looks like a computer to me, except there are NO wires in a physical brain, at least not at the physical level. Who knows what's happening at the quantum level? I guess Nature's Psyche knows, but the Human Psyche and the human brain certainly do not!

They dedicate the rest of this chapter to LTD (long-term depression), LTP (long-term potentiation), and CamKII; and then, they give these physical processes the assignment of storing our new memories and thoughts within our synapses or the AMPA receptors on the post-synaptic membrane.

LTD is a governor! It doesn't have anything to do with declarative memories. It's a refinement of motor control. It's a remapping of physical functionality. It doesn't survive the death of our physical body and physical brain, so it can't be a memory in the truest sense of the word.

Yikes!

Cerebellar LTD actually sucks in or internalizes AMPA receptors. The "amplification receptors" get internalized. It's a decrease in functionality, and has nothing to do with memories or learning. In fact, it would be a loss of memories or a loss of learning, if these AMPA receptors were actually storing our memories and learning.

LTD is re-programming or re-mapping physical functionality.

They keep employing a *category error* logic fallacy here in order to make their case. It's what the Materialists and Naturalists do, because they have never studied the Philosophy of Science; and thus, they have no idea that the Philosophy of Science falsifies Materialism and Naturalism.

They keep calling LTD and LTP "memory" and "learning". It's nothing of the sort!

We aren't consciously aware of these physical processes, so there's nothing there for the Human Psyche to remember and learn!

Furthermore, none of this LTD and LTP type of memory and learning survives the death of our physical body and physical brain, so in this case nothing was ever really learned or remembered! LTD and LTP are a modification in physical functionality, and NOT new memories!

They keep giving one example after another. The next example involves synaptic plasticity in the Hippocampus. This one involves LTP.

Once again, LTP results in the re-programming or re-mapping of our physical brain, and therefore has nothing to do with the declarative memories and learning that survive the death of our physical brain.

LTP and LDT alter brain circuitry and synapses, NOT memories. In fact, we aren't even aware when LTP and LTD take place, so there's nothing there for us to remember. Elementary!

Furthermore, these people repeatedly and erroneously call it "rewiring the brain"; but, there are NO wires within our brain, at least not at the physical level. There may be "wires" or connections at the quantum level; but if there are, they are not visible to us at the physical level. A synaptic cleft is a terminator, a dead-end, not a wire!

They keep *conflating* everything – a logic fallacy. They repeatedly keep *conflating* declarative memories with synaptic change, neuroplasticity, and remapped physical functionality. They are NOT the same thing at all, which means that these people are using a *category error* logic fallacy in order to make their case.

The problem is that declarative memories are typically ethereal, and in most humans they don't last long, at least not at the physical level. It's the "procedural memories" – the skills and habits – that tend to last a lifetime. But, the "procedural memories" created by LTP are programming or re-mapping of physical functionality, and NOT memories!

These are all physical processes that they are talking about, which won't exist after the physical body is dead and gone. They are talking about neuroplasticity and remapping physical functionality, and NOT the memories and learning which survive the death of our physical brain.

Now, they are going to try to store memories in the timing of action potentials.

They can tell you what's happening – the frequency of action potential firing is increasing, or decreasing. But, they can't tell you what it means! ONLY Nature's Psyche knows what it means! All of this is taking place outside our conscious awareness; and thus, it has nothing to do with the Human Psyche and its memories.

They tell us that NMDA receptors have a very high affinity for glutamate, so the transmitter remains bound to the receptor for tens of milliseconds – a very long time.

With high affinity, why doesn't the glutamate transmitter stay bound to the NMDA receptor for all eternity? So, who breaks the affinity or the connection; and, who chooses to break the connection at ten milliseconds and not ten seconds? Is there some kind of invisible timer built into an NMDA receptor that breaks the connection automatically at tens of milliseconds?

Whenever you have molecules that have a strong attraction or strong affinity for one another, there is NO physical process that can break that affinity or connection. This affinity or connection has to be broken at the quantum level by Nature's Psyche; otherwise, it's going to last forever.

Of course, they talk about an action potential being strong enough to awaken these dormant ion channels, but that's a pre-programmed physical function; and, again it has nothing to do with the declarative memories that survive the death of our physical body and physical brain.

ALL of this LTP and LTD looks more like computer programming or remapping of physical functionality than memory storage!

Now, they switch over to AMPA receptors, and try to show how the addition of AMPA receptors to the post-synaptic membrane "contributes to the formation of declarative memories".

YIKES!

That's the phrase they actually use.

They want you to believe that our newly created memories are being stored within these AMPA receptors; but, that's physically impossible. An AMPA receptor represents one single hardware BIT – it's OPEN or CLOSED. It's physically impossible to store megabytes of new thoughts and memories into a single hardware BIT.

They even state that "AMPA receptors in the postsynaptic membrane are continually being added and removed even in the absence of synaptic activity."

DOUBLE YIKES!

Here they are stating that AMPA receptors are being added and removed without any physical cause and without any physical reason. Who or what is telling the cell to add or remove AMPA receptors in the complete absence of synaptic activity? This doesn't make any sense from a physical perspective.

They also attribute memory consolidation to the replacement of GluR1-containing AMPA receptors with non-GluR1 AMPA receptors.

Who taught the cell how to add the right kind of AMPA receptors, and then to remove them when they are no longer needed? Why are they no longer needed? Well, they are no longer needed because apparently you have forgotten whatever memories they were tasked to store.

So, where are all of your thoughts, memories, and ideas being stored, while evolution is juggling your AMPA receptors? The only thing left is the ions and the atoms! But, ask the Materialists and Naturalists, and they will tell you that that is physically impossible. According to these people, you can't store anything within an ion or an atom, because an atom is nothing but empty space.

Do you see how the Materialists and Naturalists paint themselves into a corner that they can't get out of? Materialism, Naturalism, Darwinism, and Atheism are self-defeating, because they completely lack explanatory power. These people deliberately limit themselves to the physical level.

Now, observe what happens when we choose to take the same thing into the quantum realm or psyche realm.

At the quantum level, you can map petabytes of memories, thoughts, and ideas directly to a single ion channel, receptor, or atom, if you want to. At the quantum level, you can also theoretically store petabytes of thoughts, memories, and ideas within a single atom or ion if you want to. See how much easier it is to explain memory storage at the quantum level, than it is to try to explain it at the physical level?

What's amazing is that Quantum Neuroscience was already a proven science long before I coined the term. The concepts that I have been discussing in this book should be common knowledge to anyone who truly knows and understands how Quantum Mechanics really works.

Finally, they switch over to CamKII and try to use it, LTP, and NMDA receptors to store new memories within our hippocampus.

Animals engineered to produce too many NMDA receptors show *enhanced* learning ability in some tasks. (p. 786.)

Memory can result from experience-dependent alterations in synaptic transmission. (p. 787.)

Well, there you have it. It must be so.

But, is it really memory that we are talking about here?

Memories are eternal. Memories and real learning survive the death of our physical brain. Thoughts and memories are quantum waves.

Is LTP the kind of memory that survives the death of our physical brain; or, is it simply new functionality or the mapping of new capabilities that we are talking about here? Alas, we are once again looking at increased functionality, not memories.

Nature's Psyche maps our hippocampus to match with or model our physical surroundings. The hippocampus is a Map of Physical Functionality that is mapped by Nature's Psyche at the quantum level to match the spatial locations we use most. It has nothing to do with declarative memories!

As if they were trying to sweep it under the rug, they finally state:

CamKII and phosphorylation as a long-term memory mechanism is problematic. Pinpointing where and how a molecule contributes to learning (and memory) **can be difficult.** (*Neuroscience*, pp. 786-787.)

YA THINK?

It's not only difficult, it's physically impossible!

However, it isn't difficult to demonstrate how adding a new molecule contributes to new physical functionality.

And, it isn't difficult to demonstrate how Nature's Psyche is able to Map Physical Functionality to a specific molecule at the quantum level.

But, they persist. They want you to know that all your memories are being stored within your synapses, and no place else.

A requirement for long-term memory is the synthesis of new protein. This protein is used to assemble new synapses. These findings indicate a requirement for new protein synthesis during the period of *memory consolidation*, when short-term memories are converted into long-term ones. (p. 788.)

Now we are back to storing our long-term declarative memories in synapses; and, we are back to distributing our memories among these new synapses. Again, where are ALL those new memories, thoughts, and ideas being stored while evolution is making all of these synapses to store these new memories within? I guess evolution must be storing them in the cloud someplace.

They even dive into transcription and gene expression, trying to show how the types of proteins that the transcription factors choose to make determine the types of memories that can be stored in the synapses of our physical brain.

Again, it's new functionality that is being added by adding new synapses. It has nothing to do with memories, especially the memories that survive the death of our physical brain. Think about it logically!

Do you see how desperate they are to get your memories stored within your synapses?

Why are they so desperate?

It's because they have NOTHING else to fall back on at the physical level. They have painted themselves into a corner and can't get out of it.

Transcription can't be used to describe and produce memory storage!

Who controls the transcription process? Who is sending the transcription factors to that specific location, that specific gene, and telling the transcription factor which of the two or three possible proteins to produce this time around from that specific gene? How do the transcription factors find that specific location and gene within the genome? How do those transcription factors convert your memories to proteins, or convert these new proteins into memories? And, where are ALL of your new memories being stored, while evolution is making all of these new proteins and synapses for you, so that you can store your memories within them? Transcription factors are under the control of Nature's Psyche at the quantum level, from beginning to end!

These are ALL quantum processes taking place at the quantum level under the command and control of Nature's Psyche; therefore, it does NOTHING to explain how our memories are being stored at the physical level in our physical brain.

It's physically impossible for the addition of new synapses and new receptors to be causing and then storing the kinds of declarative memories that survive the death of our physical body and physical brain! Such a claim doesn't make logical sense – to claim that receptors, proteins, and synapses are creating and then storing all of our memories. The only way that would be possible is if Nature's Psyche maps and indexes a specific memory, event, thought, or quantum wave to a specific neurotransmitter receptor at the quantum level, and then stores that map in non-local cloud storage or God's Database. I can visualize that being possible.

But, I can't visualize shoving megabytes of new memories into a receptor that hasn't been made yet, nor deployed yet. And, you can't jump the gun at the physical level, either. If you deploy these receptors before you have any memories or experiences to store in them, then you end up adding false and fake memories to the system that you don't want and that never happened. It's a chicken and egg paradox, which can't be resolved at the physical level.

It's physically impossible to add new thoughts and memories to a receptor or a synapse at the physical level. Your thoughts only exist on the quantum level within your Human Psyche, because your thoughts and dreams can't be detected by nor recorded with our physical instruments; and, your thoughts survive the death of your physical brain.

In their desperation, they even resort to magic and oxymorons.

The switch could stay on, perhaps indefinitely, and this showed how unstable molecules could produce stable information storage. (*Neuroscience*, p. 789.)

These people are desperate to get your memories stored within your physical brain, so that your memories and their personal responsibility can cease to exist when we all die.

Okay, so I took this one out of context, which is a *strawman* logic fallacy; but, it does make an interesting point and does show how desperate they become at times.

I'm a computer scientist.

Try to imagine how a bunch of unstable iron molecules on your physical hard drive could produce stable information storage.

Keep trying to imagine it. It's only possible in your imagination.

I owned one of the original IBM PCs.

Its whopping 10 megabyte hard drive had been around for so long, that the iron molecules on the hard drive were starting to get scratched off. That thing had so many lost file clusters when I used chkdsk, that soon the whole thing was becoming non-functional.

With a hard drive, we really are talking about information storage.

However, there are NO wires in a physical brain.

These people repeatedly *conflate* switches (neurons) with RAM (memory storage). It's a *category error* logic fallacy.

A switch is either ON or OFF – like a neuron or an ion channel. A switch isn't RAM, at least NOT at the physical level. Who knows what it is doing at the quantum level?

Switches or hardware BITS make for horrible and hopeless memory storage; but, a switch makes for excellent, stable, and reliable physical functionality!

Over and over again, these people are deliberately *conflating* stand-alone switches with RAM or memory storage. They are not the same thing at all!

The kinds of complex memories and learning that survive the death of our physical body and physical brain can't be stored within a single hardware switch like a neuron, synapse, or neurotransmitter receptor, no matter how many of these switches you might choose to add to the system. It's physically impossible to store memories, ideas, and thoughts in a switch. The only thing a switch can store and process is ON or OFF.

Like a dog returning to vomit because it has nothing else to eat, they state:

Long-term memory is associated with the formation of new synapses, and forgetting is associated with a loss of these synapses. (p. 791.)

They are fixated on storing your memories in your synapses, so that your memories can cease to exist when you die.

Ask yourself how adding a new synapse or switch translates into a new memory! It's physically impossible! You can't store megabytes of new memories in a single hardware BIT!

There's NO physical entity in the synaptic cleft called "evolution," who is moving the neurotransmitters and degradation enzymes around with tractor beams at the quantum level. Evolution doesn't work at the quantum level.

Remember, a synaptic cleft was designed to scramble, randomize, and integrate every thought, memory, idea, and programming code that comes its way. It's physically impossible to store memories in a synaptic cleft.

And, how does Psyche KNOW which synapse to trigger when it wants to relive a specific memory? ALL of this is impossible at the physical level. Long-term memory storage and after-death Life Reviews are only possible at the quantum level or the psyche level.

Learning fails to occur, and the memory is wiped out, if the cerebellum is surgically removed. (*Neuroscience*, p. 793.)

They *conflate* learning and memory with physical functionality to the very end!

If you temporarily remove the functionality with some type of temporary lesion, the functionality does indeed cease, and the memories do indeed seem to get wiped out. But, both the functionality and the "memories" return when the functionality is restored. The memories were NEVER wiped out. The brain simply lost access to them for a while.

Nature's Psyche can't reprogram or remap the cerebellum or hippocampus if it doesn't exist. This has nothing to do with memory, and everything to do with brain plasticity, functionality, mapping, and synaptic change.

It's called *jumping to conclusions*.

They just *assume* that the memories are wiped out when the cerebellum is surgically removed, because the Human Psyche can no longer use its cerebellum to access those memories or Maps of Physical Functionality.

However, we KNOW for a fact that people who are born blind experience vision during their Near-Death Experiences and actually experience video during their Life Reviews. Their memories are NOT wiped out, even though they have NO video system to begin with.

The memories are NOT wiped out when the cerebellum is surgically removed, because they are being stored non-locally in the cloud by Nature's Psyche. However, removal of the cerebellum does indeed remove or eliminate any of the types of experiences and physical functionality that depend upon the cerebellum in the first place. It's always functionality that is lost, and not memories.

I have come to the conclusion that learning, memory, thoughts, and intelligence ARE a function of Psyche, and NOT the physical brain, because they survive the death of our physical brain.

They end by stating that the destruction of the hippocampus appears to impair the mechanism that "fixes" new memories in the neocortex. (p. 793.)

Well, appearances can be deceiving.

All of the Materialists, Naturalists, and Darwinists repeatedly make the claim that new memories are stored long-term in the neocortex, which technically means that all the animals that don't have a neocortex – the non-mammals – don't have any memories, which is clearly false.

I launched a beetle up in an Estes rocket a few times. By the third time he was crying and spreading his arms wide, when I tried to put him back into the payload compartment. After the fourth time, he was walking backwards. I altered the thing's physical functionality, if not its memories. After all, the thing's not supposed to have memories according to the Materialists and Naturalists, because it doesn't have a neocortex. Everything the Materialists and Naturalists say seems to be nothing but science fiction – *ad hoc just-so story-telling* at its very best.

There's NO mechanism within a neuron at the physical level for storing and processing memories. If the neurons and neurotransmitter receptors really are connected together as BYTES of programming code, data, and memories, it's all being done at the quantum level by Nature's Psyche, because there are NO visible wires connecting neurons and receptors together at the physical level.

There's NO mechanism within a neuron at the physical level for storing and processing memories. If the neurons and neurotransmitter receptors really are connected together as BYTES of programming code, data, and memories, it's all being done at the quantum level by Nature's Psyche, because there are NO visible wires connecting neurons and receptors together at the physical level.

I have come to the conclusion that learning, memory, thoughts, and intelligence ARE a function of Psyche, and NOT the physical brain, because they survive the death of our physical brain.

There's NO physical entity in the synaptic cleft called "evolution," who is moving the neurotransmitters and degradation enzymes around with tractor beams at the quantum level. Evolution doesn't work at the quantum level.

We NEED a scientific explanation for what's happening at the quantum level or psyche level! Remember, if your Interpretation of Quantum Mechanics can't explain what Nature's Psyche and the Human Psyche are doing at the quantum level and how they are doing it, then your Interpretation of Quantum Mechanics is worthless.

This is the smoking gun that explains clearly and conclusively why we need Quantum Neuroscience, Psyche, and Quantum Mechanics – so that we can explain what's happening at the quantum level. You can't use Materialism, Naturalism, Evolution, and Classical Physics to explain what's happening at the quantum level. It's physically impossible.

Technically, our environment can't make us do anything that we don't want to do; and technically, our genes can't make us do anything.

I used to be a Materialist, Naturalist, Nihilist, and Atheist; but, I got over it. In fact, though, I had to work really hard and study a lot with an open-mind in order to get over it. I had to be willing to allow ALL of the evidence into evidence.

I had an Atheist online tell me to stop making fun of her religion – to stop disparaging, debunking, and criticizing her religion. She preferred the deceptions and the lies. Skeptics don't want you being skeptical about their skepticism. It's important to learn how to doubt your doubts. Hope and faith motivate us to act and learn. I hoped to learn something new and interesting when I chose to research and study Quantum Neuroscience. I believe that I have done so. For the Atheists, Materialists, Naturalists, Darwinists, and Behaviorists, faith or trust is the enemy; and, it shows. These people never learn anything new or interesting.

I gave the Atheists, Materialists, Naturalists, and Darwinists a fair hearing for fifty years of my life. One day I decided to turn around and go the other way because I didn't like where that road was taking me. It's my right to do so, if I want to. I encourage you to do the same. My life has been getting progressively better and progressively more interesting ever since. It was a choice – a function of my Psyche. It was nothing that my Nature or my Nurture did to me or did for me. It was a choice. It was something that I, my Psyche, the REAL ME chose to do for me.

Our Psyche determines what we are going to do!

Psychic Determinism is what we are really looking at, when it comes to the quantum level – mind-over-matter. In fact, Psyche existed before there was physical matter, and it was Psyche who designed and created physical matter.

Remember, Psyche can override its Nature and Nurture at will.

We observe the Human Psyche doing so all the time!

This is how things really work, despite what the Materialists and Naturalists might claim.

All that I really wanted to know all of my life is how everything works, and now I do.

Good enough!

Applied Quantum Neuroscience

I just wanted to know how everything works, and now I do, thanks to Quantum Neuroscience. Quantum Neuroscience proved to be the answer to life, the universe, and everything.

It was worth the price of admission, which involved getting rid of My Materialism, My Nihilism, My Scientism, and My Atheism. I'd make that trade any day! I lost nothing of value, and I gained the world.

Mind-Over-Matter

God and Nature's Psyche seem to be getting functionality out of our synapses that's physically impossible. There has to be a ton of functionality taking place at the quantum level that isn't visible to our physical instruments at the physical level. Our synapses and neurotransmitters typically work way too perfectly for it all to be the result of random chance, entropy, and random diffusion.

Of course, there are the metabotropic receptors and the G-proteins, which Nature's Psyche can use to get specialized neural functionality and enhanced neural functionality and adjusted neural functionality through our synapses; but, I'm not sure that G-proteins are enough to explain all of the miraculous things that seem to be happening within our synapses when they are balanced and working right, as well as all the horrible things that seem to happen within our synapses when evolution has messed up our genes and proteins.

Nature's Psyche performs flawlessly; but, there's nothing it can do about the faulty and broken genes and proteins that we get from evolution and random mutations. Only God can fix our broken genes and proteins at the quantum level; and, He isn't always willing to do so.

The Human Psyche pays attention to what it values most; and then, Nature's Psyche maps the brain accordingly. After all of my research, I have settled upon the conclusion that Nature's Psyche takes the 100 billion neurons or switches within our brain and organizes them into Maps of Physical Functionality at the quantum level. Neurons are switches with NO wires connecting them. If they are being organized or mapped into any sort of functionality or purpose, it's being done at the quantum level, because it's physically impossible to do it at the physical level.

The goal in life is to get whatever works for you to work for you.

Any time we observe Mind-Over-Matter being used to treat and cure our physical illnesses and mental illnesses, we are in fact observing Applied Quantum Neuroscience in action.

The Placebo Effect is Applied Quantum Neuroscience, or mind-over-matter.

The Placebo Effect has been proven to exist, and it's real. It works.

Any time you are dealing with a Psychosomatic Disorder or a Personality Disorder that has NO physical cause and NO documentable genetic cause, then you are dealing with something that is best treated and cured by mind-over-matter techniques such as psychotherapy, friendship therapy, meditation, prayer, exercise, socializing at church, and the Atonement of Christ.

The mental illnesses, stress, personality disorders, and psychological disorders that have NO physical cause are best treated and most-easily cured with the Atonement of Christ.

Treating a psychosomatic disorder with prescription drugs and illicit drugs is counter-productive and contra-indicated, because these drugs can and do give a person substance-induced psychosis, which is something that I got to experience for myself for a couple of years. It's hell. You don't have to die in order to go to hell. You can do so right here, right now.

The psychiatrists literally give you another mental illness in an attempt to treat your mental illness, whenever they start pumping you full of sleeping pills and benzodiazepines. For stress, anxiety, and depression, the Atonement of Christ works best followed by cognitive therapy. If you do get on the SSRIs, pick one, get rid of everything else, and taper that ONE down to a maintenance dose that gives you ALL of the benefits of the drug with NONE of the side effects.

—

Matter-Interfering-with-Mind

The Human Psyche and its physical brain seem to be married or fused into a symbiotic whole. Thoughts and memories are quantum waves – a product of the Human Psyche. Choices are a product of the Human Psyche. These different types of quantum waves seem to exist at different frequencies. Typically, it's the Human Psyche who produces the thoughts.

However, there are times when a malfunctioning brain will produce thoughts and impulses.

During psychosis and schizophrenia, the malfunctioning brain is producing hallucinations and delusions, which are thoughts. It's as if the brain develops a mind of its own.

Obsessive-Compulsive Disorder are anxiety disorders characterized by recurring, uncontrollable, anxiety-producing thoughts and impulses. The physical brain is triggering thoughts.

Maybe that's the very definition of mental illness – when the physical brain develops a mind of its own and starts producing thoughts that aren't wanted by nor helpful to the Human Psyche. Lots of drugs can have this same effect; and, substance-induced psychosis and withdrawal symptoms such as hallucinations and delusions are the result.

Even though the thoughts or delusions are fake and faulty, the Human Psyche typically remains capable of making choices. When the ability to choose is lost, then that's the very pinnacle of mental illness.

Neuropathic pain is severe chronic pain that doesn't have a recognizable physical cause. This is the type of pain that the doctors always tell you is in your head. If it's truly all in your mind, then mind-over-matter treatments should be able to cure it, assuming of course that you find the right ones.

Schwartz, J. M. (2002). *The Mind and the Brain: Neuroplasticity and the Power of Mental Force*. New York: HarperCollins.

As documented in *The Mind and the Brain*, Jeffrey Schwartz developed a mind-over matter or psychotherapy technique for treating and curing OCD by remapping the functionality of the human brain, which has proven efficacious and has been verified through brain scans as producing noticeable results.

At the neurological level, the rationale for Refocusing is straightforward. Our PET scans had shown that the orbital frontal cortex, the caudate nucleus, and the thalamus operate in lockstep in the brain of an OCD sufferer. This brain lock in the OCD circuit is undoubtedly the source of a persistent error-detection signal that makes the patient feel that something is dreadfully wrong. By actively changing behaviors, Refocusing changes which brain circuits become activated, and thus also changes the gating through the striatum. The striatum has two output pathways, as noted earlier: direct and indirect. The direct pathway tends to activate the thalamus, increasing cortical activity. The indirect pathway inhibits cortical activity. Refocusing, I hoped, would change the balance of gating through the striatum so that the indirect, inhibitory pathway would become more traveled, and the direct, excitatory pathway would lose traffic. The result would be to damp down activity in this OCD circuit.

Some of the OCD patients, especially those willing to be treated without drugs, were recruited into the brain imaging study that Lew Baxter and I were starting, with the goal of measuring whether the positive behavioral changes we were seeing in patients were accompanied by brain changes. Our UCLA group therefore performed PET scans on eighteen drug-free OCD patients before and after they underwent ten weeks of the Four Steps, with individual sessions once or twice a week in addition to regular group attendance. The patients who signed on exhibited moderate to quite severe symptoms. What they all had in common was a willingness to be PET-scanned twice and to try a largely self-directed, drug-free treatment. Twelve of the patients improved significantly during the ten-week study period. In these, PET scans after treatment showed significantly diminished metabolic activity in both the right and the left caudate, with the right-side decrease particularly striking. There was also a significant decrease in the abnormally high, and pathological, correlations among activities in the caudate, the orbital frontal cortex, and the thalamus in the right hemisphere. No longer were these structures functioning in lockstep. The interpretation was clear: therapy had altered the metabolism of the OCD circuit. Our patients' brain lock had been broken.

This was the first study ever to show that cognitive-behavior therapy—or, indeed, any psychiatric treatment that did not rely on drugs—has the power to change faulty brain chemistry in a well-identified brain circuit. What's more, the therapy had been self-directed, something that was and to a great extent remains anathema to psychology and psychiatry. The changes we detected on PET scans were the kind that neuropsychiatrists might see in patients being treated with powerful mind-altering drugs. We had demonstrated such changes in patients who had, not to put too fine a point on it, changed the way they thought about their thoughts. Self-directed therapy had dramatically and significantly altered brain function. There are now a wealth of brain imaging data supporting the notion that the sort of willful cognitive shift achieved during Refocusing through mindful awareness brings about important changes in brain circuitry as we will see in later chapters.

(*The Mind and the Brain*).

http://publicism.info/psychology/mind/3.html

Schwartz and others used self-directed psychotherapy and the Human Psyche to treat and cure OCD; and thereby, they literally changed the patient's brain circuitry (the CPU Model) or remapped the patient's brains (Maps of Physical Functionality Model), and the change in brain functionality was documented by PET scans.

This is a documented and proven case of mind over matter. By deliberately choosing what it wants to think about and pay attention to, the Human Psyche can literally remap its brain or change its brain's functionality.

This is proof of concept where Quantum Neuroscience is concerned; and more importantly, it's also a cure for OCD that doesn't involve drugs.

—

Matter-Over-Mind

Of course, anytime we observe Matter-Over-Mind or pharmacology and genetics being used to treat and cure our physical illnesses and mental illnesses, we are observing Applied Neuroscience in action.

At the physical level, the best treatments and cures find everything that's wrong with you physically, and then produce customized treatments that treat and cure what's actually wrong with you. Throwing pills at an illness typically makes it worse, thanks to all of the side effects. Finding out exactly what's wrong with your physical body and treating that directly with the right supplement, anti-toxin, antagonist, or dietary restriction works miracles and works like a silver bullet to solve or kill what it was made for.

Reversing Alzheimer's Disease:

https://www.youtube.com/watch?v=6D5aA_-3Ip8

Tumor Necrosis Factor (TNF) Inhibition seems to be one of those miraculous physical cures for pain that everyone should know something about. They use TNF antagonists or TNF blockers to work the magic. This is neuroscience at its best.

http://www.painbreakthrough.com/what-we-treat/

Functionality Is Engineering and Programming

God does His work through us.

Why?

It's so that we can learn and grow from the experience.

But, there are limitations which apply to this rule.

—

Oh, it's happening, alright. Our brains really are processing data, running computer code, crunching numbers, transferring data, building cities, weighing decisions, and storing memories. It's just not happening at the physical level. It's all happening at the quantum level. The physical input of sensory data, and the physical output of motor commands, is the only thing happening at the physical level within the brain. It all happens at the end of an axon, where the switch is either turned ON or OFF. A synapse is a switch. Our brain is comprised of a 100 trillion switches or synapses. Our brain is also comprised of a 100 billion stand-alone switches or neurons, which are either ON or OFF. Furthermore, our brains are comprised of quadrillions of ion channels that are either OPEN or CLOSED. Everything within our brains reduces to a single BIT. There's nothing else there within our brain. Everything else is happening at the quantum level. All the important and interesting stuff is happening at the quantum level, outside our conscious awareness.

Where does brain functionality end and non-local functionality begin? It's hard to tell, because even the so-called "physical functionality" of the brain, or CPU processing, is physically impossible thanks to the synaptic clefts. It's physically impossible to transfer or process bytes of programming code through synaptic clefts, because a synaptic cleft will scramble anything that comes its way. The only thing you can get through a synaptic cleft is a single BIT that turns a post-synaptic receptor ON, opening its gate or ion channel. It's physically impossible to transfer, transmit, and process megabytes of programming code through a single hardware BIT.

When it came to the physical brain and its functionality, I had to face facts and let all the evidence teach me what it was trying to tell me, rather than making up fictional stories to try to explain it as the Materialists and Naturalists do.

Based largely upon the discoveries of Karl Lashley, I and many others have seen convincing evidence that there is no physical RAM within a physical brain, meaning that there are no physical memory engrams, no distributed memory, and no physical memory storage within a physical brain.

During my research, Lashley's discoveries prompted me to make a concentrated effort to separate memory storage (RAM) from functionality or Cerebral Processing Units (CPUs). You can make a convincing case that there is no memory storage or physical RAM within a physical brain; but, it's much harder to make a convincing case that there are no physical CPUs or no Cerebral Processing Units within our physical brains. I eventually did make such a discovery, but it only came to me through the repeated process of demonstrating to myself that there is no RAM or physical memory storage within the physical brain per se. In other words, it proved extremely fruitful during my research to make a concentrated effort to separate memory storage (RAM) from functionality or Cerebral Processing Units (CPUs).

ALL of the evidence repeatedly pointed me to the fact that our physical brains are ALL about physical functionality (Cerebral Processing Units) and not about physical memory storage. Consequently, I felt a desire to dedicate at least one essay to the physical functionality of the human brain.

So, why does that part of the brain function like a thalamus and not a cerebellum?

The answer is that someone engineered it to be that way, and then programmed it to run that way. Someone made the hardware to function that way, and then someone programmed the device to run that way.

So, who is this Someone?

Well, it can't be the Human Psyche, because we are NOT consciously aware of any of these things, their processes, and how they work. From our perspective, they just are, and they just function as designed.

It can't be evolution! Evolution of any kind cannot do engineering and programming. Even the Materialists, Naturalists, and Atheists will tell you that it's physically impossible for some invisible non-existing entity in the sky, like evolution, to design and create anything. And, I tell you that evolution (genetic change), random mutations, and natural selection didn't even exist, until after God designed and created the genes and the proteins in the first place. It has been proven scientifically, experimentally, and observationally that chemical evolution cannot make proteins and the matching genes to go along with those proteins. Evolution didn't provide us with our hardware, programming, and functionality, because it couldn't.

So, who is it that gave us our engineered brains, programming, and functionality?

According to the Orthodox Interpretation of Quantum Mechanics, from Henry P. Stapp, it is Nature's Psyche who is collapsing the wave function and making ALL of this engineering, programming, and functionality REAL.

Working off information from God's Database, Nature's Psyche deliberately and knowingly forms the glial cells and the neurons into a thalamus or a cerebellum, doing all of the engineering necessary to give these Cerebral Processing Units (CPUs) their functionality. Then, Nature's Psyche programs these devices as we go along, so that they function or run properly. The only thing that's guaranteed to mess it all up is bad genetics. During the construction of these devices, Nature's Psyche has to use proteins, and proteins come from genes. If the genes are bad, then the proteins are bad; and then, Nature's Psyche doesn't have the physical hardware necessary to create properly functioning CPUs within our brain.

All throughout their college textbooks, the Materialists, Naturalists, Darwinists, and Atheists talk about brain functions. Even when the Materialists and Naturalists use the word "function", it's automatically implied that that functionality required engineering and programming. The Materialists, Naturalists, Darwinists, and Atheists insist that evolution did all of the engineering and programming that was needed to create functional brain structures, even though that's physically impossible. Evolution of any kind can't do engineering and programming, despite what the Materialists and Naturalists claim. Evolution has NEVER been observed doing so – it has NEVER been caught in the act – because evolution can't do design, engineering, programming, and creation. Evolution doesn't have a brain, so why on earth would anyone think that evolution could design and create a brain!

Evolution can't do functionality; but, Nature's Psyche certainly can!

Quantum Neuroscience is based on the idea that Nature's Psyche makes the brain and then assigns a specific function to each CPU in the brain. Then, Nature's Psyche programs the brain after the hardware functionality is in place. Remember, according to the Orthodox Interpretation of Quantum Mechanics, from Henry P. Stapp, it is Nature's Psyche who is collapsing the wave function and making ALL of this physical engineering, programming, and functionality REAL.

After all the engineering and programming has been done, Nature's Psyche turns to specific parts of the brain which it has made in order to accomplish the physical aspects of an assignment or task that the Human Psyche has asked to be done.

The evidence suggests that Nature's Psyche does the programming and the assignment of functionality, while the brain is developing.

There are people with 90% to 97% of their brain missing, who function normally. That's physically impossible; but, completely possible if Nature's Psyche is doing the programming and the assignment of functionality, and then compensating for the missing brain structures from the Quantum Realm through Quantum Mechanics and Quantum Waves. You really don't need much of a brain, if Nature's Psyche is doing all of the work for you.

http://www.rifters.com/real/articles/Science_No-Brain.pdf

http://mypsyche.us/wp-content/uploads/2018/01/Is-Your-Brain-Really-Necessary.pdf

As long as the core motor functionality remains intact and the functionality for your five physical senses remains intact, so that your physical body moves whenever your Human Psyche wants it to move and your physical senses actually work, everything else in terms of functionality, intelligence, memories, and processing can be handled for you by Nature's Psyche non-locally in the cloud. You don't need a brain for any of that! You need parts of a brain properly "wired" or mapped for physical functionality; but, you don't need anything else! Everything else in terms of processing and memory storage can be handled by Nature's Psyche non-locally in the cloud through Quantum Waves and Quantum Mechanics.

Functionally normal people with only 3% of a brain end up being proof of concept for Quantum Neuroscience, as far as I'm concerned. They lack 97% of the physical functionality, but they are still functioning normally. That's physically impossible; but, completely possible when Nature's Psyche and Quantum Mechanics are in control of the situation.

It ALL comes down to functionality, programming, and memories; and, NONE of that seems to be controlled by our physical brain. It all seems to be developed and handled by Nature's Psyche. The only time that anything seems to go wrong is if evolution has given Nature's Psyche faulty genes, and therefore faulty proteins, to work with. Nature's Psyche can seem to compensate for everything else, except for faulty genes and faulty proteins.

Quantum Neuroscience, as well as Quantum Mechanics, is based upon mind-over-matter. Psyche, Mind, or Non-Local Consciousness exists in the Quantum Realm, or the Transdimensional Realm, or the Non-Local Realm. The Human Psyche's job is to pay attention and to make decisions and choices. Then, Nature's Psyche tries to comply and does all of its work outside our conscious awareness. All of that programming, memory storage, development, and CPU functionality is being handled by Nature's Psyche completely outside our conscious awareness. Functionally normal people with only 3% of a brain, don't know that there's anything wrong with their brain. Nature's Psyche can handle everything, except for faulty genes and faulty proteins.

Nature's Psyche commands and controls the physical realm from the quantum realm. Nature's Psyche manufactures, differentiates, programs, commands, and controls our physical functionality based upon the input or knowledge that it receives from our genome, God's Psyche, God's Blueprints, and God's Database. The only time anything goes wrong is when evolution gets in the way and mutates our genes. Nature's Psyche relies upon proteins while building functionality into our Cerebral Processing Units (CPUs). Proteins come from genes. If the genes and proteins are bad, then the physical part of the CPU is going to be glitchy or non-functional; otherwise, Nature's Psyche seems to be able to compensate. However, after the brain has been developed or constructed and functionality assigned to the different parts, if those parts of the brain get destroyed or removed, that assigned functionality seems to be lost.

NONE of this is being handled by evolution. Evolution is worthless. It is Nature's Psyche who is doing all of this brain programming for us invisibly behind the scenes, according to the Orthodox Interpretation of Quantum Mechanics. It's Nature's Psyche who collapses the wave function making these programming changes REAL and EFFECTIVE. It's not evolution's psyche doing this for us, because evolution doesn't have a psyche according to the Materialists and Darwinists.

Evolution can't implement God's plans, but evolution can certainly destroy them. Most people instinctively know that this is so; but technically, the "most people" logic fallacy cannot be relied upon to prove one's point or to make one's case. So, you are going to have to decide for yourself what it is that you want to believe. I can lead you to information, but I can't make you think. That's physically impossible.

Now, let's explore other aspects of physical functionality and observe how they are being designed, constructed, and controlled by Nature's Psyche.

Bottom-up processing is defined as neural mechanisms that involve the activation of higher cortical areas by lower cortical areas. This is feedback and sensory input that they are talking about here. There's actually a physical explanation for how this works, because it was designed to work through synapses.

However, we start to run into some serious problems whenever we start to talk about top-down processing.

Why?

It's because our neurons at the top have NO visible signs of having brains, CPUs, or RAM anywhere within them.

Top-down processing is defined as a neural mechanism that involves activation of lower cortical areas by higher cortical areas.

Can you see the problem?

It's that FIRST neuron at the top, who unilaterally fires all on its own out of the blue with NO physical input whatsoever, which ends up being the thing that's physically impossible and physically unexplainable!

Imagine that you are floating in a tank with the lights off, earplugs, nose-plugs, and you are experiencing complete sensory deprivation. Then out of the blue, you (your psyche) decides that you want to raise your arm, and your arm rises. Somehow, without any sensory input whatsoever from the physical environment, your psyche reaches out telepathically and triggers the neurons in your primary motor cortex that raise your arm. Your arm rises.

This is where top-down processing FAILS as a physical model, because the Human Psyche is at the top of the process, or the ultimate cause of that process. Unless you, your psyche, decides to raise your arm, your arm is just going to lie there and do nothing. The Materialists and Naturalists will tell you that Psyche is physically impossible; and they are right, because Psyche isn't physical in nature or origin. Then these people tell you that Psyche does not exist, and in this they are wrong. If psyche didn't exist to trigger those neurons directly and get them started and turned on, then you would just lie there like a vegetable or a rock doing nothing until you rot and die.

Let's look at a few of the brain structures and observe how the Materialists and Naturalists describe their functions and functionality. It can be extremely informative, if we let it.

A second messenger is a chemical synthesized in a neuron in response to the binding of a neurotransmitter to a metabotropic receptor in its cell membrane. This functionality is designed and programmed into the neuron at the molecular level or atomic level. Who designed and implemented all of this programmed functionality? It can't be evolution, because evolution of any kind can't do design and programming – certainly not at the quantum level or the psyche level where atomic manipulation and molecular manipulation take place. Evolution (genetic drift), random mutations, and natural selection didn't even exist, until after God designed and created the proteins and their matching genes in the first place.

The Purkinje effect is the observed function that in intense light, red and yellow wavelengths look brighter than blue or green wavelengths of equal intensity; and, in dim light, blue and green wavelengths look brighter than red and yellow wavelengths of equal intensity. Who programmed these cells to work this way?

Reciprocal innervation is the principle of spinal cord circuitry that causes a muscle to relax automatically when a muscle that is antagonistic to it contracts. Well, that required coordination and planning, didn't it?

Recurrent collateral inhibition is the inhibition of a neuron by itself. This effect is produced by the neuron's own activity via a collateral branch of its axon and an inhibitory interneuron. Only God can decide how to wire these things, so that they work right and accomplish their intended purpose.

My first gut reaction was to ask, "Who wired these things this way?" These functions looked like programming to me, until I finally realized that there are NO wires in our brain. Now my question is, "Who mapped this functionality into these things at the quantum level?" With Maps of Physical Functionality, the mapping takes place both at the quantum level and the physical level. When Nature's Psyche maps physical functionality into our brain, it's doing so at the quantum level; but then, Nature's Psyche can use those Maps of Physical Functionality to trigger or activate things for us at the physical level. Maps of Physical Functionality provide our much-needed interface or bridge between the quantum level and the physical level. It solves everything for us.

Conduction aphasia is a language deficit that is thought to be caused by damage to the neural pathway between Broca's area and Wernicke's area. In other words, the hard-line or arcuate fasciculus has been cut, and functionality has ceased. Without the functionality or the hardware, Nature's Psyche has nothing that it can command and control. That's just the way it is!

Fasciculation is the tendency of developing axons to grow along the paths established by preceding axons. It's the pioneer growth cone on the FIRST axon to forge a new pathway that is of primary interest to us at the quantum level, because such a thing requires intelligence, purpose, planning, and motivation. The "magic" or intelligence takes place in the pioneer growth cones, because an axon can go anywhere it wants to go; so, who or what is telling those growth cones where to go? Pioneer growth cones are proof of concept where Quantum Neuroscience is concerned.

The frontal eye field is a small area of prefrontal cortex that controls eye movements. Command and control require intention and intelligence. Command, control, and choice are a function of the Human Psyche. Nature's Psyche has mapped the frontal eye field of the prefrontal cortex to control eye movements. When the Human Psyche wants to move its eyes, Nature's Psyche triggers the appropriate neurons that it has mapped to control eye movements. It's simple to understand, once you choose to do so. Nature's Psyche makes Maps of Physical Functionality, and then Nature's Psyche uses those Maps of Physical Functionality to control our movements.

The fusiform face area is an area of human cortex, located at the boundary between the occipital and temporal lobes, and it is selectively activated by human faces. Who is doing the recognizing of human faces? A neuron sitting there in your head isn't going to recognize a human face! Where are all the image recognition software and parallel processing CPUs being stored in your brain? Who is doing the selective activation? Selectivity is a function of Psyche or Intelligence. Despite the Darwinists' claims to the contrary, evolution can't do selectivity or selection, because evolution isn't sentient, conscious, and aware. Whenever the Human Psyche recognizes a human face, Nature's Psyche selectively activates the fusiform face area of the human cortex. This is the simplest and most parsimonious explanation; otherwise, you have got to find a way to slam a supercomputer into the fusiform face area of your brain, which is physically impossible.

Global amnesia is defined as a loss of memory for information presented to us from all sensory modalities. Yet, these people will go on to have a full and complete Life Review. So, is it really a loss of memory, or is it a loss of functionality? Ironically, ALL of the evidence that I have encountered states that "amnesia" of any kind is a loss of functionality, and NOT a loss of memories. "Amnesia" is a misnomer. Access to the memories is lost when the functionality is lost; but, whenever the functionality is restored, the memories come back as well. So, the memories weren't really lost or wiped out after all, now were they?

This observational evidence seems to suggest that our memories aren't being stored within our brain. Instead, Nature's Psyche creates Maps of Physical Functionality within our physical brain, and whenever one of those Maps gets destroyed, then the physical functionality is lost but not necessarily the memories. Maps of Physical Functionality is the best explanation that I have encountered so far for brain function, because all of our RAM memory storage and CPU processing seems to be taking place non-locally in the cloud and not locally within our physical brain. It's physically impossible to run programming code through a synaptic cleft; so, all of that CPU processing has to be taking place someplace else besides our brain. It's physically impossible to store petabytes of RAM or even a byte of RAM within a synaptic cleft; so, all of that RAM memory storage has to be taking place someplace else besides our brain. So, what's left? Well, it's obvious and clear to me that Nature's Psyche has organized our brain into different Maps of Physical Functionality; and now, Nature's Psyche is using those Maps of Physical Functionality to get things done for us.

These observations suggest a new and better definition for memory. Memory and learning are anything that survives the death of our physical body and physical brain. If it survives the death of your physical brain and you can actually remember it, then it's truly a memory in every sense of the word, and learning has actually taken place.

Apraxia is a disorder in which patients have great difficulty enacting movements that they are asked to perform, but they have no problem performing these movements spontaneously in natural situations. It's as if the Human Psyche is having difficulty interfacing with and controlling its physical body on demand; but, the programs or routines or functionality, that Nature's Psyche has already put into place, come naturally without thinking about them.

Ataxia is loss of motor coordination. It's a loss of functionality, and NOT a loss of memories. Physical deterioration can reduce or eliminate functionality; but, these people can remember having had motor coordination earlier in their lives. Functionality can be lost; but, consolidated memories never are. It's as if our consolidated memories are being stored for us non-locally as Quantum Waves. No amount of physical damage can remove them.

The anterolateral system consists of axons in the somatosensory system, which carry signals related to pain and temperature. Sever these axons, and you can no longer feel heat, cold, or pain. Anosmia is the inability to smell. Asomatognosia is a deficiency in the awareness of the parts of one's own body, which is typically caused by damage to the parietal lobe. Astereognosia is an inability to recognize objects by touch. Aphagia is a complete cessation of eating. Adipsia is the complete cessation of drinking. Anhedonia is a general inability to experience pleasure. Alexia is an inability to read. Agraphia is an inability to write. Acquired dyslexia is caused by brain damage. Adrenogenital syndrome is a developmental disorder that masculinizes genetic females. Contralateral neglect of visual, auditory, and somatosensory stimuli typically results from damage to the right parietal lobe. These are functional and developmental abnormalities, typically caused by genes and/or physical damage of some kind.

Again, from all of this we observe that functionality or purpose is lost, but not the memories of the individual. Consolidated memories remain, as if they are being stored for us non-locally as Quantum Waves in the cloud. If functionality returns, the ability to form new memories in this department returns. Whenever physical functionality goes offline or is damaged, then there's nothing for the Human Psyche to experience in that department and nothing for Nature's Psyche to remember.

There are so many things that can go wrong, that it's miraculous that anything goes right.

Nature's Psyche and God's Psyche function perfectly; but, the physical genes and the physical structures do not. If evolution gives us genetic abnormalities, they are going to show up in our physical bodies. If we get run over by a truck, it's going to show up in our physical body.

Physical functionality is necessary in order for the Human Psyche and Nature's Psyche to be able to experience and then remember physical events. If the machinery that consolidates memories is damaged or removed, then the Psyche is unable to access and store that information, even though the physical body is still going through the motions. The physical brain is how the Human Psyche interfaces with and controls its physical body. No brain function, then no interface and no experiences; and consequently, nothing to remember from the physical realm.

It's possible for short-term functionality (what they call working memory) to remain, even though the ability to consolidate those memories and transmit them to the cloud is lost. This can get extremely complex. Remember, evolution of any kind cannot do complexity and functionality. It's physically impossible. Evolution can only do diseases that are caused by genetic mutations.

Autoimmune diseases arise when the immune system attacks healthy body cells as if they were a foreign pathogen. Autoimmune disease is a breakdown in functionality, often caused by a genetic mutation. ALL of our inherited diseases are caused by evolution or random mutations. Evolution can't design and create functionality, but evolution can certainly destroy it.

Antibodies are proteins that bind specifically to antigens on the surface of invading micro-organisms, and in so doing help to destroy the invading organism. Antibodies and B-cells identify and destroy invading organisms. What's their motivation? The functionality for destroying organisms is designed and built into them. The ability to recognize invading organisms is probably programmed and built into them as well. However, the motivation to search for and attack invading organisms would be a product of Nature's Psyche. Motivation and incentives are a function of Psyche, not an innate part of physical matter. Atoms and molecules just sit there and do nothing, without some kind of Psyche to motivate them to do things.

Cerebral commissures are tracts of axons connecting the left hemisphere with the right hemisphere. The massa intermedia is the neural structure in the third ventricle that connects the two lobes of the thalamus. The corpus callosum is the main hard-line between the left hemisphere and the right hemisphere. When the hard-line is cut, the hemispheres can no longer communicate with each other. They don't have the hardware that's necessary to do so. This is what happens to split-brain patients. It's as if they have two separate brains.

In split-brain situations, the Human Psyche has to get creative in order to accomplish communication between its two hemispheres. The physical functionality is lost, so the Human Psyche has to compensate. One such compensation is called, cross-cuing. Cross-cuing is non-neural communication between hemispheres that have been separated by commissurotomy. The helping hand phenomenon is an example of cross-cuing. In split brain patients, the Human Psyche learns that it can use the hand that knows the answer to reach out and help the other hand that doesn't know the answer. In one example, the hand that knew the answer reached over and wrote the answer on the back of the hand that was mapped to the part of the brain that needed to be able to speak the answer.

So, how do neurons communicate with each other?

Well, they aren't doing it through chemical synapses, because that's physically impossible! The synaptic cleft will scramble any message that you try to send through it! That's the way it was designed to work. The synaptic cleft is a randomizer, scrambler, and jamming device. You can't send messages through a synaptic cleft! Any complex communication through the physical brain and any complex memory storage within the physical brain is physically impossible.

So, how are neurons communicating with each other if they can't use their synapses to do so? How are our memories being stored if they can't be stored in synapses and neurons?

Well, the neurons are under the command and control of Nature's Psyche. If the neurons need to communicate with each other, then they do so through Nature's Psyche. Hook an EEG to your scalp, and what do you observe? Waves! Nature's Psyche is using Waves, or WiFi, or Quantum Telepathy to communicate with the brain that it has engineered and programmed for your use. Telepathy is WiFi at the quantum level; and, it works! Your brain is awash with waves or WiFi! Neurons communicate with each other through waves or WiFi, at the quantum level.

I think that most people will find this interesting, but the Materialists and Naturalists are going to find it annoying, because it falsifies Materialism and Naturalism. Nevertheless, the explanatory power of Quantum Neuroscience, Quantum Mechanics, and Nature's Psyche is vastly superior to anything that the Materialists and Naturalists can provide us. Using Nature's Psyche and Quantum Mechanics, we can actually explain things that are physically impossible. We can explain how they function and work.

Posttraumatic amnesia (PTA) is amnesia that is produced by a non-penetrating head injury – a blow to the head that doesn't penetrate the skull. Amnesia is defined by the Materialists and Naturalists as loss of memory, because they don't know better. However, careful observation and critical thinking will reveal that "amnesia" is the result of lost functionality, and not necessarily the loss of memories. Once the functionality is restored and the brain is healed, the memories come back. The only memories that don't come back are the memories that weren't produced and stored, while that part of the brain was offline and non-functional. Nature's Psyche can't record memories from parts of the brain that don't exist or aren't functioning properly. This is logical common sense. Nature's Psyche can't use the machinery if the machinery is damaged or missing.

Posttraumatic stress disorder has a strong psychological component and is a function of the Human Psyche. What one psyche finds stressful and unbearable, another psyche will take in stride. The physical insult was real, and the stress was real; but, the Human Psyche is the one who chooses how to deal with its after-effects. Nature's Psyche records the events for us so that we can remember them; but, it's the Human Psyche who has to learn how to deal with them.

Psychosomatic disorders are defined as psychological disorders that have no known physical cause. In other words, it's all in your head, or all in your mind. Psychosomatic disorders are physically impossible, because they are being caused by your Psyche or your Mind.

Prosopagnosia is an inability to recognize faces. When the part of the brain that recognizes faces, or objects, is damaged or removed, then Nature's Psyche can no longer use that part of the brain to recognize faces or objects. Nature's Psyche can't use lost functionality to produce memory waves from that lost functionality. It's also possible that it's the Human Psyche who actually recognizes the different faces; but, that recognition process has been mapped by Nature's Psyche to the fusiform face area at the physical level in order to produce the necessary physical functionality associated with the process. If the Map of Physical Functionality gets destroyed, then face recognition is no longer possible at the physical level; but, who knows what's happening at the quantum level or the psyche level?

According to the BioPsychoSocial Model of Mental Illness and Reality, the physical body, the Psyche, and the Society are equal contributors to our mental illnesses, diseases, psychology, well-being, and stress.

The preceding mental illnesses are examples of mind-over-matter.

In contrast, substance-induced psychosis, iatrogenic harm, psychedelic drugs, psychoactive drugs, and psychopharmacology represent matter-over-mind.

Psychopharmacology studies the influence of drugs on the brain and behavior. Psychedelic drugs alter perceptions, emotions, and cognitions. Withdrawal from prescription drugs produces psychosis, delusions, and hallucinations – false stimuli. Psychoactive drugs alter subjective experiences and change our behavior by acting on the nervous system to make these changes.

The Human Psyche can only experience what the physical brain is producing. The Human Psyche is completely unaware of what Nature's Psyche is doing, experiencing, and recording. If Nature's Psyche decides to share one of our long-term memories with us, it can download the wave of that experience into our brain, and we can re-experience it again; but, most of the time, we are left with only what the Human Psyche can remember, which isn't much in comparison to all the data that Nature's Psyche is uploading to the cloud for us.

The positive-incentive theory is the idea that behaviors such as eating, drinking, and drug addiction are motivated by their anticipated pleasurable effects. The positive-incentive theory of addiction theorizes that addictions are motivated and caused by the pleasure-producing properties of the drugs. The positive-incentive value is the value that the Human Psyche assigns to a particular action, such as taking a drug or eating a meal. Each Human Psyche is different, and it actually chooses what it finds pleasurable and what it doesn't find pleasurable. Some people find candy or chocolate more pleasurable than others do.

The postcentral gyrus functions as the primary somatosensory cortex.

The primary visual cortex receives input from the retinas, through the lateral geniculate nuclei.

These cortices were assigned to process sensory INPUT.

The precentral gyrus functions as the primary motor cortex.

Nature's Psyche uses the primary motor cortex to trigger physical actions or OUTPUT.

Who designed the functionality of these devices or Cerebral Processing Units (CPUs)?

Well, it wasn't evolution. Evolution of any kind can't do design and creation. It's physically impossible.

God's Psyche was the one who designed and created the functionality of these things.

So, who implemented this functionality and now controls this functionality.

Well, it certainly isn't evolution's psyche, because evolution doesn't have a psyche.

It is Nature's Psyche who oversees the development or construction of our physical body and physical brain, based upon information or knowledge that it accesses from God's Database. It's also Nature's Psyche who programs our brain. All of this takes place completely outside the conscious awareness of the Human Psyche and its physical brain.

Hormones are chemicals that are released by the endocrine system directly into the circulatory system.

Glucose is a simple sugar that is the breakdown product of complex carbohydrates. Glucose is the body's primary utilizable source of energy. Glucose is brain food.

Glucogenesis is the process by which proteins are converted into glucose.

Glucagon is a pancreatic hormone that promotes the release of free fatty acids from adipose tissue, their conversion into ketones, and the use of both as sources of energy.

Who designed and implemented this functionality?

Well, God's Psyche or God's Intelligence designed and created the proteins and the genes to go along with them. But, it's Nature's Psyche who decides when you need glucose rather than proteins, or ketones rather than fat. It's Nature's Psyche who collapses the wave function thereby making all of this physical functionality REAL.

I'm not going to apologize for trying to figure out how all this stuff works. I'm a scientist. That's what we are supposed to do – figure out how things work!

One of the most interesting discoveries that I made while developing Quantum Neuroscience is that it's physically impossible to store messages and memories within the synaptic cleft. The synaptic cleft will scramble any message that you try to send through it! If memories are being stored within our brain, they aren't being stored within our synapses, because that's physically impossible. So, where are our memories being stored? Where does that functionality reside?

It has been observed and estimated that our physical brain has access to petabytes of memories; but, it's physically impossible to store petabytes of RAM within our physical brain. It can't be done! So, where are our memories being stored? Where does that functionality reside?

Karl Lashley and others discovered that there are NO memory engrams or centralized RAM in our physical brain. This discovery led the Materialists and Naturalists to assume that our memories are being distributed evenly throughout ALL the neurons of our brain, which is physically impossible thanks to the synaptic cleft. Furthermore, there are NO signs of CPUs and RAM within our neurons; and, these types of things cannot exist in the synaptic cleft where random chaos rules supreme. So, where are our memories being stored? Where does that functionality reside? We are running out of options!

"The P300 wave is a positive EEG wave that usually occurs about 300 milliseconds after a momentary stimulus that has meaning for the subject." In other words, whenever the Human Psyche comes across something that it finds meaningful, Nature's Psyche will fire off a P300 wave, which is detectable by EEG.

Meaning is a product of the Human Psyche, which means that in this case, it's the Human Psyche who is developing and remembering these various "meanings", concepts, and ideas. Furthermore, the observational evidence suggests that these meanings and memories are being transmitted by waves! Waves are WiFi at the quantum level. Telepathy is also WiFi at the quantum level. Hook an EEG to your scalp, and you are going to see waves! It's Nature's Psyche who is producing these quantum waves within the neurons, because all of our brain functionality happens completely outside the conscious awareness of the Human Psyche. Memories, messages, and meaning are being stored within these waves and transmitted by these waves. That's why Karl Lashley was able to cut out any part of the brain that he wanted to cut out, and the animals still had their memories. They would lose functionality, but they didn't lose their memories.

Human beings who have chunks of their brains cut out lose the associated functionality; but, the memories and experiences that were created and consolidated by Nature's Psyche, through the physical brain, remain intact. These memories can't be destroyed because they were created as waves, and then stored as waves. However, after the functionality of the brain is destroyed, the brain no longer has anything in that department for Nature's Psyche and the Human Psyche to experience and remember. If the part of the brain that registers experiences or declarative memories is removed, then Nature's Psyche has nothing from that part of the brain that it can convert into waves or long-term memories. Nature's Psyche and the Human Psyche can't work with something that they don't have. If you are missing a hippocampus, then the Human Psyche and Nature's Psyche can't use it to create memories and waves from it. If you destroy the part of your brain that retrieves memories from non-local cloud storage, then you no longer have access to those memories, even though they still exist in God's Database.

Fascinating, is it not?

This is Quantum Neuroscience at its best, and it has infinitely more explanatory power than Materialism, Naturalism, and Classical Physics.

Using Nature's Psyche, the Human Psyche, and Quantum Mechanics, we can actually explain the physically impossible! We can actually explain how memories are being created by our brain, experienced by the Human Psyche, and then stored by Nature's Psyche within God's Database. It's happening through Waves, or WiFi, or Quantum Telepathy. We can actually explain how it's possible for our consolidated memories to remain intact, even after huge chunks of our brain have been cut out or removed. Our long-term, consolidated, declarative memories are NOT being stored within our brain, because that's physically impossible. Nothing can survive a synaptic cleft! Our consolidated memories are being stored as waves by Nature's Psyche within God's Database.

Our memories are not being stored within the Human Psyche; otherwise, we would have instantaneous access to all of them simultaneously if they were indeed stored within the Human Psyche. The Human Psyche only remembers what it chooses to pay attention to. All of the activities of Nature's Psyche take place completely outside the conscious awareness of the Human Psyche. The Human Psyche makes choices and decides what to do with the physical body and physical brain which God has given to it. Nature's Psyche makes a recording of it all, and then stores all of that data as waves in God's Database.

The Materialists, Naturalists, Darwinists, Nihilists, and Atheists aren't going to want to have anything to do with Quantum Neuroscience. That's unfortunate, because those people are willing to pay a lot of money to be told what they want to hear.

Since it is Nature's Psyche who collapses the wave function according to the Orthodox Interpretation of Quantum Mechanics, WE KNOW that it is Nature's Psyche who develops a zygote into an embryo, and then into a fetus, and finally into a newborn child, based upon the information and knowledge that it gets from our genome and God's Database.

The thing we don't know is where the genome ends, and God's Database begins.

However, based upon the estimates of the exabyte of information needed to construct a human body from start to finish, and the 750 megabytes in the human genome, the genome ends very early in the process; and, God's Database takes over from there. It's actually possible that NONE of our blueprint is stored within our genome; and, that the genome is only used by Nature's Psyche to make proteins, which is complicated enough as is.

Nature's Psyche works flawlessly during the whole construction process or developmental process; and, the ONLY thing that will mess it up is the mutated genes that evolution has given us. Bad genes equal bad proteins. Bad genes equal procedural mishaps. That's just the way it is. The physical processes can fail, but Nature's Psyche will not.

Once I finally realized and accepted the fact that it's physically impossible to store memories within a synaptic cleft and physically impossible to transfer memories and data through a synaptic cleft, I simply KNEW that our memories are NOT being distributed throughout our synapses as the Materialists and Naturalists claim, at least not at the physical level.

I concluded that Karl Lashley's discovery – that there are NO memory engrams or NO physical RAM within our brains – is a true and accurate assessment of the situation, at least at the physical level. This was not a welcome discovery for my Neuroscience teacher, and he told me that I was wrong.

Nevertheless, ALL of the evidence I encountered repeatedly suggested that our long-term consolidated declarative memories are NOT being stored within our physical brain. This discovery led me to visualize the physical brain as a collection of Cerebral Processing Units (CPUs). It became clear to me that our physical brain is all about physical functionality, and NOT memory storage. Since I'm a computer scientist and a certified computer support specialist, I found the CPU Model of brain functionality quite satisfying, because I could relate to it on a personal level.

However, NONE of the evidence was supporting the CPU Model either.

Why?

It's because CPUs and RAM are physical machinery that are wired together so that an electrical current can be passed through them.

There are NO wires in the physical brain; consequently, CPUs and RAM are an incorrect way to define and describe the physical brain.

You see, a synapse is an electrical terminator, or an electrical dead-end. At the end of the axon, the electrical current ends, and chemical transmission or neurotransmission begins. That's NOTHING like a CPU or RAM!

Not only is it physically impossible to transfer memories, data, and information through a synaptic cleft, but it is also physically impossible to transfer bytes of software and machine code through a synaptic cleft. Our brains are no more functioning as CPUs than they are functioning as RAM. If the RAM Model is successfully defeated, and it is, then the CPU Model goes down with it. They are both physically impossible to implement through synaptic clefts. It can't be done! Our brains are neither CPUs nor RAM.

That discovery was quite a disappointment, even for someone like me, who is trying to keep an open mind about all of this.

So, what are our brains, really; and, how do our brains really work?

That is the million-dollar question, isn't it?

I eventually reduced brain functionality to two summary phrases, which I have mentioned a few times in this book.

Oh, I don't have any doubt that our memories are being distributed throughout our brain among the atoms of our brain. It's just not happening at the physical level, because it is physically impossible. How would you go about putting your memories into a single atom, if I were to give you the assignment to do so? Well, you wouldn't be doing it at the physical level, because it's physically impossible. You would have to do so at the quantum level, by converting your memories into Quantum Waves, and then storing those waves within that physical atom.

Oh, it's happening, alright. Our brains really are processing data, running computer code, crunching numbers, transferring data, building cities, weighing decisions, and storing memories. It's just not happening at the physical level. It's all happening at the quantum level. The physical input of sensory data, and the physical output of motor commands, is the only thing happening at the physical level within the brain. It all happens at the end of an axon, where the switch is either turned ON or OFF. A synapse is a switch. Our brain is comprised of a 100 trillion switches or synapses. Our brain is also comprised of a 100 billion stand-alone switches or neurons, which are either ON or OFF. Furthermore, our brains are comprised of quadrillions of ion channels that are either OPEN or CLOSED. Everything within our brains reduces to a single BIT. There's nothing else there within our brain. Everything else is happening at the quantum level. All the important and interesting stuff is happening at the quantum level, outside our conscious awareness.

Our brains are processing tons of information, just not at the physical level. The only thing happening at the physical level is the opening and closing of gates and/or the triggering of individual action potentials. It ALL reduces or integrates to a single BIT of information, which is the action potential. The neuron either fires, or it doesn't.

ALL of those synapses have NO meaning at all, at the physical level. If the synapses have any meaning besides OPEN or CLOSED, it is being given to them by Nature's Psyche at the quantum level, while Nature's Psyche assigns and oversees their construction.

The synapses have NO meaning to us at the physical level. ALL of their activity takes place outside our conscious awareness. The synapses are either depolarizing the membrane potential with EPSPs or polarizing the membrane potential with IPSPs. At the physical level, ALL the synapses on a neuron are being reduced or integrated into a single BIT of information called an Action Potential, which takes place when the neuron turns ON and FIRES.

All I was ever looking for was a physical explanation for why our brain and neurotransmitter systems work. All I ever found was physical explanations why they shouldn't work.

I'm a computer scientist, and I found the brain extremely disturbing once I figured out how it really works and what it really is. There's nothing there at the physical level, except a hundred billion stand-alone BITS or switches, with NO wires connecting them. That's NEVER going to work! Not as a CPU or RAM, anyway.

With 100 trillion synapses in the adult brain and possibly up to a quadrillion synapses in a child's brain, there seems to be an invisible database that's being used by Nature's Psyche to decide where Nature's Psyche should build the synapses and what types of neurotransmitters and neuropeptides these quadrillion synapses will transmit and receive across their synaptic cleft. You certainly are not going to the petabytes of information needed for synaptic logistics within our 750-megabyte genome!

Furthermore, some neurotransmitters have a few different types of post-synaptic receptors; and, the mapping or addressing locations as well as typing or differentiation information has to be stored for these as well within this invisible database in the cloud, which I have personally chosen to call God's Database.

It's physically impossible to store the quadrillions of bytes of database – being used to map-out, type, and place our synapses within our physical brain – within the 750 megabytes of "Flash RAM" that exists within our genome. The petabytes of RAM needed to store this invisible database can't be stored at the physical level within the brain or genome, because it's physically impossible.

Remember, it's physically impossible to store quadrillions of bytes or petabytes of data within our brains or genome at the physical level. All of this invisible storage has to be taking place at the quantum level, outside our conscious awareness.

This tells us what's happening at the quantum level. All of our RAM memory storage and CPU-like functionality is taking place at the quantum level, because these types of functionality are physically impossible at the physical level, thanks to synaptic clefts.

After making these different discoveries, then I finally realized what's really going on at the physical level, where our physical brains are concerned.

Using God's Blueprint or God's Database, Nature's Psyche and God's Psyche organize our physical brains into unique physical structures that become Maps of Physical Functionality. Each unique physical structure within our brain is mapped to perform specific physical functions. Then Nature's Psyche runs or operates our brains through these Maps of Physical Functionality, which Nature's Psyche and God's Psyche have manufactured for us. Neurons are switches that trigger physical functionality (output), and registers that buffer physical senses (input). Nature's Psyche and the Human Psyche use these registers, or neurons, or Maps of Physical Functionality to perceive sensations and sensory input. Then, once the Human Psyche makes a decision or a choice, Nature's Psyche uses other switches, or neurons, or Maps of Physical Functionality to trigger a physical response or a physical action.

That's how our brains are working at the physical level.

Neurons are switches, and these 100 billion switches have been organized into Maps of Physical Functionality by Nature's Psyche and God's Psyche, for use by the Human Psyche. Nature's Psyche uses these switches to turn on physical functions for us.

In summary, the Human Psyche uses these registers, or neurons, or Maps of Physical Functionality to perceive physical sensations or physical senses. The Human Psyche makes decisions and choices based upon this physical input. Then in response to the Human Psyche's choices, Nature's Psyche uses other switches, or neurons, or Maps of Physical Functionality to trigger physical responses and physical actions by collapsing the wave function. That's how our brains are working at the physical level. It's really easy to understand, once a person chooses to allow all of the evidence into evidence.

I KNEW that I had found the correct model for brain functionality at the physical level, once I discovered that each one of our physical senses has been organized according to a specific Map of Physical Functionality, and once I discovered that our physical functionality or physical output has been mapped directly to our primary motor cortex.

When I chose to pursue Quantum Neuroscience, I chose to try to adapt Neuroscience to the Orthodox Interpretation of Quantum Mechanics, and this Maps of Functionality Model ends up being the most useful result, in my humble opinion.

https://en.wikipedia.org/wiki/Topographic_map_(neuroanatomy)

http://mypsyche.us/wp-content/uploads/2018/01/Topographic-Map.pdf

The Retinotopic Map refers to our visual system. The Retinotopic Map is a map of the retinas onto the primary visual cortex of the occipital lobes. Our primary visual cortex is organized like a Map by Nature's Psyche. It's a Map of Physical Functionality. Nature's Psyche and the Human Psyche access this Retinotopic Map telepathically at the quantum level.

The Tonotopic Map refers to our auditory system. The auditory cortex is mapped by Nature's Psyche according to sound frequency. Nature's Psyche and the Human Psyche access this Tonotopic Map telepathically at the quantum level.

The Rhinotopic Map refers to our sense of smell.

Our sense of taste tends to be included within with our Somatotopic Map.

The Somatotopic Map refers to our skin or somatosensory system, and it is represented by the somatosensory homunculus, in the primary somatosensory cortex. All of our somatosensory input is tied to a Map within our brain by Nature's Psyche. The neurons that comprise this Somatotopic Map function as registers of physical sensory input. Nature's Psyche and the Human Psyche access this Somatotopic Map telepathically at the quantum level.

Our physical senses or physical input are represented by Topographic Maps of Physical Functionality, and both the Human Psyche and Nature's Psyche have direct access to the information being produced by these various physical senses or topographic maps. Sensation or physical senses are a function of these different Topographic Maps of Physical Functionality. Perception is a product of the Human Psyche. The Human Psyche decides what to make of its physical senses.

Our Motor System or physical output is organized as a Topographic Map, which is called the motor homunculus, and comprises the primary motor cortex. All motor activity in the brain consolidates into the primary motor cortex. Get Nature's Psyche to trigger the correct neuron in the primary motor cortex, and your finger will indeed rise off the table, just like you, the Human Psyche, wanted it to.

Unlike the topographic maps of the senses, the neurons of the motor cortex are efferent neurons that exit the brain instead of bringing information to the brain through afferent connections. The motor system is responsible for initiating voluntary or planned movements (reflexes are mediated at the spinal cord level, so movements that associated with a reflex are not initiated by the motor cortex). The activation from the motor cortex travels through Betz cells down the corticospinal tract through upper motor neurons, terminating at the anterior horn of the grey matter where lower motor neurons transmit the signal to peripheral motor neurons and, finally, the voluntary muscles. – Wikipedia

A cortical homunculus is a physical representation of the human body, located within the brain. A cortical homunculus is a neurological "map" of the anatomical divisions of the body. There are two types of cortical homunculi: sensory and motor. – Uncle Google

Chosen behaviors and volition are a function of the Human Psyche, and their physical manifestation is triggered through the different Topographic Maps in the Motor Cortices within the brain by Nature's Psyche. According to the Orthodox Interpretation of Quantum Mechanics, by Henry P. Stapp, Nature's Psyche collapses the wave function thereby making the Human Psyche's requests physical, actual, and real. In contrast, reflexes are automatic programming which has been set-up by God and Nature's Psyche to take place within the spinal cord and brain stem.

Whenever the Human Psyche makes a decision or a choice to move some part of its physical body, Nature's Psyche reaches out telepathically collapsing the necessary wave functions thereby triggering the specific neurons that produce that specific physical action. The activities of Nature's Psyche take place completely outside the conscious awareness of the Human Psyche. The Materialists and Naturalists call these unconscious activities, which are being performed at the quantum level by Nature's Psyche, "procedural memories" and "implicit memories"; and, these people claim that all of our memories exist only within our physical brain.

"Memories" is a misnomer. The Materialists and Naturalists don't know what they are talking about. Unconscious activities are performed at the quantum level by Nature's Psyche; and, these unconscious activities are tied directly to Maps of Physical Functionality that Nature's Psyche has made at the quantum level. It has nothing to do with memories, and everything to do with physical functionality. Implicit memories are defined as unconscious memories; but, how can they be memories if we can't remember them?

Remember, the ultimate goal of the Materialists, Naturalists, Darwinists, Nihilists, and Atheists is to trick you into believing that you and your memories cease to exist when your physical body dies.

Relays:

Then, we have all the different structures within our brain that function as relays, consolidators, and processors of the sensory input, thereby producing physical output of their own.

The amygdala functions as our emotion chip, or a Physical Map of our different emotions. Through our amygdala, we feel our emotions at a visceral level or a physical level, particularly the emotion of fear. The Human Psyche still has feelings and emotions, after its brain is dead and gone; but, the amygdala makes the Human Psyche feel its emotions at the physical level. Without an amygdala, a person no longer feels any fear in his gut. There may or may not be fear and stress, but it isn't felt in the viscera if the amygdala is missing.

The thalamus is a relay station. It receives sensory input from many of our physical senses, and then structures, organizes, and relays that information to the specific structures in our brain that were designed to register that information. The Human Psyche learns where to go in its brain to experience the different physical sensations that are coming into its brain. The thalamus is at the anterior end of the brain stem, and its nuclei are mapped by Nature's Psyche to function as sensory relays that receive input from our physical senses and project output to the cortex.

The medial geniculate nuclei are auditory thalamic nuclei that receive input from the inferior colliculi and project or relay that information to the primary auditory cortex. The lateral geniculate nuclei are six-layered thalamic structures that receive input from the retinas and transmit their output to the primary visual cortex. The mediodorsal nuclei are a pair of nuclei in the thalamus, and damage to these is responsible for many of the memory deficits associated with Korsakoff's syndrome.

The hypothalamus seems to be a Physical Map comprised of different thermostats and regulators, pumping out hormones of different types, thereby adjusting physical functionality according to whatever Nature's Psyche deems to be necessary.

The hippocampus is a Spatial Map of our surroundings. Our external physical world gets mapped into our hippocampus by Nature's Psyche. The hippocampus is the most likely place where Spatial Maps or Spatial Memories are in fact being mapped and stored within our physical brain at the physical level, by Nature's Psyche.

There are many other structures in your brain that have been mapped by Nature's Psyche to produce specific physical functions or physical outputs for you. It's Nature's Psyche who is doing all this Mapping of Physical Functionality for you within your physical brain, completely outside your conscious awareness.

If you close your eyes and try to figure out or sense where you, your psyche, really resides and really exists within your physical body, you realize that you are not in your finger or your toe. You are not in the back of your head. You, your psyche, is located just behind the skull in your forehead, in your prefrontal cortex, in the third eye position. It's interesting, because your vision feels like it is right there in front of you, when it fact it's being processed at the back of your brain. You, your psyche, resides or is stationed in your prefrontal cortex. That's where the Human Psyche seems to reside. So, it's no coincidence that our prefrontal cortex is mapped so as to control our Executive Functions at the physical level.

Obviously, all of these different Topographical Maps of Physical Functionality have already been discovered and proven to exist; and, so has Quantum Mechanics. I'm simply the first person to combine the two into a functional whole that I have chosen to call Quantum Neuroscience. Quantum Neuroscience was already a proven science and a verified science long before I coined the term; but, through Quantum Neuroscience I demonstrate scientifically how our Human Psyche is interacting with and controlling its physical brain, with the help of Nature's Psyche and Quantum Mechanisms.

Of course, I could have cut this book down by three-fourths by going straight to this Maps of Physical Functionality Model; but, then you would no longer have the months of backstory which led me to this discovery, nor would you find it all that significant or convincing.

No, I had to explain how I got here, or it wouldn't make sense to anyone, because there's nothing like this on the planet right now, as far as I know. Quantum Neuroscience has been hiding in plain sight for a century, waiting for someone to pick it up and run with it. But, the Materialists, Naturalists, and Atheists have been avoiding and rejecting this type of science for millennia, as if it were a blight, pox, and plague. These people are never going to accept Quantum Neuroscience because it's physically impossible.

That's just the way it is. You can't get blood from a stone. It's physically impossible.

Getting the Materialists and Naturalists to accept the physically impossible is, well, physically impossible. It's going to have to be a spiritual thing for them, or they ain't gonna get it. I KNOW, because I used to be a Materialist, Naturalist, Nihilist, and Atheist, until I had a few spiritual experiences or epiphanies of my own.

Textbooks for Quantum Neuroscience

When it comes to the "Neuroscience" part of Quantum Neuroscience, I still choose the book *The Mind and the Brain: Neuroplasticity and the Power of Mental Force* by Jeffrey Schwartz as the best introduction to practical neuroscience, applied neuroscience, and clinical neuroscience that I have discovered so far. This book is Neuroscience the way it should be done! It's actually useful neuroscience, and it's completely compatible with everything that I have discovered and developed concerning Quantum Neuroscience.

When it comes to the "Quantum" part of Quantum Neuroscience, I still choose *Consciousness Beyond Life: The Science of the Near-Death Experience* by Pim van Lommel as the ultimate best introduction to Quantum Mechanics and Non-Local Consciousness of any book on the planet. This book is too good to be true; yet, there it is. It exists; and, it actually explains what's really going on when it comes to Quantum Non-Local Consciousness, Psyche, or Mind. I haven't found anything better.

With a complete understanding of those two books and a bit of intelligence, you should be able to design and develop your own version of Quantum Neuroscience that is consistent, logical, makes sense, and actually works.

When it comes to an official Interpretation of Quantum Mechanics, which matches perfectly with Quantum Neuroscience and explains everything that is going on within the mind-brain interface, the Orthodox Interpretation of Quantum Mechanics from Henry P. Stapp is the ONLY one that explains everything you need to know in a logical and useful fashion – the interplay between the Human Psyche, Nature's Psyche, and the collapse of the wave function.

Any interpretation of Quantum Mechanics that is based upon Materialism, Naturalism, and Classical Physics is absolutely worthless when it comes to Quantum Neuroscience and trying to explain how the Human Psyche interacts with, controls, and survives its physical brain; and, most of the official interpretations of Quantum Mechanics listed on the Wikipedia are based exclusively on Materialism, Naturalism, and Classical Physics; and they by definition in principle deny the existence of Psyche or Non-Local Consciousness a priori. You can't use Materialism, Naturalism, and Classical Physics to explain how the Human Psyche works; but, by using the Orthodox Interpretation of Quantum Mechanics from Henry P. Stapp, you certainly can. All you want is something that works and makes logical sense!

Henry P. Stapp was prolific. It's going to take you some time and money to acquire and process everything that he has written. Stapp tried to explain and defend his Orthodox Interpretation of Quantum Mechanics from every angle and every question that he ever encountered. It's a magnum opus, making Stapp's Orthodox Interpretation of Quantum Mechanics the most useful version and the most realistic version of Quantum Mechanics on the planet, in my humble opinion. Both Pim van Lommel and Jeffrey Schwartz mention Henry P. Stapp within their books.

Pick one of Stapp's books to begin your journey, or dive into the tons of free material that Stapp has online.

https://sites.google.com/a/lbl.gov/stappfiles/

http://www-physics.lbl.gov/~stapp/

http://www-physics.lbl.gov/~stapp/PTRS.pdf

http://mypsyche.us/wp-content/uploads/2017/10/PTRS.pdf

Books and articles from Mario Beauregard and Wilder Penfield are compatible with Quantum Neuroscience.

Penfield, W. (1975). *The Mystery of the Mind*. Princeton, New Jersey: Princeton University Press.

Beauregard, M. & O'Leary, D. (2007). *The Spiritual Brain: A Neuroscientist's Case for the Existence of the Soul*. New York: HarperCollins.

Beauregard, M. (2012). *Brain Wars: The Scientific Battle over the Existence of the Mind and the Proof That Will Change the Way We Live Our Lives*. New York: HarperOne.

On the observational and experiential side of the equation, the following three books have proven informative and essential for me personally. Theories and ideas are completely worthless, if there's no observational evidence supporting them. Unlike Materialism, Naturalism, and Atheism, Quantum Neuroscience is primarily an observational and experiential science!

Anonymous. (2013). *Teachings of the Doctrine of Eternal Lives*. Salt Lake City, UT: Digital Legend.

Buhlman, W. L. (1996). *Adventures Beyond the Body: How to Experience Out-of-Body Travel*. New York: HarperCollins Publishing.

Durham, E. (1998). *I Stand All Amazed: Love and Healing from Higher Realms*. Orem, UT: Granite Publishing and Distribution.

There are thousands of others; but for me personally, these were the books that I found most useful and interesting. They are the ones that did the trick for me and helped me to see the light. They opened me up to the possibilities of Quantum Neuroscience.

Of course, all the different books that I have written are going to be compatible with Quantum Neuroscience, because the same mind or psyche is behind them all.

http://www.amazon.com/author/science

Science 2.0 was a direct and essential precursor to *Quantum Neuroscience*.

https://www.amazon.com/dp/B0771K6WTX/

I also found an organized religion that is completely compatible with and accommodates everything that I have ever learned or discovered about Quantum Neuroscience, within the Church of Jesus Christ of Latter-day Saints and the books that the Biblical God Jesus Christ had a hand in writing and producing – the Bible, the Book of Mormon: Another Testament of Jesus Christ, the Doctrine and Covenants, and the Pearl of Great Price. Psyche, Intelligence, Non-Local Consciousness, our pre-earth existences, the eternal journey of the soul, the Doctrine of Eternal Lives, Quantum Mechanisms, and Quantum Neuroscience are all there within those books waiting to be discovered and used.

https://www.lds.org/?lang=eng

The Biblical God Jesus Christ KNEW about Psyche, Intelligence, our pre-mortal or pre-earth existences, and Quantum Neuroscience long before we started to discover these things. Jesus Christ is the Being of Light and Love, whom everyone encounters during their near-death experiences. Whenever that being identifies Himself by name, it's always Jesus

Christ whom they encounter. He's the real deal. The other gods, such as evolution, are man-made fictional gods and don't really exist.

I thought that it was totally amazing to find a revealed religion and an organized religion that is 100% compatible with Quantum Neuroscience and Quantum Mechanics. Within these scriptures (Bible, Book of Mormon, Doctrine and Covenants, and Pearl of Great Price), the Gods reveal themselves to us – tell us who they are and what they can do.

We human beings are their children, both at the quantum level and the physical level. Our spirit bodies are their offspring; and, our physical bodies are their offspring. Adam was a Son of God. The Gods are a highly advanced race of human beings. The Gods tell us that we can become like them if we prove trustworthy and reliable.

Anytime we humans encounter the Gods either in person in the flesh or during our Near-Death Experiences, there's no technology visible. It's as if the Gods don't have any technology.

Why?

Our current generation of human beings is becoming technologically advanced. We have a gadget for everything now. Why don't the Gods have more toys than we do, if they are Gods and they are trillions of years more advanced than we are?

It's because they don't need the technology. The Gods have something infinitely better – a complete mastery of Quantum Mechanics and complete control over quantum mechanisms at the quantum level. You don't need physical technology if you have mastered the quantum level, or the spiritual level. In fact, the technology tends to get in the way. Our physical bodies are the pinnacle of their physical technology. They don't need anything more.

In these scriptures, the Gods actually tell us that they use Quantum Mechanics or Quantum Mechanisms to get things done. They call it the Priesthood of God or the Power of God. It's Quantum Mechanics.

The Gods have physical bodies that don't have any physical limitations. The Gods can Quantum Tunnel their physical bodies anywhere in the universe at will. Quantum Tunneling or Teleportation involves stopping the passage of time. According to the Theory of Relativity, if you move at the speed of light or faster, time comes to a stand-still and entropy ceases to exist. The way you Quantum Tunnel a physical object is to stop the passage of time for that object or remove all of its entropy, then you can put it anywhere you want in the physical universe instantly.

The Gods can read our minds – Quantum Telepathy. The Gods can do Action at a Distance – Quantum Telekinesis. The Gods can Phase-Shift and walk through walls here at the physical level. The Gods can levitate their physical bodies. And, as we already mentioned, the Gods can teleport at will.

Most of us can't do these kinds of things, because the Gods have placed physical limitations on us. The Gods are testing us to decide what type of individuals we will choose to be. The Gods aren't going to give us all their powers if we are going to choose to use those powers to destroy everything. We are being tested here in physical mortality.

The Gods can also function flawlessly at the spiritual level or the quantum level. The Gods can reorganize the physical atoms and physical molecules at the quantum level at will, thereby making the Gods seem omnipotent. The Gods can map ALL of the knowledge and information in this universe, contained within every physical atom and within every living psyche, into their own individual personal Psyche – at the quantum level or the psyche level, thereby becoming omniscient.

Our ultimate goal is to become like the Gods, whether we realize it or not.

Once I realized what's really going on where the Gods are concerned, thanks to the Church of Jesus Christ of Latter-day Saints – the Gods are Masters and Commanders of Quantum Mechanics or the Priesthood of God – then I suddenly realized that that's a whole lot cooler and more interesting than anything we can get from Materialism, Naturalism, Darwinism, Nihilism, Behaviorism, Atheism, and Classical Physics.

We each have to start where we currently are and then go on from there.

Remember, Jesus Christ can come and get you out of hell, just for the asking. He's done so for dozens of Atheists and Heathens on record. Jesus Christ could even come and get Richard Dawkins out of hell, if Richard Dawkins were to ask. That's good enough for me. In contrast, evolution is never going to come and get you out of hell, when your psyche and spirit body land there. I learn best through comparison and contrast, which is why I use it so much in my research and writings.

You are going to have to decide for yourself if any of this is useful to you, or not. That's something that nobody else can do for you.

I used to be a Materialist, Nihilist, and Atheist. One day, a few years ago, I realized that I didn't like where that road was taking me, so I decided to turn around and go the other way. My life has been getting progressively better ever since. That is what I have experienced and observed. May your journey be merry and bright! I wish you well.

Proof of Concept – Attention Must Be Paid to This

THIS!

Schwartz, J. M., & Begley, S. (2002). *The Mind and the Brain: Neuroplasticity and the Power of Mental Force*. New York: HarperCollins.

This is the ultimate book in Applied Research and Applied Neuroscience. There is no better. It also reconciles Quantum Mechanics and Non-Local Consciousness (psyche or mind) with Neuroscience.

When I finally started doing research into Quantum Neuroscience, I eventually realized that I needed to find scientific evidence of the Human Psyche reaching out telepathically through the Quantum Zeno Effect and turning on specific neurons directly through the power of mental force or will. Until I read this book, I had no idea how much of that kind of research, and how much of that kind of scientific evidence, was already on record waiting for me to find it. I wasn't looking for it for the first fifty years of my life; therefore, I never found it until now.

Unfortunately, I can't quote the whole book, but I can submit the whole book as Proof of Concept, where Quantum Neuroscience is concerned.

People have put this book online, if you know where to look:

http://publicism.info/psychology/mind/index.html

https://issuu.com/broli/docs/sharon_begley_jeffrey_m

"Chapter 10: Attention Must Be Paid" was particularly important as Proof of Concept for Quantum Neuroscience.

http://publicism.info/psychology/mind/11.html

http://mypsyche.us/wp-content/uploads/2018/01/The-Mind-and-the-Brain.pdf

This is good stuff that everyone should have access to, in my humble opinion.

The "cocktail-party phenomenon" is the ability to unconsciously monitor the contents of one conversation while consciously focusing on another. The cocktail-party phenomenon represents mind-over-matter, and the multitasking of attention. The unconscious is a product of Psyche. Both the Human Psyche and Nature's Psyche can do unconscious processing. Attention is a function of the Human Psyche. The Human Psyche chooses what to pay attention to. Nature's Psyche and God's Psyche do most of their work outside the conscious awareness of the Human Psyche. Consciousness is a function of the Human Psyche. Choice is a function of the Human Psyche. Nature's Psyche decides how to react by collapsing the appropriate wave functions.

A paradigm is an experimental method. The Reappraisal Paradigm is an experimental method for studying emotion. Subjects are asked to reinterpret a film or photo – to change their emotional response to it – while their brain activity is recorded.

You mean they can do that?

We're not supposed to be able to do that according to the Behaviorists – change our emotional responses at will! That's forbidden.

Deliberately changing your emotional response at will is a function of the Human Psyche.

It reminds me of the movie *Equilibrium*, where they have our hero hooked up to a polygraph. He's been caught, and the waves on the lie detector are off the scale. Then suddenly everything flatlines, when he (his psyche) makes a decision or commits to a course of action; and instantly, all of his fear is gone. He shut it all down. That's the power of Psyche!

Cognitive neuroscience is a division of biopsychology that focuses on the use of functional brain imaging (PET and fMRI) to study the neural bases of human cognition. Cognition, attention, and choice are a product of the Human Psyche.

How do we know?

WE KNOW because cognition, thoughts, attention, intentions, desires, personality, memories, and choices continue to exist long after our physical brain is dead and gone, according to the empirical evidence from Near-Death Experiences (NDEs), Out-of-Body Experiences (OBEs), Shared-Death Experiences (SDEs), and after-death Life Reviews.

Through PET and fMRI, we can observe Psyche activating neurons directly and working through neurons in order to get physical tasks done.

Harper Collins permits us to make quotes in critical reviews (promotional reviews), which is what I'm now going to do.

Attention to shape and color pumps up the volume of neuronal activity in the region of the visual cortex that processes information about shape and color; attention to speed turns up the activity of neurons in the region that processes information about speed. In people, paying attention to faces turns up activity in the region whose job it is to scan and analyze faces.

If this seems somewhat self-evident, it's worth another look: the visual information reaching the brain hasn't changed. What has changed — what is under the observer's control — is the brain's response to that information. Just as visual information about the color of this book's cover reached your brain as you opened it, so every aspect of the objects on the screen (their shape, color, movements, etc.) reached the monkey's brain. The aspect of the image that monkey pays attention to determine the way its brain responds. Hard-wired mechanisms in different brain areas get activated, or not, depending on what the monkey is interested in observing. An activity usually deemed to be a property of the mind — paying attention — determines the activity of the brain.

Attention can do more than enhance the responses of selected neurons. It can also turn down the volume in competing regions. Ordinarily — that is, in the absence of attention — distractions suppress the processing of a target stimulus (which is why it's tough to concentrate on a difficult bit of prose when people are screaming on the other side of a thin wall). It's all well and good for a bunch of neurons to take in sounds at a boisterous party, but you can't make out a damn thing until you pay attention. Paying attention to one conversation can suppress the distracting ones. Neurons that used to vibrate with the noise of those other conversations are literally damped down.

Schwartz, Jeffrey M. *The Mind and the Brain: Neuroplasticity and the Power of Mental Force* (pp. 329-330). HarperCollins. Kindle Edition.

Isn't that fascinating?

Mind, or psyche, or non-local consciousness – the Human Psyche – determines the activity of the brain. Psyche is the ultimate causal agent.

Choice is a function of Psyche! Choosing what we are going to pay attention to is therefore a function of Psyche. It's the psyche or the mind who is doing the choosing. The brain is simply complying with the Human Psyche's requests and demands. We literally reach in with our psyche or mind, and simultaneously ramp up and damp down the parts of our brain that are needed to accomplish the tasks that we are interested in.

If we are willing to allow this evidence into evidence, this also explains how Concept Cells work. Nature's Psyche, completely outside of our awareness, has assigned a specific brain cell to Jennifer Aniston; and, whenever our Human Psyche decides to pay attention to Jennifer Aniston or recognizes Jennifer Aniston, that particular neuron starts firing. That specific cell doesn't need any further input from the brain itself. It just starts firing whenever the Human Psyche is paying attention to Jennifer Aniston. This is Action at a Distance or Quantum Telepathy that we are witnessing here – an integral part of Quantum Neuroscience.

Mental force, generated by the effort of directed attention, can modulate neuronal function.

When it comes to determining what the brain will process, the mind (through the mechanism of selective attention) is at least as strong as the novelty or relevance of the stimulus itself. In fact, attention can even work its magic in the total absence of sensory stimuli. If an experimenter teaches a monkey to pay attention to a certain quadrant of a video screen, then single-cell recordings find that neurons responsible for that area will fire 30 to 40 percent more often than otherwise, even when there is no there there — even, that is, when that quadrant is empty. So here again we have the mental act of paying attention acting on the activity of brain circuits, in this case turning them up before the appearance of a stimulus.

fMRIs find that activity spikes in human brains, too, when volunteers wait expectantly for an object to appear in a portion of a video monitor. Even before an object appears, attention has already stacked the neuronal deck, activating the visual cortex and, even more strongly, the frontal and parietal lobes — the regions of the brain where attention seems to originate. As a result, when the stimulus finally shows up it evokes an even greater response in the visual cortex than if attention had not primed the brain.

Schwartz, Jeffrey M. *The Mind and the Brain: Neuroplasticity and the Power of Mental Force* (pp. 330-331). HarperCollins. Kindle Edition.

It's "paying attention" that turns on or activates neurons and Cerebral Processing Units within the brain – NOT sensory input! It's Psyche who turns on the brain, not the sensory input. This is the thing that we should be paying attention to, when it comes to neuroscience. The psyche or the mind is turning on the brain and firing it up! Well, that's definitely Proof of Concept, where Quantum Neuroscience is concerned. I now have evidentiary proof that I'm on to something important, after all.

Active, focused attention to a specific attribute such as color, they discovered, ramps up the activity of brain regions that process color. In other words, the parts of the brain that process color in an automatic, "hard-wired" way are significantly and specifically activated by the willful act of focusing on color. Activity in brain areas that passively process motion are amplified when volunteers focus attention on motion; areas that passively process shape get ramped up when the volunteers focus on shape. Brain activity in a circuit that is physiologically dedicated to a particular task is markedly amplified by the mental act of focusing attention on the feature that the circuit is hard-wired to process. In addition, during the directing of such selective attention, the prefrontal cortex is activated. As we saw in Chapter 9, this is also the brain region implicated in volition or, as we are seeing, in directing and focusing attention's beam.

The following year, another team of neuroscientists confirmed that attention exerts real, physical effects. This time, they looked not for increased neuronal activity but for something that often goes along with it: blood flow. After all, blood carries oxygen to neurons just as it does to every other cell in the body. Just as a muscle engaged in strenuous aerobic activity is a glutton for oxygen, so a neuron that's firing away needs a voluminous supply of the stuff. In the 1991 experiment, some subjects were instructed to pay attention to vibrations applied to the tips of their fingers, while others were not. The researchers found that, in the subjects paying attention to the vibrations, activation in the somatosensory cortex region representing the fingertips increased 13 percent compared to activation in subjects receiving the identical stimulation but not paying attention. It was another early hint that paying attention to some attribute of the sensed world — colors, movements, shapes, faces, feels, or anything else — affects the regions of the brain that passively process that stimulus. Attention, then, is not some fuzzy, ethereal concept. It acts back on the physical structure and activity of the brain.

Attending to one sense, such as vision, does not simply kick up the activity in the region of the brain in charge of that sense. It also reduces activity in regions responsible for other senses.

Schwartz, Jeffrey M. *The Mind and the Brain: Neuroplasticity and the Power of Mental Force* (pp. 332-333). HarperCollins. Kindle Edition.

Using God's Database as a template, Nature's Psyche does the "physiological dedication" of certain brains areas and certain concept cells to certain tasks. We KNOW that this is so, because it happens completely outside our conscious awareness. The Human Psyche has NO awareness of what parts of the brain Nature's Psyche has assigned to do specific tasks.

The ONLY way to explain what's going on within these science experiments is to bring the Human Psyche, Nature's Psyche, Action at a Distance, the Human Psyche's Attention, Quantum Telepathy, and the Quantum Zeno Effect into the equation! There is NO physical explanation for what we are witnessing here! The physical parts of the brain aren't just randomly turning themselves on. Someone or something is making them turn on. There's intelligence, purpose, and intention being demonstrated here, which are all products or functions of Psyche.

Jeffrey Schwartz goes on with additional examples and evidence within this chapter, as well as throughout his whole book. He literally uses science experiments to prove that the Human Psyche exists. This IS the smoking gun that I have been looking for all of my life. This is Proof of Concept, where Quantum Neuroscience is concerned.

http://publicism.info/psychology/mind/11.html

http://mypsyche.us/wp-content/uploads/2018/01/The-Mind-and-the-Brain.pdf

http://publicism.info/psychology/mind/index.html

https://issuu.com/broli/docs/sharon_begley_jeffrey_m

These kinds of scientific discoveries are precisely the goal of Quantum Neuroscience, which is an attempt to try to find a scientific explanation for how the Human Psyche interacts with, controls, and survives its physical brain.

I'm a believer, because now I have scientific proof that this is real and true. There's no going back to my ignorance anymore. I now have Proof of Concept, where Quantum Neuroscience is concerned. Good enough!

This scientific evidence has been around for a couple of decades, but the Materialists and Naturalists have done an excellent job hiding all of this evidence from us – I have to give them that! It's what they do; and, they are very good at it. They make sure that none of this gets into our public schools. This is the stuff that they don't want you to know, and they make sure that you don't. Their ultimate goal in life is to prevent you from finding Proof of Psyche and Proof of God – by any means necessary.

But, I've seen through the ruse and found the evidence anyway, because I went looking for it. Seek and ye shall find. Knock, and it shall be opened unto you. I'm glad that I did. I love living on what everyone else calls The Fringe. Mainstream got boring. Already been there and done that! It was time for something new. There doesn't seem to be any limits to Quantum Neuroscience.

All of these brain processes take place completely outside our conscious awareness, which means that it's Nature's Psyche or God's Psyche who is doing the activation of these different Cerebral Processing Units (CPUs), and NOT the Human Psyche. The Human Psyche makes the requests or pays the attention, and Nature's Psyche does its best to comply.

As far as I can tell, Nature's Psyche has two different means or processes by which it can turn on or fire-up a specific neuron directly and intentionally.

First, Nature's Psyche can go directly to the desired neuron and open up its ion channels directly, thereby triggering one action potential after another after another, or thereby triggering a continuous release of neurotransmitters at the other end of the neuron, depending upon which set of ion channels Nature's Psyche chooses to open.

The other way that Nature's Psyche can control this process of directly activating a specific neuron is by reaching out telepathically and quantum tunneling or teleporting the correct neurotransmitters to the correct dendrites and neurotransmitter receptors on the correct post-synaptic neuron.

Quantum Mechanics actually provides a scientific explanation for these two different processes; whereas, Materialism, Naturalism, and Classical Physics do not. Materialism and Naturalism are completely worthless when it comes time to explain what's happening in the Quantum Realm, or the Non-Local Realm, or the Transdimensional Realm, or the Non-Physical Realm, or the Psyche Realm, or the Spirit Realm.

The Quantum is Psyche's Realm; and apparently, Nature's Psyche knows precisely what it is doing and how to do it. Nature's Psyche is infinitely more intelligent than we are, because it has direct access to God's Database and God's Mind. We don't, which limits us greatly. The command and control of our physical brain by Nature's Psyche takes place completely outside our conscious awareness, even making many of us believe that it isn't happening at all.

God designed it this way as a test of our agency, faith, and will-power. God intended to give us a physical experience. The test works and learning takes place because the Human Psyche is forced to make choices while imprisoned within its assigned physical body. The whole test was designed to determine what we will choose to do when left to our own devices. It works, and we learn a lot as a result. For many of us, pain is our greatest teacher. It can get our attention, when nothing else does.

Van Lommel, P. (2010). *Consciousness Beyond Life: The Science of the Near-Death Experience*. New York: HarperCollins.

The book, *Consciousness Beyond Life*, and Near-Death Experiences are also Proof of Concept for Quantum Neuroscience.

I covered this book and Quantum Mechanics from a spiritual perspective quite extensively a couple of years ago within my book, *Quantum Mechanics: From a Non-Physical Spiritual Perspective*, so I won't do so again here. Take a look at that book, if the subject interests you.

https://www.amazon.com/dp/B01J023TGU

These books are Proof of Concept, where Quantum Neuroscience is concerned.

How the Physical Brain Really Works

I'm a computer scientist. I understand CPUs and RAM. Therefore, when all the neuroscientists told me that our brains are wired together to function as CPUs and RAM, I believed them. I, like everyone else, truly believed that that is how our brain really works at the physical level – like CPUs and RAM. The neuroscientists said so. They should know, shouldn't they?

However, over the years while studying neuroscience and psychology, ALL of the evidence was telling me that there are NO memory engrams, and therefore no RAM, within our physical brain. I resisted the evidence, vacillating back and forth for months – RAM, no RAM, RAM, no RAM; and my neuroscience teachers definitely resisted the evidence; but, it was there nonetheless, and I couldn't deny it in the end. There are NO long-term memory storage modules within our physical brain.

The preponderance of the evidence finally led me to conclude that our brains are comprised of carefully designed Cerebral Processing Units (CPUs) at the physical level. A lot of that belief is still within this book, because I liked that model a lot. I'm a computer scientist after all. I could understand it and accept it. I went with that belief for months, until ALL of the evidence started telling me that it is wrong as well.

I finally found the smoking gun at the physical level that put lie to all of the materialistic and naturalistic models that the neuroscientists are trying to force-feed us. It's called Integration, or Summation. This was the straw that broke the camel's back.

Integration is the combining of ALL the synaptic input signals into ONE overall output signal. In other words, ALL of the synaptic input from the Inhibitory Post-Synaptic Potentials (IPSPs) and Excitatory Post-Synaptic Potentials (EPSPs) are being integrated, reduced, or combined into ONE output signal that we call an Action Potential.

Do you understand what this really is and what it really means?

All of the neuroscientists write the words in their books, but they don't have a clue what it really means. "Integration is the adding or combining of all the synaptic signals on a neuron into ONE overall signal called a Post-Synaptic Potential, which may or may not produce an Action Potential."

A neuron is a switch. It is either ON or OFF. That's it! That's how our brain is really working at the physical level – as a collection of 100 billion stand-alone switches.

YIKES!

My whole world just crumbled.

Can you see why?

You can't shove a gigabyte of CPU functionality into a single hardware BIT. It's physically impossible. You can't shove petabytes of RAM functionality into a single hardware BIT. It's physically impossible. In fact, you can't even process a single byte through a single hardware switch or BIT. It's physically impossible.

I finally realized that there are no CPUs and no RAM within a physical brain; and then, lots of confirmational evidence kept pouring in.

A careful study of neurotransmitter vesicles and the synaptic cleft finally made it clear to me that any message, memory, idea, or programming code that you try to send through a synaptic cleft will get randomized, scrambled, and integrated into a single post-synaptic potential; and therefore, it will cease to exist. In other words, it's physically impossible to send even a single byte of information through a synaptic cleft, because it will be scrambled and then integrated into a single BIT of information. You can't store bytes of information within a synaptic cleft nor transfer bytes of information through a synaptic cleft, because it's physically impossible.

Our brains are NOT functioning as CPUs and RAM. It's physically impossible for them to do so.

Then the final nail in the coffin came along, and killed the CPU and RAM idea dead for good. I discovered that there are NO wires in our brains! The idea that our brains are wired together is nothing but a fictional myth. It isn't true. It's a carefully crafted deception that's being produced by the Materialists and Naturalists.

The dendrites in our brain transmit or conduct post-synaptic potentials. The axons in our brain transmit or conduct action potentials. In this case, a "potential" is an electrical current comprised of ions, or electrically charged atoms. The Materialists and Naturalists call dendrites and axons "wires". They do so in order to give you the erroneous impression that your whole brain is wired together into a single functioning whole like a CPU or RAM, which technically isn't true at all. To counteract this deception, I occasionally make the statement that there are NO wires in your brain. They, of course, will disagree with this statement and point you directly to dendrites and axons in order to counter that statement.

However, thanks to the synaptic clefts and synapses that separate every neuron and astrocyte, each neuron is functionally and electrically separated from every other neuron and astrocyte by synapses, which effectively means that there are NO wires in your brain. In other words, your brain is NOT wired together into a single CPU or RAM like the Materialists and Naturalists claim that it is. Technically, your brain is NOT comprised of circuits like these people claim that it is. A wire is a single, usually cylindrical, flexible strand or rod of metal. Wires are used to bear mechanical loads or electricity and telecommunications signals. There's NOTHING like this within a physical brain. There are NO wires in your brain.

Imagine it!

Everything that the Materialists and Naturalists have been telling us about brain functionality is FALSE! There are NO wires in your brain, which means that your brain is NOT a collection of CPUs and RAM. Integration FALSIFIES their wired-brain theory, their CPU and RAM theory, and their Distributed Memory Theory. Remember, it's physically impossible to store gigabytes of programming code or petabytes of memories within a single hardware BIT such as a neurotransmitter receptor, synapse, or neuron.

I kept eliminating every myth and falsehood until hopefully ONLY the truth remained.

We have falsified and eliminated everything that the Materialists, Naturalists, and Neuroscientists preach and teach as being real and true. It's physically impossible to send even a single byte of information through a synaptic cleft and have it survive. It's physically impossible to store even a single byte of information within a single hardware BIT, like a neurotransmitter receptor, synapse, or neuron. There are NO wires, CPUs, or RAM in your physical brain.

We have falsified and eliminated everything that the Materialists, Naturalists, Darwinists, Nihilists, and Atheists have chosen to believe in; so, what's left?

The only thing left is the truth.

A neurotransmitter receptor or ion channel is a gate. It is either OPEN or CLOSED. That's it! A receptor or register is a single hardware BIT – a gate or a switch.

A synaptic cleft will randomize, scramble, and then integrate every message, memory, and idea that comes its way into a single BIT of information – OPEN or CLOSED, ON or OFF.

ALL of the synaptic input through the dendrites on a single neuron gets integrated, reduced, and combined into a single BIT of information called a post-synaptic potential, which either triggers an Action Potential or it doesn't.

A neuron functions as either a register or a switch – it's either ON or OFF. A neuron is a single hardware BIT, and nothing more! It's physically impossible to process a byte of programming code through a single hardware BIT or switch; and, it's physically impossible to store a byte of information within a single hardware BIT or neuron.

The truth that remains is that our neurons are registers and switches. That's it! There's nothing else within our physical brain besides registers and switches. They are either ON or OFF. That's it!

Our brains aren't wired together into CPUs and RAM. ALL of their Distributed Memory Theories, where they claim that our memories are being stored in synapses and being distributed in a weighted fashion among synapses, have been FALSIFIED because they are physically impossible. None of that exists at the physical level within our brains, because it's physically impossible.

In the light of actual evidence, ALL of their materialistic, naturalistic, and atheistic theories have been reduced to a single hardware BIT that we call a neuron. It's either ON or OFF. There's nothing else there in our brains at the physical level, except a collection of a 100 billion stand-alone switches that are either ON or OFF.

These are the smoking guns that FALSIFY Materialism and Naturalism at the physical level. I have exposed all the deceptions and lies so that hopefully only the truth will remain.

So, where is all of that functionality, CPU processing, and memory storage being done, if it's physically impossible for it to be done within our physical brains at the physical level?

The ONLY logical answer is that it's ALL being done at the quantum level, through quantum waves, by Nature's Psyche. WE KNOW that it has to be Nature's Psyche (or God's Psyche) who is doing all of this functionality, processing, and memory storage for us at the quantum level, because the Human Psyche isn't consciously aware of any of these processes.

Remember, through quantum waves or quantum mechanisms, Nature's Psyche can do the physically impossible at will.

Isn't it amazing what we can discover by allowing ALL of the evidence into evidence?

So, what remains after we have eliminated everything that is false and everything that has been falsified?

The truth remains.

The physical brain is comprised of Maps of Physical Functionality that have been put into place or mapped into place by Nature's Psyche at the quantum level. In other words, Nature's Psyche has organized or mapped all of those 100 billion stand-alone switches into unique and different Maps of Physical Functionality. Our brains are all about physical functionality; and, our brains have been mapped by Nature's Psyche to function as such.

This is how our brain is really working and operating at the physical level – as Maps of Physical Functionality. This is the only thing that remains after all the myths and falsehoods have been exposed and eliminated.

Oh, I severely and vigorously fought against letting go of my Cerebral Processing Units (CPUs) model of brain functionality, because I had put a lot of time into developing it and trying to prove its veracity. I liked it a lot, and I didn't want to let it go. However, I did find a way to salvage it. Eventually, I realized that all of those different Cerebral Processing Units are indeed functioning as unique and different Maps of Physical Functionality, even though they are NOT functioning as wired CPUs in the traditional sense of the word, because there are NO wires in our brains.

One-to-one mapping, or topographic maps, or topographic arrays are physically explainable.

In fact, despite all of the science fiction they have given us, the neuroscientists have also pointed us to this truth and successfully organized the structures in our brain into topographical maps or topographical arrays; and, they actually use these terms every now and then. For example, lateral inhibition is described as the inhibition of adjacent neurons or receptors in a "topographic array". Some of the neuroscientists have realized that our brain structures are organized into Topographical Maps of Physical Functionality. The Materialists and Naturalists aren't always lying to us, because the truth will remain after all the falsehoods have been eliminated.

https://en.wikipedia.org/wiki/Topographic_map_(neuroanatomy)

http://mypsyche.us/wp-content/uploads/2018/01/Topographic-Map.pdf

Notice that they haven't put a whole lot of effort into this, despite the FACT that this is indeed the way that Nature's Psyche has organized our physical brain – as Maps of Physical Functionality or Topographic Maps.

Neurons function as switches or receptors – as gates or registers. Neurons have been mapped by Nature's Psyche to do so.

This is really simple to understand, once we decide to allow all of the evidence into evidence, and then choose to pursue a preponderance of the evidence.

According to the Materialists and Naturalists, it is physically impossible to store memories or psyche within a physical atom. That also means that it is physically impossible to store memories or psyche within a molecule, a protein, a gene, a neurotransmitter, or a synapse. Memories are synonymous with Psyche; and, according to the Materialists and Naturalists, Psyche does not exist, which means that thoughts and memories do not exist.

I have observed that the Materialists, Naturalists, Darwinists, Nihilists, Behaviorists, and Atheists are almost always wrong.

Memory or thought is synonymous with Psyche, because they both survive the death of our physical body and physical brain. ALL of the evidence from Near-Death Experiences (NDEs), Out-of-Body Experiences (OBEs), Shared-Death Experiences (SDEs), and after-death Life Reviews tells us that this is so.

Science 2.0 and Quantum Neuroscience allow ALL of the evidence into evidence, and then pursue a preponderance of the evidence. ALL of the evidence is telling us that memories and psyche are the same thing, because our thoughts and memories survive the death of our physical body and physical brain.

ALL of this psyche stuff is taking place at the quantum level through quantum waves, or thoughts and memories. It's mind-over-matter. So, what's happening at the physical level?

The neurons that Nature's Psyche has mapped to our five physical senses are registers or receptors; and, the Human Psyche can learn to read and understand these registers and learn how to make sense of them. This explains our physical input. Our physical senses or physical input have been organized by Nature's Psyche as Maps of Physical Input Functionality. This is the only thing that the Materialists, Naturalists, and Neuroscientists seem to have gotten right – the Topographical Map idea.

The neurons in our motor cortices are stand-alone switches and nothing more. Nature's Psyche has organized these switches into Maps of Physical Output Functionality. Nature's Psyche uses these switches to turn ON and turn OFF physical movements for us.

Nature's Psyche has created a Map of Physical Functionality within each of the structures in our physical brain. Then, in the case of the Maps of Physical Output Functionality, Nature's Psyche uses these different switches to move our physical body around for us. Deliberate intentional practice or attention forces Nature's Psyche to improve and enhance these Maps of Physical Functionality for us.

Neurons and synapses deliberately, intelligently, and knowingly improving and enhancing their physical functionality is a sure sign of Nature's Psyche or Intelligence in action. Yes, it's Action at a Distance – a quantum function – but it's happening nonetheless, because it's physically impossible to command, control, orchestrate, and coordinate atoms and molecules at the physical level.

Nature's Psyche within each physical atom is infinitely more intelligent and knowledgeable than we are, because Nature's Psyche has direct access to God's Psyche and God's Database.

In contrast, the only thing the Human Psyche can do is to access the Maps of Our Sensory Input Functionality, and then use that information to choose what it wants to do with the physical body that God has given it. The Human Psyche makes the choice, and then Nature's Psyche carries out the command by collapsing the necessary wave functions and turning ON the appropriate switches or neurons.

This is really simple and parsimonious – really easy to understand – once a person chooses to do so. Our brain is comprised of physical switches that have been organized by Nature's Psyche into different Maps of Physical Functionality. This is how our physical brain really works.

I didn't start with Maps of Physical Functionality; but, after examining all the evidence, that's what I ended up with – Maps of Physical Functionality mapped at the quantum level by Nature's Psyche and used by Nature's Psyche to get things done for us at the physical level.

When it comes to the brain, I started out with CPUs and RAM wired together at the physical level – just like everyone else – until I learned better. It took me a while to learn better. I eventually found evidence for brain functionality at the quantum level, when I finally went looking for it. Remember, telepathy or quantum waves are WiFi at the quantum level; and remember, there are NO wires in your brain. All of the functionality within your brain is being done at the quantum level through WiFi or Quantum Waves by Nature's Psyche. That's how your brain really works.

Some people have chosen to equate Nature's Psyche with God's Psyche. They observe that the physical laws and quantum laws are a manifestation, or a function, or a creation of God's Psyche.

However, Jesus Christ was God in the flesh; and, a careful examination of His actions and abilities as documented in the New Testament reveals that God's Psyche is something completely different than Nature's Psyche. Jesus Christ had command and control of the physical elements or physical atoms at the quantum level. There were times when He even teleported to safety – Quantum Tunneling. In other words, Nature's Psyche within the different atoms and molecules has covenanted to obey ALL of God's commands; and, God has direct access to and mastery of all quantum mechanisms, because He set all this up in the first place.

Normally, without any influence or interference from God's Psyche, Nature's Psyche just goes along on autopilot following and obeying the physical laws and quantum laws that God has established, because Nature's Psyche has covenanted with God's Psyche to do so. WE KNOW instinctively and from experience how the physical laws work; and, that's why we immediately recognize the "miraculous" whenever God intervenes and uses the quantum laws to get things done for us.

God's Psyche can interact directly with Nature's Psyche at the quantum level through quantum waves, whenever God's Psyche desires to do so; but, the Human Psyche cannot. The Human Psyche isn't consciously aware of the communication taking place between Nature's Psyche and God's Psyche, or Nature's Psyche and Nature's Psyche, at the quantum level. The Human Psyche can only detect what's happening at the physical level, within all the different Maps of Physical Functionality that Nature's Psyche has mapped into our physical brain.

This explains how our brain really works at the physical level. Simple! Parsimonious!

Fascinating, is it not?

Well, at least I think it is.

The purpose of Quantum Neuroscience is to explain scientifically how the Human Psyche interacts with, controls, and survives its physical brain. I believe that I have done so. It helps that Quantum Neuroscience was already a proven and verified science before I coined the term.

This is Quantum Neuroscience at its very best. The explanatory power of Quantum Neuroscience is infinitely superior to anything that we can get from the Materialists, Naturalists, Neuroscientists, Behaviorists, and Classical Physicists, because it takes ALL of the evidence into consideration.

All you really want is the truth, unless of course you are a Materialist or Naturalist or Darwinist; and then, any fictional ad hoc story will do. These people are master story-tellers. There have been many times in my life when they have convinced me that they were right. Yet, whenever I was given sufficient evidence, the Materialists and Naturalist and Atheists have almost always been proven wrong. Materialism and Naturalism FAIL and are FALSIFIED all across the board. In fact, there are many indications that some of the Materialists, Naturalists, Darwinists, Nihilists, Behaviorists, and Atheists are deliberately and knowingly trying to trick us and deceive us. It's what they do, and they are very good at it. They've had millennia of practice.

Oh, what a tangled web we weave, when we practice to deceive.

But, all I ever really wanted is the truth. I believe that I have finally found it.

PART VIII — QUANTUM MECHANICS PROVES THAT GOD EXISTS

The observed and proven existence of Action at a Distance, Quantum Mechanics, Forces, Fields, and Non-Local Consciousness FALSIFIES Materialism and Naturalism. When those go down, Scientism, Darwinism, Nihilism, and Atheism go down with them.

My observation is that Quantum Mechanics and invisible forces and fields are where we see the Hand of God in action. Quantum Mechanics, Forces, Fields, Physical Laws, and Fine-Tuning are Scientific Proof of God's Existence. Psyche or Non-Local Consciousness is one of those invisible forces or fields – Living Light, an Infinite Singularity. This Reality was really easy for me to understand and accept once I chose to let go of my Materialism, Naturalism, Scientism, Nihilism, and Atheism.

We Materialists, Naturalists, Atheists, and Classical Physicists have a hard time understanding and accepting Quantum Mechanics because Quantum Mechanics is Supernatural in nature and origin. We have to think outside the box, before we can understand and accept what Quantum Mechanics is trying to tell us and teach us. For many of us, that's extremely hard to do. We each have to start where we currently are, and then go on from there.

There are as many interpretations or versions of Quantum Mechanics as there are physicists on this planet. However, there are three main interpretations that have floated to the top. That's the main problem of Science and the scientific methods – each theory and observation is subject to theoretically an infinite number of possible different interpretations, which can't all be true.

The three main interpretations of Quantum Mechanics are the Copenhagen Model (where nothing exists until it is observed at which point it magically springs into existence from nothing), the Parallel Worlds Model or Many-Me Model or Science Fiction Model (where there are millions of me in the multiverse each living a different life some sinning and some being good), and the Quantum Consciousness or Cosmic Consciousness Model.

The Quantum Consciousness Model or Non-Local Consciousness Model is the MOST parsimonious and logical model.

Furthermore, we KNOW from the millions of Non-Local Experiences, Near-Death Experiences (NDEs), and Out-of-Body Experiences (OBEs) on record that each non-local consciousness or psyche continues to exist and function when the physical brain is dead or offline; consequently, I choose to go with the Model of Quantum Mechanics that has the MOST observational evidence and experiential evidence supporting it, as any good scientist should. I consider myself above all else to be a scientist, and Science is Observation and Experience – or at least it should be.

By definition in principle, there's NO way to prove or observe or experience the Copenhagen Model of Quantum Physics; and, the Parallel Worlds Model or Many-Me Model has NEVER been observed or experienced. Consequently, there is really ONLY one main model of Quantum Mechanics that has ANY observational evidence or scientific evidence supporting it – the Non-Local Consciousness Model or Quantum Consciousness Model. I chose to go with the best explanation or the one that has actually met its burden of proof. There are millions of NDEs and OBEs on record; and, that reality constitutes proof of its existence! Direct observations or eye-witness accounts can indeed be used in a court of law as PROOF in order to meet one's burden of proof and to establish a preponderance of the evidence.

Go with the BEST and get rid of all the rest. That's what I chose to do.

According to the LAWS of Quantum Mechanics, the ONLY way to bring Physical Matter into existence is through CONSCIOUS OBSERVATION or what God calls the Word of Command.

Physical Matter is OSERVED into existence by some kind of Non-Local or Non-Physical Consciousness. According to Quantum Mechanics, every single particle of Physical Matter in this whole universe was OBSERVED into existence by some kind of Non-Physical or Non-Local Consciousness.

This reality and truth establishes a NEED for God's Psyche or for God's Consciousness. ONLY God's Psyche could have issued the Word of Command and provided the energy needed, which converted a portion of Dark Matter or Spirit Matter into Physical Matter during the Prime Event. There's no other logical explanation that fits the observed evidence. There are always other explanations or interpretations; but, there are NONE others that actually fit the observations or the preponderance of the evidence.

There are 13 million NDEs on record. Thousands of people have seen and conversed with the Biblical God Jesus Christ during their NDEs, OBEs, and other non-local experiences or spiritual experiences; and, Jesus Christ has come and rescued hundreds of others from hell.

The very first particle of Physical Matter was OBSERVED into existence or BROUGHT into existence by the CONSCIOUS OBSERVATION or by the CONSCIOUS DECISION of something Non-Physical or Supernatural. Not only is this reality logical, but it is also Scientifically Accurate because it matches with Reality and because it matches with the Science of Quantum Mechanics. It also matches with the Lived Experiences of the human race.

Quantum Mechanics and the Quantum Consciousness Model of Quantum Mechanics became my SECOND convincing Scientific Proof of God's Existence, after the "inability of Natural Selection, Random Mutations, Chance, and Raw Physical Matter (evolution) to design and create" proved to me that God must of necessity exist in order to have done all the science that needed to be done to produce functional genomes and the corresponding life forms on this planet.

If you believe in the Science of Quantum Mechanics, and I do; then you just KNOW that God has to exist in order to provide the Consciousness necessary to bring the first particle of Physical Matter into existence. **Without the prior existence of Consciousness, there would be NO Physical Matter whatsoever, according to the LAWS of Quantum Mechanics.** And, when it comes to Science, Quantum Mechanics is as experimentally PROVEN and as much LAW as any Science can possibly be. This reality and truth IS Scientific Proof of God's Existence.

When it comes to Quantum Mechanics, understanding Nonlocality is the KEY. What is Nonlocality? Non-Local MEANS Non-Physical. Something is Non-Local if it has NO physical location in our Physical 3D Space-Time, meaning that it is Non-Physical in nature. If something is Non-Local and Non-Physical, then that means that it is Spiritual or Transdimensional in nature. Spiritual, by definition, means Non-Local and Non-Physical. Therefore, Quantum Mechanics IS Spiritual Mechanics, or the scientific study of Spiritual Mechanisms. Nonlocality in reality MEANS Non-Physicality or Spirituality.

In summary: According to the LAWS of Quantum Mechanics, a Non-Local Non-Physical Consciousness was NECESSARY in order to bring the first particle of Physical Matter into existence. In other words, God was NECESSARY in order to bring the first particle of Physical Matter into existence — specifically, God's Consciousness was NECESSARY to bring the first Physical Particle into existence. According to the LAWS of Quantum Mechanics, every particle in this Physical 3D Space-Time was CONSCIOUSLY OBSERVED into existence or CONSCIOUSLY CHOSEN into existence by some kind of Non-Local Non-Physical Consciousness that we would rightfully call God.

If you understand Quantum Mechanics and accept Quantum Mechanics as PROVEN SCIENCE, which I do, then you just KNOW that according to the LAWS of Quantum Mechanics, God's Consciousness must exist in order to have brought the first Physical Matter into existence. It's elementary, my Dear Reader.

Once we have the Correct Definition for something like Quantum Mechanics, then The Truth of the situation is usually OBVIOUS and CLEAR.

The Theory of Evolution IS Creation by Evolution or Creation by Rocks. Once we have the Correct Definition for the Theory of Evolution and we are calling it what it really is, Creation by Evolution or Creation by Rocks, then The Truth of the situation becomes OBVIOUS and CLEAR, because most of us automatically KNOW that Evolution and Rocks cannot design and create anything at all.

Physical Matter is brought into existence and then subsequently organized by CONSCIOUS THOUGHT, not by Evolution nor Rocks.

Quantum Mechanics IS the Scientific Study of how the Non-Physical becomes Physical. Quantum Mechanics IS the Scientific Study of how the Non-Local becomes LOCALIZED in our Physical 3D Space-Time through Conscious Observation. In other words, Quantum Mechanics IS the Scientific Study of how the Spiritual becomes Physical in nature. God's Non-Local Consciousness was needed to convert a portion of Dark Matter or Spirit Matter into Physical Matter in exactly the right locations so that galaxies and stars could form.

Light is Spirit. And, Spirit is Light. The Biblical God and the out-of-body travelers have told us that Spirit is Light. Spirit is comprised of Light.

Quantum Mechanics teaches us that Physical Matter is in fact comprised of Spirit Matter. The Physical is comprised of or made out of the Spiritual. Physicist David Bohm said that Physical Matter is condensed light or frozen light.

If you slow Spirit Matter down enough by infusing space-time into it, it becomes Physical Matter. Quantum Mechanics teaches us that CONSCIOUSNESS and Conscious Observation slows down Spirit Matter or Quantum Objects to the point that they become Physical in nature or "Frozen" in nature, and thus localized in nature. Conscious Observation can grab ahold of the Spiritual, slow it down, and localize it into our 3D Space-Time, thus condensing it or making it Physical in nature. Physical Matter is condensed Spirit Matter or frozen Spirit Matter. This is what Quantum Mechanics is really trying to teach us!

For me personally, Quantum Mechanics and the Theory of Evolution and Cosmic Fine-Tuning ARE my strongest and most convincing Scientific Proofs of God's Existence. The SCIENCE of Quantum Mechanics and the FAILURES of the Theory of Evolution convinced me beyond a shadow of a doubt that God must exist.

The obvious FAILURES of the Theory of Evolution or Creation by Evolution or Creation by Physical Matter provides a Scientific Proof of God's Existence, in that God must exist in order to have done all of the design and creation and SCIENCE, which Evolution could NEVER have done. This type of Scientific Proof of God's Existence is based upon Deductive Reasoning and the elimination of the FALSE or IMPOSSIBLE PREMISES so that we end up with the right CONCLUSION.

When properly understood, THE SCIENCE of Quantum Mechanics provides convincing and irrefutable POSITIVE SCIENTIFIC PROOF of God's Existence.

According to the LAWS of Quantum Mechanics, God's Non-Local Non-Physical Consciousness must exist; otherwise, NO Physical Matter would exist right here, right now. According to Quantum Mechanics, if God did not exist, then your Physical Body would not exist and this Physical Universe would not exist. This reality is OBVIOUS and CLEAR to anyone who actually understands Quantum Mechanics.

Once I got the Correct Definition for the Theory of Evolution and for Quantum Mechanics, then THE TRUTH regarding these subjects was OBVIOUS and CLEAR.

I will discuss Cosmic Fine-Tuning in a different part of this book. For me personally, Cosmic Fine-Tuning is a very compelling and convincing Scientific Proof of God's Existence.

Evolution of any kind cannot design and create anything at all; therefore, the Theory of Evolution or Creation by Evolution is OBVIOUSLY FALSE, which means that God is the ONLY person that we know of who could have designed and created all of the genomes and life forms on this planet. For me personally, this reality has become my FIRST most solid and most convincing Scientific Proof of God's Existence. Who would have thought that the Theory of Evolution could have proven to someone that God exists? But, that's exactly what happened to me!

The Science of Quantum Mechanics makes it OBVIOUS and CLEAR that God's Consciousness was NECESSARY in order to bring the first physical particle and the first physical matter into existence, because according to the LAWS of Quantum Mechanics every single physical particle in this universe was brought into existence by CONSCIOUS OBSERVATION or the Word of Command. Quantum Mechanics, when properly understood, becomes a solid and irrefutable Scientific Proof of God's Existence!

I discuss these various Quantum realities and truths in much greater detail in my book, "Quantum Mechanics from a Non-Physical Spiritual Perspective", which can be found at this link:

https://www.amazon.com/dp/B01J023TGU

Therefore, I won't go into all of that here, thus leaving space to pursue these Quantum truths from different angles instead. In other words, this section of this book is going to be relatively short.

—

Note, the LAWS of Quantum Mechanics and my discussion on the subject also forms a CORE foundational part of my books entitled:

"The Second Comforter: Supping with Our Resurrected Lord Jesus Christ"

https://www.amazon.com/dp/B01IAKHTY6

"I Am Not a Creationist! So what am I?"

—

I mention these books just in case you want to see how I use Quantum Mechanics and Nonlocality to support my other themes and theses, because Quantum Nonlocality or Quantum Spirituality IS germane to many different topics and not just one.

Some people have asked me if I'm being paid by the word. I am verbose and prolific.

Well, actually, I'm not being paid at all. I haven't made any money from this. I've lost money or spent money – thousands of dollars. Nobody's interested in any of this.

I simply keep writing and explaining these things until I finally understand them myself. Writing is how I learn. I learn best by trying to explain these things to someone else. I also quickly figure out what I don't know and what I need to research, by trying to explain these things to others in writing. Writing forces me to learn.

I write these books for my benefit, knowing that nobody else is ever going to read them.

Others have said, "You sure like to hear yourself talk."

Well . . . no.

But, I do like reading what I have written.

While reading what I have written, I remember things that I have already forgotten. If I want to remember something, I have to get it in writing, or soon it is gone.

—

Now pay attention!

Despite the FACT that I have debunked and falsified the Theory of Evolution thousands of times in hundreds of different ways, don't ever once let me convince you that there is NO such thing as evolution or random mutations.

Evolution is REAL, very REAL. Evolution or random mutation is entropy or the second law of thermodynamics. It's REAL. Entropy is very REAL at the physical level; and, it works as advertised at the physical level.

Whenever the Materialists, Naturalists, Darwinists, Nihilists, Behaviorists, and Atheists start talking about evolution, they get most everything wrong, because evolution or entropy cannot design and create. However, these people do indeed get one thing perfectly right. Evolution, or random mutation, or entropy is indeed the CAUSE of ALL of our heritable diseases, developmental diseases, and heritable mental illnesses.

Remember, the Theory of Evolution is FALSE because random mutations or entropy cannot design and create genes, proteins, and life forms. However, evolution or random mutation or entropy is very REAL; and, it can indeed destroy genes, proteins, and life forms. Do you see how that works? It's important to understand.

The Multi-Me Theory of Parallel Worlds

Quantum Neuroscience starts by identifying the physically impossible, and then it continues by trying to find a scientific explanation for the physically impossible.

You need an Interpretation of Quantum Mechanics that explains what the Human Psyche and Nature's Psyche are doing at the quantum level. If your Interpretation of Quantum Mechanics doesn't do that, then it's worthless. You also need an Interpretation of Quantum Mechanics that explains how God's Psyche does His science and His work through quantum mechanisms. If your Interpretation of Quantum Mechanics can't do that, then it's worthless.

In Quantum Mechanics when it comes to Schrödinger's Cat Paradox, there are three main solutions or interpretations that have been presented, and many sub-variations.

The **Copenhagen Solution** where the moon, the earth, and the universe don't exist until after some human views them into existence or observes them into existence doesn't make logical sense, because it's a chicken and egg paradox. According to this model, the earth has to exist before humans or chickens can exist; but, the earth can't exist unless some human or chicken observes it into existence. The Copenhagen Model is the most-taught solution; but, it's also the most illogical solution to the Schrödinger's Cat Paradox.

The **coolest solution** from the standpoint of science fiction stories is the Many Worlds Theory, or the Parallel Worlds Theory, or the Multi-Universe Theory, or the Multi-Me Theory. This is the one that the Atheists, Naturalists, and Materialists like most because it's based upon speculation and can be made to fit with quantum theory without bringing God into the equation. There's one HUGE problem with this one, though. It IS fiction – just like Materialism, Naturalism, Darwinism, Nihilism, and Atheism – with absolutely NO observational evidence to support it. Science IS Observation; and, since NOBODY has ever lived nor experienced multiple versions of themselves or multiple earths, we KNOW for a fact that Parallel Worlds and Multi-Me's do NOT exist. The ONLY place we have ever encountered Parallel Worlds or Many Earths is within Science Fiction, which is the only place that it exists. Just because it's cool and just because it makes for the best Science Fiction, it doesn't mean that it's true or real.

In contrast, thousands of people have seen and talked with Jesus Christ during their Out-of-Body Experiences (OBEs) and Near-Death Experiences (NDEs). Hundreds if not thousands of former Atheists and Unbelievers have been rescued from hell by Jesus Christ, while they were dead and out-of-body, and then lived to tell about it. God resurrected them from the dead. In ALL of these NDEs and OBEs, Jesus Christ is the Savior; and, Satan, the demons, the evil spirits, the rebels, and the atheists or unbelievers reside in hell. In NONE of these NDEs and OBEs has Satan ever been observed as being the Savior, and Jesus Christ has never been observed as being the Grandmaster of hell.

There is ONLY ONE universe that OBErs and NDErs experience; but, there are many different dimensions, levels, or realms in our universe – each in a different phase of existence. The lowest level, or lowest frequency, or lowest phase of the Spirit World or Non-Local World is a mirror-image of our physical world; but, each higher dimension or higher frequency in the Spirit World or Non-Local World changes things; and, the non-consensus realities, or most flexible realities, or the designer realities exist at the higher dimensions or higher frequencies. You are always you, because there's ONLY ONE you; but, your Psyche or Non-Local Consciousness can function in multiple different dimensions at different phases, or different frequencies, or different native velocities including a physical reality.

This third solution to the Schrödinger's Cat Paradox, the REAL SOLUTION that has actually been observed and experienced by human beings or human psyches, IS called **Quantum Consciousness**; wherein, God, or God's Psyche, or God's Consciousness is the Ultimate Observer who observed all physical matter into existence or called all physical matter into existence.

Go with the best, and get rid of all the rest. Go with the one that is real and true – the one that has actually been observed and experienced for real.

Quantum Consciousness and the Observation of God IS Scientific Proof of God's Existence, because it's Proof of Concept. In contrast, Materialism, Naturalism, Darwinism, Nihilism, Atheism, and Parallel Multi-Me's FAIL because they have NEVER been observed nor experienced for real.

Remember, Science IS Observation. Observation or Lived Experience of any kind IS Scientific Evidence or Proof of Concept. Jesus Christ, or God, has been observed and experienced during thousands of different NDEs or OBEs while people have been "in the spirit". Remember, hundreds of Atheists have been saved from hell by Jesus Christ, just for the asking. Remember, if you ever find yourself in hell, Jesus Christ can get you out of there. He's done it before, and He can do it some more.

Remember, Non-Local means Non-Physical, or not located in our 3D Physical Space-Time Realm. Non-Local means Transdimensional. The Non-Local Realm is the Quantum Realm or the Spirit Realm.

Remember, each Physical Universe and each Physical Particle has a beginning, which means that it has a Beginner.

Remember, understanding Quantum Non-Locality is the KEY to understanding the whole of Science! If you don't understand Quantum Non-Locality, the Non-Local Realm, or the Spirit Realm, then you simply have NO clue as to how things really work.

Quantum Mechanics teaches us that conscious observation or the Word of Command from Non-Local Consciousness or Psyche IS NEEDED to convert spirit matter into physical matter. Quantum Mechanics is Spiritual Mechanics, or the study of how spirit matter really works.

In all of my books, I have chapters on Quantum Mechanics, because for me personally Quantum Mechanics IS Scientific Proof of God's Necessity, which ends up being Scientific Proof of God's Existence.

Pim Van Lommel's book, *Consciousness Beyond Life: The Science of the Near-Death Experience*, IS Scientific Proof of God's Existence, because it is scientific proof of Quantum Non-Locality or Quantum Non-Physicality.

In the Appendix of his book, *The Future of the Mind: The Scientific Quest to Understand, Enhance, and Empower the Mind*, Michio Kaku discusses Quantum Consciousness and presents the THREE MAIN solutions to the Schrödinger's Cat Paradox on pages 329 to 342.

The **first solution** is the traditional Copenhagen Model, by Bohr and Heisenberg. The Copenhagen Model never made any logical sense to me. Einstein hated it.

> To determine the state of the cat, you must open the box and make a measurement. The cat's wave (which was the sum of a dead cat and a live cat) now "collapses" into a single wave, so the cat is now known to be alive (or dead). Thus, observation determines the existence and the state of the cat. The measurement

process is thus responsible for two waves magically dissolving into a single wave. Einstein hated this. (Kaku, p. 332).

Observation is all fine and well – I keep saying that Science is Observation of both the Local Physical Realm and the Non-Local Spirit Realm. Scientific Observations, or Psyche Experiences, or Lived Experiences ARE the BEST and most convincing form of Scientific Evidence!

But, the implications of the Copenhagen Model are illogical. Under this model, the moon does not exist unless there is someone actually looking at it and observing it into existence.

The truth is that the moon exists, whether there's any life on this planet observing it or not, because God organized the moon into existence from pre-existing spirit matter that He then converted into physical matter, which He then used to organize our moon.

The **second solution** to the Schrödinger's Cat Paradox is the Parallel Universe, or Many Worlds, interpretation that was proposed in 1957 by Hugh Everett.

Under this solution, every time a choice is made, it creates a new and different parallel universe.

This is the model used in most of our Science Fiction shows on television, including *The Flash* with Earth-1, Earth-2, Earth-19, and so forth; and, the television series *Dark Matter* with their Quantum Drive or Blink Drive that can be used to teleport within our universe and can be used to phase from one physical universe into the next parallel physical universe.

The problem with the Parallel Universe Theory is that it has NEVER been experienced, has never been OBSERVED, and has never been lived or experienced for real. NOBODY has ever experienced it, outside of our Science Fiction shows on television. If it has never been experienced, then it isn't REAL!

There is NO parallel universe in which Satan was our Savior and Jesus Christ was the devil. In ALL of our NDEs, OBEs, Visions, and Spiritual Experiences as a race, Jesus Christ is always our Savior, and Satan is the dictator and slave-master in hell. Satan wants to enslave you in hell, and Jesus Christ wants to set you free from hell.

THE TRUTH will always match with REALITY, which means that it will always match with the Lived Experiences or Psyche Experiences of the human race. This reality and truth leads us to the third solution.

The **third solution** to the Schrödinger's Cat Paradox was developed by Eugene Wigner in 1967. It's called Quantum Consciousness, Non-Local Consciousness, Intelligence, or Psyche.

Cosmic Consciousness

Eugene Wigner said that **only a conscious person can make an observation that collapses the wave function**.

Since you need an infinite number of friends to collapse the previous wave function to make sure they are alive, you need some form of "cosmic consciousness", or God.

Wigner concluded: "It was not possible to formulate the laws (of quantum theory) in a fully consistent way without reference to consciousness."

In this approach, God or some eternal consciousness watches over all of us, collapsing our wave functions, so that we can say we are alive. This interpretation yields the same physical results as the Copenhagen interpretation, so this theory cannot be disproven. But, the implication is that consciousness is the fundamental entity in the universe, more fundamental than atoms. The material world may come and go, but consciousness remains as the defining element, which means that consciousness, in some sense, creates reality. (Kaku, pp. 333, 334 – bolding and editing are my doing).

In his most recent books, Michio Kaku provides all of the main solutions, and then lets you decide for yourself which one you like most. When it comes to the Schrödinger's Cat Paradox, we simply KNOW which solution matches BEST with REALITY or the Lived Experiences of the human race. As a scientist, I tend to go with the BEST, and get rid of all the rest.

Only a Conscious Psyche can collapse the wave function, when it needs to be collapsed. The very existence of the atoms that we see around us is based upon the fact that they were consciously observed into existence, or consciously called into existence by God or by God's Non-Local Consciousness or Psyche. It ALL reduces to Psyche or Non-Local Consciousness, which means that it ALL reduces to God's Psyche or God's Non-Local Consciousness, as the fundamental unit of REALITY.

This is the TRUE MODEL OF REALITY, the Ultimate Model of Reality, because it has been LIVED and EXPERIENCED for real. This is a Psyche Ontology, which is in fact the Ultimate Model of Reality. This REALITY is Scientific Proof of God's Existence. The Biblical God Jesus Christ has been lived and experienced, both in the physical body and out of the physical body by thousands of different people.

Remember, if it has never been experienced, then it isn't REAL! Lived Experience IS Scientific Evidence, because Lived Experience IS Scientific Observation or Eye-Witness Observation.

You, your Psyche, is going to have to decide for yourself whether I know what I'm talking about, or not. But, I'm sticking with my story. I used to be a Materialist, Nihilist, and Atheist, but the Scientific Evidence or Lived Experiences of the human race finally convinced me that I was wrong. It's been a few years now since I last believed in Materialism, Naturalism, Nihilism, Darwinism, and Atheism. I have been searching for and developing Scientific Proofs of God's Existence for the past couple of years now. It's now July 2017, as I write this sentence. I started searching for Scientific Proof of God's existence in the summer of 2015, after my Atheist friends formally disowned me and rejected me.

The TRUTH is that we human beings or human psyches live and experience Psyche or Non-Local Consciousness, every single day of our lives, and long after our Physical Brain is dead and gone, according to the Scientific Evidence, or Empirical Evidence, or Experiential Evidence from NDEs, SDEs, OBEs, Visions, and other types of Spiritual Experiences including the Revelations of the Biblical God Jesus Christ and the Revelations from the Biblical God Jesus Christ. There's enough Scientific Evidence in these Out-of-Body Experiences and Visions to convict these people of having had a Spiritual Experience, in a court of law. In contrast, there is NO EVIDENCE to support the philosophical claims and primary assumptions of the Materialists, Naturalists, Darwinists, Nihilists, Determinists, and Atheists.

I have chosen to go with the evidence, and to leave all of the philosophical speculation behind. That is my right as a Scientist!

Our interpretations of Quantum Mechanics, String Theory, and Origin Theories must match with REALITY and the Lived Experiences or Scientific Observations of the human race, or our interpretations of Scientific Evidence will be completely worthless. Lived Experiences of any kind, Local or Non-Local, are Scientific Observations. Materialism, Naturalism, and Atheism are completely worthless and unscientific, because they don't match with the Lived Experiences of the human race, meaning that they don't match with REALITY.

Quantum Neuroscience starts by identifying the physically impossible, and then it continues by trying to find a scientific explanation for the physically impossible.

You need an Interpretation of Quantum Mechanics that explains what the Human Psyche and Nature's Psyche are doing at the quantum level. If your Interpretation of Quantum Mechanics doesn't do that, then it's worthless. You also need an Interpretation of Quantum Mechanics that explains how God's Psyche does His science and His work through quantum mechanisms. If your Interpretation of Quantum Mechanics can't do that, then it's worthless.

Quantum Numbers and Atomic Orbitals

When it comes to physics, especially Quantum Mechanics or Transdimensional Physics, there is a common sense LAW that axiomatically states that the Smaller has the potential to dwell within or reside within, interact with, and control the Larger.

When it comes to Orbitals and the Quantum Numbers, I see Signs of Psyche and Signs of God written all over it.

https://en.wikipedia.org/wiki/Atomic_orbital

http://mypsyche.us/wp-content/uploads/2017/12/Atomic-orbital-Wikipedia.pdf

https://www.khanacademy.org/science/physics/quantum-physics/quantum-numbers-and-orbitals/a/the-quantum-mechanical-model-of-the-atom

http://mypsyche.us/wp-content/uploads/2017/12/Quantum-numbers-Khan-Academy.pdf

Who taught the Electrons to act this way? Who makes the Electrons act this way? It's too organized and too elegant for it to have happened randomly by chance. You can see God's Signature all over it! Orbitals are three-dimensional waves! That's radically advanced technology! We humans tend to visualize waves as being linear or concentric, not three-dimensional.

The Electrons don't have any strings attached. Electrons can quantum tunnel away to the other side of the universe instantaneously if they want to, so who is keeping them in orbit of a specific atom? Each individual electron can act or function any way that it wants, so who is telling each electron to submit to the Atomic Orbital System or the Orbital Way of acting, who is telling each electron to stay put in orbit of its assigned atom, who is assigning a specific orbital to each electron, and who is telling each electron to coordinate its actions with the other electrons in orbit of the nucleus of its assigned atom? Who or what is teaching the electrons to act this way and making the electrons act this way? They act differently depending upon their order in the hierarchy. This implies intelligence and coordination.

The only thing that made any logical sense to me when trying to answer these questions is God's Psyche, or God's Word of Command, or God's Physical Laws. Someone had to design the Atomic Orbital System in the first place, before the electrons could be made to conform to it. There is intelligence, design, order, and law written all over this thing, suggesting the existence of some kind of Law Giver.

So, who is making each electron coordinate telepathically with the other electrons in the atom; who assigns each electron its orbital; who is communicating this information to the electrons; and who is making each electron conform to its assigned orbital? The only thing that made logical sense to me is something smaller than the electron which can reside within and control the electron – something like Nature's Psyche. The smaller can dwell within and control the larger. In this case, we are looking at something smaller than the elementary particles of physical matter.

The very existence of something smaller than the elementary particles of physical matter – something immaterial like Psyche's command and control, or God's Laws – FALSIFIES Materialism, Naturalism, and even Darwinism, and Atheism.

The necessity of Physical Laws that are smaller than electrons, dwell within electrons, control electrons, and then coordinate the electrons telepathically demonstrates the necessity of something like Nature's Psyche and God's Psyche – the intelligence and power necessary to make the Physical Laws and then enforce the Physical Laws.

If telepathic command and control is involved, then technically God's Psyche doesn't have to reside within each electron, because He can orchestrate everything from the outside and teach each physical particle how to act when He brings it into existence. However, if there is a piece of Nature's Psyche within each quark and electron, then it makes logical sense to me that each piece of Nature's Psyche is smaller than an electron or a quark so that it can dwell within and control these elementary physical particles.

As I see it, the very existence of Atomic Orbitals, Quantum Consciousness, Physical Laws for Nature's Psyche to obey, and Quantum Mechanics IS Scientific Proof of God's Existence. The existence of Quantum Laws, Physical Laws, Order, and Organization is proof of the existence of some kind of Law Giver. There is no other logical explanation, because we all KNOW that random chance, random mutations, and random diffusion produces nothing but entropy, chaos, cancer, death, and extinction.

Someone has to be injecting and imposing syntropy, order, organization, and life into the system, because physical matter can't be doing it all by itself, because physical matter doesn't work that way. It's physically impossible for physical matter to spontaneously organize itself into something like a functional protein or a functional gene; therefore, someone else – someone intelligent – has to impose that order and structure upon it instead.

https://en.wikipedia.org/wiki/Horton_Hears_a_Who!

Horton hears a Who. Well, I'm sensing signs of a WHO and sensing the Signature of God within Atomic Orbitals and the ways that electrons have been taught to act. Of course, it took me over fifty-five years to learn about Atomic Orbitals, because I wasn't looking for such a thing and didn't know that they existed.

Likewise, it took me over fifty years to realize that every science discipline IS Scientific Proof of God's Existence, because I wasn't looking for such a thing and consequently didn't believe that such a thing was possible.

Nevertheless, there are signs and indications that there exists a Who or a Whoville within each and every electron and quark that God has ever designed, created, and brought into existence. Just because we can't see it with our physical eyes nor hear it with our physical ears doesn't mean that it doesn't exist. Observation of the Physical Laws and Quantum Laws in action tells us that it must exist. It's elementary my dear reader.

Quantum Neuroscience: this is the science and the science book that I wish I would have had access to forty years ago when I was in high school and in college the first time around. But, I had to develop it and write it for myself. I hope you enjoyed the ride. I certainly did.

Catch the Quantum Wave

Oh, it's happening, alright. Our brains really are processing data, running computer code, crunching numbers, transferring data, building cities, weighing decisions, and storing memories. It's just not happening at the physical level. It's all happening at the quantum level. The physical input of sensory data, and the physical output of motor commands, is the only thing happening at the physical level within the brain. It all happens at the end of an axon, where the switch is either turned ON or OFF. A synapse is a switch. Our brain is comprised of a 100 trillion switches or synapses. Our brain is also comprised of a 100 billion stand-alone switches or neurons, which are either ON or OFF. Furthermore, our brains are comprised of quadrillions of ion channels that are either OPEN or CLOSED. Everything within our brains reduces to a single BIT. There's nothing else there within our brain. Everything else is happening at the quantum level. All the important and interesting stuff is happening at the quantum level, outside our conscious awareness.

What's the storage capacity of the human brain?

Most computational neuroscientists tend to estimate human storage capacity somewhere between 10 terabytes and 100 terabytes, though the full spectrum of guesses ranges from 1 terabyte to 2.5 petabytes.

The math behind these estimates is fairly simple. The human brain contains roughly 100 billion neurons. Each of these neurons seems capable of making around 1,000 connections, representing about 1,000 potential synapses, which largely do the work of data storage. Multiply each of these 100 billion neurons by the approximately 1,000 connections it can make, and you get 100 trillion data points, or about 100 terabytes of information.

Neuroscientists are quick to admit that these calculations are very simplistic. First, this math assumes that each synapse stores about 1 byte of information, but this estimate may be too high or too low. Neuroscientists aren't sure how many synapses transmit at just one strength versus at many different strengths. A synapse that transmits at only one strength can convey only one bit of information — "on" or "off," 1 or 0. On the other hand, a synapse that can transmit at many different strengths can store several bits. Secondly, individual synapses aren't completely independent. Sometimes it may take several synapses to convey just one piece of information. Depending on how often this is the case, the 10-to-100-terabytes estimate may be much too large. Other problems include the fact that some synapses seem to be used for processing, not storage (suggesting that the estimate may be too high), and the fact that there are support cells that might also store information (suggesting that the estimate may be too low).

Even if we accept that the storage capacity of the brain is between 10 and 100 terabytes, estimating how much of that space is "used space" versus "free space" is very difficult — the brain is simply much more complex than an external hard drive. Not only do some parts seem to be involved in many different memories at once, but this stored data is often being corrupted and even lost. One thing is certain: The notion that humans only use 10 percent of their brain is a myth — information may be stored in every part of the brain.

These computations and estimates here are actually very good. They are in line with what I have come up with on my own.

Can you see any flaws in any of this?

Most people can't, because they aren't looking for flaws or falsehoods. The Materialists and Naturalists take all this information on blind faith as being real and true. They don't question it, and they don't give it another moment's thought!

But, I see some flaws or weaknesses. I'm a theoretician. I notice trends, and then consolidate them.

First of all, it doesn't matter how many synapses are on a neuron, because all of that information reduces to a single BIT that's called a neuron! All of the synapses consolidate into a single neuron. A neuron is either ON or OFF – it either fires, or it doesn't! A neuron is a single switch. It either lights up, or it doesn't. A neuron can transmit one single bit of information – ON or OFF. That's it! Consequently, synaptic weights are also irrelevant. They only increase or decrease the chance of a neuron firing. Everything in our physical brains reduces to that single neuron – that single bit or switch – either ON or OFF. It's physically impossible to store memories and data in synaptic weights or synaptic clefts. The synaptic weights are Black Boxes, or a convenient fiction. They can't be used as RAM or memory storage, because nothing can survive passage through a synaptic cleft, except ON or OFF – one bit of information.

The ONLY way that these synaptic weights can have any value or meaning is if Nature's Psyche is imbuing the synaptic weights it creates with meaning and then deriving some kind of value or benefit therefrom. Otherwise, all of the synapses and synaptic weights are put through temporal summation and spatial summation getting compressed into a single BIT of information that decides whether the post-synaptic neuron will fire or not.

Spatial summation is the integration of signals that occur at different sites on the neuron's membrane. Who is translating and giving purpose and meaning to these ON-OFF signals? Well, the OFF signal really doesn't have any meaning. It just means that the ion channel is CLOSED.

Temporal summation is the integration of neural signals that occur at different times at the same synapse.

ALL of the synaptic inputs and synaptic weights are being integrated into a single BIT of information that decides whether a neuron fires or not. The Materialists and Naturalists try to turn it into something magical and mystical, but it's not. ALL synapses integrate into one BIT of information, determining whether the neuron will fire or not. You can't shove petabytes of memories and data through a single BIT. It's physically impossible. You can't store petabytes of memories and data within a single BIT. It's physically impossible.

God made it physically impossible to get CPU functionality and RAM functionality out of a physical brain, because all of those stand-alone switches or neurons are not wired together in any way. According to Classical Physics, Materialism, and Naturalism, our physical brains shouldn't work; but, they do. When it come to the physical brain, there is more than meets the eye.

Since every synapse reduces to a single neuron or BIT, the most that our physical brain can have in terms of memory storage capacity is 100 billion bits of RAM, because our brain has a 100 billion switches, bits, or neurons. But, the neurons aren't connected by any wires, so our brain is comprised of 100 billion stand-alone bits, or 100 billion switches, which are either ON or OFF. With no wires connecting the neurons, you can't make RAM nor CPUs out of them! It's physically impossible. With 100 billion stand-alone switches in our brain, our brains technically have NO memory storage capacity whatsoever. Our brains also don't have any CPU functionality and can't do any processing of information, because there are no wires connecting any of it together. This is what our physical observations are trying to tell us.

Can you hear the Materialists and Naturalists showering me with vitriol right now? Well, it's happening invisibly behind the scene. But it doesn't matter, because it's physically impossible to construct RAM and CPUs out of switches with NO wiring. It can't be done! It's physically impossible to do the physically impossible.

My research and observations suggest that most of our brain is being used for processing (Cerebral Processing Units), and NOT memory storage or RAM. Karl Lashley and others discovered that our brains have NO memory engrams, which means that our brains have NO consolidated memory storage or RAM. ALL of the evidence, when critically examined and allowed into evidence, suggests that our brains have NO memory storage or physical RAM.

The writing is on the wall, for anyone who is willing to look and see.

While researching Quantum Neuroscience, it was made clear to me that nothing can survive the synaptic cleft. A neurotransmitter receptor represents a single bit of information – OPEN or CLOSED – ON or OFF. Receptors can't function as gigabytes of RAM as the Materialists and Naturalists claim. There's NO transistors or switches within our neurotransmitter receptors.

Adding a neurotransmitter receptor to a postsynaptic membrane, through Long-Term Potentiation (LTP), does absolutely nothing to add new memories to your brain. After all, where are all those memories being stored while that new receptor is being added to the postsynaptic membrane? And, how are those memories loaded into that new ion channel when nothing physical can survive the synaptic cleft? Using LTP for memory storage is synonymous with magic.

Furthermore, Distributed Memory is a myth, because it's physically impossible to distribute your memories across many different neurons through synaptic clefts. No message or memory can survive a synaptic cleft, because a synaptic cleft is a randomizer or scrambler, not RAM or memory storage! You can't store gigabytes of RAM in random chaos. It's physically impossible. You can't transmit nor distribute gigabytes of memories through random chaos. It's physically impossible. A synaptic cleft is synonymous with random diffusion or random chaos. Remember that! It's physically impossible to transmit memories and messages through a synaptic cleft, which means that it's physically impossible to store memories within a synapse.

God has deliberately designed our brains to make it physically impossible for them to function as CPUs and RAM.

The Materialists and Naturalists never stop to think about any of this, because they don't want to. The very existence of Quantum Memories or Non-Physical Memories falsifies Materialism and Naturalism, which means that these people refuse to go there.

Nevertheless, if you want to find and know the truth, you have to be willing to catch the Quantum Wave!

God has deliberately designed our brains so that it's physically impossible for them to function as CPUs and RAM.

Imagine what a former Materialist, Nihilist, and Atheist like me goes through when he first encounters such a thing! ALL of that brainwashing and conditioning was still within me and still a part of me, when I first started to study Neuroscience.

The cognitive dissonance was through the roof!

I could sense that there was something seriously wrong with our brains; but, I couldn't put it into words. It took me months to be able to identify it and put it into words.

God has deliberately designed our brains to make it physically impossible for them to function as CPUs and RAM. Nothing can survive a synaptic cleft. It's physically impossible to send memories, messages, data, information, or programming code through a synaptic cleft. If any of that is taking place within our brains, it's not taking place through our neurons and our synapses, because that's physically impossible. Nothing can get through a synaptic cleft!

Can you imagine the cognitive dissonance I was going through anytime I sat down to study Neuroscience? My brainwashing and conditioning were preventing me from seeing the light; and, ALL of my neuroscience teachers were suffering from the same malady. I kept running into one obstacle after another; and, my neuroscience teacher didn't like it whenever I brought them up. Despite his PhD in the subject, he didn't have any realistic answers either.

Remember, any message that you try to send through a synapse will be scrambled by the synaptic cleft. Therefore, any type of memory that you might try to store within a synapse will be scrambled by the synaptic cleft. It's physically impossible to transmit messages through synapses, and it's physically impossible to store memories within synapses!

It's also physically impossible to store petabytes of RAM within a physical brain. A neuron represents one single bit – ON or OFF – like a switch, except within our brain there are NO visible wires connecting these switches. Our brain is like 100 billion stand-alone switches, with no wiring connecting those switches. You can't make RAM nor CPUs out of such a thing. It's physically impossible. And, what's turning on and turning off the switches? Those switches or neurons don't have any wires on them. It's a *deus ex machina* whenever the Materialists and Naturalists claim that our memories are being created by our synapses and then stored within our synapses, because it's physically impossible. Nothing in terms of memories or information makes it through a synaptic cleft. There's no way to send video, a memory, and a complex message through a BIT, or an ion channel, or a synapse. ON or OFF – OPEN or CLOSE – is the only thing you can send through an ion channel, neurotransmitter receptor, synapse, and neuron. Everything within a physical brain reduces to or integrates into a single BIT.

Amazing, is it not?

A neuron represents a single bit, a switch. All synapses reduce to a single bit or neuron – ON or OFF. 100 billion bits or neurons is 12,500,000,000 bytes. That's 12.5 gigabytes of RAM, if our whole brain were dedicated exclusively to RAM, as memory storage. Sorry, but most of that is probably being allocated as CPU bytes and used to store and process programs or functionality, NOT memories!

Nevertheless, no matter how you choose to allocate the 12.5 gigabytes within your brain, it's physically impossible to store petabytes of memories in 12.5 gigabytes of RAM. And technically, our brains are not 12.5 gigabytes of RAM, because the bits in our brains are stand-alone switches with no wiring connecting them; therefore, in actuality, our brains as a whole represent 1 BIT of physical hardware memory storage, no matter how we choose to look at it. Our brains only do ON and OFF. Our brains are comprised of stand-alone switches.

Scientists estimate that our brains are storing petabytes of memories; but, no matter how you choose to visualize it, it's physically impossible to store petabytes of RAM in our physical brain. And despite what the Materialists and Naturalists claim, you certainly can't store petabytes of RAM in a synapse, neuron, or neurotransmitter receptor, because each one of these represents a single BIT of hardware. You can't shove petabytes into a single bit. It's physically impossible. All throughout our brains, everything reduces to a single bit – ON or OFF. That's not CPUs or RAM that we are looking at, but disconnected switches. A neuron is a stand-alone switch – a BIT of information – either ON or OFF. You can't transmit petabytes of memories through a single BIT, and you can't store petabytes of memories within a single BIT. It's physically impossible.

So, where are our memories being stored, and how are they being stored? And, how do neurons communicate with each other?

Well, the latter question is easier to answer. Hook up an EEG to your scalp, and what do you observe?

WAVES!

The waves mean something, even the simplest of them.

Neurons can communicate with each other directly, through waves. Waves are WiFi at the quantum level. Telepathy is WiFi at the quantum level. You can send messages through waves. You can also store memories holographically within waves. It can be done, and it is being done. Alas, these waves are non-physical and immaterial. Their very existence falsifies Materialism and Naturalism, which is the point that God is trying to make with all these different waves that He is using to get things done within our brains and physical bodies.

Remember, God has deliberately designed our brains to make it physically impossible for them to function as CPUs and RAM.

So, who is producing these messages, transmitting these messages, and then interpreting these messages? Who is providing all of this functionality within our brains? The only logical answer is Nature's Psyche, because all of these messages, commands, and control are taking place outside our conscious awareness. The Human Psyche isn't aware of any of these things.

It is obvious and clear from examining our brains, that there are NO wires connecting the neurons in our brains, which means that any communication taking place between the neurons is happening wirelessly through waves of some kind. The communication and "wiring", if it exists, is taking place invisibly and wirelessly through WiFi, Quantum Telepathy, or Waves. There's no other explanation for how it's being done, because it's physically impossible to send messages and memories through a synaptic cleft and have them survive.

This is proof of concept, where Quantum Neuroscience is concerned. Quantum Mechanics is needed, in order to have memories and to store memories! Quantum

Mechanics is needed to turn those 100 billion switches or bits within our brains into functional CPUs and RAM. ALL of the "wiring" within our brains is being done wirelessly through WiFi at the quantum level, by Nature's Psyche or God's Psyche.

That's how it is possible to have 100 terabytes of CPU processing taking place within our brains, and petabytes of memories being stored within our brains, because it's all being processed and stored wirelessly by Nature's Psyche using quantum mechanisms, holography or light waves, and God's Database.

In fact, if you are using Quantum Waves for CPU functionality and RAM storage, it's theoretically possible to have a petabyte of CPU functionality within a single atom, along with an exabyte of memory storage within that same atom. Psyche or Non-Local Consciousness is an infinite singularity. Psyche is theoretically capable of an infinite amount of memory storage capacity and an infinite amount of CPU functionality, within something that has NO physical size and takes up NO physical space whatsoever.

We can detect electromagnetic waves coming from our physical brain, so we KNOW that our brains and neurons are generating these kinds of waves. Again, it's our neurons that are generating all of this electromagnetic energy, and NOT our synapses or synaptic clefts! Neurons are batteries – they are either charging or discharging – they are either ON or OFF. When neurons fire, they set waves into motion. In contrast, synaptic clefts are randomizers, scramblers, and jamming devices. You are not going to get a message or memory through a synaptic cleft, unless you use Waves of some kind. Even electromagnetic waves might be able to get the job done.

However, waves at the quantum level are something completely different. They aren't detectable with our physical instruments! Quantum Waves are off the scale! Quantum Waves exist in the non-local realm or spirit world, which means that they have NO physical limitations. Quantum Waves have NO speed limits. They can and do exist at velocities greater than the speed of light. In fact, Quantum Waves can be instantaneously and simultaneously everywhere, and every-when. Everything in the quantum realm is pure syntropy. There's NO entropy in any of it. Entropy is a function of space-time. Physical matter is subject to entropy. There is NO entropy in the non-local realm where Psyche or Non-Local Consciousness resides. It's pure syntropy. Nothing ages and nothing dies. Psyche and spirit matter have NO physical limitations and NO entropy! They are endless and eternal.

According to the Quantum Law of Complementarity, if entropy exists, then syntropy must exist as well. ALL of the entropy has been loaded into the physical realm, leaving the spirit realm or non-local realm full of syntropy.

I'm not the first person to discover these truths, and I won't be the last. We all end up at the same destination, when we finally decide to allow ALL of the evidence into evidence.

The Need for a New Approach

When empirical scientific studies discover phenomena or facts that are not consistent with current scientific theories, these new facts must not be denied, suppressed, or even ridiculed, as is still quite common. In the event of new findings, the existing theories ought to be elaborated or modified and if necessary rejected and replaced. We need new ways of thinking and new forms of science to study consciousness and acquire a better understanding of the effects of consciousness.

Science Equals Asking Questions with an Open Mind

In my opinion, current science must reconsider its assumptions about the nature of perceptible reality because these ideas have led to the neglect or denial of important areas of consciousness. Current science usually starts from a reality that is based solely on perceptible phenomena. Yet at the same time we can (intuitively) sense that besides objective, sensory perception there is a role for subjective aspects such as feelings, inspiration, and intuition.

Current scientific techniques cannot measure or demonstrate the content of consciousness. It is impossible to produce scientific evidence that somebody is in love or that somebody appreciates a certain piece of music or a particular painting. The things that can be measured are the chemical, electric, or magnetic changes in brain activity; the content of thoughts, feelings, and emotions cannot be measured. If we had no direct experience of our consciousness through our feelings, emotions, and thoughts, we would not be able to perceive it.

Endless Consciousness

On the basis of prospective studies of near-death experience, recent results from neurophysiological research, and concepts from quantum physics, I strongly believe that consciousness cannot be located in a particular time and place. This is known as nonlocality. Complete and endless consciousness is everywhere in a dimension that is not tied to time and place, where past, present and future all exist and are accessible at the same time. This endless consciousness is always in and around us.

We have no theories to prove or measure nonlocal space and nonlocal consciousness in the material world. The brain and the body merely function as an interface or relay station to receive part of our total consciousness and part of our memories in waking consciousness. Nonlocal consciousness encompasses much more than our waking consciousness.

Our brain may be compared both to a television set, receiving information from electromagnetic fields and decoding this into sound and vision, and to a television camera, converting or encoding sound and vision into electromagnetic waves. Our consciousness transmits information to the brain and via the brain receives information from the body and senses.

The function of the brain can be compared to a transceiver; our brain has a facilitating rather than a producing role; it enables the experience of consciousness. There is also increasing evidence that consciousness has a direct effect on the function and anatomy of the brain and the body, with DNA likely to play an important role.

Near-death experience prompted the concept of a nonlocal and endless consciousness, which allows us to understand a wide range of special states of consciousness, such as mystical and religious experiences, deathbed visions (end-of-life experiences), perimortem and postmortem experiences (nonlocal communication), heightened intuitive feelings (nonlocal information exchange), prognostic dreams, remote viewing (nonlocal perception), and the mind's influence on matter (nonlocal perturbation).

Ultimately, we cannot avoid the conclusion that endless consciousness has always been and always will be, independently of the body. There is no beginning and there will never be an end to our consciousness. For this reason we ought to seriously consider the possibility that death, like birth, may be a mere passing from one state of consciousness into another and that during life the body functions as an interface or place of resonance.

The Near-Death Experience: Bridging Science and Spirituality

An NDE is both an existential crisis and an intense learning experience. People are transformed by the conscious experience of a dimension where time and distance play no role, where past and future can be glimpsed, where they feel complete and healed, and where they can experience unlimited wisdom and unconditional love. These transformations are primarily fueled by the insight that love and compassion for oneself, others, and nature are important conditions of life. Following an NDE, people realize that everything and everybody are connected, that every thought has an impact on oneself and others, and that our consciousness survives physical death. People realize that death is not the end.

(Van Lommel, Pim. *Consciousness beyond Life: The Science of Near-Death Experience.* pp. xv-xix.)

http://vedicilluminations.com/downloads/Consciousness-Life-After-Death/Consciousness-Beyond-Death.pdf

An Out-of-Body Experience

During an out-of-body experience, people have verifiable perceptions from a position outside and above their lifeless body. Patients feel as if they have taken off their body like an old coat, and they are astounded that despite discarding it they have retained their identity, with the faculty of sight, with emotions, and with an extremely lucid consciousness.

The out-of-body experience begins with a patient's sensation that his or her consciousness is leaving the physical body but continues to function unchanged. Sometimes this is accompanied by fear, followed by a (futile) attempt to return to the body, but patients often feel liberated and are amazed at the sight of the lifeless or seriously damaged body.

The most common vantage point is from the ceiling, and because of this unusual position some people initially fail to recognize their body. People experience their new weightless body as a spiritual or nonphysical body that can penetrate solid structures such as walls and doors. It is impossible to communicate with or touch others who are present. To their utter amazement, people go unnoticed even though they can hear and see everything. The range of vision can extend to three hundred sixty degrees, with simultaneous detailed and bird's-eye views. Blind people too have the faculty of sight while deaf people know exactly what has been said.

While this is happening, people discover that all it takes to be near someone is to think of that person.

This out-of-body experience is of scientific importance because doctors, nursing staff, and relatives can check and corroborate the reported perceptions and the moment when they were supposed to have taken place.

(Van Lommel, Pim. *Consciousness beyond Life: The Science of Near-Death Experience*. p. 19.)

http://vedicilluminations.com/downloads/Consciousness-Life-After-Death/Consciousness-Beyond-Death.pdf

All I can say is, "Buy and read the book." Van Lommel's book is essential for anyone who wants to know what's really going on.

We need a new type of science that allows ALL of the evidence into evidence, which is why I created Science 2.0. Science 2.0 allows all of the evidence into evidence, and then it pursues a preponderance of the evidence. Nothing is off-limits or out-of-bounds. Even direct experiences with the Biblical God Jesus Christ are allowed into Science 2.0. Science 2.0 isn't hiding from anything, not even God.

Psyche or Intelligence is the ultimate causal agent. Consciousness is not located in space-time. Consciousness is non-local, meaning that it's non-physical and immaterial. Psyche is living, conscious, sentient light waves. Psyche is living light. Psyche is Quantum Waves. God is called the Father of the Lights. God the Father and Heavenly Mother clothed our Psyches or Intelligences with spirit bodies made in their image, thereby making us their sons and daughters.

Quantum Waves or Thought Waves are an invisible sort of light; and, it's light that isn't limited by the speed of light. Quantum Waves are instantaneously and simultaneously everywhere. It's a whole other way of being and existing. The probability waves of Quantum Mechanics are consciousness waves, or psychic waves, or sentient waves. They have purpose and meaning, because they are alive and self-aware.

For me personally, these observations and truths became Scientific Proof of God's Necessity, because we absolutely need God if we humans want to have memories of any kind. We absolutely need God and Nature's Psyche if we want to be something more than vegetables and rocks. We absolutely need Quantum Mechanics, Supernatural Mechanisms, and Transdimensional Mechanisms if we want to have any hope of doing the physically impossible. These realities and truths also ended up being Scientific Proof of God's Existence. God is hiding in plain sight where the Materialists, Naturalists, and Atheist can't see Him and refuse to look for Him. Science doesn't get any better than this, in my humble opinion.

Original Version: 25FEB2018
Current Version: 30MAR2019

www.ingramcontent.com/pod-product-compliance
Lightning Source LLC
Chambersburg PA
CBHW052306220526
45472CB00001B/3